The Impact of Climatic Variations on Agriculture

VOLUME 1: ASSESSMENT IN COOL TEMPERATE AND COLD REGIONS

Edited by

M. L. Parry

T. R. Carter

Department of Geography, The University of Birmingham, U.K. and IIASA, Vienna, Austria

and

N. T. Konijn

Department of Soil Science, Agricultural University, Wageningen, The Netherlands

KLUWER ACADEMIC PUBLISHERS
DORDRECHT / BOSTON / LONDON

THE INTERNATIONAL INSTITUTE FOR APPLIED SYSTEMS ANALYSIS
UNITED NATIONS ENVIRONMENT PROGRAM

Library of Congress Cataloging in Publication Data

The Impact of climatic variations on agriculture.

On v. 1 t.p.: The International Institute for
Applied Systems Analysis, United Nations Environment
Program.
  Includes index.
  Contents: v. 1. Assessments in cool temperate and
cold regions.
  1. Crops and climate. 2. Climatic changes.
3. Meteorology, Agricultural. I. Parry, M. L.
(Martin L.) II. Carter, Timothy (Timothy R.)
III. Konijn, Nicolaas. IV. International Institute
for Applied Systems Analysis. V. United Nations
Environment Programme.
S600.7.C54I46   1988   630'.2'516   88-3146

ISBN 90-277-2700-7 (v.1)
ISBN 90-277-2701-5 (v. 1 pbk.)

Published by Kluwer Academic Publishers,
P.O. Box 17, 3300 AA Dordrecht, The Netherlands.

Kluwer Academic Publishers incorporates
the publishing programmes of
D. Reidel, Martinus Nijhoff, Dr W. Junk and MTP Press.

Sold and distributed in the U.S.A. and Canada
by Kluwer Academic Publishers,
101 Philip Drive, Norwell, MA 02061, U.S.A.

In all other countries, sold and distributed
by Kluwer Academic Publishers Group,
P.O. Box 322, 3300 AH Dordrecht, The Netherlands.

All Rights Reserved
© 1988 by International Institute for Applied Systems Analysis and United Nations Environment
Program.
No part of the material protected by this copyright notice may be reproduced or
utilized in any form or by any means, electronic or mechanical
including photocopying, recording or by any information storage and
retrieval system, without written permission from the copyright owner.

Printed in The Netherlands

The Impact of Climatic Variations on Agriculture

VOLUME 1: ASSESSMENT IN COOL
TEMPERATE AND COLD REGIONS

THE INTERNATIONAL INSTITUTE FOR APPLIED SYSTEMS ANALYSIS

is a nongovernmental research institution, bringing together scientists from around the world to work on problems of common concern. Situated in Laxenburg, Austria, IIASA was founded in October 1972 by the academies of science and equivalent organizations of twelve countries. Its founders gave IIASA a unique position outside national, disciplinary, and institutional boundaries so that it might take the broadest possible view in pursuing its objectives:

To promote international cooperation in solving problems arising from social, economic, technological, and environmental change
To create a network of institutions in the national member organization countries and elsewhere for joint scientific research
To develop and formalize systems analysis and the sciences contributing to it, and promote the use of analytical techniques needed to evaluate and address complex problems
To inform policy advisors and decision makers about the potential application of the Institute's work to such problems

The Institute now has national member organizations in the following countries:

Austria
The Austrian Academy of Sciences

Bulgaria
The National Committee for Applied Systems Analysis and Management

Canada
The Canadian Committee for IIASA

Czechoslovakia
The Committee for IIASA of the Czechoslovak Socialist Republic

Finland
The Finnish Committee for IIASA

France
The French Association for the Development of Systems Analysis

German Democratic Republic
The Academy of Sciences of the German Democratic Republic

Federal Republic of Germany
Association for the Advancement of IIASA

Hungary
The Hungarian Committee for Applied Systems Analysis

Italy
The National Research Council

Japan
The Japan Committee for IIASA

Netherlands
The Foundation IIASA–Netherlands

Poland
The Polish Academy of Sciences

Sweden
The Swedish Council for Planning and Coordination of Research

Union of Soviet Socialist Republics
The Academy of Sciences of the Union of Soviet Socialist Republics

United States of America
The American Academy of Arts and Sciences

Preface

Three important studies were initiated in the 1970s to investigate the relationship between climatic variations and agriculture: by the National Defense University (1980) on *Crop Yields and Climate Change to the Year 2000*, by the U.S. Department of Transportation (1975) on *Impacts of Climatic Change on the Biosphere* and by the U.S. Department of Energy (1980) on *Environmental and Societal Consequences of a Possible CO_2-Induced Climatic Change* (the CIAP study). These were pioneering projects in a young field. Their emphasis was on measuring likely impacts of climatic variations rather than on evaluating possible responses, and they focused on first-order impacts (e.g., on crop yields) rather than on higher-order effects on society.

A logical next step was to look at higher-order effects and potential responses, as part of a more integrated approach to impact assessment. This was undertaken by the World Climate Impact Program (WCIP), which is directed by the United Nations Environment Program (UNEP). The WCIP is one of four aspects of the World Climate Program that was initiated in 1979.

At a meeting in 1982, the Scientific Advisory Committee of WCIP accepted, in broad terms, a proposal from the International Institute for Applied Systems Analysis (IIASA) for an integrated climate impact assessment, with the proviso that the emphasis be on impacts in the agricultural sector. Martin Parry was asked to design and direct the project at IIASA. Funding was provided by UNEP, IIASA, the Austrian Government and the United Nations University.

This volume, and its companion volume, are the products of that project. Together they represent the work of 76 authors from 50 different scientific institutes in 17 countries. Yet they are not merely edited collections of chapters because each reports results from a *common set of impact assessments* which were designed, conducted and reported in a compatible manner by our collaborating scientists.

Discussion of the design of the assessments was initiated in September 1983 under the guidance of a steering committee, chaired by C.S. Holling, whose members were B. Bolin, W.C. Clark, M.L. Parry and P.E.O. Usher. Of particular interest were some of the ideas which had emerged that month at a meeting in Villach (Austria) on *The Sensitivity of Ecosystems and Society to Climatic Change*.

From October 1983 a core staff of three IIASA scientists (Martin Parry, Timothy Carter and Nicolaas Konijn) set about selecting case study regions (11 in all) with teams of 3 to 6 collaborating scientists in each case study. The teams came together at IIASA on two occasions, first to agree upon the details of the project design (to select impact models, develop climatic scenarios, etc.) and the second time, about six months later, to compare results and agree upon details of presentation. The "cool temperate and cold" teams (for Volume 1) met in April and October 1984, and the "semi-arid" teams (for Volume 2) met in October 1984 and April 1985. A mid-term review of the project was conducted by G.A. McKay following the October 1984 meetings.

The case study reports were submitted to IIASA over the next 18 months, and each received an international, interdisciplinary review. Individual sections of the case studies were also reviewed by specialists in appropriate fields. The reviewers for Volume 1, totaling 60 from 12 countries, were: S. Adalsteinsson, M. Andreasson, D.W. Anthony, L.M. Arthur, R.E. Bement, M. Black, M.R. Bryck, M.C. Bursa, M.G.R. Cannell, A.H.C. Carter, A.J.W. Catchpole, M. Cave, R.S. Chen, W.C. Clark, G.J.F. Copeman, M.D. Dennett, P. Driessen, S. Dunlop, J. Eadie, A. Eddy, W.R. Emanuel, K. Frohberg, H. Fukui, C.F. Green, D.S. Grundy, J. Hansen, M.N. Hough, T. Iwakiri, S.R. Johnson, R.W. Katz, W.W. Kellogg, H. van Keulen, C. van Kooten, S. Lebedeff, S. Le Duc, L. Lyles, A.J. McIntyre, H. Mansikkaniemi, T. Mela, D. Mitlin, R.E. Munn, P. Newbould, A.E.J. Ogilvie, M.S. Philip, L. Pölkki, C. Price, W.E. Riebsame, I. Saito, C.M. Sakamoto, G. Sirén, E. Siuruainen, P. Smithson, A. Smyshlaev, Y. Sugihara, J. Tarrant, J.H.M. Thornley, M. Torvella, C.C. Wallén, R.A. Warrick, and T.M.L. Wigley.

The reviewers for Volume 2 (altogether 53 from 16 countries) were: F.O. Agumba, R. Ananthakrishnan, G.F. Arkin, J.A.A. Berkhout, E. Berry, R. Bromley, I. Burton, P. Cooper, B. Currey, J.C. Dickinson, T.E. Downing, W.R. Emanuel, G. Fischer, P.A. Furley, B.E. Grandin, S. Gregory, R. Gwynne, M. Hamilton, G.H. Hargreaves, S. Hastenrath, R.L. Heathcote, C. Henfrey, B.G. Hunt, D. Jha, R.W. Katz, J.R. Kiniry, G. Knapp, V.E. Kousky, R. Lal, P. Legris, H.N. Le Houérou, M. Lipton, D.M. Liverman, J.G. Lockwood, L.O. Mearns, V.V.N. Murty, M. Nirenberg, L.J. Ogallo, J.P. Palutikof, P.W. Porter, D.A. Preston, G.W. Robertson, C. Sage, R.P. Sarker, P.L. Scandizzo, J. Shukla, K. Swindell, M. Thompson, W.I. Torry, J. Townsend, R.L. Vanderlip, F.J. Wang'ati, and A. Wood.

The reviews were then dispatched to authors, together with additional comments and general guidelines from the three volume editors, for final revisions of the case study reports.

Advice and encouragement was given by our scientific colleagues at IIASA (particularly Bill Clark, "Buzz" Holling, Jag Maini and Ted Munn) and by visitors to our project (particularly Bob Chen, Howard Ferguson, Michael Glantz, Bob Kates, Diana Liverman, Gordon McKay, Bob Watts and Dan Williams); and Peter Usher, our program officer from UNEP, was there with the right kind of support when it was needed.

Preface vii

At Birmingham University and at IIASA Martin Parry and Tim Carter edited the reports into more coherent volumes. Marilyn Brandl and Lourdes Cornelio keyboarded and corrected the text. Tim Grogan drew the figures. Bob Duis and his publications team at IIASA, in conjunction with David Larner at Reidel, did the final production work. Marilyn and Lourdes, secretaries to the Climate Project at IIASA, deserve our especial gratitude for their efficiency and dedication.

In all, more than 200 scientists and professional support staff have worked toward the publication of these two volumes. We thank them all and hope that they can take some satisfaction from the result.

Laxenburg, Austria
March, 1987

Martin L. Parry
Timothy R. Carter
Nicolaas T. Konijn

Contents

Preface   v

Part I. The Impact of Climatic Variations on Agriculture: Introduction to the IIASA/UNEP Case Studies

Contents   3
List of Contributors   5
Abstracts   7

1. The assessment of effects of climatic variations on agriculture: aims, methods and summary of results
   M.L. Parry and T.R. Carter   11
2. The choice of first-order impact models
   T.R. Carter, N.T. Konijn and R.E. Watts   97
3. Development of climatic scenarios:
   A. From general circulation models
   W. Bach   125
4. Development of climatic scenarios:
   B. Background to the instrumental record
   J. Jäger   159
5. A case study of the effects of CO_2-induced climatic warming on forest growth and the forest sector:
   A. Productivity reactions of northern boreal forests
   P. Kauppi and M. Posch   183
6. A case study of the effects of CO_2-induced climatic warming on forest growth and the forest sector:
   B. Economic effects on the world's forest sector
   C.S. Binkley   197

Part II. Estimating Effects of Climatic Change on Agriculture in Saskatchewan, Canada
G.D.V. Williams, R.A. Fautley, K.H. Jones, R.B. Stewart and E.E. Wheaton

Contents	221
List of Contributors	223
Abstract	225
1. Introduction	227
2. The effects on the agroclimatic environment	259
3. The effects on potential for wind erosion of soil	281
4. The effects on spring wheat production	293
5. The effects on the farm and provincial economy	321
6. Conclusions and implications	351
Acknowledgments	367
Appendices	369
References	371

Part III. The Effects of Climatic Variations on Agriculture in Iceland
P. Bergthórsson, H. Björnsson, Ó. Dýrmundsson, B. Gudmundsson, Á. Helgadóttir and J.V. Jónmundsson

Contents	383
List of Contributors	385
Abstract	387
1. Introduction	389
2. The effects on agricultural potential	415
3. The effects on grass yield, and their implications for dairy farming	445
4. The effects on the carrying capacity of rangeland pastures	475
5. Implications for agricultural policy	489
References	505

Part IV. The Effects of Climatic Variations on Agriculture in Finland
L. Kettunen, J. Mukula, V. Pohjonen, O. Rantanen and U. Varjo†

Contents	513
List of Contributors	515
Abstract	517

1. Introduction	519
2. The effects on barley yields	547
3. The effects on spring wheat yields	565
4. The effects on the profitability of crop husbandry	579
5. Conclusions and implications for policy	599

Acknowledgments 609
References 611

Part V. The Effects of Climatic Variations on Agriculture in the Subarctic Zone of the USSR
S.E. Pitovranov, V. Iakimets, V.I. Kiselev and O.D. Sirotenko

Contents	617
List of Contributors	619
Abstract	621
1. Introduction	623
2. The effects on agriculture and environment in the Leningrad region	639
3. The effects on spring wheat yields in the Cherdyn region	663
4. Planned responses in agricultural management under a changing climate	677
5. Results and policy implications	695
Acknowledgments	707
Appendices	709
References	719

Part VI. The Effects of Climatic Variations on Agriculture in Japan
M.M. Yoshino, T. Horie, H. Seino, H. Tsujii, T. Uchijima and Z. Uchijima

Contents	725
List of Contributors	727
Abstract	729
1. Introduction: the policy and planning issues	731
2. Development of the climatic scenarios	751
3. The effects on latitudinal shift of plant growth potential	773
4. The effects on altitudinal shift of rice yield and cultivable area in northern Japan	797
5. The effects on rice yields in Hokkaido	809

6.	The effects on the Japanese rice market	827
7.	The implications for agricultural policies and planning	853

Acknowledgment 864
References 865

Index 869

PART I

The Impact of Climatic Variations on Agriculture: Introduction to the IIASA/UNEP Case Studies

Contents, Part I

List of Contributors		5
Abstracts		7
1.	The assessment of effects of climatic variations on agriculture: aims, methods and summary of results	11
	M.L. Parry and T.R. Carter	
	1.1. Introduction	11
	1.2. Aims	12
	1.3. Alternative approaches to the problem	13
	1.4. An outline of the IIASA/UNEP approach	17
	1.5. Major findings of the present study	28
	1.6. Summary of specific results	52
	1.7. Conclusions and recommendations	68
	Appendices	77
	References	94
2.	The choice of first-order impact models	97
	T.R. Carter, N.T. Konijn and R.G. Watts	
	2.1. Introduction	97
	2.2. Climate and the crop production system	98
	2.3. Agroclimatic models	101
	2.4. Evaluation of models for estimating effects of short-term climatic variations	107
	2.5. Evaluation of models for estimating effects of longer-term climatic change	112
	2.6. Model outputs	120
	2.7. Conclusions	122
	References	122
3.	Development of climatic scenarios: A. From general circulation models	125
	W. Bach	
	3.1. Introduction	125
	3.2. Needs of decision makers	126
	3.3. Status of climatic scenario analysis for impact assessment	127
	3.4. Use of GCMs for climatic scenario analysis	129
	3.5. The climate inversion problem	148

3.6.	Outlook	148
	Acknowledgments	151
	Appendices	151
	References	153

4. Development of climatic scenarios: B. Background to the instrumental record 159
  J. Jäger
   - 4.1. Introduction 159
   - 4.2. Data 160
   - 4.3. Recent climatic variations in the northern hemisphere 161
   - 4.4. Temperature and precipitation changes in the IIASA/UNEP case study areas 175
   - Acknowledgments 179
   - References 179

5. A case study of the effects of CO_2-induced climatic warming on forest growth and the forest sector: A. Productivity reactions of northern boreal forests 183
  P. Kauppi and M. Posch
   - 5.1. Introduction 183
   - 5.2. Temperature and the productivity of boreal forests 183
   - 5.3. Data and study procedure 184
   - 5.4. Locating regions with a high ETS response 187
   - 5.5. Possible growth response 189
   - 5.6. Discussion 193
   - References 194

6. A case study of the effects of CO_2-induced climatic warming on forest growth and the forest sector: B. Economic effects on the world's forest sector 197
  C. S. Binkley
   - 6.1. Introduction 197
   - 6.2. A global forest sector model 198
   - 6.3. The ecological models 206
   - 6.4. Results 213
   - 6.5. Conclusions 216
   - Acknowledgments 217
   - References 217

List of Contributors

PARRY, Dr. Martin L.
Atmospheric Impacts Research Group
Department of Geography
University of Birmingham
PO Box 363
Birmingham B15 2TT
United Kingdom

and

International Institute for
Applied Systems Analysis
Schlossplatz 1
A-2361 Laxenburg
Austria

CARTER, Mr. Timothy R.
Atmospheric Impacts Research Group
Department of Geography
University of Birmingham
PO Box 363
Birmingham B15 2TT
United Kingdom

and

International Institute for
Applied Systems Analysis
Schlossplatz 1
A-2361 Laxenburg
Austria

KONIJN, Dr. Nicolaas T.
Department of Soil Science
Agricultural University
PO Box 37
NL 6700 AA Wageningen
The Netherlands

WATTS, Professor Robert G.
Department of Mechanical Engineering
Tulane University
New Orleans LA 70118
USA

BACH, Dr. Wilfrid
Institute of Geography
Westfälische Wilhems University
Robert-Koch-Strasse 26
D-4400 Münster
Federal Republic of Germany

JÄGER, Dr. Jill
Fridtjof-Nansen-Strasse 1
D-7500 Karlsruhe 41
Federal Republic of Germany

KAUPPI, Dr. Pekka
Ministry of Environment
PO Box 306
SF-00531 Helsinki, Finland

POSCH, Mr. Maximilian
International Institute for
Applied Systems Analysis
Schlossplatz 1
A-2361 Laxenburg, Austria

BINKLEY, Dr. Clark S.
School of Forestry and
Environmental Studies
Yale University
205 Prospect Street
New Haven, CT 06511
USA

Abstracts of Part I

Section 1

This section describes the aims and methods of the IIASA/UNEP project on the assessment of the effects of climatic variations on agriculture. It summarizes the results from case studies in cool temperate and cold regions in the northern hemisphere. Details are given elsewhere in this volume. A companion volume reports the results of studies in semi-arid regions.

An overall goal of the project was to increase our understanding not only of *first-order effects* of climatic variations on agricultural productivity, but also of *higher-order effects* on regional and national economies. The first part of this section outlines some approaches for considering these interactions. Two broad types of experiments are described: *impact experiments* and *adjustment experiments*. The former provide estimates of the first-order and higher-order effects of climatic variations on farming systems that are assumed to include no adjustments. The latter evaluate a number of adjustments available at the farm- or policy-level to offset or mitigate these effects. Experiments were conducted to examine the impacts of three types of *climatic scenario*, representing short- and medium-term climatic variations observed in the historical instrumental record, and long-term climatic changes estimated for a doubling of atmospheric carbon dioxide concentrations by the Goddard Institute for Space Studies general circulation model (the GISS $2 \times CO_2$ scenario).

The results of the case study experiments [in Saskatchewan (Canada), Iceland, Finland, in the Leningrad, Cherdyn and Central regions of the northern European USSR, and in Japan] are summarized in the second part of the section, and tabulated in appendices. Results indicate that the effects of the GISS $2 \times CO_2$ scenario on agriculture are generally greater than the effects of observed, decadal anomalies, but not always as great as the effects of observed extreme annual events. However, the frequency and magnitude of these extreme events could well increase under $2 \times CO_2$ conditions, implying that a series of on-farm adjustments (e.g., changes in crop variety, in fertilizer applications or in land allocation) and a range of policy responses (e.g., revised regional agricultural support, adjustments in agricultural infrastructure and government assistance to plant breeders) would be necessary to accommodate the changes.

The section concludes by describing a number of general research approaches that were found to be of value in these assessments, and by recommending some specific priorities for further research.

Section 2

This section introduces the first-order impact models that are used in the case studies in this volume for estimating effects of climatic variations on agriculture. The focus is on models of crop growth and yield, although other models of plant productivity or potential are also considered. Two types of agroclimatic model can be identified: (1) *agroclimatic indices* and (2) *crop-climate models*, including: (a) empirical-statistical models and (b) simulation models.

Most models were developed to accommodate present-day short-term variations in climate and few are appropriate for use in experiments to assess the effects of long-term climatic change. The advantages and limitations of using agroclimatic models for both of these purposes are illustrated with reference to a checklist of the models used in this volume. The importance of adequate model validation and sensitivity testing is emphasized. Problems of the mismatch of scales in using agroclimatic model estimates as inputs to models of higher-order effects are also examined. Finally, some alternative methods of presenting model results are discussed, and the advantages of a spatial-mapping approach for locating regions where agricultural productivity is particularly sensitive to climatic variations are described.

Section 3

The purpose of this section is to derive CO_2-induced climatic change scenarios from general circulation models (GCMs) as a basis for impact assessments conducted in the case study areas. It is stressed that such climate scenarios should neither be construed as predictions of regional climatic change, nor as estimates of forthcoming climatic events, but rather as a set of self-consistent and plausible patterns of climatic change. Data processing, model validation, and regional scenario construction are demonstrated. Model problems and improvements are discussed briefly. The selection of the GISS-model data as input for the case studies of this volume is made on the basis of model characteristics and requirements for impact analysis. It must be realized that this and all the other GCMs that have so far been used in CO_2 sensitivity studies are still plagued with many deficiencies and that the data obtained therefrom are quite uncertain. These models are, however, currently the most appropriate tools with which a future climatic change due to specific influences can be assessed. While they are constantly being improved, it is important that the methods for model-derived climatic scenarios are also developed so that they are available for impact analysis when they are needed. The ultimate goal is community preparedness, i.e., to have at hand an optimal strategy that can respond with flexibility to any potential climatic risk.

Section 4

This section examines recent changes in surface temperature, sea-level pressure, precipitation and the interannual variability of these elements in the northern hemisphere. Its purpose is to provide background information on climatic variations that have occurred in the case study areas in the recent past and which have, in some circumstances, been adopted as climatic scenarios in the climate impact assessments in those areas. Data sources are discussed with an indication of limitations of coverage and other shortcomings. The surface temperature data show that there was a warming at least over the land areas of the northern hemisphere from the turn of the century until about 1940, followed by a cooling until the mid-1960s and a subsequent warming. The spatial patterns of the temperature variations in the northern hemisphere have been analyzed using annual average data. Further studies have shown that at least in the winter season the regional temperature variations were associated with changes in the atmospheric circulation at the surface. There is no evidence for a consistent relationship between average temperature over a period and interannual variability. Long-term fluctuations in precipitation are also noted in some regions. Correlations have been found between temperature and precipitation, although the sign and magnitude of the correlation vary according to location and season.

Section 5

This section, together with the section that follows, provides a case study of the advantages and disadvantages of linking biophysical and economic models in attempts to assess the effects of climatic change. Here, the effects of changes in temperature induced by a doubling of present-day atmospheric carbon dioxide concentrations ($2 \times CO_2$) on the productivity of northern hemisphere boreal forests are assessed using a simple forest growth model. The productivity of boreal forests is closely correlated with accumulated temperatures during the year – the effective temperature sum (ETS) – this relationship forming the basis of the growth model. Monthly temperatures for a 4° latitude × 5° longitude network of grid points in the latitude band 38°N–70°N are converted to ETS both for observed (1931–1960) "present-day" conditions and for $2 \times CO_2$ conditions. Temperatures under $2 \times CO_2$ conditions are those derived from estimates of the Goddard Institute for Space Studies (GISS) general circulation model (GCM). The computed ETS values are then converted to annual growth and mapped. Several uncertainties in the analysis are acknowledged and discussed. Results indicate that the location of the boreal zone under the GISS $2 \times CO_2$ scenario would be shifted northwards by 500–1000 km, and forest growth would increase in all parts of the study area with the greatest increases occurring in northern maritime regions. The results reported here are used in Part I, Section 6, as inputs to a further set of experiments to assess the economic implications of productivity changes on the world's forest sector.

Section 6

High-latitude forests provide the raw materials for an important part of the world's forest products industry. Section 5 indicated potentially large increases in forest productivity in the boreal zone under the temperature changes implied by the Goddard Institute for Space Studies (GISS) climate model for a doubling from present levels of atmospheric carbon dioxide concentrations (the GISS $2 \times CO_2$ scenario). This section quantifies some of the economic effects of those productivity changes using an economic model of the global forest sector. The model projects production, consumption, prices and trade of 16 forest products in 18 regions which comprise the globe. Ecological effects associated with changes of climate enter the model through the forest growth and timber supply components. The model computes partial market equilibrium solutions for each of the ten 5-year periods covering 1980 to 2030, the latter date assumed to be the date at which doubled CO_2 levels are reached. The absolute values of the results are highly speculative owing to uncertainties in the climate, biological and economic models, and because of the long time period of projections. Nevertheless, assuming that relative increases in forest growth due to climate warming are greater in the boreal zone than in other regions, three results seem fairly robust:

(1) The primary benefits of the increased supply will accrue to the consumers of forest products.
(2) While some producers, both in the directly affected regions as well as in other regions, are likely to benefit from the increased timber supply, others in directly

affected regions where the increase in supply is small may actually be hurt by climate warming.
(3) Climate warming imposes significant costs on regions where productive capacity is currently marginal and where large investments have been made to grow trees.

SECTION 1

The Assessment of Effects of Climatic Variations on Agriculture: Aims, Methods and Summary of Results

Martin Parry and Timothy Carter

1.1. Introduction

There is growing evidence that increasing concentrations of carbon dioxide (CO_2) and of other radiatively active trace gases in the atmosphere may be having a long-term effect on our climate. The earth is at present experiencing a long timescale climatic warming, with global mean temperatures increasing by 0.3–0.7 °C over the last 100 years (WMO, 1986). Five of the nine warmest years in the entire 134-year global temperature record occurred after 1978, the three warmest being 1980, 1981 and 1983 (Jones et al., 1986). Although this observed increase cannot yet be ascribed in a statistically rigorous manner to the increasing concentration of CO_2 and other trace gases, its direction and magnitude lie within the predicted range of their effects. Recent assessments suggest that increases in global mean temperatures in the range of 1.5–5.5 °C are likely to occur as a result of increases in CO_2 and other trace gases equivalent to a doubling of the atmospheric CO_2 concentration, which will probably occur between 2050 and 2100 (Bolin et. al., 1986).

The magnitude of these climatic changes could be sufficient to bring about long-term changes in agricultural potential, but it is not yet clear whether these changes in potential would be reflected in spatial shifts of cropping patterns and regional changes in agricultural output. However, preliminary estimates are that in the mid- to high-latitude cereal growing zones, shifts of several hundred kilometers per °C change are possible, assuming unchanged technology and economic constraints.

In addition to potentially large-scale and long-term changes in temperature, short-term anomalies of climate continue to cause severe shocks to economies and communities around the world. In 1982 and 1983 the most pronounced El Niño event ever recorded was associated with major floods in Ecuador and Peru, persistent drought in Northeast Brazil and widespread forest fires in Australia. At the same time there occurred persistent drought in southern and eastern Africa, and in north-central India.

Droughts have also recently affected the southeastern United States (in 1986) and the US Corn Belt (in 1983). Each of these short-term events has had a substantial effect on agriculture at the regional level. The 1983 drought in the US Corn Belt, for example, reduced average US maize (corn) yields by 29% relative to the year before (Parry et al., 1985). This, combined, with a 29% reduction in planted area promoted by the US government to reduce over-production, resulted in a 50% fall in US maize production in 1983 in comparison with the previous year.

Short-term variations in temperature can also affect agricultural output, especially in northern countries where the growing season is limited by spring and autumn frosts. In Japan, for example, a cold winter in 1982–1983 was followed by a late spring and cool summer in 1983. As a result, average rice yields in the northernmost district, Hokkaido, were about 21% below normal and there was extensive damage by weather to wheat and barley both in terms of the area affected and the quantity of production lost. Details are given in the Japanese case study in Part VI of this volume.

These brief examples have, of course, greatly over-simplified the connection between climatic variations and fluctuations in agricultural output. Many other (non-climatic) factors are also at work, and it is not a simple task to unravel them. However the indications are, firstly, that short-term climatic anomalies will continue to cause short-term perturbations in output and, secondly, that future long-term changes in climate may alter the long-term agricultural potential of different regions of the earth. Taken together, these are sufficiently compelling reasons for attempting to extend our understanding of the effects of climatic variations on agriculture.

1.2. Aims

This volume considers, firstly, the range of effects that both short-term and long-term changes of climate may have on agriculture and, secondly, the array of adjustments available to mitigate or exploit these effects. The core of the volume is a number of case studies of cool temperate and cold regions in the northern hemisphere. A companion volume reports results of similar case studies in semi-arid regions (Parry et al., 1988).

The two volumes are the outcome of a research project based at the International Institute for Applied Systems Analysis (IIASA) and funded jointly by IIASA and the United Nations Environment Programme (UNEP) as part of the World Climate Impact Programme (WCIP). The WCIP is one of four components of the World Climate Programme (WCP) that was initiated by the World Meteorological Organisation (WMO) in 1979.

The purpose of the project was twofold:

(1) To investigate the effects of climatic change and variability* on agriculture.
(2) To evaluate alternative responses to these effects.

An important premise behind the study was that, while we are not yet in a position to forecast how the climate may change (either in the short- or long-term) we *can* estimate the potential consequences of each of a number of *possible* climatic changes. By considering the range of impacts from and adjustments to these possible events we can improve both our techniques of impact analysis and our armoury of potential responses. Thus, at some point in the future, when we *are* able to make reasonably accurate forecasts of climatic change, we shall also have acquired an ability both to assess their impact and to respond effectively to them.

1.3. Alternative Approaches to the Problem

The past 15 years have seen encouraging developments in the method of climate impact assessment. With the benefit of hindsight it seems that our efforts up to the mid-1970s focused mainly on the (one-way) *impact* of climate on human activity. This has recently been replaced, and hopefully improved upon, by a greater emphasis on the *interaction* between climate and human activity.

1.3.1. The emphasis on impacts

The *impact approach* is based upon the assumption of direct cause and effect where a climatic event (for example, a short-term variation of temperature) operating on a given "exposure unit" (for example, a human activity) may have an "impact" or effect [*Figure 1.1 (a)*]. Such an approach certainly characterized studies which, for example, employed regression models to seek statistical relationships between climate and agriculture with a "black-box" approach that gave little attention to understanding the nature of the relationships. This was true of some aspects of the US Department of Transportation's Climate Impact Assessment Program which sought to estimate the possible range of effects resulting from atmospheric ozone depletion (CIAP, 1975). It also characterized the approach followed by the National Defense University's study of possible effects of long-term climatic change on crop yields and agricultural production (NDU,

---

*Throughout the volume reference is made to three forms of climatic perturbation: climatic variability, climatic change and climatic variations. Following Hare (1985), *climatic variability* describes the observed year-to-year differences in values of specific climatic variables *within* an averaging period (typically 30 years). *Climatic change* describes longer-term changes *between* averaging periods either in the mean values of climatic variables or in their variability. These two descriptors form subsets of the third collective term, *climatic variations*, which embraces short-, medium- and longer-term changes with time in values of climatic variables.

a) Impact approach

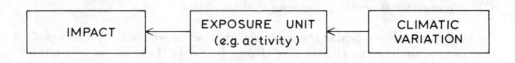

b) Interaction approach

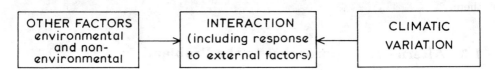

Figure 1.1. Schema of simple (*a*) impact and (*b*) interaction approaches in climate impact assessment (adapted from Kates, 1985).

1980). Here the emphasis was on the search for relatively simple connections between a climatic variation on the one hand and a possible response on the other. In reality, of course, so many intervening factors operate that it is both misleading and quite impossible to treat these three study elements (climatic variation–exposure unit–impact) in isolation from their environmental and societal *milieu*, and very few studies have followed this method with success.

1.3.2. The emphasis on interactions

More recently, the focus of attention has been to seek a better understanding of the interactions between climate and human activity by assuming that a climatic event is merely one of many processes (both societal and environmental in origin) which may affect the exposure unit [*Figure 1.1(b)*]. To illustrate, it may be argued that the magnitude of the effect of the 1930s drought on the Canadian prairies was substantially increased by the depression of farm prices and the desperate economic straits that had been reached by many Prairie farmers before the beginning of the drought. In addition, widespread ploughing of soils prone to wind erosion increased the amount of windblown dust which choked crops and reduced yields. Economics, weather and farming technology thus interacted to create a severe economic and social impact that was perhaps pre-conditioned by the Depression but triggered by drought. Likewise the effects of any change of climate in the future will be influenced by concurrent economic and social conditions and the extent to which these create a resiliency or vulnerability to impact from climatic change.

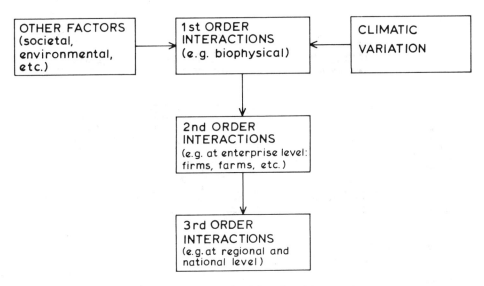

Figure 1.2. Schema of interaction approach with ordered interactions.

This *interaction approach* was adopted by the project on *Drought and Man* of the International Federation of Institutes for Advanced Study (IFIAS) (Garcia, 1981). Here the emphasis was on comprehending the syndrome of political and economic as well as meteorological events which resulted in the severe effects of drought in the Sahel in the mid-1970s.

1.3.3. Orders of interaction

Interaction models achieved greater realism by considering the "cascade" of interactions that can occur from the first-order biophysical level, through second-order levels characterized by units of enterprise (farms, corporations, etc.) to third-order interactions at the regional and national level (*Figure 1.2*). This approach was followed by the European Commission study of the socioeconomic effects of CO_2-induced climatic changes, which used outputs from general circulation models (GCMs) to generate information on possible changes in temperature and precipitation that could be used as inputs to models of runoff and biomass production. Economic interactions at the second- and third-order levels were not, however, modeled (Meinl, *et al.*, 1984).

1.3.4. The search for a fully integrated approach

Additional complexity can be introduced by studying interactions of the same order, both *within* individual sectors (such as between different farming systems) and *between* different sectors (such as between the concurrent effects of a change

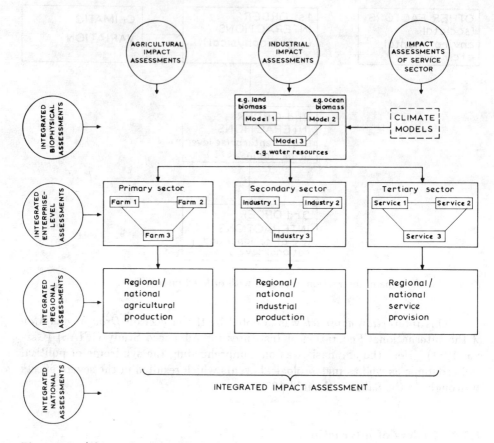

Figure 1.3. A hierarchy of models for integrated climate impact assessment.

in climate on agriculture, forestry, water resources, etc.), *and* the feedback effects operating between them. This form of fully integrated assessment has been proposed in some recent methodological studies of climate impact assessment, such as the Battelle Study (Callaway *et al.*, 1982) but has yet to be successfully implemented. In theory an integrated assessment could be performed at any level (biophysical, enterprise, or regional) and for any sector, and for any combination of these. A schema is given in *Figure 1.3*.

It will be some time before fully integrated assessments can be implemented because the full complement of systems models and our ability to connect them satisfactorily does not yet exist. Yet there are encouraging indications that some elements critical to an integrated approach have been successfully incorporated in a number of recent reviews of methods (Parry and Carter, 1984; Carbon Dioxide Assessment Committee, 1984; Warrick *et al.*, 1986; Maunder, 1986). In summary, we can say that over the past five years:

- We have moved away from studying impacts on averages (e.g., average crop yields) to considering also impacts on the range of expected effects (e.g., anomalously high or low yields).
- There has been some progress in linking models of different orders and of different sectors to produce more integrated assessments.
- Some recent assessments have experimented with a range of adjustments, and have thus begun to answer the question: "How might agriculture best respond to climatic change?"
- We have begun to make assessments over real (geographic) space not simply at a few arbitrarily-chosen points.
- There has been some attempt to match the spatial scales of impact assessment to the spatial scales at which policy-makers and planners operate.

1.4. An Outline of the IIASA/UNEP Approach

The approach adopted in the IIASA/UNEP project has three general characteristics. Firstly, it attempts a partially integrated assessment of impacts. Secondly, it adopts both a direct and an adjoint research method. Thirdly, it reports two types of experiment: impact and adjustment experiments.

1.4.1. A partially integrated approach

Three elements in the IIASA/UNEP project characterized its integrated approach:

Firstly, the project sought to assess the effect of climatic variations on agriculture by using a *hierarchy of models* that included the following:

- Models of climatic variation (based on outputs from general circulation models, on analysis of the instrumental record, or a combination of these).
- Biophysical models of first-order relationships, that is those between certain climatic variables (e.g., temperature, precipitation, insolation, windspeed, etc.) and biophysical supply or demand (e.g., biomass productivity, energy demand, etc.). In particular these considered the effect on crop growth, livestock production, livestock health, etc.
- Economic models of second-order relationships, that is those at the enterprise level (e.g., of farms, firms, agencies, institutions, etc.). These considered the effects (*inter alia*) of changes in farm-level production on farm profitability, farm purchases, etc.
- Economic models of higher-order relationships, for example those between farm profitability and regional employment or gross domestic product.

Secondly, it was considered important to consider the effects of climatic variations and their interactions with other physical systems, distinguishing between:

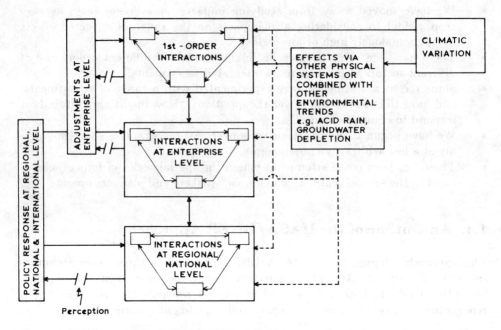

Figure 1.4. Schema of the IIASA/UNEP project's approach: an interactive approach to climate impact assessment with ordered interactions, interactions at each level, and some social and physical feedbacks.

- Those in which the effects of the climatic variation are transmitted through other physical systems (e.g., by pests and diseases; by changes in soil structure, soil nutrients, soil erosion, salinization, etc.).
- Those in which the effects of the climatic variation are themselves affected by other concurrent environmental trends (such as acid deposition, groundwater depletion, etc.).

Thirdly, the project considered two types of response to climate impacts:

- Adjustments at the enterprise level (which at the farm level might include changes of crops, increased irrigation, changes in fertilization, etc.).
- Policy responses at the regional, national and international level.

Figure 1.4 gives a schema of the project's study method, which incorporates the three elements outlined above.

1.4.2. The use of direct and adjoint methods

The scientific method most frequently adopted in climate impact assessment is the *direct method* in which, for example, the effects of a change in an input variable (such as a change in temperature) are traced, in a number of steps, along

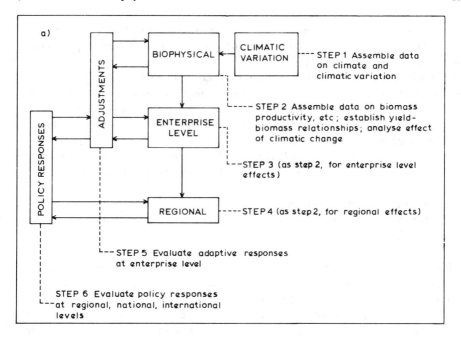
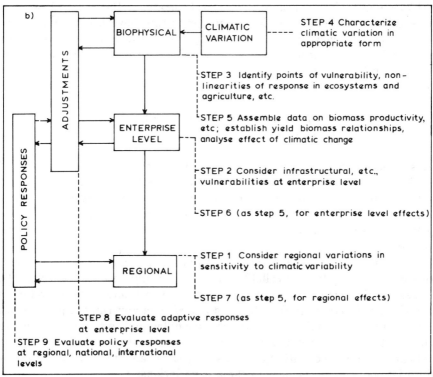

Figure 1.5. (a) Direct and (b) adjoint methods of climate impact assessment. Both approaches were adopted in the IIASA/UNEP project.

several pathways (e.g., temperature → crop biomass productivity → forage level → carrying capacity → livestock production → meat and milk supply, etc.) [Figure 1.5(a)]. The question is thus posed: "For a given variation in climate, what is the effect on, for example, ecosystems, economy and society?" An advantage of this approach is that assessments can be made even if the number of climatic scenarios is restricted (for example, in the case of projected possible future climates). The analysis is conducted on the basis of the character of the climatic changes rather than their likely impacts.

An alternative or *adjoint* method (Parry and Carter, 1984) focuses first on the sensitivity of the exposure unit and addresses the questions:

- To what aspects of climate is the exposure unit especially sensitive?
- What changes in these aspects are required to perturb the exposure unit significantly?

Climatic scenarios can then be characterized partly on the basis of these detected sensitivities and partly along lines adopted in the direct approach described above. The steps are illustrated in *Figure 1.5(b)*. An advantage of this approach is that it can help identify sensitivities independently of state-of-the-art climatic scenarios and can allow climatic changes to be expressed in the form which has direct meaning for the exposure unit. To illustrate, in the Saskatchewan study in this volume, the effects of climate on agriculture have been characterized in terms of the frequency of days of blowing dust because a sensitivity study of Prairie agriculture identified such events as critical to the sustainability of cereal production in the region.

1.4.3. The conduct of impact experiments and adjustment experiments

In this project two broad types of response experiment were conducted (cf. *Figure 1.7*). In a number of *impact experiments* each set of models in the hierarchy simulates a limited number of feedbacks within its own sub-systems. The form of analysis is therefore essentially sequential, estimates of effects being based largely on assumed and essentially static sets of agronomic and economic responses.

Adjustment experiments involve altering some of these assumptions to evaluate various options available to offset or mitigate the effects. These responses might occur at the enterprise level (for example, a farm-level decision to switch to a different crop) or through a change in government policy (such as a change in farm subsidies). Experiments for different crops, varying amounts of fertilizer, etc., can enable a new set of impact estimates to be generated which can then be compared with the initial estimates, and thus help in evaluating appropriate policies of response to climatic variations.

In the studies reported here it has frequently proved useful to perform these experiments in a sequence, firstly, conducting impact experiments on the (unrealistic) assumption that the economic and social systems will undergo no change and secondly, conducting adjustment experiments which incorporate an increasingly complex pattern of assumed responses.

1.4.4. Some integrating concepts

Three broad concepts provided a common focus for the study: the evaluation of climatic changes as changes in risk, the mapping of impact areas, and the emphasis on vulnerability of marginal systems to climatic change.

Climatic change as a change in risk

Government interest in impacts from climate is often a short-term concern (e.g., the effects of droughts, floods, cold spells, etc.). Long-term climatic changes, such as that which might result from increasing atmospheric CO_2, are frequently not perceived by governments to be of immediate importance. This suggests, therefore, that a useful way to express long-term climatic change for the policy maker is as a change in the *frequency* of short-term anomalies. For example, estimates from this study indicate that in Saskatchewan the frequency of drought would increase from 3% of all months at present to about 9% under a doubled CO_2 climate. One advantage of this type of approach is that the change can be expressed as a change in the *risk of impact* (Parry, 1985a). This would enable governments to devise programs that accommodated specified tolerable levels of risk, and adjust policies to match the new risk levels.

Delimiting impact areas

In order to conduct useful impact analyses, it is important to be specific about the exposure unit being studied and its location. One method of identifying areas that can be affected by climatic variations focuses on the shift of limits or margins representing boundaries between arbitrarily defined classes (Parry, 1985b). The classifications may be of land use, agroecological potential, vegetation, crop yields, etc., and the boundaries delimit zones on maps that may shift in response to a change in climate, thus defining "impact areas". This method can be combined with the change-in-risk approach to help identify agricultural areas that are particularly sensitive to climatic variations. Examples reported in this volume include spatial shifts of thermal resources, of zones of net primary productivity, and of agricultural potential.

Vulnerability at the margin

Underlying the identification of spatial shifts of climatic resources is the idea that sensitivity to climatic variability may be more readily observed towards the margins of tolerance of an organism or activity, or at the margins of comparative advantage between two competing activities. We can define three types of "marginality":

(1) *Spatial or geographical marginality*, which describes the edge of a specified region defined in biophysical or economic terms.
(2) *Economic marginality*, where returns on an activity barely exceed costs.
(3) *Social marginality*, where an underdeveloped population becomes isolated from its indigenous resource base and is forced into marginal economies that contain fewer adaptive mechanisms for survival (Baird *et al.*, 1975).

These marginal areas or groups do not necessarily coincide on the ground. Marginal populations, for example, are not always found in areas that are poorly endowed with resources. But it is reasonable to suggest that, whatever the type of marginality, it is characterized by a special sensitivity to changes in resource availability (such as changes in climate, which can itself be regarded as a resource). The margins can be mapped and their shifts can be used to designate areas of impact from climatic change or climatic variability.

Marginality, risk and the spatial shift of these are, thus, three concepts that underlie the studies in this volume: The focus is on regions near the northern limit of agriculture where levels of agricultural risk are already high; close attention is given to changes in risk that may result from changes of climate; and there is emphasis on mapping these effects in order to identify areas of possible impacts.

1.4.5. The use of case studies

In order to develop and test the methodology in a manageable research context the project focused on a number of case studies. The case studies fall into two groups: cool temperate and cold regions, and semi-arid regions.

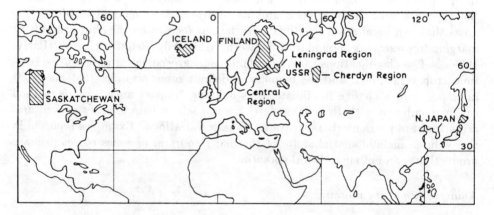

Figure 1.6. Location of the case study regions: Saskatchewan (Canada), Iceland, Finland, the Leningrad, Central and Cherdyn regions (USSR) and northern Japan.

Cool temperate and cold regions

Consideration of CO_2-induced climatic changes is a prominent feature in these studies, partly because CO_2-induced climatic warming is expected to be more pronounced at higher latitudes, and partly because agriculture in these more economically developed regions is less vulnerable to short-term climatic variability. Results of these case studies are reported in this volume.

Four criteria determined the choice of case studies, namely that they should:

(1) Be near the northern limit of agriculture and thus, *prima facie*, be sensitive to variations in temperature.
(2) Include a variety of geographical locations, farming types, and political structures (i.e., both centrally planned and market-oriented economies).
(3) Have available for use a variety of crop-climate and economic models already developed and validated for the region.
(4) Have scientists and policy advisers with sufficient interest and time to make a commitment to research on the project.

The regions chosen as case studies were: Saskatchewan (Canada), Iceland, Finland, the Leningrad, Central and Cherdyn regions of the northern USSR, and northern Japan (*Figure 1.6*).

Semi-arid regions

Vulnerability to drought is symptomatic of some agricultural systems in these regions, where a high degree of sensitivity to present-day climatic variability is a major source of system disruption. Case studies were conducted in the central Sierra of Ecuador, Northeast Brazil, central and eastern Kenya, dry tropical India, in the Stavropol and Saratov regions of the southern USSR, and in southeast Australia. Results of these are reported in Volume 2 (Parry *et. al*, 1988).

Organization

In each case study a team of between 4 and 7 scientists and policy advisers performed the experiments and submitted results. The teams met together at IIASA at the beginning of the project to establish common goals, agree upon climatic scenarios and upon methods of impact assessment, and to determine the intended outputs. They then returned to their home regions to complete the research, which was reported in draft form at a second meeting at IIASA. Omissions, overlaps and errors were discussed at that meeting, and results were submitted to IIASA for external reviewing and subsequent revision.

1.4.6. A common set of experiments

In order to establish a degree of comparability which would enable subsequent synthesis and the generalization of results, the case studies adopted similar climatic scenarios, similar model hierarchies and similar analytical approaches (*Figure 1.7*).

Types of climatic scenario

Each case study was asked to consider the effects of three scenarios of climatic variation:

(1) A single, anomalous weather year, taken from the region's historical instrumental record.
(2) An extreme weather decade or period of weather-years, also taken from historical data.
(3) A long-term change of climate that might result from a doubling of atmospheric CO_2. Data for this scenario were derived from grid-point outputs from the Goddard Institute for Space Studies (GISS) GCM (Hansen *et al.*, 1983, 1984). It is referred to in the following pages as the "GISS $2 \times CO_2$ scenario". It should be noted that the "direct" effects of increased atmospheric CO_2 on plant growth and water-use efficiency were *not* considered in the impact experiments (except for a single example in the USSR study).

Thus short-term, medium-term and long-term climatic variations were all considered. Details of the development of scenarios are given in *Appendix 1.1* and also in Sections 3 and 4.

Linking the models

In each case study a variety of models of crop-climate relationships were used to estimate first-order effects. Descriptions and comparisons of these can be found in Section 2. Three main types of model were used:

(1) *Agroclimatic indices* that combine into a single term those meteorological variables that most influence plant growth.
(2) *Empirical-statistical models* that relate a sample of annual crop yield data to weather data for the same time period and area using statistical techniques such as regression analysis. These were most valuable for estimating the impacts of short-term climatic anomalies, the magnitude of which were *within* the range of conditions upon which the model was based (thus not requiring extrapolation of model relationships). Assessments using these models were most successful in areas where crop yields are highly sensitive to a *single* climatic variable.
(3) *Simulation models* that are based on an understanding of the relationships between the basic process of plant and crop growth and environmental factors. Despite their general requirements for quite detailed meteorological and physiological data, simulation models are more firmly based on experimental observation than statistical models, and were therefore more suited to conducting adjustment experiments in this project (particularly when considering longer-term effects of changes in climate).

Aims, methods and summary of results 25

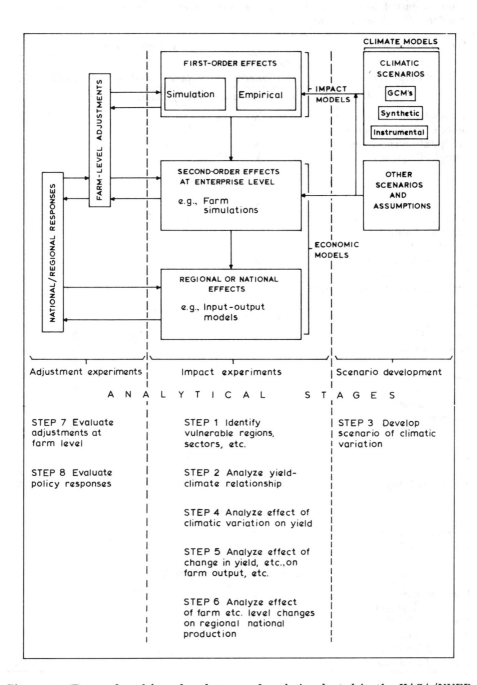

Figure 1.7. Types of model used and stages of analysis adopted in the IIASA/UNEP project.

Table 1.1. Checklist of experiments on the effects of climatic variations on agricultural production. Experiments in Part A assume *ca.* 1980 technology. Experiments in Part B assume adjusted technology. In each case study three or more climatic scenarios were considered (for details, *see* text).

Type of experiment	Case study regions				
A. Impact experiments	Saskatchewan (Canada)	Iceland	Finland	N. USSR	Japan
Effects on derived parameters of climate					
Accumulated temperatures	X		X		X
Onset of growing season		X			
Precipitation effectiveness	X				
Drought frequency	X				
Effects on potential					
Biomass potential	X				X
Biomes		X			
Cropping limits					X
Forest limits			X		
Wind erosion of soil	X				
Index of soil fertility				X	
Probability of crop maturation			X		
Effects on crop yields					
Spring wheat	X		X	X	
Winter wheat				X	
Barley			X	X	
Oats			X	X	
Winter rye				X	
Corn for silage				X	
Potatoes				X	
Vegetables				X	
Hay		X		X	
Pasture		X			
Rice					X
Yield variability			X		X
Effects on livestock					
Carrying capacity		X			
Carcass weight		X			
Higher-order economic effects					
Farm income	X		X		
Farm employment	X				
Provincial GDP	X				
Additional costs				X	
National food supply					X
National food stocks					X

Table 1.1. continued.

Type of experiment	Case study regions				
B. Adjustment experiments	Saskatchewan (Canada)	Iceland	Finland	N. USSR	Japan
Changes in crop type					
From spring to winter varieties	X				
From early- to late-maturing varieties			X	X	X
Changes in fertilizer applications					
To increase or optimize yields				X	
To stabilize 1980 yields		X			
Changes in soil drainage					
To prevent waterlogging				X	
Changes in land allocation					
To stabilize yield	X				
To optimize production			X	X	
Changes in bought-in food					
To supplement fodder supply		X			

Outputs from these were used as inputs to models of second-order effects where these models were available. In some case studies detailed farm simulations were available, but elsewhere farm-level responses needed to be described in a non-quantitative fashion. To illustrate, the case study in Finland used a hierarchy largely of regression models: first-order effects of climate on cereal yields were estimated using regression models, and farm-level profitability was then investigated as a second-order effect of the climate-induced yield changes by balancing the gross returns per unit of production against farmers' expenditures. Average net return was thus obtained on a regional basis assuming present-day price and cost levels. Finally, the implications of changes in crop yields and production for agricultural policy were examined, both at the regional level (in relation to income equalization policies within Finland) and at the national and international levels (with respect to national average farm income, self-sufficiency objectives and overseas trade).

In contrast, the Saskatchewan study used a mixed hierarchy of simulation and regression approaches. A simulation model estimated altered yield levels of spring wheat that were used as inputs to farm-level models, converted to production figures and then aggregated by farm size and soil zone to give provincial production and commodity changes. These were then used as inputs to a regional input–output model which considered impacts both on the agricultural and the non-agricultural sectors. Finally, changes in output levels for various economic sectors were translated into changes in employment using a third, employment, model (for details see Part II).

Analytical stages

Where possible the case studies adopted the adjoint approach, considering first the varying sensitivities of different regions, sectors and farming systems to different types of climatic variation and thus identifying, a priori, points of special vulnerability that should be the focus for study. Scenarios of climatic variation were then constructed using those climatic variables (or derived variables, such as effective temperature sum, onset of growing season, etc.) considered most appropriate for the particular impact study. Various orders of effect and response were then considered (*Figure 1.7*).

Table 1.1 is a checklist of the impact experiments conducted for each case study. It includes:

(1) Effects on derived climatic parameters (e.g., effective temperature sum, precipitation effectiveness, etc.).
(2) Effects on agricultural potential (e.g., shift of altitudinal cropping limits, changes in probability of ripening, etc.).
(3) Effects on crop yields.
(4) Effects on livestock.
(5) Higher-order effects such as on farm income, farm employment or on national food stocks.

These were followed by adjustment experiments which considered the efficacy of various on-farm adjustments as responses to variations in climate, including:

(1) Changes of crop type or cultivar to suit the expected weather.
(2) Changes in the amount of treatments applied, especially fertilizer applications.
(3) Changes in the timing of farming operations to suit the weather (e.g., the timing of soil preparation, planting and fertilizing).

More general policy responses at the regional and national level were also considered. These included:

(1) Changes in regional land-use allocations.
(2) Changes in national agricultural support policies.

1.5. Major Findings of the Present Study

In this subsection we report some general conclusions about the effects of climatic variations. Consideration is given to types of first-order effect, types of higher-order effect, potential farm-level adjustments and potential national policy responses. We then discuss the project results in more detail, prior to making recommendations about adjustments, policy responses and future research directions.

1.5.1. Spatial complexities and non-linearities

The geography of climatic change and its effect

It should first be emphasized that it is not always a straightforward matter to make valid generalizations about the effects of climatic variations, largely because they can exhibit such a complex pattern in both time and space. Not only are there likely to occur varying degrees of *absolute* change in climatic variables at different locations, but the effect of these changes is very much a function of the change in climate *relative* to the existing (baseline) conditions. For example, while effective temperature sums (ETS) in southern Finland were found to increase by 33% under the GISS $2 \times CO_2$ scenario, in northern Finland (where they were already one-tenth lower than in the south) the increase is 43%. Thus much of the variation in yield responses is a function of the geography of existing potential.

While that geography is often complex, there are, however, some general spatial patterns which can be detected. For example, because agricultural potential generally decreases northward in most of the areas studied, the same *absolute* increases of temperature have greater *relative* effects on crop potential at higher than at lower latitudes.

In addition, because the magnitude of temperature changes expected to result from increases in atmospheric CO_2 is greater at higher latitudes, we can reasonably expect substantial effects on crop potential in far northern regions. To illustrate, we may draw upon results from the Japanese case study in this volume. These indicate that mean July–August temperatures increase more in northern Japan (+3.5 °C in Hokkaido) than in central Japan (+3.2 °C in Tohoku) under the GISS $2 \times CO_2$ scenario. The effect of these different absolute changes on different regional climates means that effective accumulated temperature increases significantly more in Hokkaido (+35%) than in Tohoku (+29%). As a result, the estimated increases in rice yields in Hokkaido are twice that in Tohoku. Similar differential effects are found in other northern regions. This has profound implications for the support policies that at present bolster farm incomes in northern regions (*see* Subsection 1.5.8 below).

The non-linear effects of climatic change

Because the relationship between climatic change and its effect on agriculture is frequently non-linear, small climatic changes can have large effects. Consider, for example, the relationship between temperature and spring wheat yield in Saskatchewan. Here temperatures of around 1 °C less than average are generally beneficial for yields because the crop's moisture requirement for transpiration is decreased. However, reduced temperatures also imply a shortening of the

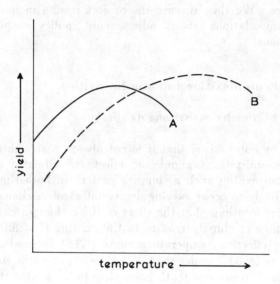

Figure 1.8. Hypothetical yield-temperature response curves for two crops (A and B) in the same region.

potential growing season, and at temperatures 2 °C less than average the crop will generally fail to ripen [and yields are thus nil, cf. *Figure 1.20* (a)]. In this instance a 1 to 2 °C decrease in growing season temperature is a *critical threshold* beyond which quite small variations of climate can have a dramatically increased effect. One of the present priorities in climate impact assessment is to identify critical thresholds such as these, and estimate whether they are likely to be exceeded under certain scenarios of climatic change.

In general the relationship between crop yield and climate exhibits a greater degree of linearity than that exemplified above. Normally we might not, therefore, expect a small increase of temperature to have a dramatic effect on yields. In the core grain growing regions of North America and Western Europe, for example, a 2 °C warming with no change in precipitation is estimated to reduce average yields by 10±7% (Bolin, *et al.*, 1986). However, the *differences in response* by different crops may be critical. Consider, for example, the yield response to average temperature of two crops in the same region (*Figure 1.8*). If we assume that these crops are otherwise equally competitive, then the cross-over of the curves indicates the point at which the crops are equally profitable. To the left of the cross-over it would make sense, *ceteris paribus*, to grow crop A, and to the right crop B. In this simple example a small climatic change would result in a major change in choice of crop and thus of land use in the region. In reality patterns of cropping and livestock production throughout the world are the result of the intersection of similar (though more complex) functions. A small change in one of these functions, perhaps due to a change of climate, can cause a radical change in regional production patterns.

In addition we should note the strongly non-linear effects on the frequency of extreme events such as droughts and cold spells that can occur as a result of changes in mean climate. For example it has been demonstrated that, in northern Europe near the present altitudinal limit of cereal cropping, a 0.6 °C increase in mean annual temperature can decrease the frequency of "cool summer" crop failure by a factor of 14 (Parry and Carter, 1985). Similarly, an increase of 1.7 °C in mean maximum temperatures in the US Corn Belt increases by three times the likelihood of a run of 5 consecutive days with damaging high temperatures for maize (Mearns *et. al.*, 1984).

Effects can be multiplicative and cumulative

The ultimate effects of a climatic anomaly on agriculture may often be greater than simply the first-order effects on yield. In some cases, two or more first-order effects may induce a combined effect that is *multiplicative*. For example, in Iceland during a cold decade, hay yields are reduced but the fodder requirement for sheep is increased because there is reduced grazing available and a shorter-than-normal growing season. The combined effect is to reduce carrying capacity quite sharply and to increase the requirement for "bought-in" feed. Conversely, higher temperatures, higher grass yields and a longer grazing season would reduce the requirement both for winter stall-feeding in particular, and for winter housing for stock in general, thus radically altering the present-day allocations of capital to these expensive items.

In addition, apparently minor annual effects of climatic anomalies on agriculture can, if repeated over successive years under a changing climate, *accumulate* to produce significant long-term effects. Some first steps at modeling these effects have been taken in the Japanese case study where the GISS $2 \times CO_2$ climatic scenario has been introduced into a regional econometric rice model, not as an instantaneous change in climate, but as a linear transition between present-day and $2 \times CO_2$ temperatures over a 17-year simulation period. Although this is, of course, a very rapid and somewhat "unreal" climatic change, the results indicate that the annual increment in national rice supply due to progressively higher temperatures is still relatively small. However, over the 17-year period, these small increases in estimated supply would, assuming unaltered government policies, lead to more than a doubling of rice stocks and thus produce a major national rice surplus. Appropriate changes in national agricultural policies to avoid this outcome are considered in Subsection 1.5.8 below.

1.5.2. Climatic change *versus* present-day anomalies

How can we judge the magnitude of effects of possible future changes of climate? One way is to use as a bench-mark the effects of *present-day* climatic extremes. An advantage is that we can measure the effects of these events and also have some idea of the more appropriate ways of responding to them; and these may provide some clue to effects and responses in the longer-term. However, caution is necessary because we are comparing part of the interannual variability around the mean with changes in the mean. We consider these complications and make detailed comparisons later in this subsection (1.6.3). At this point it is useful to summarize our conclusions.

Understandably, the relative magnitude of present-day extremes and mean conditions under the GISS $2 \times CO_2$ scenario depends greatly on the geography of existing variability and of changes predicted by the GISS GCM. The former is characterized by large differences between temperate and continental locations, differences that are reflected, for example, in the present-day location of boreal forests (*see* Section 5). The latter are as much the product of present limitations of GCMs as they are a reflection of the future geographical pattern of climatic change. These limitations are discussed in *Appendix 1.1* and in Section 3.

In these comparisons of the effects of short-term and long-term climatic variations we assume that management and technology are at 1980 levels. It should thus be emphasized that we are comparing simulated, not actual effects. Appropriate adjustments in farming systems could well reduce the levels of impact. With these caveats in mind, it appears that:

- In Saskatchewan an *average* weather-year under the GISS $2 \times CO_2$ climate would be broadly similar in its scale of effect on wheat yields and wheat production to that estimated for the most anomalous five-year period on the local instrumental record – the extremely warm and dry period 1933–1937. Individual extreme years, such as 1961, have exceeded this level of anomaly.
- In Iceland temperatures under the GISS $2 \times CO_2$ scenario exceed any of those experienced in individual years over the past century. Effects on yields of hay and carrying capacity of sheep are about *four times* those of the 10 warmest years since 1931.
- In Finland the effects on crop yield of higher temperatures assumed under the $2 \times CO_2$ scenario are roughly *double* those estimated to occur in weather typical of the most extreme warm periods that have occurred this century (e.g., 1931–1940 and 1966–1973).
- In the northern USSR, the effects on crop yield of increased temperatures and precipitation under the GISS $2 \times CO_2$ scenario are negative, and *opposite* to the effects of warmer-than-average conditions characteristic of the present-day climatic variability.
- In Japan the effects on rice yields of higher temperatures assumed under the GISS $2 \times CO_2$ scenario are of a *similar magnitude* to those estimated for one of the warmest individual years in the instrumental record (1978).

1.5.3. First-order effects of climatic variations on agriculture

In this and the following subsections we describe in general the types of first-order and higher-order effects of variations of climate. Specific results of impact experiments are summarized in Section 1.6.

Changes in length of growing season and period of crop growth

Two of the major implications of CO_2-induced temperature changes for the growth of agricultural crops at present cultivated in cool temperate and cold regions are: (a) a lengthening of the potential growing season, and (b) an increase in plant growth rates and thus a shortening of the required growing period. For example, under the GISS $2 \times CO_2$ scenario the growing season in southern Saskatchewan is lengthened by 4-9 weeks, while the estimated maturation period for spring wheat is reduced by 4-14 days (*see* Part II, Section 4). One strategy to cope with the resulting mismatch between crop requirements and the thermal resources available for growth would be the substitution of crop varieties with greater thermal requirements (*see* Subsection 1.5.7). However, other characteristics of the growing season may place constraints on the choice of substitutes. For example, in southern Saskatchewan moisture constraints due to high evapotranspiration rates would restrict potential plant growth during the summer period (July and August), splitting the effective growing season into two parts. Crops such as winter wheat or maize, with higher thermal requirements and a greater ability to withstand dry summer conditions, might be suitable as substitutes for spring wheat under these conditions.

Changes in mean crop yield

Except where moisture is a limiting factor, yields of most crops in cool temperate and cold regions display a marked positive relationship to growing season temperatures. This relationship varies, of course, from crop to crop. It is demonstrably stronger in the north than in the south of some of the areas we have studied, and clearly indicates the marginal conditions of these northerly regions. The results also suggest that the cultivars grown in these areas, while bred to survive the cool conditions, are rarely able to obtain yields approaching the full potential.

Crop yields estimated for the GISS $2 \times CO_2$ scenario are generally above the present-day average, a result that can be attributed to large predicted increases in precipitation offsetting the increased water demand from crop transpiration caused by markedly higher temperatures. Under these conditions, the low-temperature constraints on achieving potential yields are removed. However, the temperature increases are in some cases so great that crop cultivars are ill-adapted to the longer and more intense growing season and to the assumed range of (higher) temperatures under the GISS $2 \times CO_2$ scenario. As a result, mean yield levels are usually no greater than those obtained during anomalously warm periods at the present-day.

Some crops can be expected to do better under the GISS $2 \times CO_2$ scenario than others. For example winter wheat would probably give higher yields than spring cereals in Saskatchewan and in the Central (Moscow) Region of the USSR. Also, in the latter region crops with greater thermal requirements such as maize would show a greater increase in yield than cold-region crops such as potatoes and oats. The *differences in yield* between various crops can thus be expected to alter. This will affect the relative profitability of different crops and thus the extent to which they are grown in different regions. We provide a more detailed discussion of yield responses under various climatic scenarios later in this section (1.6.2).

Changes in the variability of yields

Since we at present know little about changes in the interannual variability of temperature and precipitation which might accompany a change of climate, in the experiments reported here this was generally assumed to be equal to the present-day variability. However, a number of the "extreme period" instrumental scenarios indicated that past changes in mean climate, though more modest than the $2 \times CO_2$ changes, were accompanied by significant changes in year-to-year variability. Moreover, several experiments indicated that whatever changes in climatic variability are assumed, these are not necessarily reflected in equivalent changes of crop yield variability. The rice model results for Hokkaido illustrate two contrasting cases:

(1) During a 1957 to 1966-type "stable" climate the coefficient of yield variation is about one-third lower than during a 1974 to 1983-type "unstable" climate although mean temperatures and mean yields for the two periods are very similar.

(2) Comparison of the GISS $2 \times CO_2$ scenario results with those for the baseline climate (1974–1983) indicates a reduction of over a quarter in interannual yield variability and a small increase in mean yields, *in spite of* an assumption that there is no change in interannual climatic variability. This is due to a lower frequency of cold summer damage under the $2 \times CO_2$ climate.

The second point emphasizes that the effect of a change in mean climate and/or year-to-year variability on mean yield can be different from its effect on the above- or below-average yields. Furthermore, there is an added spatial dimension if we attempt to identify patterns of regional variability. *Figure 1.9* illustrates these points by showing how the relative effects of the cool- and warm-period climatic scenarios and the GISS $2 \times CO_2$ scenario on spring wheat yields in Finland differ, both between highest-yielding (5 percentile), average-yielding (50 percentile) and lowest-yielding (95 percentile) years, and between central Finland and southern Finland. The former differences can be explained by contrasts in climatic variability between scenarios; the latter are largely attributable to the greater susceptibility in southern than in northern Finland of spring wheat to moisture stress during warmer-than-average years.

Changes in the probability of yields

A corollary of the effect on yield variability is the effect on the probability or frequency of a given yield level. Given the strongly non-linear relationship between variations in average climate and the frequency of extreme events, discussed

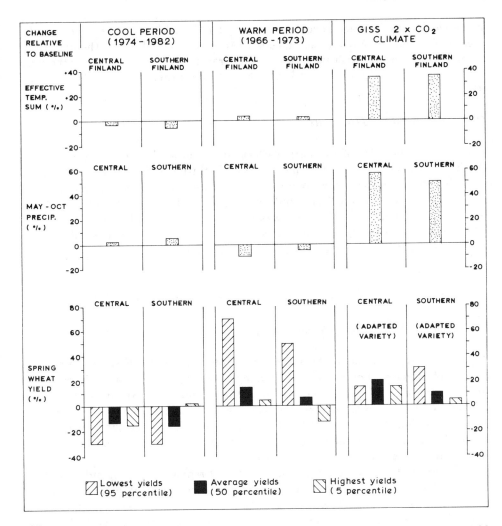

Figure 1.9. The influence of climatic variations on different levels of spring wheat yield in central and southern Finland. Three yield levels are represented: lowest yields (95 percentile), average yields (50 percentile) and highest yields (5 percentile). Yields are relative to the 1959–1983 baseline. (Based on information in *Figures 3.6–3.8*, Part IV.)

above in Subsection 1.5.1, it is clear that climatic variations can be especially important in marginal farming areas where the risk of failure is already significant. In some of the experiments reported here, the level of risk is referred to as the level of *crop certainty*. An example of this is presented in the case study in Iceland, where the potential area of successful barley ripening is expressed in terms of the number of lowland meteorological stations at which barley would ripen in 60% of years, and in 90% of years (*Appendix 1.3*).

Changes in yield quality

While yields may increase in some regions under the GISS $2 \times CO_2$ scenario, the *quality* of output may decrease. This is the case in Iceland where the digestibility of grass decreases with increases in temperature, thus reducing the otherwise marked gains in output on both hayfields and unimproved rangelands (*see* Part III).

Changes in sensitivity to inputs

One of the difficulties of estimating effects of climatic change on agriculture is that the sensitivity of yield to inputs such as fertilizers and pesticides also varies with climate. In Iceland, for example, it is evident that hay yield would be more responsive to altered fertilizer inputs during anomalous climatic periods if the mean level of nitrogen fertilization was lower than at present. This means that adjusting levels of fertilization can be an effective stabilizing response in extreme years, a point that is considered further in Subsection 1.5.7).

1.5.4. Effects on the spatial pattern of agricultural potential

Spatial shifts of comparative advantage

Different crops growing in the *same* region often respond differently to a given change of climate. For example, assuming no change in rainfall, winter wheat yields in the Central (Moscow) Region of the European USSR are estimated to increase under a 1.5 °C increase in temperature while spring barley and oats yields are reduced (*Appendix 1.5*). In order to optimize agricultural output from the region there might therefore occur a substantial switch of crops, an adjustment which is explored further in Subsection 1.5.8.

Likewise, the *same* crop grown in *different* regions may respond differently to a similar change of climate. In northern Finland, for example, barley would benefit from a higher ETS without moisture shortage, whereas in southern Finland yields could actually decrease as a result of the same change in temperatures due to early summer moisture stress (*Appendix 1.4*). Comparable sensitivity analyses of spring wheat yields in the northern and southern parts of the European USSR indicate similarly contrasting responses: at Cherdyn (west of the Ural Mountains) yields increase with increasing temperature, while near Volgograd (some 1000 km farther south) they decrease (*Appendix 1.5* and Parry *et. al.*, 1988).

The combination of these two complexities of response (by different crops or varieties in the same region and by the same crops or varieties in different regions) would bring about substantial shifts in the comparative advantage which one crop or mix of crops would have over another. In the Central Region (USSR), for example, a 1.5 °C warming would suggest a strong move away from barley, oats and grassland and into winter wheat and maize for silage (*Appendix 1.5*).

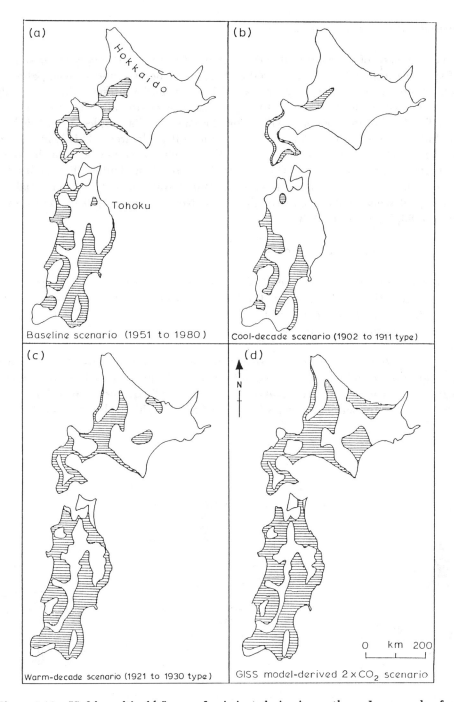

Figure 1.10. "Safely cultivable" area for irrigated rice in northern Japan under four climatic scenarios (from Part VI). The safely cultivable area (shaded) is defined by the minimum level of accumulated temperatures during the growing season required for the crop to complete its normal life cycle.

Spatial shifts of crop potential

A consequence of the factors described above, together with many other sources of spatial complexity, is that climatic change can be expected to bring about a spatial shift of crop potential. Areas which are, under present climatic conditions, judged to be most suited to a given crop or combination of crops or to a specified level of management will change location. In its simplest form this kind of shift can be seen as a shift in limits of the cultivable area. This is illustrated in *Figure 1.10* for northern Japan where the "safely cultivable" area for irrigated rice under the GISS $2 \times CO_2$ climate is more than double that under the present-day climate. Similar large-scale climate-related shifts of potential for maize and wheat have been investigated in North America (Blasing and Solomon, 1983; Rosenzweig, 1985).

Regional analogues as indicators of impact

Analogue regions for each case study area, based upon changes in temperature and precipitation assumed under the GISS $2 \times CO_2$ scenario, are shown in *Figure 1.11*. Regional analogues of this kind are useful for interpreting the GISS $2 \times CO_2$ scenario for several reasons:

(1) As effective illustrations of climatic change. As illustrations alone, regional analogues serve to highlight the magnitude of the future change in climate *within* regions in terms of the present-day differences in climate *between* regions. For example, the experiments reported elsewhere in this volume indicate that

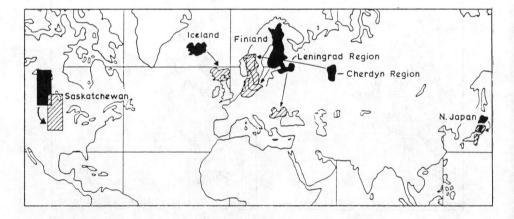

Figure 1.11. Present-day regional analogues of the GISS $2 \times CO_2$ climate estimated for the case study regions: Saskatchewan, Iceland, Finland, Leningrad and Cherdyn regions (USSR) and Hokkaido and Tohoku districts (Japan).

Iceland's climate under the GISS $2 \times CO_2$ scenario is similar to the climate of northeast Scotland today.

(2) As indicators of likely agricultural adaptations. Since agricultural activities have been developed and often adjusted by many generations of farmers to match local climatic resources, it is reasonable to assume that agriculture is, in most cases, closely tuned to present-day climate. Exceptions occur where economic, social and political forces have pushed certain farming types into regions which are climatically marginal to them. Elsewhere, however, present-day farming types in analogue regions are a useful indicator of the adaptive strategies required to retune agriculture to altered climatic resources (e.g., by the substitution of new crops, the altered allocations of land to various uses, adjustments in management practices, etc.). To illustrate, rice varieties at present grown in central and southern Japan, which was identified as an appropriate regional analogue for the GISS $2 \times CO_2$ climate in northern Japan (Hokkaido), were used in adjustment experiments with crop simulations for Hokkaido to evaluate their adoption as an appropriate adaptive response. These rice varieties yielded 18–21% more under the GISS $2 \times CO_2$ climate than did present-day Hokkaido varieties (*see* Part VI).

(3) As indicators of potential productivity. Current agricultural productivity in the analogue region may also be of a similar magnitude to that attainable under a changed climate in the study area. This possibility offers a new (often unique) opportunity to verify model estimates of crop response to climatic change. For example, a linear relationship has been found between national hay yields in Iceland during the period 1901–1983 and a temperature index based on seasonal temperatures. The average temperature index for the period 1931–1960 is 1.0, while the highest observed annual index is about 1.2. Under the GISS $2 \times CO_2$ scenario, however, the mean index rises to about 1.5, approaching the average present-day value in Northern Ireland (which is 1.74). If the yield/temperature relationship for Iceland is also valid for conditions in Northern Ireland, we could expect average hay yield under similar management to be about 74% higher in Northern Ireland than in Iceland. In fact, yields in Northern Ireland are in good agreement with this estimate, thus providing some justification for extrapolating the hay yield index to the $2 \times CO_2$ scenario conditions in Iceland (*see* Part III).

There are, of course, many difficulties with using regional analogues, due generally to regional variations in environmental, management and technological factors and in particular to differing responses to climatic variations by these various factors. Perhaps the most important general difficulty with respect to crop growth is the variation of *daylength* with latitude. Crop varieties at high latitudes are bred for short growing seasons with high photoperiods. Under $2 \times CO_2$ conditions, while substitution by varieties at present grown at lower latitudes would make sense from a climatic standpoint, these varieties might not be suited to the long daylengths of their new environment. For this reason, geographical shifts in agricultural potential under a CO_2-induced climatic change may well have important implications for plant breeding (*see* Subsection 1.7.5).

1.5.5. Complexity due to other concurrent effects

Further complexity stems from the relationship between climate and other environmental processes. There are, indeed, numerous other paths by which changes in climate can have an indirect effect on agriculture, largely through concurrent effects on other physical systems.

Effects on water resources

Major changes in water resources may result from changes in climate and have serious implications for agriculture. To illustrate, changes in precipitation in Japan assumed under the GISS $2 \times CO_2$ scenario may involve increases in winter snowfall, increased spring snowmelt and thus changes in water availability for rice irrigation (*see* Part VI). Changes in groundwater supply and frequencies of flooding are also sources of major indirect effects of climatic changes on agriculture considered in detail in the case studies (*see* Parts V and VI).

Effects on soil fertility

Increases in precipitation estimated for some regions under the GISS $2 \times CO_2$ scenario are likely to lead to increased soil leaching and a consequent reduction in soil nutrient status. In the Leningrad region of the USSR an index of soil fertility indicates falls in soil productivity of more than 20% by the year 2035 as a result of precipitation increases and assuming trend fertilizer applications. Large increases in fertilizer applications would be necessary to restore productivity levels (*Appendix 1.5*).

Effects on soil erosion

Changes in climate involving changes in the frequency of dry spells may affect rates of wind erosion. In Saskatchewan the frequency of moderate and extreme droughts is estimated to increase three-fold under the GISS $2 \times CO_2$ scenario (and thirteen-fold if increases in temperature are not accompanied by increases in precipitation). Estimates of changes in the potential for wind erosion vary from −14% to +29% according to the assumed changes in precipitation (*Appendix 1.2*).

Effects on the incidence of pests and diseases

Variations of climate can cause spatial shifts in the survival and development of pests and diseases, somewhat similar to the spatial shifts of agricultural potential discussed above. However, factors affecting the transport, migration and dispersal of pests and diseases require more extensive research before these effects can be effectively modeled. For this reason little attention has been given to them in the case studies reported here. Where, however, their effects have been estimated it is clear that the consequences are serious. For example, while blight and rust do little damage to potatoes and the few cereals that are grown in Iceland's cool climate at present, they could be expected to reduce yields much more under the warmer climate assumed for the GISS $2 \times CO_2$ scenario. This could make Iceland's climate similar to that in northeast Scotland today where cereal losses from diseases amount at present to 10–15% (*see* Part III).

1.5.6. Higher-order effects of climatic variations on agriculture

Effects on farm profitability

Assuming for the present that there is no change in costs of inputs, we can estimate the effect of yield changes on farm profitability by analyzing the effect on net returns per hectare. In southern Finland, for example, net returns to barley cultivation under the GISS $2 \times CO_2$ scenario are estimated to increase by about three-quarters and net returns of oats about three-fold (*Appendix 1.4*). Yet present-day returns to farming in Finland, as in much of Europe, are heavily dependent on government support, and it is unrealistic to believe that this level of support would continue, in particular because of the production surpluses that would ensue. Changes in government agricultural support as an appropriate response to climatic change are considered in Subsection 1.5.8.

Effects on regional production costs

Changes of yield may also affect production costs, particularly in centrally planned economics where regional production targets are fixed and where levels of inputs are often adjusted to counter weather-related variations of yield. In the Central (Moscow) Region of the USSR, for example, a $+1\,°C$ temperature change can reduce production costs of winter wheat and corn by 22% and 6%, respectively, while increasing production costs for cool-summer crops such as barley, oats and potatoes (*Appendix 1.5*).

Effects on regional and national food production

The aggregate effect of altered yields on regional or national food production is the outcome of an extremely varied set of interactions, because yields are altered by differing amounts according to crop type, soil type, level of input and type of management. Owing to the complexities and uncertainties of modeling these relationships it has thus far been possible only to provide estimates for single crops. The results from experiments in the present study indicate that in Saskatchewan provincial wheat production under the GISS $2 \times CO_2$ scenario falls by 18%, while in Japan annual rice production increases by about 5% (*Appendices 1.2 and 1.6*).

Effects on regional farm income

By using aggregated data on crop yield as inputs to farm simulation models, changes in regional farm income can be estimated. To illustrate, five farm simulation models were developed in the Saskatchewan study, one for each of five farm types. These were used to simulate commodity changes, net income and cash flow under different levels of crop production (estimated by a crop-climate model for the different climatic scenarios). These effects were then aggregated according to the type, size and dominant soil zone of farms, to give provincial totals representative of 65 000 farms in Saskatchewan. In this way, provincial farm incomes were estimated to be reduced by 78% in an extreme dry year (1961-type), by 25% under a dry period scenario (1933 to 1937-type), and by 7% under the GISS 2 × CO_2 scenario (*Appendix 1.2*). Varying margins of error accompany these estimates, details of which are given in Part II.

Effects on regional farm income disparities

Government policies of response in reducing price support for agriculture would need to be sensitive to the *regional* pattern of climatic effects. For example, in Finland under the GISS 2 × CO_2 scenario yields of barley and oats are estimated to increase both in the north and in the south of the country, but farmers in the south obtain greater increases in net return than farmers in the north. Moreover, the year-to-year variability of net return is lower in the south than in the north (*Appendix 1.4*). Thus, *ceteris paribus*, the south of Finland would further increase its advantage over the north for growing barley and oats. In contrast, the regional differences in estimated relative changes of spring wheat yields (higher in the north than in the south) suggest that the comparative advantage of cultivating spring wheat in the south may be reduced. This has major implications for regional planning because the low-productivity northern regions are at present heavily subsidized by the government (*see* Part IV). Adjustments in regional policies are considered in Subsection 1.5.8.

Effects on over-supply

In many of the cool temperate and cold areas covered in this volume agriculture is heavily and expensively subsidized by governments in order to maintain farm incomes and boost domestic food output. Without large-scale reductions in this support, it is likely that massive surplus production would occur in these areas under a warmer climate. In Japan, for example, the national rice stock is estimated to double under the GISS 2 × CO_2 scenario. That would leave the government with large surpluses of a rice variety that has a very limited market outside Japan and is at present produced at a cost well above the world level. It is clear, then, that government policy might need to be altered radically to reduce price support for rice farmers or to enforce the "set-aside" of farmland in Japan (*see* Part VI).

Higher-order effects on the wider economy

Changes in farm profitability can be expected to affect other, non-agricultural sectors of the regional economy. Continuing with the Saskatchewan example, an input–output model was used to estimate the effects of provincial agricultural commodity changes (determined by the farm simulations, described above) on other sectors of the provincial economy (e.g., marketing, transportation and provincial household purchasing power). Reasonable estimates of effects under scenarios of short-term climatic variations are possible using this model and assuming present-day relationships between sectors of the economy, but these become unreliable under scenarios of longer-term climatic change. For example, purchasing power falls by 38% under the extreme dry year scenario, by 13% under the dry period scenario and by 3% under the GISS 2 × CO_2 scenario (*Appendix 1.2*). Likely future adjustments in province-wide economic activity, which clearly would affect the last of these estimates, are considered in Subsections 1.5.7 and 1.5.8.

Effects on employment

Changes in agricultural production and profitability and the effects of these on the wider economy are likely, in addition, to affect the level of employment. In the Saskatchewan study, present-day statistical relationships are used to simulate changes in provincial employment as a function of changes in economic activity estimated for 12 sectors by the input–output model under different climatic scenarios. In an extreme dry year, agricultural employment is estimated to fall by about 9% and total provincial employment by 6%, whereas under the dry period scenario (1933 to 1937-type) the reductions are 3% and 2%, respectively (*Appendix 1.2*). Quite large margins of error accompany these estimates and these are likely to be greatly magnified for the estimates of effects under the GISS 2 × CO_2 scenario (reductions of 0.8% in agricultural employment and 0.5% in total employment). Substantial changes in employment structure can be expected to occur over this time-scale that are not considered in the simulations. An impact experiment of this kind merely gives an indication of the sensitivity of employment to climate, not a prediction of likely effects over the long-term. Further details are given in the Saskatchewan case study (Part II).

Parallel effects on other sectors

The focus of this volume is on the responses to climatic variations of a single economic sector, agriculture. But while the results of first-order "biophysical" impact experiments at the plant or crop level are largely separable from the concurrent first-order effects on other sectors (e.g., on forests, water resources, transportation, etc.), the higher-order effects and their various implications are not. For instance, the study of potential effects of the GISS 2 × CO_2 scenario on the productivity and economic profitability of northern boreal forests (Sections 5 and 6) indicates that the traditional balance between forestry and agriculture in many northern regions could be significantly altered under a 2 × CO_2 climate.

Specifically, the enhanced growth rates of boreal forests estimated by a simple biological model for the GISS $2 \times CO_2$ scenario would lead to increased supply of forest products on the world market. The economic benefits of this (estimated by an economic model of the global forest sector) would accrue primarily to the consumers of forest products, through lower prices. Some producer-countries would also benefit (e.g., Finland) while others, despite the increased productivity, may lose (e.g., Sweden).

From results such as these we can begin to bring together some of the hitherto disparate strands of climate impact assessment. For example, by weighing the estimated benefits of the GISS $2 \times CO_2$ scenario for Finnish forestry against the estimated effects of the same scenario on agriculture, some indications can be gained of the wider implications of CO_2-induced climatic change for national- and regional-level policies in Finland.

1.5.7. Potential farm level adjustments to climatic variations

The effects that have been summarized above assume no changes in technology, management or in background economic conditions (product prices, labor supply, etc.). This assumption is unrealistic, at least with respect to long-term changes in climate, because experience tells us that these factors are likely to change substantially over the kind of time-scale estimated for atmospheric CO_2 doubling. The estimates are not, therefore, a prediction of future effects but a *sensitivity analysis* of present-day agriculture, enabling identification of those aspects and areas which may be especially vulnerable to climatic variations. This form of analysis also provides a basis for the next step: the evaluation, firstly, of potential adjustments at the farm level and, secondly, of potential policy responses at the regional or national level. The former are considered here, the latter in Subsection 1.5.8.

The farm level adjustments that have been considered are largely those that could be put in place now, since this enables us to parameterize them and use them as inputs to the linked models. Assumptions about uncertain future developments in technology, demand and prices are much less easy to specify. The adjustments are of five types: of crop variety, of soil management, of land allocation, of harvesting efficiency and of purchases to supplement production.

Changes in crop variety

(1) Change from spring to winter varieties. Under the GISS $2 \times CO_2$ scenario, yields of spring-sown crops (e.g., wheat, barley and oats) would be reduced in some regions, owing to the increased frequency of moisture stress early in the growing period. However, a switch to winter wheat or, in some areas, winter rye would reduce damage by high evapotranspiration rates and, in addition, take advantage of the longer potential growing season (assuming that snow cover remained sufficient to protect the crop against winter kill). In Saskatchewan experiments with a switch of 10% of the cropped area from spring wheat to winter wheat indicated reduced overall wheat output in a normal year, but

increased output in a drought year (*Figure 1.12*). A next step would be to evaluate the effects over (say) 5 or 10 years of variable weather, to see which "mix" of varieties offer the highest aggregate production or (alternatively) the most stable income over a given period. Different planting strategies could then be devised to meet different farming goals.

(2) Change to varieties with higher thermal requirements. A logical way of exploiting longer and warmer growing seasons at high latitudes is to use later-maturing varieties with higher thermal requirements. In many cases the estimated yields of present-day varieties of crops such as spring wheat in N. USSR and early-maturing rice in northern Japan derive little benefit from a warmer 2 × CO_2 climate, and may even be reduced. Yet several experiments have shown that by substituting varieties with higher thermal requirements higher yields can be achieved (*Figure 1.13*). For example, in the Cherdyn region of the USSR, spring wheat varieties with thermal requirements 50 and 100 growing degree-days (GDD) greater than current varieties gave yields 16% and 26% above the baseline under the GISS 2 × CO_2 climatic scenario. These estimated yield increases were further enhanced when direct effects of atmospheric CO_2 on plant photosynthesis and water use were also simulated.

In central and southern Finland, the response of spring wheat varieties with thermal requirements 120 GDD above the present-day varieties was tested for the GISS 2 × CO_2 scenario. Again, average yields were increased significantly both in central Finland (by about 20%) and in southern Finland (by approximately 10%).

In northern Japan, simulated yields under the GISS 2 × CO_2 scenario of middle- and late-maturing rice were well above baseline values (+23% and +26%, respectively) and much improved on the yield increase estimated for present-day early-maturing varieties (+4%) (*Figure 1.13*).

(3) Change to varieties giving less variable yields. As discussed above, we know little about what changes could occur in the interannual variability of temperature and precipitation as a result of increasing atmospheric CO_2. Any increase could have a profound effect on yield security. But, even if we assume the same degree of variability under an altered climate, the effects on crop yields are considerable. It is possible, however, to test a number of varieties for stability of yield in order to reduce these effects. Our data from Hokkaido indicate that later-maturing varieties of rice would give higher *and* less variable yields under the GISS 2 × CO_2 climate (*Figure 1.13*).

The responses of different varieties to cool or warm summers under the changed climate are also of interest. In Japan, anomalously cool summers under the GISS 2 × CO_2 scenario would be warmer than their present-day counterparts and hence, with present varieties, they would provide higher yields, exceeding even the 2 × CO_2 average yields. However, later-maturing varieties with higher thermal requirements would not respond favorably to the coolest summers, even under the warmer 2 × CO_2 climate. In contrast, an anomalously warm summer in the 2 × CO_2 climate would actually lead to below-average

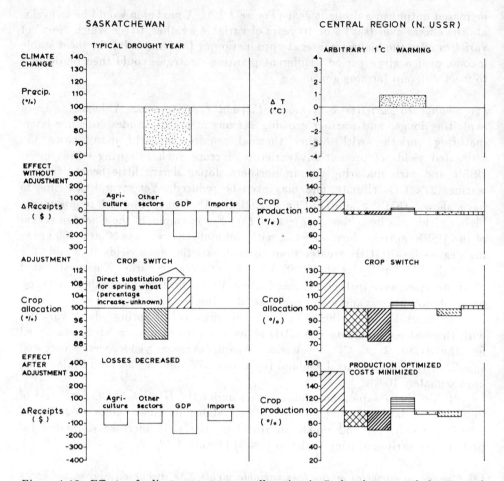

Figure 1.12. Effects of adjustments to crop allocation in Saskatchewan and the Central Region (N. USSR) and to fertilizer applications in Iceland, on agricultural receipts and productivity under different climatic scenarios (continued on facing page).

yields of present-day varieties, while the later-maturing varieties would be better equipped to exploit the warm conditions, giving above-average yields (*Figure 1.13*).

Similar results are presented for central and southern Finland in *Figure 1.13*. Yields of an adapted spring wheat variety would exceed those of present-day varieties across a range of conditions both in central and in southern Finland.

Changes in fertilizing and drainage

(1) Altered fertilizer applications. The present study includes two types of adjustment experiment with fertilizers. The first, for the Leningrad region (USSR), increased levels of fertilizer application to optimize winter rye yields

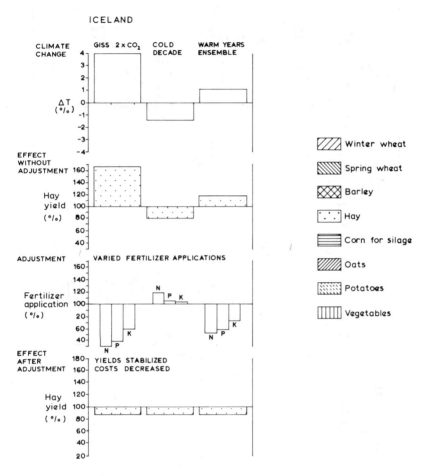

Figure 1.12. continued

under the GISS $2 \times CO_2$ scenario. While estimated yields for increased fertilizer applications based on projections of the present-day trend are lower than forecasted trend yields in the absence of a climatic change, a 50% increase in fertilizer applications would increase yields to levels above the trend (*Figure 1.13*). The second, for Iceland, varied fertilizer applications in order to maintain constant production levels. This indicated that it would be possible, by decreasing applications under warmer conditions and increasing applications in cool conditions, both to stabilize hay yields at a level only slightly lower than the present-day average and to reduce costs significantly (because long-term fertilizer use would be greatly diminished) (*see Figure 1.12*). Variable applications of fertilizer to stabilize yields by offsetting the effects of anomalously cool or warm summers are at present being tested for feasibility by the Icelandic government.

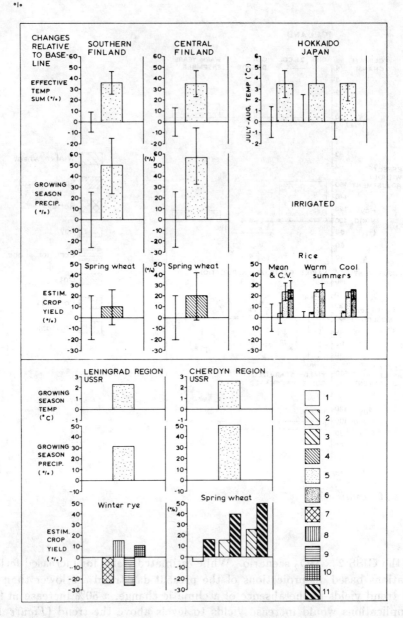

Figure 1.13. Estimated crop yields under the GISS 2 × CO_2 scenario for present-day and for adjusted crop varieties and management conditions. Model experiments in Finland, northern USSR and northern Japan. 1 = present variety; 2,3,4 = varieties with thermal requirements 50, 100 and 120 growing degree-days higher than present; 5,6 = newly introduced middle-maturing (Koshihikari) and late-maturing (Nipponbare) varieties with transplanting date 25 days earlier than present; 7 = present variety with technology trend projected to the year 2035: values relative to trend yields without climatic change; 8 = present variety with fertilizer applications 50% above those in 7; 9 = present variety with drainage activity increased by 2 km/km² above that in 7; 10 = combination of 8 and 9; 11 = includes "direct" effects of carbon dioxide.

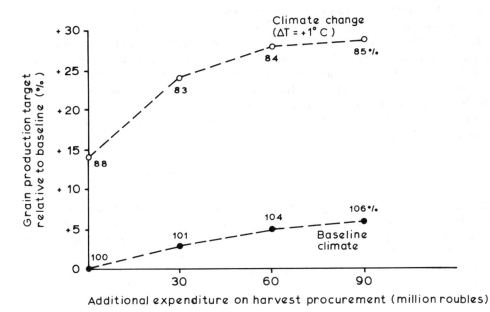

Figure 1.14. Cost of increased grain production in the Central region, N. USSR. Changes in total grain production are plotted in percentages relative to the baseline (vertical axis) as a function of additional expenditure (roubles) on harvesting capacity, equipment and labor (horizontal axis). The resultant costs per ton are shown on the graph as percentage values relative to the baseline for (*a*) the baseline climate (1931–1960) and (*b*) an arbitrary increase in mean annual temperature of +1 °C (Source: *Tables 4.5* and *4.7*, Part V).

(2) Improvements in soil drainage. Increased precipitation predicted in the GISS $2 \times CO_2$ experiment might be expected to lead to increased soil erosion, thus offsetting both the beneficial effects of a warmer climate and technological improvements. Improvements in soil drainage are an adjustment which was evaluated in the Leningrad region, N. USSR (*Figure 1.13*). The effect of these measures is to reduce yields slightly, an effect that would have to be weighed against reduced erosion and more efficient disposal of nitrate pollutants in the region, in order fully to assess overall benefit (for full details, *see* Part V).

Changes in farm expenditure

Even if a climatic change leads to higher crop yields, there may not be sufficient capacity to collect and store the greater output. Under these circumstances, increased expenditure on new equipment, additional labor and increased storage facilities may be economically justified in the light of the higher production obtained. For example, the cost effectiveness of increasing the efficiency of grain production in the Central Region around Moscow (USSR) is illustrated in *Figure 1.14* for baseline and warmer (+1 °C) climatic conditions. Under the baseline climate an increased expenditure on infrastructure of 30 million roubles (1 rouble =

ca. US$ 1 in 1986) could increase the harvested grain production by about 3%, but the average cost per unit of this increased production would rise by about 1%. Under the warmer climate, production simulated for baseline expenditure would increase by about 14% and would increase by another 10% with an additional investment of 30 million roubles. In this instance the average cost per ton of production would actually fall (from 88% to 83% of the baseline value). Thus the improved production potential fully justifies an increased expenditure under the warmer climate, while the equivalent increase in expenditure under present climatic conditions would not be justified.

Changes in off-farm purchases required to supplement farm production

Off-farm purchases, which represent both real costs to the farmer and opportunity costs to the agricultural sector, can also be affected. For example, the cost of bought-in feed required to offset the reductions in herbage yield estimated for a cool decade in Iceland would be, on average, 29–38% of a farmer's net income. In a warmer climate, the deployment of this "saved" expenditure elsewhere offers one range of adjustments that has yet to be fully explored (*see* Part III).

1.5.8. Potential regional and national policy responses to climatic variations

In addition to the farm level adjustments in management and technology described above, a number of potential regional and national policy responses were evaluated.

Changes in regional land-use allocation

(1) Changes of land use to optimize production. Because different crops respond differently to changes of climate and to varying levels of fertilizer application under those climates, any attempt to maximize output of each crop while minimizing production costs is likely to identify quite different allocations of land to alternative crops under different climates. For example, in the Central Region (USSR), an "optimal" land use for a climate that is on average 1 °C warmer than the present one would have an expanded area under winter wheat, maize and vegetables and a reduced area under northern temperate-zone crops such as spring-sown barley, oats and potatoes (*Figure 1.12*). This pattern of land use begins to resemble that at present found farther south in the USSR and points again to the value of using regional analogies to identify possible responses to climatic change (*see* Subsection 1.5.4 above).

Broadly similar experiments in Iceland have investigated the increase in carrying capacity of both improved grassland and unimproved rangeland under the GISS 2 × CO_2 scenario. Assuming unaltered technology, carrying capacity increases by two-thirds or more. Given that inputs are unchanged, estimates show that 12% increases in carcass weight of sheep could be obtained at the 1980 stocking levels, or alternatively the 1980 capacity levels could be maintained with feed requirements reduced by 53% (*see Appendix 1.3*).

(2) Changes of land use to stabilize production. Some experiments, still in progress in Saskatchewan and not reported in this volume, consider the efficacy of removing marginal cropland from production as a means of drought mitigation. Wheat crops on this land tend to be profitable in years of normal or above-normal rainfall but can cause major losses in dry years. When the marginal cropland is "converted back" in modeling experiments from wheat to pasture for beef cattle the result is a reduction in the interannual variability of provincial farm income due to weather, with only small losses in overall production when averaged over a span of weather-years.

Changes in national agricultural policy

(1) Changes of policy to maintain national food security while avoiding oversupply. The effect of price support policies in many countries is to encourage home production often at prices well above that of the world market. The case studies of Finland and Japan in this volume indicate that major national surpluses of foodgrains can be expected to result from increases in growing season temperatures. These would be an unnecessary expense on the national exchequers and create a problem of disposal. Radical changes of policy would be needed to avoid these circumstances, in particular the reduction of price support for cereal farmers and the introduction of set-aside programs to encourage farmers to take cropland out of production (*see* Parts IV and VI).

(2) Changes of policy to maintain equitable regional farm incomes. Because variations of climate have differing degrees of effect in different regions (*see* Subsection 1.5.1 above), the present regional pattern of farm incomes is likely to alter. As a result, government policies designed to reduce regional discrepancies may need substantial revision in order to maintain a level of equitable support. In general, where northern regions (such as north Finland), which at present enjoy a high level of government support, would experience marked increases of agricultural potential as a result of increased growing season temperatures, farm support would need to be reduced accordingly.

(3) Changes of policies supporting farm inputs. Where national farm policies tend to encourage inputs such as fertilizers and improved drainage these can be retuned to encourage new levels of input appropriate for the altered levels of agricultural potential. In addition, further support may be needed in traditional areas of agricultural extension such as in land management (e.g., instituting new soil management practices to control erosion), water management (e.g., improving efficiency of water use) and pest control (e.g., adopting resistant crop varieties and cultivation practices).

1.6. Summary of Specific Results

Figures 1.15–1.19 summarize, for Canada, Iceland, Finland, N. USSR and Japan, estimates of the effects of three types of climatic variation: an extreme year, an extreme decade and a long-term change (here represented by outputs from the GISS GCM 2 × CO_2 experiment). *Appendices 1.2 to 1.6* list the more detailed results. In both figures and appendices results are arranged according to the hierarchial structure of the experiments adopted in the project. From top to bottom they contain values for the regional climatic scenarios, for estimated first-order effects, and for estimated higher-order effects. Estimates are relative to the baseline period, (i.e., to the baseline climate 1951–1980, or to crop yields simulated for that climate, or to actual yields, production, income, etc., for that period). In the impact experiments management and technology are held constant at *ca.* 1980 levels, although they are varied in the adjustment experiments.

1.6.1. The impact experiments: A summary of estimated effects without adjustments

We summarize here the results of the impact experiments conducted for the five case studies. This set of experiments assumes that the effects occur without adjustment in the agricultural system. The range of adjustments available for reducing negative effects or enhancing positive ones has been outlined in Subsection 1.5 above. Further details of the results are given in *Figures 1.15–1.19* and in *Appendices 1.2–1.6*. Short definitions of some of the measures of impact are given before the appendices. For a full assessment of the experiments the reader is referred to the individual case study reports in this volume. Unless otherwise stated all changes are relative to the 1951–1980 baseline.

Southern Saskatchewan (see Figure 1.15)

(1) *Under the extreme dry year (1961-type) scenario:* growing season precipitation totals are 36 to 71% below the 1951–1980 baseline (depending on location); growing season temperature is 1.3°C to 3.5°C above the baseline; effective temperature sum (ETS) increases 12 to 18%; precipitation effectiveness decreases 23 to 40%; wind erosion potential varies from −15% to +367%, according to location; the climatic index of agricultural potential decreases 53 to 84%; spring wheat yields decrease 74 to 79%, according to soil zone; total provincial wheat production falls 76%; average farm household income falls 78%; total employment in the province falls 6%; total provincial household purchasing power decreases 38%.

(2) *Under the extreme dry period (1933 to 1937-type) scenario:* growing season precipitation varies from 47% below to 7% above the 1951–1980 baseline (depending on location); growing season temperatures are 0.0°C to 2.3°C above the baseline; ETS increases 3 to 16%; precipitation effectiveness decreases 21 to 27%; wind erosion potential varies from −57% to +123%,

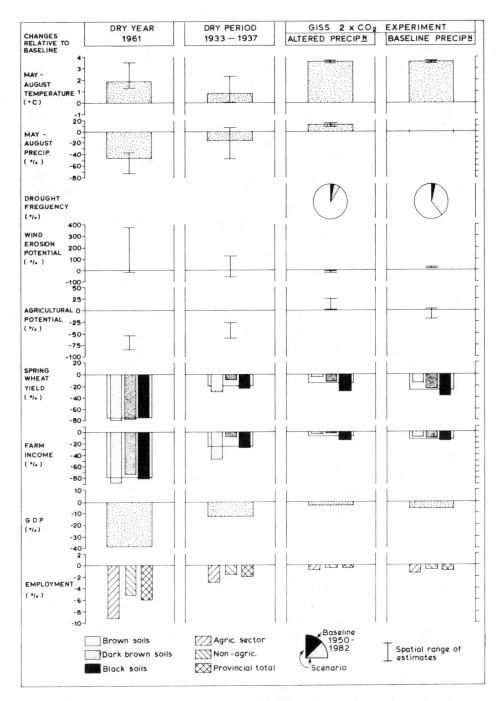

Figure 1.15. Estimated effects of climatic variations on agricultural production in Saskatchewan. Baseline climate is 1951–1980, unless otherwise indicated. Broad bars represent provincial mean values. (Sources: *Appendix 1.2* and Part II.)

according to location; the climatic index of agricultural potential decreases 26 to 60%; spring wheat yields decrease 9 to 29%, according to soil zone; total provincial wheat production falls 20%; average farm household income decreases 25%; total employment in the province falls 2%; total provincial household purchasing power decreases 13%.

(3) *Under the GISS 2 × CO_2 scenario with temperature and precipitation adjusted:* growing season precipitation totals are 9 to 14% above the 1951–1980 baseline; growing season temperatures are 3.5 °C to 3.6 °C above the baseline; ETS increases 48 to 53%; precipitation effectiveness increases 5 to 12%; drought frequency increases ×3; wind erosion potential decreases 3 to 14%; the climatic index of agricultural potential increases 1 to 30%; spring wheat yields decrease 4 to 29%, according to soil zone; total provincial wheat production decreases 18%; average farm household income decreases 7%; total employment in the province falls 0.5%; total provincial household purchasing power decreases 3%.

(4) *Under the GISS × CO_2 scenario with only temperature adjusted:* growing season temperatures are 3.5 °C to 3.6 °C above the baseline; ETS increases 48 to 53%; precipitation effectiveness decreases 10 to 12%; drought frequency increases ×13; wind erosion potential increases 24 to 29%; the climatic index of agricultural potential varies from −18% to +3%; spring wheat yields decrease 15 to 37%, according to soil zone; total provincial wheat production falls 28%; average farm household income decreases 12%; total employment in the province falls 1%; total provincial household purchasing power falls 6%.

Iceland (see Figure 1.16)

(1) *Under the cold-decade (1859 to 1868-type) scenario:* mean annual temperature is 1.3 °C below the 1951–1980 baseline; onset of growing season is delayed 13 days; the area with potential for tree growth is almost nil; hay yields decrease 19%; pasture yields decrease 16 to 17%, according to the level of fertilization; sheep carrying capacity of improved grassland decreases 34%; of rangeland decreases 19 to 21%; average lamb carcass weight decreases 4%; sheep feed requirements increase 22%.

(2) *Under the 10 warmest years scenario:* mean annual temperature is 1.1 °C above the 1951–1980 baseline; onset of growing season is brought forward 7 days; hay yields increase 18%; pasture yields increase 16 to 17%, according to the level of fertilization; sheep carrying capacity of improved grassland increases 40%; of rangeland increases 18%; average lamb carcass weight increases 3%; sheep feed requirements decrease 16%.

(3) *Under the GISS 2 × CO_2 scenario:* mean annual temperature and precipitation are 4.0 °C and 15% above the 1951–1980 baseline; onset of growing season is brought forward 48 days; hay yields increase 66%; pasture yields increase 49 to 52%; sheep carrying capacity of improved grassland increases 253%; of rangeland increases 64 to 66%; lamb carcass weight increases 12%; sheep feed requirements decrease 53%.

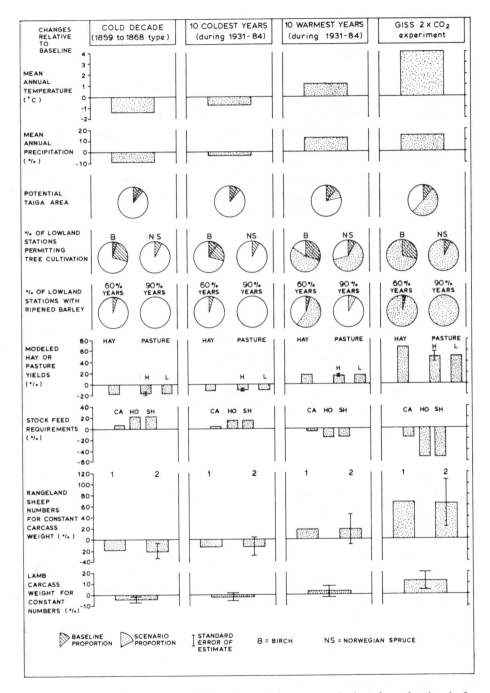

Figure 1.16. Estimated effects of climatic variations on agricultural production in Iceland. Baseline climate is 1951–1980. H = high input (120 kg/ha nitrogen); L = low input (80 kg/ha); Ca = cattle; Ho = horses; Sh = sheep; 1 = from national model; 2 = from refined national model. (Sources: *Appendix 1.3* and Part III.)

Finland (see Figure 1.17)

(a) *North/Central Finland*

(1) *Under the cold-period (1974 to 1982-type) scenario:* mean growing season precipitation is 2% above the 1959–1983 baseline; ETS decreases 2%; barley yields decrease 1%; spring wheat yields decrease 12%.
(2) *Under the warm-period (1966 to 1973-type) scenario:* mean growing season precipitation is 8% below the 1959–1983 baseline; ETS increases 3%; barley yields increase 3%; spring wheat yields increase 15%.
(3) *Under the GISS $2 \times CO_2$ scenario:* ETS and growing season precipitation are 35% and 59% above the 1959–1983 baseline; barley yields increase 14%; spring wheat yields increase 20% (using new varieties)

(b) *Southern Finland*

(1) *Under the cold-period (1974 to 1982-type) scenario:* mean growing season precipitation is 7% above the 1959–1983 baseline; ETS decreases 5%; barley yields increase 4%; spring wheat yields decrease 15%.
(2) *Under the warm-period (1966 to 1973-type) scenario:* mean growing season precipitation is 3% below the 1959–1983 baseline; ETS increases 2%; barley yields decrease 8%; spring wheat yields increase 10%.
(3) *Under the GISS $2 \times CO_2$ scenario:* ETS and mean growing season precipitation are 37% and 50% above the 1959–1983 baseline; barley yields increase 9%; spring wheat yields increase 10%.

(c) *All Finland*

(1) *Under the cold-decade (1861 to 1870-type) scenario:* net return on barley decreases 31% in the south; national farm income falls by 7 to 10%, according to the impact model used.
(2) *Under the warm-decade (1931 to 1940-type) scenario:* net return on barley increases 42% in the south and decreases 18% in the north; national farm income varies from −7 to +16%, according to the impact model used.
(3) *Under the GISS $2 \times CO_2$ scenario:* net return on barley increases 77% in the south and 47% in the north; national farm income increases 12 to 27%, according to the impact model used.

N. USSR (see Figure 1.18)

(a) *Leningrad Region*

(1) *Under the $+0.8°C$ scenario:* mean growing season precipitation is 5% above the 1951–1980 baseline; winter rye yields increase 31% (but only 4% relative to trend).

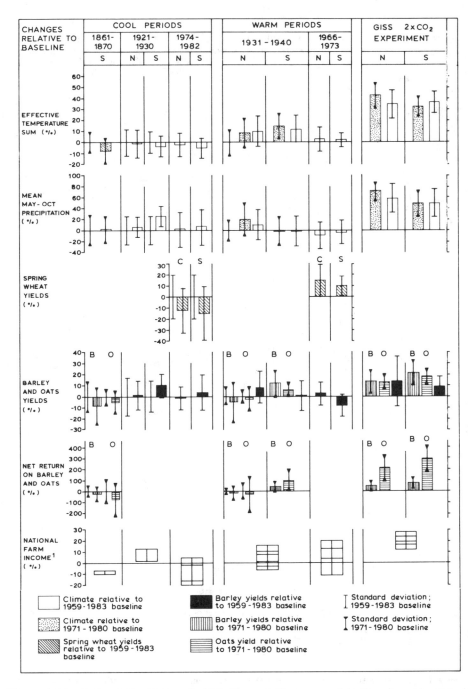

Figure 1.17. Estimated effects of climatic variations on agricultural production in northern (N), central (C) and southern (S) Finland. Baseline climate is 1959–1983 or 1971–1980, as shown. 1 = estimated farm income based on yield estimates from all regions. (Sources: *Appendix 1.4* and Part IV; J. Mukula, personal communication.)

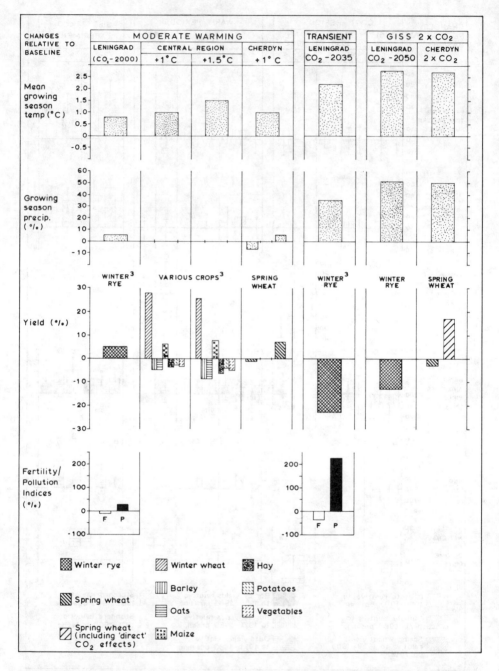

Figure 1.18. Estimated effects of climatic variations on agricultural production in the northern European USSR. Baseline climate is 1951–1980 for the Leningrad and Cherdyn regions and 1931–1960 for the Central Region. 1 = climatic change estimated using an empirical method (Vinnikov and Groisman, 1979); 2 = assumed date of CO_2-doubling; 3 = relative to technology trend. (Sources: *Appendix 1.5* and Part V.)

(2) *Under the GISS $2 \times CO_2$ scenario:* mean May–October temperature and precipitation are 2.2 °C and 36% above the 1951–1980 baseline; winter rye yields increase 6% (but decrease 22% relative to trend).

(b) *Cherdyn Region*

(1) *Under the $+1°C$ scenario:* growing season precipitation is unaltered from the 1951–1980 baseline; spring wheat yields increase 3%.
(2) *Under the GISS $2 \times CO_2$ scenario:* mean May–September temperature and precipitation are 2.7 °C and 50% above the 1951–1980 baseline; spring wheat yields decrease 3%.

(c) *Central Region*

(1) *Under the $+1°C$ scenario:* annual precipitation is unaltered from the 1931–1960 baseline; trended winter wheat yields increase 33%; trended barley yields decrease 1%.
(2) *Under the $+1.5°C$ scenario:* annual precipitation is unaltered from the 1931–1960 baseline; trended winter wheat yields increase 30%; trended barley yields decrease 4%.

Japan (see Figure 1.19)

(a) *North Japan*

(1) *Under the "cool-year" (1980-type) scenario:* mean July–August temperature is 1.6 °C below the 1951–1980 baseline; effective accumulated temperature (EAT) decreases 14%; rice cultivation limit falls 250 m; rice yield index decreases 40%; district rice supply falls 36%.
(2) *Under the "warm-year" (1978-type) scenario:* mean July–August temperature is 2.5 °C above the 1951–1980 baseline; EAT increases 25%; rice cultivation limit rises 400 m; rice yield index increases 15%; district rice supply increases 18%.
(3) *Under the GISS $2 \times CO_2$ scenario:* mean July–August temperature is 3.5 °C above the 1951–1980 baseline; EAT increases 35%; rice cultivation limit rises 560 m; rice yield index increases 5%; district rice supply increases 17%.

(b) *Central Japan*

(1) *Under the "cool-year" (1980-type) scenario:* mean July–August temperature is 2.6 °C below the 1951–1980 baseline; EAT decreases 21%; rice cultivation limit falls 410 m; rice yield index decreases 48%; district rice supply falls 18%.
(2) *Under the "warm-year" (1978-type) scenario:* mean July–August temperature is 2.3 °C above the 1951–1980 baseline; EAT increases 20%; rice cultivation limit rises 370 m; rice yield index increases 8%; district rice supply increases 10%.

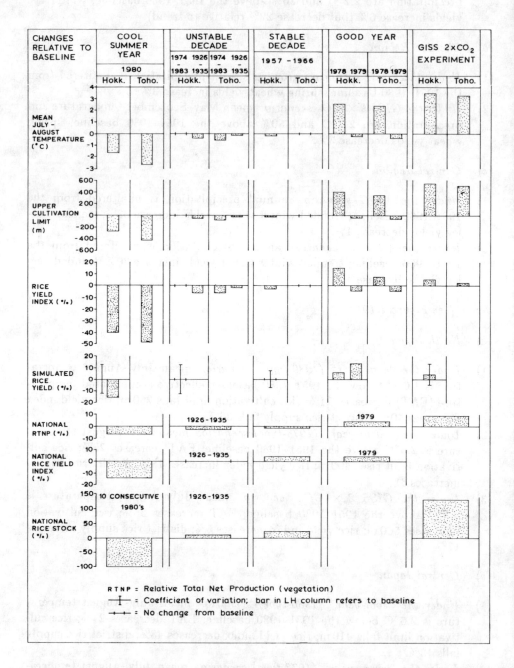

Figure 1.19. Estimated effects of climatic variations on agricultural production in Hokkaido (Hokk.), Tohoku (Toho.) and all Japan. Baseline climate is 1951–1980, unless otherwise indicated. (Sources: *Appendix 1.5* and Part VI.)

(3) *Under the GISS 2 × CO_2 scenario:* mean July–August temperature is 3.2 °C above the 1951–1980 baseline; EAT increases 29%; rice cultivation limit rises 510 m; rice yield index increases 2%; district rice supply increases 7%.

(c) *All Japan*

(1) *Under the "cool-year" (1980-type) scenario:* net biomass production decreases 7%; rice yields decrease 13%; national rice supply falls 15%; over 10 consecutive occurrences of a similarly cool year, national rice stocks are exhausted.
(2) *Under the "warm-year" (1979-type) scenario:* net biomass production increases 3%; rice yields increase 3%; national rice supply is unchanged.
(3) *Under the GISS 2 × CO_2 scenario:* net biomass production increases 9%; national rice supply increases 6%, over a 16-year period national rice stock triples.

1.6.2. Analysis of the results

In this subsection we analyze the results of the first-order impact experiments, particularly the assessments of crop yield responses to climatic variations. The results can be divided into two types:

(1) Those illustrating the sensitivity of agriculture to present-day climatic variations (based on the instrumental scenarios and on a range of model sensitivity tests).
(2) Those indicating the possible effects of selected longer-term climatic changes that are beyond the range of recorded variations (based on the GISS 2 × CO_2 scenario).

In the following subsection (1.6.3) we compare the magnitude of effects of the 2 × CO_2 climate with that of present-day climatic variations, including a discussion of higher-order effects where these have been estimated.

The sensitivity of agriculture to present-day climatic variations

The results of individual scenario experiments are depicted for each case study in *Figures 1.15–1.19*. In addition, to highlight the sensitivity of crop yields to variations in temperature (a key variable, given the northerly location of the case study regions), the graphs in *Figure 1.20* compare the modeled responses of different crop types to changes in growing season temperatures alone (with other factors such as precipitation and solar radiation assumed fixed at their baseline values). Some general observations can be noted:

(1) *A positive yield/temperature relationship.* Yields of most of the crops studied display a positive relationship with growing season temperature within

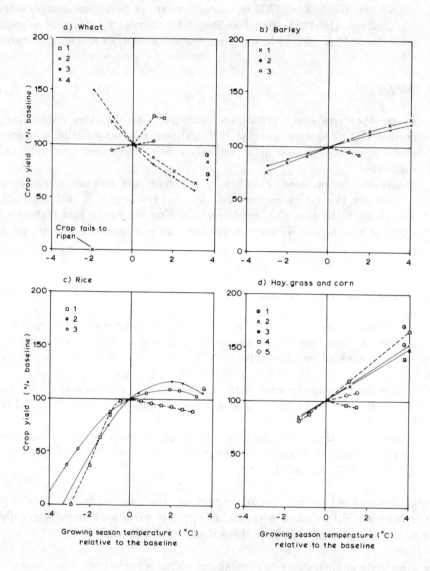

Figure 1.20. Modeled crop yield responses to temperature with other climatic variables assumed fixed at baseline values. Each plot represents estimates by a particular model for a specific geographical region. (a) Wheat: 1. Winter wheat, Central Region (USSR); 2. Spring wheat, Cherdyn Region (USSR); 3,4. Spring wheat, Saskatchewan (crop districts 6b and 7a). (b) Barley: 1. Northern Finland; 2. Southern Finland; 3. Central Region (USSR). (c) Rice: 1. Hokkaido (simulation model); 2. Hokkaido (yield index); 3. Tohoku (yield index). (d) Hay, grass and corn: 1. Hay, Iceland (national); 2,3. Pasture, Iceland (high and low inputs); 4. Hay, Central Region (USSR); 5. Corn (maize) for silage, Central Region (USSR). Solid curves represent empirical relationships that are described by continuous functions. Dashed curves are inferred relationships from individual model runs. The letter G adjacent to plotted points denotes responses to temperature changes under the GISS $2 \times CO_2$ scenario.

the normal range of interannual temperature variations (especially for below-average temperatures. This emphasizes the limiting effects of temperatures in many of these northern marginal environments.

(2) *The importance of other climatic factors.* In reality, climatic variations involve more than just the temperature changes depicted in *Figure 1.20*. Indeed, several of the plots could be construed as potentially misleading if considered in isolation. For example, rice yields in Japan are extremely sensitive to variations in solar radiation, while precipitation variations have a significant effect on yields of barley in Finland, on spring wheat yields in the Cherdyn region, USSR, and (overridingly) on spring wheat yields in Saskatchewan.

(3) *Differences between model estimates.* Different models may give apparently contradictory results for the same crop and region. This is illustrated for rice yields in Hokkaido [see *Figure 1.20(c)*]. According to a simple index relating interannual temperature variations to rice yields, above-average temperatures are beneficial to rice yields over a wide range of conditions. In contrast, results from a simulation model imply that positive departures from the same 1951–1980 baseline temperatures are detrimental to yields. The discordant results here are a consequence of (a) the inter-correlation between a modeled climatic variable in the yield index (temperature) and an unmodeled variable (solar radiation), and (b) the smoothing of long-term average climatic data used in the simulation model. In other examples, discrepancies may be less easily explained.

(4) *The influence of non-climatic factors.* Even if the location and climatic characteristics of two regions are similar, other non-climatic factors may interact with climatic variations to produce markedly different yield responses to climate. For example, a higher relative sensitivity to temperature is observed for pasture yields in Iceland at lower levels of fertilizer application [*Figure 1.20(d)*]. This has important implications for using varied fertilizer applications to stabilize annual yields and to maintain stocking levels.

(5) *Carry-over effects.* It is important to examine the effects of extreme weather-years in the context of the climate in the preceding years. One example is the case of two consecutive years (1936 and 1937) with similar warm, dry conditions in southern Saskatchewan (*Appendix 1.2*). Estimated average spring wheat yields are reduced relative to the baseline in both years, but the reduction is nearly three times as large in 1937 (55%) as in 1936 (20%). This is because the modeled soil moisture reserves in early 1936 (following a relatively cool and wet year in 1935) are considerably greater than in early 1937 (following the 1936 drought).

Estimated effects under the GISS $2 \times CO_2$ scenario.

Results of experiments with the GISS $2 \times CO_2$ scenario are summarized in *Figure 1.21*. As noted above, the direct effects of doubled atmospheric carbon dioxide concentrations were not considered in this project. It is also important to restate that the GISS model estimates are crude in spatial and temporal

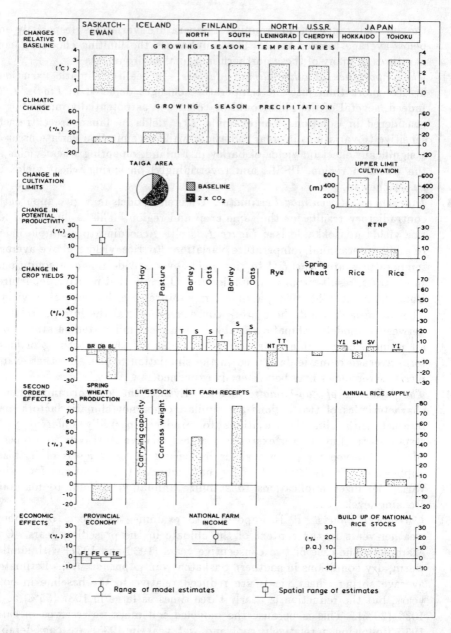

Figure 1.21. Effects under the GISS $2 \times CO_2$ scenario: regional changes in climate and their estimated effects on agricultural potential, on crop yields, on livestock and on the regional economy (relative to the baseline). RTNP = relative total net production of vegetation (national); BR = brown soils; DB = dark brown soils; BL = black soils; T = temporal model; S = spatial model; NT = no trend; TT = technology trend; SV = simulation model for variable climate; SM = simulation model for mean climate; YI = yield index; CC = carrying capacity; CW = carcass weight; FI = farm income; FE = farm employment; G = GDP; TE = total employment. For information on baselines, see *Appendices 1.2–1.6*.

resolution, unrealistic in their representation of present-day climate and thus highly uncertain in their representation of a $2 \times CO_2$ climate. They are regarded here not as predictions, but as plausible scenarios of future climate that are useful in extending our understanding of the *methods* of estimating biophysical and economic effects of climatic change.

The climatic changes are different from region to region as indicated by the temperature and precipitation plots. These differences reflect the differential regional climatic responses to a global atmospheric CO_2 doubling simulated by the GISS GCM, as well as different methods of interpreting the GISS model outputs in each region (*see Appendix 1.1*). Nevertheless, in *all* regions the $2 \times CO_2$ scenario projections are of substantially higher temperatures accompanied in most regions by significant increases in precipitation relative to the 1951–1980 baseline period. No other climatic variables are adjusted in the scenario, so where they are required for impact experiments, they are usually fixed at baseline values (*see* Section 2 for further details).

The main effects on crop yields to be noted from *Figure 1.21* and the appendices are as follows:

(1) *Southern Saskatchewan.* Spring wheat yields are reduced on average by 18%, owing to water stress induced by the higher temperatures that is not fully offset by increased precipitation. The impacts vary between soil zones, an effect that is largely conditioned by the baseline climatic conditions rather than the soil properties themselves, for the latter are not modeled explicitly in this application. The effects of the GISS $2 \times CO_2$ scenario assuming baseline precipitation are more severe, with an average yield decrease of 28% estimated (*see Figure 1.15*).

(2) *Iceland.* Hay and pasture yields increase markedly under the GISS $2 \times CO_2$ scenario, indicating the high sensitivity of these crops to temperature conditions in this marginal environment. Precipitation effects are not accounted for in the model experiments because these are of minor significance for herbage yields in most years under present climatic conditions, and because the scenario estimates imply a large precipitation increase (21%) accompanying the 3.7°C temperature increase. This indicates that even under a warmer climate moisture stress may be a limiting factor only in exceptional years. Some justification for this assumption, and for the magnitude of the estimated yield changes, is provided by information on grass growth in northern Britain.

(3) *Finland.* Barley and oats yields increase but not as substantially as they might with lower precipitation amounts. There are no consistent differences in percentage increases between north and south – these depend on the models used. However, the model results all indicate that the coefficient of variation of barley and oats yields would increase in N. Finland but decrease in S. Finland (*Appendix 1.4*).

(4) *N. USSR.* Winter rye yields in the Leningrad region decrease, both for present technology (relative to baseline yields) and for trend fertilizer applications under the changed climate (relative to yields for trend fertilizer applications under the baseline climate) (*Figure 1.18*). The reduced yields

are mainly attributable to high precipitation totals under the GISS 2 × CO_2 scenario, both through short-term waterlogging and harvesting difficulties, and because of longer-term impoverishment of soil fertility. Spring wheat yields in the Cherdyn region (N. USSR) also decrease slightly (probably owing to moisture stress and to the premature development of the crop under the increased temperatures). Although outside the general scope of this volume, the direct effects of doubled concentrations of atmospheric carbon dioxide on yields were also considered along with the climatic effects. Estimates suggest a substantial (17%) increase in spring wheat yields when both factors are combined.

(5) *Northern Japan.* Rice yields increase slightly under the GISS 2 × CO_2 scenario. Estimates were for average 2 × CO_2 temperatures using a yield index for Hokkaido and Tohoku districts, and for year-to-year variations in 2 × CO_2 temperatures using a simulation model, also for Hokkaido. In spite of differences between the two models, the results undoubtedly reflect the reduced risk of cool summer damage to rice that would probably be associated with increased average temperatures. This effect is highlighted by a 27% reduction in the coefficient of variation of yields estimated by the simulation model for Hokkaido.

1.6.3. 2 × CO_2 *versus* present-day climatic anomalies: A comparison of their effects on agriculture

A fair degree of caution is necessary when comparing impact estimates for different climatic scenarios. In particular, an important distinction should be drawn between the *extreme year* and the *period-averaged* climatic scenarios. The former are part of the interannual variability around a given mean condition while the latter represent changes in the mean condition, the meteorological mechanisms of which may be quite different from those causing year-to-year fluctuations. For example, the reasons for the change to warmer conditions under the GISS 2 × CO_2 scenario in Saskatchewan are certainly different from those explaining the extremely warm, dry year 1961 even though locally the magnitude of change is similar. Moreover, altered average conditions are accompanied by their own extreme year events, which are important to consider because their individual effects on agricultural productivity may, when averaged, be quite different from the effects of a period-averaged change in climate.

Nevertheless, a comparison of estimated impacts under the GISS 2 × CO_2 scenario with those in recently experienced anomalous years or periods can give us some idea of the possible scale of future impacts and whether they lie within the range of recent experience. In particular it would indicate (a) how far present-day farming systems, which are resilient to some of the existing short-term climatic variability, can accommodate possible *future* longer-term changes in climate, and (b) how radical need to be the adjustments to farming systems in order to accommodate the future changes. Such a comparison yields the following conclusions:

(1) *In Saskatchewan* first-order effects under the GISS 2 × CO_2 scenario are broadly similar to those averaged for an extreme dry period (1933 to 1937-type), while higher-order effects are less severe. Total provincial wheat production under the 2 × CO_2 scenario falls by 18% (in the dry period by 20%), provincial employment by 7% (25%), and provincial purchasing power by 3% (13%) (*Appendix 1.2*). In contrast, effects under the extreme dry year (1961-type) are between three and eleven times as great as for the 2 × CO_2 scenario. A crude conclusion is that an *average* weather-year under the GISS 2 × CO_2 climate would be broadly similar in the scale of its first-order effect to that typical of the *anomalously* dry 1933 to 1937-type weather. The magnified higher-order effect of the 1933 to 1937-type weather relative to the GISS 2 × CO_2 conditions is probably attributable to the different spatial pattern of the yield changes under the two scenarios (*see Figure 1.15*). This analysis has, of course, assumed in both instances that management and technology are at 1980 levels. Appropriate adjustments in farming systems could well reduce the levels of impact. Once again we should emphasize that variability of weather around the GISS 2 × CO_2 average climate would almost certainly bring more severe effects in occasional extreme years.

(2) *In Iceland* effects under the GISS 2 × CO_2 scenario exceed those estimated for anomalously warm years in the instrumental record. This is due to the large degree of warming (an increase of 4 °C in mean annual temperature) estimated by the GISS model and also to the relatively low present-day variability inherent in Iceland's highly maritime climate. Thus, while variations in hay yields range from −13% to +18% for the 10 coldest years and 10 warmest years since 1931, under the GISS 2 × CO_2 climate they are estimated at +66% (*Appendix 1.3*). Similarly, while rangeland carrying capacity at present varies from −13% to +18% and sheep carcass weight from only −2% to +3%, under the GISS 2 × CO_2 climate they would increase by about 64% and 12%, respectively. With additional increases attainable from the introduction of new crops with high thermal requirements (e.g., wheat, sugar beet, vegetables), it is clear that a warming of the scale envisaged by the GISS model would confer substantial benefits in Iceland well beyond those experienced in anomalously warm years today.

(3) *In Finland* effects under the GISS 2 × CO_2 scenario exceed those for anomalously warm periods under present-day climate, though the difference is less marked than in Iceland. The latter is due partly to a weaker temperature/yield relationship for grain crops in Finland than for grass crops in Iceland, partly to the greater sensitivity of Finnish grain yield to precipitation changes, and partly to the greater interannual variability of temperatures in Finland than in Iceland. To illustrate, while effective temperature sums are 12 to 15% above normal in the south of the country in an exceptionally warm period (1931 to 1940-type), they are estimated to increase by 33 to 37% under the GISS 2 × CO_2 scenario (*Appendix 1.4*). Barley yields increase by 1 to 12% in the warm period, and by 9 to 21% under the 2 × CO_2 scenario. The percentage changes in

farmers' net returns (+42% in the warm period and +72% in the $2 \times CO_2$ climate) are greater than the yield changes because of the intrinsically low level of net return obtained at present. In contrast, the inferred levels of aggregated national farm income closely reflect the yield changes (*Figure 1.17*). In comparative terms, however, and as a rough approximation, all of the effects under the $2 \times CO_2$ scenario are about double those currently experienced during an abnormally warm run of years.

(4) *In the USSR* both winter rye yields in the Leningrad region (+4%) and spring wheat yields in the Cherdyn region (+7%) are above normal for small increases in growing season temperature and precipitation. However, conditions under the GISS $2 \times CO_2$ scenario appear detrimental. They may either be be too wet, leading to nutrient depletion and waterlogging (in the Leningrad region winter rye yields even before the assumed date of CO_2-doubling are 22% below trend yields), or they may be too warm, leading to premature crop ripening (in the Cherdyn region spring wheat yields are 3% below the baseline).

(5) *In Japan* effects under the GISS $2 \times CO_2$ scenario are of a similar magnitude to those estimated for an anomalously warm year (1978-type). Effective accumulated temperatures in Hokkaido in the warm year are 25% above normal, and under the $2 \times CO_2$ scenario, +35%. Effects on rice yields are +6% and +4%, respectively. Economically, however, the effects of the GISS $2 \times CO_2$ scenario are quite different from those of an anomalously warm year. The former implies a prolonged period of oversupply with a continuous build-up of rice stocks. In contrast, the latter represents a single year of oversupply that would normally be balanced against other years of deficit (*Figure 1.19*). Once again, therefore, it is misleading to compare the effects of single-year anomalies with those of long-term changes in mean climate. Not only would the new mean climate be characterized by its own range of anomalous years but, unlike the single years, it represents a new average level of agro-climatic resource to which farming systems can be retuned. Thus the introduction of later-maturing rice varieties, which would take advantage of the warmer and longer summers, are estimated to enable an increase of yields by more than a quarter (*Appendix 1.6*).

1.7. Conclusions and Recommendations

The results summarized here have been drawn from five substantial case studies in Saskatchewan (Canada), Iceland, Finland, the northern European USSR and Japan. For the accuracy, meaning and full context of the results, reference should be made to the case studies themselves.

The estimates reported in this volume are *not predictions* of future effects. Present-day uncertainties and inaccuracies in simulating the behavior of the world's climate and in evaluating the agricultural implications of climatic change do not permit realistic predictions to be made. Furthermore, we cannot forecast what technological, economic and social developments in agriculture will occur over the next half century. The estimates should therefore be considered as measures of the *present-day sensitivity* of agriculture to climate change. Moreover, as significant as the *numbers* presented in these estimates are the *methods* developed to achieve them, and the *insights* that have thus been gained with respect to the range of on-farm adjustments and larger-scale policy responses most appropriate for matching agriculture to an altered climate. The estimated effects, on-farm adjustments and potential policy responses are summarized below.

1.7.1. Possible effects of climatic variations (without farm level adjustments or policy responses)

In summarizing these effects we are assuming, for the present, that technology and management do not alter in response to variations of climate. Subsequently we shall relax this assumption. Many of the unrealities of the GISS $2 \times CO_2$ climate scenario, discussed in *Appendix 1.1*, should also be stressed.

Spatial complexities and non-linearities

For a number of reasons the effects of climatic variations are likely to vary considerably in their magnitude:

- The patterns of effect of climatic variations are likely to be geographically complex, firstly because the amounts of *absolute* change in, for example, temperature and precipitation will vary considerably from region to region, and secondly because their effect on agriculture is a function of change *relative* to the existing (baseline) climate which itself is also geographically varied.
- The relationships between climatic variations and crop growth, and especially between climatic variations and farming decisions, may be strongly *non-linear*. In certain regions small climatic variations can exceed a critical *threshold* that would require farming systems to alter radically.
- Relatively small changes in mean climate can bring about major changes in the frequency of anomalous (e.g., unusually warm or dry) years, even if there is no change in interannual variability around the mean. This can have a substantial effect on the stability of agricultural production.
- The effects of climatic variations on agriculture can be (a) *multiplicative*, where different elements in a farming system are each adversely (or each positively) affected, such that the combined result is substantial and (b) *cumulative*, where carry-over effects from one year to the next (such as in soil moisture or production surpluses) have an incremental influence on profitability.

Changes in yield and yield response

Variations in climate can affect agricultural output through

- Changes in length of the potential growing season and changes in plant growth rates leading to changes in the required growing period. This may result in a mismatch between crop requirements and climatic resources and require the substitution of crop varieties.
- Changes in mean yield. These vary from crop to crop, resulting in changes in their relative profitability which may lead to crop substitution.
- Changes in the variability of yields, which can occur even when no change in interannual climatic variability is assumed to accompany a change in mean climate.
- Changes in the level of crop certainty (i.e., the probability of a given yield), which may result from quite small changes of mean climate owing to the strongly non-linear relationship between changes in mean climate and the frequency of anomalous weather-years.
- Changes in yield quality, which may be unrelated to changes in the yield amount.
- Changes in the sensitivity of plants to differing levels of application of fertilizers, pesticides and herbicides.

Changes in the spatial pattern of agricultural potential

In addition to changes in yield at a given location, a long-term climatic change will induce varied responses in different plants and animals in different regions. This is likely to result in substantial spatial shifts of agricultural potential, particularly:

- Spatial shifts of comparative advantage, resulting in the widespread replacement of present-day crop types and varieties in certain regions.
- Spatial shifts of crop potential, where the physiological limits to certain crops may contract or be extended as a result of changes of climate.

The use of *analogue regions* to observe the consequences of such spatial shifts may be rewarding, particularly in order to indicate likely levels of productivity and likely agricultural adaptations.

Other first-order effects

The present study has focused largely on the effect of climatic variations on rain-fed agriculture. It has not considered, for example, what influence changes in the composition of the atmosphere might have on plant growth. Neither has it examined fully the indirect effects of climatic variations on agriculture, usually transmitted through concurrent effects on other environmental systems. Four paths of impact are likely to be important:

Aims, methods and summary of results 71

- Effects through changes in water resources that may assume a greater importance under a changed climate for irrigation or flood control.
- Effects through changes in soil fertility, particularly as a result of altered rates of leaching resulting from precipitation changes.
- Effects through changes in rates of soil erosion, again largely as a result of altered rates of precipitation.
- Effects through changes in the incidence of agricultural pests and diseases.

Higher-order effects

If we assume, for the present, that technology and management do not alter in response to a long-term change of climate, we may expect the most pronounced effects to be:

- Changes in farm profitability, owing to altered yield levels.
- Changes in regional production costs, particularly in centrally planned economies with regional production targets.
- Changes in regional and national food production, being the aggregate outcome of varying yield responses by different crops on different soil types, under different systems of management.
- Changes in regional farm income, as a result of altered levels of individual farm profitability.
- Changes in regional farm income disparities, also as a result of changes in the regional pattern of productivity and profitability.
- Changes in rates of over-supply due to regional changes in agricultural potential.
- Changes in regional economic activity, stemming from changes in production, incomes and investment in agriculture that affect the vitality of other sectors of the economy.
- Changes in employment that result from altered economic activity in both the agricultural and non-agricultural sectors.
- Changes in other economic sectors, which occur in parallel to changes in agriculture and should be examined alongside the agricultural effects in order to assess overall impact.

1.7.2. Some potential adjustments at the farm level

A number of "adjustment experiments" were conducted as part of the IIASA/UNEP project to identify those adjustments that represent the most appropriate responses to possible long-term changes of climate. These adjustments may be summarized as follows:

(1) *Changes in crop variety*, viz:
- Changes from spring to winter crop varieties, to reduce losses due to early summer moisture stress.

- Changes to varieties with higher thermal requirements, to increase yields in warmer growing seasons.
- Changes to varieties giving less variable yields, to reduce losses in anomalous weather-years and increase yield security.

(2) *Changes in fertilizer applications and drainage*, viz:
- Increasing fertilizer application, to maximize yields in warmer growing seasons.
- Decreasing fertilizer applications, to minimize input costs under more beneficial climatic conditions.
- Varying fertilizer applications to reduce interannual yield variability due to climate.
- Improving soil drainage, to reduce soil erosion that may result from altered levels of precipitation.

(3) *Changes in farm expenditure*, to match investment in machinery, storage facilities, labor, etc., to altered levels of production that are expected to result from variations of climate.

(4) *Changes in off-farm purchases*, required to supplement farm production that may alter as a result of variations of climate.

1.7.3. Some potential policy responses at the regional and national levels

In addition to the farm level adjustments described above, a number of large-scale changes in agricultural policy may also be necessary as responses to long-term changes in climate. These could include:

(1) *Changes in regional land allocations*, viz:
- Changes of land use to optimize production, in particular policies to encourage regional-scale substitutions of crop types in order to match actual cropping patterns to new spatial patterns of agricultural potential.
- Changes of land use to stabilize production, to reduce the interannual variability of production caused by short-term variations of climate.

(2) *Changes in national agricultural policy*, viz:
- Changes of policy to maintain national food security in the context of variations of climate, while avoiding costly over-supply.
- Changes of policy to maintain equitability of regional farm incomes in the context of inequities resulting from variations of climate.
- Changes to policies supporting farm inputs and land management, such as support policies for fertilizers, erosion control, water management, etc., to encourage appropriate on-farm adjustments to variations of climate.

1.7.4. Recommended research approaches

Five general research approaches have emerged from this study as valuable, complementary ways of assessing the possible effects of climatic variations:

(1) An *integrated approach* to climate impact assessment is needed in order to comprehend (a) the interactions between climatic variations and other changes in the physical environment (e.g., soil acidification, groundwater pollution, etc.) and (b) the interconnections between effects on various sectors of society and economy (e.g., on agriculture, energy, industry, etc.). This wide-ranging comprehension is necessary if we are to increase our understanding of the range of appropriate economic and social responses to climate.

(2) An *iterative approach*, usually within a model framework, is particularly useful, both for evaluating the first-order effects of climatic variations (e.g., for a number of different climatic scenarios) and for testing a range of adaptive responses to impacts (e.g., adjusted crop planting dates, altered crop varieties, varied fertilizer applications, altered land allocations, etc.). Such an approach is important for considering the comparative costs and benefits of different response options (e.g., the costs of increasing fertilizer applications weighed against improved returns from higher crop yields).

(3) An *adjoint approach* to the study of effects of climatic variations can be as valuable as the *direct approach*. The former focuses on the intrinsic sensitivities of economy and society to climate and then asks, "What type or degree of change in climate would cause a significant perturbation of them?" The latter asks the question: "For a given climatic change, what is the impact on (for example) ecosystems, economy and society?" Both approaches are valid, indeed complementary, lines of inquiry. An advantage of the first is that it can assist in the identification of *thresholds* or *distinct non-linearities* in response to climatic change, helping us recognize what kinds of climatic change are critically important and what are not.

(4) The *spatial analysis* of effects is an essential part of climate impact assessment. There are numerous examples given in this volume of the value of specifying changes in climate as spatial shifts of climatic types, or vegetation zones, or levels of agricultural potential, etc.

(5) The *use of analogue regions* is a particularly valuable aspect of spatial analysis, helping to describe an altered climate, characterize its possible effects and evaluate potential adjustments through the study of present-day farming types.

Within each of the broad approaches summarized above, a number of more specific tasks can be recommended. These are considered below.

1.7.5. Research priorities

This study has highlighted a number of priorities for future research. In summarizing these we shall refer to *Figure 1.22* which depicts the approach adopted in the IIASA/UNEP project and identifies points that warrant further attention. These research needs are summarized below, together (in italics) with examples specifically related to effects of climatic variations on agriculture:

(1) More appropriate data on climatic variations for impact assessment:
- greater spatial and temporal resolution and representativeness of climatic data
    e.g., to provide suitable inputs for crop-climate models
- a greater range of climatic data
    e.g., information on variations in specific variables such as windspeed and solar radiation, measures of variability as well as means, etc.
- more information on likely rates of change
    e.g., to enable consideration of required rates of adaptation (such as the development or substitution of new crop varieties, etc.)

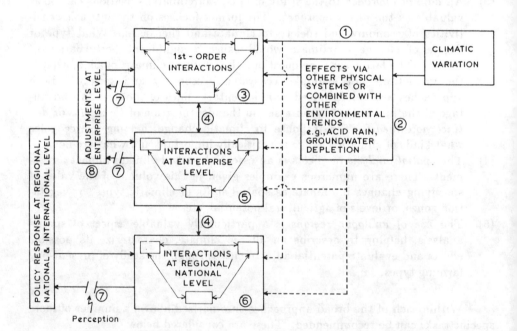

Figure 1.22. Research priorities in the assessment of effects of climatic variations: The schema of the IIASA/UNEP project's approach, with priority tasks identified by numbers that are described in the text.

(2) Improved understanding of the relationship between changes of climate and changes in other physical systems – specifically, climate and:
- soil nutrients, soil erosion and salinization
    e.g., to enable consideration of effects of precipitation changes on nutrient leaching
- pests and diseases, and their vectors
    e.g., to enable analysis of changes in crop and livestock losses

(3) Improved understanding of the relationship between climatic variations and first-order impacts in agriculture; specifically:
- climatic variations and production potential
    e.g., effects on indices of agricultural potential
- climatic variations and agricultural yields
    e.g., by empirical or simulation modeling of crop-climate and livestock-climate relationships
- direct impacts of climate on input requirements
    e.g., changes in fertilizer requirements

(4) Greater precision in tracing the higher-order effects of climatic change in the secondary and tertiary sectors (i.e., second to nth-order impacts); specifically:
- by using hierarchies of linked climate, impact, and economic models to trace the pathways of impacts
    e.g., to trace effects of altered climate on crop yields, on gross production, on farm profitability, on regional competitiveness, etc.

(5),(6) Greater precision in simulating impacts at the enterprise level and regional level
    e.g., by farm simulations to estimate altered farm incomes, by regional agricultural planning models to estimate altered optimal regional land-use allocations, etc.

(7) Improved understanding of the role of perception of climatic change (its actual or potential impacts, and the array of potential responses):
- as an expression of the direct impacts
    e.g., changes in mean, in frequency of extreme events, in risk levels
- as an expression of the costs and benefits of impacts
    e.g., changes in mean levels of farm income, in the stability of farm income, in regional employment, etc.
- as an expression of the choice range of potential or actual adjustive mechanisms; their costs and benefits
    e.g., changes in comparative advantage between crops, in cost effectiveness of improved harvesting capacity, in sensitivity of crops to fertilizer applications

(8) Improved understanding of adaptive responses at the enterprise level, especially of:
- firm and farm-level decision making, particularly in the face of risk, uncertainty, and constraints on choice

e.g., the range of on-farm adjustments of management and technology (fertilizing, soil drainage, crop-livestock combinations, etc.) at present available
- the resilience of economic systems
  e.g., the relative resilience of mono-cropping, mixed cropping and mixed crop-livestock farming to climatic variations

(9) Strategy formulation at the regional, national, and international level, especially with respect to:
- the range of possible policy responses
  e.g., government price support and crop insurance schemes
- the process of policy formulation
  e.g., centralized planning of production targets and land allocation, strategies of regional development, domestic consumer requirements, export earnings, etc.
- the priority areas for resource allocation
  e.g., plant breeding, agricultural extension services, fertilizer trials, etc.
- the constraints on choice (with respect to other resource needs and problems)
  e.g., competing land uses, limits on expenditure, self sufficiency objectives, etc.

1.7.6. The structure of this volume

It is clear from the list of research requirements presented above that much still needs to be done. We have emphasized that the assessments presented here, and in the companion volume on semi-arid areas, are preliminary. The present task is to improve our methods of impact assessment so that, when usable predictions of climatic change are available, we have sound means of acting upon them in terms of evaluating possible effects and determining appropriate responses. There are, however, encouraging signs in this volume that appropriate methods of assessment are now emerging.

The remainder of Part I of the volume considers, firstly, the choice of impact models and the development of climatic scenarios for the case study experiments. There then follows some further consideration of the advantages and difficulties involved in linking biophysical and economic models, using as an example a case study of the possible effects of CO_2-induced climatic warming on forest growth and the forest sector. The remaining five parts of the volume contain the results of the case studies.

Appendix 1.1

Development of the Climatic Scenarios

In order to conduct experiments to assess the impacts of climatic variations on agriculture in the different regional case study areas, it was first necessary to obtain a quantitative representation of the changes in climate themselves. Since climatic changes cannot yet be accurately predicted, the approach adopted as an alternative was to specify a set of *plausible future conditions* that we label "scenarios".

The climatic scenarios were selected to provide climatic data that were:

- Meteorologically realistic, comprising combinations of variables that were both physically plausible and geographically consistent.
- Freely available or easily derivable, and
- Suitable as inputs to impact models.

As a common thread running through all five of the case studies in this volume, climatic scenarios were selected to reflect issues of particular concern in these regions, characterizing (a) typical short-term anomalies and their relation to (b) possible longer-term CO_2-induced climatic change. To represent the first of these, two broad types of *instrumental* scenario were chosen:

(1) A single, anomalous weather-year, taken from the historical instrumental record.
(2) An extreme weather decade or period of weather-years, also from historical data.

The second issue of CO_2-induced climatic change was considered through the selection of a third, *general circulation model (GCM)-based* scenario:

(3) The climate simulated by the Goddard Institute for Space Studies (GISS) GCM for doubled concentrations of atmospheric carbon dioxide (Hansen et. al., 1983, 1984), representing a physically realistic, globally-consistent simulation of a future $2 \times CO_2$ climate that was common to all study areas. For details, *see* Section 3.

In addition, several more arbitrary changes in climate or *synthetic* scenarios were chosen in some regions. Finally, in order to provide a reference against which to compare various scenario experiments, the period 1951–1980 was adopted (wherever data and methods permitted) as the standard climatological *baseline*.

By examining *first* the nature and sensitivity of the exposure unit (i.e., the crop or livestock unit being affected by climate), usually through the use of an impact model (cf. step 2, *Figure 1.7*), it was possible to identify those important climatic variables (in terms of their potential impacts) that needed to be included in the scenario. In light of the high latitude location of the agricultural activities being studied, seasonal temperatures formed the basis of the scenario construction in four out of five of the case study regions, the only exception being Saskatchewan, where precipitation was the dominant criterion for scenario development.

The instrumental scenarios

In general, these scenarios were selected to reflect recorded short-term historical events that could occur again in the future. As such they represent regional events for which combinations of climatic data are available (from existing meteorological observing stations) that are suitable as inputs to impact models. Moreover, in many cases, they have been chosen not purely on the basis of a climatic anomaly *per se*, but also with the recorded impacts in mind. For example, the 1929–1938 "dry decade" scenario in

Saskatchewan, represents the 10 consecutive years with the greatest predicted average shortfall of wheat yields (relative to trend) in the 53-year period, 1928–1980. This *diagnostic* approach to scenario selection (i.e., diagnosing the weather-related impact before selecting the scenario) contrasts with the more *prognostic* features of GCM-based scenarios, discussed below.

The types of instrumental scenario selected in each region are summarized in *Table 1.2* and illustrated in *Figure 1.23*. Their selection is described in detail in each case study. Most were developed to assess the impacts at the present-day (with current agricultural systems, technologies, crops, etc.) of past, observed climatic anomalies that might be expected to occur again even in the absence of longer-term climatic change (e.g., the 1860s cold period in Iceland and Finland, the cold year 1980 in Japan and dry year 1961 in Saskatchewan). Other anomalies served as "analogue" data sets for possible CO_2-induced warming (e.g., the ensemble of the ten warmest years from 1931–1984 in Iceland). While most scenarios were selected to reflect anomalies in *average* climate, several (e.g., in the Japan study) were based on the interannual *variability* of climate.

The GISS $2 \times CO_2$ scenario

There is a detailed discussion in Section 3 of the reasons for selecting the GISS GCM experiment as a basis for the $2 \times CO_2$ scenario. The GISS GCM produced outputs for an 8° latitude × 10° longitude network of grid points showing the simulated change in monthly averaged climatic variables between $1 \times CO_2$ (present-day) and $2 \times CO_2$ (future) equilibrium conditions. Statistical smoothing techniques were used to transfer the model-generated data to a finer resolution (4° × 5°) grid system and to filter out large amplitude temporal fluctuations in the data.

Values of mean monthly temperature (°C), precipitation rate (mm/day) and cloud cover (proportions) were provided for a 4° × 5° resolution grid point coverage representing each case study area. Five types of information were available:

(1) Observation data (1931–1960 mean values).
(2) GISS model generated $1 \times CO_2$ equilibrium values.
(3) GISS $1 \times CO_2$ – Observations (used for validation of the GISS model).
(4) GISS $2 \times CO_2$ equilibrium values.
(5) GISS $2 \times CO_2$ – GISS $1 \times CO_2$ (i.e., the change in equilibrium climate for a doubling of CO_2).

In order to apply the GISS $2 \times CO_2$ data to the particular climatic and agricultural conditions at present observed at a regional level, it was first necessary to evaluate the performance of the GISS model in simulating present-day climatic conditions. At a global scale, it is normally sufficient to compare the GISS $1 \times CO_2$ data with the 1931–1960 interpolated grid point observation data (*see* Section 3, *Figure 3.4*). However, the impact experiments at regional level are usually conducted for specific areas or sites with climatic records that reflect local conditions which are not always observable in the gridded data sets.

Therefore, in each regional case study, the GISS $1 \times CO_2$ data were compared instead (by interpolating between grid points) with 1951–1980 mean data for sites considered representative of the agricultural regions being investigated. The poor correspondence of real and modeled climates in many areas had implications for the development of the $2 \times CO_2$ scenario, discussed below. This weak validation of the GISS model, along with its coarse spatial and temporal resolution, mean that very little confidence can be attached to model estimates of regional climatic change. The estimates should be regarded, rather, as one possible scenario of future climate that is useful for exploring the methods of impact assessment, preparing for the time when more accurate model predictions of climatic change become available.

Table 1.2. Types and characteristics of climatic scenarios selected in the IIASA/UNEP Project, Volume 1.

			Climatic Scenarios				
			Instrumental			Synthetic	GCM/Empirical
Case study	Major climatic constraint on agriculture	Baseline period	Extreme event	Anomalous period	Other	Incremental changes	$2 \times CO_2$
Saskatchewan	P	1951–80	P–	P–		T±,P±	GISS (T_d,P_r)
Iceland	T	1951–80	T–	T–	T±	T±	GISS (T_d,P_r)
Finland	T	1959–83 1971–80		T±		T±,P±	GISS (T_d,P_d)
N. USSR:							
Leningrad	T	1951–80					V/G & GISS Transient (T_d,P_d)
Cherdyn	T,P	1951–80				T±,P±	GISS (T_d,P_d)
Central Region	T	1931–60				T+	
Japan	T	1951–80	T±	T±,S,U		T±,H±	GISS (T_d,P_d)

T = Temperature; P = Precipitation; H = Hours of sunshine; S = Stable; U = Unstable; + = Above average; – = Below average; d = Adjustment by "differences"; r = Adjustment by "ratios"; V/G = Vinnikov/Groisman empirical method (*see* Part V). Note that two different baseline periods were adopted in different sections of the Finland study (*see* Part IV).

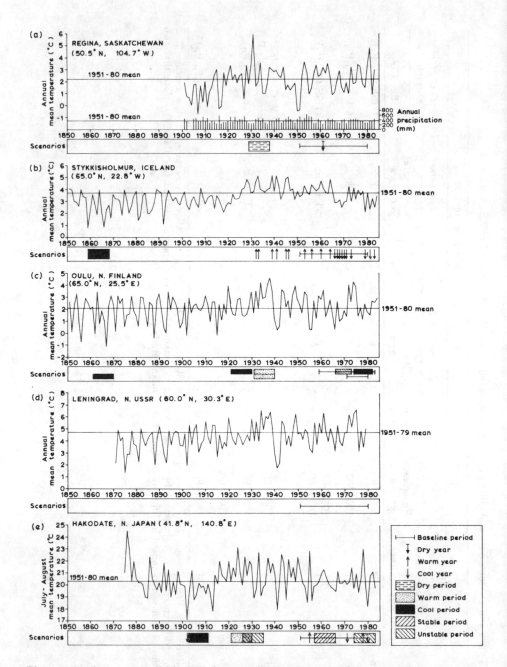

Figure 1.23 Instrumentally-based climatic scenarios, selected for the five case studies. Scenarios are depicted in horizontal bars beneath long-term climatic time series from meteorological stations within each case study region. The 1951–1980 mean climate (used as the baseline throughout this volume) is marked as a horizontal line through the plotted points on each graph.

The use of the GISS data depended on the nature of the impact models employed. All of the agroclimatic models required information on temperature conditions, most required precipitation data, and several also needed data on solar radiation, evaporation and windspeed. The GISS outputs of monthly temperature and precipitation proved to be adequate for adjusting 1951–1980 observation data to the $2 \times CO_2$ levels, but the cloud cover data were not suitable for simulating changes in solar radiation. As a result, the radiation term in some models, along with other variables not available from the GISS model, was not adjusted from 1951–1980 mean values (except where empirical methods were used for approximating these; see Section 2). Where models required temperature or precipitation inputs at a sub-monthly (e.g., daily, weekly, etc.) time resolution, the monthly GISS adjustments were applied to all values within each month.

It was assumed throughout the different studies that, while the GISS model may not reproduce the observed present-day climate very closely, the change between $1 \times CO_2$ and $2 \times CO_2$ equilibrium conditions was representative of the difference between the present (1951–1980) climate and a future $2 \times CO_2$ climate. In most cases, temperatures and precipitation were adjusted simply by adding the *difference* between $1 \times CO_2$ and $2 \times CO_2$ values to the 1951–1980 mean values. However, in some drier regions this led to an exaggerated change in precipitation, so that an alternative technique employing *ratios* rather than differences was adopted instead (see Part II, Section 1 for a full explanation of this).

In most applications, adjustments were made for individual station locations by interpolating between the gridded GISS model values, either manually from the printed data sets, or using computer interpolation routines, or by interpolating mapped values to the station locations by eye.

Finally, the time component of a CO_2-induced climatic change was incorporated in two of the case study experiments: in the Leningrad region (USSR) and in Japan. Although the GISS model only simulated climatic conditions for $1 \times CO_2$ and $2 \times CO_2$ conditions, the transition between these two equilibrium states was approximated by assuming a linear rate of change up to an assumed CO_2-doubling date. In the Leningrad study, this approach was combined with an empirical method of specifying regional climatic change based on changes in northern hemisphere temperatures, which was applied to the period up to 2005 (following Vinnikov and Groisman, 1979 – see Part V).

Synthetic scenarios

Synthetic scenarios comprise climatic data that are generated arbitrarily to simulate a climatic change. These were found to be useful in some regions for testing the sensitivity of impact models. A popular approach is to alter the climatic input data to a model by systematic increments above and below the baseline values (e.g., at intervals of 0.5 °C for temperatures and 10% for precipitation), and to study crop responses to these changes. Several of these sensitivity experiments are illustrated in this volume as two- or three-way plots of climatic variable(s) versus crop yield.

In the absence of time-dependent "transient" GCM estimates of climatic changes over the period of increasing atmospheric carbon dioxide concentrations, and since most GCM "equilibrium" climatic change experiments are for doubled or quadrupled concentrations of CO_2, synthetic scenarios may also be used to represent conditions intermediate between the present and $2 \times CO_2$ climates. This procedure was adopted for the Central Region of the European USSR, where the two impact models that were employed (to estimate the effects of climatic change on agricultural production, costs and land allocation) were unable satisfactorily to accommodate the large changes in climate estimated for the GISS $2 \times CO_2$ scenario. Instead, more modest arbitrary changes in temperature (alone) of +0.5, +1.0 and +1.5 °C were assumed.

Appendices 1.2–1.6

The following appendices summarize the results of experiments conducted to assess the effects of climatic variations on agricultural production in the five case studies reported in this volume. Short definitions of some general measures of effect are listed below, while full details are given in the case studies themselves. The rows in the appendices are organized according to the hierarchical structure of the experiments, containing (from top to bottom) values of the regional climatic scenarios, first-order impacts, higher-order and economic impacts, and adjustment experiments (where appropriate). Baseline values are shown in the first column and, unless otherwise noted, estimates are relative to the 1951–1980 climate and to productivity simulated for that climate, with management and technology at *ca.* 1980 levels. Adjustment experiments are indicated with a solid black circle.

Definitions of some general measures of effect used in the appendices

Most of the terms used in the appendices to describe the effects of climatic variations are either self-explanatory or are specific to individual case studies. For the latter, reference to sources of definitions is given in the appendix footnotes. More general terms are defined below.

Climatic index of agricultural potential. An index that indicates the potential for biomass productivity, reflecting the length of growing season and the temperature and moisture resources available to plants during the season (*see* Part II).

Drought frequency. This is based on the frequency of months with a Palmer Drought Index lower than −4.00 (indicating drought or severe drought) during a period of several decades (*see* Part II).

Effective accumulated temperature. The annual accumulated sum of air temperatures above 0 °C during those days when daily mean air temperature is greater than a specified baseline (10 °C in most applications).

Effective temperature sum. The annual accumulated sum of air temperatures above a specified threshold temperature that represents a minimum requirement for the commencement of active plant growth (5 °C is commonly used in this volume).

Household purchasing power. Average provincial gross domestic product (GDP) per household (*see* Part II).

Potential for tree growth. Qualitative description based on three quantitative measures used in the Iceland study (Part III): two giving the total lowland area with temperatures (in growth units) sufficient to sustain birch and Norwegian spruce, the third indicating the total potential taiga area, according to the location of the 10 °C mean July isotherm.

Precipitation effectiveness. An index that characterizes the annual moisture regime of a region on the basis of monthly precipitation and air temperature. Low values (below 32) indicate arid and semi-arid conditions, high values (above 64) indicate humid and wet conditions (*see* Part II).

Trended yields. Model estimates of future yields under present-day or changed climate and under certain assumptions about projected technology.

Wind erosion potential. An index combining windspeed (representing the erosive force on soil particles) with soil moisture (providing a cohesive force resisting erosion – represented by the *precipitation effectiveness index*). High values denote a high potential for wind erosion of soil; low values denote a low potential (*see* Part II).

Appendix 1.2. The estimated effect of climatic variations on agricultural production in Saskatchewan. Baseline values are shown in the first column and, unless otherwise noted, estimates are relative to the 1951–80 climate and to productivity simulated for that climate, with management and technology at ca. 1980 levels. Dual entries reflect the spatial range of impacts. Source: Part II.

CANADA (Southern Saskatchewan)[a]	Baseline (1951–80)[b]	Dry year (1961-type)	Scenarios Dry period (1933 to 1937-type)[b]						GISS $2 \times CO_2$ experiment	
			(1933-type)	(1934-type)	(1935-type)	(1936-type)	(1937-type)	(Period-mean)	(T,P)[c]	(T)[c]
Climate										
Mean May–August temperature[d]	14.4 to 16.4°C	+1.3 to +3.5°C	−0.1 to +2.4°C	−0.4 to +2.0°C	−2.0 to +0.5°C	+0.7 to +3.9°C	+0.0 to +3.0°C	+0.0 to +2.3°C	+3.5 to +3.6°C	+3.5 to +3.6°C
Mean May–August precipitation[d]	177 to 249 mm	−71 to −36%	−49 to +21%	−49 to +28%	−31 to +57%	−62 to −9%	−52 to −6%	−47 to +7%	+9 to +14%	No change
Effective temperature sum[e,f]	1440 to 1880 GDD[g]	+12 to +18%						+3 to +16%[h]	+48 to +53%[i]	+48 to +53%[i]
Precipitation effectiveness index[f,j]	28 to 46[k]	−40 to −23%						−27 to −21%[h]	+5 to +12%[l]	−12 to −10%[l]
First-order effects										
Drought frequency (PDI$\leq$ −4)[m]	3.0%								× 3	× 13
Wind erosion potential[n]	6.6 to 77.0	−15 to +367%						−57 to +123%[h]	−14 to −3%	+24 to +29%
Climatic index of agricultural potential[f,p]	8.1 to 14.9[q]	−84 to −53%						−60 to −26%[h]	+1 to +30%[r]	−18 to +3%[r]
Spring wheat yields[s]:										
Brown soils	1034 kg/ha	−79%	−59%	+5%	+3%	−33%	−67%	−29%	−4%	−15%
Dark brown soils	1190 kg/ha	−77%	−31%	−5%	+55%	−8%	−59%	−9%	−12%	−24%

Aims, methods and summary of results

Higher-order effects

Provincial wheat production[t]	10.2M tons	−76%	−37%	−14%	+26%	−21%	−58%	−20%	−18%	−28%

Farm household income[t]:									
Brown soils	$540M	−88%	−26%	−40%	−6%	−16%	−53%	−47%	−3%	−9%
Dark brown soils	$800M	−72%	−12%	−20%	+31%	+36%	−26%	−9%	−1%	−10%
Black soils	$950M	−79%	−12%	−31%	−14%	−3%	−36%	−26%	−14%	−15%
Small sized farms	$480M	−74%	−14%	−35%	−11%	−6%	−34%	−28%	−10%	−14%
Medium sized farms	$1210M	−82%	−12%	−27%	+9%	+14%	−35%	−21%	−7%	−12%
Large sized farms	$600M	−75%	−22%	−28%	+3%	+5%	−41%	−30%	−6%	−11%
Provincial total	$2290M	−78%	−15%	−29%	+3%	+7%	−36%	−25%	−7%	−12%

Employment[u]:									
Agricultural sector	88 000 m.y	−9.1%	−1.8%	−3.4%	−1.0%	+0.8%	−4.1%	−3.0%	−0.8%	−1.4%
Other sectors	335 000 m.y	−5.1%	−1.0%	−1.9%	−0.5%	+0.5%	−2.3%	−1.7%	−0.5%	−0.8%
Provincial total	423 000 m.y	−5.9%	−1.2%	−2.2%	−0.6%	+0.5%	−2.7%	−2.0%	−0.5%	−0.9%
Provincial household purchasing power	$6260M	−38.3%	−7.5%	−14.2%	−4.2%	+3.5%	−17.2%	−12.7%	−3.4%	−5.8%

a Estimates for a northern station, Uranium City, are footnoted where appropriate.
b Unless otherwise noted.
c Adjusted precipitation (P) and/or temperatures (T).
d Range for all crop districts; *Table 4.2*, Part II.
e Above a 5°C base: *Figure 2.1*, Part II.
f Baseline values are for representative stations. Scenario estimates for five long-period stations.
g Uranium City: 1145 GDD.
h Dry period scenario is 1929 to 1938-type.
i Uranium City: +50% (T,P), +50% (T); *Table 2.5*, Part II.
j Units arbitrary; *Figure 2.2*, Part II.
k Uranium City: 40.
l Uranium City: +26% (T,P), −10% (T); *Table 2.5*, Part II.
m Source: *Table 6.1*, Part II. Baseline period: 1950–82.
n As note f but for four long-period stations. High values indicate high potential for erosion, units arbitrary; *Table 3.3*, Part II.
o Three long-period stations.
p Uranium City; *Figure 2.3*, Part II.
q Uranium City: 7.8.
r Uranium City: +74% (T,P), +17% (T); *Table 2.5*, Part II.
s Source: *Table 4.7*, Part II.
t Sources: *Tables 5.5–5.7*, Part II.
u Sources: *Tables 5.8–5.9*, Part II and R. Fautley (personal communication).

Appendix 1.3. The estimated effect of climatic variations on agricultural production in Iceland. Baseline values are shown in the first column and, unless otherwise noted, estimates are relative to the 1951–80 climate and to productivity simulated for that climate, with management and technology at *ca.* 1980 levels. Adjustment experiments are indicated with a solid black circle. Values in parentheses are standard errors of estimates. Percentages in square brackets are absolute. Source: Part III.

			Scenarios		
Iceland	Baseline (1951–80)[a]	Cold decade (1859 to 1868 -type)	10 coldest years (during 1931–84)	10 warmest years (during 1931–84)[a]	GISS $2 \times CO_2$ experiment
Climate[b]					
Mean annual temperature	3.7°C	−1.3°C	−0.8°C	+1.1°C	+4.0°C
Mean annual precipitation	704 mm	−9%	−3%	+12%	+15%
Onset of growing season	4 May	17 May	9 May	27 April	17 March
First-order effects[c]					
Theoretical area of taiga	11000 km²	−27%	−27%	+100%	+482%
Percentage stations permitting:					
Birch cultivation	[29%]	[4%]	[10%]	[85%]	[100%]
Norwegian spruce cultivation	[8%]	[0%]	[0%]	[71%]	[100%]
Percentage stations with ripened barley:					
60% years	[4%]	[0%]	[0%]	[60%]	[100%]
90% years	[0%]	[0%]	[0%]	[8%]	[100%]
Improved grassland hay yields[d]	4620 kg/ha	−19%	−13%	+18%	+66%
Pasture yields:[e,f] high input	5260 kg/ha	−16% (3%)	−10% (2%)	+16% (2%)[g]	+49% (6%)
low input	4671 kg/ha	−17%	−10%	+17%[g]	+52%
Effects on livestock					
Feed requirements:[c,h] cattle	100	+5%	+3%	−5%	−16%
horses	100	+22%	+14%	−17%	−53%
sheep	100	+22%	+14%	−16%	−53%

Aims, methods and summary of results

Effects on livestock (continued)

Carrying capacity:					
Improved grassland:[c,h]	cattle	100	−23%	−16%	+24%	+98%
	horses	100	−34%	−24%	+42%	+253%
	sheep	100	−34%	−24%	+40%	+253%
Rangeland:	sheep[c]	847 000	−19%	−13%	+18%	+66%
	sheep[j]	866 400	−21% (17%)	−13% (19%)	+18% (23%)	+64% (26%)
Lamb carcass weight[i]		14.05 kg	−4% (3%)	−2% (3%)	+3% (5%)	+12% (7%)
Adjustment experiments						
Fertilizer requirement for yield of 4000 kg/ha on improved grassland:[j]	nitrogen	110 kg/ha	+55% •	+31% •	−24% •	−55% •
	phosphorus	31 kg/ha	+48% •	+29% •	−19% •	−45% •
	potassium	79 kg/ha	+27% •	+15% •	−10% •	−25% •
Adjustments to maintain baseline milk production:[k]						
Pasture area:[e,f]	high input	28 600 ha	+19% •	+11% •	−14% [g] •	−33% •
	low input	32 200 ha	+20% •	+12% •	−14% [g] •	−34% •
Feed concentrates:[e,f]	high input	46 000 tonnes	+52% •	+32% •	−52% [g] •	−100% •
	low input	46 000 tonnes	+54% •	+33% •	−54% [g] •	−100% •
Changes in yields due to altered fertilization relative to changes due to temperature:[e,f]	high input[l]		33% •	56% •	55% [g] •	21% •
	low input[m]		48% •	81% •	72% [g] •	27% •

[a] Unless otherwise noted.
[b] Sources: *Table 1.5* and *Table 5.3*, Part III.
[c] Source: *Table 2.12*, Part III.
[d] Nitrogen application: 160 kg/ha.
[e] Sources: Equation 3.2 and *Tables 3.9–9.11*, Part III.
[f] Input levels (nitrogen): 120 kg/ha (high); 80 kg/ha (low).
[g] Scenario temperatures +1.3°C relative to baseline.

[h] Assuming 1941–49 management levels.
[i] Source: *Table 4.2*, Part III. Results calculated for 1966–83 mean carcass weight (13.92 kg).
[j] Source: *Table 2.7*, Part III.
[k] 100 million liters per year.
[l] Change in nitrogen application of 30 kg/ha.
[m] Change in nitrogen application of 26.6 kg/ha.

Appendix 1.4. The effect of climatic variations on agricultural production in Finland. Baseline values are shown in the first column. Unless otherwise noted, estimates are relative to either the 1959–83 or the 1971–80 climate (as indicated), and to productivity simulated for those climates, with management and technology at ca. 1980 levels. Adjustment experiments are indicated with a solid black circle. Values in parentheses are standard deviations (scenario values relative to baseline values). Source: Part IV.

FINLAND	Baseline (1959–83/ 1971–80)	Scenarios					GISS $2 \times CO_2$ experiment
		Cool periods		Warm periods			
		(1861 to 1870-type)	(1921 to 1930-type)	(1974 to 1982-type)	(1981 to 1940-type)	(1966 to 1973-type)	
Climate[a]				*Southern Finland*			
Effective temp. sum:							
Baseline 1959–83	1263 GDD (121)		−4% (+2%)	−5% (−4%)	+12% (+29%)	+2% (−38%)	+37% (+0%)
Baseline 1971–80	1231 GDD (126)	−8% (+20%)			+15% (+23%)		+33% (+0%)
Mean May–Oct precip.:							
Baseline 1959–83	342 mm (87)		+26% (−26%)	+7% (+29%)	−1% (+6%)	−3% (−14%)	+50% (+0%)
Baseline 1971–80[b]	342 mm (99)	+2% (−17%)			−1% (−7%)		+50% (+0%)
First-order effects							
Barley yields:							
Temporal model[c]	3075 kg/ha (428)		+10% (−25%)	+4% (+17%)	+1% (−3%)	−8% (−28%)	+9% (−37%)
Spatial model[d]	2315 kg/ha (329)	−8% (+26%)			+12% (−13%)		+21% (−11%)
Oats yields[d]	2386 kg/ha (174)	−5% (+36%)			+6% (−16%)		+18% (+12%)
Spring wheat yields[e]	2300 kg/ha (450)			−15% (+22%)		+10% (−56%)	+10% (−11%)[f]
Higher-order effects							
Net return (barley)[g]	616 FIM/ha (316)	−31% (+26%)			+42% (−14%)		+77% (−11%)
Net return (oats)[g]	134 FIM/ha (153)	−80% (+35%)			+101% (−16%)		+286% (+12%)

Northern/Central Finland

	Baseline						
Climate[a]							
Effective temp. sum:							
Baseline 1959–83	1080 GDD (134)		−1% (−1%)	−2% (−15%)	+10% (+12%)	+3% (−10%)	+35% (+0%)
Baseline 1971–80	1094 GDD (134)	−2% (−4%)			+9% (+12%)		+43% (+0%)
Mean May–Oct precip.:							
Baseline 1959–83	313 mm (81)		+6% (−27%)	+2% (+19%)	+11% (+7%)	−8% (−5%)	+59% (+0%)
Baseline 1971–80[b]	288 mm (54)				+20% (+63%)		+72% (+0%)
First-order effects							
Barley yields:							
Temporal model[h]	2560 kg/ha (429)		+1% (−22%)	−1% (−39%)	+8% (−16%)	+3% (−39%)	+14% (+37%)
Spatial model[i]	2359 kg/ha (195)				−5% (+131%)		+14% (+39%)
Oats yields[i]	2386 kg/ha (110)				−2% (+135%)		+13% (+68%)
Spring wheat yields[j]	2000 kg/ha (400)			−12% (+0%)		+15% (−25%)	+20% (+25%)[f]
Higher-order effects							
Net return (barley)[g]	657 FIM/ha (187)				−18% (+131%)		+47% (+39%)
Net return (oats)[g]	134 FIM/ha (97)				−34% (+134%)		+210% (+67%)

Whole of Finland

	Baseline						
Economic effects							
National farm income[l]	6764 FIM/kg	−10 to −7%	+1% to +13%	−20 to +5%	−7 to +16%	−11 to +20%	+12 to +27%

Sources: Table 5.1, Part IV, and J. Mukula, personal communication.

[a] Sources: *Table 5.1*, Part IV, and J. Mukula, personal communication.
[b] May–September precipitation (not shown) used in model experiments.
[c] Relative to 1959–83 baseline; Helsinki region; *Table 2.2*, Part IV.
[d] Relative to 1971–80 baseline; Helsinki (Kaisaniemi); *Table 4.3*, Part IV.
[e] Relative to 1959–83 baseline; values derived from mapped yields for Helsinki area. Sources: *Figures 3.6, 3.7 and 3.8*, Part IV.
[f] New variety with thermal requirements 120 GDD greater than present.
[g] Relative to 1971–80 baseline, based on yield estimates (note d); *Table 5.2*, Part IV.
[h] As note c, but for Ähtäri station (south-west of Oulu).
[i] As note d, but for Oulu station.
[j] As note e, but for central Finland (approximate latitude 64°N).
[k] As note g, but based on yield estimates according to note i.
[l] Based on the range of modeled mean yield estimates; following Section 5, Part IV.

Appendix 1.5. The estimated effect of climatic variations on agricultural production in the northern European USSR. Baseline values are shown in the first column and, unless otherwise noted, estimates are relative to the 1951–80 climate and to productivity simulated for that climate, with management and technology at *ca.* 1980 levels. Adjustment experiments are indicated with a solid black circle. Dates assumed for changes are indicated in square parentheses. Source: Part V.

NORTHERN EUROPEAN USSR	*Baseline climate (1951–80)*[a] *1980 technology*	*Baseline climate (1951–80)*[a] *Technology trend for given date*	*Scenarios:* *Moderate warming by given date; technology trend where specified*			*Scenarios: GISS 2 × CO_2 experiment* *1980 management*	*Scenarios: GISS 2 × CO_2 experiment* *Adjusted management*
				Leningrad region			
Climate[b]				[*2000*]		[*2035*][c]	
May–Oct temperature	12.2°C	0.0°C		+0.8°C		+2.2°C	+2.2°C
May–Oct precip.	351 mm	+0 mm		+5%		+36%	+36%
First-order effects[d]							
Soil fertility index[f]	10.7[e]	+22% [*2000*] +33% [*2035*]		+18%		−22%	+17%[g] • −25%[h] •
Surface water pollution by nitrogen[f]	0.152 g/l	+18% [*2000*] +25% [*2035*]		+58%		+314%	+501%[g] • +136%[h] •
Winter rye yields[f]	2.0 t/ha	+26% [*2000*] +36% [*2035*]		+31%		+6%	+57%[g] • +0%[h] • +50%[i] •
				Cherdyn region			
Climate[j]				[*Arbitrary date*]		[*Arbitrary date*]	
May–Sept temp.	13.4°C		+1.0°C	+1.0°C	+1.0°C	+2.7°C	+2.7°C
May–Sept precip.	298 mm		−20 mm	+0 mm	+20 mm	+50%	+50%
First-order effects							
Spring wheat yields[k]	1.15 t/ha		−1%	+3%	+7%	−3% (+17%)	+16% (+39%)[l] • +26% (+49%)[m] •

Aims, methods and summary of results

	Central region		
Climate[n]	[1995]	[1995]	[1995]
May–Oct temperature	12.5°C	+1°C	+1.5°C
May–Oct precip.	376 mm	+0 mm	+0 mm
First-order effects[o]			
Crop yields:[f]			
Winter wheat	20.3 q/ha +3%	+33%	+30%
Barley	17.0 q/ha +4%	−1%	−4%
Oats	15.5 q/ha +6%	+1%	−2%
Corn for silage	240.0 q/ha +1%	+7%	+9%
Hay	44.0 q/ha +7%	+3%	+0%
Potatoes	132.0 q/ha +6%	+3%	+1%
Vegetables	328.0 q/ha +2%	−1%	−3%
Higher-order effects & adjustments			
Production costs (allocations)[p]:	[1980 prices (areas)]		GISS 2 × CO_2 scenario not considered
Winter wheat	15.0 r/q (150 000 ha)	−22% (+29%)	(+26%) ● ● ●
Barley	13.0 r/q (130 000 ha)	+5% (−23%)	(−20%) ● ● ●
Oats	13.0 r/q (50 000 ha)	+5% (−27%)	(−25%) ● ● ●
Corn for silage	2.8 r/q (130 000 ha)	−6% (+5%)	(+3%) ● ●
Hay	4.5 r/q (450 000 ha)	+4% (+0%)	(+0%) ● ●
Potatoes	15.0 r/q (120 000 ha)	+3% (−6%)	(−4%) ● ●
Vegetables	9.2 r/q (25 000 ha)	+3% (+3%)	(+5%) ●

[a] Unless otherwise noted.
[b] Values for 1931–60; source: Müller (1982).
[c] 2 × CO_2 climate assumed to occur in 2050.
[d] Sources: *Figures 2.7–2.9*, Part V.
[e] Units are arbitrary.
[f] Scenario values include technology trend.
[g] 50% increase in fertilizer applications.
[h] Improved drainage activity.
[i] 50% increase in fertilizer application combined with improved drainage activity.
[j] Station: Kirov (1931–60); *Figure 3.5*, Part V.
[k] Simulated yields incorporating "direct" effects of CO_2 on crop are in parentheses; *Table 3.2* and Section 3.4, Part V.
[l] New middle-maturing variety.
[m] New late-maturing variety.
[n] Station: Moscow (1931–60); Müller (1982).
[o] Sources: *Tables 4.4, 4.6* and *4.8*, Part V.
[p] Area of crop required to minimize total expenditure.

Appendix 1.6. The effect of climatic variations on agricultural production in Japan. relative to the 1951–80 climate and to productivity simulated for that climate, with with a solid black circle. Values in parentheses are standard deviations or coefficients of

			Scenarios	
		Cool year	Unstable decades	
JAPAN[a]	Baseline (1951–80)[b]	(1980-type)[b]	(1926 to 1935-type)	(1974 to 1983-type)
HOKKAIDO				
Climate				
Mean Jul–Aug temp.[c]	20.6°C	−1.6°C	−0.3°C	+0.0°C
Effective accumulated temperature[d]	3025 GDD	−14%	−3%	+0%
First-order effects				
Rice cultivation limit	250 m	−254 m	−48 m	+0 m[e]
Rice yield index	100	−40%	−6%	+0%[f]
Simulated rice yields:				
Average climate & Present cultivar	5.05 t/ha	−24%		−11%
Annual weather &:				
Present cultivar[h]	4.48 t/ha (12%)	−15%		+0% (+0%)
Middle-maturing c.[h]		+18%[i] ●		
Late-maturing c.[h]		+17%[i] ●		
Higher-order effects				
District rice supply[j,k]	0.85 M t (22%)	−36% (+4%)[l]		
TOHOKU				
Climate				
Mean Jul–Aug temp.[c]	22.8°C	−2.6°C	−0.1°C	−0.5°C
Effective accumulated temperature[d]	3676 GDD	−21%	−1%	−4%
First-order effects				
Rice cultivation limit	350 m	−413 m	−16 m[e]	−79 m[e]
Rice yield index	100	−48%	−1%[f]	−6%[f]
Higher-order effects				
District rice supply[j,k]	3.05 M t (11%)	−18% (−1%)[l]		
WHOLE OF JAPAN				
First-order effects				
Relative total net prod'n (vegetation)	380 M t/year	−7%	−2%	
Actual rice yields	100	−13%	+0%	
Higher-order effects				
National rice supply[j,k]	12.27 M t (11%)	−15% (−2%)[l]	−4% (−0%)	
National rice stock[j]	8.34 M t	−100%[l]	+11%	

[a] Sources: *Tables 2.1, 4.3* and *7.1*, Part VI, unless otherwise noted.
[b] Unless otherwise noted.
[c] Interannual variability assumed unchanged from baseline.
[d] Above 0°C (when mean temperature exceeds 10°C); equation (3.9), Part VI.
[e] Estimates from Section 4.3, Part VI.
[f] Estimates from equations (4.2) and (4.3), Part VI.
[g] Source: *Table 5.2*, Part VI.

Aims, methods and summary of results

Baseline values are shown in the first column and, unless otherwise noted, estimates are management and technology at *ca.* 1980 levels. Adjustment experiments are indicated variation. Source: Part VI.

Stable decade (1957 to 1966-type)	Scenarios (continued)		GISS $2 \times CO_2$ experiment
	"Good years"		
	(1978-type)	(1979-type)	
−0.1°C	+2.5°C	−0.2°C	+3.5°C[c]
−1%	+25%	−2%	+35%
−16 m	+397 m	−32 m[e]	+556 m
−2%	+15%	−4%[f]	+5%
−11%	−6%	+0%[g]	−7%
+0% (−33%)	+6%	+13%[g] •	+4% (−27%) •
	+26% •	+32%[g,i] •	+23% (−30%) •
	+31% •	+35%[g,i] •	+26% (−33%) •
	+18%	+6%	+17% (−6%)[m]
			+14% (−8%)[n]
+0.0°C	+2.3°C	−0.3°C	+3.2°C[c]
+0%	+20%	−3%	+29%
+0 m[e]	+365 m	−48 m[e]	+508 m
+0%[f]	+8%	−3%[f]	+2%
	+10%	+5%	+7% (−3%)[m]
			+6% (−3%)[n]
−1%		+3%	+9%
+3%		+3%	
−3% (−2%)	+3%	−0%	+6% (−3%)[m]
			+5% (−3%)[n]
+20%			+191%[m]
			+133%[n]

[h] Relative to 1974–83 baseline.
[i] Scenarios adjusted relative to GISS $2 \times CO_2$ mean climate.
[j] Relative to 1966–82 baseline.
[k] Estimates from *Figures 6.4–6.9*, Part VI.
[l] 10 consecutive occurrences of 1980-type climate from 1973–1982.
[m] $2 \times CO_2$ equilibrium climate.
[n] Transition climate from present (in 1966) to $2 \times CO_2$ (in 1982).

References

Baird, A., O'Keefe, P., Westgate, K. and Wisner, B. (1975). *Towards an Explanation and Reduction of Disaster Proneness*. Occasional Paper 11, University of Bradford, Disaster Research Unit.

Blasing, T.J. and Solomon, A.M. (1983). *Response of the North American Corn Belt to Climatic Warming*, DOE/NBB-004. Prepared for the United States Department of Energy, Office of Energy Research, Carbon Dioxide Research Division, Washington, D.C.

Bolin, B., Döös, B.R. Jäger, J. and Warrick, R.A. (1986). *The Greenhouse Effect, Climatic Change and Ecosystems*, SCOPE 29, Wiley, Chichester, 542 pp.

Callaway, J.M., Cronin, F.J., Currie, J.W. and Tawil, J. (1982). *An analysis of methods and models for assessing the direct and indirect impacts of CO_2-induced environmental changes in the agricultural sector of the US economy*, PNL-4384, Pacific Northwest Laboratory, Battelle Memorial Institute, Richland, Washington.

Carbon Dioxide Assessment Committee (1983). *Changing Climate*, National Academy of Sciences, Washington, D.C., 496 pp.

Climate Impact Assessment Program (CIAP) (1975). *Impacts of Climatic Change on the Biosphere*, Monograph 5, US Department of Transportation, Washington, D.C.

Garcia, R. (1981). *Drought and Man: The 1972 Case History. Vol. 1: Nature Pleads Not Guilty*, Pergamon, New York.

Hansen, J., Lacis, A., Rind, D., Russell, G., Stone, P., Fung, I., Ruedy, R. and Lerner, J. (1984). Climate sensitivity: analysis of feedback mechanisms. In J. Hansen and T. Takahashi (eds.), *Climate Processes and Climate Sensitivity*, Maurice Ewing Series, **5**, pp. 131-63, American Geophysical Union, Washington, D.C.

Hansen, J., Russell, G., Rind, D., Stone, P., Lacis, A., Lebedeff, S., Ruedy, R. and Travis, L. (1983). Efficient three-dimensional global models for climate studies: Models I and II. *Mon. Wea. Rev.*, **111**, 609-662.

Hare, F.K. (1985). Climatic Variability and Change. Ch. 2 in R.W. Kates, J.H. Ausubel and M. Berberian (eds.), *Climate Impact Assessment: Studies of the Interaction of Climate and Society*, SCOPE 27, Wiley, Chichester, pp. 37-68.

Jones, P.D., Wigley, T.M.L. and Wright, P.B. (1986). Global temperature variations between 1861 and 1984, *Nature*, **322**, 430-434.

Kates, R.W. (1985). The Interaction of Climate and Society, Ch. 1 in R.W. Kates, J.H. Ausubel and M. Berberian (eds.), *Climate Impact Assessment: Studies of the Interaction of Climate and Society*, SCOPE 27, Wiley, Chichester, pp. 3-36.

Maunder, W.J. (1986). *The Uncertainty Business: Risks and Opportunities in Weather and Climate*, Methuen, New York.

Mearns, L.O., Katz, R.W. and Schneider, S.H. (1984). Extreme high-temperature events: changes in their probabilities with changes in mean temperature, *J. Clim. Appl. Meteor.*, **23**, 1601-1613.

Meinl, H., Bach, W., Jäger, J., Jung, H.-J., Knottenberg, H., Marr, G., Santer, B., Schwieren, G. (1984). *Socio-economic impacts of climatic changes due to a doubling of atmospheric CO_2 content*, Commission of the European Communities, Contract No. CLI-063-D.

Müller, M.J. (1982). *Selected climatic data for a global set of standard stations for vegetation science*, Junk, The Hague, 306 pp.

National Defense University (NDU) (1980). *Crop yields and climatic change to the year 2000*, Vol. 1, Washington, D.C., Fort Lesley J. McNair.

Parry, M.L. (1985a). Estimating the Sensitivity of Natural Ecosystems and Agriculture to Climatic Change – Guest Editorial, *Climatic Change*, **7**, 1–3.

Parry, M.L. (1985b). Assessing the impact of climatic change on marginal areas. Ch. 14 in R.W. Kates, J.H. Ausubel and M. Berberian (eds.), *Climate Impact Assessment: Studies of the Interaction of Climate and Society*, SCOPE 27, Wiley, Chichester, pp. 351–367.

Parry, M.L. and Carter, T.R. (eds.) (1984). Assessing the Impact of Climatic Change in Cold Regions, *Summary Report SR-84-1*. International Institute for Applied Systems Analysis, Laxenburg, Austria, 42 pp.

Parry, M.L. and Carter, T.R. (1985). The effect of climatic variations on agricultural risk, *Climatic Change*, **7**, 95–110.

Parry, M.L., Carter, T.R. and Konijn, N.T. (1985). Climatic change. How vulnerable is agriculture? *Environment*, **27**, 4–5, 43.

Parry, M.L., Carter, T.R. and Konijn, N.T. (eds.) (1988). *The Impact of Climatic Variations on Agriculture. Volume 2, Assessments in Semi-Arid Regions*, Reidel, Dordrecht, The Netherlands.

Rosenzweig, C. (1985). Potential CO_2-induced effects on North American wheat producing regions, *Climatic Change*, **7**, 367–389.

Vinnikov, K.Ya. and Groisman, P.Ya. (1979). An empirical model of present-day climatic changes. *Meteorologia i Gidrologia*, **3**, 25–28 (in Russian). English translation available in: *Soviet Meteorol. Hydrol.*, **3**.

Warrick, R.A. and Gifford, R.M., with Parry, M.L. (1986). CO_2, climatic change and agriculture. Assessing the response of food crops to the direct effects of increased CO_2 and climatic change. In B. Bolin, B.R. Döös, J. Jäger and R.A. Warrick (eds.), *The Greenhouse Effect, Climatic Change and Ecosystems. A Synthesis of the Present Knowledge*, SCOPE 29, Wiley, Chichester, pp. 393–473.

WMO (1986). *Report of the International Conference on the Assessment of the Role of Carbon Dioxide and of Other Greenhouse Gases in Climate Variations and Associated Impacts*, Villach, Austria, 9–15 October 1985. WMO-No. 661, World Meteorological Organization, Geneva, 78 pp.

SECTION 2

The Choice of First-Order Impact Models

Timothy Carter, Nicolaas Konijn and Robert Watts

2.1. Introduction

The quantitative estimation of crop responses to climatic variations is a fundamental requirement for the studies reported in this volume. It is the procedure that translates a climatic perturbation (described by a climatic scenario) into a tangible first-order, biophysical effect (on crop production). Such a result can subsequently be used to evaluate further higher-order effects on agriculture and the wider economy. The quantitative tools that are used to evaluate the influence of climate on crops are termed *agroclimatic models.*

In this Section we offer an introduction to the agroclimatic models that are employed in the IIASA/UNEP case study experiments. We will focus primarily on agricultural crop models since nearly all the models described in the volume are of this type. However, much of our discussion is equally applicable to other plant models (such as the forest growth model reported in Part I, Section 5).

First, we describe briefly the complexities of the crop production systems to be modeled, showing how effectively such systems are adapted to present-day climatic variations and considering how certain components of a system might respond to changes in climate. Secondly, we look at the types of agroclimatic models that have been used to represent the crop production system. A model checklist is also introduced that summarizes features of the models used in this volume. The checklist is used, thirdly, to illustrate some advantages and limitations of using agroclimatic models for simulating the effects of present-day climatic variations. Fourthly, the additional problems (and some solutions) associated with estimating the effects of longer-term climatic change are discussed, also with reference to the checklist. Finally, to show how agroclimatic models have been used in this volume, we point to some different methods of presenting model outputs.

2.2. Climate and the Crop Production System

2.2.1. Environmental factors affecting crop growth and yield

To be able to show the effects of climatic variations on the physical processes determining crop growth, a model should respond to at least one of the variables that help to describe the climate. No matter where crop production takes place the following conditions determine the final production level: the radiative and temperature regimes and the soil water available for plant growth. There are also other environmental effects that are important for crop growth, and are themselves climate-sensitive, including the availability of plant nutrients and the interference of pests and diseases. These can be considered as indirect effects of the climate. When any of the above factors is not considered in a model, it often means that under the actual local physical environment that particular variable

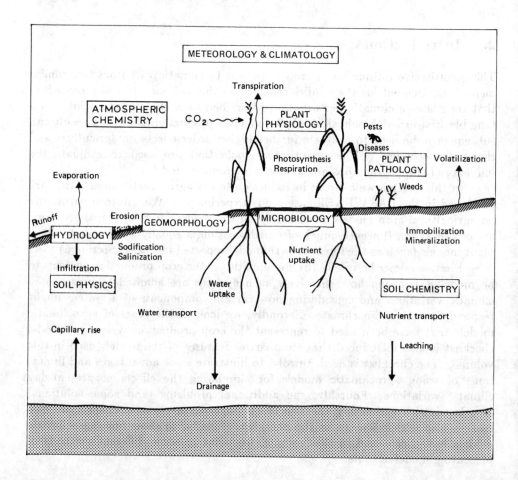

Figure 2.1. Physical processes in plant production and some research disciplines that study them.

does not induce short-term responses sufficiently large to affect crop yields. Such an omission may become significant when longer-term climate changes are considered. This is the case for atmospheric pollutants, particularly carbon dioxide, which are likely to have an important bearing on plant productivity in the future, independent of their effects through changes in climate. However, these issues are not within the scope of this volume.

The more important processes are illustrated in *Figure 2.1* and have been grouped according to their parent research discipline. While the major concern throughout this volume is to evaluate the effects on agriculture of changes in *climate* (the uppermost component depicted in *Figure 2.1*), the diagram reminds us that such investigations are at the interface of many research specialisms. Thus they require an interdisciplinary approach, as is evidenced by the authorship of each component in the case studies.

2.2.2. Constructing a crop calendar

Some agroclimatic models relate only climatic factors to crop yield, but we will take a broader view here, recognizing that other variables such as farm management activities not only contribute to the level of the crop yield, but can also be decisive in whether one obtains a yield at all. We can illustrate the full range of aspects that can be incorporated in a model by referring to a crop calendar (*Figure 2.2*). Spring wheat is taken as our example since it is an important crop in several of the regions described in this volume. Similar calendars could be constructed for other crops, however, based on information contained in the case studies and from elsewhere in the literature (e.g., FAO, 1978; Duckham, 1963). The models we are interested in can be compared with the appropriate calendar to see whether certain important aspects are included. Subsequently we will use a checklist to match appropriate models to particular tasks.

In *Figure 2.2*, various development stages of spring wheat are recognized which are related to the crop's special requirements. The following requirements are distinguished: radiation, water, nutrients, weed control, pest control, and soil conservation. These are represented over time as vertical lines within horizontal bars; the closer the lines in the bar, the more important are the requirements. In addition, each plant has its typical optimum and extreme temperature ranges that can vary according to the stage of growth. Similarly, a crop's sensitivity to sporadic extreme weather events such as frosts, hailstorms or strong winds, changes during its development.

For most of the crop requirements the farmer has tools to reduce any adverse effects of the environment. Thus, each of the requirements indicated in *Figure 2.2* is accompanied by a shaded horizontal bar that shows how farmers can act, given that they have the means and sufficient know-how. For example, soils may not be capable of meeting the nutrient requirements, and require the application of fertilizers at the right time and in an appropriate manner to increase production. To what degree these applied nutrients become available to the plant, however, is also a function of the weather. Moreover, although the farmer can counteract to a certain degree the negative effects of the weather,

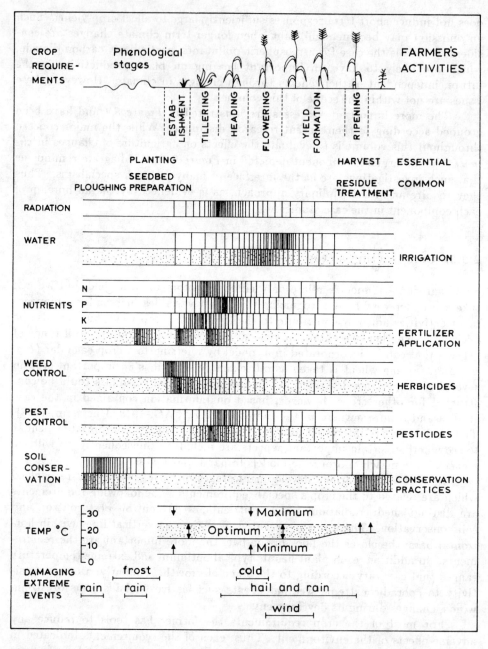

Figure 2.2. A crop calendar for spring wheat. The timing of important crop requirements (listed in the left hand column) and the farmer's activities to help meet these requirements (listed to the right of the diagram) are indicated within the accompanying horizontal bars (unshaded bars for plant requirements; shaded bars for management activities) as vertical lines. The closer the lines in the bars, the more important are the requirements or activities. The optimum range of temperatures and the sensitivity of a crop to extreme weather events are depicted in the lower part of the diagram.

losses in nutrients may still occur due to leaching, immobilization or volatilization.

2.2.3. Long-term effects of crop production

The crop calendar in *Figure 2.2* illustrates some of the crop requirements and farming activities that affect crop production on a short-term basis. Most present-day applications of agroclimatic models consider these factors and their interannual variations. However, other factors may become more important over the longer term. For example, soil properties such as organic matter content change naturally from year to year, partly as a function of climate, although anthropogenic activities tend to mask these background effects. Over the longer-term, however, man's activities can disguise cumulative adverse changes to soils such as depleted organic matter content that may be accelerated by climatic change and contribute to reduced crop production through processes such as soil erosion.

Additionally, while farmers may not currently be able to fulfill some of the crop requirements depicted in the crop calendar, developments in plant breeding may come to their aid in the future. Plant breeders may develop better adapted varieties, such as short-straw varieties where there is a risk of lodging, or quick-maturing varieties where growing seasons are short. The modeling of such adjustments in crop varieties as a response to climatic change is discussed below.

2.3. Agroclimatic Models

Having described some of the factors that influence crop production and yield, we now consider how these various factors can be isolated and their impacts quantified within a formal model framework. We can view a model as a set of input variables that lead to a set of output variables through a number of mathematical relationships. In an agroclimatic model, the output variable of primary interest is usually some measure of crop productivity (e.g., yield, biomass potential, land suitability). Input variables can include any combination of climatic variables such as temperature, precipitation, windspeed, and solar radiation, other environmental variables (e.g., soil properties), and various management variables (e.g., irrigation and the application of fertilizers and pesticides).

We can distinguish between two general techniques for examining the response of agricultural crops to climatic variations. The first provides relative measures of crop suitability through the use of agroclimatic indices. The second facilitates estimates of potential crop productivity by means of a crop-climate modeling procedure. Both techniques, summarized below, have been widely employed in the case study experiments reported in this volume.

Table 2.1. Some measures used to describe temperature conditions for plant growth.

Index	Critical temperature	Data requirements	Calculation technique	Summation period	Region of use and reference to part/section in this volume	Purpose of use
Effective temperature sum (ETS)	5°C	Daily mean temperatures	Summation of temperatures above critical temperature	Monthly	Finland IV/2 and 3	Development phases of barley and spring wheat
					Finland IV/4	Regression models for barley and oats
					N. USSR V/1 and 2	Characterization of thermal resources
Effective temperature sum (ETS)	5°C	Monthly mean temps and st. deviation of daily mean temps about the monthly mean	Approximate method	Annual	Northern hemisphere zone (38–70° N) I/5	Regression model of forest growth
Growing degree-days (GDD)	5°C	Long-term mean and st. deviation of monthly temps	"Thom" method (Thom, 1954a,b)	Annual	Saskatchewan (Canada) II/2	Characterization of thermal resources and potential for maize cultivation
Warmth index	5°C	Monthly mean temperatures	12 month summation of temps above critical temp. × no. days in month	Annual	Japan VI/3	Characterization of thermal resources during the effective growing season

Name	Critical temp	Variable	Computation	Period	Region	Purpose
Coldness index	5 °C	Monthly mean temperatures	12 month summation of temps *below* critical temp. × no. days in month	Annual	Japan VI/3	Characterization of relative severity of temperatures during the winter half-year
Effective accumulated temperature	0 °C	Daily mean temperatures	Summation of temperatures *above* critical temperature	Period when daily mean temperatures are above 10 °C	N. USSR V/1 and 2 Japan VI/1 and 3 N. Japan VI/4	Characterization of thermal resources Regional evaluation of average and variability of thermal resources Evaluation of cultivable area for rice
Forest growth units	1.6 °C	Monthly mean temperatures	Computation of monthly temperature index and summation over 12 months	Annual	Iceland III/2	Assessing the probabilities of successful tree growth
ETS (requirements for barley)	3 °C	Daily mean temperatures; latitudes; total May–Sep precipitation	Summation of temps *above* critical temp. adjusted for latitude and precip.	Annual	Iceland III/2	Assessing the probabilities of barley ripening
Cooling degree-days	22 °C	Daily mean temperatures	Summation of temps *below* critical temp.	Early reproductive phase of rice development	Hokkaido (Japan) VI/5	Approximate equation for evaluating sterility of rice spikelets

2.3.1. Agroclimatic indices

In essence, an agroclimatic index is a derived variable that is defined either by manipulating values of a meteorological variable into a different form or by combining variables with empirically-derived coefficients into a composite term. For the impact analyst such indices offer two advantages. Firstly, an index constitutes a single term to which crop growth and development is found to be particularly responsive. Secondly, the statistical problem of collinearity amongst meteorological variables is minimized and the number of degrees of freedom reduced (Robertson, 1983).

Indices have been developed to examine various aspects of the agroclimate, either singly or in combination. For example, the most common derived variable used to describe the thermal agroclimate is Effective Temperature Sum (ETS), usually measured in units of growing degree-days. ETS is frequently adopted to estimate the timing of crop development stages, and is calculated as the integrated excess of temperature above a fixed datum (base temperature) over a period required for a specific phase of development (Parry and Carter, 1984).

Given such an apparently straightforward concept, it is perhaps surprising that there exist so many different (and potentially confusing) names and computational methods for what is essentially the same measure. Nine different approaches have been identified in this volume alone. They are summarized in *Table 2.1*. Several authors calculate ETS using daily mean temperatures (e.g., for the Finland study using temperatures above a base temperature of 5 °C). Others utilize monthly mean data to compute degree-days, either with no consideration of within-month temperature variability (e.g., for Japan temperatures above a 5 °C base) or employing special techniques to simulate the daily temperature distribution (e.g., for the latitude band 38–70 °N above a 5 °C base). For both the USSR and Japan, effective accumulated temperatures are computed above a 0 °C base for daily mean temperatures during the period when daily mean values are above 10 °C, and a base of 1.6 °C is used to calculate Forest Growth Units in Iceland, a measure analogous to ETS.

Rather than calculating temperature sums *above* a threshold value, some studies utilize the same procedure to derive cooling degree-days by integrating temperature *below* a given threshold. These measures have been used in the Japanese study for evaluating the sensitivity of rice to cool temperature conditions (*see Table 2.1*).

The duration of the effective growing period for crops can also be represented by various measures, such as the Frost-Free Period in the northern USSR (Part V, Section 2) or the Effective Period in Japan (Part VI, Section 3). Furthermore, moisture conditions are very important in influencing crop growth and also in affecting the cohesive properties of some soils. Two moisture indices have been used in the Saskatchewan Study (Part II, Section 2): the Palmer Drought Index and Thornthwaite's Precipitation Effectiveness Index, the latter forming a component of the wind erosion model reported in Section 3 of that report. Another, the Relative Dryness Index (RDI), has been correlated to world vegetation types, and forms one component of the Chikugo model used in the Japan case study (Part VI, Section 3).

Finally, two indices are reported in the volume that combine several different aspects of the plant environment in a single term to provide an indicator of potential productivity. The first, used in the Saskatchewan case study, combines a "heliothermic factor" (characterizing thermal and radiative conditions) with a "moisture factor" in a "climatic index of agricultural potential" (*see* Part II). The index is used as a relative measure of potential biomass productivity in different parts of Saskatchewan for different climatic scenarios, and also provides useful information on agroclimatic conditions within a growing season that can indicate which cropping practices might be suitable under a given climatic regime.

The second, already mentioned, is the Chikugo model for estimating net primary productivity. This relates the potential productivity of natural vegetation to local radiation conditions and to the ability of plants to exploit these conditions, a function of moisture conditions represented by the RDI. The model is used to estimate potential shifts in zones of net primary productivity under different climatic scenarios in Japan (*see* Part VI). Since the Chikugo model estimates plant productivity in absolute terms, it bears a close resemblance to certain types of models used for estimating crop yields. Thus, we now turn from more general estimates of agroclimatic and biomass potential, to consider models for specific crops.

2.3.2. Crop-climate models

We have adopted the term "crop-climate model" to represent a range of models that relate crop variable(s) to meteorological variable(s), irrespective of time scale (WMO, 1985). Thus, unlike some other common descriptors (e.g., "crop-weather" or "crop-yield"), this one focuses explicitly on responses of crop yields to *climatic* variations, but over a *range of time-scales*, which is convenient from the point of view of the different climatic scenarios employed in the model experiments in this volume.

There are numerous reviews of crop-climate modeling in the literature (e.g., Baier, 1977, 1982; Sakamoto, 1981; Robertson, 1983; Nix, 1985; WMO, 1985) and all attempt some kind of model classification. For our purposes, two broad classes of model can be distinguished: empirical-statistical models and simulation models.

Empirical-statistical models are developed by taking a sample of annual crop-yield data together with a sample of weather data for the same area and time period, and relating them through statistical techniques such as multiple regression analysis. This procedure is sometimes labeled a black-box approach since it does not easily lead to a causal explanation of the relationships between climate and crop yield [*Figure 2.3(a)*].

This description should not imply, however, that these models are developed blindly or indiscriminately. The most effective empirical-statistical models are usually the product of a careful and well-informed selection of suitable explanatory variables, based on close understanding of basic crop physiology.

a) Empirical/statistical

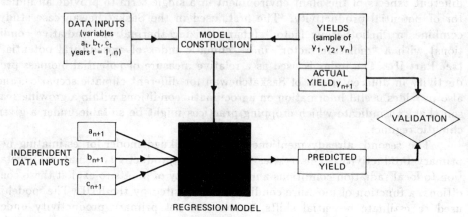

b) Simulation

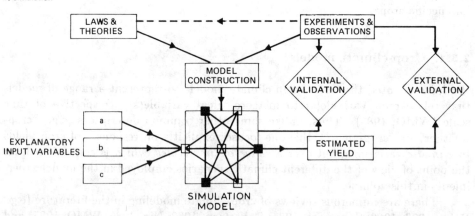

Figure 2.3. The construction, operation and validation of (*a*) an empirical-statistical (black-box) model and (*b*) simulation model (schematic).

Simulation models generally treat the dynamics of plant or crop growth over the growing season through a set of mathematical expressions tying together the interrelationships of plant, soil and climate processes. Some of these relationships are well enough understood to be regarded as accepted laws of physics, chemistry and biology and are often referred to as deterministic functions (Lyons, 1982). Other processes which are either poorly understood or of secondary interest to the modeler are frequently represented by empirical functions.

Thus, no simulation model can be described as truly deterministic since all incorporate at least some empirical (black-box) elements. However, they differ from empirical crop models in their development and operation. In the simulation approach, plant growth processes are prespecified, and output data are *generated internally* by the model, as illustrated in *Figure 2.3(b)*.

An intermediate type of approach employing simple deterministic mathematical functions in "mechanistic schemes" has also been identified in the literature (e.g., WMO, 1985) but since they, like more sophisticated computer simulation models, are process-based and operate over the development phases of a crop, we have subsumed them into the simulation model category.

Both types of model are used in the case studies in this volume to estimate crop responses to climatic variations of two kinds:

(1) Short-term anomalies that have been observed in (or are within the range of) the instrumental climatic record.
(2) Longer-term climatic change estimated for future doubled concentrations of atmospheric carbon dioxide.

While some models were not developed with the explicit intention of examining effects of climate or weather on crops, most are able to account for present-day climatic conditions, including short-term variations of climate mentioned above. The advantages and limitations of using different models for this application are evaluated in Subsection 2.4. The additional problems of using such models for analysis of crop responses to longer-term changes in climate that are outside the historical range of climatic observations are discussed in Subsection 2.5. For many of the model characteristics examined, the reader is referred to the corresponding column heading in a model checklist (*Figure 2.4*). This is used in the following sections to provide a broad assessment of the crop-climate models used in this volume.

2.4. Evaluation of Models for Estimating Effects of Short-Term Climatic Variations

We can evaluate a model according to the following general criteria: objectives, data requirements, assumptions and sources of error, and validation.

2.4.1. Objectives

In general, models are constructed in order to satisfy certain objectives. These condition what levels of detail and explanation are required in developing the model. As a simplification, empirical statistical models are built for the sole purpose of yield prediction; simulation models offer a means to analyze plant processes as well as to estimate yield. Model objectives are categorized under "intended outputs" in *Figure 2.4*. Differences in modeling objectives will become apparent in the following discussion, but it is important to emphasize here that few models have been developed with the explicit objective of estimating crop responses to climatic change.

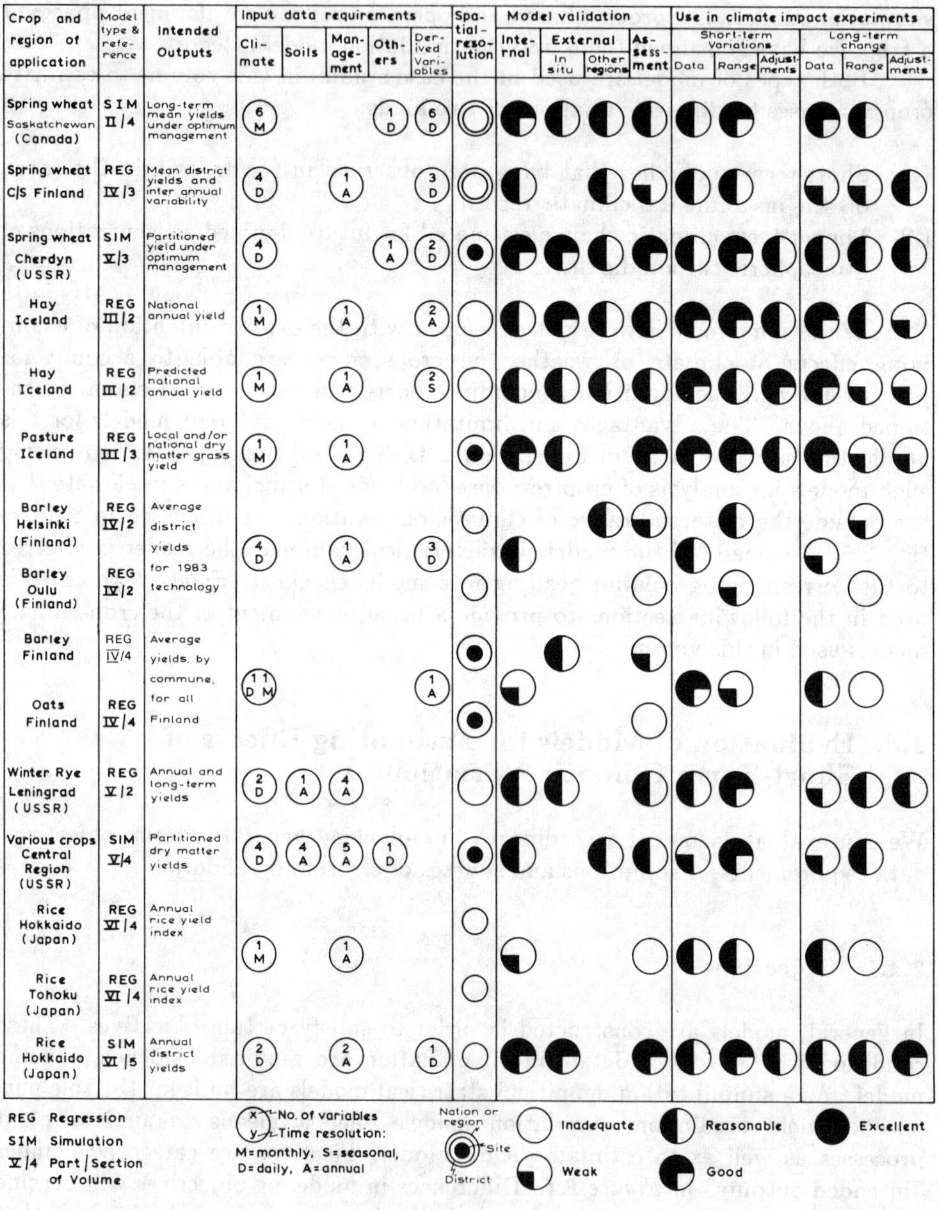

Figure 2.4. A checklist of the crop-climate models employed in case study experiments (for explanation, *see* text).

2.4.2. Data requirements

A model is only as good as the data set upon which it is based. Depending on the level of sophistication of a model, its data requirements may include information on climate, crop yields, soils, drainage, other site characteristics, management practices, and other factors, each at a certain time resolution and representing a particular site or region. Adequate data are required first, for constructing the model algorithms, second, for running the model (i.e., as inputs) and third, as independent objective measures against which to validate model outputs.

In general, fewer and more simple model variables allow use of a longer time series of data and permit a more straightforward and rapid calculation process. Most empirical-statistical models are constructed to meet this objective, both for the purpose of simplicity and to avoid the statistical problem of inflated correlation coefficients due to a high number of explanatory variables (Sakamoto, 1981). Continuing this line of reasoning, it is no surprise that some of the most effective statistical models have been developed in areas where variations in crop growth and yield are governed by a *single* major weather factor (e.g., temperature and hay yields in Iceland, *see* Part III).

For an empirical treatment of a more complex system, data bases may become prohibitively large, and in this situation the explanatory, simulation approach may be preferable since some of the variables can be generated internally by the model. However, data requirements are normally greater for simulation models because of the short time-steps frequently employed; and such data are seldom available over long time periods. This imposes a formidable restriction on the applicability of many simulation models for the analysis of climatic variations over more than a few years.

An additional logistical advantage of most empirical models lies in their prediction of yields for quite large regions. Thus, climatic data are often regional averages derived from a number of sites, any one of which need not necessarily provide a continuous record. Conversely, simulation models tend to require a single, unbroken run of data for all variables at a specific site.

Details about the type, number and temporal resolution of model input data are accommodated under the heading "input data requirements" in the model checklist (*Figure 2.4*). The practical availability and manageability of these data for conducting climate impact experiments are assessed separately under the heading "data".

2.4.3. Assumptions and sources of error

The effectiveness of a model, for whatever purpose it is designed, depends to a large extent on the nature and validity of its assumptions. Although each model has its own unique set of assumptions, some are common to most models.

Most of the weaknesses of empirical models can be attributed to their statistical assumptions. Firstly, in classical linear regression analysis it is assumed that variables are independent. Unfortunately, few of the climatic variables that affect crop yield are *not* related to each other and many are also related to themselves over time (i.e., are autocorrelated). These problems of *multicollinearity* usually result in regression coefficients with large, unstable variances. Secondly, the relationships between crop yields and climatic variables are rarely linear. Where they are modeled as such this can also lead to inflated variance, and the resulting coefficients may be unstable to the extent that their sign can change with the addition of new observations (Biswas, 1980).

One potential solution for reducing collinearity is to combine intercorrelated variables into an *agroclimatic index* (see above). Surrogate variables such as these have been used to represent soil moisture information or radiative thermal conditions and can be related directly to potential agricultural productivity (e.g., the climatic index of agricultural potential used in the Saskatchewan case study, *see* Part II).

A third, simplifying assumption that pervades most models but may have a greater influence in the empirical-statistical type, concerns the *interpolation* and *extrapolation* of values. Since most models are developed and "tuned", using functions pertaining to an observed range of present-day conditions, it is likely that interpolation will be the dominant procedure employed in model runs conducted under "normal" conditions. However, altering the climate may introduce values outside the modeled range. The problems of extrapolating existing model relationships are examined in Subsection 2.5.1.

A fourth assumption, which affects both model types but in different ways, involves the inclusion of *management practices and technology* as explanatory variables. In simulation models these are either included as explicit input variables (a separate category in *Figure 2.4*), or are specified as constants. In the latter case, this assumption may restrict the model's effectiveness in estimating long-term effects of technological change.

Empirical-statistical models often treat technological change as a time trend based on yield statistics over a relatively long period of time. This procedure has two disadvantages. First, it relies upon the subjective separation of technological trends from those that can be attributed to the environment (including climate). Secondly, as in the case of simulation models, a future scenario requires some assumptions about the extrapolation of the technology trend.

Finally, *sporadic climatic events* such as hail storms, floods, late or early frosts can often have devastating effects on crops. While simulation models generally make allowance for these episodic events through the use of critical tolerance levels, empirical-statistical models often do not because of their (usually) short duration, and this can have implications for the analysis of longer-term climatic change.

We have given short descriptions here of some of the more important model assumptions and sources of error. (More complete discussions of these and others can be found in the reviews by Sakamoto, 1981; Baier, 1982; Biswas, 1980; Katz, 1979 and WMO, 1985.)

2.4.4. Validation

The ultimate test of any model is to assess how closely its predictions correspond to actual measured observations. The whole credibility of a model's forecast of crop response rests on the rigorous testing of its sensitivity and verification of its outputs against independent observations. These aspects thus form an important component of the model checklist (*Figure 2.4*).

A *sensitivity analysis* can tell us a lot about the inherent stability of a model, its effective range of operation and its potential applicability to climatic change experiments. In essence, a sensitivity analysis seeks to evaluate a model's responses to incremental changes in magnitude of each input variable (both singly and in combination). These adjustments should be distributed such that the extreme values lie well outside the observed natural range. Similar procedures are employed for those mechanisms, including feedback processes, that are generated internally, and for the range of uncertainty surrounding parameter values. In this way, we can gain insights into:

(1) The physical realism of the modeled relationships over a wide range of conditions.
(2) The relative importance of the various types of external forcing.
(3) The relative influence of each modeled variable in determining model outputs.
(4) Some of the model's limitations, including information concerning:

   (a) the conditions under which the model breaks down.
   (b) which of its coefficients are potentially unstable, under what circumstances and with what effect (e.g., see Katz, 1979).

In order to appraise its performance, it is necessary to *validate* a model's estimates against observed data. This procedure has a number of attendant difficulties, not least the problem of securing adequate data for its implementation. The types of data required for validation depend to a large extent on the nature of the model's predictions.

As a minimum requirement, the outputs from an empirical-statistical model should be verified against at least ten years of independent crop yield data, reflecting a variety of conditions. Ideally, these should be selected from a different area than that for which the model was constructed but for the same time period, in order to test the model's common applicability (referred to as "other regions" in the "external" validation category – *Figure 2.4*). However, because of the area-specific nature of most statistical models (due in part to their general neglect of variables such as soils, drainage and management that are often highly heterogeneous over space) this form of cross-validation is rare. Model results are usually compared with observations from the same area but for a different time period [*Figure 2.3(a)* and the heading "in situ" in *Figure 2.4*]. Further validation techniques are also available as alternative or (preferably) as complementary tests of model reliability (e.g., Draper and Smith, 1981), but these have seldom been used for such purposes (and not at all in this volume).

For simulation models, the validation requirements are slightly different. Firstly, the data requirements are usually very much greater than for statistical models. A proper validation procedure should include not only a comparison of modeled outputs with real conditions, but also the verification of each and every internally derived value [*Figure 2.3(b)* and the category "internal" in *Figure 2.4*]. In some instances, however, it may be either impractical or even technically impossible to measure a particular variable.

Secondly, the model structure should be scrutinized for internal consistency and the model's behavior tested under a range of different possible conditions (WMO, 1985). This is one of the main functions of a sensitivity analysis (discussed above).

Thirdly, in validating any type of model care should be taken to ensure that the independent observations being used are totally compatible with the model estimates. Efficient scrutiny of the data may uncover unforeseen peculiarities to which a model may or may not be sensitive (such as the introduction of irrigation to an area, the replacement of a crop hybrid, or a change in the total area under the crop).

In order to provide an approximate judgment on the demonstrated validity of the models used in this volume, we have examined the validation procedures and results reported for each model, and entered our findings under the category "assessment" in *Figure 2.4*.

2.5. Evaluation of Models for Estimating Effects of Longer-Term Climatic Change

For the purposes of estimating crop responses to future climatic change, it is tempting simply to substitute information on the new climatic conditions into existing agroclimatic models and to study the model outputs accordingly. However, even if a model performs satisfactorily under present-day conditions according to the criteria outlined in the previous section, the utmost caution is required in simulating crop responses to a long-term change in the climate. Three categories of problem need to be overcome in conducting these studies:

(1) Problems relating to the validity of model relationships.
(2) Difficulties concerning model linkages with respect to scenarios of climatic change.
(3) Problems of linking agroclimatic models to models of higher-order effects.

2.5.1. Model relationships

There are four aspects that may affect the validity of model relationships: the range and applicability of model functions, the importance of natural long-term environmental processes, the future influence of management and technology on agriculture, and the effects of other environmental factors.

Model functions

The extrapolation of a model's functional relationships outside the range for which they were developed and validated is one of the most common errors in conducting climatic change experiments. In the model checklist (*Figure 2.4*) we have included a category labeled "range" to assess the level of confidence that can be placed in model estimates of crop response to the range of both short- and long-term climatic variations. Not being process-based, empirical-statistical models are especially likely to contain functions that require reevaluation for the changed conditions. Simulation models may be able to accommodate certain features of a climatic change, however. For instance, a change in climate may alter the importance of critical threshold values for crop growth that are often included in simulation models. Alternatively, if a crop is especially vulnerable to the effects of episodic events, then it is of critical importance that such events should be modeled effectively, for any change in their frequency may have damaging consequences (Rosenberg, 1982).

Both of the above cases can be illustrated with reference to changes in frequency of sporadic low- and high-temperature damage to rice in northern Japan. Using an empirical rice yield index, the effects of these weather anomalies can be inferred from an implicit relationship between damage and mean monthly temperatures (the input data for the model). In contrast, these short-term temperature anomalies can be input directly and their effects modeled explicity in a simulation model for the Hokkaido region (*see* Part VI, Sections 4 and 5).

Another weakness exhibited by many empirical-statistical models stems from their omission of important explanatory variables that are correlated with modeled variables. Under a changed climate the relative influence of such variables on crop response may change thus invalidating the original relationships. Further comparison of the two rice yield models serves to illustrate this. The strong relationship between rice yield and temperature under present climatic conditions in the rice yield index is partly an effect of above-average temperatures being associated with above-average solar radiation conditions (a relationship implicit in the index). However, while temperatures increase significantly under a doubling of atmospheric carbon dioxide concentrations according to the estimates of the Goddard Institute for Space Studies (GISS) general circulation model (GCM), levels of solar radiation change little. The temperature-yield relationship as used in the yield index would thus be weakened, but since both variables are incorporated in the simulation model that model can accommodate the climatic change.

One method of justifying the extrapolation of a model relationship is discussed in the Iceland case study, involving the use of a regional analogue. Under the GISS $2 \times CO_2$ scenario, the climate of Iceland resembles the present-day climate of northern Britain. By inferring from this that future hay yields in Iceland might attain similar levels to current yields in Scotland and northern Ireland, present-day temperature-yield relationships in Iceland have been tentatively extrapolated to the $2 \times CO_2$ climatic conditions (*see* Part III).

Long-term environmental processes

Over short time periods, in situations where the rates of change of certain environmental processes are relatively slow, their effects may be imperceptible. Thus, for the purposes of modeling short-term crop response, they can often be fixed or omitted completely. However, in a consideration of longer term climatic changes over periods ranging from decades to centuries, these processes may exert an influential and possibly dominant role in crop growth. It is therefore important to incorporate them in models, at least on an annual updated basis. Examples include changes in soil properties (*see* Subsection 2.2.3), and the efficiency of field drainage. In the model checklist, the ability of models to account for these longer-term changes can be inferred from the nature of the input variables that are required (usually on an annual basis – *Figure 2.4*).

Management and technology

Some of the assumptions that are made in model experiments about past trends in management and technology were discussed in Subsection 2.4.3. Projecting these trends with any confidence into the near future is very difficult, however, and over the longer term (beyond about 5–10 years) almost impossible, particularly in highly developed agricultural systems where technological advancement is a function of research and innovation, and where management practices are subject to world market forces (both highly unpredictable entities).

In agroclimatic models where management and technology factors are represented, two approaches have been adopted to handle the effects of climatic change. The first specifies a time horizon for the climatic change (say 50 years) and extrapolates the management/technology trend over that period. For example, for projecting the effects of climatic change over the period 1980–2035 in the Leningrad region (northern USSR), it was assumed that regional fertilizer applications (a model input) would increase linearly from observed 1980 levels to levels currently recorded on experimental plots by the year 2035 (the experimental plot data thus placing an upper limit on the extrapolation – *see* Part V, Section 2).

The second (more common) approach, sidesteps the issue altogether by estimating the effects of climatic change under present-day management and technology (i.e., fixed at *ca.* 1980 levels). However, even when this approach is adopted, if a management factor can be modeled explicity (as in the Leningrad case), it may be instructive to study a crop's sensitivity to changes in management, not as a scenario of future management conditions but rather as a potential adjustment to mitigate the effects of climate. *Adjustment experiments* of this sort have been conducted in several case studies. For example, modeled fertilizer applications have been adjusted to simulate how temperature effects on hay yields may be offset in Iceland (Part III), and altered planting dates of newly introduced rice varieties have been input to a rice yield simulation model for northern Japan to study the effects on annual yields (Part VI). Columns have been included in the checklist to indicate to what extent and with what degree of success models have been used in adjustment experiments (*Figure 2.4*).

Other environmental factors

Attendant on, and possibly contributing to, climatic change are likely to be other factors that influence agricultural productivity, such as atmospheric, soil and water pollution. One attempt, within a model framework, to simulate the combined direct and indirect effects of doubled carbon dioxide concentrations on crop yield is reported for the Cherdyn region (northern USSR), using a spring wheat simulation model (Part V). Although strictly outside the scope of this volume, such model experiments are, nonetheless, rarely conducted and merit further study (Warrick et al., 1986).

2.5.2. Linking first-order impact models to scenarios of climatic change

A first consideration for modelers wishing to simulate the effects of future climatic change is how best to quantify that change. In this volume we consider three basic types of scenario of future climate:

(1) Instrumental scenarios, using past observed climatic anomalies as analogues of future conditions.
(2) Synthetic scenarios representing arbitrary adjustments to values of climatic variables.
(3) General circulation model (GCM) scenarios of climate under doubled concentrations of atmospheric carbon dioxide.

The temporal and spatial resolution of these scenarios varies widely, and it is this consideration which is a major determining factor in the selection of an appropriate scenario as an input to a particular model. With the first two types of scenario, data are often available at the level of detail appropriate for direct input into a model, but if not, the kinds of difficulties encountered are likely to be similar to those found with the GCM scenarios. In the following, therefore, we address three main problems encountered in using GCM outputs as inputs for agroclimatic models: the matching of data, the specification of a baseline, and the consideration of time-dependent changes.

Matching of data

The GISS GCM provided the estimates for the $2 \times CO_2$ climatic scenario used by impact modelers in this volume (Part I, Section 3). In common with the outputs from other GCMs, the GISS model-derived data were of a very restricted nature. Values for only three climatic variables were provided (mean air temperature, precipitation rate and cloud cover) at a monthly time resolution, and for a 4° latitude × 5° longitude network of grid points over each study area. Some solutions to these difficulties are described below.

Missing variables. Several methods were adopted to account for climatic variables not provided but required as inputs to the agroclimatic models. In one method the missing variables were fixed at present-day, baseline values. For example, baseline values of solar radiation or hours of sunshine were used in experiments with the spring wheat model in the Saskatchewan study (Part II), with the rice simulation model and with the Chikugo model for net primary productivity in the Japan study (Part VI). A second method employed empirical techniques for computing missing variables on the basis of other climatic data. For example, in the spring wheat model used in the northern USSR study (Part V), sunshine hours and vapor pressure were both estimated (using regression equations) from temperatures that were available from the GISS model. Thirdly, where maximum and minimum temperatures were required as model inputs, it was usually assumed that the magnitude of change in these variables between baseline and $2 \times CO_2$ conditions corresponded to the changes estimated by the GISS model for mean temperatures (e.g., for the spring wheat model experiments in Finland – see Part IV).

The second, regression method may provide quite reliable estimates of missing variables for climatic scenarios within (though not outside) the range of instrumental observations. However, none of these three alternative procedures is an effective substitute for observed or GCM-generated data, where the changes in different variables are (or should be) internally self-consistent (WMO, 1985).

Temporal resolution. Although the monthly time step of GISS model outputs matched the input requirements of some agroclimatic models (e.g., the hay yield and pasture yield models used in Iceland, *see* Part III; and the rice yield indices used in Japan, *see* Part VI), many models required input data at a higher temporal resolution (*Figure 2.4*). One method of improving the compatibility of time scales where the time resolution of the predicted climatic variables is too coarse for input into an agroclimatic model, is to generate a synthetic set of finer-scale data. For example, in the Saskatchewan study a sine curve interpolation technique is used to generate smoothed daily temperatures from monthly means, and weekly precipitation is adjusted by allocating fixed proportions of the weekly total to particular days (Part II). These techniques are particularly useful for processing GCM-generated mean data, for the spring wheat model was developed to estimate long-term average yields. However, the model has also been used to estimate yields under "extreme year" scenarios, the character of which are most accurately represented by *observed* daily data. This explains the higher rating in *Figure 2.4* accorded the (less demanding) data requirements of this model for experiments with long-term changes than with short-term variations in climate.

Alternatively, instead of improving the time resolution of the input data, another compromise is to run the impact model at a coarser resolution. A third method is to assume that the values for changes in monthly mean climate provided by the GISS model could be used in adjusting all daily or weekly values within the appropriate month (i.e., an assumption of unchanged within-month variability between baseline and $2 \times CO_2$ conditions). This method was used in adjusting the input data to the rice yield simulation model for Japan (Part VI).

Spatial resolution. The spatial resolution of the GISS model-derived gridpoint data (already interpolated from the original 8° latitude × 10° longitude GCM outputs – see Part I, Section 3) is too coarse to provide the level of local detail required in many of the agroclimatic models indicated in *Figure 2.4*. To illustrate, four adjacent grid points straddling latitude 60° N enclose an area of about 125 000 km^2 (greater than the entire surface area of Iceland, or about one-third of the area of Finland). It was thus necessary to employ spatial interpolation procedures to evaluate the GISS model-derived climatic changes for the locations or regions specified for agroclimatic model experiments. Techniques include computer interpolation routines, manual interpolation from mapped isolines of the climatic changes, and weighted interpolation procedures based on the distance of the site location from adjacent GISS model grid points. Clearly, these interpolation procedures may disguise significant sub-grid-scale variations in climate that can occur both under present-day and (undoubtedly) under future conditions. But until finer-scale GCM predictions are available, information on future local climatic variations must be based on observed present-day conditions viz, the climatological baseline.

Specifying a baseline

In order to provide reference points with which to compare both scenarios of future climate and their modeled effects on agricultural productivity, two types of "baseline" condition need to be specified. Firstly, the baseline *climate* should be representative of present-day climatic conditions in the study region. For experiments reported in this volume, the climatological baseline period adopted was 1951–1980, and data from this period (when available) were used as inputs to agroclimatic models. These data provide a fairly accurate representation of the local-scale climate under which the modeled crops are cultivated. This is in contrast to the GISS model estimates of present-day climate which are neither local scale (see above) nor accurate (*see* Part I, Section 3). If the GISS model estimates of 2 × CO_2 climate are judged to be similarly inaccurate, then, rather than using these estimates directly to specify the 2 × CO_2 climate, an alternative approach can be employed. Throughout this volume it is assumed, therefore, that the *change* between GISS modeled 1 × CO_2 and 2 × CO_2 equilibrium conditions (interpolated to the study locations, as described above) can be used to describe the change between the present-day (baseline) and a future 2 × CO_2 climate.

In most crop-climate models, it is necessary to specify a second baseline condition that represents the present-day levels of *management and technology*. With empirical-statistical models that include a time trend (*see* Subsection 2.4.3) it is customary to specify a point on the trend line to represent the baseline condition (for example, the estimates of barley yield in Finland under each climatic scenario were projected to a 1983 baseline on a time trend, *see* Part IV, Section 2). With some empirical-statistical models and most simulation models, management factors comprise independent model variables, for which baseline values can be specified (for instance, 1950–1983 baseline fertilizer applications at experimental stations were used in scenario experiments with the pasture yield model in the Iceland case study, see Part III, Section 3). Finally, where models incorporate neither a trend term nor independent management variables, crop yields modeled for the baseline climate may need to be adjusted to correspond to actual crop yields over the baseline period (for example, simulated rice yields in Hokkaido are adjusted by a factor of 0.84 to match recorded yields during the baseline period, 1974–1983, see Part VI, Section 5).

Time-dependent changes

A third set of problems encountered in linking scenarios of climatic change to agroclimatic models concerns the time development both of the climate and of the simulated effects. Most $2 \times CO_2$ estimates from GCMs (including the GISS model estimates used in this volume) represent an equilibrium state of the modeled climate that is reached following a step-like (instantaneous) doubling of CO_2. No information is provided about either when the doubling will occur, or the nature of the transition between present-day and $2 \times CO_2$ equilibrium climates – the *transient* response (Part I, Section 3). In the absence of this information, it is common to use the $2 \times CO_2$ estimates as inputs to agroclimatic models directly, as a *step-like* climatic perturbation. A model is thus simulating a long-term change in average climate as if it were part of the year-to-year variability. With only one agroclimatic model, in the USSR study, is an attempt made to account for transient climatic change (Part V, Section 2). Annual climatic input data are obtained by interpolating linearly between the mean baseline climate (1951–1980), the climate in 2005 (empirically-derived) and the GISS $2 \times CO_2$ scenario climate (assumed to occur in 2050). The model simulates both the direct effects and the indirect effects (through changes in soil conditions) of this climatic change on average winter rye yields during the period 1980–2035.

2.5.3. Linking first-order impact models to models of higher-order effects

Several examples are presented in this volume of model linkages where output data from an agroclimatic model have been used as input data to other models that assess higher-order effects of a climatic scenario. The latter models include, for example, models of farm income, of land allocation, of employment or of food stocks. Each has its particular data requirements that are not always matched by the outputs from agroclimatic models. Four "mismatches" are illustrated here.

Requirements for yield estimates for more than one crop-type or variety

As an example of this, the farm simulation model used in the Saskatchewan study requires information on grain yields of four crops under different climatic scenarios, although estimates are only available for spring wheat yields from an agroclimatic model (Part II). In this instance, the percentage changes in spring wheat yields relative to baseline yields are assumed to be applicable to the other three crops also, in order to conduct the experiment. Otherwise, four separate agroclimatic models would be required to estimate the independent responses of each crop to the climatic scenarios. Alternatively, as demonstrated in the northern USSR case study, a single model may be capable of simulating the responses of different crops, given that certain model parameters specific to each crop can be determined. In this example, yield estimates were provided by a crop production model for seven types of crop as inputs to a regional agricultural planning model (Part V).

Requirements for crop yield estimates under different management practices

These can sometimes be simulated directly by an agroclimatic model, but where this is not possible, information about the ratios of reported yields under each management practice may need to be applied to modeled yield estimates to provide adequate input data for the second-order model. For example, the ratios of spring wheat yields on fallow to yields on stubble observed in 1979–1980 in Saskatchewan are used to estimate fallow yields from modeled stubble yields (Part II).

Requirements for aggregated data

Agroclimatic models rarely produce estimates of productivity or yields for the same spatial units as are required in modeling higher-order effects. The former models, even if they have large-area applicability, usually consider smaller subunits, such as crop districts or communes (*see Figure 2.4*). The latter models, in contrast, are commonly developed to characterize conditions at administrative district, economic unit, provincial or even national scale. Aggregation from one scale to another is therefore necessary; for example, from mean crop district yields to mean yields by soil zone, and subsequently to yields by farm type that were needed as inputs to the farm simulation model in Saskatchewan (Part II).

Requirements for time-dependent data

Many economic models operate over a simulation period of several years, in order to accommodate realistic interannual variations in exogenous driving variables. Thus, annual estimates of agroclimatic productivity are often required as inputs to such models. Some agroclimatic models can, of course, supply such information (e.g., the spring wheat yield model for Saskatchewan estimated yields for each year of the scenario period 1933–1937, with these 5 annual values being introduced as years 3–7 in the 10-year sequence required for operating the farm simulation model, *see* Part II). In the absence of modeled estimates, it is common to use *recorded* annual data to represent the baseline condition within the simulation period (often data used in calibrating the economic model) against which the scenario estimates can then be compared. It is noteworthy that some economic models incorporate management responses in their relations, which raises some interesting questions about further linkages that can be explored between economic models and agroclimatic models. For example, if an agroclimatic model estimates a decrease in crop yields for a given climatic scenario, and if the standard management response to a yield decrease (simulated in an economic model) is an increase in fertilizer applications, then this represents an important feedback into the agroclimatic model for estimating crop yields in the subsequent year (a type of approach advocated in the Iceland case study for examining strategies to stabilize annual national fodder supply, *see* Part III).

2.6. Model Outputs

After conducting a scenario experiment with an agroclimatic model, it is important to present the results (e.g., estimates of biomass productivity, of dry matter potential or of crop yield) in a clear, informative and unambiguous manner. However, while each of these criteria may be satisfied within an individual study, it is useful to highlight some differences in presentation between studies that pose problems in attempting to compare model results. Two broad methods have been adopted for representing model outputs: statistical or graphical methods, and a spatial mapping approach.

2.6.1. Statistical/graphical methods

Scenario estimates are usually compared with estimates for the baseline climate to indicate the magnitude of change induced by a climatic anomaly. Summary statistics (e.g., mean, standard deviation, coefficient of variation, etc.) are commonly used to describe these changes in productivity (often depicted graphically on bar charts), but these can be subject to misinterpretation. Measures of "mean" yield over a given time period, for example, may be quite different depending on the techniques used to compute them. To illustrate, model estimates of mean rice yield in Hokkaido based on the period-averaged climate (1951–1980) are significantly higher than the averaged estimates of rice yield during either of the decades 1957–1966 and 1974–1983 based on annual weather, the latter estimates accounting for the (non-linear) effects of annual weather anomalies on yields (Part VI, Section 5). Furthermore, changes in interannual yield variability under different climatic scenarios can be considered either in absolute terms (e.g., as standard deviations or percentiles) or in relative terms with respect to mean yields (e.g., as coefficients of variation). For example, while the standard deviation of estimated oats yields in southern Finland increased by 21 kg/ha relative to the baseline under the GISS $2 \times CO_2$ scenario, a concurrent rise in mean yields (by 435 kg/ha) resulted in the coefficient of variation (standard deviation/mean) actually decreasing (by 5%) for the same scenario (Part IV, Section 4).

These measures of interannual variability are often depicted graphically as range bars on either side of the mean. They should not be confused, however, with range bars that are sometimes used on graphs to show how estimates vary over space within a region (although mapping procedures are often more effective for presenting this information, as discussed below). Nor should range bars be confused with confidence intervals, another statistical and graphical device widely employed.

Given the large number of uncertainties related to the scenarios, models and data, confidence intervals should be attached to *all* model estimates. The calculation of these can be a very time-consuming procedure, however, requiring rigorous sensitivity testing (*see* Subsection 2.4.4) and screening of data for errors. The development of the pasture yield model in the Iceland study is a good illustration of the many tests that are required to assess the validity and the uncertainty of even the most simple linear relationship (Part III, Section 3).

2.6.2. The spatial mapping approach

One focus of the model experiments reported throughout this volume has been on assessing not only the magnitude but also the *geographical pattern* of the effects of climatic variations on agriculture. Where model estimates are available for a number of locations within a region, it is possible to represent many of the summary statistics (described above) as geographical isolines on a map. One advantage of this approach is that the modeled effects of different climatic scenarios can be represented in terms of geographical shifts in, for example, levels of agricultural potential, of crop yield or of biomass productivity.

The approach is particularly useful for analyzing the effects of longer-term climatic changes, for large-scale shifts in vegetation zones (biomes) or cropping systems are only likely to occur as a long-term response to an enduring change. However, maps of year-to-year shifts in productivity, with the local short-term effects they imply (e.g., crop failures, increased growth rates, frost damage, etc.) can also give an indication of the inherent stability and resilience of a regional ecosystem or farming system. There are many examples throughout this volume of maps showing modeled climate-induced shifts:

(1) In vegetation zones (e.g., shifts in the boreal forest zone in the northern hemisphere under the GISS 2 × CO_2 scenario based on a forest growth model, *see* Part I, Section 5).
(2) In net primary productivity (e.g., shifts in levels of NPP of natural vegetation in Japan under five climatic scenarios computed using the Chikugo model, *see* Part VI).
(3) In agricultural potential (e.g., mapped values calculated using the climatic index of agricultural potential for locations in southern Saskatchewan, *see* Part II).
(4) In crop yields (e.g., model simulations of spring wheat yields mapped across crop districts in southern Saskatchewan for four climatic scenarios, *see* Part II).

(5) In crop yield variability (e.g., maps of simulated changes in the 5, 50 and 95 percentile values of spring wheat yields in Finland under three climatic scenarios, see Part IV).

2.7. Conclusion

In this section we have attempted to outline some of the techniques that have been employed in this volume to assess the first-order effects of climatic variations on crop production. We have particularly concentrated on describing and evaluating the use of agroclimatic models, developing a model checklist to help in this assessment.

We can conclude that agroclimatic models are far from perfect: many are inadequately validated or tested, and several are used for experiments that they appear to be unsuited to tackle. However, in many of the studies that follow, such analyses represent the *first attempt* to model the effects of climatic change on agriculture. Despite many difficulties (highlighted throughout this section) models *have* been developed or refined for climate impact studies, techniques *have* been devised for validating and testing these models and their outputs, and system feedbacks and adjustments *have* been incorporated in model experiments. Finally, linkages *have* been developed between these first-order models and both climate models and higher-order impact models. The next two sections consider the first of these linkages, namely the provision of input data for agroclimatic models through the development of climatic scenarios.

References

Baier, W. (1977). Crop-weather models and their use in yield assessments. *WMO Technical Note No. 151*, World Meteorological Organization, Geneva, 48 pp.

Baier, W. (1982). Agroclimatic modeling: An overview. In D.F. Cusack (ed.), *Agroclimatic Information for Development: Reviving the Green Revolution*, Westview, Boulder, Colorado, pp. 57-82.

Biswas, A.K. (1980). Crop-climate models: A review of the state of the art. In J. Ausubel and A. Biswas (eds.), *Climatic Constraints and Human Activities*, Volume 10, IIASA Proceedings Series, Pergamon, Oxford, pp. 75-92.

Draper, N. and Smith, H. (1981). *Applied Regression Analysis*. 2nd Edition, Wiley, New York.

Duckham, A.N. (1963). *The Farming Year*. Chatto and Windus, London.

FAO (1978). *Crop Calendars*. FAO Plant Production and Protection Paper No. 12, Food and Agriculture Organization, Rome.

Katz, R.W. (1979). Sensitivity analysis of statistical crop-weather models. *Agric. Meteorol.*, **20**, 291-300.

Lyons, T.C. (1982). Deterministic models for the ecological simulation of crop agricultural environments. In G. Golubev and I. Shvytov (eds.), *Modeling Agricultural-Environmental Processes in Crop Production*, IIASA Collaborative Proceedings Series, CP-82-S5, International Institute for Applied Systems Analysis, Laxenburg, Austria.

Nix, H.A. (1985). Agriculture. In R.W. Kates, J.H. Ausubel and M. Berberian (eds.), *Climate Impact Assessment: Studies of the Interaction of Climate and Society*, SCOPE 27, Wiley, Chichester, pp. 105–130.
Parry, M.L. and Carter, T.R. (eds.), (1984). *Assessing the Impact of Climatic Change in Cold Regions*. Summary Report, SR-84-1, International Institute for Applied Systems Analysis, Laxenburg, Austria.
Robertson, G.W. (ed.), (1983). *Guidelines on Crop-Weather Models*. Task Force on Crop-Weather Models, World Meteorological Organization, Geneva.
Rosenberg, N.J. (1982). The increasing CO_2 concentration in the atmosphere and its implication on agricultural productivity. II. Effects through CO_2-induced climatic change. *Climatic Change*, 4, 239–254.
Sakamoto, C.M. (1981). Climate-crop regression yield model: An appraisal. In A. Berg (ed.), *Application of Remote Sensing to Agricultural Production Forecasting*, Balkema, Rotterdam, pp. 131–138.
Thom, H.C.S. (1954a). The rational relationship between heating degree days and temperature. *Mon. Wea. Rev.*, **82**, 1–6.
Thom, H.C.S. (1954b). Normal degree days below any base. *Mon. Wea. Rev.*, **82**, 111–115.
Warrick, R.A. and Gifford, R.M., with Parry, M.L. (1986). CO_2, climatic change and agriculture. Assessing the response of food crops to the direct effects of increased CO_2 and climatic change. In B. Bolin, B.R. Döös, J. Jäger and R.A. Warrick (eds.), *The Greenhouse Effect, Climatic Change, and Ecosystems*, SCOPE 29, Wiley, Chichester, pp. 393–473.
WMO (1985). *Report of the WMO/UNEP/ICSU-SCOPE Expert Meeting on the Reliability of Crop-Climate Models for Assessing the Impacts of Climatic Change and Variability*, WCP-90, World Meteorological Organization, Geneva, 31 pp.

SECTION 3

Development of Climatic Scenarios:
A. From General Circulation Models

Wilfrid Bach

3.1. Introduction

This section evaluates the advantages and disadvantages of a number of general circulation models (GCMs) in deriving climatic scenarios for subsequent assessments of climate impact in the IIASA/UNEP project's case studies in Saskatchewan, Iceland, Finland, the northern USSR and Japan. It begins by outlining the nature of the problem and the information needs of decision makers. There then follows some comparison of the model-based and analogue approaches to scenario development. The main body of this section is concerned with the choice of GCM experiment that is most appropriate for the impact assessments which follow. The chosen scenario is then described in some detail.

Fossil fuel combustion, synthetic chemicals production, biomass burning, forest and soil destruction are the main activities through which mankind has set in motion a global experiment. Worldwide monitoring gives concrete evidence that the chemical composition of our atmosphere is undergoing a major change. Until recently the main concern was with the constantly increasing concentration of atmospheric carbon dioxide (CO_2). Now, after a closer look, some 30 additional radiatively-active trace gases have been identified – and this number is expected to increase in the future (Ramanathan et al., 1985). Their contribution will significantly enhance the CO_2-effect due to their greater growth rates, longer residence times in the atmosphere, and higher radiation absorption efficiencies.

An important consequence of the modification of the atmosphere's composition is a change in the radiation budget, and hence the climate, implying a general warming at the surface of the earth with potentially far-reaching impacts on global ecosystems, agriculture and water resources in the near term, and melting of ice and sea-level changes in the longer term. Physical reasoning and model simulations strongly suggest that a substantial anthropogenic influence on climate has already been effected, although it is not yet possible to prove it in a scientifically rigorous way (Hansen et al., 1985; Wigley and Schlesinger, 1985). Moreover, many of the climate-influencing factors, acting either individually or in combination, may result in synergistic effects, whose consequences we have barely begun to investigate (Bach, 1983, 1984).

All of this must be seen in connection with the pressure of the constantly increasing population and the limited environmental and food resources which will further aggravate the vulnerability of society to climatic change. It has been said that in a future warmer world there would be winners and losers, implying that the positive and negative impacts of the ongoing climatic change and variability would balance out. It is further maintained that any negative effects through climate on food production could be compensated through genetic technology by breeding novel strands of grain with higher yield, and through foreign trade. Unfortunately, experience from the recent past does not justify any such optimism. On the contrary, the extremely vulnerable hybrids developed under the so-called Green Revolution require the application of large amounts of pesticides and artificial fertilizer, thereby putting a heavy strain on the environment and the economy, especially of developing countries. Moreover, any grain surplus is first and foremost exported to those countries that can pay for it. The vulnerability of the hunger-stricken countries continues to be large, because the potential for yield increases is rather limited owing to capital, energy, sociopolitical, environmental and climatic constraints. To reduce this vulnerability by determining the range of potential climatic changes and the extent to which they might be able to disrupt the socioeconomic fabric of society remains an important task.

Finally, to reduce the risk of climatic impacts to our life-support systems requires community preparedness, especially among those who have to make the decisions and those who are charged with providing the necessary information. The first part of this section therefore considers the information needs of the decision makers and their policy advisers and to what extent these can be satisfied at present by scientists. The remainder emphasizes the use of general circulation models for impact analysis and makes some suggestions for improvements.

3.2. Needs of Decision Makers

Experiments with climate models indicate that the response of the atmosphere to an increase of CO_2 and other trace gases will be uniform neither in space nor in time but will show rather pronounced regional and seasonal heterogeneities. Therefore, to be useful in impact analysis, decision makers need information on the likely distribution of the regionally and seasonally varying climate parameters in enough detail to detect also small-scale anomalous behavior. In addition, it is important to get some idea of possible changes in the frequency and severity of such weather phenomena as droughts, storms, floods, etc. – changes that can be expected as part of the changed climate patterns. An indication of which areas would be most affected is essential.

Ideally the information should be as "surprise-free" and as reliable as possible. This means that, besides CO_2, it is very important that the other climate-influencing factors, and in particular the other greenhouse gases (e.g., CFMs, CH_4, N_2O, O_3, etc.) plus aerosol and land-use changes also be considered. Since decisions deal with problems as they evolve in time, information to influence such decisions should also describe the transitional phases of climate, and not simply an equilibrium situation at some future time. Decision-making also requires some indication of the degree of uncertainty or likely error associated with the change of each climate parameter. The information should be supplied in a form suitable for the respective user.

Finally, policy advisers and scientists must engage in dialogue so that priorities can be set which ensure that useful information is available when it is needed. Some of these needs and the status of the fulfillment of such needs are discussed below.

3.3. Status of Climatic Scenario Analysis for Impact Assessment

Climate impact studies analyze the interactions between climate and society. The purpose of such studies is to assess the consequences for society of climatic change and climatic variability caused advertently or inadvertently by mankind's activities. They are systems analyses which provide a synthesis of all relevant factors so that decision makers can reach more rational decisions. In cases where it is difficult to describe future climatic conditions, it has become customary to make use of climate scenarios. These are not meant to be predictions but rather internally consistent specifications of climatic conditions over space and time to be used for analysis of societal responses to climatic change.

There are two main approaches for assessing the patterns of climatic change in a future, warmer world. One is the analogue method in which the regional and seasonal patterns of past warm climates (based either on proxy or instrumental data) are used to construct warm-world scenarios as analogues for a future CO_2-induced warm climate (Lough et al., 1983). The second is the physical method which uses climate modeling (general circulation models) to construct climate scenarios with the purpose of quantitatively assessing the regional and seasonal patterns of climatic changes induced by CO_2 and other trace gases (Bach et al., 1985). To give an idea of the potential usefulness of these methods in the decision-making process, their respective advantages and disadvantages are briefly summarized.

The main advantage of analogue scenarios is that they are based on warm climates that have actually existed. The main disadvantage is that the causes of warm climates are not necessarily related to effects of CO_2 and other trace gases, but that they may be due to changes in other boundary conditions such as land-sea distribution, composition of the atmosphere, orography and topography. According to Palutikof et al. (1984), who have used instrumental records of the past century to construct warm-world analogues, another disadvantage of this method is that the observed warm-cold differences in the instrumental data are much smaller than the most conservative temperature increase for a CO_2-doubling; so that they can only be taken as indicative of conditions during the early stages of a CO_2-doubling and, hence, are indicative only of conditions during the early stages of a CO_2-warming. Moreover, Pittock and Salinger (1982), Pittock (1983) and Jäger and Kellogg (1983) have pointed out that the use of individual years in some studies to represent warm and cold climate periods is open to serious criticism since a one-year transient response may have quite different characteristics compared with a warm period of several decades. Thus a future CO_2-induced warmer climate may differ considerably from scenarios based on warm climates of the past (Schlesinger, 1983).

The main advantage of scenarios based on climate models is that they can be used to describe, in a physically consistent manner, not only past and present climates but also how climate could change in response to a change in some external forcing such as through CO_2 or other greenhouse gases. The main disadvantage is the inherent inability to construct a model that is a perfect replica of the actual climate system. To what extent this can be improved is probably not only a matter of understanding and modeling the physical processes but also one of computer capacity and cost.

The available climate models can be applied in two basically different ways:

(1) To study the response of the model climate to a time-dependent continual increase in CO_2 and/or other greenhouse gases in the atmosphere (these are the transient response studies).
(2) To assess the sensitivity of the model climate to the perturbation, for example, of a prescribed, time-independent doubling of CO_2 or other trace gases (these are the equilibrium response experiments).

In the real world the concentrations do not increase as a step function, such as a CO_2-doubling, but increase rather continually. Schneider and Thompson (1981) and Thompson and Schneider (1982a,b) question, therefore, the extent to which the results from equilibrium models can be applied to the real world. Their investigations indicate that the ocean can delay the effects up to several decades and that the spatial patterns of the equilibrium response may differ significantly from those of the transient response. Bryan *et al.* (1982), on the other hand, conclude from their findings that sensitivity studies of climatic equilibrium can give at least a rough indication of the prediction of the zonal distribution of sea surface temperature trends. Palutikof *et al.* (1984) state that even the most sophisticated climate models are not yet realistic enough to provide the necessary spatial detail with a sufficient degree of reliability. On the other hand, a committee of the United States National Academy of Sciences (CDAC, 1982) had concluded earlier that climate models are useful for constructing scenarios that can at least suggest the scales of spatial and temporal climatic changes.

To summarize the present state-of-the-art methodology of scenario analysis for use in climate impact assessment, it is fair to say that both the analogue and the physical methods are still beset with many difficulties. The use of these methods and the interpretation of the results derived from them must, therefore, be approached with caution.

The purpose of this introductory section is to discuss the second method, namely the construction of model-derived climate scenarios for use in impact assessment. This would ideally require models that have both a transient response and a high spatial/temporal revolution. Such models are not yet available. Among the available hierarchy of climate models it is only the three-dimensional general circulation models (GCMs) that are capable of generating data at high spatial and temporal resolutions. Up to now GCMs have only been used in time-independent equilibrium response experiments – an approach that is not fully realistic. The following sections demonstrate the methodology of deriving climate scenarios from currently available GCMs for the purpose of impact assessment.

3.4. Use of GCMs for Climatic Scenario Analysis

3.4.1. Characteristics of GCMs

A GCM consists of a prognostic and diagnostic system of equations incorporating the physical and dynamical processes that control the climate (Manabe, 1983). Generally, it has three main components, namely a model of the atmosphere, a heat and water balance model of the continental surface, and a scheme to incorporate the ocean.

The processes of the climate system are described by the governing equations such as the horizontal momentum equation, the continuity equations for mass and water vapor, the thermodynamic energy equation, the hydrostatic equation, and the equation of state. The first three of these equations are used to determine the rates of change of the prognostic variables including principally surface pressure, surface and soil temperature, horizontal velocity, as well as water vapor. Additionally, GCMs have many diagnostic variables, of which vertical velocity, clouds and surface albedo are the more important ones (Schlesinger, 1983).

Several of the small-scale physical processes important to climate are not explicitly resolved due to the limited spatial resolution of GCMs. The usual procedure is to parameterize them, i.e., to relate them either statistically or empirically to the scale of those variables which are resolved. The small-scale processes which are parameterized include transfer of solar and terrestrial radiation, turbulent transfer of heat, moisture and momentum between the earth's surface and the atmosphere and within the atmosphere, as well as condensation of water vapor (i.e., clouds). The processes and parameterization schemes differ between the various GCMs which accounts for some of the differences in the results (Schlesinger, 1983).

The simulation of climatic change with a GCM requires the prescription of certain parameters and boundary conditions such as the solar constant and orbital parameters, land/sea distribution and topography, total atmospheric mass and composition, etc. Moreover, the earth's climate cannot be successfully simulated without coupling the atmosphere with the ocean, the ice and the biosphere. The ocean has a significant influence on climate because of its large heat storage capacity. Special attention is required for the computation of sea surface temperature (SST) since it is determined by ocean circulation and the energy exchange between ocean and atmosphere. So far most of the present GCMs have only a rudimentary incorporation of the ocean effect.

After a GCM's parameterization and boundary conditions have been set, the rate of change of the prognostic variables are determined for a global network either at discrete grid points, as in gridpoint or finite difference models, or by a finite number of prescribed mathematical functions, as in spectral models. The variables at the grid points or the mathematical functions are then integrated forward in time in discrete steps starting from some given initial value. The choice of the time steps ranging from 10 to 40 minutes depends on the grid size, the speed of the fastest moving disturbance, and the integration method. Depending upon the specific model and its initial conditions, integration may take from a few months (without the ocean effect) to several years (with atmosphere-ocean coupling) before the model reaches its equilibrium climate, i.e., when its time-averaged statistics are no longer perceptibly changing. The statistics of such a control integration (the control run) define the model's climate. This can be compared with the observed climate, such as in model validation (*see* Subsection 3.4.5), or with a perturbed climate, such as for a CO_2 doubling (see the CO_2-induced climatic change scenarios in Subsection 3.4.6).

3.4.2. Model criteria for scenario analysis

To be useful scenarios for the evaluation of potential impacts of future climatic change, GCMs should:

(1)   Be based on a realistic geography and topography.
(2)   Have a high spatial resolution.
(3)   Have an adequate temporal resolution.
(4)   Incorporate a coupled model of the atmosphere-ocean circulation.
(5)   Simulate realistically the patterns of the observed climate.

Table 3.1 summarizes some of these criteria for all those GCMs that have so far been used in CO_2 sensitivity studies. An examination of this list can help us determine the suitability of the available GCMs for impact analysis in subsequent case studies.

Geography and topography

In the earliest CO_2 sensitivity experiments a GCM was used with a horizontal domain extending over only 120° longitude and from the equator to 81.7°N (GFDL, No. 1) or from the equator to the pole (GFDL, No. 2). One half of these sectors is land and the other half sea. This is termed idealized geography in *Table 3.1*. Furthermore, there is no topography in either model, which means that the surface elevation is uniform everywhere. Results from such models would not be useful in impact analysis. All other GCMs have a global domain extending from pole to pole together with a realistic land/sea geography and topography. It should be noted, however, that even in those cases termed realistic, the topography has a uniform elevation and surface roughness within each grid area. This is considered in more detail for the GISS model in Subsection 3.4.3.

High spatial resolution

The analysis of regional climatic impacts requires a high geographical resolution. In GCMs the spatial resolution is a compromise between the demand for high coverage and limitations in computer memory and time. At present none of the GCMs goes below a spatial resolution of 4° latitude by 5° longitude which, at the equator, amounts to approximately 400 × 500 km. Thus, much regional climatic detail cannot yet be resolved. The GISS model has an even coarser grid spacing which is prompted by limitations in available computer capacity and by a desire to perform long-term experiments. Hansen et al. (1983) have made test runs with various resolutions which indicate that this may not be a critical issue. This is demonstrated in more detail in Subsection 3.4.3. All CO_2 experiments, except those at OSU and UKMO, were conducted with nine-layer GCMs that include both the troposphere and the stratosphere. The UKMO model has only five vertical layers, and the OSU GCM uses a two-layer model of only the troposphere.

Adequate temporal resolution

Especially for agricultural impact analysis it is mandatory that the GCMs simulate an annual cycle and also a diurnal insolation cycle. Except for the GFDL (Nos. 1, 2), the OSU (No. 5) and NCAR (No. 6) models, all other GCMs have an annual cycle so that monthly and seasonal means can be generated (*Table 3.1*). Furthermore, except for the GFDL (Nos. 1, 2, 3), OSU (No. 5) and NCAR (Nos. 6, 7) GCMs, all other models consider also the effects of the important diurnal cycle. A prerequisite for a good modeling of the diurnal cycle is the realistic treatment of the planetary boundary layer and surface energy transfer processes. Moreover, for the simulation of the annual cycle an adequate parameterization of the ocean's effect on climate must be taken into account.

Atmosphere–ocean schemes

The way the ocean is incorporated affects not only the temporal resolution of a model, but it can also alter considerably the reaction of the model to a change in atmospheric boundary conditions. In GCM experiments with a climatological ocean model (OSU No. 4; UKMO Nos. 8 and 9), in which the sea surface temperature (SST) and the sea ice cover are prescribed, the ocean acts as if it were an infinite heat sink, i.e., as the atmosphere warms, there is no corresponding warming of the ocean, and hence no return of energy to the atmosphere (*Table*

Table 3.1. Criteria for and characteristics of general circulation models used in CO_2-induced climatic change studies.

Model/Researcher	Geography	Topography	Spatial resolution	Annual cycle/diurnal cycle	Ocean/sea ice	CO_2-simulation	Temperature increase (K) due to CO_2[a]
(1) GFDL Manabe, Wetherald (1975)	Idealized land/ocean	None	V: 9 layers H: Eq. to 81.7° λ = 120° sector ($\Delta\Phi$ = variable $\Delta\lambda$ = 6°)	No/no	Swamp (no heat capacity or transport)/ice cover predicted	$2 \times CO_2$	2.9
(2) GFDL Manabe, Wetherald (1980)	Idealized land/ocean	None	V: 9 layers H: Eq. to Pole λ = 120° sector ($\Delta\Phi$ = 4.5°, $\Delta\lambda$ = 5°)	No/no	Same as (1)	$2 \times CO_2$	3.0
(3) GFDL Manabe, Stouffer (1980); Manabe, Wetherald, Stouffer (1981)	Realistic	Realistic	V: 9 layers H: Pole to Pole, λ = 360° (spectral)	Yes/no	68 m mixed layer (no horizontal heat transport)/ice thickness predicted	$4 \times CO_2$	4.0
(4) OSU Gates, Cook, Schlesinger (1981)	Realistic	Realistic	V: 2 layers H: Pole to Pole, λ = 360° ($\Delta\Phi$ = 4°, $\Delta\lambda$ = 5°)	Yes/yes	"Climatological ocean", sea surface temp. prescribed/ice cover prescribed	$2 \times CO_2$	0.2
						$4 \times CO_2$	0.4
(5) OSU Schlesinger (1983)	Realistic	Realistic	Same as (4)	No/no	Swamp (no heat capacity or tranport)/ice cover predicted	$2 \times CO_2$	2.0
(6) NCAR Washington, Meehl (1983)	Realistic	Realistic	V: 9 layers H: Pole to Pole, λ = 360° (spectral) ($\Delta\Phi$ = 4.5°, $\Delta\lambda$ = 7.5°)	No/no	Same as (5)	$2 \times CO_2$	1.3[b]
						$2 \times CO_2$	1.4[c]

Climatic scenarios: A. From general circulation models

Model	Geography	Topography	Resolution	Ocean	Diurnal/Seasonal cycle	Experiment	ΔT
(7) NCAR Washington, Meehl (1984)	Realistic	Realistic	Same as (6)	50 m mixed layer (no ocean heat transport)/ice cover predicted	Yes/no	2 × CO$_2$	3.5
(8) UKMO Mitchell (1983)	Realistic	Realistic	V: 5 layers H: Pole to Pole, λ = 360°, 330 km	"Climatological ocean", sea surface temperature prescribed + 2 K/ice cover prescribed	Yes/yes	2 × CO$_2$	0.2
						2 × CO$_2$	2.2
(9) UKMO Mitchell, Lupton (1984)	Realistic	Realistic	Same as (8)	Sea surface temperature prescribed + lat. varying increment, ice cover predicted	Yes/yes	4 × CO$_2$	4.5
(10) GISS Hansen et al. (1984)	Realistic	Realistic	V: 9 layers H: Pole to Pole, λ = 360° (ΔΦ = 8°, Δλ = 10°)	60-70 m mixed layer with prescribed seasonal depth and prescribed horizontal transport/ice thickness predicted	Yes/yes	2 × CO$_2$	4.2
(11) OSU Gates, Potter (1985)	Realistic	Realistic	Same as (4)	Variable-depth mixed layer	Yes/yes	2 × CO$_2$	1.5
(12) OSU Gates et al. (1984) Schlesinger et al. (1985)	Realistic	Realistic	Same as (4)	Complete ocean GCM	Yes/yes	2 × CO$_2$	1.5

V = vertical resolution; H = horizontal resolution; Φ, λ: latitude, longitude. GFDL: Geophysical Fluid Dynamics Laboratory, Princeton, USA; GISS: Goddard Institute for Space Studies, New York, USA; NCAR: National Center for Atmospheric Research, Boulder, USA; OSU: Oregon State University, Corvallis, USA; UKMO: Meteorological Office, Bracknell, UK. Source: Adapted from Schlesinger (1983) and updated.

[a] Global mean equilibrium temperature at the surface (some of the models may not have run long enough to have reached equilibrium).
[b] Clouds (prescribed).
[c] Clouds (predicted).

3.1). Thus, the response of these GCMs is restricted to the atmosphere and the land surfaces. The surface air temperature changes estimated from such GCMs, which are typically about an order of magnitude less than those from more realistic models, can, at best, give only a rough idea of the lower end of the potential impacts. Therefore, Mitchell (1983) modified this approach by enhancing the SST at all ocean grid points by 2K for the $2 \times CO_2$ case (UKMO No. 8), and later Mitchell and Lupton (1984) adjusted their $4 \times CO_2$ experiment by a latitudinally varying increment (UKMO No. 9). This kind of parameterization does not, however, remove the basic deficiency of climatological ocean models, namely the incomplete simulation of the feedbacks between atmosphere and ocean. They are therefore not very useful for impact analysis.

In another scheme the ocean is treated as a swamp, i.e., like perpetually wet land, of zero heat capacity without any heat transport and with an infinite water vapor source. It can simulate neither annual nor diurnal cycles because heat cannot be stored in the oceans, and thus, only annual averages can be obtained. Therefore, the use of results from this model type for agricultural impact is also severely restricted. The GFDL GCMs Nos. 1 and 2, the OSU GCM No. 5 and the NCAR GCM No. 6 belong to this category (*Table 3.1*).

Another approach is the fixed-depth mixed layer ocean model. It incorporates some of the surface layer-atmosphere feedback, but it still has neither horizontal heat advection by means of ocean currents, nor vertical heat transfer between the various ocean layers such as in upwelling and downwelling areas. To this category, which is presently one of the most advanced ocean models used in CO_2 studies, belong the GCMs developed at GFDL (No. 3), NCAR (No. 7), and GISS (No. 10).

Finally, CO_2 experiments are under way with a variable-depth mixed layer ocean model at OSU (No. 11 by Pollard, 1982; Gates and Potter, 1985), and with an atmosphere GCM coupled to a complete ocean GCM at GFDL (Manabe et al., 1979; Spelman and Manabe, 1984), at OSU (No. 12 by Gates et al., 1984; Schlesinger et al., 1985), at NCAR, at the Max Planck Institute for Meteorology in Hamburg, and perhaps elsewhere. The latter approach could eventually yield more realistic results.

Summary of evaluation

None of the present GCMs shown in *Table 3.1* meets all of the requirements listed above. But there are differences in the currently available models that make them more or less useful for impact analysis. A somewhat realistic topography and geography with a global domain is considered a prerequisite in impact studies. Except for the GFDL models (Nos. 1 and 2) this criterion is more or less fulfilled by all the other models. Particularly for agricultural impact analysis both an annual and a diurnal cycle is essential so that the swamp models developed by OSU (No. 5) and NCAR (No. 6), and the mixed layer ocean models used by GFDL (No. 3) and NCAR (No. 7) were also disregarded. The climatological ocean models of OSU (No. 4) and UKMO (No. 8) with prescribed SST and sea ice cover for a CO_2 doubling result in a surface temperature increase that is about an order of magnitude lower than that obtained from more realistic modeling with feedbacks between the atmosphere and the ocean, and were therefore not considered. The UKMO model (No. 8) with the enhanced SST by 2 K at all ocean grid points was also rejected because it does not solve the basic deficiency of climatological ocean models, namely the neglect of ocean-atmosphere feedbacks. The GFDL (No. 3) and UKMO (No. 9) GCMs were only run for a CO_2 quadrupling. Since in the following case studies emphasis is on the near-term impacts (i.e., over the next 50 years or so), $4 \times CO_2$ experiments were not considered as realistic. At the time when the data were needed for the case studies, the most advanced OSU models (Nos. 11 and 12) had not been run long enough to have reached equilibrium so that their data were not available as input for scenario analysis.

Thus, based on this type of scrutiny, it was decided that currently the GISS model (No. 10) might be the best choice to supply the input data for the impact models to be used in the $2 \times CO_2$ case studies as described in the various chapters of this book. While this is as good a choice as can at present be made, it should not be overlooked that extensive sensitivity experiments have clearly shown the desirability of using higher spatial resolution and the need for improvements in modeling the various physical processes (Hansen et al., 1984). Hydrological processes at the surface, moist convection, clouds, and boundary layer transport processes, each critically influencing the climate simulations, are examples of the present crude treatments. More of these difficulties are discussed in the following and the final section.

3.4.3. Description of the GISS model

Physical processes within a gridbox

The GISS GCM is a gridpoint model (*see* Hansen et al., 1983, for a detailed description). The model structure at a gridbox is shown in *Figure 3.1*. Each gridbox has the appropriate fractions of ocean, ocean ice, land, and ice sheets on land. Snow may occur on land, land ice and sea ice, as determined by the model. Snow depth is computed, and snow albedo includes the effects of snow age and masking by vegetation. Ground temperature computations include diurnal variation and seasonal heat storage (*see* z_1 and z_2 in *Figure 3.1*). This is very important because of the highly non-linear dependence of latent and sensible heat fluxes on temperature. Ground moisture is assessed as a function of precipitation and evaporation, and it also incorporates a water-holding capacity appropriate for the root zone of the respective vegetation. Evaporation is computed and runoff is dependent upon precipitation rate, ground wetness, soil type, and vegetation.

Convection is a very important sub-grid-scale process responsible for the vertical transport of moisture, sensible heat, and momentum. Moreover, it determines the development of clouds and precipitation, thereby coupling dynamical and radiative processes in the atmosphere. Cloud cover and vertical distribution are computed and can thus respond and feed back on a CO_2-induced climatic change. Convective clouds are specified to be proportional to the mass

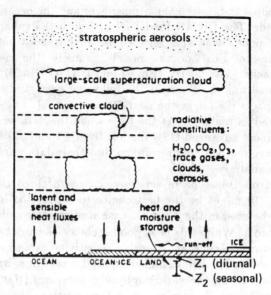

Figure 3.1. Schematic illustration of the GISS model structure at a single gridbox. (Source: Hansen et al., 1983.)

of the rising saturated moist air. A large-scale cloud cover is assigned, when the whole air layer is saturated. Large-scale clouds amount to about 80% of the global cloud cover generated by the GCM. Cloud type, altitude and thickness are the basis for the cloud radiative properties. Absorption of radiation by cloud particles and by water vapor within the cloud are included. Furthermore, the cloud radiative properties are wavelength-dependent.

The radiation term comprises heating by absorption of solar radiation and heating (or cooling) by absorption and emission of terrestrial radiation. It is difficult to model this accurately throughout the troposphere and stratosphere and to have it properly respond to climatic changes in the temperature-pressure profile, cloud distribution or atmospheric composition. This is due to the fact that accurate integration over complex and overlapping absorption bands and the proper inclusion of the effects of multiple scattering pose a major problem in GCMs. All significant gaseous absorbers such as H_2O, CO_2, O_3, O_2 and N_2O are included in the solar radiation term. Also absorption and scattering by aerosols are considered. The thermal radiation calculations are based on both the major absorption bands of H_2O, CO_2 and O_3 and the weaker bands of CO_2, O_3, N_2O and CH_4. Complete radiation calculations are made only once every 5 hours. Sensitivity tests show, however, that this does not affect much the monthly mean diurnal cycle.

Land coverage and elevation

Beside the treatment of the physical processes within a gridbox, other criteria important to impact analysis are the inclusion of a realistic geography and

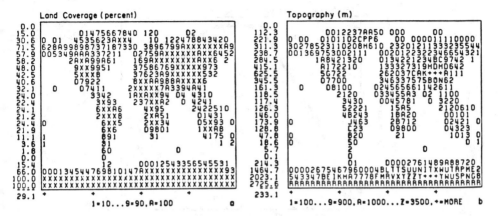

Figure 3.2. Digital maps of land coverage and topography for the 8° × 10° GISS model. A blank on either map is identical to zero. (*a*) For land coverage 0 is 0–5%, 1 is 5–15%, 2 is 15–25%, etc., 9 is 85–95%, A is 95–100% and X is exactly 100%. (*b*) For topography 0 is 0–50 m, 1 is 50–150 m, 2 is 150–250 m, etc., 9 is 850–950 m, A is 950–1050 m, B is 1050–1150 m, etc., Z is 3450–3550 m and + is more than 3550 m. Longitudinal averages are on the left above the area-weighted global average. (Source: Hansen *et al.*, 1983.)

topography. *Figure 3.2(a)* shows for the 8° latitude by 10° longitude model version the percent of area covered by land (*see* the legend). Each digit corresponds to a gridbox. About 29% of the globe is covered by land. *Figure 3.2(b)* shows the elevation range for each gridbox.

Effects of different resolutions

Another important criterion in impact analysis is the grid spacing of a model. For this purpose Hansen *et al.* (1983) conducted a series of experiments with 4° × 5°, 8° × 10° and 12° × 15° horizontal resolutions. Comparisons were made for January or the three winter months, depending on the available observed data. Tests were made for a large number of meteorological elements, including temperature and precipitation, and were represented either as zonal means or global maps. In *Figure 3.3(a)* the calculated zonal mean surface air temperature for three different resolutions is compared with observations. In general, the results of all three resolutions are in good agreement with observations, except near the North Pole, where the temperature for the 12° × 15° resolution is several degrees too low, and near the South Pole, where the modeled temperature for all resolutions is too high. The agreement between the results of the different resolutions and observations is not as good for precipitation as it is for temperature. But, as *Figure 3.3(b)* shows, there are major improvements in going from 12° × 15° to 8° × 10°, perhaps also to 4° × 5° resolution. Hansen *et al.* (1983) conclude from this that the main features of global climate can be realistically simulated with a resolution as coarse as 1000 km.

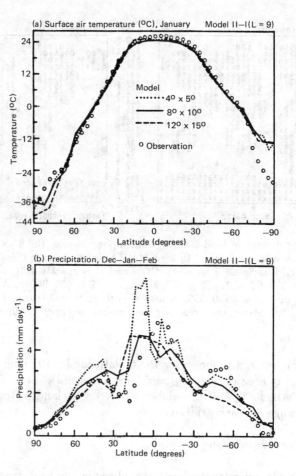

Figure 3.3. Zonal mean surface air temperature (*a*) and precipitation (*b*) for three horizontal resolutions of the GISS model. Model results are means from a five-year run for 8° × 10° resolution, but from a single month or season for the other resolutions. Observations for (*a*) are from Crutcher and Meserve (1970) and for (*b*) are from Schutz and Gates (1971). (Source: Hansen *et al.*, 1983.)

Model feedbacks

There are many physical processes responsible for, for example, a temperature increase. Feedback analysis is an important procedure for identifying those physical mechanisms that exert the greatest influence. Hansen *et al.* (1984) have used a one dimensional (1-D) radiative convective model (RCM) to study the relationship of surface temperature changes with individual feedback processes that might also occur in 3-D GCMs. There is, however, *a priori* no guarantee that these changes will be similar, because the simple global and annual averages of the 1-D RCM cannot account for the non-linear nature of the physical processes and their 2-D and 3-D interactions.

The 3-D GISS model yields a warming of about 4K for a CO_2 doubling. This corresponds to a net feedback factor of $f = 3$–4. The latter is defined by $f = \Delta T_{eq}/\Delta T_0$, where ΔT_{eq} is the equilibrium change of global mean surface air temperature and ΔT_0 is the change of surface temperature that would be required to restore radiative equilibrium if no feedbacks occurred. The main positive feedback processes in the model are changes in atmospheric water vapor, clouds, and snow/ice cover, and the feedback factors calculated for these processes are 1.6, 1.3 and 1.1, respectively. Several other potential feedbacks, such as land/ice cover, vegetation cover, and ocean heat transport, were held fixed in these tests. Finally, it should be noted that at present information on cloud processes is not adequate to allow confirmation of the cloud feedback.

3.4.4. Data processing

A necessary step prior to model validation is to transfer the model-generated data to the grid system of the observed data. The observed temperature data as compiled by Schutz and Gates (1971; 1972a,b; 1973a,b; 1974a,b) and the precipitation data as compiled by Jäger (1976) were available as averages for the 30-year period of 1931–1960 on the OSU grid system (Gates, personal communication, 1983). This has a spacing of 4° latitude by 5° longitude resulting, for example, for the European area [cf. *Figure 3.5(a)*], in 165 grid points (i.e., 30°N, 34°N,...,70°N; 30°W, 25°W,...,40°E). A spatial smoothing technique was used to transfer the model-generated GISS data to the OSU grid and temporal smoothing was applied to simulate better the mean climate (*see Appendix 3.1*).

3.4.5. Model validation

Before a climate model can be used with any degree of confidence it must be validated. The most common validation procedures include the comparison of simulated data with observed data in their spatial and temporal variations, the reproduction of past climates, comparison of different model simulations, and model tuning.

It is standard procedure to tune model parameters such that they satisfy observational constraints. But it must be realized that the resulting agreement of model-generated data with observed data does not really validate the model's performance, since the data had already been used to develop the model and thus do not allow an independent model check.

Many of the GCMs agree in their basic model structures but differ in the way they parameterize the physical processes. Comparison of differences and similarities between different model simulations can help reduce disagreements and improve model performance. Extensive pattern analysis, i.e., comparing the areas of agreement or disagreement as simulated by different GCMs, has been performed by Meinl *et al.* (1984) and Bach *et al.* (1985).

Another validation procedure is to look at the extent to which past climates can be reproduced by a model. GCM simulations have been conducted:

(1) For the Cretaceous period about 65–220 MYBP (i.e., million years before present), when it was 10–20 K warmer than today (Barron and Washington, 1984).
(2) For the last Glacial Maximum about 18 000 years BP, with global mean surface air temperatures about 4K lower than today (Hansen et al., 1984).
(3) For the Holocene post-glacial epoch about 6000–9000 years BP, when it was 1–2 K warmer than today (Kutzbach and Otto-Bliesner, 1982).

It should be noted, however, that climate sensitivity inferred from paleoclimatic data does not provide much direct model validation due to the presence of additional feedbacks on very long time scales (Dickinson, 1986).

The most direct approach to model validation is to compare the model-generated data with recent observations over a variety of space and time scales. For example, Manabe and Stouffer (1980), using the GFDL model No. 3 in *Table 3.1*, were able to simulate quite well the large seasonal variation in surface air temperature changes over continents of the northern hemisphere, and Hansen et al. (1984), using their GISS model, were able to reproduce well the observed annual mean sea/ice cover in both hemispheres. So far very little validation has been done for the simulation of the diurnal cycle.

Impact analysis is only useful on a regional scale. Therefore, to be used in regional impact studies, climate models must be validated also on a regional scale. On this scale differences between model-generated and observed climate parameters can be caused not only by the imperfect performance of the model, but also by the differences in the grid systems of the different data sets, and by factors related to observed data collection.

To demonstrate this, model validation is carried out for areas of different areal extent using the GISS model; i.e., one for the hemispheric belt of 38°N to 70°N covering the case study areas of Saskatchewan, Iceland, Finland, the northern USSR and Japan (*see* the stippled areas in *Figure 3.4*, and another smaller area covering most of Europe and North Africa from 30°W to 40°E and 30°N to 70°N in *Figure 3.5*). This is done in order to show the model performance over different regions. *Figure 3.4(a)* shows the annual mean temperature differences between the model-generated and the observed data for the hemispheric belt. Underestimates (of up to 3 K) and overestimates of equal magnitude are found over large parts of the area. Over southern Greenland and Siberia the model shows large overestimates of 6 K and greater. This is probably due to the inadequate modeling of the ice/snow-albedo feedbacks. *Figure 3.4(b)* shows the annual mean precipitation differences. Over most of the study areas the errors are about 1 mm/day, rising to over 2 mm/day over areas off the northwest coast of the USA. These are very large errors frequently amounting to 50% of the observed precipitation.

Figure 3.5 shows a similar comparison of the observed with the modeled data on a regional scale for Europe and North Africa. In the western half of the study area the differences between the measured and observed temperature data

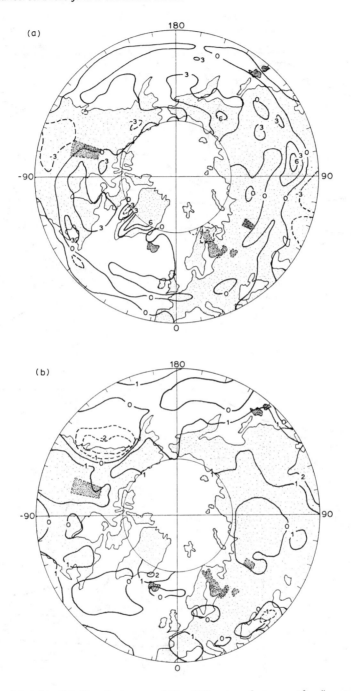

Figure 3.4. Model validation in terms of the differences between the "control" climate generated by the GISS GCM and the observed climate for the northern hemispheric belt 38°N to 70°N for: (a) mean annual temperature (K) (Schutz and Gates observation data), and (b) mean annual precipitation rate (mm/day) (Jäger observation data). Case study regions adopted in this volume are indicated by heavy stipple.

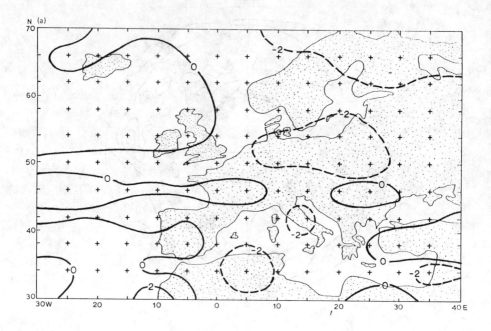

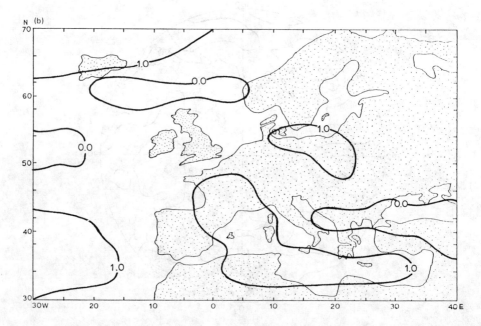

Figure 3.5. Model validation for the European region in terms of the difference between the "control"' climate generated by the GISS GCM and the observed climate for: (*a*) mean annual temperature (K) (Schutz/Gates observation data), and (*b*) mean annual precipitation rate (mm/day) (Jäger observation data). Also shown in (*a*) is the 4° latitude × 5° longitude grid system adopted in this study.

are close to zero, implying almost perfect agreement. In the eastern half, however, there are considerable underestimates (ca. 2 K) by the model except for some areas over the Balkans and Turkey [*Figure 3.5(a)*]. In contrast, the simulated annual mean precipitation rate shows considerable overestimates (ca. 1 mm/day) over large areas [*Figure 3.5(b)*].

Therefore, all of these results must be interpreted with caution. *Table 3.1* shows that the GISS model has a spatial resolution which is much coarser than that of the observed data. Therefore, the model cannot produce information in the same detail as that provided by the observed distributions. Also, at such a coarse grid spacing, large areas are covered which include a rather diverse topography. Moreover, climate conditions at a specific location within a grid may differ markedly from the average for the grid. Therefore, all of this may not only hamper a reliable representation of the physical processes that determine the weather and climate, but it may also render the interpretation of the model results more difficult (Rind and Lebedeff, 1984). Nevertheless, CO_2 experiments based on the areas which have been validated are still useful since they can give at least a general idea of the changes that are possible in a CO_2-warmed earth, even though details of changes at any given location are presently beyond the capability of this or any other available GCM (Wilson and Mitchell, 1984).

3.4.6. Construction of CO_2-induced climatic change scenarios

With the above caveats in mind we can now look at seasonal temperature and precipitation change scenarios due to a CO_2 doubling developed with the GISS model for Europe. *Figure 3.6* shows that a doubling of the atmosphere's CO_2 content in the GISS model raises the surface air temperature in Europe between 3 K and 7 K. Temperature increases in excess of 5 K are found over northern and northeastern Europe in winter and over North Africa in all other seasons. The large temperature increase over northeastern Europe may be related to the reduction in snow cover from autumn to spring. The strong temperature rise over North Africa is probably due to decreases both in soil moisture and in evaporation, making more heat available for surface air warming. The significance of the temperature changes was calculated from monthly data of the control and perturbed experiments using the Chervin method described in *Appendix 3.2*. The changes are significant at the 5% level over the whole study area in all seasons.

Figure 3.7 shows the regional distribution of the precipitation rate changes by season for a CO_2 doubling in the model. Decreases are indicated by dashed isohyets. According to these scenarios, a precipitation reduction is possible in southwestern parts and precipitation increase in northern regions in all but the autumn seasons. Noteworthy are the precipitation rate decreases in spring and in summer over the Mediterranean and southwestern Europe where it is already dry, and the strong precipitation increases over northeastern Europe where it is already wet. If such changes should come about they would not be without practical importance to agriculture. In contrast to temperature, precipitation rate changes, with their greater inherent variability, show smaller areas of

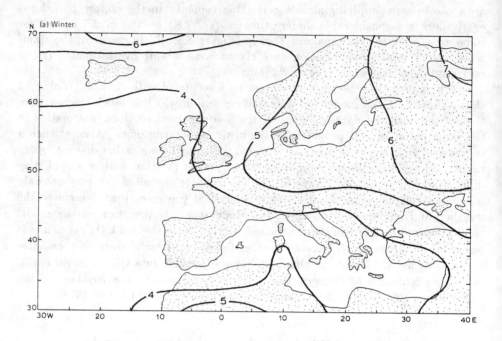

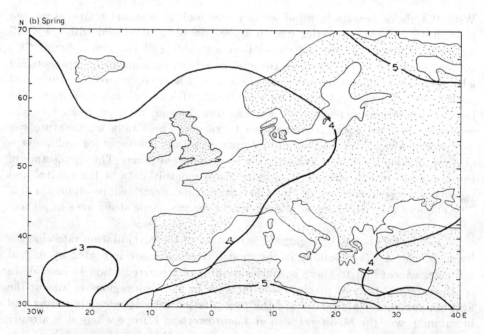

Figure 3.6. Regional distribution of the average surface temperature change (K) between the GISS model $1 \times CO_2$ (control) and $2 \times CO_2$ experiments in the European region by season. All values are statistically significant at the 5% level.

Climatic scenarios: A. From general circulation models   145

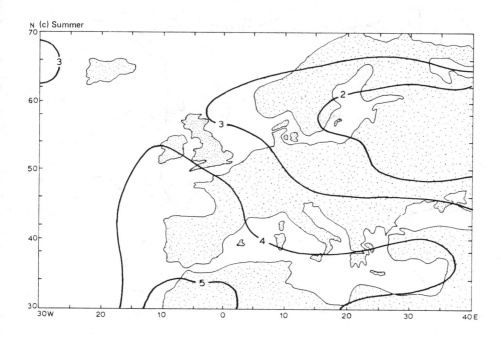

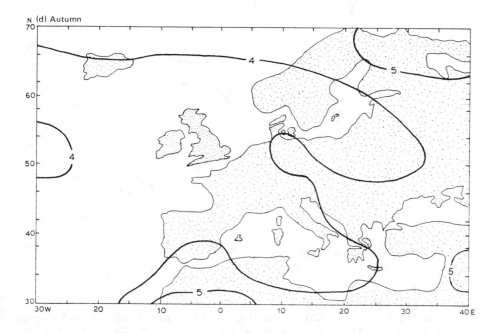

Figure 3.6 continued

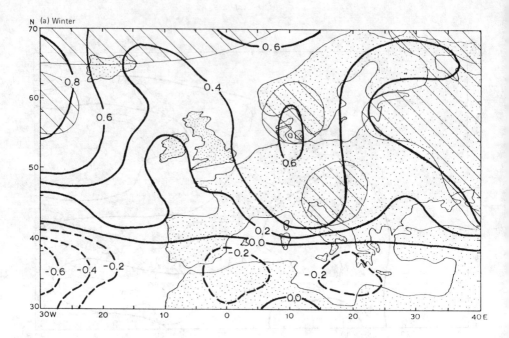
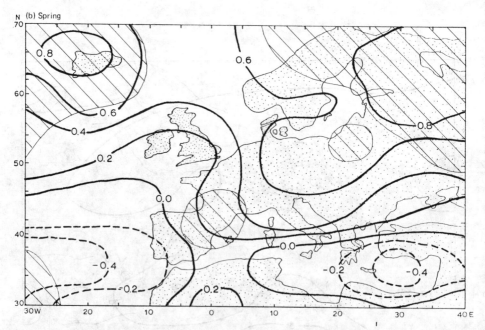

Figure 3.7. Regional distribution of the change in average precipitation rate (mm/day) between the GISS model $1 \times CO_2$ (control) and $2 \times CO_2$ experiments in the European region by season. Shading indicates changes that are statistically significant at the 5% level.

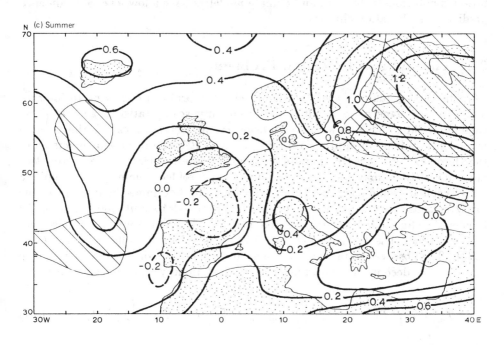

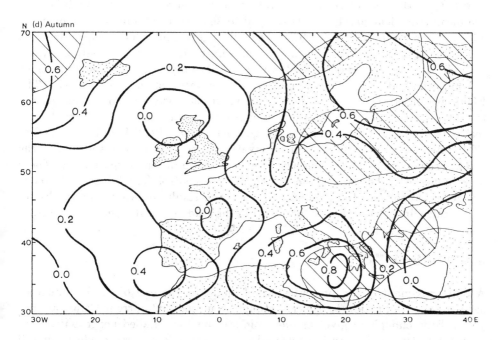

Figure 3.7 continued

statistically significant changes. It should be pointed out that the simulation of precipitation is notoriously difficult, and climate modelers have a low level of confidence in predicted precipitation change.

3.5. The Climate Inversion Problem

Climate impact analysts are often faced with the problem of having to interpolate climate changes on small scales from large-scale statistics generated by current GCMs. This downscale transformation has been called the climate inversion problem, an appropriate term, because it is the reverse or inversion of the familiar problem of parameterization (Gates, 1985). There are two approaches that may be used to solve this problem, namely to couple a local/regional dynamical model with a GCM, or to determine the most probable local/regional structure of a climate variable from GCM simulation based on the large-scale climate. The first method has not yet been applied. The second method, which has been applied for Oregon, is briefly as follows (Kim et al., 1984). The observed data, such as the mean temperature or precipitation for the month m are decomposed into their respective stationary and transient parts:

$$T_m = \bar{T} + T'_m; \quad \bar{T} \equiv \frac{1}{M} \sum_m T_m; \quad P_m = \bar{P} + P'_m; \quad \bar{P} \equiv \frac{1}{M} \sum_m P_m \qquad (3.1)$$

where T_m is the spatial mean temperature for month m; $\bar{T}$ is the deviation from the mean (transient part); and M is equal to 360, the total number of months in a 30-year record from 1948 to 1977. The notation is similar for precipitation (P).

Next, the variations at each station are examined and the relationship for each climatic variable is quantified in terms of the empirical orthogonal functions (EOFs) which define the dominant spatial patterns of local variations. The EOFs are based on the covariance matrix. In this way the large-scale results from GCMs can be systematically assigned or distributed on a regional/local scale. Using a set of 49 stations in Oregon, Kim et al. (1984) found that the first EOF determined from 30 years of observations accounts for 78% and 81% of the total variance of the local temperature and precipitation, respectively. These high values may be due to the strong surface control (coastal area, Cascade Range in Oregon) on the regional climate. The results indicate that the use of large-scale climate models may be useful in the study of local climatic impacts.

3.6. Outlook

A fair assessment of the present state of affairs at the end of the introductory section and as a preview to the case studies would be that current GCMs are not yet realistic enough to give reliable estimates on the detailed geographical and seasonal scales required in impact analysis. Nevertheless, these models can still be used as a tool to construct scenarios of possible climatic change, and they can serve as an educational vehicle to suggest the methods for both climate model improvement and coupling of climate with impact models.

The development of GCMs and their use in sensitivity experiments is certainly a very expensive affair and can therefore be conducted at only a few research centers around the world. To reproduce something as complex as the climate system and to make the simulations come as close to the real world as possible is extremely complicated and not straightforward. It is important to develop a variety of GCMs in order to enable comparisons between them. These comparisons can help identify the different parameterization schemes and the different sub-grid-scale treatments. Gates (1985) has pointed out that any specific parameterization requires a certain resolution and that realization of this will accelerate progress in the improvement of GCMs.

An important problem with present GCMs relates to their treatment of clouds (Dickinson, 1986). The current approach is to simulate cloud formation whenever moist convection occurs or whenever a critical humidity is reached. The model-generated cloud climatology is then compared with the observed clouds. This meets with great difficulties because there is such a wide variety of cloud types whose radiative properties are quite different. Of special importance to modeling the radiative properties of clouds are, for example, liquid water content, drop sizes, fractional cover, height, and spatial scale. Most cloud properties are of sub-grid-scale. Cloud behavior is very complex and its effect on the earth's heat budget is especially difficult to model. Therefore, improved cloud prediction schemes and a more comprehensive data base are urgently needed.

Another important problem is the coupling of the atmosphere to the ocean in a way which allows realistic thermal and dynamic interaction. Current GCMs simulate poorly the temporal and spatial variations in sea surface temperature (SST), a parameter that is so vital in affecting climate patterns. The variations of SST depend upon the complex interrelations between the spatially varying net surface heating and such dynamical processes as small-scale vertical mixing and large-scale horizontal transport by ocean currents, which are both responsible for the redistribution of the absorbed solar energy. Difficulty in understanding ocean circulation means that it is not ordinarily modeled as a dynamic process. Neglecting ocean dynamics could mean that both the modeled latitudinal and regional distributions of a climatic change, induced by the increasing trace gas concentrations in the atmosphere, will be in error (Rind and Lebedeff, 1984). Mesoscale eddies perform a significant role in oceanic transport of kinetic energy, but hitherto their dynamic nature is not well understood, and current ocean general circulation models have resolutions that are too coarse to resolve these eddies. Vertical mixing is another poorly understood process which so far is only crudely incorporated in the GCMs. Modeling the entire ocean circulation remains an important task.

Modeling the boundary layer processes is another area plagued with many deficiencies. There are still many unrealistic features, especially in the vicinity of mountains, which implies that the simulation of the dynamical flow over and around obstacles requires much improvement. A more accurate simulation of regional climates also requires improvements in the modeling of the surface hydrological/biological interactions. This is important because plants affect the water-holding capacity of the soil-vegetation system, the transport of water from the soil to the atmosphere, the surface roughness (by increasing turbulence), and the surface albedo, thereby changing the stability and the convective processes (Rind, 1984). Dickinson (1986) has pointed out that the parameterization for the calculation of land evapotranspiration may require the assessment of a realistic diurnal cycle of surface temperature and evapotranspiration for use in mesoscale or global climate models. All of this is hampered by the complexity and detail of the small-scale surface processes as well as the lack of an adequate data base.

Moreover, there are many other factors which can influence the climate system. For example, under ideal conditions, increasing emissions from fossil fuel use may stimulate photosynthesis, thereby acting as a sink for atmospheric CO_2 (Kohlmaier et al., 1985). However, fossil fuel use has also been related to acid rain, photochemical oxidants, and heavy metals considered by some to be major ingredients for the rapidly proceeding forest dieback in mid-latitudes (Bach, 1985). Ongoing tropical deforestation (Sinclair, 1985) is already now a major additional biospheric source of CO_2; forest dieback in temperate and polar regions may eventually add to this source. The combined effects on the regional and transient climate could be considerable and would therefore require more attention.

In addition to the CO_2 perturbation experiments, similar studies also involving the other radiatively-active trace gases would be very important. The combined radiative forcing effect of the recently identified trace gases is already equal in magnitude to that of CO_2 due to the higher growth rates, residence times and radiation efficiencies. Pioneering work pointing out these additional effects has been done by Flohn (1981). Most of our knowledge stems from one-dimensional radiative-convective models. Ramanathan et al. (1985) give a present-state-of-the-art account. Dickinson and Chervin (1980) have made one of the rare attempts to study the sensitivity of trace gases other than CO_2 using a GCM. The perturbation involved an input of 10 ppbv of chlorofluoromethanes which would correspond to an increase of the current $CFCl_3$ and CF_2Cl_2 concentrations by 55 and 35 times, respectively. The results indicate mean surface temperature and precipitation change patterns similar to those obtained from CO_2 perturbation experiments. In general, results from such model simulation support once more the need for a fully coupled atmosphere-ocean GCM.

Finally, it is important that, beside the time-independent equilibrium response experiments, more emphasis is placed on modeling the transient or time-dependent climatic effects of the increasing trace gas concentrations in the atmosphere (Schneider, 1984). It is clear that from a decision-making standpoint it is more important to get information on climatic change patterns as they are evolving in time, rather than at some point in the future as is obtained from some arbitrary forcing such as a doubling of CO_2. To be meaningful, such transient projections into the future would have to be conducted for a variety of energy (Bach, 1986) and trace gas scenarios (Ramanathan et al., 1985) as well as different aerosol types and concentrations (Kondtratyev and Prokofiev, 1984). So far most of this transient work has been conducted with one- and two-dimensional models (e.g., Hansen et al., 1984; Thompson and Schneider, 1982b). However, transient simulations using the more complete three-dimensional GCMs are now under way for the period of 1958 to 2030 at the Goddard Institute for Space Studies, and perhaps elsewhere. Such studies should receive more emphasis.

Acknowledgments

I wish to thank W.L. Gates, Oregon State University, for making available the observed temperature (Schutz/Gates) and precipitation (Jäger) data plus model-generated data; J.E. Hansen, Goddard Institute for Space Studies; S. Manabe, Geophysical Fluid Dynamics Laboratory; J.F.B. Mitchell, British Meteorological Office, and W.M. Washington, National Center for Atmospheric Research, for supplying further model-generated data. The support of the Commission of European Communities (Contract No. CLI-063-D) and of the Deutsche Forschungs- und Versuchsanstalt für Luft- und Raumfahrt (Contract No. V 30501-0004/81) is gratefully acknowledged. Some of this material is part of a major study on the socioeconomic impacts of climate changes due to a doubling of the atmospheric CO_2 content, carried out in cooperation with Dornier System, Friedrichshafen. Special thanks go to my assistant H.J. Jung for carefully reading this manuscript.

Appendix 3.1. Smoothing Techniques

Spatial smoothing

The purpose of spatial smoothing is to remove small-scale fluctuations from the data while at the same time conserving the large-scale deviations of interest. In this study spatial smoothing was used to interpolate data from a grid system with latitude φ_i and longitude λ_i to a given point (φ,λ) on the earth's surface. This was done using a Gaussian filter of the form:

$$Y'(\varphi, \lambda) = \frac{\sum_i \exp(\delta_i/\alpha)^2 \ Y(\varphi, \lambda_i)}{\sum_i \exp(\delta_i/\alpha)^2} \qquad (A3.1)$$

where δ_i is the distance (degrees) between point (φ_i, λ_i) and point (φ, λ); $Y(\varphi_i, \lambda_i)$ is input data; and $Y'(\varphi,\lambda)$ is the interpolated value for point (φ,λ). Tests have shown that the best choice for the filter width α is 3° (i.e., approximately 330 km in Europe).

Temporal smoothing

The data for the control $(1 \times CO_2)$ and the perturbed $(2 \times CO_2)$ GCM experiments show considerable monthly fluctuations in all climate parameters. Therefore, temporal smoothing was applied to better simulate the mean climate. Using a variety of filters it was found that a fourth-order binomial filter gave the best results:

$$Y'_{i+N/2} = \frac{1}{2^N} \sum_{k=0}^{N} \frac{N!}{k!\,(N-k)!} \ Y_{i+k} \qquad (A3.2)$$

where Y_i is input for month i; Y'_i is output for month i, and i is the cyclical index modulo 12, i.e., $i \pm 12 = i$; N is the order of filter (in this case 4).

Appendix 3.2. Statistical Significance of Model Results

The differences between the results of the control $(1 \times CO_2)$ and perturbed $(2 \times CO_2)$ experiments are caused by changes in boundary conditions. Such differences in the simulated data, just as in the observed data, can be masked because there is an inherent variability in any time-averaged quantity. The natural variability or noise inherent in the model is compared with the signal obtained from the differences between the control and perturbed runs. This signal-to-noise ratio gives an indication of the statistical significance of the changes between the two model runs, and it must first be established in order to avoid simply comparing the various noise levels of the models. Statistical significance can be improved and noise levels reduced if the models are integrated over sufficiently long time periods (Schlesinger, 1983).

Since the time series simulated by GCMs are highly correlated in space and time the standard technique of statistical inference cannot be used. Therefore, in this study the statistical significance of the GCM-generated geographical distributions of surface temperature and precipitation differences was tested with a method described by Chervin (1981). First, sampled ensembles of the model's control and perturbed climate are assembled from sets of independent, finite time-span realizations of GCM simulations over several annual cycles. These samples, derived from both the control and the perturbed cases, are used in order to obtain reasonable estimates of the ensemble average and the inherent variability (i.e., the ensemble standard deviation). Then the null hypothesis that there is no difference between the control and the perturbed experiment is tested at each gridpoint by the first-moment test variate:

$$r_1 = \frac{<m_C> - <m_E>}{(\sigma_C^2 + \sigma_E^2)^{1/2}} \tag{A3.3}$$

where $<m_C>$ and $<m_E>$ are the estimate ensemble time averages for the control $(1 \times CO_2)$ and the experiment $(2 \times CO_2)$; σ_C^2 and σ_E^2 are the ensemble variances. The numerator is the signal and the denominator is an estimate of the variance or noise, so that the quantity r_1 is a measure of the signal-to-noise ratio. Since the equality of ensemble averages cannot always be assumed, the traditional t-test for the equality of the first moments cannot be applied in this case. Therefore, a modified test statistic is used which evaluates significance via an approximation by:

$$r_1 = \frac{t'}{\sqrt{N}} \tag{A3.4}$$

where the probability distribution of t' is rather similar to that of Student's t-distribution, and N is the number of independent samples from the $1 \times CO_2$ and $2 \times CO_2$ cases. It should be noted that in this procedure N is the same for both cases. The acceptance regions for r_1 for the null hypothesis that the population means are equal, such that:

$$r_1(\min) < r_1 < r_1(\max) \tag{A3.5}$$

where $r_1(\min)$ and $r_1(\max)$, are determined from tabulated values of the t-distribution with N-1 degrees of freedom. The same technique has also been used by Mitchell (1983) and Washington and Meehl (1984).

This test procedure has been criticized because it involves many decisions at the gridpoint level instead of only one decision at the global level (Hasselmann, 1979; Storch, 1982). The reason for this is that, in general, a global decision cannot be replaced by a sum of local decisions. Neglect of the spatial correlation in meteorological fields results in overestimates of the area with significant changes. Livezey and Chen (1983) have devised a strategy with which the significance of the differences between control and perturbed GCM experiments can be evaluated.

There are other test procedures. Katz (1982) uses a parametric time series model to estimate the variance of time averages. In this case the variance is evaluated by fitting an autoregressive process to the data. The statistical significance is represented in terms of confidence intervals as used, for example, by Schlesinger (1983). Hayashi (1982) has introduced a reliability ratio such that the significance and reliability of the signal is assessed by comparing it with the confidence interval. For further detail see Bach et al. (1985).

Finally, it is important to note that the statistical analysis must always be supplemented by explanations of the physical processes inducing a climatic change (Gilchrist, 1983).

References

Bach, W. (1983). Carbon dioxide/climate threat: fate or forebearance? In W. Bach et al. (eds.), *Carbon Dioxide: Current Views and Developments in Energy/Climate Research*. Reidel, Dordrecht, pp. 461–509.

Bach, W. (1984). *Our Threatened Climate: Ways of Averting the CO_2-Problem through Rational Energy Use*. Reidel, Dordrecht.

Bach, W. (1985). Forest dieback: extent of damages and control strategies. *Experientia*, **41**, 1095–1104.

Bach, W. (1986). Global air pollution: modeling the combined greenhouse effect of CO_2 and other trace gases. *Proceedings of the 7th World Clean Air Congress & Exhibition*. Sydney, pp. 330–340.

Bach, W., Jung, H.J. and Knottenberg, H. (1985). Modeling the influence of carbon dioxide on the global and regional climate. Methodology and results. *Münstersche Geogr. Arbeiten*, **H.21**, Schöningh, Paderborn.

Barron, E.J. and Washington, W.M. (1984). The role of geographic variables in explaining paleoclimates: results from cretaceous climate model sensitivity studies. *J. Geophys. Res.*, **9**, 1267–1279.

Bryan, K., Kamro, F.G., Manabe, S. and Spelman, M.J. (1982). Transient climate response to increasing atmospheric carbon dioxide. *Science*, **215**, 56–58.

CDAC (Carbon Dioxide Assessment Committee) (1982). *Carbon Dioxide and Climate: A Second Assessment*. National Research Council, National Academy Press, Washington, D.C.

CDAC (1983). *Changing Climate*. National Research Council, National Academy Press, Washington, D.C.

Chervin, R.M. (1981). On the comparison of observed and GCM simulated climate ensembles. *J. Atmos. Sci.*, **38**, 885–901.

Crutcher, H. and Meserve, J. (1970). *Selected Level Heights, Temperatures and Dew Points for the Northern Hemisphere.* NAVAIR 50-1C-52, US Government Printing Office, Washington, D.C.

Dickinson, R.E. (1986). How will climate change? The climate system and modelling of future climate. In B. Bolin, B.R. Döös, J. Jäger and R.A. Warrick (eds.), *The Greenhouse Effect, Climatic Change and Ecosystems.* Wiley, Chichester, pp. 206–270.

Dickinson, R.E. and Chervin, R.M. (1980). Sensitivity of a general circulation model to changes in infrared cooling due to chlorofluoromethanes with and without prescribed zonal ocean surface temperature change. *J. Atmos. Sci.,* **36**, 2304–2319.

Flohn, H. (1981). *Major Climatic Events Associated with a Prolonged CO_2-Induced Warming.* ORAU/IEA-81-8 (M), Institute of Energy Analysis, Oak Ridge Associated Universities, Oak Ridge, USA.

Gates, W.L. (1985). The use of general circulation models in the analysis of the ecosystem impacts of climatic change. *Climatic Change,* **7(3)**, 267-284.

Gates, W.L. and Potter, G.L. (1985). The response of a coupled atmospheric GCM and mixed layer ocean model to doubled CO_2. Abstract for Third Conference on Climate Variations: Symposium on Contemporary Climate 1850–2100, Los Angeles, Jan. 5–11, p. 132.

Gates, W.L., Cook, K.H. and Schlesinger, M.E. (1981). Preliminary analysis of experiments on the climatic effects of increased CO_2 with an atmospheric general circulation model and a climatological ocean. *J. Geophys. Res.,* **86**, 6385-6393.

Gates, W.L. Han, Y.J. and Schlesinger, M.E. (1984). *The Global Climate Simulated by a Coupled Atmosphere–Ocean General Circulation Model: Preliminary Results.* Climatic Research Institute, Report No. 57, Oregon State University, Corvallis.

Gilchrist, A. (1983). Increased carbon dioxide concentrations and climate: the equilibrium response. In W. Bach *et al.* (eds.), *Carbon Dioxide: Current Views and Developments in Energy/Climate Research.* Reidel, Dordrecht, pp. 219–258.

Hansen, J., Lacis, A., Rind, D., Russell, G., Stone, P., Fung, I., Ruedy, R. and Lerner, J. (1984). Climate sensitivity: analysis of feedback mechanisms. In J. Hansen and T. Takahashi (eds.), *Climate Processes and Climate Sensitivity.* Geophysics Monograph **29**, Maurice Ewing Vol. 5, American Geophysics Union, Washington, D.C., pp. 130–163.

Hansen, J., Russell, G., Lacis, A., Fung, I., Rind, D. and Stone, P. (1985). Climate response times: dependence on climate sensitivity and ocean mixing. *Science,* **229**, 857–859.

Hansen, J., Russell, G., Rind, D., Stone, P., Lacis, A., Lebedeff, S., Ruedy, R. and Travis, L. (1983). Efficient three dimensional global models for climate studies: models I and II. *Mon. Wea. Rev.,* **110**, 609–662.

Hasselmann, K. (1979). On the signal-to-noise problem in atmospheric response studies. In *Meteorology of Tropical Oceans.* Royal Meteorological Society, pp. 251–259.

Hayashi, Y. (1982). Confidence intervals of a climatic signal. *J. Atmos. Sci.,* **39**, 1895–1905.

Jäger, J. and Kellogg, W.W. (1983). Anomalies in temperature and rainfall during warm arctic seasons. *Climatic Change,* **5(1)**, 39–60.

Jäger, L. (1976). Monatskarten des Niederschlags für die ganze Erde. *Berichte des Deutschen Wetterdienstes,* **18(139)**, 38 pp.

Katz, R.W. (1982). Statistical evaluation of climate experiments with general circulation models: a parametric time series modeling approach. *J. Atmos. Sci.*, 39, 1446–1455.

Kim, J.-W., Chang, J.-T., Baker, N.L., Wilks, D.S. and Gates, W.L. (1984). The statistical problem of climate inversion: determination of the relationship between local and large-scale climate. *Mon. Wea. Rev.*, 112, 2069–2077.

Kohlmaier, G.M., Plöchl, M., Keeling, C.D. and Revelle, R. (1985). Modeling efforts and experimental evidence of a CO_2 stimulation effect on different types of vegetation. In *Atmospheric Carbon Dioxide, its Sources, Sinks and Global Transport*. Kanderstag Conference, 2–6 Sept. 1985, pp. 158–168.

Kondratyev, K. Ya. and Prokokiev, M.A. (1984). Typifying atmospheric aerosol for assessments of its climatic impact. *Phys. of the Atm. and Ocean*, 5, 339–349.

Kutzbach, J.E. and Otto-Bliesner, B.L. (1982). The sensitivity of the African–Asian monsoonal climate to orbital parameter changes for 9000 years BP in a low-resolution general circulation model. *J. Atmos. Sci.*, 39, 1177–1188.

Livezey, R.E. and Chen, W.Y. (1983). Statistical field significance and its determination by Monte Carlo techniques. *Mon. Wea. Rev.*, 111, 46–59.

Lough, J.M., Wigley, T.M.L. and Palutikof, J.P. (1983). Climate and climate impact scenarios for Europe in a warmer world. *J. Clim. Appl. Meteor.*, 22, 1673–1684.

Manabe, S. (1983). Carbon dioxide and climatic change. In B. Saltzman (ed.), *Advances in the Theory of Climate*. Advances in Geophysics, Vol. 25, Academic Press, New York, pp. 39–82.

Manabe, S. and Stouffer, R.J. (1980). Sensitivity of a global climate model to an increase of CO_2 concentration in the atmosphere. *J. Geophys. Res.*, 85 (C 10), 5529–5554.

Manabe, S. and Wetherald, R.T. (1975). The effects of doubling the CO_2 concentration on the climate of a general circulation model. *J. Atmos. Sci.*, 32, 3–15.

Manabe, S. and Wetherald, R.T. (1980). On the distribution of climatic change resulting from an increase in CO_2 content of the atmosphere. *J. Atmos. Sci.*, 37, 99–118.

Manabe, S., Bryan, K. and Spelman, M.J. (1979). A global ocean–atmosphere climate model with seasonal variation for future studies of climate sensitivity. *Dyn. Atmos. Oceans*, 3, 393–426.

Manabe, S., Wetherald, R.T. and Stouffer, R.J. (1981). Summer dryness due to an increase of atmospheric CO_2 concentration. *Climatic Change*, 3, 347–386.

Meinl, H., Bach, W., Jäger, J., Jung, H.-J., Knottenberg, H., Marr, G., Santer, B. and Schwieren, G. (1984). Socioeconomic impacts of climatic changes due to a doubling of atmospheric CO_2 content. *Research Report to CEC/DFLR*. Dornier-System, Friedrichshafen.

Mitchell, J.F.B. (1983). The seasonal response of a general circulation model to changes in CO_2 and sea temperatures. *Quart. J. Roy. Meteor. Soc.*, 109, 113–152.

Mitchell, J.F.B. and Lupton, G. (1984). A $4 \times CO_2$ integration with prescribed changes in sea surface temperatures. *Progr. in Biomet.*, 3, 353–374.

Palutikof, J.P., Wigley, T.M.L. and Lough, J.M. (1984). Seasonal climate scenarios for Europe and North America in a high CO_2, warmer world. *TRO 012*. US Department of Energy, Washington, D.C.

Pittock, A.B. (1983). Recent climatic change in Australia: implications for a CO_2-warmed earth. *Climatic Change*, 5(4), 321–340.

Pittock, A.B. and Salinger, M.J. (1982). Towards regional scenarios for a CO_2-warmed earth. *Climatic Change*, **4(1)**, 23–40.

Pollard, D. (1982). The performance of an upper ocean model coupled to an atmospheric GCM: preliminary results. *Report No. 31*, Climatic Research Institute, Oregon State University, Corvallis.

Ramanathan, V., Cicerone, R.J., Singh, H.B. and Kiehl, J.T. (1985). Trace gas trends and their role in climatic change. *J. Geophys. Res.*, **90 (D3)**, 5547–5566.

Rind, D. (1984). The influence of vegetation on the hydrologic cycle in a global climate model. In J.E. Hansen and T. Takahashi (eds.), *Climate Processes and Climate Sensitivity*. Geophysics Monographs **29**. Maurice Ewing Vol. 5, American Geophysics Union, Washington, D.C., pp. 73–91.

Rind, D. and Lebedeff, S. (1984). *Potential Climatic Impacts of Increasing Atmospheric CO_2 with Emphasis on Water Availability and Hydrology in the United States*. Report prepared for the US Environmental Protection Agency, Washington, D.C., 96 pp.

Schlesinger, M.E. (1983). *Simulating CO_2-Induced Climatic Change with Mathematical Models: Capabilities, Limitations and Prospects*. III.3–III.139, US DOE 021, Washington, D.C.

Schlesinger, M.E., Han, Y.J. and Gates, W.L. (1985). The role of the ocean in CO_2-induced climatic change: a study with the OSU coupled atmosphere–ocean general circulation model. *Climatic Research Institute Report No. 60*, OSU, Corvallis.

Schneider, S.H. (1984). On the empirical variation of model-predicted CO_2-induced climatic effects. In J.E. Hansen and T. Takahashi (eds.), *Climate Processes and Climate Sensitivity*. Geophysics Monographs **29**, Maurice Ewing Vol. 5, American Geophysics Union, Washington, D.C., pp. 187–201.

Schneider, S.H. and Thompson, S.L. (1981). Atmospheric CO_2 and climate: importance of the transient response. *J. Geophys. Res.*, **86(64)**, 3135–3147.

Schutz, C. and Gates, W.L. (1971). *Global Climatic Data for Surface, 800 mb, 400 mb: January. R-915-ARPA*. The Rand Corporation, Santa Monica, CA, 173 pp.

Schutz, C. and Gates, W.L. (1972a). *Supplemental Global Climatic Data: January. R-915 1-ARPA*. The Rand Corporation, Santa Monica, CA, 41 pp.

Schutz, C. and Gates, W.L. (1972b). *Global Climatic Data for Surface, 800 mb, 400 mb: July. R-1029-ARPA*. The Rand Corporation, Santa Monica, CA, 180 pp.

Schutz, C. and Gates, W.L. (1973a). *Global Climatic Data for Surface, 800 mb, 400 mb: April. R-1317-ARPA*. The Rand Corporation, Santa Monica, CA, 192 pp.

Schutz, C. and Gates, W.L. (1973b). *Supplemental Global Climatic Data: January. R-915/2-ARPA*. The Rand Corporation, Santa Monica, CA, 38 pp.

Schutz, C. and Gates, W.L. (1974a). *Global Climatic Data for Surface, 800 mb, 400 mb: October. R-1425-ARPA*. The Rand Corporation, Santa Monica, CA, 192 pp.

Schutz, C. and Gates, W.L. (1974b). *Supplemental Global Climatic Data: July. R-1029/1-ARPA*. The Rand Corporation, Santa Monica, CA, 38 pp.

Sinclair, L. (1985). International task force plans to reverse tropical deforestation. *Ambio*, **14(6)**, 352–353.

Spelman, M.J. and Manabe, S. (1984). Influence of oceanic heat transport upon the sensitivity of a model climate. *J. Geophys. Res.*, **89**, 571–586.

Storch, H.V. (1982). A remark on Chervin-Schneider's algorithm to test significance of climate experiments with GCMs, *J. Atmos. Sci.*, **39**, 187–189.

Thompson, S.L. and Schneider, S.H. (1982a). Carbon dioxide and climate: has a signal been observed yet? *Nature*, **295**, 645–646.

Thompson, S.L. and Schneider, S.H. (1982b). Carbon dioxide and climate: the importance of realistic geography in estimating transient temperature response. *Science*, **217**, 1031–1033.

Washington, W.M. and Meehl, G.A. (1983). General circulation model experiments on the climatic effects due to a doubling and quadrupling of carbon dioxide concentration. *J. Geophys. Res.*, **88**(C 11), 6600–6610.

Washington, W.M. and Meehl, G.A. (1984). Seasonal cycle experiments on the climate sensitivity due to a doubling of CO_2 with an atmospheric general circulation model coupled to a simple mixed layer ocean model. *J. Geophys. Res*, **89**, 9475–9503.

Wigley, T.M.L. and Schlesinger, M.E. (1985). Analytical solution for the effect of increasing CO_2 on global mean temperature. *Nature*, **315**, 649–652.

Wilson, C.A. and Mitchell, J.F.B. (1984). *The 5-Layer Model Climate over Western Europe and the Frequency of Occurrence of Extreme Values of Temperature, Precipitation and Wind for Selected Grid Boxes: The Changes with $4 \times CO_2$ and Prescribed Sea Surface Temperatures.* MET O 20 Tech. Note II/224. Bracknell.

SECTION 4

Development of Climatic Scenarios:
B. Background to the Instrumental Record

Jill Jäger

4.1. Introduction

Climate statistics are an important source of descriptive information about the climate. They are obtained by averaging measured climatic elements such as temperature, precipitation and pressure over a time period that is considerably longer than the limit of predictability for atmospheric motions (about two weeks with current forecasting methods). Often statistics are obtained by averaging data over a large number of years and 30 years of data are usually used to define "normals". A description of the climate of an averaging period requires, in addition to measures of central tendency (means or totals), estimates of the variability during the averaging period. Suitable measures of variability include the standard deviation of elements such as temperature and pressure, and the frequency spectrum of observed variations. It is common to refer to short-term climatic changes lasting only a few decades as climatic fluctuations. The term trend is usually used to denote fluctuations on time scales of around 20 years or more.

A more rigorous definition of climate is obtained by using an ensemble average rather than a time average (Dickinson, 1986). An ensemble average is made over a hypothetical infinite set of earths with the same external influences (i.e., the same solar input, the same atmospheric composition, etc.) but different detailed weather patterns. Weather noise is the uncertainty in the ensemble average arising from the sampling of unpredictable weather fluctuations, that part of climatic variability which arises from day-to-day weather variations.

This section looks at the characteristics of recent climatic fluctuations in the northern hemisphere. Its purpose is to provide background information on these fluctuations as a preliminary step toward the selection of some of them as climatic scenarios in the case study impact assessments. Their selection is described in more detail, case study by case study, in subsequent parts of this volume. Since, for the purposes of the assessment of climatic impacts on agriculture, the climatic data for the surface are of importance, only the fluctuations in the surface elements (temperature, sea-level pressure, precipitation) are considered here. The variations in other elements, such as temperature and pressure in the upper atmosphere, are necessary for a more complete description of climate and for analysis of the nature and causes of climatic change, but are not as useful, given the present state of knowledge, for the present study. In addition, a complete description of climate would require consideration of the variations in other components of the climate system (oceans, ice and snow, the biota and the land surface), but such analysis will not be made here.

Although there have been improvements in the data coverage of the northern hemisphere during the last 100 years, it must be noted that the coverage, even for variables such as surface air temperature, is still far from complete and there has been a tendency for investigators to make hemispheric generalizations on the basis of records from a limited geographical area without consideration of the hemispheric representativeness of the area under study.

The following sections describe the data availability and the fluctuations of temperature, sea-level pressure and precipitation observed during the last 100 years. The causes of the observed fluctuations are generally unexplained although there are some clear associations between regional temperature changes and circulation changes. The precise causal mechanisms of climatic change are not well established and no mechanisms have been unequivocally and quantitatively associated with fluctuations observed in the recent past.

4.2. Data

Most attempts that have been made to combine station surface air temperature data into an average for the northern hemisphere have used essentially the same data source, *World Weather Records* (WWR) (*see* WWR, Smithsonian Institution, 1927, 1934, 1947 and WWR, US Weather Bureau 1959–1982). These have been supplemented with additional data from published and manuscript material in meteorological archives (Bradley *et al.*, 1985; Jones *et al.*, 1985) and the enhanced data set used to compile a gridded data set (Jones *et al.*, 1986). These authors have assessed the effects of the two main sources of possible error in estimating land-based surface air temperature in the northern hemisphere: errors in the individual station records used to compile hemispheric averages and changes in spatial coverage. They have interpolated the correct and corrected stations onto a regular 5° latitude by 10° longitude grid by an objective method for all months from 1851 to 1984 and estimate that the hemispheric temperature series is probably reliable on a year-to-year basis after 1875 (Jones *et al.*, 1986).

Other compilations of the surface air temperature data have been made recently by Vinnikov *et al.* (1980), Hansen *et al.* (1981) and Yamamoto (1981). These studies are largely based on land station data and the results are briefly compared in Subsection 4.3.

A few studies have also been made of ocean or ocean-plus-land temperature changes (Folland et al., 1984; Barnett, 1978, 1984; Chen, 1982; Paltridge and Woodruffe, 1981; Fletcher, 1984). It is, however, very difficult to obtain a strictly homogeneous time series of ship-based data due to changes in instrumentation and problems related to instrument exposure. Either sea surface temperatures (SSTs) or ship-based (marine) air temperatures (MATs) may be used, but both types of data require adjustments to produce a homogeneous record.

For analysis of changes in the atmospheric circulation at the surface, gridded monthly average sea-level pressure data are available. Several analyses have used the data set available from the US National Center for Atmospheric Research (NCAR) (Jenne, 1975), which extends from 1899 to the present. The data are for a 5° latitude-longitude grid from 20°N to 85°N. From 1899 to 1939 the data are from the US *Historical Map Series*. Since 1939 several sources have been used. Another gridded set of sea-level pressure data is produced by the UK Meteorological Office and extends from 1873 to the present, with coverage limited to the sector 100°W to 90°E between 1881 and 1898.

Even though long records do exist, the analysis of long-term fluctuations in precipitation amount is more difficult than the analysis of, say, surface air temperature. Precipitation is a discontinuous process and two closely located measurement stations may record significantly different amounts of rain. The areal averaging of precipitation data is difficult because of sparseness of coverage in some places and because of the effects of orography on precipitation. There are also problems in measuring rainfall amount and it is generally accepted that rain gauges tend to underestimate the amount of precipitation reaching the ground (Corona, 1978). Factors such as wind, evaporation, splashing out of the gauge and height of the gauge above the ground affect the amount of rainfall measured. Since one can assume that such errors will be nearly the same each year at an individual station, an analysis of long-term fluctuations is possible, provided that homogeneous station records are used (e.g., records with no significant changes of instrumentation or station location). One precipitation data set is provided by NCAR and extracted from the World Monthly Surface Climatological Data. For example, Corona (1978) used precipitation data from about 1300 stations over the northern hemisphere. The homogeneity of these data is uncertain. Tabony (1981) looked at another data set, comprising rainfall totals for 185 stations in Western Europe that had homogenized records (*see* Subsection 4.3.4).

4.3. Recent Climatic Variations in the Northern Hemisphere

4.3.1. Changes in surface air temperature

As discussed in the previous subsection, there have been a number of studies of the surface air temperature fluctuations of the northern hemisphere. Three of the four studies based on land stations give very similar results, which is not surprising since they rely on the same basic data source. Minor differences exist owing to small differences in the data sources and the different methods of producing area averages (Wigley *et al.*, 1986). In the case of Yamamoto (1981),

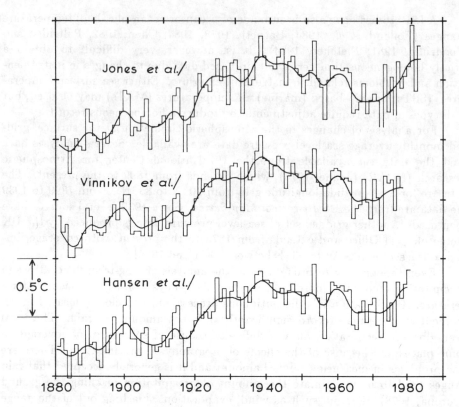

Figure 4.1. Comparison of three estimates of northern hemisphere temperature changes. The top curve is from Jones *et al.* (1982); the middle curve is from Vinnikov *et al.* (1980); the lower curve is based on the data of Hansen *et al.* (1981) for the whole hemisphere. The curves show both annual mean values and values smoothed with a 10-year Gaussian filter (padded at the ends to cover the whole period of record). (Source: Wigley *et al.*, 1986.)

averages were based on the incorrect assumption that no changes had occurred in regions with no data, so these results are not directly comparable with the others. *Figure 4.1* compares three estimates of northern hemisphere temperature changes based on land stations. The three curves are highly correlated. Each analysis shows the general trends of warming to around 1940, cooling to the mid-1960s and warming since about 1970. The curves shown in *Figure 4.1* have often been used as indicators of global mean change, but a true global average can only be obtained by including data from the ocean areas and all of the southern hemisphere. Recent comparisons of land and ocean data and of southern and northern hemisphere data have shown remarkable parallels (Wigley *et al.*, 1985).

To date no comprehensive synthesis of the land-based and marine temperature records has been published. A somewhat *ad hoc* synthesis has been produced, however, to give a global mean temperature curve that is at least an improvement on the "global" curves based on land stations (Wigley *et al.*, 1986).

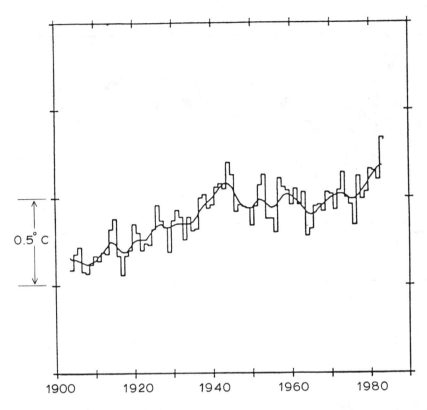

Figure 4.2. Estimated global mean annual surface temperature changes since 1904. Smooth curve shows 10-year Gaussian filtered values. (Source: Wigley *et al.*, 1986.)

The resulting global mean annual surface temperature changes since 1904 are shown in *Figure 4.2*. The curve shows the temperature increase until about 1940, followed by a slight cooling and a renewed temperature increase after the early 1960s.

Figure 4.3 shows the surface air temperature fluctuations in the northern hemisphere in winter (December, January, February) and summer (June, July, August) calculated using the data compiled by Jones *et al.* (1986). The values for winter show large year-to-year variability before 1900. After 1900 there was a general temperature increase until 1940, an irregular temperature decrease until about 1970 and an increase since about 1970. The average temperatures for the land stations in the northern hemisphere in the summer months also show a temperature increase from 1900 until 1940, but this was followed by decreasing temperatures until the end of the 1970s. The year-to-year variability in surface temperature at the land stations in the northern hemisphere is smaller in the summer than in the winter and the amplitude of the fluctuations is also smaller.

Annual and seasonal changes in air temperatures over high latitudes of the northern hemisphere have also been examined for the period 1881–1980 (Kelly *et*

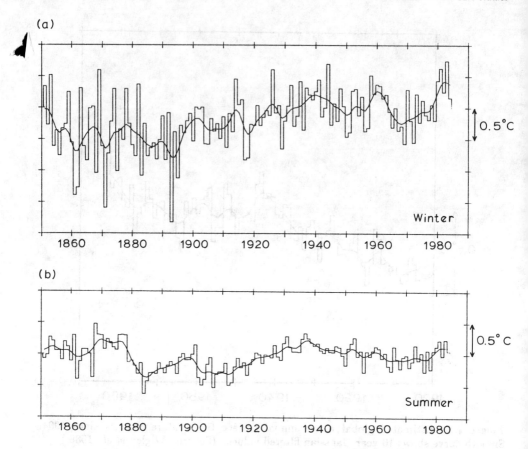

Figure 4.3. Surface air temperature anomalies from the 1951 to 1970 mean for the northern hemisphere for (a) winter and (b) summer. (Source: Jones et al., 1986.)

al., 1982). The trends in Arctic temperatures were broadly similar to those for the northern hemisphere during the study period, although the Arctic fluctuations were greater in magnitude and more rapid. *Figure 4.4* shows the annual temperature anomalies computed for the Arctic ($65°N-85°N$). These averages are most representative of the conditions over the land masses of the region. Data were not available for major oceanic areas or for the central Arctic for most of the 1881–1980 period.

The correlation between the annual temperature data for the northern hemisphere and for the Arctic is high (about 0.8), but the range of variations in the Arctic data is greater by a factor of 3 (Kelly et al., 1982). The overall range of the long-term fluctuations in temperature has been greatest in winter. The spatial pattern of changes in Arctic temperatures was also analyzed using principal component analysis. Trends were found to be generally consistent over most of the Arctic, but certain regions were particularly sensitive to long-term fluctuations, most notably northwest Greenland and around the Kara Sea (Kelly et al., 1982).

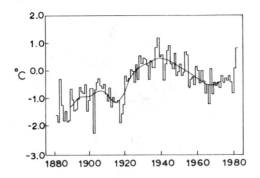

Figure 4.4. Annual temperatures (°C) as departures from the reference period 1946–1960 averaged over the Arctic (65°N–85°N). (Source: Kelly et al., 1982.)

The spatial patterns of northern hemisphere temperature change during three distinct periods – strong warming, 1917–1939; cooling, 1940–1964; warming, 1965–1980 – have been analyzed (Jones and Kelly, 1983). A linear trend was fitted at each grid point to the annual temperature data for each of these periods. The trends (expressed in °C/year $\times$ 10^{-1}) are shown in *Figure 4.5(a)* 1917–1939, (b) 1940–1964, and (c) 1965–1980. The shaded areas indicate the grid points at which the slope of the trend is significant at the 5% level although, as the authors point out, the statistical test is invalidated by the preselection of the three periods as ones of maximum change.

During the warming phase of 1917–1939 [*Figure 4.5(a)*] maximum warming occurred over Greenland, the Barents Sea, northwestern Siberia, northern Europe, the USA and the Far East. Cooling occurred in central Canada and central Asia. Jones and Kelly (1983) suggest that this pattern implies changes in the number of mid-latitude cyclones entering the Arctic and variations in the strength of the continental anticyclones.

The distribution of annual temperature changes during the 1940–1964 cooling phase [*Figure 4.5(b)*] shows major cooling over the Kara Sea, northern parts of the USSR, Alaska and northwestern Canada. Warming occurred over the Ukraine and Kamchatka. The cooling during this period was not as marked as the earlier warming, and the spatial patterns were not as coherent.

The recent warming, 1965–1980 [*Figure 4.5(c)*], was strongest over the Greenland Sea, the Barents Sea and northern Scandinavia, most of the USSR, Alaska, northwestern Canada, the southwestern USA and northern Africa. Cooling occurred over the Canadian Arctic islands and northwest Greenland and, to a lesser extent, over the Mediterranean region. The spatial pattern was as coherent as that of the previous cooling, and the magnitudes of the trends were similar (Jones and Kelly, 1983).

In all three periods the magnitudes of the trends were greatest in the Arctic and in the interior of the two main continents. The pattern of the cooling phase (1940–1964) was not opposite to the pattern of the earlier warming (1917–1939), and the two patterns for the warming phases were different. However, the recent

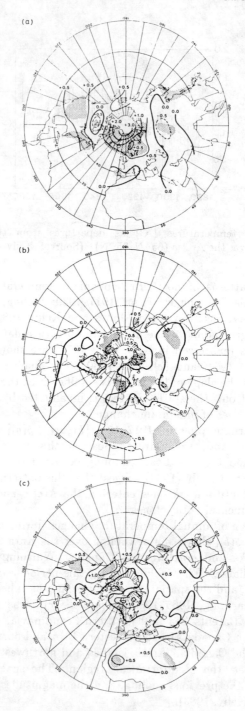

Figure 4.5. Linear trend of gridpoint annual temperatures (°C/year × 10^{-1} over the periods (a) 1917–1939, (b) 1940–1964, (c) 1965–1980. Shaded areas are significant at the 5% level. (Source: Jones and Kelly, 1983.)

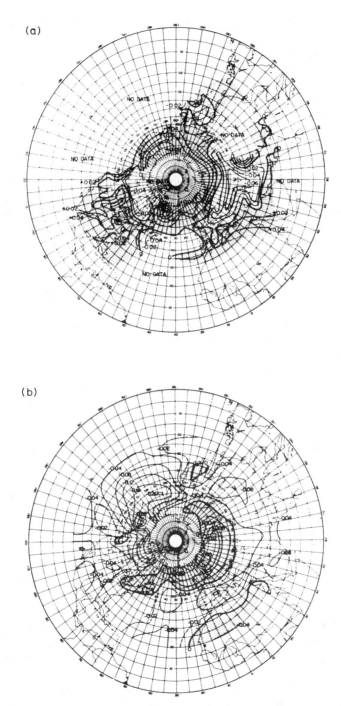

Figure 4.6. Isopleths of the slope of the regression line fitted to trends in winter over the period 1900–1941 of (a) surface temperature (in °C/year) and (b) sea-level pressure (in mb/year). (Source: van Loon and Williams, 1976a.)

warming (1965–1980) was almost opposite to the pattern of the cooling phase, and this suggests that the recent warming has been returning surface temperatures to levels similar to those experienced in the 1940s.

The work by Jones and Kelly (1983) considered the linear trends only in the annual data. It is important for our understanding of the nature of climatic change to look at the changes as a function of season and possibly as a function of month. A series of earlier studies has examined the regional distributions of surface temperature changes in winter, summer, spring and autumn in the northern hemisphere (van Loon and Williams, 1976a,b; Williams and van Loon, 1976). The station data collected for *World Weather Records* were used for the analysis of surface temperature changes between 1900 and 1972. Unlike the studies reported above, the data were not gridded. In addition to studying arbitrary 15-year periods, van Loon and Williams examined the period of "warming" from 1900 to 1941 and the period of "cooling" from 1942 to 1972.

Figure 4.6(a) shows the isopleths of the slope of the regression line of winter mean surface temperature for 1900–1941 (in °C/year). Although the northern hemisphere as a whole had a trend of increasing temperatures during this period (*see Figure 4.1*), it is clear from *Figure 4.6(a)* that not all regions had increasing temperatures during this period. In particular, the temperature decreased over northern Canada, eastern Europe and central Asia. The largest temperature trends were in the polar latitudes (up to 0.18 °C/year over the 40-year period).

4.3.2. The connection between trends of mean temperature and circulation

In several cases it has been shown that surface temperature trends are associated with changes in the atmospheric circulation. In addition to computing linear trends in the surface temperature data, van Loon and Williams (1976a,b) and Williams and van Loon (1976) computed linear trends in the gridded sea-level pressure data. The pressure data were also used to calculate the south-north and west-east components of the geostrophic wind, and regression lines were also computed for these components.

Figure 4.6(a) shows the linear trend in surface temperature in the winter season for the period 1900–1941, and the trends in sea-level pressure during the same period are shown in *Figure 4.6(b)*. North of 50°N the pressure increased over almost half of the circumference and fell over the other half. It was also demonstrated that the winter surface temperature changes from 1900 to 1941 were largely consistent with the changes in circulation. For example, the cooling over central Europe and the central USSR is clearly related to increased easterly flow into these areas. The areas of warming centered over Greenland and the Barents Sea are related to the increased southerly flow into these areas (van Loon and Williams, 1976a). For the summer season, in contrast, it was found that the regional temperature trends in the northern hemisphere were not so clearly associated with changes in advection as those in winter (van Loon and Williams, 1976b).

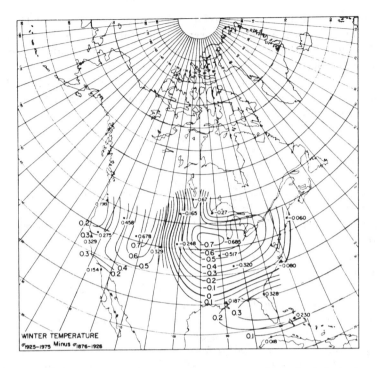

Figure 4.7. The difference, 1925–1975 minus 1876–1926, of standard deviation of winter (DJF) temperature (°C) in North America. (Source: van Loon and Williams, 1978.)

4.3.3. Changes in interannual variability

The question of whether climate is more variable in cool periods has been examined using station data (for which there was a complete 100-year record until 1975) from Europe and North America and reported in *World Weather Records* (van Loon and Williams, 1978). The station data were not necessarily homogeneous. It has been pointed out subsequently that data inhomogeneities, caused by urbanization in particular, are important in some regions and especially in the USA (Jones et al., 1986). Standard deviations for the temperature data and precipitation data were computed after the linear trend had been removed since the focus was on variability from year-to-year and not on the slow changes expressed by the long-term trend (van Loon and Williams, 1978).

Figure 4.7 shows the change of standard deviation of winter temperature from 1876–1926 to 1925–1975 in North America. There is a large-scale pattern with decreasing variability in the midwest and northeast and increasing variability in the south and west. The change from the first 51-year period to the second was found to be statistically significant. However, no single connection was found between the trend of the mean and the trend of variability of both temperature and precipitation (van Loon and Williams, 1978).

An examination of surface pressure anomalies over much of the northern hemisphere and over a more limited area near Britain over a period of about 100 years yielded little evidence for any unusual variability of climate in recent years on annual, one-month and five-day time scales (Ratcliffe et al., 1978). A circulation change in the east Atlantic–British Isles sector was noted around 1940, but this did not result in any detectable increased variability of monthly temperature or rainfall over Britain.

4.3.4. Changes in precipitation

In contrast to the studies of surface temperature, there have been relatively few studies of hemispheric averages of precipitation. More studies have looked at the characteristics of the rainfall record at individual stations or for geographical regions (e.g., Wigley et al., 1984). Investigations of long-term precipitation records have generally uncovered noticeable fluctuations on a decadal timescale but no significant long-term trends.

In the USA monthly temperature and precipitation data have been used to calculate the Palmer Drought Severity Indices from January 1895 to April 1981 for a 60-point grid over the USA (Karl and Koscielny, 1982). The authors found no statistically significant cycles of drought during the past 86 years. Moreover, visual inspection of the time series for the nine regions indicated that if there were any long-term trends in the data, they were rather small and were not likely to be statistically significant.

In another study, precipitation data from 150–300 stations in North America were analyzed for the period 1935–1974 (Corona, 1978, 1979). Apart from the large peaks in precipitation for several winters between 1937 and 1941, no significant trend was found in the rainfall data, although an approximately 22-year cycle in summer precipitation for North America was suggested (Corona, 1978).

The main patterns of European rainfall anomalies have been examined using principal component analysis of 182 homogenized rainfall series from 1861 to 1970 (Tabony, 1981). The network was divided into 15 regions. *Figure 4.8* shows the decadal means of annual precipitation for the 15 regions in Europe. Increases in precipitation are evident in a belt stretching from southwest France throughout northwest Germany to southwest Scandinavia. This increase was mainly a feature of the winter halfyear in France and the summer halfyear in Germany.

Trends in northern hemisphere precipitation have been analyzed for the period 1891–1979 for the months of January and July (Gruza and Apasova, 1981). The data for the period 1891–1975 were taken from maps prepared at the Principal Geophysical Observatory in Leningrad and at the USSR Hydrometeorological Center for 1976–1979. The values were converted to percentages of the long-term mean to provide smoother and more coherent fields. However, the data are probably not homogenized. The long-term record of anomalies in Eurasia showed an upward trend in January over the entire period (linear trend: 1.2% per year) and a very small negative trend in July (-0.1% per

Climatic scenarios: B. Background to the instrumental record   171

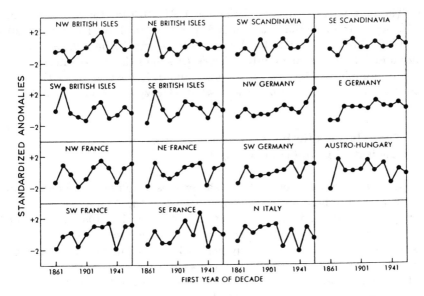

Figure 4.8. Decadal means of annual rainfall over 15 regions of Europe. (Source: Tabony, 1981.)

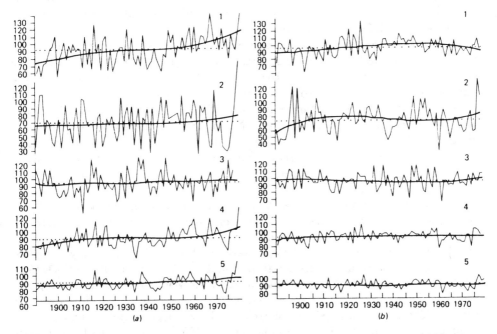

Figure 4.9. Long-term changes in precipitation anomalies in (*a*) January and (*b*) July by region: 1 = northern Asia, 2 = Africa, 3 = America, 4 = Eurasia, 5 = northern hemisphere. (After Gruza and Apasova, 1981; source: Houghton, 1984.)

year). The long-term trends for the period 1891–1979 for Europe were found to be 0.4% per year in January and 0.15% per year in July. During the 30-year period 1946–1975, the trends for Europe were found to be -2.0% per year in January and 1.1% per year in July. *Figure 4.9* shows the long-term series of precipitation anomalies computed for January and July for regions of the northern hemisphere and for the hemisphere as a whole (Gruza and Apasova, 1981).

4.3.5. The relationship between temperature and precipitation fluctuations

A number of studies have looked at the anomalies of temperature and rainfall in the northern hemisphere for seasons or years when the Arctic was warmer than average (e.g., Williams, 1980; Wigley *et al.*, 1980; Namias, 1980; Jäger and Kellogg, 1983). There were two main reasons for focusing on warm years or seasons in the Arctic. Firstly, it is generally believed that an increased carbon dioxide concentration would lead to a globally averaged warming (*see* Part I, Section 3). Secondly, model and observational studies have suggested that the Arctic is more sensitive to climatic changes than the hemisphere as a whole. The study of the warm Arctic seasons showed that large and coherent anomalies of temperature occurred elsewhere in the hemisphere when the Arctic was warm, but these anomalies were not always positive. In addition, it was found that because of changes in the atmospheric circulation, there were also large and coherent anomalies in the rainfall distribution. The magnitudes and distribution of the anomalies were different in the different seasons.

Figure 4.10 shows the differences between average winter precipitation during the five warmest consecutive seasons in the Arctic (1934–1938) and the five coldest consecutive seasons in the Arctic (1966–1970). Over Eurasia the differences of precipitation had a general pattern of increased precipitation north of 60°N, decreased precipitation between 40°N and 60°N, and increased precipitation between 30°N and 40°N. Over North America there was increased precipitation along the west coast and in a band south of the Great Lakes. The differences shown in *Figure 4.10* are an example of the scale of precipitation changes associated with temperature changes; other examples are discussed by Jäger and Kellogg (1983).

In a further study, changes in mean seasonal temperature and total seasonal precipitation have been analyzed using monthly averages of temperature and precipitation for each of the contiguous states of the USA (Diaz and Quayle, 1980). Three periods were selected based on time series of areally weighted winter and annual mean temperature for the contiguous USA: 1893–1920, 1921–1954, 1955–1977. No systematic relationship was found between changes in mean temperature and standard deviation from period to period and season to season. There was some positive correlation between precipitation and variance of precipitation. During the warm phase (1921–1954), winter mean temperatures increased most over the eastern USA, and, at the same time, precipitation increased throughout the south and northward to the Great Lakes. Generally, the eastern half of the USA received more precipitation while its western half

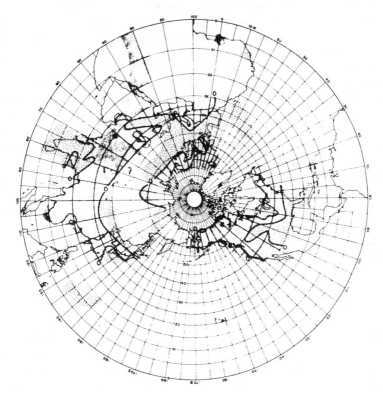

Figure 4.10. Differences between average winter precipitation during the five warmest consecutive Arctic seasons and the five coldest consecutive Arctic seasons. (Source: Jäger and Kellogg, 1983.)

received less. In the cooling period that followed, the greatest drop of mean temperature occurred over those areas that had the greatest increases during the warming period. There were gains in mean annual precipitation in the eastern half of the USA while the Rocky Mountain region and California received less. Therefore, in terms of integrated precipitation over the USA, there was greater aridity during the middle warm period compared with the two cooler periods. The drier conditions developed primarily over the southwestern USA.

The correlation between temperature and precipitation over the USA and Europe has also been examined (Madden and Williams, 1978). Time series of seasonal temperature means and precipitation totals were determined for North American and European stations using data from the *World Weather Records* for the period 1897–1960; 72 stations in North America and 26 stations in Europe were considered. In the winter season over North America the plains states showed a significant negative correlation, that is, during the 64-year period, cold winters were usually wet winters and vice versa. In two areas, however, there was a significant positive correlation: in the northern California and Oregon coast area and in an area from New England southwestwards to the lower Mississippi valley. It was suggested that the distribution of positive and negative correlations was associated with the tracks of cyclones and with the relationship of temperature and precipitation within the various sectors of the cyclone. In the summer, only one small area on the Pacific coast had a positive correlation, so that over a large part of the USA cool summers are usually wet summers and vice versa. Over Europe in winter, the majority of the area considered had a significant positive correlation. The pattern was therefore somewhat different from that over North America where a large area of negative correlation occurred over the middle of the continent. In summer the area of significant negative correlation was larger (cool summer usually meant wet summer and vice versa), and covered most of Europe except the Mediterranean. Thus, Madden and Williams showed that over most areas in North America and Europe summer temperatures and precipitation tended to be negatively correlated, and in other seasons patterns of correlation were more complex with large areas of both positive and negative correlation. They also found that these patterns occur on all time scales since the normalized cospectra, showing the contribution to the total correlation from low-, medium-, and high-frequency variations, in general had the same distributions as the correlations.

4.3.6. Summary

This review has considered fluctuations of surface temperature, sea-level pressure and precipitation over the northern hemisphere during the last 100 years. Although there have been improvements of the data coverage and quality during this period, it must be noted that the coverage, even for variables such as surface temperature, is still far from complete, especially over the ocean areas. The surface temperature data show that over the land areas of the northern hemisphere there was a general warming from the turn of the century until about 1940, followed by a cooling until the mid-1960s and a subsequent warming. The timing of these warming and cooling phases varied somewhat according to season. The spatial patterns of trends during these periods have also been investigated. During the warming phase of 1917–1939 maximum warming occurred over Greenland, the Barents Sea, northwestern Siberia and northern Europe, the USA and the Far East. Cooling occurred in central Canada and central Asia. During the cooling phase of 1940–1964, major cooling occurred over the Kara Sea, northern parts of the USSR, Alaska and northwestern Canada, while warming occurred over Kamchatka and the Ukraine. The recent warming (1965–1980) was strongest over the Greenland Sea, the Barents Sea and northern Scandinavia, most of the USSR, Alaska, northwestern Canada, the southwestern USA, and northern Africa. Cooling occurred over the Canadian Arctic islands and northwest Greenland. At least for the winter season, the regional surface temperature fluctuations have been demonstrated to be associated with atmospheric circulation changes and related changes in advection.

Analysis of precipitation records is hampered by sampling and measurement problems, the discontinuity of the precipitation process and relatively large natural variability. No studies of the spatial and seasonal distributions of linear trends in precipitation data have been made, so that the precipitation record has not been treated in the same detail as the surface temperature record. Some evidence of long-term trends has been found in homogeneous European rainfall records, whereas analyses of records from the continental USA have found no significant long-term trends and large decadal fluctuations in the precipitation totals.

4.4. Temperature and Precipitation Changes in the IIASA/UNEP Case Study Areas

The following parts of this volume consider the impact on agriculture and forestry of changes in climate in five case study areas: Canada (Saskatchewan), Iceland, Finland, the northern USSR and Japan. The preceding discussion has shown that there have been spatial differences in the temperature and precipitation fluctuations in the northern hemisphere during the past hundred years. In this subsection the temperature and precipitation records at grid points or meteorological stations in the case study areas are examined.

Figure 4.11(a) shows the winter temperatures at four grid points near to the case study areas from 1850 to 1983. The grid points are 50°N, 110°W (southwestern Saskatchewan, Canada); 65°N, 20°W (west-central Iceland); 60°N, 30°E (close to the case study areas of southern Finland and Leningrad, USSR), and 40°N, 140°E (northwestern Tohoku, Japan). For the location of the case study areas see Part I, Section 3, especially Figure 3.4. The data are from the data set described by Jones et al. (1986). The values shown in Figure 4.11 can be compared with the averages for the northern hemisphere shown in Figure 4.2. The magnitude of the variations is larger for the grid points than for the northern hemisphere. In particular, the gridpoints at 60°N, 30°E and 50°N, 110°W show large variations. The temperatures in the northern hemisphere showed an increase from the beginning of the 1900s until about 1940, a slight drop from 1940 to 1960 and a subsequent increase. In the northern USSR, Finland and Saskatchewan case study areas, long-term fluctuations of this kind are not visible in the winter temperature record. In Iceland, however, there was a warming from the end of the 1800s until about 1940 followed by a cooling. The winter temperatures in Japan show a fluctuation essentially opposite to that of the northern hemisphere average with a period of lower temperatures before 1940 and a period of higher temperatures afterwards. None of the curves in Figure 4.11(a) shows a warming since 1970 as found for the northern hemisphere average.

The temperature fluctuations at the same grid points in summer are shown in Figure 4.11(b). The curves for 50°N, 110°W (southwestern Saskatchewan); 65°N, 20°W (west-central Iceland) and 60°N, 30°E (southern Finland, Leningrad region) show an increase of summer temperatures from 1900 until about 1940, in common with the northern hemisphere average. The Iceland curve shows a steady decrease of summer temperatures after 1940, and a decrease also occurred at 60°N, 30°E, although not as steady as in Iceland. The Japan summer temperatures show no major long-term trends since about 1910; previous to this there was a slight warming.

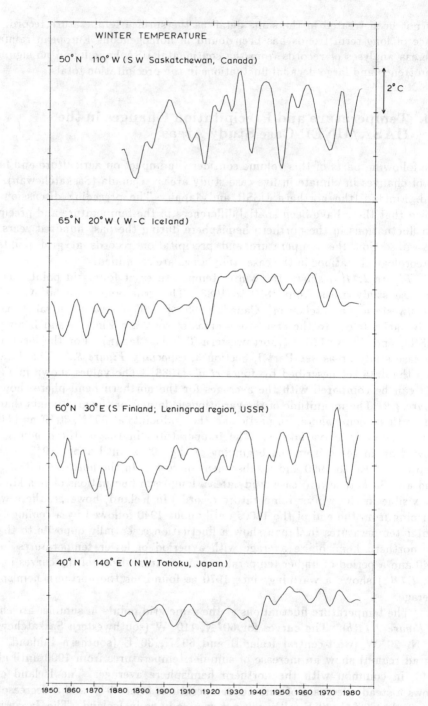

Figure 4.11. (a) Winter temperatures as departures from the reference period (1951–1970) mean at four grid points located in the case study areas. (Source: Climatic Research Unit, University of East Anglia, Norwich, UK.)

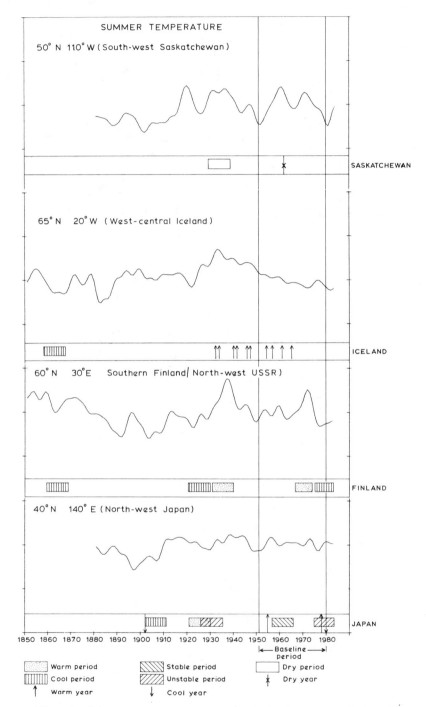

Figure 4.11. (*b*) As for (*a*): summer temperatures (with the instrumental scenarios used in each region marked below each curve – note that no instrumental scenarios were chosen in the USSR study).

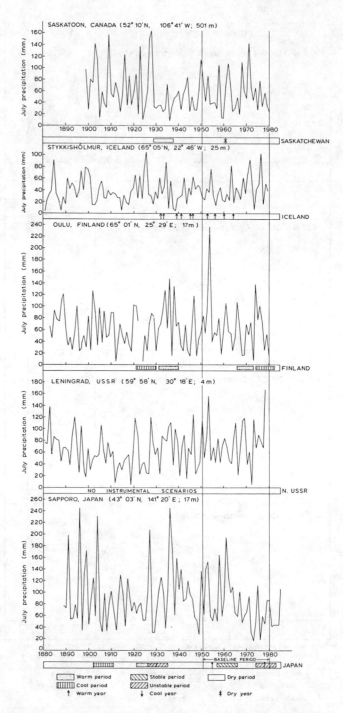

Figure 4.12. July precipitation totals (mm) at meteorological stations in the five case study areas, 1881–1980. Instrumental scenarios selected in four of the regions are marked below their respective curves.

The curves of winter and summer temperature shown in *Figure 4.11* show that there have been long- and short-term fluctuations in temperature in the case study areas with no general similarity to the average for the land-based northern hemisphere. Indeed, given the nature of temperature fluctuations over the hemisphere during the last hundred years, with large-scale regional temperature trends varying according to period and location, there would be no reason to expect that the temperature records of individual grid points would mirror the changes computed for the hemisphere as a whole. The gridpoint fluctuations are of larger magnitude than the hemispheric ones, as would be expected as a result of the averaging process.

Figure 4.12 shows the July precipitation totals (mm) at five meteorological stations in the case study areas (Saskatoon, Stykkishólmur, Oulu, Leningrad, and Sapporo). There is some evidence of long-term precipitation fluctuations at these stations. At Stykkishólmur, Iceland, for example, there was an increase of July precipitation amount from a minimum at the end of the 1930s until the end of the 1970s. At Leningrad, USSR, there was a decrease between 1880 and 1920 and an increase between 1920 and 1955. The dry years of the 1930s are clear in the data from Saskatoon, Canada, as is the period of large interannual variability between 1900 and 1930. At Sapporo (northern Japan) there has been a decrease of the July precipitation since the end of the 1930s and relatively low interannual variability during the last 20 years. More detailed descriptions of the instrumental climatic record for each of the case study regions are to be found in the following parts of this volume. These data formed the basis for the selection of instrumental climatic scenarios, which involved the identification of individual years and periods of years with anomalous weather. The most important scenarios are indicated in *Figures 4.11* and *4.12*. The reference period for impact experiments is 1951–1980.

Acknowledgments

This section was written while the author was visiting the Climatic Research Unit, Norwich, UK. I would like to thank Tom Wigley and Phil Jones of the Climatic Research Unit for providing me with data used in the preparation of this section and for their valuable comments on earlier versions.

References

Barnett, T.P. (1978). Estimating variability of surface air temperature in the northern hemisphere. *Mon. Wea. Rev.*, **106**, 1353–1367.

Barnett, T.P. (1984). Long-term trends in surface temperature over the oceans. *Mon. Wea. Rev.*, **112**, 303–312.

Bradley, R.S., Kelly, P.M., Jones, P.D., Diaz, H.F. and Goodess, C. (1985). A climatic data bank for the northern hemisphere land areas. 1851–1980. *DOE Tech. Report No. TR017*, US Dept. of Energy, Carbon Dioxide Research Division, Washington D.C.

Chen, R.S. (1982). Combined land/sea surface air temperature trends, 1949–1972. MSc Dissertation (Massachusetts Institute of Technology).

Corona, T.J. (1978). The interannual variability of northern hemisphere precipitation. *Environmental Research Paper No. 16*, Colorado State University, Fort Collins, Colorado, 27 pp.

Corona, T.J. (1979). Further investigations of interannual variability of northern hemisphere continental precipitation. *Environmental Research Paper No. 20*, Colorado State University, Fort Collins, Colorado, 20 pp.

Diaz, H.F. and Quayle, R.G. (1980). The climate of the United States since 1895: spatial and temporal changes. *Mon. Wea. Rev.*, **108**, 249–266.

Dickinson, R.E. (1986). How will climate change? The climate system and modelling of future climate. In B. Bolin, B.R. Döös, J. Jäger and R.A. Warrick (eds.), *The Greenhouse Effect, Climatic Change and Ecosystems*. Wiley, Chichester, pp. 206–270.

Fletcher, J.O. (1984). Clues from sea–surface records. *Nature*, **310**, 630.

Folland, C.K., Parker, D.E. and Kates, F.E. (1984). Worldwide marine temperature fluctuations 1856–1981. *Nature*, **310**, 670–673.

Gruza, G.V. and Apasova, Ye.G. (1981). Climatic variability of the northern hemisphere precipitation amounts (in Russian). *Meteorologia i Gidrologia*, **5**, 5–16.

Hansen, J.E., Johnson, D., Lacis, A., Lebedeff, S., Lee, P., Rind, D. and Russell, G. (1981). Climatic impact of increasing atmospheric carbon dioxide. *Science*, **213**, 957–966.

Houghton, J.T. (ed.) (1984). *The Global Climate*. Cambridge University Press, 234 pp.

Jäger, J. and Kellogg, W.W. (1983). Anomalies in temperature and rainfall during warm Arctic seasons. *Climatic Change*, **5**, 39–60.

Jenne, R. (1975). *Data Sets for Meteorological Research, NCAR TN/1A-111*. National Center for Atmospheric Research, Boulder, Colorado.

Jones, P.D. and Kelly, P.M. (1983). The spatial and temporal characteristics of northern hemisphere surface air temperature variations. *J. Climatol.*, **3**, 243–252.

Jones, P.D., Raper, S.C.B., Bradley, R.S., Diaz, H.F., Kelly, P.M. and Wigley, T.M.L. (1986). Northern hemisphere surface air temperature variations: 1851–1984. *J. Clim. Appl. Meteor.*, **25**, 161–179.

Jones, P.D., Raper, S.C.B., Santer, B.D., Cherry, B.S.G., Goodess, C., Bradley, R.S., Diaz, H.F., Kelly, P.M. and Wigley, T.M.L. (1985). A grid point surface air temperature data set for the northern hemisphere. 1851–1984. *DOE Tech Report*. US Dept. of Energy, Carbon Dioxide Research Division, Washington D.C.

Karl, T.R. and Koscielny, A.J. (1982). Drought in the United States: 1895–1981. *J. Climatol.*, **2**, 313–329.

Kelly, P.M., Jones, P.D., Sear, C.B., Cherry, B.S.G. and Tavakol, R.K. (1982). Variations in surface air temperatures: Part 2, Arctic Regions, 1881–1980. *Mon. Wea. Rev.*, **110**, 71–83.

Madden, R.A. and Williams, J. (1978). The correlation between temperature and precipitation in the United States and Europe. *Mon. Wea. Rev.*, **106**, 142–147.

Namias, J. (1980). Some concomitant regional anomalies associated with hemispherically averaged temperature variations. *J. Geophys. Res.*, **85**, 1580–1590.

Paltridge, G. and Woodruffe, S. (1981). Changes in global surface temperature from 1880 to 1977 derived from historical records of sea surface temperature. *Mon. Wea. Rev.*, **109**, 2427–2434.

Ratcliffe, R.A.S., Weller, J. and Collison, P. (1978). Variability in the frequency of unusual weather over the approximately the last century. *Quart. J. Roy. Meteor. Soc.*, **104**, 243–255.
Smithsonian Institution (1927, 1935, 1947). *World Weather Records*, Miscellaneous Collections, Volumes 79, 90, and 104, Smithsonian Institution, Washington D.C.
Tabony, R.C. (1981). A principal component and spectral analysis of European rainfall. *J.Climatol.*, **1**, 283–294.
US Weather Bureau (1959–1982). *World Weather Records*, 1941–1950, 1951–1960, 1961–1970, US Department of Commerce, Washington D.C.
van Loon, H. and Williams, J. (1976a). The connection between trends of mean temperature and circulation at the surface: Part I. Winter. *Mon. Wea. Rev.*, **104**, 365–380.
van Loon, H. and Williams, J. (1976b). The connection between trends of mean temperature and circulation at the surface: Part II. Summer. *Mon. Wea. Rev.*, **104**, 1003–1011.
van Loon, H. and Williams, J. (1978). The association between mean temperature and interannual variability. *Mon. Wea. Rev.*, **106**, 1012–1017.
Vinnikov, K.Ya., Gruza, G.V., Zakharov, V.F., Kirillov, A.A., Kovneva, N.P. and Ran'kova, E.Ya. (1980). Current climatic changes in the northern hemisphere (in Russian). *Meteorologia i Gidrologia*, **6**, 5–17.
Wigley, T.M.L., Angell, J.K. and Jones, P.D. (1985). Analysis of the temperature record. In M.C. MacCracken and F.M. Luther (eds.), *Detecting the Climatic Effects of Increasing Carbon Dioxide, DOE/ER-0235*. US Department of Energy, Washington D.C.
Wigley, T.M.L., Jones, P.D. and Kelly, P.M. (1980). Scenarios for a warm, high-CO_2 world. *Nature*, **283**, 649–652.
Wigley, T.M.L., Jones, P.D. and Kelly, P.M. (1986). Empirical climate studies: warm world scenarios and the detection of climatic change induced by radiatively active gases. In B. Bolin, B.R. Döös, J. Jäger and R.A. Warrick (eds.), *The Greenhouse Effect, Climatic Change and Ecosystems*. Wiley, Chichester, pp. 271–322.
Williams, J. (1980). Anomalies in temperature and rainfall during warm Arctic seasons as a guide to the formulation of climate scenarios. *Climatic Change*, **2**, 249–266.
Williams, J. and van Loon, H. (1976). The connection between trends of mean temperature and circulation at the surface: Part III. Spring and Autumn. *Mon. Wea. Rev.*, **104**, 1591–1596.
Yamamoto, R. (1981). Change of global climate during recent 100 years. *Proceedings of The Technical Conference on Climate – Asia and Western Pacific*. WMO Report No. 578, World Meteorological Organization, Geneva, pp. 360–375.

SECTION 5

A Case Study of the Effects of CO_2-Induced Climatic Warming on Forest Growth and the Forest Sector: A. Productivity Reactions of Northern Boreal Forests

Pekka Kauppi and Maximilian Posch

5.1. Introduction

The purpose of this section is to evaluate the effects of changes in climate under the GISS 2 × CO_2 scenario (considered in Section 3) on the productivity of boreal forests. Some of the results reported here will be used as inputs to a further set of experiments concerned with the effect of productivity changes on forestry as an economic activity (*see* Section 6). Together Sections 5 and 6 provide a case study (at a hemispheric scale) of the advantages and limitations of linking biophysical and economic models in attempts to assess the effects of climatic change.

5.2. Temperature and the Productivity of Boreal Forests

Cool climate – particularly in spring, summer and autumn – restricts the productivity of boreal forests. Low winter temperatures may not be so crucial, however, because plants are adapted to tolerate low temperatures in their dormant stage. Low productivity occurs partly because of a short growing season due to spring and autumn frosts which force plants to adapt by remaining dormant over a large part of the year. In addition, low growing season temperatures tend to restrict the rate of biological processes and in this way also to decrease productivity. It appears that the productivity of boreal ecosystems is rather closely correlated with the amount of heat accumulated during the year, as obtained by subtracting the plants' threshold temperature from the mean temperature of the day and summing up over the year. The value obtained is termed the annual effective temperature sum (ETS) measured in degree-day units.

Two essentially different factors, the magnitude of the warming and the potential of ecosystems to react to it, determine how ecosystems will respond to a possible climatic warming. Under the GISS $2 \times CO_2$ climatic scenario the largest warming would take place in continental regions such as North America and continental Siberia. The GISS $2 \times CO_2$ climatic scenario represents future averaged equilibrium climatic conditions, based on simulations by the Goddard Institute for Space Studies (GISS) general circulation model (GCM) for a doubling of the present concentrations of atmospheric carbon dioxide. (For details, see Section 3.) Smaller increases in temperature are estimated in maritime regions. Also, the potential of ecosystems to react to the warming can be assumed to vary in time and space. Within the boreal zone maritime ecosystems may conceivably be potentially more responsive (Kauppi and Posch, 1985). Continental ecosystems were estimated to be potentially less responsive. In this way, it appears that the two factors – climatic warming and ecological response – tend to have regional distributions with opposite effects. However, in our earlier analysis, we did not examine which of the two factors would dominate. This study focuses on the possible long-term increase in the productivity of boreal forests and combines the estimated increase in temperature together with the potential of forests to react to this increase.

Before proceeding with this, however, some important caveats are in order:

(1) Increased atmospheric CO_2 concentrations can themselves cause enhanced photosynthetic activity (and hence net productivity) and also increase the water utilization efficiency of plants (Kramer, 1981; Kauppi, 1987).
(2) Changes in other climatic variables – such as precipitation, insolation and windspeed – are also likely to accompany CO_2-induced temperature changes.
(3) Several other factors are important to tree growth, for example, the incidence and activity of pests and diseases, the rates of nutrient cycling and other processes in the soil and ground litter, and the susceptibility of trees to frost damage.

These factors, while important, are *not* considered in this study.

5.3. Data and Study Procedure

In line with experiments conducted in individual case study regions, this analysis is based on the GISS GCM-generated results for estimating temperature conditions for $1 \times CO_2$ and $2 \times CO_2$ environments (*see* Section 3). In order to generate a scenario for a climatic warming the $2 \times CO_2$ results were compared with the $1 \times CO_2$ results and to gridded observed data (source: Schutz and Gates –

see References, Section 3) for describing the existing climate. All data used in this study are given on a grid covering the northern hemisphere between 28°N and 70°N. The grid size is 5° in an east-west and 4° in a north-south direction; this gives a total of $9 \times 72 = 648$ data points.

Observed mean monthly temperature data were converted into ETS values for each grid point, applying a threshold temperature of $T_0 = +5\,°\text{C}$. Calculating ETS from the temperature time series would require data on daily observations, yet only monthly mean temperatures were available, and a method was developed for approximating ETS from these data. It was assumed that the daily temperatures are normally distributed around the monthly mean, $\bar{T}$. This assumption was tested statistically using Finnish data from 20 meteorological stations, 10 years and 12 months each year. For these data the assumption was found to be valid with only a few exceptions. The standard deviation σ varied from 3.0°C to 5.5°C in winter and from 1.5°C to 4.0°C in summer. Therefore a value of $\sigma = 3\,°\text{C}$ was selected to describe that distribution, and on this basis the monthly ETS was computed as follows (N is the number of days per month):

$$\begin{aligned} ETS &= \int_{T_0}^{\infty} (T - T_0) \frac{N}{\sqrt{2\pi}\sigma} \exp\left[-\frac{(T-\bar{T})^2}{2\sigma^2}\right] dT \\ &= \frac{N\sigma}{\sqrt{2\pi}} \exp\left[-\frac{(\bar{T}-T_0)^2}{2\sigma^2}\right] + \frac{N}{2}(\bar{T}-T_0)\left[1 + \text{erf}\left(\frac{\bar{T}-T_0}{\sqrt{2}\sigma}\right)\right] \end{aligned} \quad (5.1)$$

where

$$\text{erf}(x) = \frac{2}{\sqrt{\pi}} \int_0^x e^{-t^2} dt$$

Monthly ETS values were summed up to give the annual ETS. Note that equation (5.1) is of an identical form to an equation used to compute ETS in an earlier paper (Kauppi and Posch, 1985). However, the integration in equation (5.1), in contrast to the earlier equation, applies to *monthly* (rather than annual) mean temperature and requires standard deviations of daily mean temperature around the *monthly* (rather than the annual) mean.

ETS was calculated in the same way for a climatic scenario referring to doubled CO_2 conditions (assuming the same standard deviation). The scenario was constructed using GISS model results by first subtracting the computed $1 \times CO_2$ (reference) temperature from the computed $2 \times CO_2$ temperature for each grid point. The difference was added to the Schutz and Gates observed temperature at each grid point to correct for errors in the modeled reference level. Scenario estimates for mean monthly temperatures (and thus, of course, for ETS) were higher than the corresponding reference temperatures at all grid points.

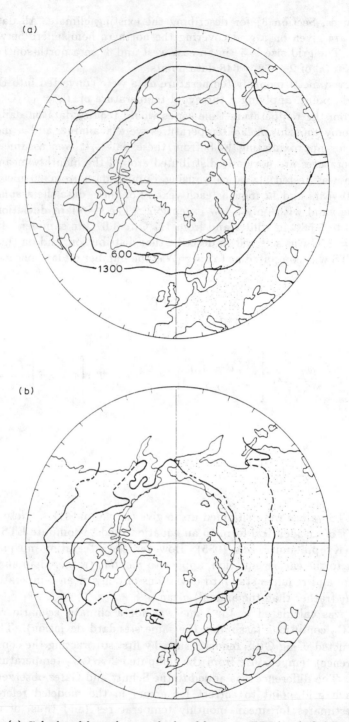

Figure 5.1. (*a*) Calculated boreal zone: the band between ETS isopleths of 600 and 1300 degree-days. (*b*) Observed boreal zone, redrawn from Hämet-Ahti (1981).

5.4. Locating Regions with a High ETS Response

To test whether or not the annual ETS concept is applicable for describing the boundaries of the boreal zone, isopleths of ETS were compared with the vegetation map compiled by Hämet-Ahti (1981). Based on the situation in the Fennoscandian peninsula, ETS isopleths of 600 and 1300 degree-days were used to describe the northern and southern boundaries of the zone. These calculated boundaries were compared with those observed (*Figure 5.1*). In the Fennoscandian peninsula the values 600 and 1300 degree-days quite accurately bound the central parts of the biome, i.e., the northern boreal, middle boreal and southern boreal subzones as classified by Hämet-Ahti (1981). However, in certain areas of the Pacific region the calculated and observed locations of the zone tend to depart. The calculated northern boundary fails to include some parts of the boreal biome in Alaska and Kamchatka. The match is again quite reasonable from British Columbia eastwards to the Atlantic Ocean and further east towards western Siberia, that is, over at least two-thirds of the zone.

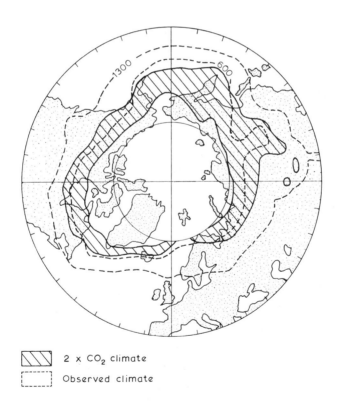

Figure 5.2. Calculated boreal zone for the GISS 2 × CO_2 climate scenario relative to the calculated present-day zone [from *Figure 5.1(a)*].

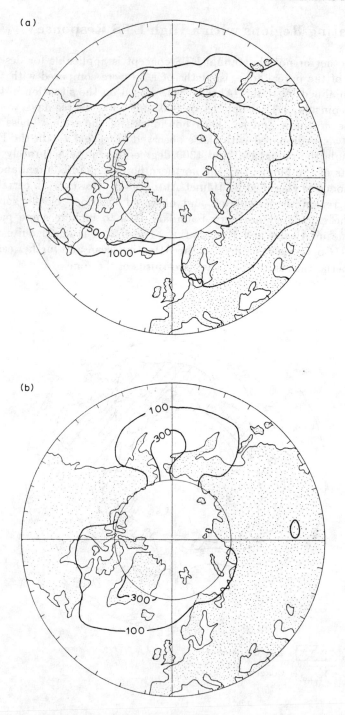

Figure 5.3. Estimated change of ETS for the GISS $2 \times CO_2$ climate scenario (a) in absolute units (degree-days), (b) in relative units with isopleths indicating doubling (100% increase) and quadrupling (300% increase) of ETS.

One must bear in mind that the two sets of data are not fully independent. The region is so large that the vegetation map has been constructed not only on the basis of botanical field observations but also using supplementary temperature data (similar to those used in estimating the calculated zone) as a proxy for botanical information.

The 600 and 1300 degree-days ETS boundaries were calculated in the same way for the GISS $2 \times CO_2$ scenario. According to this scenario the zone would shift northwards by 500–1000 km (*Figure 5.2*). Emanuel et al. (1985a) reported even larger estimates for such shifts based on a different climate model, namely that of Manabe and Stouffer (1980) and the Holdridge classification of vegetation zones; although in response to criticisms of their methods (Rowntree, 1985) these estimates have since been revised downwards (Emanuel et al., 1985b).

A comparison was made between the estimated ETS for the doubled CO_2 scenario ETS_2 and that representing the observed climate using the Schutz and Gates temperature data ETS_0. The increase of ETS was calculated in absolute units, $ETS_2 - ETS_0$, as well as in relative units, $100(ETS_2 - ETS_0)/ETS_0$. The relative increase indicates how dramatic the change would be in relation to the historical reference level. An interesting finding was that highest increases, at a given latitude, were estimated over maritime regions. This result was obtained with both indicator variables, that is, both in absolute and in relative units (*Figure 5.3*).

5.5. Possible Growth Response

Effective temperature sum is a variable which has been applied in the boreal zone both in agriculture and in forestry investigations. It has relevance especially in the northern part of the zone where soil moisture is almost always sufficient even to the extent that excess water and the subsequent peat formation frequently limit forest productivity (Mikola, 1950). Significant moisture deficiency appears only towards the southern part of the zone. Here drought may limit agricultural crop yields and, hence, the ETS variable is only weakly correlated to agricultural crop yield (Mukula et al., 1978; see Part IV). The ETS variable alone is not particularly useful in studies which focus on any specific ecosystem, but it may have some value in rough regional comparisons. Therefore, for demonstrative purposes, we convert the ETS results to estimates of potential forest productivity.

Average tree growth varies in Finland from $1.2 \, m^3 \, ha^{-1} \, yr^{-1}$ in northern regions to $5.0 \, m^3 \, ha^{-1} \, yr^{-1}$ in southern regions and is rather closely correlated with ETS (*Figure 5.4*). Each data point in *Figure 5.4* represents one "Forestry Board" district. The size of the districts varies from 4000–51 000 km^2. ETS data were average values for the period 1931–1960, and the growth data were inventory results from 1951–1953. The inventory was based on a random sample from all forests, and in this way the data represent the average case rather than the conditions of any specific ecosystem. The national forest inventory has been repeated seven times between 1920 and 1980. The regional pattern of tree growth as measured in the third inventory (1951–1953) is representative for the

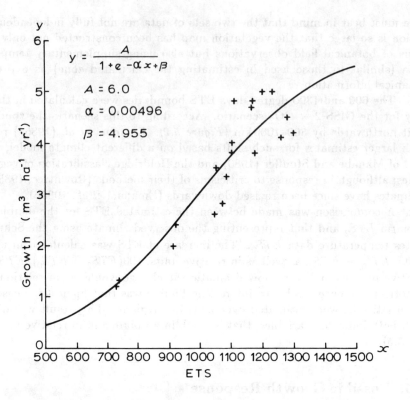

Figure 5.4. Empirical relationship between annual effective temperature sum (ETS) and annual mean tree growth.

whole 60-year period (*Yearbook of Forest Statistics*, 1983). Finnish forests are managed for timber production and are low-land forests with relatively sheltered exposures, their growth figures being generally higher than those of forests elsewhere in the boreal zone. Thus conditions for growth that pertain to Finnish conditions, if applied to other parts of the boreal zone, tend slightly to overestimate productivity in that zone (as discussed in Section 6).

The relation between ETS (denoted as x) and tree growth (y) was assumed to be adequately represented by a logistic function:

$$y = \frac{A}{1 + e^{-\alpha x + \beta}} \tag{5.2}$$

where A is the maximum growth rate (assumed = $6.0 \, \text{m}^3 \, \text{ha}^{-1} \, \text{yr}^{-1}$), and α and β are coefficients (0.00496 and 4.955, respectively).

Data from Finland were used to determine the parameters of the function (*see Figure 5.4*). The leveling-off of the function for high ETS values takes into account the drought limitation of growth in high ETS conditions. This functional relationship between ETS and growth was used to map productivity

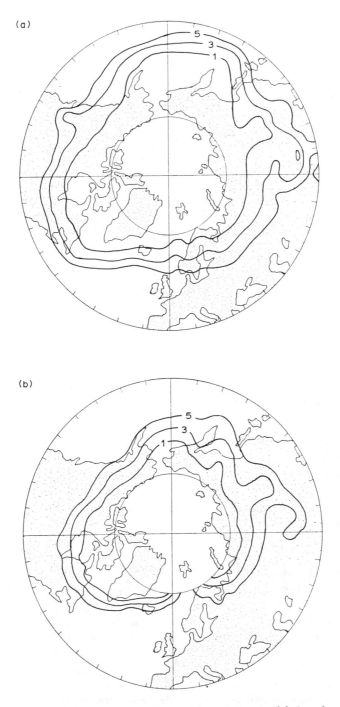

Figure 5.5. Estimated productivity regions of boreal forests (a) for observed (Schutz and Gates) climatic data, (b) for the GISS $2 \times CO_2$ climate scenario. Values refer to the regional average of potential stemwood productivity in $m^3 ha^{-1} yr^{-1}$.

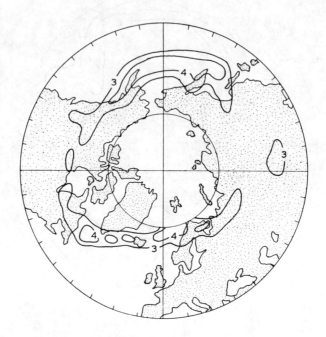

Figure 5.6. Estimated change in potential productivity from baseline level to doubled CO_2 conditions (in $m^3 ha^{-1} yr^{-1}$).

regions for the observed climate [*Figure 5.5(a)*] and for the GISS $2 \times CO_2$ climate [*Figure 5.5(b)*].

The changes of growth from observed climate to doubled CO_2 climate were estimated in absolute units ($m^3 ha^{-1} yr^{-1}$). The greatest growth changes were estimated in maritime regions in the northern parts of the observed zone (*Figure 5.6*). Several factors affect this result. In the northernmost regions even the doubled CO_2 climate was estimated to be too cold to maintain any major productivity. In southern regions observed climate is currently rather favorable, but under doubled CO_2 conditions it was assumed that moisture stress would restrict growth (leveling-off in *Figure 5.4*). During the winter period, with temperatures below the threshold, ecosystems are viewed as being dormant and ETS, by definition, does not respond to temperature variations. So, while ETS does not respond at all to a large rise in winter temperature from, say, $-40\,°C$ to $-15\,°C$, in contrast it responds rather strongly to a modest temperature rise from, say, $+3\,°C$ to $+7\,°C$. Such temperatures (around the threshold temperature) occur quite frequently in maritime regions within the boreal zone, and this is the reason why the response is estimated to be higher in these regions.

An earlier hypothesis, mentioned above, stated that maritime areas of the boreal zone would be particularly susceptible to impacts from a possible climatic change (Kauppi and Posch, 1985). This new result suggests that despite the counteracting effect of relatively lower estimated climatic warming over maritime regions, maritime ecosystems might nevertheless respond more strongly than continental ecosystems. Highest responses were estimated in Labrador, southern Greenland, Iceland, northern Fennoscandia and around the Bering Strait (*see* Figure 5.6).

5.6. Discussion

If, indeed, the greenhouse effect of CO_2 changed the climate as much as is supposed above, quite substantial changes could be expected between the boreal biome and the neighboring biomes – tundra to the north and the temperate zone to the south. It is almost impossible to assess which ecological, social or economic consequences would have the greatest impact. In this section, however, we have tried to specify regions within the boreal forest zone which would respond more strongly than others from the standpoint of productivity. This has been done by using rather simple indicator variables: effective temperature sum and forest productivity. The largest increase in growth is estimated to occur in maritime regions of the northern parts of the biome.

Several different sources of uncertainty in this kind of analysis have emerged from the study. First of all there is uncertainty associated with the input data. Observed climate, for example, necessarily contains approximations because the area is very large and it is only partially covered by sufficiently dense networks of meteorological stations. Further uncertainties are associated with the modeling of climatic changes due to doubled CO_2 conditions, related both to the simplifying assumptions and inaccuracies of the climate model itself and to the omission in the analysis of changes in other climatic variables such as precipitation and evapotranspiration, although these would perhaps have fewer consequences in the boreal zone than elsewhere in the world. Moreover, while it is reasonably clear that atmospheric CO_2 concentrations will continue to increase in the next few decades even if the anthropogenic sources were "switched off" immediately (due to the ocean inertia effect – *see* Section 3), it is not certain that CO_2 concentrations will ever grow to values twice as high as the reference level.

Our method of calculating ETS is an approximation, especially the assumption that daily temperatures are distributed normally around the monthly mean, which has to be checked. A comparison of the results by assuming different distributions (including the rather unlikely case of an uniform distribution around the monthly mean) showed, however, that *ETS* values are rather insensitive to the particular shape of the distribution as long as they are unimodal. We also compared our method with the widely used formula $ETS = N(\bar{T} - T_0)$ for $\bar{T} > T_0$, and here we found a deviation in the *ETS* of 30–40 degree-days per month for $\bar{T} \simeq T_0$. That simple formula is equivalent to assuming that the daily temperature is equal to the monthly mean, a rather rough approximation we tried to overcome by using our formula.

A basic assumption of the calculations is that boreal forests are in equilibrium both with the present and with the $2 \times CO_2$ climate. The approach is valid in the long run assuming that the atmospheric concentrations of CO_2 and other greenhouse gases will stabilize at a new, relatively constant level. Time will be needed after such a stabilization for the climate to reach a new equilibrium (see, for example, Hansen et al., 1985) and for the forests to become adapted to the new climate. The above results thus refer to the possible conditions in the future when essentially all equilibria have been reached.

If the rate of a climatic change is low, forests can change smoothly; first due to the phenotypic acclimatization, and then due to genetic adaptation. The latter process is rather slow in forests, where individual trees live 100–200 years and where the minimum time period between two successive generations is roughly 30 years. In this time perspective the observed and expected rates of the increase of the atmospheric CO_2 concentration are rather high. One would not anticipate a smooth ecological transition unless the climatic response times are very long. In case of a rapid climatic change, forests would be susceptible to, for example, pests or extreme climatic events. In addition, as mentioned earlier, it is necessary to consider the so-called fertilizing effect of increased atmospheric CO_2 on photosynthesis and plant growth, a matter which is outside the scope of this section (see Shugart et al., 1986). At the present state of knowledge it is very difficult to assess what actually would happen during such an obviously quite troublesome transition period. Nonetheless, with these caveats in mind, Section 6 considers some of the economic implications of possible changes in forest productivity.

References

Emanuel, W.R., Shugart, H.H. and Stevenson, M.P. (1985a). Climatic change and the broad-scale distribution of terrestrial ecosystem complexes. *Climatic Change*, 7, 29–43.
Emanuel, W.R., Shugart, H.H. and Stevenson, M.P. (1985b). Response to comment: climate change and the broad-scale distribution of terrestrial ecosystem complexes. *Climatic Change*, 7, 457–460.
Hämet-Ahti, L. (1981). The boreal zone and its subdivision. *Fennia*, 159, 69–75.
Hansen, J., Russell, G., Lacis, A., Fung, I., Rind, D. and Stone, P. (1985). Climate response times: Dependence on climate sensitivity and ocean mixing. *Science*, 229, 857–859.
Kauppi, P. (1987). Forests and the changing chemical composition of the atmosphere. In M. Kallio, D.P. Dykstra and C.S. Binkley (eds.), *The Global Forest Sector: An Analytical Perspective*. J. Wiley and Sons, Chichester.
Kauppi, P. and Posch, M. (1985). Sensitivity of boreal forests to possible climatic warming. *Climatic Change*, 7, 45–54.
Kramer, P.J. (1981). Carbon dioxide concentration, photosynthesis and dry matter production. *Bioscience*, 31, 29–33.
Manabe, S. and Stouffer, R.J. (1980). Sensitivity of a global climate model to an increase of CO_2 concentration in the atmosphere. *J. Geophys. Res.*, 85, 5529–5554.
Mikola, P. (1950). On variations in tree growth and their significance to growth studies. *Commun. Inst. For. Fenn.*, 38, 5.

Mukula, J., Rantanen, O. and Lallukka, U. (1978). Crop certainty of oats in Finland 1950-1976. *Report Nr. 10*, Finnish Agricultural Research Center, Department of Plant Husbandry, Jokioinen (in Finnish).

Rowntree, P.R. (1985). Comment on "Climatic change and the broad-scale distribution of terrestrial ecosystem complexes" by Emanuel, Shugart and Stevenson. *Climatic Change*, 7, 455–456.

Shugart, H.H., Antonovsky, M. Ya., Jarvis, P.G. and Sandford, A.P. (1986). CO_2, climatic change and forest ecosystems. Assessing the response of global forests to the direct effects of increasing CO_2 and climatic change. In B. Bolin, B.R. Döös, J. Jäger and R.A. Warrick (eds.), *The Greenhouse Effect, Climatic Change and Ecosystems*, SCOPE 29, Wiley, Chichester, pp. 475–521.

Yearbook of Forest Statistics (1983). *Official Statistics of Finland*, Vol. XVIIA, No.14, Folia Forestalia 550.

SECTION 6

A Case Study of the Effects of CO_2-Induced Climatic Warming on Forest Growth and the Forest Sector: B. Economic Effects on the World's Forest Sector

Clark S. Binkley

6.1. Introduction

The preceding section pointed to potentially significant increases in the productivity of high-latitude forests under the GISS 2 × CO_2 climatic scenario (for details of the GISS 2 × CO_2 scenario, *see* Section 3). These high-latitude forests provide the raw material for an important part of the world's forest products industry. In Finland and Sweden, expansion of mill capacity in the forest products industry is limited by the growth of timber needed to supply the mills. Concern has been expressed both in eastern and in western Canada about the adequacy of forest growth to support the extant industry. Increasing forest growth in those regions could conceivably have significant economic impacts on forest products production, prices and trade throughout the world.

This section quantifies some of the economic effects of the changes in high-latitude forests associated with the productivity changes described in Section 5. The complexities of both the economic and the ecologic systems involved renders any such assessment tentative at best. The first subsection discusses the economic model used in the analysis. The next subsection explains the ecologic model and describes how the ecologic and economic models are linked. The third subsection presents the simulated adjustments in the world's forest sector in response to climate warming. The final subsection comments on the results and presents summary estimates of the economic gain to the forest sector associated with climate warming.

6.2. A Global Forest Sector Model

The global forest sector model developed at the International Institute for Applied Systems Analysis (IIASA) was used in this assessment (Kallio *et al.*, 1987, especially Parts V and VI). The model projects production, consumption, prices and trade of 16 forest products in 18 regions which comprise the globe. The model is integrated in the sense that it includes components describing all aspects of forest products production: forest growth, timber supply, processing facilities and final demand. The ecological effects associated with changes of climate enter the model through the forest growth and timber supply components of the model.

The model computes a partial market equilibrium for the world's forest sector using a mathematical programming approach first suggested by Samuelson (1952). Market equilibrium occurs when prices are such that the quantity of each product which is supplied to the market just equals the quantity of the product demanded at that price. The point of market equilibrium possesses two important kinds of consistency: firstly, prices equal costs for the last unit produced and, secondly, material flows balance in such a way that, for example, enough timber is harvested to produce all of the products that are consumed. The equilibrium solutions are partial in the sense that many factors that influence forest products prices and consumption levels, such as the costs of labor and energy, are exogenous to the model. Because the model is designed to examine long-run structural change in the sector, these partial market equilibrium solutions are computed for each of the ten 5-year periods covering 1980 to 2030.

Although the model solutions are linked between time periods, intertemporal market equilibrium is not explicitly modeled. This limitation is important to the timber supply model because timber owners can choose when to harvest their trees: if prices are too low in one period, harvests are readily deferred in anticipation of prices rising in another period. Because timber prices influence the costs of all forest products, this limitation is not confined to one component of the model. Instead of using a complex intertemporal equilibrium model of timber supply, we adopt a simple procedure which captures many of the adjustments that would be expected in a more complete presentation of the sector (Binkley and Dykstra, 1987).

Figure 6.1 illustrates the dynamic structure of the model. The first large rectangle represents the market equilibrium model for a particular period. It is composed of submodels for each of the regions which are linked together by a series of equations describing trade.

The solution for one 5-year period indicates the quantity of timber harvested for that period, prices, production and consumption in each region, and trade flows of raw materials and finished products among regions. This information is used to project timber-growing stocks, processing costs, mill capacities and other factors that are used to solve the market equilibrium problem for the next time period. The process is iterated until the specified projection horizon has been reached. All of the simulations reported here are initialized with 1980 values.

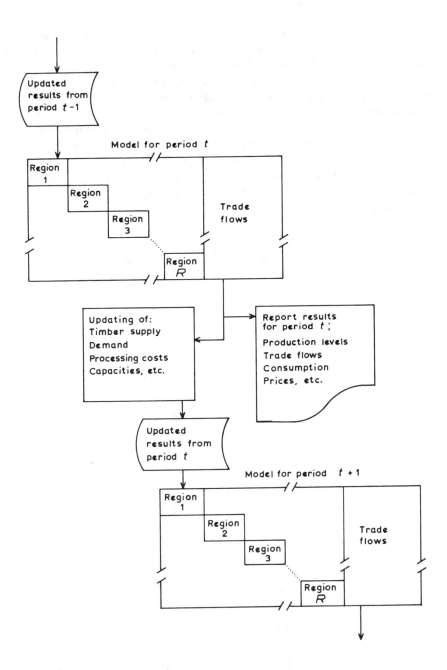

Figure 6.1. The IIASA forest sector model. This links a series of static models through time without imposing intertemporal market equilibrium criteria. The solution from one period is used to update production capacity, timber supply and trade constraints which are used as inputs to solve for market equilibrium in the next period.

6.2.1. Mathematical formulation

Market equilibrium assumes that each producer and trade agent involved in the forest sector is a profit maximizer (this assumption is relaxed for the centrally planned economies; see Subsection 6.2.2) and that each consumer purchases from the producer or trader who offers the lowest price. Given prices P_{ik} for each region i and commodity k, profit maximization results in a certain level of supply (comprising domestic production and net imports) of each commodity in each region. If, for all products and regions, such supply equals demand at that price as computed from the demand functions, then P_{ik} is an equilibrium price.

As a convenient analytical device, the equilibrium price, consumption levels and trade flows can be found by solving the following optimization problem (the time subscript is omitted). The problem is to find q_{ik}, y_{im} and e_{ijk} for all $i, j, k,$ and m to maximize

$$\sum_{ik} \int_0^{q_{ik}} P_{ik}(q_{ik}) \, dq_{ik} - \sum_{im} \int_0^{y_{im}} C_{im}(y_{im}) \, dy_{im} - \sum_{ijk} D_{ijk} e_{ijk} \tag{6.1}$$

subject to

$$q_{ik} - \sum_m A_{ikm} y_{im} + \sum_j (e_{ijk} - e_{jik}) = 0 \qquad \text{for all } i \text{ and } k \tag{6.2}$$

$$0 \leq y_{im} \leq K_{im} \qquad \text{for all } i \text{ and } m \tag{6.3}$$

$$L_{ijk} \leq e_{ijk} \leq U_{ijk} \qquad \text{for all } i, j \text{ and } k \tag{6.4}$$

where indices i and j refer to regions, k to products, m to production activities, and:

q_{ik} = Consumption of product k in region i.

$P_{ik}(q_{ik})$ = Price of product k in region i associated with a particular level of consumption.

y_{im} = Level of annual production in region i for production activity m.

$C_{im}(y_{im})$ = Marginal cost of production by process m in region i associated with a particular level of production y_{im} (assumed to be a non-decreasing function of activity level y_{im}).

D_{ijk} = Unit cost of transporting commodity k from region i to region j (including possibly tariffs which are proportional to quantity).

e_{ijk} = Quantity of commodity k exported from region i to region j.

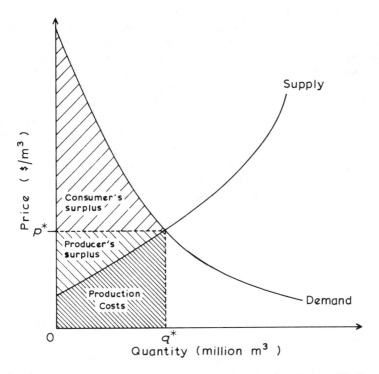

Figure 6.2. Equating supply and demand to determine the partial equilibrium price and quantities. The area under the demand curve up to a point q represents the benefits to consumers associated with consuming q units. The area under the supply curve represents total production costs. At market equilibrium, producers charge p^*/m^3 and sell q^* million m³; their gross revenues are p^*q^*. Profits equal gross revenues minus production costs and are labeled producers' surplus in the figure. If output expands beyond q^*, the costs of the extra production exceed the benefits of that production. Consequently the point where supply and demand are equal maximizes the sum of benefits to producers and consumers.

A_{ikm} = Net output of product k per unit of production for process m in region i.

K_{im} = Production capacity associated with process m in region i.

L_{ijk}, U_{ijk} = Lower and upper bounds on trade flows in product k between regions i and j.

The objective function, equation (6.1), is the sum of consumers' and producers' surplus as illustrated in *Figure 6.2*. Consumers' surplus, represented by the first term in equation (6.1), represents the total amount consumers are willing to pay for a specified quantity of a product. Producers' costs, represented by the second term of equation (6.1), are the total costs to society (including timber, labor, energy, and so on) of a certain level of output. The difference between benefits to consumers and costs to producers represents the net benefit to society of a particular level of consumption. Because we include trade, transportation costs must be deducted from this sum to compute net social benefit. Transportation costs are represented by the third term in equation (6.1). The point at which net social benefit is maximized defines the equilibrium point, that is, the point at which supply and demand are balanced.

The economic model is completed with equations (6.2–6.4). Equation (6.2) represents material balances: in each region for each commodity consumption equals production minus net exports. Equation (6.3) constrains production levels to lie at or below levels of industrial capacity in each region. Inequality equation (6.4) reflects limitations on trade between regions arising from quotas, agreements and trade inertia caused by institutional restrictions on rapid changes in trade patterns.

6.2.2. Products and regions

Computational limitations proscribe separate regional submodels for each country in the world. Instead, countries were aggregated into regions on the basis of the structural similarities of their forest sectors and their importance to the world markets in forest products. Similarly, wood is used in a very wide array of industrial and consumer products, and it is not feasible to model each end use. *Table 6.1* lists the 18 regions used in the present assessment along with the estimated share of world production for each of the products included in the analysis.

The table also distinguishes final products (F) which are consumed outside the forest sector, and intermediate products (I) which are consumed within the forest sector. The possibilities for the flow of wood among products and regions can be illustrated by considering the case of coniferous timber. The harvesting of coniferous trees produces coniferous logs, coniferous pulpwood and fuelwood. The logs can be converted into lumber or plywood in the region where the timber harvest took place, or the logs can be exported to other regions for processing. The lumber and plywood can either be consumed in the region where it is produced, or exported to another region for final consumption.

It is technically possible to convert only part of a log into lumber and plywood; some of the waste material is suitable for making pulp and paper products. These wastes, as well as the pulpwood resulting from the timber harvest, may be exported to another region or may be used within the region for the production of pulp and paper. The pulp can be exported or used domestically for the production of paper. Production of some paper products, printing and writing papers, for example, requires both coniferous and nonconiferous pulp.

Although most of the world trade in forest products passes between market economies, a significant and growing fraction of this trade is associated with countries with centrally planned economies. For these regions the market assumptions that underlie the model formulation discussed above do not apply. Instead, we assume that production levels are fixed according to a central plan, projected along a trend of the past 20 years. Consumption targets are set for each product assuming that the relationship between consumption and production during the past 20 years will hold in the future. We assume that countries with centrally planned economies will sell exports at the highest possible prices and will purchase imports at the lowest possible prices, and will minimize the deviations from consumption targets. A function estimated on historical data models the trade-off these countries face between gains in export revenue and the losses associated with not meeting consumption targets. Fedorov et al. (1987) discuss this component of the model in more detail.

6.2.3. Production process and trade

Production is Leontief: a unit of output requires fixed proportions of inputs. Conversion factors indicate the quantity of logs, pulpwood and fuelwood produced from trees, and the quantity of logs and pulpwood required to produce specified quantities of lumber, panels, pulp and paper of various types.

For each product, up to three production technologies are included, each with different conversion factors. One technology describes "old" capacity (more than 10 years old), one describes "modern" capacity (less than 10 years old) and the third is "new" capacity which may be added if it is profitable enough to do so. The conversion factors for old and modern technology are fixed at levels that reflect the actual capacities of the extant forest products industry in each region.

The conversion factors for the new capacity improve over time to reflect technological progress. For each product we define a "best" technology which represents the maximum technical efficiency engineers believe is possible by 2030. The conversion coefficients for new capacity in each region trend toward the values for this best technology. Thus we assume worldwide convergence of technical production efficiency.

In each period production is constrained by the level of installed capacity. Industrial capacity changes between periods in response to profitability. If production of a particular commodity in a particular region is profitable, then capacity is added. Capacity is lost through depreciation, so total capacity can decline if losses through depreciation exceed additions triggered by profitability. Dykstra and Kallio (1987) describe the production component of the forest sector model in more detail.

Each production process possesses a constant cost per unit of output which shifts over time in response to changes in real costs and exchange rates. In addition, a transportation cost (Wisdom, 1987), including tariffs, is assessed on any material which is traded between regions.

Trade between regions is constrained to change only slowly over time (Kornai and Kallio, 1987). This reflects the fact that many institutional factors tend to limit the rate of change in trade.

Table 6.1. 1980 regional production shares (in %) and total world output[a] of selected forest products[b].

	Conif. logs (I)	Nonconif. logs (I)	Conif. pulpwood (I)	Nonconif. pulpw. (I)	Fuel-wood (F)	Conif. sawnw. (F)
Western Canada	10.9	1.2	8.2	5.1	–	8.4
Eastern Canada	6.2	–	9.3	6.6	–	4.8
Western USA	14.6	–	10.3	1.1	1.2	10.5
Eastern USA	10.7	13.8	24.6	27.3	1.4	5.9
Brazil	1.6	8.2	1.6	7.0	3.0	–
Chile	–	–	1.2	–	–	–
Rest of Latin America	–	4.9	1.7	1.4	5.8	–
Finland	3.5	–	5.7	3.8	–	3.3
Sweden	3.7	–	6.7	3.4	–	3.6
Western Europe	7.9	7.1	9.5	15.3	2.0	10.0
USSR	21.8	7.0	8.3	5.5	5.8	27.6
Eastern Europe	5.5	7.7	4.9	5.5	–	5.7
Africa	–	5.8	–	–	28.3	–
China	3.7	4.9	2.1	2.3	11.9	4.3
Japan	2.9	1.7	1.1	11.7	–	9.6
Southeast Asia	–	22.5	–	–	12.9	–
Australia/New Zealand	1.2	2.3	2.5	1.6	–	–
Rest of World	3.4	10.9	–	1.2	26.4	2.6
TOTAL[a]	604.7	284.2	386.7	144.3	1350.9	311.6

	Nonconif. sawnw. (F)	Veneer & plyw. (F)	Composite panels (F)	Conif. white pulp (I)	Nonconif. w. pulp (I)
Western Canada	–	4.0	2.2	13.4	5.0
Eastern Canada	–	2.2	1.2	9.3	11.4
Western USA	–	17.4	5.1	5.4	–
Eastern USA	15.8	19.0	14.3	27.2	32.7
Brazil	6.7	2.6	2.4	–	8.4
Chile	–	–	–	2.6	–
Rest of Latin America	5.4	1.8	1.5	–	–
Finland	–&	1.5	1.9	6.7	5.4
Sweden	–	–	3.1	9.9	5.0
Western Europe	8.9	6.2	32.4	5.1	4.5
USSR	10.6	5.5	13.8	6.7	2.5
Eastern Europe	7.6	3.5	13.1	4.5	1.5
Africa	4.9	1.8	–	–	–
China	6.9	3.5	–	3.2	2.5
Japan	6.2	18.3	3.4	2.2	17.3
Southeast Asia	10.4	6.8	–	–	–
Australia/New Zealand	2.0	–	1.7	1.0	–
Rest of World	12.6	5.5	1.5	–	–
TOTAL[a]	115.0	45.3	58.7	31.3	20.2

Table 6.1 continued

	News-print (F)	Printing & writing paper (F)	Household & sanitary paper (F)	Packaging & board (F)	Recycled paper (I)
Western Canada	4.7	–	–	–	–
Eastern Canada	29.2	2.9	3.4	2.5	2.3
Western USA	4.7	2.4	8.0	5.8	4.4
Eastern USA	11.3	28.8	37.5	35.5	27.3
Brazil	–	1.5	2.3	2.5	2.3
Chile	–	–	–	–	–
Rest of Latin America	–	2.2	4.5	2.7	4.0
Finland	5.4	4.4	1.1	2.6	–
Sweden	5.4	1.5	2.3	2.7	1.1
Western Europe	11.3	24.6	20.5	18.7	26.7
USSR	5.8	8.4	2.3	4.1	4.7
Eastern Europe	1.6	2.2	3.4	4.5	2.8
Africa	–	–	1.1	–	–
China	3.9	5.7	1.1	–	2.1
Japan	10.5	9.0	10.2	12.3	16.3
Southeast Asia	–	–	–	–	–
Australia/New Zealand	1.9	–	1.1	1.4	1.5
Rest of World	1.6	3.7	1.1	2.2	2.3
TOTAL[a]	25.7	45.5	8.8	84.5	47.2

[a] Round and solid wood products in million m^3, pulp paper products in million tons, – indicates less than 1% of total world production.
[b] F and I denote final producer and intermediate products, respectively.

6.2.4. Demand

Final product demand in a region is a function of that region's price, income and population level, and of a time trend reflecting technological shifts. Wibe and Kallio (1987) used data from the FAO *Yearbook of Forest Products* (FAO, 1984) to estimate the demand equations used in the results reported below.

Demand shifts through time in response to changes in population and income levels. The absolute level of demand and, hence, the level of production and prices is quite sensitive to these two variables. FAO population projections were used with some modifications, but the comparable FAO projections of gross domestic product give levels of demand which are much higher than indicated by other studies of long-term developments in the sector. Consequently, the empirical negative relationship between percentage growth in GDP and GDP level is used to modify the FAO exponential GDP growth rates (Wibe and Kallio, 1987).

6.2.5. Timber supply

The ecological processes of the forest are linked to the economic processes of the market through timber supply. While elaborate optimization models of timber supply have been constructed, we work with a simple formulation which captures many of the important market adjustments one observes in the more elaborate models (Binkley and Dykstra, 1987).

In a given period, the marginal cost of timber removals is an increasing function of the annual removal quantities. The growing stock is divided into four classes: large and small coniferous trees, and large and small nonconiferous trees. Large trees produce saw and veneer logs, pulpwood and fuelwood, whereas small trees produce pulpwood and fuelwood but are not suitable for lumber or veneer production. A separate timber supply equation was estimated for each growing stock class in each region.

The timber supply curves are shifted through time in response to changes in the level of growing stock. As the inventory of growing stock declines, it becomes more costly to sustain a given level of removals. Access and harvesting costs are apt to be higher, and the value of non-timber benefits greater per unit of timber output, when smaller inventories of timber are held. We assume that the supply elasticity with respect to timber inventory is unity. That is, a 1% increase in the inventory of forest growing stock produces a 1% increase in harvest, all else equal. Thus, timber costs (and therefore product prices, and production and consumption levels) depend on forest growth. Ecological considerations enter the model by determining raw material costs.

6.3. The Ecological Models

Two ecologic models are used in this assessment. The first is a model used to project the levels of timber inventory in the global forest sector model. The second describes how forest growth may be affected by CO_2-induced climate warming.

6.3.1. Timber inventory projection

The timber inventory projection model strikes a balance between biological reality and simplicity necessitated by the global scale of the analysis. Despite its simplicity, the model seems to mimic closely the results of more elaborate and more biologically realistic models (Binkley and Dykstra, 1987).

The inventory projection model begins with a basic accounting identity

$$I_{t+1} = I_t - h_t + g_t \qquad (6.5)$$

The starting inventory I_0 is an initial condition of the model. The harvest level h_t is part of the model solution for period t. Forest growth g_t completes the model.

Forest growth depends on numerous edaphic, environmental and genetic factors which are exogenous to the model. The principal endogenous factor is the density of the forest or "stocking level". We assume that the relationship between forest growth and forest density is nonlinear. Denote forest density $x_t = I_t/A_t$ (in m^3/ha) where A_t is the forest area in period t. Let $\dot{x} \equiv dx/dt$ be forest growth per unit area (m^3/ha/yr). At $\dot{x} = 0$, x is zero because no timber growth can take place in the absence of trees. There also exists some maximum level of stocking (the "carrying capacity") above which mortality exceeds productivity and net growth ceases.

It is well known that any arbitrary function can be approximated with a power series, so

$$\dot{x}_t = b_0 x_t + b_1 x_t^2 + b_2 x_t^3 + \ldots \tag{6.6}$$

Note that this relationship is constrained to pass through the origin on the $(x, \dot{x})$ plane.

Dividing equation (6.6) by x_t expresses forest growth on a percentage basis

$$\frac{\dot{x}_t}{x_t} = b_0 + b_1 x_t + b_2 x_t^2 + \ldots \tag{6.7}$$

This model was estimated for each region (in Binkley and Dykstra, 1986, *Table 21.5* gives the estimates of b_0, b_1, and b_2 for each region). In no case were more than three terms of equation (6.7) used. The regions affected by CO_2-induced climate warming were all modeled with two parameters, so $b_2 = 0$ in these regions.

Binkley and Dykstra (1987) compared the results of this simple growth model with those from more elaborate and more biologically realistic growth models. The deviations were small. The worst case occurred in projecting the coniferous timber inventory in Finland. The simple model described above deviated from the more complex model by 11.1% after 40 years, for an average annual deviation of about 0.25% per year.

The growth term g_t in equation (6.5) is modeled as

$$g_t = \dot{x}_t A_t \tag{6.8}$$

In the base scenario of the global forest sector model, forest land area A_t is changed over time to account for trends in afforestation and deforestation in some of the developing regions.

6.3.2. Forest growth and CO_2-induced climate warming

In Section 5, Kauppi and Posch estimated the effects on boreal forest productivity of the temperature increases estimated under the GISS $2 \times CO_2$ scenario. Their results were used to modify the forest growth functions and area of forest land used in the forest sector model, and thereby to simulate the economic effects of CO_2-induced climate warming. Before describing precisely how the Kauppi–Posch results were incorporated into the forest growth model, some caveats are in order (Kauppi, 1987; Smith, 1985; and Shugart et al., 1986, provide good reviews of the kinds of effects elevated CO_2 levels and climate warming might have on forests).

Some limitations of the analysis

The boreal forests alone are considered in this analysis because:

(1) The projected temperature increases are larger in this region than in any other forest area.
(2) The effect on plant growth of an increase in temperature is very significant in this forest type (Shugart et al., 1986).

While forests in these areas are likely to be the most affected, those in other regions would probably be affected as well (Emanuel et al., 1985). However, any changes outside the boreal forests are ignored.

Changes in precipitation patterns are likely to accompany changes in global temperature patterns, and these effects are ignored in our analysis. While some parts of the boreal forest suffer excess moisture, changes in precipitation could offset all or part of the temperature-induced increase in growth.

In laboratory and greenhouse settings (with all other growth requirements optimized), increases in CO_2 concentrations have been shown to enhance net productivity directly. The efficiency of water utilization also improves and consequently growth would increase in drier regions. These "fertilization effects" are not considered in the following analysis.

Increases in temperature lengthen the growing season but may also alter the probability of frost damage to young shoots, with consequent effects on growth (Cannell and Smith, 1985). Other possible impacts on tree growth due to climate warming include changes in the rates of litter decomposition, and therefore on nutrient cycling, and changes in the range and activity of microbial pathogens and harmful insects (Smith, 1985).

Climate changes are likely to affect agricultural lands as well as forest. At least in the USA, most forest land use changes occur at the agricultural margin (Binkley, 1983). Different changes in agricultural and forest productivity are likely to displace the margin of comparative advantage between these two activities. Such shifts in the margin between forestry and agriculture have not been included in this assessment.

In short, we have focused on only one of the many possible effects on forests of increased levels of atmospheric CO_2.

The time scale of forest response is also important to consider (Shugart et al., 1986). Growth responses due to increased temperature and CO_2 levels are apt to be quite rapid because the photosynthetic machinery is in place to take advantage of better growing conditions. Changes in forest land area, on the other hand, require trees to colonize new areas. Shugart et al. (1986) cite evidence that boreal forests can advance quite rapidly at their altitudinal and northern latitudinal limits in response to even small increases in temperature. On the other hand, invasion of other plant communities into the current boreal forest will probably require the death or significant decline of the extant forest before a forest better adapted to the warmer conditions can colonize. Over the 50-year projection horizon of the model, we assume that the boreal forest persists but becomes increasingly more productive as temperature increases.

The Kauppi-Posch model

In Section 5 Kauppi and Posch use a simple statistical model to simulate the effect of these temperature changes on forest productivity. They regressed observed levels of forest growth in 19 forest districts in Finland on effective temperature sum (ETS, the annual accumulated sum of daily mean temperatures above an effective growth threshold of 5 °C). This regression model was used in conjunction with temperature levels derived from estimates by the GISS general circulation model of climate for doubled concentrations of atmospheric CO_2 (*see* Section 3) to estimate forest productivity at 648 grid points (at intervals of 4° latitude × 5° longitude) covering the northern hemisphere between 38° N and 70° N.

Incorporating climate effects in the forest growth model

Two kinds of biological effects were identified in this procedure: the increase in growth on extant forest lands, and the increase in the extent of forest land as tundra areas become warm enough to support forest ecosystems. The first effect was estimated by comparing estimated growth rates in the current climate and in the 2 × CO_2 climate. The second effect was estimated as the percentage increase in the area with estimated growth greater than $0.5 \, m^3 \, ha^{-1} \, yr$, the limit used in most statistical series to define "exploitable forest land". Both calculations were performed by imposing a map of political boundaries on the Kauppi-Posch grid system, and by computing countrywide averages from the growth estimates at each of the grid points. Sweden and Finland, two of the regions of interest, each contain only a few grid points, so linear interpolation of the relevant grid points was used to improve the precision of the estimates.

In both calculations certain areas were excluded from consideration. These include the desert and semidesert areas of the USSR south of 50° N, and the grainlands in the southwestern USSR and south-central Canada (Olson and Watts, 1982). Without significant increases in moisture, the former are unlikely to support forests irrespective of temperature level. The high value of the latter

Table 6.2. Impact of the GISS 2 × CO_2 scenario on forest growth and area.

	Increase in forest area (%)	Growth		
		Current[a]	2 × CO_2[a]	Increase (%)
Canada:				
East	22.2	1.89 (1.60)	4.47	136.5
West	36.8	2.21 (2.13)	3.53	59.7
Finland	25.0	3.15 (3.18)[b]	5.45	73.0
Sweden	16.7	3.90 (3.01)	5.90	51.3
Norway	0.0	2.51 (2.40)	5.63	124.3
USSR	85.0	1.82 (1.41)	3.20	75.8

[a] m^3 ha^{-1} yr^{-1}, based on the Kauppi–Posch growth model (see Section 5) and temperature data interpolated on a 4° × 5° latitude/longitude grid by Schutz and Gates (see Section 3).
[b] Estimates of current growth, m^3 ha^{-1} yr^{-1} (Binkley and Dysktra, 1987, Table 21.5).

lands for agricultural production probably excludes their use for timber production unless, of course, this value changes in response to the effects of a doubling CO_2 (see, for example, Part II).

Table 6.2 shows the results of these calculations. With the exception of the USSR, the percentage increase in growth on extant forest areas exceeds the increase in forest area.

Forest land increases

To simulate the economic effects of these temperature changes, we apply these percentage increases in forest land area to the current "exploitable" forest area [the forest land where industrial cuttings have occurred or could occur periodically under the current level of technical and economic development in the region (ECE, 1976)]. Forests excluded from this category include those in which industrial cutting is prohibited or severely restricted by law, or in which the physical productivity is too low or the delivered wood cost to the nearest markets too high to warrant periodic industrial harvests.

In no case does the calculated increase in exploitable forest area under climate warming exceed the current total forest area of the country. To the extent that land is excluded from the "exploitable" category due to low productivity, our model reasonably portrays the land area shifts which might be expected.

This view of forest area expansion differs somewhat from that of Emanuel *et al.* (1985). They argue that with elevated temperature, the extent of the boreal forest will decline sharply, replaced primarily by cool temperate steppe and by cool moist forest [under the Holdridge (1964) classification]. The boreal forest is pushed northward into areas now occupied by tundra. Their results are driven by the increased water stress associated with higher temperature and assume no change in precipitation due to climate warming and no increase in the efficiency of water use by trees due to higher atmospheric CO_2 concentrations: "For example, with warming, Cool Temperate Moist Forest replaces Boreal Wet Forest, but if an associated increase in average annual precipitation also occurs, the change would be to Cool Temperate Wet Forest" (Emanuel et al., 1985; p. 39).

Increases in forest growth

The figures in parentheses in *Table 6.2* give the best available estimates of the average forest growth in each region (*see* Dykstra and Binkley, 1987, for a discussion of how these estimates were developed). In general the growth levels estimated from the Kauppi–Posch model using the current climate are reasonably close to those derived from field surveys. The Kauppi–Posch model was estimated using data from Finland. The very close agreement between the two estimates of growth in that country give a measure of confidence in the averaging process used to interpolate the growth data from the grid-based information.

The maximum deviation between the two estimates of growth occurs in Sweden – a surprising result given the geographical proximity and similarity of forest types in Sweden and Finland. The growth estimates based on temperature alone tend to overstate the actual levels of growth, suggesting that on a countrywide basis, site conditions are poorer or the forests not so intensively managed as in Finland. This latter consideration surely obtains in Canada and in the USSR.

Carter (unpublished observations) examined the Schutz and Gates interpolations of observed (1931–1960) temperature used by Kauppi and Posch to estimate ETS and growth. His calculations, intended to test whether exaggerated temperatures were the source of the high growth estimates in Sweden, suggest rather, that the interpolation procedure may *understate* actual temperatures, and hence ETS values. Consequently, if more representative temperature data were used, the Kauppi–Posch model would probably overstate forest growth in Sweden to a larger degree than indicated in *Table 6.2*.

To simulate the economic effects of the $2 \times CO_2$ scenario, the timber inventory projection component of the global forest sector model is modified using the results in *Table 6.2*. To simulate the increase in forest area, the initial forest area is incremented by a constant amount each 5-year simulation period until the increase shown in *Table 6.2* is reached by the year 2030. *Table 6.3* shows the results of this calculation.

To simulate the increase in forest growth, the intercept b_0 in equation (6.7) is increased over time, but the slope coefficient b_1 is held constant. The increase in forest growth is assumed to occur at all stocking levels, and the maximum stocking level also increases. To calculate the change in the intercept term, we assume that the increase in forest growth shown in *Table 6.2* will occur at the current stocking level.

Figure 6.3 illustrates the procedure. The curve labeled "Base" depicts the relationship between the percentage growth of the forest ($\dot{x}/x$) and forest stocking (x) in the base scenario without climate warming. This is equation (6.7).

Table 6.3. Parameters for the forest growth model under the GISS $2 \times CO_2$ scenario. Only the intercept b_0 of the growth equation (6.7) is changed in response to climate warming. The change between 1980 and 2030 is assumed to be linear. For the full growth equations, *see Table 21.5* in Binkley and Dykstra (1987).

Region	Growth equation intercept (b_0)		Forest land area (mill. ha)	
	1980	2030	1980	2030
Western Canada			92.5	126.5
Coniferous	0.0458	0.0560		
Nonconiferous	0.0220	0.0348		
Eastern Canada	0.0300	0.0566	98.6	120.5
Finland	0.0700	0.1010	19.5	24.3
Sweden	0.0700	0.0855	22.2	25.9
Soviet Union	0.0300	0.0385	534.5	988.8

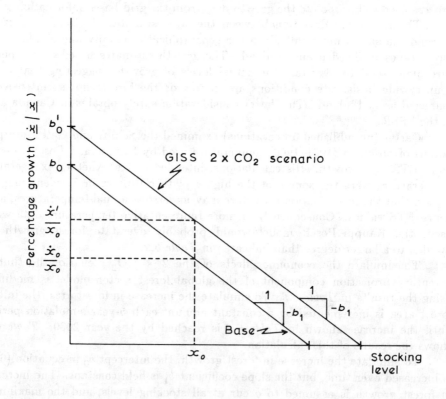

Figure 6.3. The forest growth equation (6.7) shifts upward at all levels of forest stocking to model the increased growth associated with climate warming; x_0 refers to the current stocking level of forest density ($m^3 ha^{-1}$) and $\dot{x}_0$ to the current level of forest growth ($m^3 ha^{-1} yr^{-1}$). Climate warming increases growth to $\dot{x}'$. To model this effect a new intercept term b_0' is calculated so the growth function passes through the point (x_0, $\dot{x}'/x_0$) but retains the original slope b_1.

The point x_0 refers to the current stocking level, and $\dot{x}_0$ refers to the current level of forest growth. To simulate the effect of climate warming, we assume that the growth curve moves upward at all stocking levels. The size of the displacement is calculated so the new curve goes through the point $(\dot{x}_0, \dot{x}'/\dot{x}_0)$; $\dot{x}'$ is calculated by applying the percentage increase in growth due to climate warming (last column in *Table 6.2*) to the estimate of current growth (*Table 6.2*). *Table 6.3* shows the values of the intercept term b_0' which results from this procedure. Again we assume a linear increase in this parameter from its current level in 1980 to the maximum level in 2030.

6.4. Results

To assess the effects of CO_2-induced climate warming on the forest sector, the forest growth component of the global forest sector model was modified as described above, and the model was solved for a 50-year projection horizon. Before describing the simulation results, the treatment of timber removals in the USSR requires discussion.

Recall that timber harvests in the USSR are set on the basis of trends in the past central plans. Consequently, increasing the timber inventory alone will have no impact on that region's production, prices or trade. The simulations of climate warming assume that the USSR may increase timber harvests to the point that the ratio of harvests to timber inventory is the same as in the base scenario without climate warming. This upper bound on removals may not be reached if the extra timber is not needed to meet domestic consumption targets, and if the transportation costs make it unprofitable to export.

For the GISS $2 \times CO_2$ scenario, the assumed upper bounds for annual Soviet timber production rise gradually from their 1980 level of 359.5 million m^3 to 633.2 million m^3 in 2030 (versus 430 million m^3 in the base scenario). Current net annual growth of Soviet forests is about 880 million m^3, and annual growth of timber considered to be economically exploitable is about 640 million m^3 (Fedorov et al., 1987). Thus the assumed upper bounds on Soviet timber harvest levels seem to be economically feasible.

Throughout this discussion, results are reported:

(1) Only for the simulation year 2030.
(2) Primarily for coniferous products.
(3) Generally with reference to a "base" scenario.

Each of these points requires some elaboration. First, because climate warming and the associated impacts on forest growth are long-term processes, it seems most appropriate to report results for this time period. The effects of climate warming are phased in linearly, and the economic effects are roughly proportional to the biological effects. Suffice it to note that the economic effects of climate warming appear by the turn of the century.

Second, the boreal forest is primarily a coniferous forest, so the increases in forest growth associated with climate warming will largely occur in coniferous volume. In the simulations reported below, the 2030 coniferous timber inventory for the world increases 29% over the base case level whereas the nonconiferous inventory increases only 5%. (In total the inventory of merchantable timber increases by about 50 billion m^3, or 16%.) Thus the direct effects of climate warming occur primarily with coniferous products. Because some products require both coniferous and nonconiferous inputs (e.g., printing and writing paper), increased production of coniferous products induces changes in nonconiferous markets.

Third, the results are discussed in relation to base scenario values. Many of the parameters of the global forest sector model are uncertain. Precisely stated, a scenario is defined by setting values for the exogenous parameters. The base scenario reflects the best statistical estimates for parameters which are amenable to statistical estimation, and values intermediate in the range of controversy for those, such as future levels of gross domestic product in each region, where statistical estimates are very apt to be unrealiable. Dykstra and Kallio (1987) give the details of this base scenario.

The changes in the forest growth model used in the climate warming scenario affect the economic results indirectly. In the regions that are directly affected by climate warming, the higher levels of forest growth lead to an increase in the timber inventory. This increase in timber inventory translates into an increase in timber supply. Harvests in those regions increase, and timber prices decline. Lower timber costs make the industry in those regions more profitable, so processing capacity shifts into those regions (and out of others) to take advantage of the lower wood costs. As processing capacity shifts out of a region, timber demand declines and timber prices will fall. Thus an increase in timber supply in some regions means that timber prices will be lower in all regions.

In the regions not directly affected by climate change, timber harvests will decline in response to lower timber prices. Timber owners in those regions will clearly be hurt: both the amount of timber they can sell and the prices they will receive decline. In the directly affected regions, the increase in harvest helps offset the reduction in stumpage prices.

Table 6.4 shows the changes in prices and harvest levels for coniferous sawlogs and pulpwood, and the change in timber revenues (including revenues from the sale of nonconiferous timber). Adjustments in harvest levels are characterized by large increases in the affected regions with small decreases in the rest of the world. The price changes for pulpwood are much larger than those for sawtimber. This occurs because wood costs are only a small fraction of cost of the paper products which use pulpwood, while sawtimber represents a large part of the cost of lumber and plywood. As a consequence, the demand for pulpwood is much less elastic than is the demand for sawtimber. Hence an increase in pulpwood supply has a bigger impact on pulpwood prices than a comparable increase in sawtimber supply has on sawtimber prices.

In general, timber producers in the affected regions benefit by the warmed climate, but the size of the benefit is small except in Finland. In Sweden, harvest levels do not increase enough to offset the reduction in prices, so timber owners are actually worse off with a faster growing forest. The main

Table 6.4. Price, harvest and sale changes due to climate warming (%). Changes are from the base scenario for 2030. – Indicates less than 1 million m^3 of production.

	Coniferous sawlogs		Coniferous pulpwood		Income for timber sales
	Price	Harvest	Price	Harvest	
Directly affected					
Western Canada	−16.2	+25.0	−48.5	+20.9	+1.9
Eastern Canada	−22.9	+48.0	−40.6	+33.8	+6.5
Finland	−8.6	+63.2	−31.9	+51.4	+22.4
Sweden	−14.9	+24.5	−34.1	+11.8	−7.6
USSR	−13.8	+48.3	−45.6	+28.5	+1.3
Other regions					
Western USA	−15.3	−3.2	−52.4	+8.5	−25.5
Eastern USA	−20.5	−6.0	−38.3	−4.4	−19.6
Brazil	−9.3	0.0	−40.1	−5.8	−8.8
Chile	−20.7	0.0	−38.9	+0.5	−24.5
Western Europe	−16.5	−4.4	−30.5	−12.5	−19.5
Eastern Europe	−18.2	0.0	−29.9	−41.5	−22.7
Africa	−10.5	−	−31.6	−	−6.8
China	−13.0	−3.1	−42.6	+27.3	−17.8
Japan	−11.5	−4.2	−44.6	+2.4	−15.2
Southeast Asia	−19.7	−	−40.6	−	−6.8
Australia/ New Zealand	−15.4	0.0	−53.6	−1.3	−24.8

beneficiaries of the increased timber harvest are the consumers of forest products who, because of lower cost timber, pay less for the forest products they purchase.

During the last two decades, Brazil, Chile and Australia/New Zealand have invested heavily in coniferous plantations. In this scenario, the timber from those plantations is harvested to avoid the high transportation costs of using the increased harvest from the higher latitudes. Yet in these regions timber prices and total timber receipts are sharply reduced owing to the interregional adjustments that take place. While it may be economic to harvest mature plantations in this scenario, the lower prices associated with the climate warming call into question the profitability of establishing the plantations in the first place. Analysis of the issue lies beyond the scope of the present inquiry, but is nonetheless important.

Product prices will fall as a consequence of lower timber prices. Lower product prices imply higher consumption. How will the higher level of consumption be met?

Market equilibrium requires that the increase in consumption just equals the increase in timber supply. Part of the increased timber supply in the warmed regions will be processed and consumed within the region. Part of it will be processed and exported in the form of final products to regions where timber supplies did not increase. And finally, part of it will be exported in the form of raw materials to be processed and consumed in other regions. Much of the economic adjustment to climate warming is of this latter sort.

In western Canada, eastern Canada and the USSR, most of the increase in timber supply is exported in the form of unprocessed raw materials. In western Canada, 81.0% of the increase is exported as logs to either the USA or Japan. Virtually all of the increase in Soviet timber supply is exported as logs to China. The USSR's market for logs in Japan is almost completely lost to exports from western Canada. In eastern Canada, a small part of the increased supply is processed into either lumber or newsprint which is then exported. Most of the increased supply is exported as sawlogs or pulpwood to the USA for processing and consumption. As a consequence, production of lumber, plywood, particle board and packaging paper in the eastern USA actually increases despite the reduction in timber harvests (by 11.9%, 5.1%, 11.8% and 15.1%, respectively.)

The adjustments in Finland and Sweden are more complex. All of the increased timber supply in Finland is processed into products. Lumber production increases by about 12 million m^3 annually, or by about two-thirds. Virtually the full amount is exported. Production of printing and writing papers increases by 40%. As a consequence, demand increases for both the coniferous and nonconiferous white pulp needed to produce that line of papers. The increase in coniferous timber supply is adequate for the production of the needed coniferous pulp, but the requirement for nonconiferous pulp is met by increased imports from the eastern USA.

Finland's extra printing and writing paper production is exported, primarily to western Europe. Consequently, the production of this type of paper in western Europe plummets (from 9.4 to 2.2 million tons annually), and with it, so does the demand for and the price of both coniferous and nonconiferous pulpwood.

Because of the significant increase in Finnish exports of printing and writing paper to western Europe, Sweden loses some of its market there. This loss is partly offset by a 25% increase in coniferous lumber production, all of which is exported.

As a consequence of all of these adjustments, annual turnover (excluding timber and pulpwood exports) in the Finnish forest products industry increases by about US$21 billion (1980 dollars). Despite the increase in timber supply, the annual turnover for the Swedish forest products industry declines by about US$2 billion and in eastern Canada by about US$3 billion. In contrast, western Canada gains about US$3 billion in annual sales. The industry in western Europe suffers a loss in turnover of about US$15 billion annually. The impact on the forest sector in other regions is negative but lower than these amounts.

6.5. Conclusions

This section has linked a simple biological model of the effects of climate warming on boreal forest productivity to an economic model of the world's forest sector. The absolute values of the results are highly speculative owing to uncertainties in the biological, economic and climate models, and the long time period required for the effects to arise. Assuming that climate warming enhances the growth of the boreal forests (and changes forest growth in other regions by only a small amount), three results seem fairly robust. First, the primary benefit of the increased supply will accrue to the consumers of forest products, either individuals or the industries which use forest products as inputs. Second, some producers, both in the directly affected regions as well as in other regions, are likely to benefit from the increased timber supply. Producers in directly affected regions where the increase in timber supply is small may actually be hurt by climate warming (e.g., Sweden). Third, climate warming imposes significant costs on regions where productive capacity is currently marginal (e.g., western Europe), and where large investments have been made to grow trees (e.g., Brazil, Chile and Australia/New Zealand).

The structure of the economic model makes it comparatively simple to measure the net benefits to forest sector producers and consumers due to the additional timber supply attending climate warming. Recall that the objective function of the programming model, the sum of producers' and consumers' surpluses, measures the net benefit to society of production and consumption in the forest sector. From the solutions to the base and climate warming scenarios, the change in total surplus can be measured. These differences can be discounted and summed over the ten analysis periods to estimate the net present benefit to society (with respect to the forest sector) of climate warming. The calculations result in a figure of about US$150 billion (1980 dollars, discounted to 1985 dollars) if a real interest rate of 4% is used, and about US$25 billion if a real interest rate of 10% is used. While the changes in social surplus are significant, the differential impacts on various segments of the sector seem more critical.

Acknowledgments

The work reported in this section was completed while the author was on leave from Yale University working with the Forest Sector Project at the International Institute for Applied Systems Analysis, Laxenburg, Austria. Several people assisted in this work: Dennis Dykstra, Miloslav Lenko and Gábor Kornai organized and made the computer runs of the IIASA Forest Sector Model which were reported here. Maximilian Posch assisted in the interpolation and analysis of the changes in forest growth associated with climate warming. Timothy Carter provided many useful comments on several drafts of this manuscript as did William H. Smith, William R. Reifsnyder and Paul Waggoner on the penultimate version. Of course, any errors of analysis or interpretation are the responsibility of the author alone.

References

Binkley, C.S. (1983). Private forest landuse. In J.P. Royer and C.D. Risbrudt (eds.), *Nonindustrial Private Forests: A Review of Economic and Policy Studies*. Duke University School of Forestry and Environmental Studies, Durham, NC.
Binkley, C.S. and Dykstra, D.P. (1987). Timber Supply. In M. Kallio, D.P. Dykstra and C.S. Binkley (eds.), *The Global Forest Sector: An Analytical Perspective*. John Wiley & Sons, London.
Cannell, M.G.R. and Smith, R.I. (1985). *Climate Warming, Spring Phenology and Frost Damage on Trees*. (Unpublished observations.)

Dykstra, D.P. and Kallio, M. (1987). Base scenario. In M. Kallio, D.P. Dykstra and C.S. Binkley (eds.), *The Global Forest Sector: An Analytical Perspective*. John Wiley & Sons, London.
ECE (1976). European timber trends and prospects, 1950 and 2000. Supplement to Volume XXIX, *Timber Bulletin for Europe*, United Nations Economic Commission for Europe, Geneva.
Emanuel, W.R., Shugart, H.H. and Stevenson, M.P. (1985). Climatic change and the broad scale distribution of terrestrial ecosystem complexes. *Climatic Change*, 7, 29–43.
FAO (1984). *Yearbook of Forest Products, 1971–1982*. Food and Agricultural Organization of the United Nations, Rome.
Fedorov, V., Kallio, M., Reteium, A. and Smyshlayev, A. (1987). Planned economies. In M. Kallio, D.P. Dykstra and C.S. Binkley (eds.), *The Global Forest Sector: An Analytical Perspective*. John Wiley & Sons, London.
Holdridge, L.R. (1964). *Life Zone Ecology*. Tropical Science Center, San José, Costa Rica.
Kauppi, P. (1987). Forests and the changing chemical composition of the atmosphere. In M. Kallio, D.P. Dykstra and C.S. Binkley (eds.), *The Global Forest Sector: An Analytical Perspective*. John Wiley & Sons, London.
Kornai, G. and Kallio, M. (1987). Trade and inertia barriers. In M. Kallio, D.P. Dykstra and C.S. Binkley (eds.), *The Global Forest Sector: An Analytical Perspective*. John Wiley & Sons, London.
Olson, J.S. and Watts, J.A. (1982). *Carbon in Live Vegetation of Major World Ecosystems*. ORNL 5862. Oak Ridge National Laboratory, Oak Ridge, TN.
Samuelson, P.A. (1952). Spatial price equilibrium and linear programming. *Amer. Econ. Rev.*, 42, 283–303.
Shugart, H.H., Antonovsky, M.Ya., Jarvis, P.G. and Sandford, A.P. (1986). CO_2, climatic change and forest ecosystems. Assessing the response of global forests to the direct effects of increasing CO_2 and climatic change. In B. Bolin, B.R. Döös, J. Jäger and R.A. Warrick (eds.), *The Greenhouse Effect, Climatic Change, and Ecosystems*, SCOPE 29, Wiley, Chichester, pp. 475–521.
Smith, W.H. (1985). Forest quality and air quality. *J. For.*, 83, 82–94.
Wibe, S. and Kallio, M. (1987). Demand functions for forest products. In M. Kallio, D.P. Dykstra and C.S. Binkley (eds.), *The Global Forest Sector: An Analytical Perspective*. John Wiley & Sons, London.
Wisdom, H. (1987). Transportation cost model. In M. Kallio, D.P. Dykstra and C.S. Binkley (eds.), *The Global Forest Sector: An Analytical Perspective*. John Wiley & Sons, London.

PART II

Estimating Effects of Climatic Change on Agriculture in Saskatchewan, Canada

Contents, Part II

List of Contributors		223
Abstract		225
1.	Introduction	227
	1.1. Aims	227
	1.2. The study area	229
	1.3. Agroclimatic impacts and societal responses	237
	1.4. Review of previous analyses of actual or simulated climatic events and impacts	242
	1.5. Selecting climatic scenarios for this case study	245
	1.6. Impact models used	252
2.	The effects on the agroclimatic environment	259
	2.1. Effects on growing degree-days	260
	2.2. Effects on the precipitation effectiveness index	262
	2.3. Effects on the climatic index of agricultural potential	264
	2.4. Effects on drought frequency using the Palmer Drought Index	270
	2.5. Discussion of agroclimatic change estimates	278
3.	The effects on potential for wind erosion of soil	281
	3.1. Modeling wind erosion	283
	3.2. Comparison of the wind erosion climatic factor with observations	286
	3.3. Model sensitivity analysis	288
	3.4. Impact analysis and interpretation	288
	3.5. Discussion: wind erosion and climatic change	292
4.	The effects on spring wheat production	293
	4.1. Methodology	293
	4.2. Comparison of model results with observations	297
	4.3. Model sensitivity	298
	4.4. Effects of climatic changes on growing season agroclimate	301
	4.5. Effects of climatic changes on spring wheat maturity	305
	4.6. Effects of climatic changes on yields and production	308

5. The effects on the farm and provincial economy 321
   5.1. Economic and employment impacts 321
   5.2. Methodology 323
   5.3. Model input data and simulations 329
   5.4. Provincial economy impacts 338
   5.5. Effects of adjustments made in crop rotations in response to adverse climatic conditions 342
   5.6. The role of economic models in impact studies 348

6. Conclusions and implications 351
   6.1. The scope of this study 351
   6.2. Summary of results 352
   6.3. Implications for climatic impact analysis 354
   6.4. Implications for policy 360
   6.5. Recommendations 363

Acknowledgments 367
Appendices 369
References 371

List of Contributors

WILLIAMS, Mr. G.D.V.
R. R. No. 4, Coldwater, Ontario
Canada L0K 1E0
(Formerly with Canadian Climate
Centre, Atmospheric Environment
Service, Environment Canada,
Downsview, Ontario)

FAUTLEY, Mr. R.A.
Prairie Farm Rehabilitation
  Administration
Agriculture Canada
Motherwell Building
1901 Victoria Avenue
Regina, Saskatchewan
Canada S4P OR5

JONES, Mr. K.H.
Atmospheric Environment Service
Central Region, Environment Canada
PO Box 4080
Air Terminal Building
Regina, Saskatchewan
Canada S4S 4Z8

STEWART, Dr. R.B.
Resources and Environment Section
Agricultural Development Branch
Agriculture Canada
Central Experimental Farm
Ottawa, Ontario
Canada K1A OC5

WHEATON, Ms. E.E.
Environment Sector
Saskatchewan Research Council
15 Innovation Boulevard
Saskatoon, Saskatchewan
Canada S7N 2X8

Abstract

The Canadian province of Saskatchewan is one of five case studies in cool temperate and cold regions considered in the IIASA/UNEP project to assess the impacts of climatic change and variability on food production. Saskatchewan agriculture, which is confined to the southern part of the province by the rocky land of the north, is constrained by long cold winters and limited moisture. Nonetheless, provincial production of wheat accounts for no less than one-eighth of the total traded on international markets.

In common with the other case studies in high-latitude areas (Iceland, Finland, northern European USSR and Japan), experiments have been conducted to simulate the impacts of several different climatic scenarios representing:

(1) An extreme period of weather-years (the dry 1930s in Saskatchewan).
(2) A single anomalous year (1961); this and (1) both taken from the historical instrumental record.
(3) The climate simulated by an atmospheric general circulation model (GISS) for doubled concentrations of atmospheric CO_2.

The standard climatic normal period (1951–1980) was used as the reference for comparing the results of the scenario experiments.

The data on the changes in climate implied in the various scenarios were translated, using several models, into estimates of "first-order" impacts on the thermal and moisture resources for agriculture, the potential for biomass productivity and for soil erosion by wind, and yields of spring wheat. To trace the downstream effects of the yield changes, these were first input to farm simulation models where results were aggregated to give provincial production and commodity changes. These were used as inputs to a regional input-output model which provided estimates of changes in output levels for various economic sectors. Finally, the changes in output were, in turn, translated into changes in employment using an employment model.

The results suggest that, with the climate as in recent decades and with agricultural technology and management at present levels (1980), Saskatchewan can expect occasional drought years with moisture resources so reduced that the potential for wind erosion is doubled, spring wheat production is only about a quarter of normal and losses to the agricultural economy brought about by this production shortfall exceed Can$1800 million and 8000 person-years. The impact on other sectors of the economy includes a provincial GDP reduction of Can$1600 million and a further reduction in jobs of 17 000 person-years.

Substantial adverse effects were also indicated for the extreme 5- or 10-year periods that recur occasionally. During these periods of warmer than normal growing seasons and subnormal moisture resources, biomass dry matter production may be reduced by nearly half and spring wheat production by about one-fifth, the latter effect leading to a loss to agriculture of about 2600 person-years and an annual economic loss of nearly Can$600 million. In the broader economy, provincial GDP is reduced by more than Can$500 million and 5600 more person-years are lost.

A shift to a warmer long-term climate, even if precipitation increases as predicted by the GISS $2 \times CO_2$ experiment, would reduce spring wheat yields by about 16%, causing annual losses to agriculture of over Can$160 million and 700 person-years unless there were major adaptive adjustments in agriculture. Such a climate might reduce wind erosion potential and increase average potential biomass productivity, but at the same time droughts could become more frequent and severe. With climatic warming without increased precipitation, all the impacts would be generally adverse and more intense, particularly drought frequency and severity.

One potential adaptive response to climatic change was tested: the switching of 10% of the cropped area in the Dark Brown Soil Zone from spring to winter wheat. In the economic analyses it was found that, in drought years, losses were considerably lower if part of the cultivated area was under winter wheat rather than all being under spring wheat, but the reverse was true in normal years.

In spite of the difficulties of adequately validating the impact models used in the experiments, and the apparent poor fit of the GISS general circulation model estimates to present Saskatchewan climatic conditions, the results have demonstrated that it is possible to make useful translations of climatic scenarios into estimates of likely effects on agricultural production, on the environment and on the economy. In addition, a set of recommendations are made with respect to:

(1) Developing the analytical methods and tools: including improving the linkages between different impact models, testing and refining the performance of impact models, estimating impacts for a wider range of crops and varieties, cataloging the results of impact experiments for different climatic scenarios for possible use by agricultural planners and seeking analogue areas with present climates comparable with the scenario climate in the study area.

(2) Suggesting appropriate policies of response: in particular, through intensified support for impacts research, policy exercises that address the problem of translating impact information into policies of response, a reassessment of agricultural planning strategies (e.g., the development of new crops and management techniques more suited to a changing environment) and the expansion of agricultural extension activities to help farmers respond to changing climate.

SECTION 1

Introduction

1.1. Aims

Canada was a logical choice as one of the cold-margin case studies for the climate impacts project of the International Institute of Applied Systems Analysis (IIASA) because it is a cold country with an agricultural industry that is quite sensitive to climatic fluctuations and has considerable experience in analyzing that sensitivity. It was considered impractical to use the whole country for this analysis, however. Instead, the province of Saskatchewan (*Figure 1.1*) was selected as the Canadian case study area. This province is a major agricultural producer in Canada, and its agriculture has been, and is likely to continue to be, particularly sensitive to climatic fluctuations. For example, it was estimated that nearly nine-tenths of the difference between the 573 kg/ha Saskatchewan wheat yield in 1961 and the 1861 kg/ha yield in 1966 was due to differences in the climate of the two years (Williams, 1973). The degradation of agricultural soils, a major problem in the province (Rennie and Ellis, 1978), is also quite sensitive to climate. Saskatchewan's markedly continental climate, with agriculture strongly constrained by moisture limitations as well as by cold, is in contrast to the situations in case study areas such as Iceland and Finland. In this respect the province is fairly representative of the northern part of the North American Great Plains.

Within the context of the IIASA/UNEP climate impacts project, the objectives of this case study are:

(1) To illustrate the applicability of several models to the analysis of the impacts on Saskatchewan agriculture of climatic changes or fluctuations implied in historical climatic data and in general circulation model (GCM) simulations.

(2) To attempt to quantify the impacts of specified climatic changes or fluctuations.

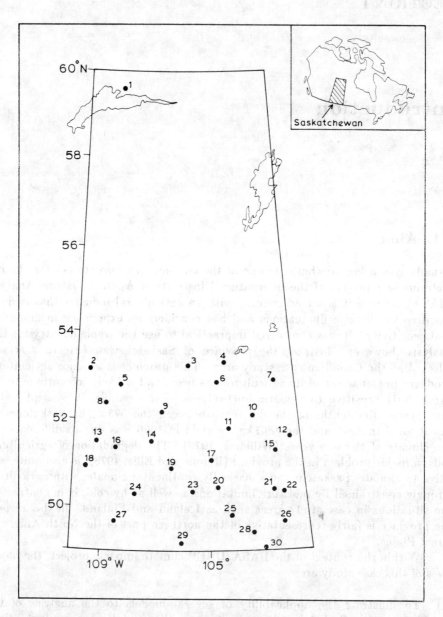

Figure 1.1. Saskatchewan, climatological stations. Station numbers: (1) Uranium City, (2) Waseca, (3) Prince Albert, (4) Lost River, (5) North Battleford, (6) Melfort, (7) Hudson Bay, (8) Scott, (9) Saskatoon, (10) Lintlaw, (11) Wynyard, (12) Kamsack, (13) Kindersley, (14) Outlook, (15) Yorkton, (16) Bad Lake, (17) Strasbourg, (18) Leader, (19) Tugaske, (20) Regina, (21) Broadview, (22) Whitewood, (23) Moose Jaw, (24) Swift Current, (25) Yellow Grass, (26) Carlyle, (27) Shaunavon, (28) Midale, (29) Rockglen, (30) Estevan.

(3) To consider some of the more appropriate policy responses to reduce the effect of negative impacts and take full advantage of positive ones.

In line with other case studies in the IIASA/UNEP project, the effects of three types of climatic fluctuations are considered:

(1) A single anomalous year.
(2) An extreme period of weather-years.
(3) The climate simulated by an atmospheric general circulation model for doubled concentrations of atmospheric CO_2.

Also in common with other case studies is the integrated use of a hierarchy of models to estimate various levels of impact: "first-order" impacts on the agroclimatic environment, effects on crop yields and "downstream" effects of these yield changes on farm production, employment and the provincial economy. This introduction provides background information on the case study area, discusses some relevant policy issues and outlines the reasoning behind the selection of the various scenarios and impact models.

1.2. The Study Area

Saskatchewan's location (*Figure 1.1*) is in a similar latitude range to that of the British Isles. Its northern border (60°N) is at the same latitude as the southern tip of Finland, or of Leningrad in the USSR. The province is about 640 km across from west to east at its southern border (49°N), and occupies 652 000 km^2, or 6.4% of the area of Canada. The northern part of the province is dominated by the rough rockland of the Pre-Cambrian or Canadian Shield (Clayton *et al.*, 1977; Richards and Fung, 1969) which is largely unsuitable for agriculture (*Figure 1.2*). Soils limitations would therefore preclude much northward expansion of agriculture, even if the climate were favorable. Any change in the thermal climate would be likely to have disadvantages as well as advantages for Saskatchewan agriculture.

The southern area of the province is part of the Interior Plains of North America. This southern, agricultural area can be divided into three soil zones. Using the Canadian soil classification system (Clayton *et al.*, 1977), these zones can be described as follows: Zone I, Brown Chernozemic, in the southwest (*Figure 1.2*); bordered by Zone II, Dark Brown Chernozemic; with Zone III, in the northeastern part of the agricultural area, composed mainly of Black and Dark Gray Chernozemic and Gray Luvisolic soils. Both Chernozemic and Luvisolic soils, developed mainly on glacial till or fluvial or lacustrine deposits, are weakly to moderately calcareous and are dominantly loamy. The Chernozemic soils developed under grass vegetation, while the Luvisolic soils developed under forest. Zones I and II have significant areas of Solonetz soils, and Zone III of organic soils.

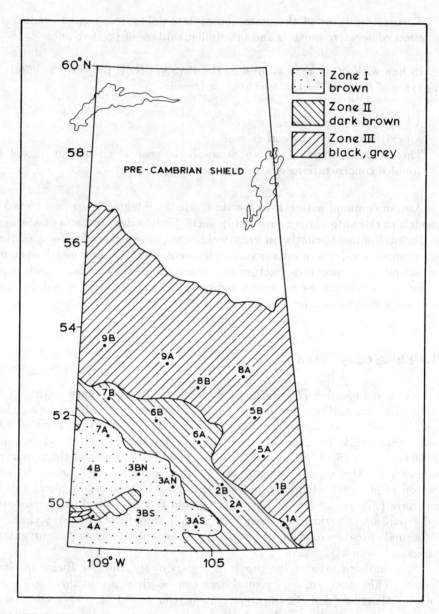

Figure 1.2. Saskatchewan crop districts and soil zones.

The terrain of the Dark Brown, and Black and Gray Soil Zones (II and III) is mainly in the 100–600 m altitude range, and "the total aspect is one in which relief and vegetation show little change over great distance," except for some entrenched river valleys and various minor features related to glacial action, such as spillways, dune areas and moraines (Richards and Fung, 1969). In Zone I (Brown Soils) the terrain is rougher and the altitudes range from 600 m to 1500 m.

Zones III and II have been found to be most productive and Zone I least productive per hectare, both because of the climates associated with these zones, and for reasons related to the soils themselves. In an analysis of crop district data, cereal yields were lowest in Zone I, and a simple moisture deficit for May to July, which was negatively correlated to yield and explained a substantial part of the yield variations, was 142 mm for Zone I, 106 mm for Zone II and 43 mm for Zone III (Williams et al., 1975). It was also found in analyses that first accounted for the influence of weather and soil texture and then examined other macroscale soil effects, that the Brown Soils generally had poorer crop yields than did the dominant soils in Zones II and III (Sheppard and Williams, 1976).

The soils of the agricultural area, while dominantly loamy, range in texture from light sands to heavy clays. Sands are least productive as they have the least capacity to retain water and plant nutrients, and crops on sand tend to suffer most from drought. At the other extreme clays tend to be quite productive as they hold more water and nutrients. Crops grown on clays are likely to suffer relatively less from drought, but on the other hand these soils, because they dry out more slowly, will be slower to warm up in the spring, so their growing season is effectively shortened.

Saskatchewan's pronounced continental climatic regime is a result of its location to the lee of the Rocky Mountains, which obstruct the movement of moist air from the Pacific, and its great distance from other maritime influences. The high latitude and proximity to the source of cold continental Arctic air to the north, without any significant barrier in that direction, is a dominant influence on the province's climate, especially in winter when continental Arctic air covers the province most of the time. The overall result is a continental climate of short, hot summers, long and very cold winters, and meager precipitation.

Temperature ranges between summer and winter are large (*Figure 1.3*), and the spring and autumn transition seasons are short. Contrasts between day and night temperatures are great, which means that crops may suffer damage from freezing even when average temperatures are relatively warm. Month-to-month and year-to-year variability are also rather high (*Figures 1.4–1.6*). For example, monthly mean temperatures may differ by 10°C or more from one month to the next in the winter and transition seasons. Standard deviations of 1.4–1.7°C are typical for June, July and August mean temperatures. The restricted thermal resources constitute a serious agroclimatic constraint: the average season free of temperatures below 0°C is between 90 and 120 days long for most Saskatchewan climatological stations (Environment Canada, 1982). Thus, depending on location, only one-quarter to one-third of the year is usually warm enough for significant crop growth. This freeze-free season is quite variable from year to year, typically ranging from as short as 50 days in some years, to as long as 150 in others.

Average annual precipitation ranges between 300 mm and 500 mm, with lowest amounts in southwestern parts of the province. Precipitation is greatest during the growing season, particularly in June (*Figure 1.3*). This one month typically accounts for between one-sixth and one-fifth of the annual precipitation

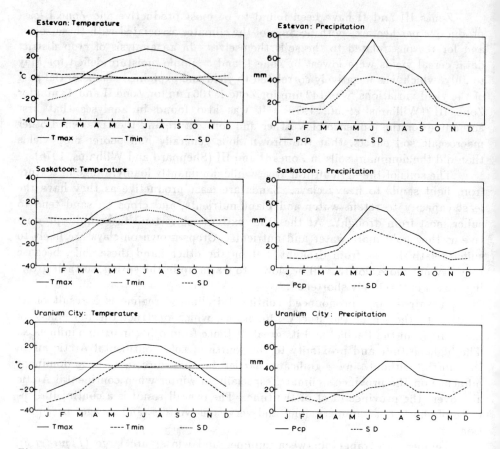

Figure 1.3. Normal monthly mean daily maximum and minimum temperature (*T*max, *T*min), total precipitation (Pcp) and standard deviations (SD) of monthly precipitation and mean temperature, for three locations.

in southern Saskatchewan. Variability of precipitation in southern Saskatchewan from year to year and month to month is quite pronounced (*Figures 1.4–1.6*). The southwestern area of the province that would be classified as "Dry" in the Köppen climatic classification system varies greatly from year to year, and examples for different years showing the boundary between this and the more humid zone to the northeast have been mapped (Richards and Fung, 1969). A considerable degree of moisture limitation, however, is characteristic of most of the agricultural area of the province. The interannual variation in moisture is the main cause of the great variability in crop yields. This has been recognized for many years. For example, prairie wheat yields were found to be related to the sum of April–July plus the preceding August–September precipitation (Williams, 1962) when the data were graphed for the years 1885–1961.

Saskatchewan has nearly 40% of Canada's farmland and grows a major portion of the country's cereal grain production (*Table 1.1*). The province's wheat is domestically very important, accounting for about 60% of Canada's

Introduction

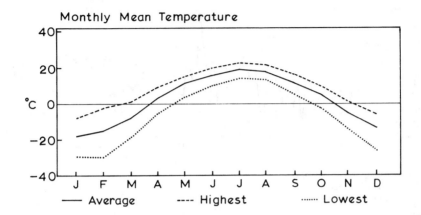

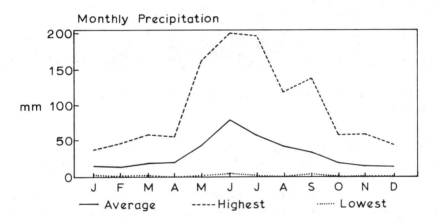

Figure 1.4. Regina average, highest and lowest monthly mean temperature and precipitation, 1884–1983.

wheat production. It is also quite important internationally because, although Saskatchewan produces only about 3% of the world's wheat, the fact that more than 70% of Canadian wheat is exported results in 1 ton of every 8 tons in the international wheat trade being from Saskatchewan.

Wheat production and summer fallow are the dominant cultivated agricultural land uses in Saskatchewan (*Figure 1.7*). In 1980 there were 7 million ha of land in wheat, about the same amount in fallow, and 1.3, 1.0 and 0.8 million ha in barley, oilseed crops and hay, respectively [for further information, *see*, for example, Statistics Canada (1982)]. The remaining agricultural land is chiefly unimproved land, used mainly for native hay and pasture. In the drier areas of the province it is common to leave the land fallow in alternate years, particularly in the context of wheat production. In 1980 about three-quarters of Saskatchewan's wheat crop was "fallow-sown", i.e., it was planted on land that

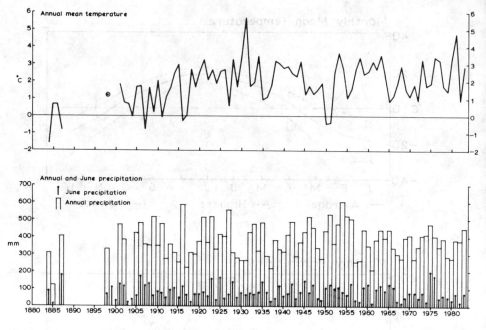

Figure 1.5. Trends in annual temperature and precipitation, and June precipitation for Regina, 1884–1983.

had lain fallow the previous year. The remaining one-quarter was "stubble-sown", that is it was planted on the "stubble" or residue from a crop grown on the land the year before.

Ostensibly, summer fallow is for moisture conservation since the stored moisture available to the crop at the beginning of the growing season will be greater after a fallow year than after a crop year. In fact, the main purpose of fallow now may be to facilitate the release of soil nitrogen for plant use and reduce crop surpluses (Rennie and Ellis, 1978). While the practice of summer fallow produces larger yields per harvested hectare, it leaves the land fallow without crop cover for a year. This is conducive to soil degradation and it also hastens the exhaustion of soil nitrogen.

Saskatchewan farms are typically highly mechanized commercial enterprises of 200–300 ha each, and farms of more than 1000 ha are not uncommon. The total number of farms of all types in Saskatchewan in 1980 was about 67 000. There are approximately 11 000 cereal farms in Zone I (Brown Soils), 16 500 in Zone II (Dark Brown) and 25 000 in Zone III (Black, Gray). Mixed farms, which have both grain and livestock, usually cattle, number 1150 in Zones I and II, and 2100 in Zone III. The remaining farms included 7500 cow/calf or feeder operations, 3282 hog farms and 870 irrigation farms; 45% of the beef cattle are in Zone III, where only about 4 ha of native pasture are needed per head, while 26% are in Zone II and 29% in Zone I. In Zone I, 9 ha of such pasture are required per head, reflecting the poorer climatic and soil resources of Zone I that

Introduction

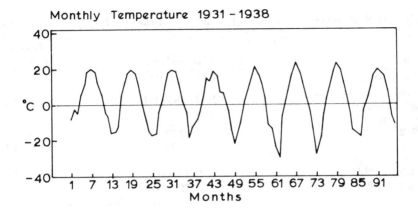

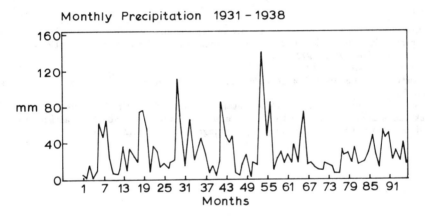

Figure 1.6. Illustration of month to month changes in temperature and precipitation: examples for 96 consecutive months, January 1931–December 1938, Regina.

have been identified elsewhere in the analysis of cereal production (Williams *et al.*, 1975; Sheppard and Williams, 1976).

Saskatchewan farming is becoming increasingly specialized; for example, about: 600 farmers produce all the dairy products, 300 farmers all the poultry and eggs, 3% of the beef farmers produce over 20% of the beef, and one-fifth of Saskatchewan farmers account for over half of the total agricultural sales value. There is a trend to increased emphasis on grain production, and this is likely to lead to use of lighter soils for grain and hence to greater susceptibility to climatic adversity. More intensive cultivation is also being practiced, and this is making the soil more prone to degradation. In addition to the trend to more specialization of farm enterprises, there is also increasing geographic concentration of crops, e.g., an increasing proportion of the province's barley and oilseed is being produced in Zone III.

Table 1.1. Saskatchewan in relation to Canada and the world[a].

	Saskatchewan	Percentage of Canadian total
Agricultural land[b]	26 328 000 ha	38
Average annual farm cash receipts, 1973–1978[c]	Can$2 153 941 000	22
Wheat crop area:		
1982[d]	7 932 000 ha	63
1983[e]	8 377 000 ha	61
Wheat production:		
1982–1983[d,f]	16 438 000 t	61
1983–1984[e,f]	15 296 000 t	58

	Saskatchewan values as a percentage of the world total	
	1982–1983[d,f]	1983–1984[e,f]
Wheat production	3%	3%
Wheat trade	14%	13%

[a] Prepared for illustration purposes only, using readily available recent data. For more comprehensive information, refer to sources such as Statistics Canada publications.
[b] *Canada Year Book, 1976–1977.*
[c] *Canada Year Book, 1983–1984.*
[d,e] Statistics Canada *Field Crop Reporting Series* (1983) No. 8 and (1984) No. 6 (data on "all wheat").
[f] Agriculture Canada Market Commentary, Sept. 1984. We have used the data on Saskatchewan's, contribution to Canada's wheat production, and Canada's share of world trade in estimating the Saskatchewan percentage of world wheat exports.

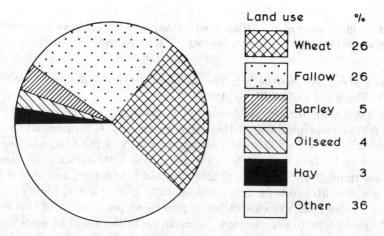

Figure 1.7. Principal agricultural land uses in Saskatchewan, 1980. (Source: Statistics Canada, 1982.)

A particularly interesting recent development in Saskatchewan agriculture is the increased emphasis on winter wheat. Because the winters have been considered too severe for the survival of fall-planted wheat, wheat has been almost exclusively a spring-sown crop in Saskatchewan in the past. However, in the early 1980s, winter wheat expanded rapidly. In 1982, only 18 000 ha were seeded to winter wheat, but over 400 000 ha of winter wheat were seeded in 1984, according to the Saskatchewan Crop Insurance Board. Winter wheat, which is sown to emerge and establish root systems some time before the onset of winter, is better able to exploit the fall and winter precipitation and to avoid damage from midsummer drought and from frosts that come at a time when spring wheat would be still immature and subject to cold injury. The expansion of winter wheat reflects changing market conditions, efforts by farmers to spread their workload by growing fall-sown as well as spring-sown crops, a perceived need to minimize spring wind erosion of soil by having a crop that covers the ground in spring, and perhaps an absence of severe winters in the early 1980s. Shewchuk (1984) has noted an apparent trend toward increasing temperatures at Saskatoon in the past two decades, particularly in winter.

It is clear that Saskatchewan agriculture is quite sensitive to climate and will probably be strongly affected by future climatic fluctuations and changes. This probability gives rise to several policy considerations which are discussed in the next section.

1.3. Agroclimatic Impacts and Societal Responses

The adaptation of Saskatchewan agriculture to the province's agroclimate and land resources and to the "economic climate" is an ongoing process which will continue in response to environmental and economic changes. A policy question of particular interest is whether society should attempt to anticipate changes and implement coping strategies in advance or should simply react to changes as they occur. We assume the former, and in particular this case study relates to the anticipation of climate-related changes. Anticipating such changes requires several types of knowledge. One of these is the understanding of how specified climatic changes are likely to affect agriculture, and that is the focus of this report. This case study does not attempt to predict climatic change; rather, it demonstrates how predictions or scenarios of climatic change can be translated into estimations of impacts on Saskatchewan agriculture, using models available from other lines of research.

In this section we review the highlights of agroclimatic impacts in Saskatchewan since 1929 and examine some of the particular policy issues relevant to climatic impact analyses. These issues relate to such matters as direct effects on agricultural production or the natural environment that might be generated by agroclimatic changes, and agriculture's response through management practices involving changing crops, systems, rotations, cultivation, pest control, etc. Also considered are the implications for farmers' financial situations, crop storage, government support of agriculture, and the interactions with domestic and world trade.

1.3.1. Historical background

A series of drought years from 1929 to 1938, which occurred at the same time as an economic depression, markedly reduced crop yields, thus contributing to severe local hardship and social upheaval. Massive drought-related farm abandonment took place. The number of abandoned farms in Saskatchewan increased fourfold from 1926 to 1936: from 4907 in 1926, to 5193 in 1931, to 12 831 in 1936 (Britnell, 1939). By the fall of 1937 two-thirds of the rural population were living on government relief payments because of drought-related crop failures that year (Britnell, 1939). Soil degradation, resulting from a combination of climatic stress and inappropriate management, was a major problem. At times 100 000–200 000 ha in the Canadian prairies were blowing out of control, i.e., they were completely denuded of topsoil or in the process of rapidly becoming so (Gray, 1978). Soil and crop yield losses are discussed in Sections 3 and 4.

Government response in the 1930s, in addition to direct relief payments to families in drought areas, included such measures as financial assistance in resettling the migrating farm families in areas less affected by drought, payments to doctors to maintain medical services for the rural population, and intensive research and education efforts to try to reduce soil losses. Other responses included the creation of a new government agency, the Prairie Farm Rehabilitation Administration, and the promotion of the development of large irrigation projects (Gray, 1978; Spence, 1967), to help combat some of the drought-related agricultural problems.

Over the years since the 1930s, technological and demographic changes and the continued evolution of economic structures have altered the character of the impacts of climatic fluctuations and the responses to them. The most severe single-year drought episode in the Canadian prairies in the past five decades, 1961, was used as one of the scenarios in this case study, as discussed in Subsection 1.4. By 1961, because of the increasingly complex economic system with its interdependencies and government involvement, the effects were much more diffused throughout the economy and society than in the 1930s. The impacts were felt more through difficulties of meeting grain marketing commitments and through adverse effects on Canada's trade balance, than through the local hardship that had characterized earlier drought impacts. Although there were several drought years from 1957 to 1968, the 1961 drought was somewhat isolated from the others, so its impact may have been less than if there had been two consecutive very dry years like 1936 and 1937. This emphasizes the importance of carefully selecting scenario years for a study such as this.

It has been estimated that under 1980 economic conditions, a period with frequent droughts like 1929–1938 or 1957–1968 in the Canadian prairies would result in drought-related wheat production losses totaling over Can$2000 million (Williams, 1983; Berry and Williams, 1985), or 5% or 6% of the total crop for the period. Economic impacts are explored further in Section 5.

In 1974 delayed planting due to cool wet spring weather, hot dry conditions in midsummer, and an early fall frost contributed to reduced production of

Saskatchewan wheat. However, climate was only part of the complex of causes for reduced wheat production that year, and since the year coincided with a period of sharply rising wheat prices, the impact was not commonly recognized (Glantz, 1979).

Some of the possible responses to impacts of future climatic fluctuations can be inferred from the 1984 experience as described at the time in an article on farm income (Tung, 1984). Drought in 1984 in southern Saskatchewan and other parts of the Canadian prairies was expected to reduce prairie grain production by 16% from the 1983 level despite a 2.5% increase in the crop area, with economic impacts such as the resulting farm income reductions spread over the 1984-1985 crop year. Response to these impacts included crop insurance payments, which were expected to exceed Can$300 million in 1984 due to the drought. In addition, an interim payment of Can$100 million under the Western Grain Stabilization Act was expected to help cushion the blow. The drought also contributed to feed grain and hay shortages for livestock producers, who were expected to reduce their beef cow herds and sell some calves that would otherwise have been kept over winter. A response to this situation has been a government-sponsored program to provide financial assistance to help prairie farmers retain their herds instead of selling them under such circumstances. For 1984 it seems likely that the various assistance and insurance programs, and an increase from the earlier anticipated levels in initial payments for grain, would tend to balance the negative drought impacts on farm income (Tung, 1984).

The year 1984 was also a particularly bad one for soil drifting, and public awareness of this was heightened by temporary closures of roads due to poor visibility in several dust storms and at least one associated traffic death (*Star Phoenix*, June 1, 1984). Soil erosion, together with cultural practices that fail to add organic matter, have contributed to a 40-50% decline in the organic matter content of Saskatchewan soils since 1900, and this reduction is continuing (Rennie and Ellis, 1978). Organic matter helps maintain a favorable soil structure for crop production, and it is quite effective in retaining soil moisture and nutrients for use by plants, so the organic matter losses reduce the crop productive capacity of the soil, particularly during droughts.

The frequency and intensity of wind erosion occurrences in the 1980s (*see* Section 3) are a reflection of the inadequacy of society's response to soil degradation problems. There has been a preoccupation with increased productivity and far too little attention to the conservation of the land resource.

Comparisons of the more recent years with the 1930s may give the impression that agriculture's coping mechanisms for dealing with climatic adversity are improving. However, the adverse effects may simply be less apparent at the local level. Trends towards more frequent cultivation may have largely negated the benefits of improved knowledge of soil conservation. In any event, continued vigilance and ongoing research are needed to deal with the impacts of future agroclimatic changes, both directly on productivity and on the land resource.

1.3.2. Implications for policy; project objectives

If the frequency of adverse conditions, such as those described above, was substantially altered by a change in Saskatchewan's climate, we could expect concomitant changes in the quantity, quality and variability of agricultural production. This is considered in Sections 4 and 5 of this report. It would also have major implications for the maintenance of the land resource (considered in Section 3) and would affect the welfare of the producer, the requirements for transportation and storage, and international trade. For example, if levels of production became more variable, more storage might be required.

One of the possible changes in Saskatchewan's agroclimate being suggested by climatic modeling results is an increase in summer dryness resulting from increased atmospheric CO_2 (Manabe et al., 1981). This could have important effects on Saskatchewan agriculture. Periods of drought lasting a year or more have large socioeconomic impacts in the province. If an intensification of these recurrent negative effects can be anticipated through impacts research, better coping strategies can be developed than if it happens unexpectedly. Strategies to take advantage of potentially beneficial climatic changes can also be more effective if they can be prepared in advance.

Climate-induced changes in farm production levels have ripple effects in other sectors of the economy as well as social implications. Reduction in productivity can lead to movement of people, both out of agriculture to other industries and out of agricultural districts. Sectors linked to agriculture would be affected by changes in agriculture even without such movements. The retail trade, the farm machinery industry, the development and maintenance of services such as roads, schools and hospitals, and the viability of small urban centers, would be affected by such changes in agriculture. Thus there may be a need to consider climatic impacts on agriculture, even in formulating policies with respect to economic and social matters not directly related to agriculture.

As we shall consider in Section 3, climatic changes could lead to conditions that would be more conducive to several important soil degradation processes which would further reduce the ability of crops to withstand climatic adversity. It is estimated that 160 million tons of soil are lost in the Canadian prairies annually due to wind erosion and 117 million tons by water erosion, and that the total measurable annual costs to prairie agriculture at present average about Can$368 million, or Can$12.31/ha (Sparrow, 1984; Prairie Farm Rehabilitation Administration, 1983; Bircham and Bruneau, 1985). Erosion can reduce wheat yields by 100–200 kg/ha/cm soil loss (Sparrow, 1984), which is equivalent to about a 5–10% wheat yield reduction in Saskatchewan. While there is considerable knowledge of appropriate techniques of soil conservation (see Section 3), there is also a need for information on the efficiency and cost effectiveness of such measures and for demonstrating their applicability at the farm level (Sparrow, 1984). The expectation of climatic changes would have implications for policies regarding support for research and extension relating to soil conservation. Another serious soil degradation problem in Saskatchewan is salinization, which already affects perhaps a million ha of Saskatchewan farmland, 160 000 ha severely, and is continuing to spread (Fairbairn, 1984). Salinization has reduced crop yields in affected areas by 10–75% despite farmers' efforts to compensate by increasing fertilizer applications; resulting crop losses exceed Can$260 million annually for the prairie provinces (Sparrow, 1984).

A number of other questions have also been important in framing this project. For example:

(1) What is the relative probability of any particular climatic scenario?
(2) Will climatic warming cause more frequent droughts and will these tend to last longer?
(3) Would the direct benefits of CO_2 to plants compensate for this?
(4) Should we also be considering cooling scenarios, for instance, related to increased dust pollution or volcanic activity?
(5) Which will better equip us to deal with future climatic fluctuations and changes: study of past agroclimatic impacts, or study of the possible impacts associated with scenarios such as those derived for CO_2 doubling?

These questions have research and policy implications. Other related questions are:

(1) What should society be prepared to pay to develop coping strategies and mechanisms as protection against a variety of possible climatic changes?
(2) How do the costs to society of ensuring that agriculture keeps in step with climate-related changes compare with the costs of failing to keep in step?
(3) What sort of policies do we need to help Saskatchewan agriculture not only minimize adverse effects of future climatic changes, but also benefit, where possible, from such changes?

The difficulties to be expected in dealing with such questions are exemplified by those encountered in estimating the potential value of climatic forecasts for agriculture (Glantz, 1979), but we believe it is important that they be addressed. This case study is directed toward providing information and methodology to help solve such problems.

We do not attempt in this report to assess the relative probabilities of different climatic scenarios, but we do demonstrate techniques for translating scenario information into assessments of the associated impacts on agriculture. These techniques will make it possible for planners and those who formulate policy to maintain an awareness of the implications for agriculture of the scenarios that seem most likely according to the latest climatological research. The techniques are also important for comparing scenarios for several reasons, including the consideration that a fairly improbable scenario may nevertheless be of considerable interest if its impact is likely to be large. Furthermore, such translations would be needed in any prediction of the combined direct and climate-related effects of CO_2 increases. This study illustrates applications for both historically based scenarios and scenarios that have been developed using climate model simulations. It provides only a few examples of the many possible climatic impact analyses that could be undertaken to provide information for use in policy formulation, but these examples consider some of the most important climate-related problems affecting Saskatchewan agriculture. Effects on a particular crop (Section 4), and hence on the provincial economy (Section 5), as well as on biomass productivity in general (Section 2) are estimated. Likely impacts on drought severity and frequency (Section 2) and on an important soil degradation process (Section 3) are also calculated. As a prelude to these analyses, changes in the thermal and moisture conditions are examined (Section 2). The remainder of Section 1 reviews previous climate impact work in the Canadian prairies and deals subsequently with the selection of scenarios for this case study and the tools available for the impact analyses.

1.4. Review of Previous Analyses of Actual or Simulated Climatic Events and Impacts

1.4.1. Analyses of historical climatic events

There is an extensive literature that deals with climatic fluctuations and change and their agricultural impacts in Canada's prairie provinces. One of the earliest analyses that mentions particular anomalous years is that by Unstead (1912). He commented, for example, on the drought of 1899–1900 which, at Indian Head, Saskatchewan (east of Regina), "almost ruined the crop", reducing wheat yields to 360 kg/ha on stubble, although yields were 1060 kg/ha on fallow. Several more recent reviews have listed a number of climatically extreme years and the associated impacts on 19th- and early 20th-century Canadian prairie agriculture (Strange, 1954; Beltzner, 1976; Allsopp, 1977 and the Canada Committee on Agrometeorology, 1977). Gray (1978) and Spence (1967) have discussed climatic events of the 1930s and societal responses.

Reviews of temperature trends have been made for 1921–1970 in southern Saskatchewan (Magill, 1980) and 1964–1983 at Saskatoon (Shewchuk, 1984). Climatic trends have been analyzed for a number of years prior to 1960 in northern Alberta (Carder, 1962; Edmonds and Anderson, 1959), and for 1940–1974 for all regions of Canada including the prairies (Thomas, 1975). Villmow (1956) examined the nature of the "Dry Belt" of southwestern Saskatchewan and southeastern Alberta and discussed the climatic conditions of some of the driest years.

Treidl (1978) discussed the impacts of climatic variability on prairie wheat production, and Glantz (1979) reviewed the effects on Saskatchewan wheat production of the short growing season and drought of 1974. The delineation of drought areas affecting spring forage in the prairies has been demonstrated for several of the years 1976–1982 by Dyer et al. (1982). Street and McNichol (1983) used soil moisture estimation techniques in analyzing droughts for the 1928–1981 period at several prairie locations.

Britnell (1939) made a quite comprehensive study of the Saskatchewan wheat industry and included analyses of the socioeconomic impacts of weather-related crop failures during the 1930s. Of particular interest were data on dollar returns from different crop districts. For example, there were no returns in 1937 in crop districts 2, 3, 4, 6 and 7 (districts are shown in *Figure 1.2*); they were 51% of the break-even returns (Can$13.10/ha) in district 1, 85% in district 9, 122% in district 5 and 180% in district 8. Since the economic depression would have had similar effects in the different districts, the differences among the districts may broadly be attributed to the climatic conditions.

Relevant early empirical analyses involving wheat yields and weather data include those by Hopkins (1935) and by the then Dominion Bureau of Statistics (1941). Robertson (1974) analyzed 1923–1972 experimental plot wheat yields at Swift Current in relation to weather and obtained weather-based estimates that agreed fairly well with the observed yields. The observed wheat yields were well below the 50-year average from 1929 to 1938, as were the weather-based estimates.

The purpose of the analysis by the Dominion Bureau of Statistics (1941) of 1921–1940 wheat and weather data was to develop large area crop prediction techniques for estimating prairie wheat yields prior to harvest. Further work on such techniques was reported by Williams and Robertson (1965), who analyzed 1952–1963 crop district wheat yields in relation to precipitation, and by Williams (1969b, 1973), who analyzed 1952–1967 crop district wheat yields, with precipitation and estimated potential evapotranspiration as explanatory variables. The procedures were expanded to include soil factors and trend terms as explanatory variables and to provide estimates for oats and barley as well as wheat in analyses of 1961–1972 data (Williams et al., 1975; Sheppard and Williams, 1976).

The analytical procedures that had been developed for yield prediction purposes were subsequently adapted for identifying drought years from the standpoint of wheat production (Williams, 1983; Berry and Williams, 1985). Regression analyses were performed on annual wheat-yield data in relation to data for the nine 10-day intervals in May to July from the two-layer water budget of Street and McNichol (1983). Those data were on the basis of a 100 km grid, and 72 grid points covered the prairie wheat-growing area. The crop yield analyses (Williams, 1983; Berry and Williams, 1985) were performed for 1928–1980, using spatial groupings of the grid points; for example, in an analysis of a group of 10 grid points there would be $53 \times 10 = 530$ cases, one for each point for each year. The resulting regression equations were then employed to make water-budget-based yield estimates for each of the 53 years and 72 grid points.

Canadian prairie wheat yields appear to have been trending upwards over the past 50–60 years. It may be postulated that up to this point technological advances have more than compensated for soil degradation and losses in native fertility (Rennie and Ellis, 1978). In any event, any crop-weather analysis including long historical series of the yields obtained by farmers in a region should take such trends into account. In the study of the 1928–1980 wheat data (Williams, 1983; Berry and Williams, 1985), no attempt was made to "de-trend" the yield data prior to analysis, as this might have resulted in removing some weather-related effects. Instead, for the purposes of regression analyses, trend was considered as a cause of yield variability, just as weather was. It was found useful to assume a linear trend from 1928 to 1950 and another from 1950 to 1980, and to incorporate these trend terms, along with weather terms, in the regression equation. Use of a more complex trend term was avoided because

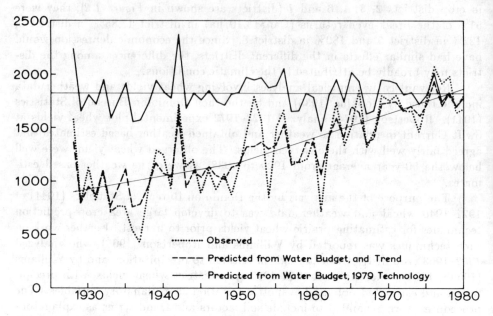

Figure 1.8. Comparison of Canadian prairie wheat yields 1928–1980 (dotted line), yield estimates based on weather and trend (dashed line), and estimates based on weather and assumed 1979 technology. (Source: AES Drought Study Group, 1986.)

with such a term one runs the risk of incorporating more of the climatic variability in the trend term than is necessary.

After deriving regression equations that enabled one to make estimates for every grid point for every year from 1928 to 1980, these equations were used, with trend variable values corresponding to 1979, to estimate the wheat yields that could have been expected with 1979 technology but with weather as in each of the years 1928–1980. In effect the "de-trending" was done after, rather than before, the regression analysis. These de-trended weather-based yield estimates enabled one to use any year or period of years as a scenario to obtain, for example, answers to questions such as: "What wheat production can we expect this year if our weather is like that of 1961; or in the next decade if the weather is like that of 1957–1966 (assuming 1979 technology)?"

Attempts at answering some of these questions have been published (Williams, 1983; Berry and Williams, 1985), and a number of other such questions are considered in a report that has now been issued by the Canadian Climate Centre (AES Drought Study Group, 1986). *Figure 1.8*, based on that information, shows the observed prairie wheat yields (dotted curve), the yields predicted from water budget and trend (dashed curve), and the "de-trended" estimates, i.e., the predictions made using the trend level for 1979 but with the weather as in each of the years indicated (solid curve).

1.4.2. Analyses including simulated climates

There have been several studies in which the effects of simulated changes in climate in the prairie provinces or in Canada as a whole have been analyzed. For example, climates were simulated on a year-by-year basis for the prairies for 1952-1967 by multiplying the original instrumentally observed precipitation data for June and July by 1.1, and by 1.3, in order to investigate the likely impacts of 10% and 30% increases in precipitation on wheat yields (Williams, 1970). A wheat-yield model (Williams, 1969b) was applied to the original and simulated climate data, and it was concluded that the increased precipitation would benefit prairie wheat production (Williams, 1970).

A similar analysis with climates simulated for each year from 1961 to 1972 and using wheat, oat and barley models (Williams et al., 1975) to assess the impacts, concluded that wheat yields would be increased 3% by a 10% increase in precipitation for all months, and reduced 4% by a 10% precipitation decrease, while oats and barley would be less sensitive to such changes (Williams, 1975).

Using a crop phenology model with climatic scenarios obtained by adding or subtracting simulated temperature changes to adjust 30-year climatic normals, it was concluded that 1 °C cooling would reduce the area of Canada suitable for barley production by one-seventh and for wheat production by one-third (Williams and Oakes, 1978). In a study area in the Peace River region of Alberta, it was estimated that the part of the area unsuitable for ripening barley was 6% with the 1931-1960 climate, and would increase to 35% with 1 °C cooling and 70% with 2 °C cooling (Williams et al., 1980).

Using a crop growth model with climates altered by adjusting temperatures by specified amounts and adjusting precipitation by various percentages, Bootsma et al. (1984) assessed the sensitivity to climatic change of several types of crops in several areas across Canada, including two study areas in Saskatchewan. They found that warming without a precipitation increase would generally reduce yields, cooling accompanied by shorter growing season lengths would also reduce yields and increasing precipitation would generally increase yields. Under a warmer climate, yields would be particularly sensitive to precipitation increases.

Adopting a similar set of climatic adjustments, it was estimated that biomass productivity would tend to increase substantially in northern Alberta but decrease somewhat in the south of that province with climatic warming and would generally increase with increasing precipitation (Williams, 1985). The same study concluded that the warming associated with CO_2 doubling could be expected to increase biomass productivity for Alberta as a whole by 16% if precipitation remained unchanged, by 33% if precipitation increased 10%, and to reduce provincial biomass productivity slightly if precipitation decreased 10%.

1.5. Selecting Climatic Scenarios for this Case Study

Climatic scenarios were selected both to reflect issues of particular concern in the region being studied and to enable comparisons to be made with results in other case studies in this volume. Scenarios were chosen from three broad types:

(1) A single anomalous weather-year taken from the historical instrumental record.
(2) An extreme weather-decade or period of weather-years, also from historical data.
(3) A climate simulated using GISS (Goddard Institute for Space Studies) atmospheric general circulation model (GCM) results for doubled concentrations of atmospheric CO_2 (Hansen et al., 1984).

In common with the other case studies and in order to provide a reference against which to compare results of analyses involving these scenarios, the use of a standard climatic scenario based on the 1951–1980 period was adopted and is referred to as scenario HIST1 throughout the remainder of this case study. This section discusses the selection of scenarios for this case study.

1.5.1. Selection of extreme weather-years and decades

The most severe drought year from the standpoint of the weather effect on prairie wheat production was 1961 (*see Figure 1.8*). The observed prairie wheat yield was farthest below trend that year and the relative severity of the 1961 prairie drought has also been noted elsewhere (Thomas, 1961; Williams, 1962, 1969b; Williams et al., 1975). In bar charts of Canadian wheat scenarios selected from the period since 1930 (Hinckley, 1976), the value for 1961 is the smallest single-year value.

The year 1961 has also been noted as a severe drought year with specific reference to Saskatchewan (Williams and Robertson, 1965; Williams, 1973). Also, in the analysis of yields at Swift Current, Saskatchewan (Robertson, 1974), the 1961 weather-based yield expectation was the smallest from 1942 to 1972 and the observed yield that year the smallest from 1950 to 1972. The climate of 1961 was therefore selected as the extreme-year scenario for this case study and is referred to as HIST2. It should be noted that for some of the analyses, the crop year from August 1, 1960, to July 31, 1961, was used instead of the calendar year 1961.

The second and third lowest values of de-trended yield estimates (*see Figure 1.8*) were for 1936 and 1937, and these have been characterized as the second and third worst prairie drought years (Williams, 1983; Berry and Williams, 1985). Britnell (1939) described the Saskatchewan situation in those years as follows: "In 1936 the excellent crop prospects of June were converted by heat and drought into another failure and the lowest wheat yields on record. In 1937, drought brought the most complete crop failure in the history of settlement over the entire prairie wheat belt." There were serious prairie droughts in seven years of the 1929–1938 decade: 1929, 1931, 1933, 1934 and 1936–1938. From the standpoint of weather reducing wheat yields, that decade was the most severe of any 10 consecutive years in the 53-year period 1928–1980 (*see Figure 1.8*). The climate of the decade 1929–1938 was therefore selected as the extreme decade scenario in this case study. It is identified as scenario HIST3.

Introduction

Further examination of *Figure 1.8* and the associated data identifies 1933–1937 as the worst five consecutive years for the prairies and probably for Saskatchewan, and data on dollar returns at the time, tabulated by Britnell, (1939), tend to confirm this for Saskatchewan. The climate of the 1933–1937 period was selected as the extreme 5-year period scenario. For reasons explained in Section 4, this 5-year extreme was used for analysis in Sections 4 and 5 rather than the extreme decade. The 1933–1937 climate is designated as scenario HIST4. Results are presented both for the individual HIST4 years and for the 5-year average.

For the drought frequency analysis (*see* Section 2), weather data for every year from 1950 to 1982 were assessed individually (HIST5).

1.5.2. Selection of scenarios to simulate climate for $2 \times CO_2$

Data for $1 \times CO_2$ and $2 \times CO_2$ GCM-derived equilibrium climates based on the GISS model (Hansen *et al.*, 1984) were obtained for grid points including the case study area (*see* Part I, Section 3). In this study these equilibrium climates will be referred to as GISSE (for $1 \times CO_2$) and GISSC (for $2 \times CO_2$).

Comparisons were made between the monthly GISSE ($1 \times CO_2$) data for the case study area and instrumentally observed climatic normals (HIST1) for nearby climatological stations. It was found that the GISSE temperatures were significantly higher in winter and lower in summer than observed data; in fact, even the GISSC ($2 \times CO_2$) temperatures were lower than the observed temperatures in summer. An example is shown here for 50°N latitude, 105°W longitude with HIST1 data for Regina (*Figure 1.9*). For precipitation, the GISSE data indicated a much wetter regime than the observed climate. Although the general annual precipitation pattern was similar, with an early summer peak, monthly amounts for GISSE were substantially higher, particularly in winter when they were more than twice the observed climatic normal amounts (*Figure 1.10*).

This comparison reveals that the equilibrium $1 \times CO_2$ climate depicted by the GISS data for Saskatchewan bears little resemblance to the present climate of that province. In fact, with its much flatter annual temperature curve and greater precipitation, it is rather more similar to the present climate of the northern part of the island of Newfoundland off Canada's Atlantic coast. Nevertheless, although the GISSE climate appears unrealistic for Saskatchewan, for the purposes of this study, it is assumed that the *changes* from GISSE to GISSC correspond to the changes that could be expected in Saskatchewan climate with CO_2 doubling. This assumption, while questionable, is not unreasonable, since both the GISSE and GISSC data are derived from simulations of the global climate that are both physically realistic and globally consistent. Notwithstanding any doubts surrounding the validity of the GISS model estimates, it was not the intention of this case study to *predict* future climate and its impacts. Rather, the climatic scenarios were constructed to provide a feasible representation of future climate that could be used for testing the methods of impact assessment.

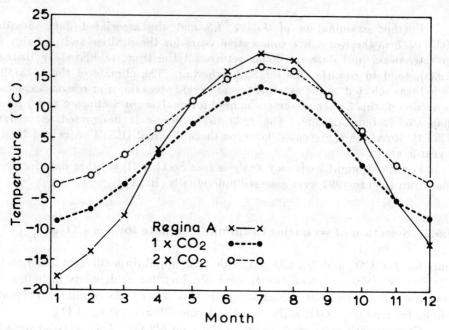

Figure 1.9. Comparison of Regina 1951–1980 normal (HIST1) mean monthly temperature with GISS $1 \times CO_2$ (GISSE) and $2 \times CO_2$ (GISSC) results for 50°N latitude, 105°W longitude.

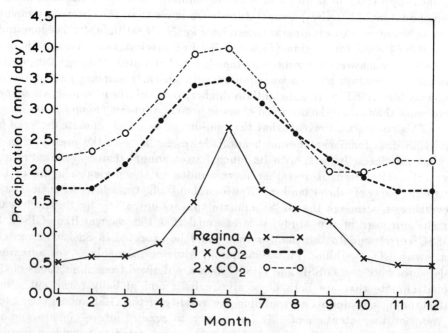

Figure 1.10. Comparisons of Regina 1951–1980 normal (HIST1) monthly precipitation rate with GISS $1 \times CO_2$ (GISSE) and $2 \times CO_2$ (GISSC) results for 50°N latitude, 105°W longitude.

Introduction

There are several ways of making use of GCM results for impact analyses. It seemed more reasonable, for analyzing impacts of expected climatic effects of CO_2 doubling in Saskatchewan, to use the present climate of Saskatchewan as a starting point, rather than using a climate corresponding to the present climate of a cool maritime area. We therefore used HIST1 as the basic reference scenario and simulated climates for CO_2 doubling by applying to HIST1 data adjustments based on the changes from GISSE to GISSC.

The adjustment of instrumentally observed data to simulate a changed climate is analogous to the problem of adjusting climatic normals at a long-term station to "simulate" such normals at a short-term station, as discussed in texts such as Conrad and Pollak (1950). For example, if Ta30 is the 30-year normal January temperature at station A, and Ta5 and Tb5 are 5-year averages of January temperature at stations A and B, then one can use the difference between the 5-year normals at A and B to estimate the 30-year normal at B: Tb30 − Ta30 = Tb5 − Ta5; Tb30 = Ta30 + (Tb5 − Ta5). In this example we simulated the January normal temperature at location B, given the corresponding normal at A and the adjustment difference, Tb5 − Ta5. In the same way one can take an historically based January temperature for location C (Tch) and simulate the corresponding temperature for a 2 × CO_2 climate (Tcc) given the adjustment difference Tc2 − Tc1, where Tc1 and Tc2 are the GCM equilibrium climate temperatures for 1× CO_2 and 2× CO_2, respectively. The simulated temperature for 2 × CO_2 is Tcc = Tch + (Tc2 − Tc1). Using the terminology of Conrad and Pollak (1950), this procedure can be called the "method of differences".

The method of differences is used for temperature because of the "quasi-constancy" of temperature differences, but for precipitation the "method of ratios" is employed because for precipitation it is the ratios, rather than the differences, that are "quasi-constant" (Conrad and Pollak, 1950). Thus, to estimate the normal January precipitation at short-term station b, one can use the relationship (Pb30/Pa30) = (Pb5/Pa5); Pb30 = Pa30 × (Pb5/Pa5). Likewise, one can simulate a 2 × CO_2 January precipitation value at location C (Pcc) from an historical January precipitation value (Pch) and the ratio of the 2 × CO_2 to 1 × CO_2 precipitation amounts (Pc2/Pc1). Thus Pcc = (Pc2/Pc1) × Pch. The example in *Table 1.2*, involving average daily precipitation data for GISSE (Pc1) and GISSC (Pc2) at 50°N, 105°W, compares the effects of applying the methods of differences and of ratios in adjusting precipitation. Regina precipitation values (Prc) for three months are derived for 2 × CO_2 by both methods using the historically based (Prh) normals and 1952 observed precipitation.

For the normals in this example June precipitation is presently more than three times April precipitation, and this relationship is maintained for CO_2 doubling when ratios are used for the calculations, but not when differences are used. With differences, the values derived for CO_2 doubling have less relative variability than the historical values. This is much more pronounced in application to a particular year such as 1952 for Regina (*see Table 1.2*). This has important implications for any analyses involving data on a year-by-year basis. For example, the use of differences for simulating precipitation amounts for 2 ×

Table 1.2. An example to compare methods of adjusting mean monthly precipitation totals under the GISS scenario.

Scenarios and adjustments	April	May	June
GISSE Pc1 (mm/day)	2.8	3.4	3.5
GISSC Pc2 (mm/day)	3.2	3.9	4.0
Pc2 − Pc1 (mm/day)	0.4	0.5	0.5
Pc2/Pc1	1.14	1.15	1.14
Normal Prh (mm/day)	0.79	1.50	2.65
Prc by differences	1.19	2.00	3.15
by ratios	0.90	1.73	3.02
1952 Prh (mm/day)	0.04	0.41	3.31
Prc by differences	0.44	0.91	3.81
by ratios	0.05	0.47	3.77

CO_2 would tend to make it appear that CO_2 doubling would greatly reduce drought frequencies while using the method of ratios might give a different result. This question of relative variability was further examined by computing the coefficient of variation (CV) for Regina May precipitation amounts for 1952–1961. CV was 65% for the observed precipitation data and for the 2 × CO_2 amounts derived by the method of ratios, but was only 49% for the 2 × CO_2 data derived by the method of differences.

In consequence of the above considerations, the method of differences was applied to HIST1 (normal) temperature data and the method of ratios to the HIST1 precipitation data to derive a climatic scenario for 2 × CO_2, designated here as GISS1 and also abbreviated below as 2 × CO_2 TP. The differences were obtained by subtracting GISSE from GISSC temperature data, and the ratios by dividing GISSC by GISSE precipitation data (*Figure 1.11*).

Examination of the GISSC and GISSE precipitation data revealed that the GISS model results indicated general increases of precipitation in Saskatchewan (*Figure 1.11*) accompanying the increases in temperature for 2 × CO_2. However, Villmow (1956) noted that, historically, low precipitation and high temperatures tended to occur in the same summers. Robertson (1974) found maximum temperatures negatively correlated with growing season rainfall throughout the major part of the 1923–1972 period at Swift Current. Manabe et al. (1981) have postulated drier summers in middle latitudes with increasing CO_2. We thought it reasonable to assume that CO_2 warming might occur without being accompanied by increased precipitation. To facilitate analysis of this possibility, use was made of an alternative scenario in which temperatures were increased as with GISS1, but precipitation was not increased (abbreviated below as 2 × CO_2 T). This scenario is designated GISS2.

For the drought frequency analyses, the data for the individual years from 1950 to 1982 (HIST5) were used with adjustments derived from the changes from GISSE to GISSC to obtain 2 × CO_2 scenarios. The resulting scenario incorporating both the temperature and precipitation changes is GISS3. The scenario for 1950–1982 year by year that reflects only the temperature changes is GISS4.

Introduction

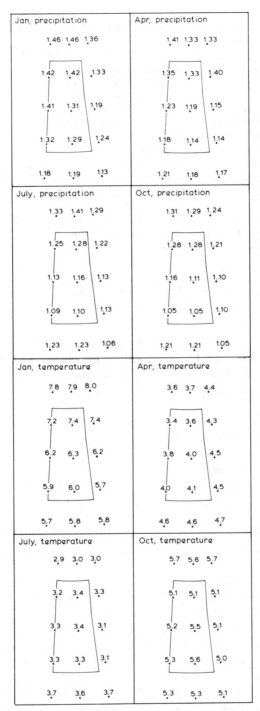

Figure 1.11. Precipitation ratios (*Pcp*adj) and temperature differences (*T*adj) for adjusting historical climatic data to derive $2 \times CO_2$ (GISS) scenarios: values for selected months for 15 grid points.

Table 1.3. Scenarios and variables in the Saskatchewan case study.

Scenario	Climatic variables		Section
HIST1: 1951–1980 normals	$T\text{max}h1$ $T\text{min}h1$ $Pcph1$		2–5
HIST2: 1961	$T\text{max}h2$ $T\text{min}h$ $Pcph2$		2–5
HIST3: 1929–1938 averages	$T\text{max}h3$ $T\text{min}h3$ $Pcph3$		2,3
HIST4: 1933–1937 year-by-year	$T\text{max}h4$ $T\text{min}h4$ $Pcph4$		4,5
HIST5: 1950–1982 year-by-year	$T\text{max}h5$ $T\text{min}h5$ $Pcph5$		2
GISSE: GCM output for $1 \times CO_2$.	$To1$ $Pcpo1$		2
GISSC: GCM output for $2 \times CO_2$.	$To2$ $Pcpo2$		2
GISS1: $T\text{adj} = To2 - To1$; $Pcp\text{adj} = Pcpo2/Pcpo1$;	$T\text{max} = T\text{max}h1 + T\text{adj}$, $T\text{min} = T\text{min}h1 + T\text{adj}$, $Pcp = Pcph1 \times Pcp\text{adj}$		2–5
GISS2: $T\text{adj} = To2 - To1$;	$T\text{max} = T\text{max}h1 + T\text{adj}$, $T\text{min} = T\text{min}h1 + T\text{adj}$, $Pcp = Pcph1$.		2–5
GISS3: $T\text{adj} = To2 - To1$; $Pcp\text{adj} = Pcpo2/Pcpo1$;	$T\text{max} = T\text{max}h5 + T\text{adj}$, $T\text{min} = T\text{min}h5 + T\text{adj}$, $Pcp = Pcph5 \times Pcp\text{adj}$		2
GISS4: $T\text{adj} = To2 - To1$;	$T\text{max} = T\text{max}h5 + T\text{adj}$, $T\text{min} = T\text{min}h5 + T\text{adj}$, $Pcp = Pcph5$.		2

All are monthly data. T = temperature, Pcp = precipitation, max = mean daily maximum, min = mean daily minimum, $T\text{mean} = (T\text{max} + T\text{min})/2$, adj = adjustment, h = historical, o = GCM output. Section = part of case study in which scenario is used. See text for details.

The scenarios used in this case study, both those based on historical data and those derived using GCM results, are summarized in *Table 1.3*. Temperature and precipitation data for the May to August period are given by crop district for the various scenarios in Subsection 4.5.

1.6. Impact Models Used

In climatic impacts analysis one typically adapts or borrows models derived for other purposes. For example, models originally developed for wheat-yield prediction (Williams *et al.*, 1975) were adapted for analyzing the impacts of extreme years such as 1937 (Berry and Williams, 1985). Likewise, a crop growth model developed for land evaluation purposes (Dumanski and Stewart, 1981) was employed in assessing impacts associated with various simulated climates (Bootsma *et al.*, 1984).

While it is desirable that such models should have been tested for the purposes for which they were originally developed, it is often not possible to test adequately the validity of models for impacts work. For instance, an impacts model (AES Drought Study Group, 1986) indicated that a year with weather like 1935 or 1954 in the Canadian prairies should be rather favorable for wheat assuming 1979 technology (*Figure 1.8*, solid curve). Technology in those years was not able to prevent severe losses due to wheat rust, so the observed yields (dotted curve, *Figure 1.8*) were very low, but those results may be quite irrelevant for validating impacts given 1979 technology. These are extreme cases, but they illustrate the severe limitations of using historical crop data for model validation in impact studies.

Introduction

For a scenario of a simulated future climate, validation may thus be quite difficult, even if good analogues for the simulated climate are found in the historical climatic record. If past climatic records for the region do not provide such analogues, the validation issue may be even more problematic. The use of analogues from other parts of the world may be helpful in this respect, but even then there are differences associated with changed locations, such as differences in daylength and soil quality, that may detract from their usefulness in validation. In any discussion of climatic impacts models it is therefore important to keep in mind that stringent validation of such models for impacts work is generally not possible, and that results of validating such models for applications other than impacts work may not be too relevant.

Some of the impacts models that have been used in Canada have been reviewed earlier (Subsection 1.4). Several other models suitable for impact studies have been applied in unpublished work in the agencies represented in this case study or elsewhere in Canada.

The selection of impact models for use in this case study was based on the following considerations:

(1) Relevance to analysis of the impacts on Saskatchewan's agricultural productivity, the stability of the agroclimatic resource, or the conservation of the soil, associated with the available climatic scenarios.
(2) Illustration of a broad spectrum of impact analysis capabilities. Both in the selection of models and in the particular procedures followed, we attempted to demonstrate a variety of capabilities, and to avoid, for example, using two models that would provide answers on essentially identical climatic impacts.
(3) Suitability of the model for the particular application. The model had to have demonstrated applicability to macroscale analysis and sensitivity to changes in the variables being analyzed.
(4) The existence of the model in a practical form. There was not time to develop models that had been postulated but never made operational.
(5) Capacity to provide useful answers rather quickly with the available resources and data. If a previously used model would require many person-months of time to make it operational again, or to perform the required analyses, it was not considered appropriate for this case study.

As spring wheat production is the major agricultural enterprise in Saskatchewan, the likely impacts of climatic changes or fluctuations on that crop were of particular interest, and a model relating the crop to climate was required. As has been noted elsewhere, the empirical yield–weather equations often used for such analyses tend not to reflect effects such as changes in the length of growing season (Williams, 1975; Hinckley, 1976). A crop growth model was selected (Section 4) which avoided this deficiency. The model is one of several available for different important Canadian crops (Dumanski and Stewart, 1981), so the methods demonstrated here for spring wheat could readily be applied for those other crops in future work.

Climate-related changes in wheat yields would in turn have a number of consequences, not only directly on the wheat growers, the transportation industry, etc., but also indirectly on other sectors. For analyzing these types of impacts, economic models, such as those used by the Prairie Farm Rehabilitation Administration, are needed. PFRA's models had been constructed for the purpose of enhancing the understanding of the susceptibility of the Saskatchewan economy to recurrent droughts, and they are quite appropriate for analyzing climatic impacts in Saskatchewan. One of these models was therefore selected to translate the wheat-yield impacts of Section 4 into economic terms (*see* Section 5).

The climatic changes implied in the GISS scenarios would undoubtedly be followed by changes in the crops that would be grown in Saskatchewan. In a much more comprehensive investigation, growth models for a number of crops could be applied and the various anticipated shifts in crops could be taken into account in an integrated manner. This was beyond the scope of this pilot study, except for some limited consideration relating to winter wheat (Section 5). Another approach to the problem is to consider that as climate changes, yields of presently well adapted crops will be reduced, but overall crop production may be maintained or even increased, with climatic change being accompanied by a shift in crop mix to crops better adapted to the new conditions. A model that estimates the potential total dry matter or biomass productivity in general but is not specific to any particular crop was selected (Section 2) to provide useful information on the likely change in crop productivity and to complement the estimates for spring wheat.

Recurrent droughts are a major problem in Saskatchewan, as is evident from the discussion of previous sections. If we were viewing climate as stable in the long term but with short-term fluctuations, then an analysis such as that of *Figure 1.8*, which relates to the most important crop, would be appropriate for obtaining a general picture of the effects of drought on Saskatchewan agriculture. Droughts during the months of August to April, as in 1976–1977 (Street and McNichol, 1983), which had serious effects, for example on forage crops, might not be considered as very important as they would not seriously affect spring wheat or other spring-sown cereals.

If the climate is considered to be dynamic and changing over the centuries, then a drought analysis tied to a single major crop is inappropriate. For instance, if climate changed to the extent that winter wheat became the major crop in Saskatchewan, droughts in the August to April period would probably have much more substantial impacts on the agricultural industry than is presently the case.

To obtain a general overview of the likely frequency and severity of droughts under a changed climate, an approach was adopted that was not related to a particular crop. A drought index which has been widely used in North America and was considered appropriate for this purpose was computed for each month for 33 years for changed climates to analyze the impacts on droughts (Section 2).

All of the models discussed to this point relate either to agricultural productivity or to conditions that directly affect it, but they do not reflect impacts

Table 1.4. Impact models and scenarios used in the Saskatchewan case study.

Impact model	HIST:	1	2	3	4	5	GISS: E	C	1	2	3	4	Part II, Section
Growing degree-days		x	x	x				x	x	x			2
Precipitation effectiveness		x	x	x						x	x		2,3
Palmer Drought Index					x							x	x	2
Climatic index of agricultural potential		x	x	x						x	x		2
Wind erosion potential		x	x	x						x	x		3
Spring wheat yield		x	x		x					x	x		4
Provincial agricultural economy		x	x		x					x	x		5
Provincial employment		x	x		x					x	x		5

on the sustainability of production. In particular, if climatic changes affect the rate of soil degradation, then there will be long-term indirect effects on productivity due to these soil-related effects. Also, continuing soil degradation can reduce agriculture's capacity to cope with climatic adversity; for instance, loss of the finer soil particles and organic matter can make crops more susceptible to drought damage (Rennie and Ellis, 1978). Agriculture can be considered analogous to having a bank account (Bursa, personal communication, 1984): the agricultural production is the "interest" from the account, but the soil is the "principal". Soil degradation, which reduces the "principal", ultimately will reduce the productivity.

Soil degradation is a very serious problem in Saskatchewan. In view of the implications of soil degradation in a changed climate, it was considered essential in this case study to demonstrate methods for assessing the impacts of climatic change on soil degradation. Wind erosion is a major component of Saskatchewan's climate-related soil degradation. A model which had been applied successfully in mapping soil erosion potential in adjacent parts of the United States was considered appropriate for this work and was therefore employed in this case study (Section 3).

The impact models and the scenarios used in the model experiments are summarized in *Table 1.4*. We are well aware that all the impacts models used in this case study have important shortcomings. Chief among these, as already noted, is the difficulty of validating models for climatic impacts work. Particular deficiencies of each of the models are discussed in subsequent sections of this report. We have demonstrated the successful linking of different models, with the wheat-yield and economic analyses of Sections 4 and 5. We have not pursued some other rather obvious linkages between models, for example, to translate changed drought frequencies (Section 2) to changed winter wheat yields, or altered winter wheat production patterns to effects on wind erosion potential. With the limited time and resources available for the case study, priority was given instead to demonstrating a variety of what we consider to be quite useful models for estimating climate change impacts in Saskatchewan.

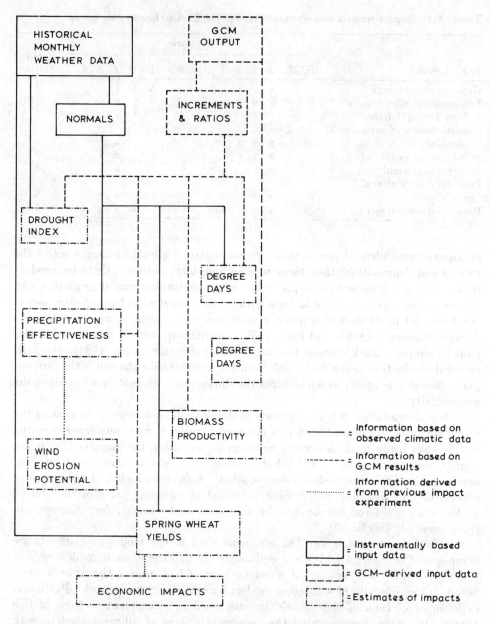

Figure 1.12. Schematic diagram of the Saskatchewan case study.

More comprehensive studies in the future should examine inter-model linkages in detail.

The flow of data in this study is illustrated in *Figure 1.12*, which distinguishes between information based on observed climatic data, that derived from the GISS general circulation model results and that related to impacts. Monthly data for individual years are used directly in the drought analysis (Section 2), in some of the wind erosion analyses (Section 3), and in some of the wheat-yield analyses (Section 4), while 1951–1980 climatic normals (HIST1) are employed with other analyses. The GCM results, consisting of a temperature and a precipitation value for each month and location for the equilibrium climate (GISSE) and for $2 \times CO_2$ (GISSC), were used directly in some preliminary analyses, including a set of computations for growing degree-days. For all subsequent analyses, the data employed in the models were obtained from historical instrumental observations, either unadjusted (HIST scenarios), or adjusted by data from the GCM results (GISS scenarios).

For the purposes of this case study, all the agriculturally relevant changes we have computed are considered as "impacts" (.__. outline, *Figure 1.12*). These include changes in heat and moisture conditions for crops (growing degree-days, precipitation effectiveness, drought frequency), in the potential for environmental degradation (wind erosion), in the potential dry matter production (biomass), in yields of a specific crop (spring wheat) and in the farm and provincial economy (economic impacts).

Since policy makers will require information on the types of agricultural impacts likely to result from future climatic change in an area such as Saskatchewan, this case study demonstrates how such information can be generated for each of several important types of impact. Climatic scenarios considered include those based on past extremes, the 1961 drought and the 1929–1938 decade, and those simulated for a $2 \times CO_2$ atmosphere by using GISS GCM results to adjust instrumentally observed climatic data. In Section 2 we consider impacts on the agroclimatic environment (thermal and moisture effects, a potential biomass index which combines these, and a drought index used to analyze moisture variability), in Section 3 the soil environment (wind erosion potential), in Section 4 spring wheat yields, and in Section 5 the economic implications. Section 6 comprises the case study conclusions and recommendations.

SECTION 2

The Effects on the Agroclimatic Environment

In Section 1 several scenarios were selected that were based on instrumentally observed historical climatic data (HIST) and on results from a general circulation model (GISS). In this section we employ indices or models to analyze the impacts associated with these scenarios on the thermal climate (growing degree-days), the moisture climate (precipitation effectiveness), the combined thermal–moisture effects (climatic index of agricultural potential which reflects biomass potential), and on drought frequency.

This section of the case study differs from Sections 3, 4 and 5 in that it does not examine specific processes (e.g., wind erosion), crops (e.g., spring wheat), or the impacts on the economy. Instead it considers more general effects on the agroclimatic environment. In a changed climate, crops and processes may be altered so much that results of applying a particular model may be misleading when considered by themselves. Section 2 not only provides a useful prelude to the subsequent parts by describing the changes in the agroclimatic environment, but it also provides some general indications, for instance, with respect to biomass productivity and drought frequency, that are useful in the interpretation of the more specific results.

The GISS model results provided were on a grid point basis. The following are the latitudes and longitudes of the six points used to represent southern Saskatchewan in this part of the study: 50°N 100°W, 50°N 105°W, 50°N 110°W, 54°N 100°W, 54°N 105°W, 54°N 110°W (*see Figure 1.11*). The temperature differences for obtaining the $2 \times CO_2$ scenario temperatures for stations in southern Saskatchewan were obtained by interpolating to the station locations from the values for these six grid points, using a Gaussian filter procedure (as described in Part I, Section 3). The same method was used to obtain the ratios for adjusting precipitation to the $2 \times CO_2$ scenarios. For Uranium City, the adjustment increments and ratios were obtained by manual interpolation from nearby grid points.

2.1. Effects on Growing Degree-Days

Growing degree-days (GDD) have long been in use in Canada for characterizing the thermal climate for crops. GDD have sometimes been referred to as "day-degrees", or "accumulated temperatures" (Unstead, 1912). Other terms that have been applied include "heat units" and "effective growth-heat"; and the quantity required for a crop to reach maturity has been called the "summation constant" or the "remainder index" (Holmes and Robertson, 1959). The term "growing degree-days", which has been commonly used in recent work such as that by Edey (1977), will be employed in this case study. Elsewhere in this volume, however, the measure is termed "effective temperature sum".

Growing degree-days are the accumulation of temperature above some threshold or base temperature, b. For $b = 5\,°C$, on a day with mean temperature of 19°C there are $t - b = 14$ degree-days above b, where t = the mean temperature for the day = (maximum + minimum for day)/2. To compute GDD for a month, these excesses above the base are summed: that is, GDD for the month = the sum of $(t - b)$ for every day in the month for which t is greater than b. Each plant species would have a different threshold, but 5°C is widely used for such calculations (and is quite close to the 42°F base most often used for growing degree calculations in the United States).

The longer the growing season and the higher the temperatures, the greater the annual GDD. Thus GDD provides an indication of the combined effects of growing season length and temperature (Chapman and Brown, 1978). Degree-days are ordinarily computed each day and accumulated through the season. Computation of long-term degree-day normals can be performed with the daily data, but if one wants to compute monthly normals of degree-days from monthly temperature normals, some adjustment is desirable to obtain comparable values, particularly for those months where the base temperature is close to the monthly mean. For example, a monthly mean just below the base would give zero degree-days without such adjustment, but such a month would normally include days where the mean temperature exceeded the base temperature, thus contributing to the degree-day summation. The method of Thom (1954a,b) for computing degree-days from monthly mean temperature normals and standard deviations of monthly mean temperatures overcomes this problem. This method was therefore used with both the simulated $2 \times CO_2$ (GISS) and instrumentally based (HIST) scenarios. Holmes and Robertson (1959) have demonstrated the application of Thom's methods to growing degree-days. Standard deviations were derived from those published for stations in the area for 1951–1980 (Environment Canada, 1982).

It should be noted that the use of standard deviations for 1951–1980 mean monthly temperatures in the degree-day computations implies that we are asking the question: What is the impact on the thermal climate of the scenario temperature changes if the temperature variability remains unchanged? It seems likely that the sensitivity of annual GDD to changes in the standard deviation would be quite low, particularly for assessments of the effects of a warmer climate. In future research this sensitivity could be tested, and scenarios of changed variability could also be assessed.

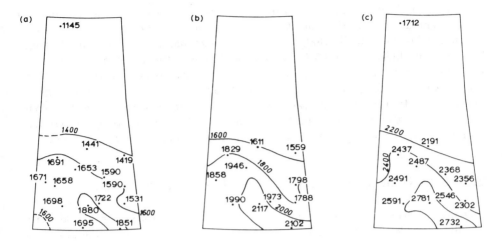

Figure 2.1. Growing degree-days for (a) 1951–1980 (HIST1), (b) 1960–1961 (HIST2) and (c) 2 × CO_2 (GISS1). (Data for HIST3 were insufficient for mapping but are included in *Table 2.5*.)

In preliminary analyses, growing degree-days were computed using data for the 1 × CO_2 and 2 × CO_2 GCM equilibrium climates, GISSE and GISSC, for the 15 Saskatchewan area grid points formed by the intersections of the 46°, 50°, 54°, 58° and 62°N latitude and 100°, 105° and 110°W longitude lines. It was found that degree-days above 5°C were about 80–95% higher for GISSC than for GISSE, without any very significant spatial pattern except that the increases along the 100°W meridian (to the east of the province) were generally lower, in the 70–80% increase range, than those along the 105°W and 110°W meridians.

When the calculations were made for the 1951–1980 climatic normals (HIST1), and the method of adding temperature increments to normals for simulating a 2 × CO_2 climate (GISS1) was used, degree-day computation results indicated an increase of about 50% in growing degree-days with a doubling of CO_2 (*Figure 2.1*). In view of the large difference between the present climate of the region as indicated by HIST1, and the GISSE climate, as discussed in Section 1, the difference in degree-day impacts between those derived using the GISSE and GISSC climates (more than 80% degree-day increase) and those based on HIST1 and GISS1 (about 50% increase in degree-days) is not surprising.

The extreme year scenario (HIST2) (*Figure 2.1*) indicated growing degree-day totals between 10% and 20% higher than normal, and the scenario for the extreme decade (HIST3) indicated summations 3–16% above normal (cf. *Table 2.5*). On the basis of these calculations in comparison with the increases of about 50% for GISS1, it would seem that the growing season warmth associated with a doubled CO_2 atmosphere would be much greater than that of extreme years or decades in the recent past. In fact, the estimates suggest that for the growing season, the increase in thermal resources due to CO_2 doubling would be at least three times larger than that experienced during recent extreme periods.

The implications of this warming can be illustrated by the example for maize. The annual GDD value at Windsor, in an important maize-growing area of Ontario, is 2533, while that at Morden, Manitoba, at the cold climate fringe for maize production, is 1908 (Environment Canada, 1982, Vol. 4). If the growing season thermal resources on a long-term basis increased to the levels indicated by the GDD results for HIST2 and HIST3, it would be warm enough for maize production in some central and southeastern parts of Saskatchewan. With the much greater warming implied in the GISS1 scenario, thermal resources would be suitable for maize throughout practically all of southern Saskatchewan.

2.2. Effects on the Precipitation Effectiveness Index

The precipitation effectiveness index of Thornthwaite (1931) can be calculated readily from long-term climatic normals of monthly mean temperature and monthly precipitation. It is useful for characterizing spatial patterns of the moisture regime for mapping the climate of a continent (Thornthwaite, 1931), and also for other applications requiring a simple moisture index, such as the assessment of the potential of climate for wind erosion of soil (Lyles, 1983).

The precipitation effectiveness index, I, is calculated as follows from the data for all months:

$$I = \sum_{i=1}^{12} 115 \left[(P_i/25.4)/(1.8 T_i + 22)\right]^{1.11} \qquad (2.1)$$

where P is normal monthly precipitation (mm), T is normal monthly mean temperature (°C), and i indicates months.

The calculations are made for each month and the results added to obtain I, as indicated by the equation. Where monthly precipitation is less than 12.7 mm, P is set to 12.7, and where the monthly temperature is less than -1.7 °C, T is set to -1.7 °C. The temperature threshold reflects the fact that at temperatures much below freezing there is very little evaporation; the precipitation restriction is needed in connection with wind erosion applications (Section 3), where I is in the denominator in a formula and should not be allowed to approach zero, and because wind erosion is probably not sensitive to monthly precipitation totals between 0 and 12.7 mm (Lyles, 1983). For the purposes of this case study we have also defined a monthly precipitation effectiveness index, I_m, to be the value of the index for a single month, i as indicated by the quantity to the right of the summation sign. I_m is used for some of the analyses in Section 3.

Examination of the precipitation effectiveness normals (*Figure 2.2*) suggests a dry belt running across the province from the Kindersley-Bad Lake area, which has values at or near the top of the semiarid range (28 and 31), through Regina and Moose Jaw (35) to Broadview in the east (36). This agrees fairly well with spatial moisture patterns obtained in other agroclimatic analyses (Agriculture Canada, 1976, 1977).

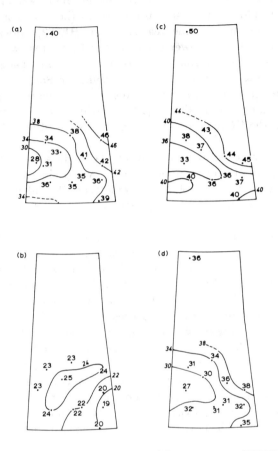

Figure 2.2. Precipitation effectiveness I for (a) 1951–1980 (HIST1), (b) 1960–1961 (HIST2), (c) $2 \times CO_2 TP$ (GISS1) and (d) $2 \times CO_2 T$ (GISS2). (HIST3 data are included in Table 2.5.) Thornthwaite (1931) defined vegetation classes as follows:

A   (wet) Rain forest, $I = 128$ or greater.
B   (humid) Forest, $I = 64$–127.
C   (subhumid) Grassland, $I = 32$–63.
D   (semiarid) Steppe, $I = 16$–31.
E   (arid) Desert, $I =$ less than 16.

Computed precipitation effectiveness values with the GISS1 scenario exceeded those for HIST1 (the 1951–1980 normal) by 1–13% in southern Saskatchewan and by over 25% at Uranium City in the north (Figure 2.2). For GISS2 the index values were 10–12% below the HIST1 values. This suggests that the increased moisture stress resulting from warming as implied in the GISS1 scenario for $2 \times CO_2$ would be more than offset by the increased precipitation, so that the long-term climatic moisture regime would be somewhat improved. Without the increased precipitation, the warming associated with

CO_2 doubling would apparently lead to a significantly drier climate. Under these conditions, the dry belt from the Kindersley-Bad Lake area east to Regina and Moose Jaw would be in the semiarid category (i.e., I less than 32).

The results for the extreme year scenario (HIST2) indicated precipitation effectiveness values 18–53% lower than the HIST1 results, with 50% reductions typical of eastern parts of southern Saskatchewan and general reductions of 30–40% further west. In such a year, all of southern Saskatchewan would be in the semiarid category (*Figure 2.2*). This is in general agreement with the findings of Richards and Fung (1969), who showed most of southern Saskatchewan in the Köppen dry climate category in 1961, and some small areas as very dry.

For the five stations with readily available data for the extreme decade analysis, HIST3, the precipitation effectiveness was reduced by 21–27% from the HIST1 values. As in the case of the extreme year, the calculations suggest that in such an extreme decade all of southern Saskatchewan would be in the semiarid category.

A comparison of the results suggests that with a future climate like that of the past half century, the precipitation effectiveness index could be expected to be 30–50% below normal in an extreme year and 25% below in an extreme decade. With CO_2-related climatic warming, the values for the climate over the much longer term would be about 10% below the recent normals (GISS2), while with both the higher temperatures and precipitation indicated by GISS for CO_2 doubling, the moisture regime would be improved somewhat, as indicated by the higher precipitation effectiveness values in the GISS1 results.

It may be noted that while the magnitude of the estimated changes in thermal climate as indicated by growing degree-days was much less for the scenarios based on historical data than for CO_2 doubling, the reverse was true for moisture. This is perhaps to be expected since the extreme year and decade were selected on the basis of effects that were largely moisture-related (Section 1).

2.3. Effects on the Climatic Index of Agricultural Potential

The climatic index of agricultural potential (Turc and Lecerf, 1972), CA, reflects both thermal and moisture resources and is computed by summing the 12 monthly products of heliothermic and moisture factors as follows:

$$CA = \sum_{i=1}^{12} HT_i * Fs_i \tag{2.2}$$

where HT is a heliothermic factor reflecting both temperature and a solar factor ($HT = Ft \cdot Fh$), Fs is a soil moisture factor computed using precipitation and temperature in a climatic soil moisture budgeting procedure, and i indicates months.

Fs is 0 for very dry conditions and increases to 1.0 for plentiful moisture; excesses of moisture are ignored. HT has only a lower limit; it is 0 where the mean daily minimum temperature for the month is 0 °C or lower, and at higher temperatures it increases with increasing temperature. Thus CA reflects the length of growing season and the amount of heat and moisture available to plants during the season. It can be taken as indicative not only of the agricultural potential of the climate but also of the potential for biomass productivity in general, including forest productivity (Turc and Lecerf, 1972).

The term biomass productivity is used here as equivalent to total harvestable dry matter, which would include the weight, after drying, of all plant material from a crop. Turc and Lecerf (1972) suggest that each unit of CA is equivalent to a dry matter production of about 0.6 t/ha. For example, in an area of southern Sweden where CA = 21, total dry matter production for maize is 12 t/ha, or 0.57 t/ha per unit of CA. Because it does not relate to a specific crop but to agricultural productivity in general, CA is useful in analyzing the impacts of climatic changes that are of sufficient magnitude that crop selections are likely to change. Such temporal climatic changes may be seen as analogous to moving on the earth to a region with a different climate − and as Turc and Lecerf (1972) point out − different regions would require different crops to exploit the agricultural potential. For the purposes of this study we use CA as a relative index only, rather than trying to interpret results using some conversion factor. In addition to the annual CA values, annual sums of Fs and HT were also computed.

Climatological data for the available stations were used to compute CA for the various scenarios. These data included monthly averages of daily mean temperature and of daily minimum temperature, and monthly total precipitation. Average monthly global solar radiation simulated by computer models for 1967–1976 (McKay and Morris, 1985) and monthly averages of day length for the nearest whole degree of latitude were also used. The mean temperature and mean daily minimum temperatures (Tmean, Tmin) were adjusted for the GISS scenarios as explained previously (*see Table 1.3*). Solar radiation data were not adjusted, which implies that the questions being asked in this analysis are of the form: What is the impact on potential biomass productivity of the scenario temperature and precipitation changes if the solar radiation remains unchanged? The mathematical basis of CA is discussed further in *Appendix 2.1*, and the computations have been explained in detail previously (Williams, 1985).

In investigating the impact of any scenario climate, one can examine the effect on annual CA (*Figure 2.3*), which reflects the combined effects of the changing thermal situation and altered moisture stress due to temperature changes together with the effects of the changed precipitation. Also one can examine the effects on a month-by-month basis for any location.

Simulated CO_2 warming, accompanied by increased precipitation (GISS1 scenario), markedly increased the heliothermic resources as indicated by HT (*Figure 2.4*), while for moisture (Fs) there was some improvement at over half the stations and deterioration at others. The balancing of these effects resulted in an increase of potential biomass production (as expressed by CA), ranging

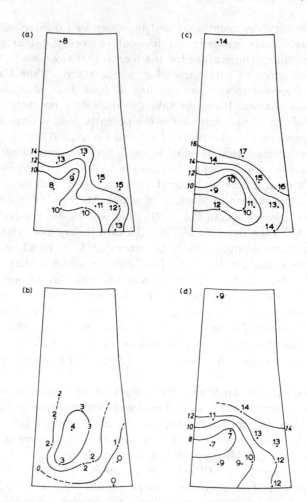

Figure 2.3. Climatic index of agricultural potential (CA) for (*a*) 1951–1980 HIST1, (*b*) 1960–1961 HIST2, (*c*) GISS1, and (*d*) GISS2. (HIST3 data are included in *Table 2.5.*)

from only slight at some southern Saskatchewan locations, up to a 30% increase at Prince Albert (*Figure 2.4*), and a 74% increase at Uranium City in the north.

With simulated CO_2 warming without the increased precipitation (GISS2 scenario), CA still increased 17% at Uranium City, reflecting the advantage of warming at a cold, northern location; there was also a 3% increase at Prince Albert; but generally in southern Saskatchewan, the calculated increase in moisture stress more than balanced the heliothermic benefits and CA was reduced by between 1% and 19% (*Figures 2.3 and 2.4*).

For the extreme year scenario (HIST2), estimated biomass productivity in southern Saskatchewan as indicated by CA ranged from nil or a few percent of normal (HIST1) at eastern locations to 47% of HIST1 at Saskatoon (*Figure 2.3*). For the extreme decade (HIST3), values around half the normal were typical.

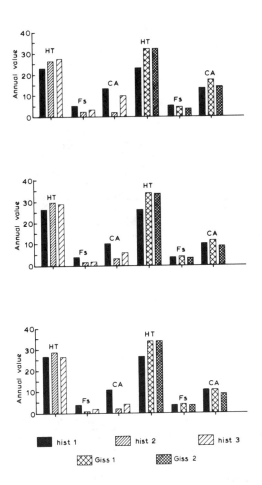

Figure 2.4. Impact on CA (biomass potential) and moisture (Fs) and heliothermic (HT) components, of (*a*) 1961 (HIST2), (*b*) 1929–1938 (HIST3), (*c*) $2 \times CO_2TP$ (GISS1), and (*d*) $2 \times CO_2T$ (GISS2) in comparison with HIST1 for Prince Albert (top), Swift Current (middle), and Regina (bottom).

The results suggest that with no fundamental change in the climate, occasional years can be expected with climatic conditions that can support only a quarter of normal biomass productivity in southern Saskatchewan generally, and virtually nil productivity at some locations. Periods as long as 10 years with only half the normal productivity can also be expected. Climatic warming without increased precipitation (GISS2) could be expected to result in long-term conditions that could support only about 90% of the productivity that the present climatic resources (HIST1) would support. The climate resulting from such warming accompanied by increased precipitation (GISS1) could be expected to support productivity averaging perhaps 7% or 8% higher than the HIST1 levels.

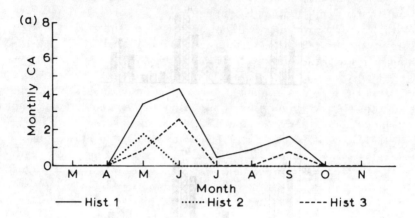

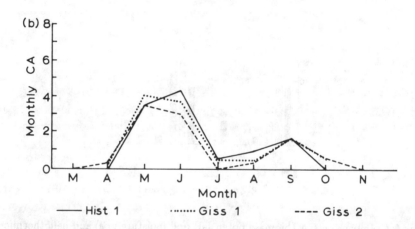

Figure 2.5. Distribution of CA on a month-by-month basis for Regina for (*a*) 1951–1980 (HIST1), 1960–1961 (HIST2), and 1929–1938 (HIST3), and (*b*) 1951–1980 (HIST1), 2 × CO_2TP (GISS1) and 2 × CO_2T (GISS2). (November–March period not shown as CA = 0 for those months.)

What is particularly interesting is the change in character of the growing season. This can be illustrated by examination of the month-by-month results for Regina. At Regina, the normal (HIST1) percentage contribution to annual CA is greatest in June (40%) as a result of the combination of Fs at a fairly high value and the second highest monthly HT value. There is a secondary peak of 15% for September (*Figure 2.5*). The results indicated that with a simulated 2 × CO_2 climate (GISS1), the September peak would be retained but the early summer peak would become earlier as June Fs decreased. From early June for HIST1, the peak would move back to late May for GISS1, and there would be a small April contribution. The secondary, late-season peak of monthly CA tended to come somewhat later with GISS1, with the September peak reduced and a significant October contribution developing.

With $2 \times CO_2$ warming without increased precipitation (GISS2), some increases in the September contribution were indicated at Regina, and the estimated monthly CA for July was reduced to zero.

For the extreme decade scenario (HIST3), June CA at Regina was 60% of normal (HIST1), there was a much greater reduction in the May and September values, and the CA contributions of all other months were eliminated completely, so that June accounted for the major part of the annual CA. With the scenario based on the August 1960 to July 1961 period (HIST2), the only contribution to CA at Regina was a much-below-normal May 1961 value. In the other 11 months the CA contribution was nil. CA was zero for July and August with HIST3 and for July with GISS2. The July and August CA values were also lower for GISS1 than for HIST1. In fact, the results for all the climatic change scenarios considered here implied reduced biomass productivity in July and August at Regina.

The lengthening of the season and relative strengthening of the earliest part for the GISS scenarios was fairly general for most locations. This indication of a tendency toward two distinct growing seasons, one in spring/early summer, the other in early fall, separated by a dry midsummer period, was much accentuated if simulated CO_2 warming was not accompanied by increased precipitation. In that case the absence of a precipitation increase markedly weakened the computed July biomass potential. The estimates for GISS2 indicated July values of zero at Swift Current as well as Regina, and much reduced July values at Saskatoon.

At all the locations analyzed, the May–June total contribution to CA was less for the extreme decade scenario (HIST3) than normal (HIST1). For HIST3, as in the case of the GISS scenarios, the July/August contribution to CA was reduced in comparison with the HIST1 results. The HIST3 results, however, in contrast to the GISS scenario results, did not suggest any tendency for the May/June peak to shift to an earlier date. In fact, the tendency was the opposite, with the CA contribution becoming increasingly concentrated in June and less in May.

At Prince Albert, representing the northern part of Saskatchewan's agricultural area, there was no midsummer minimum for CA with HIST1; instead, CA became smaller from month to month from the peak in June through to September, becoming zero in October, and the September value was considerably smaller than the August value. In contrast, the HIST2, HIST3 and GISS scenarios all resulted in larger CA values in September than in August for Prince Albert, with a distinct August or July/August minimum. At Uranium City, representing the northern part of the province, there were CA peaks in June and August with the HIST1 scenario. Data were not available for the historically based scenarios at Uranium City, but with the GISS scenarios there was the same tendency as in the south for the early summer CA peak to shift to an earlier date. In this case, May became the peak CA month for GISS1 and GISS2, and there was a June minimum for CA, especially with GISS2. August remained the late season CA peak month at Uranium City with the GISS scenarios, but there was a considerable increase in September CA.

To recapitulate the effects on the seasonal distribution of CA: a common tendency among GISS1, GISS2 and HIST3 scenarios was toward more distinctly separated early and late season peaks. For HIST3, this tendency was expressed by low CA values in July or August or both. For the GISS scenarios, in addition to this tendency to reduced CA values in July and August (or in June at Uranium City), there was also a shift of the early season peak to an earlier date and some shift toward a later date for the late season peak. In the case of the extreme year scenario (HIST2) the CA contribution became quite undependable with nil values in most months.

2.4. Effects on Drought Frequency Using the Palmer Drought Index

This section addresses the question of assessing how the frequency and duration of droughts might be expected to alter as a result of a doubling of atmospheric CO_2.

2.4.1. Methods

Some measure of the increased soil moisture stress as it would vary from year to year can be obtained by applying the $2 \times CO_2$ temperature increments and precipitation ratios to past data to find the change in moisture conditions. The change in drought frequency and duration can be measured using a drought index and a water balance model. In this study, the Palmer Drought Index (Palmer, 1965) was chosen to measure the frequency and duration of droughts. Palmer designed the index to have internal consistency and temporal and geographic comparability, which makes it ideal for research on climatic change and fluctuation (Whittmore et al., 1982). He defined drought as an interval of time, generally of the order of months, during which the soil moisture supply at a given place consistently falls short of the climatically expected, or climatically appropriate, moisture supply. The severity of drought is a function of both duration and magnitude.

Since the Palmer Drought Index employs only meteorological data for its input parameters, it lends itself to use in the analysis of the changes in drought frequency and severity for various CO_2 scenarios. Precipitation data, and potential evapotranspiration estimated using temperature data, are used in computing the index. Thus, the model may be used to simulate impacts of specified changes in the temperature and precipitation climate.

Palmer Drought Index (PDI) coefficients were calculated over the 1950–1982 period (HIST5). The actual calculation of these coefficients is described in somewhat more detail in *Appendix 2.2*.

Table 2.1. Temperature increases (°C) and precipitation ratios used in developing 2 × CO_2 scenarios at lat. 50° N long. 105° W.

Month	Temperature increase (Tadj)	Precipitation ratio (Pcpadj)
JAN	6.0	1.29
FEB	5.6	1.31
MAR	4.8	1.26
APR	4.1	1.17
MAY	3.7	1.15
JUNE	3.3	1.15
JULY	3.3	1.12
AUG	3.8	1.05
SEPT	4.7	0.98
OCT	5.3	1.09
NOV	5.7	1.27
DEC	6.0	1.29

The coefficients derived from the historic period (HIST5) represent the climate appropriate for this time. Each Palmer Drought Index value represents a departure from the long-term average of the historic data set. The PDI values for the 1950–1982 period are essentially Z-scores with a mean of zero and a standard deviation of one.

It is important to realize that these coefficients are used to find the departures of Z-scores from the historic normals. The coefficients should not be recalculated for the 2 × CO_2 scenario (GISS3 or GISS4) set of data. If they were, the resultant Z-scores would then measure the departure of moisture conditions from these normalized 2 × CO_2 scenarios. This would, in effect, be using the GISS results as the historic baseline period against which to compare those same GISS scenario results, which would be meaningless. For this analysis, HIST5 is the historic baseline period, and the coefficients for assessing the effects associated with GISS3 or GISS4 must be those computed from the HIST5 data. This then gives a Z-value which is a departure from the historic normals, and it allows the Z-values to be used in estimating the impacts associated with the GISS scenarios.

As in other parts of this case study, differences between 2 × CO_2 and 1 × CO_2 equilibrium climates from the GISS model were used to derive the monthly increments for temperature. These temperature increments were then added to the historical mean temperatures for each month throughout the 1950–1982 period (HIST5).

Table 2.1 illustrates the temperature differences and precipitation ratios used for stations in Saskatchewan in the PDI analysis. The differences in increments or ratios among different locations in Saskatchewan are quite small compared with the variations among months, therefore, only one location is shown in *Table 2.1*, but it is approximately representative of all 11 stations used in the PDI analysis.

The simulated climate data for 2 × CO_2 for each month in the 1950–1982 period were then used as input for the PDI model. The two CO_2-doubling scenarios were employed, one using the increased temperature and precipitation as indicated by the GISS model (GISS3), the other using only the temperature increase (GISS4). Palmer Drought Index values for the GISS3 and GISS4 scenarios were compared with those that were computed from the 1950–1982 climatic data (HIST5).

2.4.2. Results of PDI frequency analysis

Various Palmer Drought Index values can be characterized, in terms of the deviation from normal climate, as follows:

Greater than +6	severe wet spell
+4 to +6	extremely wet spell
+2 to +4	wet spell
+2 to −2	near normal
−2 to −4	dry spell
−4 to −6	drought
less than −6	severe drought

A value of −4 to −6 over a period of several months generally is reflected in lower crop yields and water supply problems. A severe drought (PDI less than −6) has serious economic consequences because of water shortages and crop failures due to drought.

The results indicate, for all the stations analyzed, a definite shift in the water balance, to a more drought-prone regime under the GISS $2 \times CO_2$ scenarios (*Table 2.2*). This shift appears to take place at the expense of the midrange (near-zero) frequencies rather than of the wet period frequencies. Positive and negative PDI values were fairly equally distributed for HIST5, but with the GISS3 scenario negative PDI values occurred 61.2% of the time for southern Saskatchewan. (An average value of the PDI-class percentages for the 11 stations given in *Table 2.2* was assumed to be representative for the southern Saskatchewan region.) The occurrence of severe drought months increased from 0.1% of the months to 0.9%. Drought months (below −4) increased in frequency from 3.0% to 9.1%. It appears that during the relatively dry periods, the increase in precipitation indicated for this region by the GISS $2 \times CO_2$ scenario would not be enough to offset the increased evapotranspiration caused by the increase in temperature.

Despite the increase in drought frequency, however, there is no general decrease in the estimated frequency of extreme or severe wet spells (above +4) with $2 \times CO_2$ (*Table 2.2* and *Figure 2.6*). The increase in evapotranspiration computed for the GISS3 scenario, when accompanied by the increased precipitation of that scenario, would not have reduced soil moisture during past periods of plentiful precipitation, and when the GISS3 scenario adjustments were applied to the data for the years 1950–1982, the wet spells were maintained. In fact, at some stations, the frequency of PDI values over +4 and +6 actually increased (e.g., Regina, Kindersley).

Table 2.2. Frequency of PDI values (in %) above or below index values under 1950–1982 (HIST5) and doubled CO_2 (GISS3 and GISS4) scenarios.

		PDI value							
Station	Scenario	Below −6	Below −4	Below −2	Below 0	Above 0	Above 2	Above 4	Above 6
Yorkton	HIST5	0.3	3.5	23.2	50.8	49.2	17.7	8.6	0.3
	GISS3	1.5	10.4	39.4	59.1	40.9	14.9	5.6	0.5
	GISS4	20.5	51.8	79.3	86.9	13.1	0.5	0.0	0.0
Kindersley	HIST5	0.0	1.3	23.5	52.8	47.2	21.1	4.3	0.3
	GISS3	0.0	8.1	33.1	60.9	39.1	17.9	4.5	1.3
	GISS4	4.5	32.3	62.4	85.1	14.9	4.3	0.3	0.0
Swift Current	HIST5	0.0	2.3	22.2	56.1	43.9	24.5	4.8	0.8
	GISS3	0.5	6.8	34.8	60.6	39.4	23.0	5.6	0.8
	GISS4	6.3	37.9	63.6	81.8	18.2	3.3	0.0	0.0
Moose Jaw	HIST5	0.3	4.8	20.7	55.8	44.2	19.2	7.6	0.8
	GISS3	0.3	7.1	27.5	62.1	37.9	18.9	7.8	0.3
	GISS4	6.8	29.8	66.2	82.3	17.7	2.5	0.0	0.0
Regina	HIST5	0.0	3.8	29.0	57.8	42.2	23.5	10.6	4.6
	GISS3	0.8	10.6	40.4	62.9	37.1	22.2	11.9	4.8
	GISS4	6.6	39.4	64.6	79.8	20.2	10.1	0.5	0.0
Prince Albert	HIST5	0.0	1.3	22.0	48.2	51.8	20.7	11.1	3.0
	GISS3	0.5	9.6	29.5	59.1	40.9	19.7	11.6	2.8
	GISS4	9.1	36.4	70.7	83.1	16.9	5.6	0.0	0.0
North Battleford	HIST5	0.0	1.0	17.4	59.9	40.1	15.4	9.3	2.5
	GISS3	0.5	3.5	28.8	64.4	35.6	15.7	8.1	2.8
	GISS4	4.3	37.6	77.0	86.9	13.1	5.3	0.0	0.0
Saskatoon	HIST5	0.0	0.8	23.7	54.6	45.4	22.0	4.8	0.0
	GISS3	0.0	5.6	29.5	59.1	40.9	17.9	6.8	0.0
	GISS4	4.5	31.1	66.4	87.9	12.1	3.5	0.0	0.0
Hudson Bay	HIST5	0.0	3.8	19.4	49.5	50.5	19.4	4.3	0.0
	GISS3	1.5	11.1	37.1	66.2	33.8	15.2	5.1	0.0
	GISS4	29.3	61.6	82.1	93.2	6.8	0.0	0.0	0.0
Broadview	HIST5	0.3	6.1	24.8	56.1	43.9	23.7	11.6	5.3
	GISS3	2.5	14.9	38.6	64.6	35.4	19.9	9.6	5.8
	GISS4	15.2	45.2	69.9	81.8	18.2	8.1	4.0	0.5
Estevan	HIST5	0.3	6.6	24.8	46.0	54.0	23.7	4.0	0.5
	GISS3	1.8	12.4	30.8	54.0	46.0	20.2	3.8	0.5
	GISS4	11.1	32.3	62.1	88.6	11.4	4.5	0.0	0.0
Area mean	HIST5	0.1	3.0	22.8	52.9	47.1	21.0	7.4	1.6
	GISS3	0.9	9.1	33.6	61.2	38.8	18.7	7.3	1.8
	GISS4	10.8	39.6	69.5	85.2	14.8	4.3	0.4	0.05

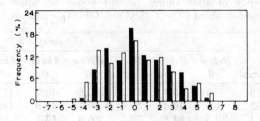
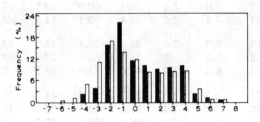
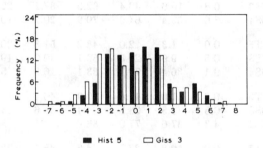

Figure 2.6. Percent drought frequency for Saskatoon (top), Swift Current (middle) and Yorkton (bottom) for 1950–1982 (HIST5) and $2 \times CO_2$ TP (GISS3). The PDI numbers shown are upper limits for classes indicated, i.e., the -8 to -7 class is marked -7, and the 5 to 6 class is marked 6.

The results indicate that CO_2 doubling as simulated by the GISS model would lead to a more drought-prone climate (due to increased evapotranspiration) but one that maintains the variability of extreme dry and wet spells.

Ordinarily southern Saskatchewan has a net soil moisture loss for the months May to September due to evapotranspiration exceeding precipitation. This moisture deficit is increased considerably with warmer temperatures. Not only does more evapotranspiration occur because of the higher temperatures, but the evapotranspiration season is extended. The PDI analyses for GISS3 suggests that the increase both in the intensity of evaporation and in its duration for $2 \times CO_2$ would more than offset the increase in precipitation for the area. This result could be a reflection of some characteristics of the procedures for calculating potential evapotranspiration in the PDI analyses and therefore must be

Table 2.3. Longest consecutive period of PDI value below -4 over 1950–1982 (HIST5) and expected length under $2 \times CO_2$ TP (GISS3) scenario.

	1950–1982 (HIST5)			$2 \times CO_2$ (GISS3)		
Station	Starting month	Year	Length (months)	Starting month	Year	Length (months)
Yorkton	June	1961	7	June	1961	8
Kindersley	May	1980	3	May	1980	3
Swift Current	May	1980	5	May	1980	6
Moose Jaw	May	1980–81	17	May	1980–81	20
Regina	May	1981	6	June	1980–81	18
Prince Albert	June	1964	3	May	1964	9
North Battleford	July	1959	2	March	1959	7
Saskatoon	August	1981	1	May	1981	12
Hudson Bay	July	1963–64	12	June	1962–64	22
Broadview	July	1961	10	June	1961	11
Estevan	June	1958	9	May	1958	13
Mean			6.8			11.7

treated with caution. It does, however, reveal how sensitive southern Saskatchewan is to temperature. The historic drought years of 1934, 1961 and 1981 all had mean annual temperatures above the station normals. Regina, for instance, has a mean annual temperature of 1.8 °C for its period of record. The mean annual temperature for 1934 was 3.5, for 1961 it was 3.2 and for 1981 it was 4.7 °C. This higher-than-normal mean annual temperature in drought years is also reflected in the records for other southern Saskatchewan stations.

The length of the longest consecutive period of months below -4 for each station for 1950–1982 (HIST5), and the effects of the change to GISS3, were examined (*Table 2.3*). In some cases a significant number of drought months is added to the length of the drought period by the change from the HIST5 to GISS3, while in other cases the length of the drought period is either not increased, or is increased in length by only one month. The major part of the increase in drought-month frequency does not occur, however, in isolated cases, but tends to be before or after a significant drought period. Thus it appears that droughts under a GISS3 scenario would be longer and more severe than with HIST5 despite the increased precipitation.

The results of using the GISS increase in temperature without the increase in precipitation (GISS4) yielded substantial increases in drought frequency. The 11 station average of frequencies of severe drought (PDI less than -6) increased from 0.1% to 10.8%. For drought, PDI less than -4, there was an increase from 3.0% to 39.6% (*Table 2.2*). Large increases in severe drought frequencies were indicated in the easternmost agricultural areas of the province where there were increases ranging from 11% to 29% for PDI less than -6. Areas in the west did not show nearly as large an increase in drought frequency because they are already quite dry under normal climate conditions.

Table 2.4. Theoretical (Gumbel) return periods (years) for PDI values of −4, −6 and the lowest value in 33 years for the HIST5 (1950–82) and GISS3 (2 × CO_2 TP) scenarios.

Station	33 year lowest PDI				−6 (severe drought)		−4 (drought)	
	Lowest value	Year	HIST5	GISS3	HIST5	GISS3	HIST5	GISS3
Yorkton	−6.28	1961	22.0	11.5	19.0	10.0	7.0	4.0
Kindersley	−4.22	1958	11.5	6.5	35.0	17.0	10.0	5.6
Swift Current	−4.93	1980	14.5	8.2	28.0	15.0	8.6	5.3
Moose Jaw	−6.10	1981	25.0	14.5	23.5	13.5	8.0	5.0
Regina	−5.08	1981	11.0	7.0	27.0	10.0	6.6	4.2
Prince Albert	−5.88	1964	23.0	13.0	24.0	13.5	8.5	5.3
North Battleford	−5.46	1959	22.5	12.5	30.0	16.0	9.5	5.7
Saskatoon	−4.40	1980	11.5	7.5	31.0	17.5	9.5	6.0
Hudson Bay	−5.63	1964	26.0	11.0	34.0	13.5	10.0	4.4
Broadview	−6.04	1961	15.5	8.6	15.0	8.5	6.5	4.1
Estevan	−6.18	1958	21.5	12.5	19.0	11.2	6.8	4.4

It appears that at present during the summer, western parts of the region lose most of their soil moisture, so that with the GISS4 climate little more evaporation can take place due to the lack of available soil moisture in summer. Eastern areas are not nearly as dry under normal conditions and therefore lose much more water from evaporation under the increased temperatures with 2 × CO_2.

2.4.3. Return periods

The estimation of extreme values is important for agricultural planning. If the frequency of drought conditions increases significantly with a warmer climate, then the existence of farming becomes threatened due to increased economic stress.

For the calculation of return periods, it is necessary that the data be independent of each other. PDI monthly values are, of course, by design, not independent since the PDI was constructed so as to take into account the length of the drought up to the month under consideration.

While monthly PDI values are not independent, yearly extreme values show no obvious periodicity, and, in fact, appear to be stochastic (Ramachandra and Padmanabhan, 1984). Return periods were calculated using Gumbel distributions, as explained by Kendall (1959). The return periods for PDI values of −4 and −6 were calculated (*Table 2.4*).

In interpreting them it should be kept in mind that the return periods are based on the lowest PDI value for each year, so a value of −4 for a year may only reflect a single isolated drought month. A value of −6, however, is much

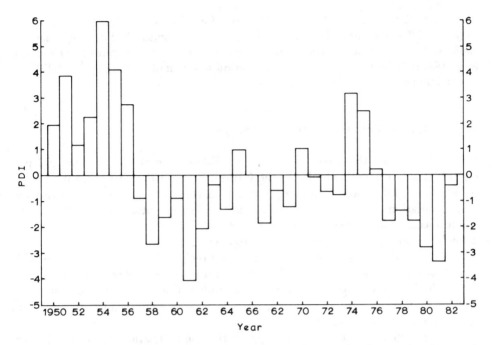

Figure 2.7. Average PDI values for 11 stations in southern Saskatchewan for the month of October.

more indicative of drought occurrence, as −6 cannot be reached without at least a couple of months less than −4.

Historically, the northern agricultural areas have suffered the least from a drastic water reduction. Stations representative of these areas (Saskatoon, Hudson Bay, North Battleford) all show a return period of 30 years or greater for PDI = −6. Southeastern sections, on the other hand, have been most prone to drastic changes in water supply, with Broadview, Estevan and Yorkton all showing a return period of less than 20 years for a −6 PDI.

Calculations using the 2 × CO_2 climatic scenario (GISS3) shortened the return period of severe water reduction for all stations. In the southeast, the return period of −6 PDI occurrences is reduced to 8–12 years rather than the historical 15–19 years. Northern stations also show return periods significantly shorter, with HIST5 return periods of over 30 years for −6 becoming 14–18 years under GISS3 conditions.

PDI values for an area are much less variable than for a point (Palmer, 1965). The areal average of PDI values for the agricultural part of the province reached −4.09 in 1961 for October, the only instance when the areal average fell below −4 for the time period studied (*Figure 2.7*). The pronounced crop-yield reduction associated with the 1961 drought has been discussed earlier (Section 1). Two stations reached their lowest PDI value in that year: Yorkton (−6.28) and Broadview (−6.04). Five other stations recorded 1961 as the second or third lowest for PDI values.

The historical worst case PDI value for Yorkton (−6.28) has a return period of 22 years under HIST5, but only 11.5 years with GISS3. For most stations, in fact, when the historical worst case Palmer Drought Index (*Table 2.4*) is analyzed, it is found to have a return period about half as long under GISS3 as under HIST5.

2.4.4. Conclusions from the PDI analysis

We have assumed that the variability and distribution of temperature and precipitation would remain similar in the future to that of the past three decades as little reliable information on likely changes in these characteristics is available. Our calculations, made on this assumption, indicate that the $2 \times CO_2$ climatic conditions simulated here on the basis of GISS3 and GISS4 would bring a more drought-prone climate to southern Saskatchewan.

The results suggest that the increase in precipitation indicated by the GISS model would not be enough to offset the increase in evapotranspiration caused by the higher temperatures. With CO_2 doubling, the climate would apparently shift to a generally more droughty regime, with increases in length and frequency of droughts due to increased soil moisture stress.

Although the PDI analyses suggest that soil moisture stress would generally increase with the GISS3 scenario, the assumption made here, that the variability of precipitation would be just as great as in the past, leads to the conclusion that there would still be some years that would be considered wet. The increased frequency of dry periods would be at the expense of normal periods; the frequency of wet periods would remain essentially unchanged.

Results of the return periods analysis suggests that what would now be the most severe drought at a location in one or two decades, would occur twice as often with GISS3. For example, a drought of severity such that it had a return period of 20 years under present climatic conditions should be expected to have a return period of 10 years with CO_2 doubling.

The frequency analysis showed that if the climate warms according to the GISS CO_2 doubling scenario but without the accompanying increase in precipitation (GISS4), a marked shift to increased drought frequency could be expected.

2.5. Discussion of Agroclimatic Change Estimates

This subsection summarizes results for selected stations from the various models employed in this part of the case study (*Table 2.5*) and draws some inferences about the effects of the different scenarios on Saskatchewan's agroclimatic environment. In interpreting these results it is, of course, important always to keep in mind the limitations of the models.

The results suggest that with no fundamental change in the climate, Saskatchewan could nevertheless expect to experience, at least once every two or three decades, an unusually warm year (HIST2) with an agroclimatic environment so dry as to reduce agricultural production by 75%. At least once every 50

Table 2.5. Summary of estimated changes in Saskatchewan's agroclimate corresponding to various climatic scenarios.

Annual % increase (+) or decrease (−) from 1951–80 normal period (HIST1)

Scenario	HIST2 Aug 1960 –July 61	HIST3 1929–38	GISS1 $2 \times CO_2 TP$ (Temp and Pcp)	GISS2 $2 \times CO_2 T$ (Temp only)
Degree-days above 5°C				
Uranium City	–	–	+50%	+50%
Saskatoon	+18%	+3%	+50%	+50%
Swift Current	+17%	+16%	+53%	+53%
Regina	+15%	+6%	+48%	+48%
Yorkton	+13%	–	+48%	+48%
Precipitation effectiveness				
Uranium City	–	–	+26%	−10%
Saskatoon	−23%	−21%	+12%	−10%
Swift Current	−33%	−27%	+9%	−12%
Regina	−37%	−26%	+5%	−11%
Yorkton	−53%	–	+6%	−10%
CA – agricultural potential				
Uranium City	–	–	+74%	+17%
Saskatoon	−53%	−48%	+8%	−18%
Swift Current	−67%	−42%	+14%	−9%
Regina	−84%	−60%	+1%	−12%
Yorkton	−93%	–	+5%	−9%

Monthly contribution (%) to CA for Uranium City (UC), Swift Current (SC), and Regina (RG)

	HIST1 1951–80			HIST2 1961		HIST3 1929–38		GISS1 $2 \times CO_2 TP$			GISS2 $2 \times CO_2 T$		
	UC	SC	RG	SC	RG	SC	RG	UC	SC	RG	UC	SC	RG
Apr	0	0	0	0	0	0	0	0	5	3	0	6	3
May	0	32	32	24	100	9	21	28	35	37	42	36	37
June	34	40	40	0	0	49	62	13	36	34	2	31	33
July	21	2	5	0	0	0	0	18	0	4	12	0	0
Aug	34	9	8	76	0	13	0	27	4	4	27	2	4
Sept	10	17	15	0	0	29	18	14	14	14	16	17	18
Oct	0	0	0	0	0	0	0	0.1	7	5	0.1	8	6
CA	8	10	11	3	2	6	4	14	12	11	9	9	10

Drought frequency (%) in Southern Saskatchewan (Palmer Drought Index)

	Less than				Greater than			
PDI	−6	−4	−2	0	0	2	4	6
HIST5 (1950–82)	0.1%	3.0%	22.8%	52.9%	47.1%	21.0%	7.4%	1.6%
GISS3 $2 \times CO_2$ TP	0.9%	9.1%	33.6%	61.2%	38.8%	18.7%	7.3%	1.8%
GISS4 $2 \times CO_2$ T	10.8%	39.6%	69.5%	85.2%	14.8%	4.3%	0.4%	0.05%

Table 2.5. *continued*

Return period in years for PDI

	−4		−6	
	HIST5 1950–82	GISS3 $2 \times CO_2 TP$	HIST5 1950–82	GISS3 $2 \times CO_2 TP$
Saskatoon	9.5	6.0	31.0	17.5
Swift Current	8.6	5.3	28.0	15.0
Regina	6.6	4.2	27.0	10.0
Yorkton	7.0	4.0	19.0	10.0

or 60 years, a decade (HIST3) with unusually warm growing seasons and a marked midsummer, no-growth period could be expected with moisture so limited that biomass productivity for the decade would be reduced by 50% from normal.

With the climatic changes predicted for a $2 \times CO_2$ atmosphere (GISS1), Saskatchewan could expect a 50% increase in growing season thermal resources, and some increases in the long-term moisture resources and potential biomass productivity, but more frequent and severe droughts. Results indicated that normal return periods of about 8 years for droughts and 26 years for severe droughts would be reduced by 40% and 50%, respectively. If the climatic warming predicted by the GISS model occurred without the predicted changes in precipitation (GISS2), a 50% increase in thermal resources could still be anticipated, but the province could expect decreases in moisture resources, decreases in potential biomass productivity except in the north, a more pronounced no-growth period between the early and late season growth periods, and substantial increases in drought frequency and severity. The implications of these and other results for agricultural policy and planning in Saskatchewan are discussed in Section 6.

The results for GISS1 and GISS3 imply a change to an agroclimate that has, respectively, both higher average soil moisture, as indicated by the precipitation effectiveness, and more frequent drought as calculated using PDI. This would be possible if long-term soil moisture increased but became more variable from year to year. Perhaps the computational procedures using temperatures adjusted by the method of differences and precipitation by the method of ratios leads to an effect equivalent to increasing soil moisture variability, and such an implied increase might or might not correspond to reality. The fact that in the CA calculations F_s decreased at 5 of the 11 southern Saskatchewan stations, although precipitation effectiveness increased at them all for GISS1, suggests that apparent contradictions may be partly due to differences among models. The relative weight given to temperature effects in the different models could be quite important. An assessment of whether the agroclimate under GISS1 would be likely to be both more drought-prone and have higher long-term, average soil moisture will have to await further research comparing the sensitivity and behavior of the various models.

SECTION 3

The Effects on Potential for Wind Erosion of Soil

Just as drought severity and frequency and biomass productivity are affected by climatic change (Section 2), so is soil degradation. Wind erosion of soil is considered here as it is a major soil degradation process in Saskatchewan. About 60% of the total annual soil loss due to wind and water on the Canadian Prairies can be ascribed to wind erosion (PFRA, 1983). Considering that soil is one of the principal basic resources for agriculture, such a threat to the soil is of great concern. The objective of this study is to evaluate the effects of climatic changes on the potential for wind erosion of soil. Effects, controls, and modeling of wind erosion of soil are discussed, a comparison with records of blowing dust is made, and a sensitivity analysis is presented. Then the wind erosion potentials of the climatic scenarios are examined. Suggested research directions and policies can be found in Section 6.

The effects of wind erosion are numerous, ranging from the physical to the economic (*Table 3.1*). The physical effects range from soil and crop damage to traffic fatalities. Not only is the more productive soil removed from the field (Bennett, 1982), but increased erosion results in the risk of the saline layer found in some soils coming closer to the surface (Mermut et al., 1983). Fertilizers or herbicides incorporated into the soil may also be lost from the field or harmfully redistributed in the soil, air and/or water. Fields can also be damaged by being buried by drifts of coarse textured soil (Brady, 1974).

The economic and sociological impacts on the farming community of drought and wind erosion have been severe (Section 1). Farm abandonment during the 1930s was common and wind erosion of soil contributed to this migration. Soil drifting even rated ahead of drought for crop destruction in at least one year of the "dirty thirties" (Gray, 1978).

There are many techniques available for attempting to control the problem. They range from the use of shelterbelts, conservation tillage, strip farming, nurse crops, fall-seeded crops and crop rotations to more esoteric methods such as the

Table 3.1. Some physical and economic effects of wind erosion and deposition.

Physical effects	*Economic consequences*
Soil damage	**Soil damage**
(1) Fine material, including organic matter, may be removed by sorting, leaving a coarse lag.	(1–4) Long-term losses of fertility give lower returns per hectare.
(2) Soil structures may be degraded.	(3) Replacement costs of fertilizers and herbicides.
(3) Fertilizers and herbicides may be lost or redistributed.	
(4) Soil may be buried by deposits of poorer quality soil.	
Crop damage	**Crop damage**
(1) The crop may be covered by deposited material.	(1–7) Yield losses give lower returns.
(2) Sandblasting may cut down plants or damage the foliage.	(1–3) Replacement costs, and yield losses due to lost growing season.
(3) Seeds and seedlings may be blown away and deposited in hedges or other fields.	(5–6) Increased herbicide and pesticide costs.
(4) Fertilizer redistributed into large concentrations can be harmful.	
(5) Soil-borne disease, weed seeds and pests may be spread to other fields.	
(6) Herbicides and pesticides redistributed into large concentrations can be harmful and losses from original location reduces effect.	
(7) Rabbits and other pests may inhabit dunes trapped in hedges and feed on the crops.	
Other damage	**Other damage**
(1) Soil is deposited in ditches, hedges, along fences, on roads, in reservoirs, lakes and streams.	(1) Costs of removal and redistribution.
(2) Fine material is deposited in houses, on washing and cars, etc.	(2,3) Cleaning costs.
(3) Farm machinery, windshields etc. may be abraded, and machinery "clogged".	(4) Loss of working hours and hence productivity declines.
(4) Farm work may be held up by the unpleasant conditions during a "blow".	(5) Costs of a decrease in transportation and communication efficiency – costs of accidents and fatalities.
(5) Visibility is decreased – transportation and communication interruptions and accidents can result.	(6) Adverse effects on human health from dust inhalation – costs of health care. Non-point source pollution of water bodies – costs of water treatment. Environmental degradation from air pollution.[a]
(6) Air pollution.	
(7) Changes in the earth-atmosphere energy budget such as increases in the rate of radiative cooling of the atmosphere.[b]	

Adapted from Wilson and Cooke (1980, p. 218) with additions by E. Wheaton.
[a] Heathcote (1980, p. 43).
[b] Guedalia *et al.* (1984).

use of chemicals. Each method has its advantages and disadvantages. These methods are well documented in several publications (e.g., FAO, 1960; Agriculture Canada, 1966; Canessa, 1977; Moldenhauer et al., 1983; US Department of Agriculture, 1983; PFRA, 1983; Bircham and Bruneau, 1985). It must be emphasized that unless these techniques are rigorously applied and are proven economically viable, wind erosion will most likely continue to be a major problem.

3.1. Modeling Wind Erosion

Erosion is initiated when windspeed is greater than the threshold value required to lift and transport vulnerable soil particles (*Figure 3.1*). The most susceptible particles are those large enough to protrude into the wind field, but light enough to be easily moved. It should be noted that even though a certain size range of particles is most susceptible to wind erosion, with strong winds a large range of soil particle sizes is susceptible (Zachar, 1982). So with greater windspeeds, soil texture becomes less important in determining wind erosion.

In addition to wind, the other climatic variables that affect wind erosion directly are precipitation and temperature. Precipitation and temperature affect soil moisture which provides a cohesive force to bind the particles together making them more resistant to the force of wind. On the other hand, raindrop impacts and wetting (producing unequal swelling) can cause breakdown of aggregates and thus increase their susceptibility to wind erosion.

Wind, precipitation and temperature have been combined in a model by Chepil et al. (1962) to provide an index to describe quantitatively the relation between climate and the potential level of wind erosion. The other main parameters affecting wind erosion of soil are vegetative cover, soil texture and structure, topography and fetch. These parameters are used with the climatic variables to estimate actual soil loss. Only the climatic elements are modeled for the purposes of this study. Other available models for calculating wind erosion potential (e.g., Schwab et al., 1966; Pasak, 1978) are very similar, being a function of the cube of the windspeed and the inverse square of a soil moisture index, but the Chepil model appears to be the best available.

The process of wind erosion can be accounted for using the fundamental laws of classical physics (Novak and van Vliet, 1983). The amount of soil moved is related to the strength of the forces acting on the soil particles. The forces acting on the soil particles to initiate movement are wind pressure and the impact of saltating particles. [Saltation is the bouncing, jumping motion of particles driven by the wind (Bagnold, 1941)]. The restraining forces acting on the particle are the cohesive force of water (if the soil contains water) and gravity.

The pressure of the wind is proportional to the square of the wind velocity (Blackwood et al., 1967; Bagnold, 1941; Chepil and Woodruff, 1963) and the impact force of saltating particles is proportional to the windspeed (Bagnold, 1941). Therefore, the mass of soil moved is proportional to the product of the wind and impact forces, which is the cube of the wind velocity.

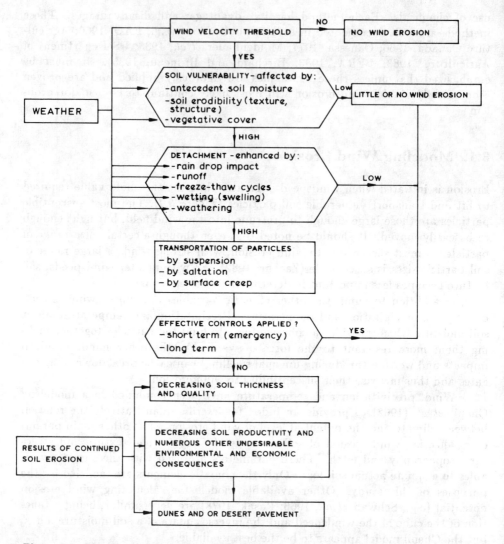

Figure 3.1. Elements in the process of wind erosion.

The force of cohesion is provided by adsorbed water films surrounding the soil particles. The resisting forces of cohesion and gravity are equal to the square of the effective moisture (Chepil and Woodruff, 1963).

The wind erosion climatic factor is based on the principle of erosion being directly proportional to the cube of the mean annual windspeed and inversely proportional to the square of the effective soil moisture. The precipitation effectiveness index (Section 2) is used to represent the soil moisture.

The equations for the wind erosion climatic factor as given by Lyles (1983) were used for this study. That source provided one of the best documentations of the model. The computation is as follows:

The effects on potential for wind erosion of soil 285

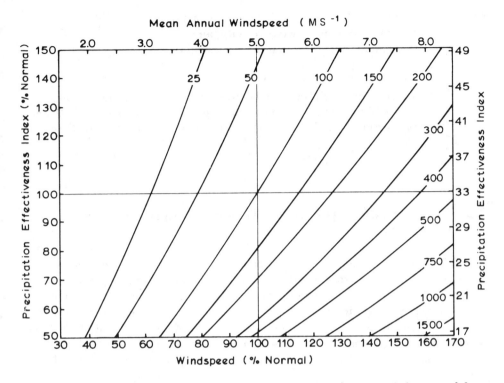

Figure 3.2. Sensitivity of the wind erosion climatic factor (percent of the normal for a representative station) to changes in windspeed and precipitation effectiveness index [equation (3.1) in text]. The sample station, near Saskatoon, is representative of the central study area. HIST1 values of mean annual windspeed (5.0 m s^{-1}) and annual precipitation effectiveness index (33) give a normal climatic factor of 44.3. Note that the highest windspeed value (8.5 m s^{-1}) is equal to the mean monthly windspeed from the west-northwest direction for Swift Current in December. It is the highest mean monthly windspeed for any direction.

$$C = 386\,(\bar{u}_z)^3/(I)^2 \qquad (3.1)$$

where: C is the annual wind erosion climatic factor,
$\bar{u}_z$ is the average annual windspeed (m s^{-1}) at a height z of 9.1 m, and
I is the annual precipitation effectiveness index.

Typical values for the wind erosion climatic factor for Saskatchewan stations for the 1951–1980 normal period are given below in *Figure 3.4(a)*.

The climatic factor can also be computed on a monthly basis. Woodruff and Armbrust (1968) provide the following method:

$$C_m = 34.48\,(\bar{u}_m)^3/(I)^2 \qquad (3.2)$$

where: C_m is the monthly wind erosion climatic factor,
$\bar{u}_m$ is the monthly mean wind velocity, and
I is the annual precipitation effectiveness index.

A disadvantage of equation (3.2) is that only the windspeed is determined on a monthly basis; the precipitation effectiveness index is an annual value. This was solved by the simple substitution of the monthly precipitation effectiveness index (I_m) (see Section 2) and the monthly windspeed into equation (3.1) to give

$$C_m = 386\,(\bar{u}_m)^3 / (I_m)^2 \qquad (3.3)$$

3.2. Comparison of the Wind Erosion Climatic Factor with Observations

It is not easy to validate the wind erosion climatic factor because there are few measurements of wind erosion rates across the Canadian Prairies (de Jong, 1984; PFRA, 1983). However, the wind erosion climatic factor may be compared with blowing-dust or dust-storm data that are available. The climatic factor was compared with the annual number of dust storms in Kansas (Chepil et al., 1963). A high incidence of dust storms was found to be associated with high values of wind erosion climatic factors (correlation coefficient r of 0.68 and highly significant).

In the present study the relationship of the wind erosion climatic factor (C) and the number of days with blowing dust or sand (D) was examined for Saskatchewan stations. Blowing dust or blowing sand is defined as "dust or sand, raised by the wind to moderate heights above the ground", and the visibility at eye level may be reduced to 1 km (Environment Canada, 1977). The number of days with blowing dust has been recorded for several Saskatchewan stations from 1977 to the present (Environment Canada, 1977–1983).

Most of the 14 stations with available data had at least one day with blowing dust every year and one station reported 19 days with blowing dust in one year. The mean annual value for the 7-year period of record (1977–1983) ranged from 5.0 days in the south to about half that number in the northern agricultural area.

The year-to-year variation of the annual wind erosion climatic factor [equation (3.1)] and annual number of days with dust have been compared for a number of stations (for example, see Figure 3.3). There appears to be only a weak association between the two, with the year 1980 for Regina providing a noticeable exception, where the high climatic factor is not reflected in a high incidence of days with blowing dust. Further analysis from thirteen stations for the seven years (1977–1983) showed that the association between C and D appears to improve with an increase in the number of stations reporting blowing dust and a greater number of days with dust.

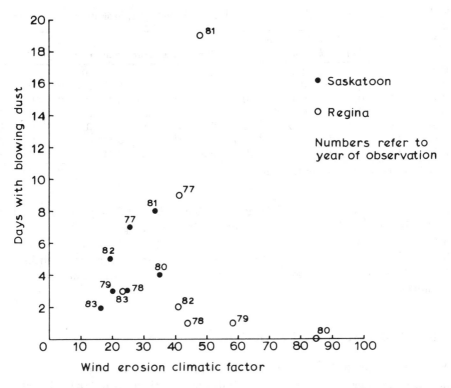

Figure 3.3. Scattergraph of days with blowing dust against annual wind erosion climatic factor for Saskatoon and Regina, 1977–1983.

The monthly variation of C [equation (3.3)] and D was also examined as there is a definite seasonal pattern. The primary peak of D occurs in the spring to early summer for all but one of the years of record (*Table 3.2*). One of the reasons for this seasonality is that the greatest mean monthly wind velocity occurs in the spring (March to May) for most locations (10 out of the 11). Another important reason is that the vegetative cover is sparse in the spring. Precipitation is not usually at a maximum in the spring and unless snowmelt contributes significantly, dryness of the top layer of soil is also a likely factor.

Even though the climatic potential for wind erosion is relatively high in September and October, the actual amount of blowing dust may be low owing to restraining factors such as vegetative cover. If this assumption is correct, it has great significance in terms of the control of wind erosion, as it suggests that high climatic potentials for erosion coincide with the exposure of soil after harvest. It also follows that a better vegetative cover at the time of primary risk (spring) as provided by crops such as winter wheat, would reduce wind erosion.

This pattern of D, with a pronounced spring maximum, probably contributes to a decrease in the degree of correlation of annual C and annual D. For example, if spring is wet but the rest of the year is dry, the annual climate factor will be high. The wet spring will dampen the likelihood of high annual D

Table 3.2. Average monthly number of days with dust (D), Saskatchewan stations.

	No. of Stations	Jan	Feb	Mar	Apr	May	June	July	Aug	Sep	Oct	Nov	Dec	Year
1977	13	0	0	1.1	2.4[a]	0.8	0	0	0.2	0	0.1	0.1	0	4.7
1978	7	0	0	0	0	0.1	0.6[a]	0.3	0	0	0.1	0.4	0	1.5
1979	9	0	0	0.1	0	0	0.7[a]	0.1	0.1	0.7	0	0	0	1.7
1980	12	0	0	0	0	1.8[a]	0.2	0.2	0	0.2	0.3	0.2	0.1	3.0
1981	12	0	0	0	4.8[a]	3.3	0.1	0.2	0	0.7	0	0	0.1	9.2
1982	11	0	0	0.1	1.6[a]	0.5	0.5	0	0.2	0.1	0	0	0	3.0
1983	12	0.1	0	0	0	0.1	0.3	0	0.3	0.6[a]	0	0	0	1.4

[a] Maximum monthly value for the year.

however, and the resulting relationship of C and D would be poor. The opposite could also occur. A dry, windy spring, resulting in a high annual D, could occur in a year that is otherwise wet (i.e., with a low C). This combination did occur in 1981. The apparent sensitivity of annual D to spring climatic conditions thus emphasizes the need to examine conditions at a monthly rather than annual resolution.

3.3. Model Sensitivity Analysis

A sensitivity analysis of the model was conducted to assess the relative importance of windspeed (U_z) and the precipitation effectiveness index (I) in determining the wind erosion climatic factor. It was particularly important to study model sensitivity to changes in wind velocity because wind anomalies were not available as GISS model outputs for scenario experiments, and windspeed has a major effect on the model results because the third power is used. The climatic factor was calculated using equation (3.1) for windspeeds between 70% above and 50% below the mean windspeed for a sample station in Saskatchewan and for values of I ranging from 50% above to 50% below the mean (*Figure 3.3*). For example, by keeping I at the normal value, a decrease in wind velocity of 20% resulted in a substantial decrease of the climatic factor to only 51% of its original value, while a 20% increase led to a tremendous increase, to 173% of the original value. Thus, without increases in I (which would tend to offset the increases in the climatic factor), even slight increases of wind with future climatic changes would tend to exacerbate the wind erosion risk.

3.4. Impact Analysis and Interpretation

The wind erosion potential of climate (as given by the wind erosion climatic factor) was calculated for some of the climatic scenarios listed in Section 1. The climatic factors for the 1951–1980 normal period are shown in *Figure 3.4(a)* for comparison. *Figure 3.4(b–d)* gives the wind erosion climatic factors by station for the scenarios and includes these results expressed as percentages of the normal climatic factor.

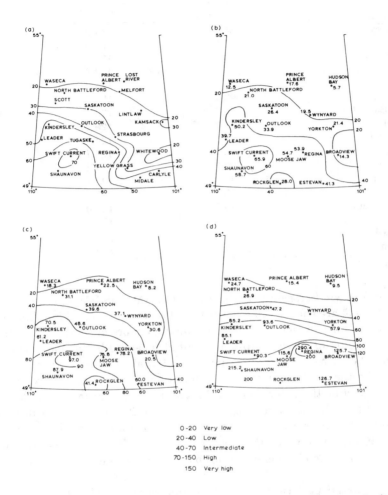

Figure 3.4. Wind erosion climatic factors based on: (a) 1951–1980 normals (HIST1), (b) 2 × CO_2TP (GISS1), (c) 2 × CO_2T (GISS2) scenario, and (d) 1961 values (HIST2).

3.4.1. The 2 × CO_2 scenario (GISS1)

The doubling of atmospheric carbon dioxide scenarios were based on the GISS global climatic model (Section 1). Normal wind data (Environment Canada, 1982) were used for these scenarios as wind anomalies were not available as GISS outputs.

Although the climatic scenario conditions simulated by the GISS1 model are warmer than the normal climate, greater than normal precipitation amounts are also predicted. As indicated previously, the wind erosion potential in a warmer climate is expected to increase owing to the decrease in soil moisture.

Table 3.3. Wind erosion climatic factor for the climate scenarios expressed as a percentage of the HIST1 climatic factor.

Station	GISS1	GISS2	HIST2
Broadview	58.8	84.3	496.8
Estevan	96.0	139.5	294.7
Hudson Bay	86.4	124.2	143.9
Kindersley	87.5	122.8	148.4
Leader	81.7	125.9	175.1
Moose Jaw	92.7	128.5	195.9
North Battleford	82.7	122.4	105.9
Outlook	88.1	126.2	243.1
Prince Albert	97.2	124.3	85.1
Regina	86.7	125.7	466.9
Rockglen	90.9	134.4	–
Saskatoon	86.3	129.4	154.2
Shaunavon	86.1	128.9	315.5
Swift Current	85.6	126.0	117.3
Waseca	83.9	122.8	165.8
Wynyard	77.1	146.6	–
Yorkton	84.5	123.4	233.5

– data unavailable.

However, the wind erosion potentials based on the GISS1 scenario [*Figure 3.4(b)* and *Table 3.3*] are considerably lower than the levels expected for climatic warming alone, and are in fact less than the climatic factors for normal conditions.

The reduction in wind erosion potential is caused by the increase in precipitation predicted by the GISS1 scenario counterbalancing and reversing the effect of the increase of temperature (precipitation effectiveness index, Section 2). Most of the climatic factors [*Figure 3.4(b)*] are in the range from 80% to 90% of normal. Precipitation increases in the order of 20% of normal will apparently offset temperature increases of 3–4 °C and a reduction in wind erosion potential results. It should be noted that total soil erosion will not necessarily decrease as the increase in precipitation would probably cause an increase in water erosion of soil.

The spatial pattern of climatic factors for the GISS1 scenario is generally similar to that of the normal period. The Swift Current–Shaunavon area in the southwest is the core of the greatest erosion risk (greater than 60%) and the northeast has the least risk (less than 20%). The isopleths have the general northwest to southeast orientation exhibited for the normal period. The shift in isopleths from the normal period to the GISS1 scenario is slight, but noticeable. The isopleths shift to the southwest by about 10–50 km. An increased frequency of hot dry periods (noted in Section 2) would increase the potential for severe wind erosion.

3.4.2. The $2 \times CO_2$ scenario with no change in precipitation (GISS2)

This scenario was particularly important here because of the sensitivity of the wind erosion model results to precipitation. Most of the wind erosion climatic factors for this scenario are in the range from 22% to 47% above the 1951–1980 normal [*Figure 3.4(c)* and *Table 3.3*]. Furthermore, the decrease in biomass estimated using this scenario (Section 2) could contribute to an increased wind erosion owing to the reduced protective cover. On the other hand, a change in crop, for example to winter wheat, could improve protection, especially during the spring.

In terms of spatial pattern, the Swift Current–Shaunavon area in the southwest retains the core of the highest values [*Figure 3.4(c)*]. Nevertheless, the area with wind erosion potential classified as high has expanded into a large area of southwestern Saskatchewan south of a Regina to Kindersley line.

The shift in isopleths that would result, as compared to the normal period conditions, is about 70 km northeastward. The greatest shift of isopleths occurs in the central portion of the region and the smallest shift occurs in the northeast.

3.4.3. Extreme decade – 1929–1938

This time period is often called the "dirty thirties" because of the many dust storms that occurred. Badly blowing years had returned in 1929 after a series of wet years in the 1920s. In 1934 the Saskatchewan Wheat Pool recorded 157 reports of serious crop damage from soil drifting. In 1935 the rains came back, but the soil still blew. In 1936 there were 3–4 times as many reports of soil drifting as in 1934. In 1937 the dust began to blow early, strongly and frequently (Gray, 1978). Thus the 1929–1938 decade was considered appropriate for analysis. Unfortunately, only four of the stations used had adequate data available for 1929–1938 and alternative procedures were not developed. The results for Saskatoon suggest a very high wind erosion potential for the period, but the three other available stations indicate the opposite. These lower values were apparently due to the lower mean annual windspeeds during that period. No doubt the technology of the time contributed to the dust storms of the 1930s. At the moment we can only speculate; the results are inconclusive and further analysis is needed. In further work it would be of interest to examine the impact on wind erosion of a decade with temperature and precipitation as in 1929–1938, but with normal windspeeds.

3.4.4. Extreme years

To explore further past instances of severe wind erosion events as useful indicators of the future, climatic factors were calculated for those years considered to have had severe wind erosion events. The wind erosion risks for years during which little or no wind erosion occurred were also determined in order to estimate the range of risks.

The year 1961 was certainly a severe drought year and most of the wind erosion potentials were very high [*Figure 3.4(d)*], but there was no independent evidence that the author is aware of, of severe dust storms that year. There were severe dust storms in the spring of 1981, but the wind erosion potentials were not above normal that year for most locations because of the masking effect of the later portion of the growing season (Section 3.2).

3.5. Discussion: Wind Erosion and Climatic Change

The GISS1 scenario indicates a decrease in wind erosion potential from the normal in southern Saskatchewan (14% decrease on average). The wind erosion potential would be rated as low to intermediate over most of agricultural Saskatchewan. In contrast, the scenario of warming without the precipitation increase (GISS2) indicates average wind erosion potentials of about 26% higher than normal.

If the extreme years examined in this study are considered as an indication of range, then variations from average long-term climate scenarios can be assessed. For example, although the long-term climate of the GISS1 scenario indicated reduced wind erosion potentials, occurrences of years as extreme as 1961 would still produce situations conducive to severe wind erosion. The more frequent, longer and more intense dry periods estimated with GISS1 conditions (Section 2) further emphasize the possibility of periods of increased wind erosion potentials.

SECTION 4

The Effects on Spring Wheat Production

In this section the impacts on a specific crop, spring wheat, are discussed. As noted previously (Section 1), spring wheat production is the major crop production activity in Saskatchewan and, therefore, is a logical choice for study in impact analyses related to the agricultural industry.

In this study a crop growth model is used to estimate the changes in spring wheat yields that could be expected to accompany the changes in climate outlined in *Tables 1.3* and *1.4*. In the following subsections details of the crop growth model and its sensitivity to changes in various climatic parameters are outlined. The impact of the different climatic scenarios on spring wheat yields and production are then discussed. Finally, the derived yield changes are incorporated into farm and provincial level input–output models to examine the possible impacts on the economy of Saskatchewan. The results of this application are presented in Section 5.

4.1. Methodology

4.1.1. The yield and phenological models

The procedures used to estimate spring wheat phenology and yields are briefly described in the following; a detailed discussion of the yield model is provided by Stewart (1981), while Robertson (1968) and Williams (1974) provide a more detailed discussion of the phenological model. Calculation of spring wheat yields is based on the methodology developed by the FAO (1978) and utilizes tabulated results from the de Wit (1965) photosynthesis model to compute "constraint-free" yields. Yield estimates assume a sigmoidal cumulative growth curve with development incremented up to the number of days required to mature the crop. Net biomass production (B_n) is calculated as a function of the gross biomass production (B_{gm}) capacity of the crop, determined by its photosynthetic response to temperature and radiation, a maintenance respiration coefficient (C_T) and the number of days required to reach maturity (N). This relationship is expressed as:

$$B_n = 0.36 B_{gm}/(1/N + 0.25 C_T) \tag{4.1}$$

To estimate N, a phenological model developed by Robertson (1968) is used. Robertson's model describes the phenological development of spring wheat as a function of temperature and photoperiod in the form:

$$\sum_{i=S_1}^{S_2} \left[a_1(L_i - a_0) + a_2(L_i - a_0)^2 \right] \left[b_1(T\max_i - b_0) + b_2(T\max_i - b_0)^2 \right.$$
$$\left. + b_3(T\min_i - b_0) + b_4(T\min_i - b_0)^2 \right] = 1 \tag{4.2}$$

where: L_i is the photoperiod (duration of daylight in hours) on day i, $T\max_i$ is the maximum air temperature on day i, $T\min_i$ is the minimum air temperature on day i, S_1 is the date of a phenological stage in the development of wheat toward maturity and S_2 is the next stage, and a_0-a_2 and b_1-b_4 are coefficients.

In the model, five phenological phases are considered: planting to emergence, emergence to jointing, jointing to heading, heading to soft dough and soft dough to ripe. For each stage a different set of a and b coefficients are used. The reader is referred to Robertson (1968) for a detailed description of the model and for a more precise definition of each of the crop phenological stages.

If the commencement date for a phase is known, then by using equation (4.2) with the appropriate set of coefficients, the date this phase ends and the next begins can be estimated. This is accomplished by summing the value calculated by equation (4.2) from day to day until a value of 1 is reached. The date the crop ripens or matures is derived by continuing the summation from the planting date through all five phases. N is then derived as: $N = \text{IEND} - \text{ISTART} + 1$, where ISTART and IEND are the Julian dates that the crop is planted and reaches maturity, respectively.

The planting date is calculated as the date the smoothed mean minimum air temperature first exceeds 5 °C in the spring. This represents, with a 50% probability, the average date for the last spring and first autumn "killing frosts" (-2.2 °C) when using averaged 30-year climatic normals data (Sly and Coligado, 1974). In determining this date, the monthly temperature data are first converted to daily values using the Brooks (1943) sine-curve technique. The planting date is then derived by computer interpolation of the first day the minimum air temperature reaches 5 °C. Similar criteria are used to determine the end of the growing season in the autumn. If an estimated autumn frost occurs before the crop reaches maturity, the crop is assumed killed and the yield component set equal to zero.

Crop dry matter yield (B_y) is then derived as:

$$B_y = B_n \times H_I \tag{4.3}$$

where H_I is the harvest index, defined as that fraction of the net biomass production that is economically useful; i.e., the grain component.

The experimental work of Major and Hamman (1981) at Lethbridge, Alberta, for Neepawa wheat is utilized to calculate harvest index values. Using their data, it was found that the harvest index was inversely related to moisture availability. That is, if moisture is limited, more of the crop biomass in terms of percent is converted into yield than if moisture is not limited. This relationship is expressed by using the ratio of actual evapotranspiration to potential evapotranspiration (AE/PE). On the basis of the Major and Hamman (1981) results, in the yield calculations, if the value of AE/PE is greater than 0.75, the value of H_I is set equal to 0.35. As AE/PE decreases below 0.75 the value of H_I is increased linearly to a maximum value of 0.52. This value is reached when AE/PE has declined to approximately 0.36.

In this study PE is calculated using the Penman (1963) method and AE is derived using a combination soil moisture budgeting–split canopy evapotranspiration model. The former involves using the techniques described by Baier et al. (1979) and the latter the work of Ritchie (1972). Details of the procedures used to calculate both AE and PE are provided by Stewart (1981).

Values of B_y computed by equation (4.3) are constraint-free or genetic potential yields and neglect the effects of yield-reducing factors such as moisture stress; weeds, pests and diseases; climatic effects on yield components, yield formation, or quality of produce; and field workability. For the purposes of this study, values of B_y were corrected by a moisture stress yield-reducing factor (MSF) to give values of estimated dry matter yield (B_{ye}), in the form:

$$B_{ye} = B_y \times MSF \tag{4.4}$$

All other yield reducing factors are assumed negligible.

Moisture stress is derived using an expression relating the relative yield decreases to the relative evapotranspiration deficit in the form:

$$MSF = [1 - k_y (1 - AE/PE)] \tag{4.5}$$

where k_y is an empirically derived coefficient for crop yield response to moisture deficit. For spring wheat the value of k_y is set to 1.15 based on the work of Doorenbos and Kassam (1979).

The procedures for estimating dry matter yields presented above are designed to evaluate the long-term crop production capability under optimum management conditions on a continental scale from standard climatic information. The input data required include long-term monthly averages of temperature, precipitation, incoming global solar radiation, windspeed and vapor pressure. These data are normally available from observation networks, or alternatively can be derived using simple empirical equations.

4.1.2. Historical and $2 \times CO_2$ climatic scenarios

For analysis of climate in relation to spring wheat, the data must be on the same geographic basis for climate as for wheat production. Crop yield and production data are reported annually for the 20 Saskatchewan crop districts (CD) (1a to 9b, *Figure 1.2*) by the provincial government. The climatic data were subsequently converted to the basis of these districts for comparative purposes (i.e., yield model estimates versus commercial yields).

The 1951–1980 climatic normals for climatological stations (Environment Canada, 1982) had been converted by computer interpolation to a 100×100 km equal area grid system (LeDrew et al., 1983). These grid data were employed to compute the climatic normals (the HIST1 scenario described in Section 1) for the 20 crop districts using procedures described by Stewart (1981).

Monthly values of crop district mean daily maximum and minimum temperatures and total precipitation (*T*max, *T*min and *P*cp, *Table 1.3*) used in the analysis for 1961 (HIST2 scenario) and 1933–1937 (HIST4 scenario) were obtained from Mack (1982). The 5-year period, 1933–1937, was used rather than the 1929–1938 period because weather data in crop district form were only available for these years.

Temperature and precipitation data for the GISS1 and GISS2 scenarios were derived by adjusting the climatic normals (HIST1), as explained in Section 1. The estimation of phenological and biomass parameters in the crop model require *T*max and *T*min. Since the GISS model output only provided monthly mean temperature values, it was assumed that the effect on *T*max and *T*min was the same (i.e., if *T*adj = 3 °C, then 3 °C was added to both *T*max and *T*min).

As discussed in Section 1, the GISS model outputs for use in simulating $2 \times CO_2$ climatic scenarios were provided for grid points covering the case study area. For the spring wheat analysis, these data from the nine points of intersection of the 50°, 54° and 58° N latitude and 100°, 105° and 110° W longitude lines were used. The temperature differences and precipitation ratios were mapped and values interpolated by hand to obtain the *T*adj and *P*cpadj adjustments at the centroid of each crop district. The centroids are labeled by crop district number in the maps outlined in the subsequent sections.

Climatic data for the GISS1 and GISS2 scenarios were simulated using the HIST1 temperature and precipitation data with the adjustments for these scenarios. For other climatic data required for the crop model, including solar radiation, wind speed and vapor pressure, normals for the 1951–1980 period were used.

Daily information for all climatic parameters except precipitation was generated from the monthly data using the sine-curve interpolation technique of Brooks (1943), as in previous work (Williams, 1969a; Williams and Oakes, 1978; Dumanski and Stewart, 1981). Precipitation data were converted to weekly values and these values were distributed among days, assuming that 60% would be received the first day, 30% the second, 10% the third, and 0% for each of the remaining 4 days. Daily information was used in simulating the water balance for the crop model.

4.2. Comparison of Model Results with Observations

Results for the phenological model have been compared to data from other sources in studies relating to wheat and barley on the Canadian Great Plains (Williams, 1971, 1974). Despite the limitations discussed in those papers, the model can be applied in most potential agricultural areas of Canada or in other areas with continental climates and long summer day lengths. However, as we have already noted (Section 1), such findings may not necessarily be too helpful in assessing the applicability of the methods to climatic impact analysis, especially for scenarios of climate as different as that anticipated with a doubled level of CO_2 in the atmosphere.

The crop growth model used in this study was designed to evaluate long-term potential crop yields under optimum management conditions. However, there are no long-term experimental results in Canada available for validating this model because of continual technological change in crop varieties. However, some relevant short-term experimental data are available. Model estimates were compared with experimental work carried out by Major and Hamman (1981) at Lethbridge, Alberta, and by Onofrei (personal communication, 1984) at six locations in Manitoba.

To compare model estimates with observations at the Alberta and Manitoba sites, climatic data including the mean (Tmean), maximum (Tmax), and minimum (Tmin) air temperatures, precipitation totals, and plant available soil moisture data observed for each year were obtained. For the Alberta site only monthly observations were available while for the Manitoba sites daily data were obtained. Results of the comparison of estimated yields to observed values are given in *Table 4.1*. Results show that the model is within 15% of observed values.

These results are remarkably good considering that the model was not designed for use with data for individual years but, rather, for applications employing monthly data averaged over several years. The equations used to compute crop biomass and yield employ averaged growing season information as opposed to the actual day-to-day values that would be used in a model designed

Table 4.1. Comparison of model estimates with observed experimental yields in Alberta and Manitoba.

			Dry matter Yield (kg/ha)		
Year	Location	No. of sites	Model	Observed	Model/ observed
1976	Alberta[a]	1	3495	3098	1.13
1977	Alberta[a]	1	2454	2623	0.94
1982	Manitoba[b]	6	3547±659	3967±440	0.89
1983	Manitoba[b]	6	2666±607	3098±763	0.86

[a] Data from Major and Hamman (1981) at Lethbridge, Alberta, for Neepawa Wheat.
[b] Data from Onofrei (personal communication, 1984) for six sites in Manitoba (Beausejour, Winnipeg, Woodmore, Marapolis, Bagot and Teulons) for Glenlea

for real time application. For this reason the model does not simulate plant growth and photosynthetic activity on a daily basis as would be the case with sophisticated models, nor was it intended to do so. It was instead designed to simulate what happens to the crop biomass productivity for averaged growing season conditions. Consequently, it is recognized that many environmental factors affecting crop growth and productivity are unaccounted for in the model framework.

A more detailed physiological model, such as Cornell University's Soil–Plant–Atmospheric Model (Shawcroft et al., 1974), can simulate daily crop growth processes. A model of this sort, however, is designed for site-specific applications requiring considerable data input which can only be obtained for individual sites or crop canopies. These data, in most cases, cannot be extrapolated to the large-scale applications. Furthermore, it is prohibitively expensive for a model of this sort to be used for large area application, and even if it could be used in this way, it has no real advantage because in most cases it has been found to be no more accurate than much simpler models. It was decided, therefore, that a detailed physiological model would not be appropriate for this study.

The model used in this study was selected, despite its inherent weaknesses, because it is physically based and is able to respond to various environmental stimuli in much the same way as a plant would respond. It requires input data that are readily available, and it is inexpensive to use for continental-scale applications. In addition, it responds reasonably well to the sorts of environmental stresses being addressed in this study and currently is the only model of this sort available for use in Canada.

It is emphasized that the comparison outlined in Table 4.1 includes experimental yields which represent the potential or maximum yields that can be obtained under optimum management practices. They do not represent the yields obtainable under current commercial conditions that are considerably less than the potential. For example, reported commercial yields were 59% and 60% of the values given in Table 4.1 for experimental wheat yields, respectively, for 1982 and 1983 in Manitoba. Similarly, in Alberta the ratio was 76% and 70%, respectively, for 1976 and 1977. For this reason, all scenario yield estimates are expressed in terms of percent of normal, where normal represents the model estimate derived using the climatic data averaged for the 1951–1980 period (HIST1). It is assumed that the effects on yields of variations in climatic conditions are the same for both commercial and experimental spring wheat production.

4.3. Model Sensitivity

The sensitivities of modeled yields, phenology and growing season lengths to changes in temperature and precipitation were analyzed. Results are provided for departures from normal temperature of from -2 to $+3$ °C and precipitation values from 60% to 130% of normal, using the 1951–1980 normal period as the base data (Figures 4.1 and 4.2). With 1951–1980 (HIST1) climatic conditions, the growing season length (GSL) in Saskatchewan averages 115 ± 6 days. Much

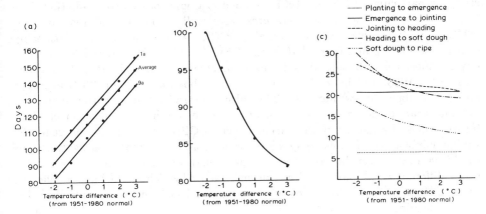

Figure 4.1. Effects of temperature deviations from 1951–1980 normal (HIST1) on: (a) changes in growing season length for crop districts 1a, 9a and a provincial average for Saskatchewan, (b) changes in maturation time for spring wheat, and (c) changes in spring wheat phase development lengths.

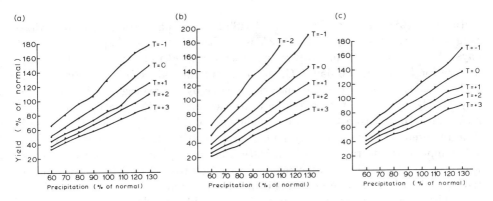

Figure 4.2. Sensitivity of model yields (% of normal) to changes in precipitation (% of normal) and temperature (°C) for crop districts (a) 1a, (b) 6b and (c) 7a in Saskatchewan.

of the variation in GSL can be explained by the latitude and by the higher elevation of the terrain in the southwest corner of the province. In general the effect on the growing season length of increasing or decreasing the temperature [*Figure 4.1(a)*] is to increase or decrease the GSL by about 10 days/°C which is generally split equally between the start and end of the growing season.

It should be emphasized that these figures represent statistical results derived from long-term climatic averages. When records for specific years are examined, no relationship between the beginning and ending dates of the growing season is evident. Thus, the fact that temperatures are above normal does not necessarily mean that the growing season length will be longer. On the contrary, the growing season can be shorter than normal in some warm years and longer than normal in some cold years. From the historical data used in this study, in particular, 1935 and 1961 are good examples of this contradiction. The former was a year when temperatures were well below normal (as much as $-2\,°C$), the latter well above normal ($+3\,°C$). According to the relationship shown graphically in *Figure 4.1(a)*, the GSL should have averaged 92 and 145 days, respectively. In actual fact, the GSL averaged only about 8 days below normal for 1935, and instead of being 30 days above normal for 1961, averaged about 3–5 days below normal. This also illustrates the fact that using an extreme year, or set of years, as an analogue to obtain an indication of likely future values of an agroclimatic variable such as GSL, given a return to generally similar climatic conditions, may be quite misleading in that exactly the same condition may never be repeated. A particular average value can represent many different combinations of conditions, each of which could have a different impact. History, from this point of view, can be construed as potentially being very misleading.

4.3.1. Sensitivity of phenological development

With 1951–1980 climatic conditions (HIST1), the maturation time for wheat ranges from 86 to 98 days in Saskatchewan with an average of approximately 90 days. Decreasing the temperature increases the length of time needed to reach maturation by about 5–6 days/$°C$, whereas, as temperature increases by 1, 2 and 3 $°C$, the effect is to reduce the maturation time by, on average, 4, 7 and 9 days, respectively. The effect of a temperature decrease on the spatial variation of maturation time within the region, as shown, would be to increase the range in time required to reach maturity among crop districts from 12 to 19 days for a 1 $°C$ temperature decrease. As temperature increases, however, the overall effect would be to reduce the average regional variation from 9 to 6 to 5 days, respectively, for incremental temperatures increases of 1, 2 and 3 $°C$ [*Figure 4.1(b)*].

The effect of temperature variations on the five development phases of spring wheat is presented in *Figure 4.1(c)*. Results indicate that the first two phases are virtually unaffected by temperature variations. Only after the crop has reached the jointing stage does temperature have an effect. Overall, the effect of temperature is translated into an increase in phase length as temperature decreases and a reduction in phase length as temperature increases.

As temperature changes the length of the growing season, it also changes the placement of the beginning and end of this period, or window, in relation to the summer solstice (period of maximum light intensity). As a consequence of the higher temperature, the planting date takes place further in advance of the solstice, and crop growth and development occurs during a period of higher incoming solar radiation. The shift also enhances the effect of daylight on maturity. On the other hand, as also shown, a decrease in temperature would have the opposite effect.

4.3.2. Sensitivity of model yields to temperature and precipitation

Any change in the length of time required for a crop to mature ultimately has a corresponding effect on crop yields. The sensitivity of model yields to seasonal temperature departures from normal of −2 to +3 °C and precipitation variations from 60% to 130% of normal (HIST1) are outlined in *Figure 4.2* for crop districts 1a, 6b and 7a. These crop districts represent a southeast to northwest transect across the study area. The data presented reveal some interesting features including:

(1) Model yields for a specific temperature are linear in relation to precipitation changes, although the slope of this relationship differs slightly from one crop district to another. This difference can be attributed to the difference in base precipitation and temperature from place to place (*Table 4.2*).

(2) The effect of temperature on yield is nonlinear for a given precipitation level. This is primarily the response of the model to changes in the length of time required to mature spring wheat which ranges from 81 to 101 days for temperature departures of +3 to −2 °C, respectively. In general, for the range of temperatures considered here, the effect of decreasing temperatures from the current level would apparently increase yields, whereas increasing temperature would decrease yields. This is primarily associated with the change in moisture stress resulting from changes in crop transpiration demand. Decreasing temperature in the absence of any change in precipitation would reduce moisture stress, while increasing temperature has the opposite effect.

(3) The model clearly shows that for any temperature change a corresponding change in precipitation is required to maintain current yield levels. For example, if temperatures were to decrease by about 1 °C, the model projects that existing yield levels could be maintained with corresponding precipitation decreases of from 12% to 14%. Similarly, if temperatures were to increase by +3 °C, as now projected by some of the current GCM models for a doubling of atmospheric CO_2 concentration, the model suggests that a precipitation increase of about 40% would be required to maintain existing spring wheat yield levels.

4.4. Effects of Climatic Changes on Growing Season Agroclimate

Average growing season climatic conditions for the 1951–1980 normals (HIST1) period in Saskatchewan are illustrated in *Figure 4.3(a)–(c)*. *Figure 4.3(a)* outlines the growing season length available for crop growth; *Figure 4.3(b)*, the thermal resources available during the growing season, expressed in degree-days above 5 °C; and *Figure 4.3(c)*, the available moisture in terms of precipitation received during this growing period. These figures basically define the growing season as being relatively short, warm and dry. The average GSL varies from

Table 4.2. Temperature (°C) and precipitation (% of normal) deviations from the HIST1[a] normals averaged over the period May to August for the HIST2, HIST4 and GISS1 scenarios.

Crop District	Baseline HIST1	Climatic scenarios							
		HIST2	GISS1	HIST4	1933	1934	1935	1936	1937
Temperature (°C)									
1a	15.9	1.8	3.5	0.2	0.7	−0.1	−1.7	1.2	1.1
1b	16.0	1.4	3.5	0.0	0.4	−0.3	−2.0	0.9	1.2
2a	16.0	1.9	3.6	1.0	1.2	0.7	−1.0	2.1	1.9
2b	16.3	1.6	3.5	0.2	0.6	−0.2	−1.7	1.2	1.2
3an	16.4	1.3	3.5	0.5	0.7	0.3	−1.6	1.3	1.6
3as	15.7	2.5	3.6	1.3	1.4	1.0	−0.7	2.9	2.0
3bn	15.8	2.1	3.5	0.9	1.4	0.7	−1.1	1.6	1.8
3bs	15.3	2.5	3.6	1.2	1.7	0.9	−0.8	2.4	1.8
4a	14.9	1.8	3.6	1.4	1.1	1.1	−0.5	3.3	1.9
4b	14.7	3.5	3.5	2.3	2.4	2.0	0.3	3.9	3.0
5a	15.6	1.4	3.6	0.1	0.7	−0.4	−1.7	0.7	1.1
5b	14.8	1.9	3.5	0.5	1.0	0.0	−1.0	1.2	1.5
6a	16.0	1.6	3.5	0.2	0.8	−0.2	−1.6	0.9	1.0
6b	16.1	1.8	3.6	0.1	0.5	−0.2	−1.5	0.9	0.8
7a	15.8	1.8	3.6	1.0	1.5	0.6	−0.9	2.4	1.6
7b	15.7	1.3	3.6	0.1	0.5	−0.4	−1.3	1.4	0.4
8a	14.4	1.7	3.5	1.0	1.1	−0.2	0.0	1.6	2.3
8b	15.5	1.5	3.6	1.8	1.5	1.2	0.5	2.3	2.3
9a	14.4	1.8	3.6	1.8	2.1	1.2	0.4	2.8	2.6
9b	14.4	1.8	3.5	0.3	−0.1	1.8	−0.9	0.9	0.0
Mean	15.5	1.9	3.6	0.8	1.1	0.5	−0.9	1.8	1.6
Precipitation (% of baseline)									
1a	248.7	38	109	84	77	74	157	47	64
1b	248.8	32	110	79	90	66	146	46	48
2a	225.7	42	110	89	121	79	133	64	49
2b	219.8	37	111	93	118	79	147	72	51
3an	199.9	44	110	88	101	81	122	79	57
3as	209.8	47	111	76	89	70	115	56	51
3bn	192.0	52	111	100	122	90	135	91	64
3bs	195.5	50	112	88	96	88	113	77	66
4a	188.5	64	112	79	119	71	83	58	62
4b	176.5	57	113	78	119	91	72	60	48
5a	237.1	29	111	91	113	69	142	75	54
5b	237.4	36	112	96	113	88	129	82	67
6a	212.7	46	112	93	101	81	153	76	53
6b	209.9	52	113	82	79	81	112	73	63
7a	195.8	45	113	53	51	51	69	38	62
7b	210.4	64	114	75	67	100	83	50	76
8a	240.4	34	113	82	83	91	93	71	73
8b	222.9	40	112	107	97	128	135	83	94
9a	238.4	40	113	72	71	84	89	49	67
9b	247.7	61	114	71	77	70	72	54	81
Mean	217.9	46	112	84	95	82	115	65	63

[a] HIST1 values are actual 1951–1980 averages, T(°C), P(mm).

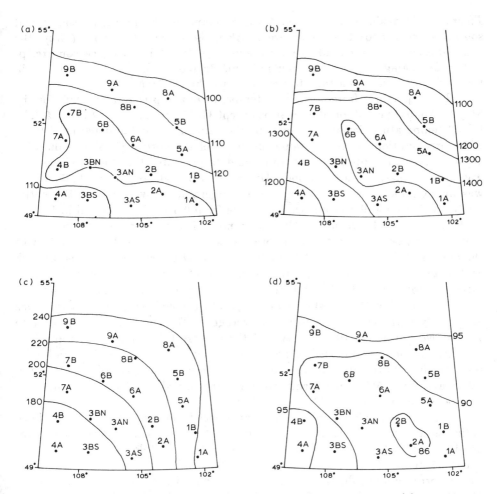

Figure 4.3. The 1951–1980 (HIST1) climate in southern Saskatchewan. (*a*) Variation in growing season length (days), (*b*) variation in growing season degree-day totals over 5 °C, (*c*) variation in growing season precipitation totals (mm), and (*d*) variation in spring wheat maturation time (days).

100 to 120 days generally decreasing in a south to north direction. The exception to this pattern is in the southwest corner of the study area where the higher elevation of the terrain in general, and particularly in the Cypress Hills, results in the GSL being of similar duration to that of the more northerly agricultural area of the province. The quantity of heat available for crop growth (GDD), as presented in *Figure 4.3(b)*, follows the growing season length pattern with the greatest amount of heat available in the central part of the study area (1400 degree-days) and the least in the north (1100) and southwest corners (1200). It may be noted that when GDD for the year as a whole are considered (*Figure 2.1* – HIST1 map), including the early spring and late autumn contributions beyond the ends of the growing season as defined here, the totals are 300–400 units higher. The spatial patterns, however, are quite similar to those presented in

Figure 4.3(b) again with the lowest values found in the north, northeast and southwest corner of the study area. Also, as shown in *Figure 4.3(c)*, the southwest corner of the study area is the driest with total growing season precipitation averaging slightly less than 180 mm. In moving both north and east in the study area, values increase to slightly more than 240 mm.

The departures from normal for temperature and precipitation data for the May to August period for all scenarios except GISS2 are given in *Table 4.2*. The GISS2 temperature deviations would be the same as those tabulated for GISS1, while precipitation for GISS2 would be 100% of the normal (HIST1) values for all districts. It is quite apparent from the data presented that the HIST4 period was an extremely variable period. During this period, except for 1935, which was an extremely cool wet year, mean temperatures (averaged for the entire study area) were well above normal, ranging from 0.5 °C above normal in 1934 to +1.8 °C in 1936. Extremes of 3 °C above normal in temperatures in individual crop districts occurred in the areas of greatest deviation in 1936 and 1937, with the greatest being 3.9 °C in 1936. In 1933 and 1934 the extremes ranged as high as 2.4 and 2.0 °C respectively. In the middle of this warm sequence was an extremely cold year (1935) with temperatures averaging 0.9 °C below normal.

Coincident with the abnormal temperature conditions, precipitation was also highly variable, ranging from an average of 37% below normal in 1937 to 15% above normal in 1935. In two of these years (1936 and 1937) significant decreases in precipitation below normal occurred, amounting to 35% and 37%, respectively. On the other hand, during the extremely cold year of 1935, precipitation averaged 15% above normal with a range as high as 57% above normal to 31% below normal for individual crop districts.

The high variability in weather experienced between years in the HIST4 period is also very evident. For example, the year-to-year fluctuation in temperature within the study area ranged from a 1.5–2.0 °C decrease between 1934 and 1935, to a 2.5–4.0 °C increase between 1935 and 1936. Coincident with this was a corresponding fluctuation in precipitation with a 33% increase between 1934–35 and a 50% decrease from 1935 to 1936. Add to this the two extremely warm dry, consecutive years that occurred in 1936 and 1937, and it becomes quite evident why this period is undoubtedly one of the most variable 5-year periods on record in western Canadian history. The effect of this variability on the agricultural system was devastating in relation to the technology available at that time and, as will be shown in following subsections, a return to similar conditions in conjunction with 1980 levels of technology, although not as devastating, would nevertheless be significant in terms of the reduction in yields and the subsequent effects on the economy.

The difference in climatic conditions for the HIST2 (1961) extreme year are given in *Table 4.2*. As shown, the temperature deviation, while averaging 1.9 °C above normal, ranged from +1.3 to +3.5 °C throughout the study area. The extreme nature and extent of the regional deviation in climate from normal is also quite apparent with every crop district being at least 1.3 °C above normal. As in the HIST4 drought period, precipitation was well below normal, averaging about 46% of normal for the entire agricultural area; the range varied from a low of 29% to a high of 64%.

In all cases the historical years tend to represent extremes in variability for both temperature and precipitation. The GISS1 scenario, on the other hand, tends to represent an extreme in temperature only with temperatures averaging 3.5–3.6 °C above the existing normals. As well, the projected climatic change is virtually identical over the entire study area, whereas historically the data show considerable regional variation in temperature deviations from normal. The combination of warmer temperatures and increased precipitation for the GISS1 scenario also contrasts markedly to the historical experience where warmer conditions have usually coincided with below normal precipitation. The GISS1 scenario suggests that precipitation levels would be increased by slightly more than 10% above the normal (HIST1) levels.

All of these climatic scenarios have important impacts on crop yields and subsequently on the economy. In the case of an extreme year the following subsections describe how short-term perturbations in climate might affect spring wheat yields. With the GISS scenarios we attempt to assess the likely effects on spring wheat of the implied shifts in long-term average climate. It should be recognized that spring wheat is not the only crop that would be affected by these perturbations and shifts. We recognize this fact in the present study; however, it was beyond the scope of this project to evaluate the impact on all crops.

In discussing the impacts of the various climatic scenarios on spring wheat yields, the following sections outline the effect of the temperature changes on the ability of spring wheat to mature, the effect of temperatures and precipitation in terms of their impact on yield, and subsequently the impact on provincial crop production. The derived wheat yield values are then used in Section 5 as input into an input–output economic model to determine the economic impacts.

4.5. Effects of Climatic Changes on Spring Wheat Maturity

This subsection focuses on the part of the growing season from spring wheat planting to maturity. The average time required to mature spring wheat, as determined by the biometeorological time scale, for HIST1 ranges from 86 to 98 days (*Table 4.3*) with the lower values observed in the southeast and central crop districts [*Figure 4.3(d)*]. Comparing these maturity requirements with the available GDDs [*Figure 4.3(b)*] shows the close correlation of the length of time required to reach maturity and the total heat available. Ignoring the effect of day length and comparing the thermal resources over the growth period of spring wheat from planting to ripening revealed that this crop generally requires from 1000 to 1100 degree-days in Saskatchewan. These results, together with the sensitivity analyses presented in Subsection 4.3, indicate that the amount of heat required to mature wheat is basically the same throughout the agricultural area in Saskatchewan. The key factor affecting wheat development is the rate of heat accumulation, and as can be seen from examining *Figures 4.3(b)* and *4.3(d)*, the areas with the longest maturation time requirement correspond to the coolest areas. Conversely, the warmer the temperature the faster spring wheat matures. Of course, as shown in *Figure 4.1(b)*, the effect of temperature on maturity is nonlinear. The results of the sensitivity analysis suggests that spring wheat

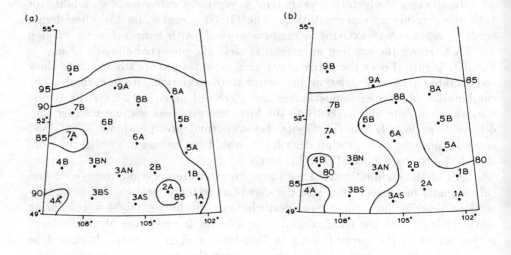
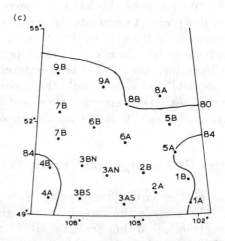

Figure 4.4. Variations in spring wheat maturation time (days) for: (a) 1933–1937 (HIST4) scenario, (b) 1961 (HIST2) scenario, and (c) 2 × CO_2TP (GISS1) scenario.

maturation time would level off somewhere between 70 and 80 days for a temperature increase somewhat greater than 3°C above the HIST1 level.

Derived maturation requirements for the HIST4 (1933–1937), HIST2 (1961) and GISS1 (2 × CO_2TP) scenarios are shown in *Figure 4.4*; the differences from normal for individual years within the HIST4 period are given in *Table 4.3*. The data indicate that climatic warming would generally reduce maturation time. For example, the average maturation requirements for the HIST4 scenario differ from normal by +2 to −12 days, while for HIST2 the difference between crop districts ranges from −5 to −18 days. There is considerable variation among the individual years making up the HIST4 average. However, the largest decrease in maturation time generally coincides with the greatest increases in temperature

Table 4.3. Difference in maturation days between HIST1 (1951–1980 normal), HIST2, HIST4 and GISS1 scenarios.

Crop district	HIST1 (1951–1980)	HIST4 individual years					HIST4 (Mean)	HIST2 (1961)	GISS1
		1933	1934	1935	1936	1937			
1a	88	−2	8	10	−3	−4	2	−8	−4
1b	87	−2	9	10	−2	−5	2	−7	−4
2a	86	−4	6	3	−5	−7	−1	−7	−4
2b	86	−2	10	5	−2	−5	1	−7	−4
3an	87	−5	7	4	−2	−8	−1	−6	−5
3as	88	−6	6	4	−3	−8	−1	−9	−6
3bn	89	−7	7	0	−7	−8	−3	−7	−7
3bs	92	−9	5	1	−8	−10	−4	−11	−10
4a	96	−10	3	12	−11	−11	−4	−9	−12
4b	98	−18	−2	−8	−15	−18	−12	−18	−14
5a	89	−3	9	10	−1	−5	2	−7	−5
5b	92	−4	9	7	−2	−8	0	−9	−11
6a	87	−1	11	7	0	−7	2	−7	−4
6b	87	−2	10	3	−1	−6	1	−7	−5
7a	88	−8	7	0	−7	−11	−4	−8	−6
7b	89	−2	11	1	−4	−4	0	−5	−7
8a	93	−3	13	−6	−6	−12	−3	−10	−14
8b	87	−2	11	0	−6	−6	0	−7	−7
9a	95	−10	2	−9	−11	−15	−8	−10	−13
9b	94	4	−9	7	−4	19	3	−9	−12
Mean (±SD)	90 (±4)	−5 (±5)	7 (±5)	3 (±6)	−5 (±4)	7 (±7)	−2 (±4)	−8 (±3)	−8 (±4)

(*Tables 4.2* and *4.3*). Crop district 4b, which for all historical scenarios has the largest increases in temperature from normal, has temperature departures of +2.4, +2.0, +0.3, +3.9 and +3.0 °C, for individual years from 1933 to 1937 respectively (average +2.3 °C), and +3.5 °C for HIST2. These correspond to the largest decreases in maturation requirements, averaging 12 days for HIST4 and 18 days for HIST2. At the same time, lower temperatures for parts of the study area in some years in the HIST4 period coincide with increases in the length of time required to mature the crop. This is particularly evident in 1935 when, with temperatures averaging 0.9 °C below normal, there was an increase of about 3 days in maturation time.

Comparison of the estimated effects of warming for the GISS1 scenario [*Figure 4.4(c)*] to those observed for the historical scenarios reveals both similarities and differences. The predicted increase in Tmean for GISS1, averaging 3.5 °C above the HIST1 level for the May to August period (*Table 4.2*), is comparable with that obtained for some crop districts for HIST2 and HIST4. The effect of the large-scale warming associated with GISS1 is to reduce spring wheat maturation time to the 79 to 84-day range. However, unlike the historical scenarios where the longest requirements remain in the northern part and southwestern corner of the agricultural area, the time required to reach maturity tends to be reversed. That is, the northern region has the shortest requirement, 79–80 days, as opposed to 82–84 days in the south and central parts. This may be attributable to the planting date advances of 2–4 weeks, and the coincident greater increases in the northern districts' day lengths.

Comparison of the changes in maturation time (*Table 4.3*) with those for May to August temperatures (*Table 4.2*) indicates that the GISS1 decreases are less than might be expected from a historical perspective. The tendency for the increase in impact on maturation time to level off at higher temperatures [*Figure 4.1(b)*] could only account for part of this effect. For example, the increase in May to August temperature was considerably less for HIST2 than for GISS1 for all but one of the crop districts, but in 12 of the districts the reduction in time to mature was greater for HIST2 then GISS1. Also, in most historical cases where the district temperature departures for a year were between +2.4 and +3.5 °C, the reduction in time to maturation was greater than that for GISS1, for which the departure was about +3.5 °C.

Consideration of growing season start and end and growing season length, May to August temperatures, and days to maturity, for a cross section of districts (1a, 4b and 9b, *Table 4.4*), can be used to explain this contradiction. The reduced impact for GISS1 is attributable to the fact that the temperature increase for this scenario is distributed fairly uniformly among the growing season months and equally between Tmax and Tmin. The distribution, however, is quite nonuniform for the historical scenarios, as is illustrated for HIST2 for district 4b (*Table 4.5*). In this example the temperature deviation is much greater for HIST2 than for GISS1 in June. As well, the fact that the positive departure in mean temperature in May is reflected nearly exclusively in the maximum temperature may also be relevant. Greater heat during certain critical periods may account for the greater impact of the temperature increases on maturation time for the historical scenario than for GISS1; however, further research would be needed to confirm this.

An interesting point revealed in comparing the HIST2 and GISS1 scenarios is the fact that in both cases the results indicate the greatest impacts in the northern and southwestern districts. Reductions in maturation time of at least 10 days were estimated for crop districts 3bs and 4b in the southwest and 8a and 9a in the northeast for HIST2, and in 3bs, 4a and 4b in the southwest and 5b, 8a, 9a and 9b in the north with GISS1. It also appears that for warming conditions of a degree or more the difference in maturation time between crop districts would tend to disappear, with the region becoming much more homogeneous; i.e., ranges in maturation times are 86–98 days for HIST1, 79–86 days for HIST2, and 79–84 days for GISS1.

4.6. Effects of Climatic Changes on Yields and Production

In this subsection the impact of the various historical and model-generated scenarios on spring wheat yields and production are outlined. The relative areas of wheat grown on stubble and fallow land are shown in *Table 4.6*. The percentage of crop grown on stubble and fallow varies for each soil zone. Approximately 87% of the wheat is grown on fallowed land in the brown soil zone, 80% in the dark brown zone and 75% in the black soil zone.

Table 4.4. The difference in spring wheat planting date, growing period temperature (°C) and maturation time between HIST1, and the HIST2 and GISS1 scenarios for crop districts 1a, 4b and 9b.

Crop	GSS	GSE	GSL	Tmean		Tmax	Tmin	Maturity
				HIST1 (1951-80)				
1a	140	256	117	17.1		24.0	10.2	88
4b	137	258	122	15.9		24.6	7.3	98
9b	145	253	109	15.7		22.3	9.2	94
				HIST2 (1961)				
1a	144	286	143	19.4	(+1.8)	28.1	10.8	80
4b	135	283	149	18.6	(+3.5)	27.0	10.3	80
9b	138	258	121	17.1	(+1.8)	24.6	9.7	85
				GISS1 ($2 \times CO_2 TP$)				
1a	113	292	180	18.7	(+3.5)	25.5	12.0	82
4b	124	276	153	17.6	(+3.5)	25.4	9.9	84
9b	126	284	132	17.7	(+3.5)	24.4	11.1	83

GSS = Growing season start; GSE = Growing season end; GSL = Growing season length.

Table 4.5. Saskatchewan crop district 4b monthly mean, maximum and minimum temperature deviations from HIST1 for the HIST2 and GISS1 scenarios.

	May			June			July			August		
	Tmean	Tmax	Tmin	Tmean	Tmax	Tmin	Tmean	Tmax	Tmin	Tmean	Tmax	Tmin
HIST1	9.4	14.4	4.3	14.7	23.7	5.8	18.1	27.3	9.0	16.7	26.2	7.2
HIST2	2.5	4.4	0.7	4.9	4.5	5.2	1.6	1.0	2.0	5.0	5.2	4.9
GISS1	3.6	3.6	3.6	3.3	3.3	3.3	3.3	3.3	3.3	4.0	4.0	4.0

4.6.1. Assumptions concerning effects on stubble and fallow yields

The yield model described in Subsection 4.3 can estimate the yields from either cropping system since the basic difference between the two is the soil moisture reserves at the time of planting. Generally these reserves, and consequently yields, are higher on those soils that have been fallowed the previous year (*Table 4.6*). In this study yields have been estimated for what is assumed to be a continuous cropping (stubble) system since there is no satisfactory way of determining what the actual moisture reserves would be on the planting date for a fallow crop other than actual measurement. A simple procedure has been used to derive the soil moisture reserves based on the annual ratio of precipitation to potential evaporation. Results were very favorable on comparing this approach with that for calculating total moisture reserves on May 1st developed by Sly (1982).

The absolute yield differences between the stubble and fallow yields, although showing a considerable range within each of the soil zones (*Table 4.6*), appears to be quite similar when the averages for the three zones are compared. Similarly, the ratio of stubble to fallow yields appears to be fairly consistent throughout Saskatchewan. In the economic impact analyses presented in Section

Table 4.6. Reported stubble, fallow and all spring wheat areas and yields by crop district and major soil zone in Saskatchewan, averaged for the period 1961–1979.

Crop district	All spring wheat		Stubble		Fallow		S/F (%)	Area stub (%)	Fal-stub yield diff (kg/ha)
	Area (ha)	Yield (kg/ha)	Area (ha)	Yield (kg/ha)	Area (ha)	Yield (kg/ha)			
Brown soil									
3as	460 956	1480	67 216	1044	393 740	1553	67	15	509
3an	246 741	1458	24 970	1028	221 771	1515	69	10	487
3bs	332 321	1383	36 841	921	295 480	1445	64	11	524
3bn	456 522	1480	62 084	1029	394 438	1564	66	14	535
4a	162 568	1232	17 167	798	145 401	1275	63	11	477
4b	254 288	1443	23 788	887	230 500	1503	59	9	616
7a	419 392	1678	65 542	1211	353 850	1768	68	16	557
Av. yield		1478		1034		1547	67	12.8	513
Total area	2 332 788		297 608		2 035 180				
Dark brown soils									
1a	306 060	1555	63 660	1175	242 400	1660	70	21	485
2a	308 341	1553	56 026	1148	252 315	1618	71	18	470
2b	421 332	1719	101 102	1308	320 230	1843	71	24	535
6a	577 485	1582	112 189	1184	465 296	1689	70	19	505
6b	436 298	1526	83 737	1094	352 561	1661	66	16	567
7b	311 155	1700	51 310	1195	259 845	1820	66	16	625
Av. yield		1602		1190		1715	69	19.8	525
Total area	236 071		468 024		1 892 647				
Black soils									
1b	217 027	1679	39 793	1252	177 234	1754	71	18	502
5a	419 413	1695	81 228	1282	338 185	1789	72	19	507
5b	375 040	1760	98 728	1317	276 312	1877	70	26	560
8a	190 300	1789	59 636	1432	130 664	1915	75	31	483
8b	280 207	1768	95 287	1410	184 920	1994	71	34	584
9a	283 747	1655	70 125	1206	213 622	1824	66	25	618
9b	191 196	1713	51 473	1330	139 723	1972	67	27	642
Av. yield		1721		1323		1861	71	25.4	538
Total area	1 956 930		496 270		1 460 660				
Province									
Av. yield		1595		1169		1691	71	23.4	522
Total area	6 650 389		1 261 902		5 388 487				

S/F = Stubble/fallow yield ratio.

5 the data presented here are used to adjust the model-generated stubble yields to derive fallow system and combined fallow/stubble system yields and total production estimates for Saskatchewan.

In using the approach described above it is assumed that the climatic impact on fallow and stubble crop yields would be identical. Historically this has not been true. The production records show generally that as conditions have become drier the ratio of stubble to fallow yields has tended to decrease. This point is illustrated in *Table 4.6* where the stubble/fallow ratio tends to be higher in the black and dark brown soil zones (71% and 69%), and lower in the brown soil zone (67%) coinciding with drier conditions. Overall, the ratio varies from 59% in the dry southwest to 75% in the more moist northeast. This point can be further demonstrated by reference to an individual year (1961) and a single crop district (4a). For this year the ratio of reported stubble to fallow yields was 38%, while, as shown in *Table 4.6*, the 1961–1979 average was 63%. The available soil moisture reserves at planting contribute greatly to the yield variations and these vary from year to year depending on the climate, soil type and land use. Without actual measurements, they are virtually impossible to estimate adequately. For this reason, although the assumption of equality of climatic effects on stubble and fallow yields is not a good assumption, it was necessary considering that there is no alternative for this case study.

4.6.2. Summary of effects

Reported average wheat yields for stubble crops in Saskatchewan for the 1961–1979 period are presented in *Table 4.6*. As shown, the range varies from less than 800 kg/ha in the southwest to more than 1400 kg/ha in the north and east. Generally, the spatial difference in yield between major soil zones averages 155 kg/ha in moving from the brown to dark brown soil zones and 135 kg/ha from the dark brown to black zone. Comparing the patterns for reported yields with that for historical precipitation [*Figure 4.3(c)*] indicates clearly the strong link between yields and precipitation.

Effects of the HIST4 (1933–1937) scenario

Reported average spring wheat yields provide the basis against which to compare the derived scenario yields (*Table 4.7*). Results clearly show a high degree of variability of weather impacts on yields throughout Saskatchewan, both spatially and temporally, with no two individual years having the same impact patterns (*Figure 4.5*). However, there is some tendency for the southwest corner and the northern parts to be more adversely affected. It appears that drought conditions are likely to reduce yields more in these areas than in the eastern and central districts. These latter districts are affected by the drought impact in some years while in others they are not. For example, for crop district 6a – a centrally located district – the yield was more than 25% below normal in 1933, while for 1935 it was more than 25% above normal. On the other hand, crop district 4a in the southwest corner and 8a in the northeast are consistently below normal for HIST2 and all HIST4 years. The southeast corner, for the average of the HIST4 years, indicates that yields are only slightly below normal, whereas in the southwest and northern portions average yields are 40–50% below normal [*Figure 4.6(a)*].

Another important feature demonstrated in the results is the influence of the previous year's moisture on current year yields. This feature is clearly illustrated using the three consecutive years 1935, 1936 and 1937. As previously mentioned, 1935 for the study area was an extremely cool wet year with temperatures averaging approximately 0.9°C below normal and precipitation 15%

Table 4.7. Derived stubble spring wheat yields (kg/ha) for the HIST2, HIST4 and

Crop district	Reported stubble yield[a]	% all yield	HIST4 period				
			1933	1934	1935	1936	1937
			Brown				
3as	1044	71	372	1070	1240	1069	311
3an	1028	71	462	1036	1389	1147	411
3bs	921	67	442	1042	1422	1018	286
3bn	1029	70	595	1148	1012	405	469
4a	798	65	380	543	584	512	287
4b	887	61	757	871	558	592	139
7a	1211	72	182	1291	939	320	334
Mean	1034		424	1082	1069	696	339
% norm	100	69	41	105	103	67	33
			Dark brown				
1a	1175	76	810	1069	2150	826	828
2a	1148	75	986	872	1998	1010	457
2b	1308	76	1177	1163	2167	1787	348
6a	1184	75	797	1118	1762	1430	432
6b	1094	72	404	1195	1348	661	323
7b	1195	70	661	1318	1601	617	790
Mean	1190		818	1127	1839	1090	490
% norm	100	74	69	95	155	92	41
			Black				
1b	1252	75	1018	764	2275	992	726
5a	1282	76	1017	910	2038	1399	531
5b	1317	75	1106	979	1563	1277	670
8a	1432	80	1071	1057	1090	1011	636
8b	1410	80	857	1239	1575	1048	677
9a	1206	73	444	792	678	367	362
9b	1330	78	892	660	1325	444	879
Mean	1323		917	950	1493	983	627
% norm	100	77	69	72	113	74	47
			Province				
Mean	1206		760	1037	1520	953	507
% norm	100		63	86	126	79	42

above normal over the May to August period. On the other hand, 1936 and 1937 were very warm dry years with temperatures averaging +1.8 and +1.6 °C above normal and precipitation 65% and 63% of normal, respectively (see Table 4.2). These years represent essentially the case of an extremely good year followed by two consecutive drought years. It should be noted, as well, that they are similar to HIST2 with respect to the departures of temperature from normal (+1.9 °C for HIST2), but not as severe in terms of below normal precipitation (46% for HIST2). The results indicate that although the 1936 deviations of temperature and precipitation were quite similar to those for 1937, the estimated effect on crop yields was significantly less, with an average 21% reduction in 1936 as compared with 58% in 1937. The effect of the drought in 1936 magnified the drought impact in 1937 basically by reducing the plant available soil moisture at the

GISS scenarios by crop district and major soil zone in Saskatchewan.

HIST1 (mean)	HIST2 (1961)	GISS1 $(2 \times CO_2)$ (TP)	GISS2 $(2 \times CO_2)$ (T)
		Brown	
812	115	999	901
889	258	1061	953
842	202	908	813
726	305	1014	895
461	182	641	560
583	148	726	626
613	255	1173	1017
735	215	992	877
71	21	96	85
		Dark brown	
1137	283	1021	955
1065	230	1085	936
1328	158	1061	903
1108	247	1090	963
786	247	927	754
997	647	1076	930
1084	274	1043	910
91	23	88	76
		Black	
1155	292	1047	973
1179	81	1001	931
1119	219	886	794
973	567	914	736
1079	345	1048	867
529	451	796	698
840	658	907	795
994	346	941	830
75	26	71	63
		Province	
965	289	989	868
80	24	82	72

[a] All scenario yields are based on the average 1961–1979 reported stubble yields.

beginning of the 1937 growing season. The available soil moisture in 1936 was, on the average, 28% above normal on the crop planting date, in comparison with 40% below normal at the beginning of 1937. Had the soil moisture reserves in 1937 been at or near normal levels, the effect on crop yields of the shortage of growing season precipitation would have been significantly different. This is evident, in that the overall production in Saskatchewan, as estimated by the yield model, would be about 20% below normal for 1936, while for 1937 the combined effect of the dry previous year and growing season drought would be to reduce yields by 60% from the normal. This demonstrates clearly the importance of available soil moisture reserves at planting time to Saskatchewan wheat production, particularly for drought years.

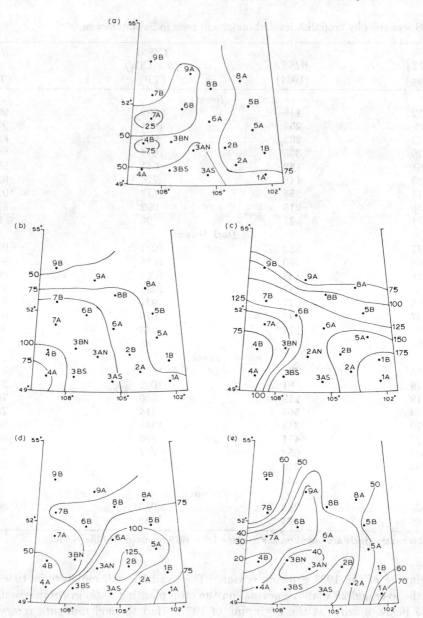

Figure 4.5. Simulated variation in Saskatchewan spring wheat yields (percentage of 1961–1979 average): (a) 1933, (b) 1934, (c) 1935, (d) 1936, and (e) 1937.

As shown, below-normal precipitation has a greater impact when moisture reserves are below normal. The above-normal moisture reserves tend to reduce the potential impact of both below-normal precipitation and/or higher temperature. These are important features that tend to be overlooked when historical analogues are used to represent a potential future condition. The result is that

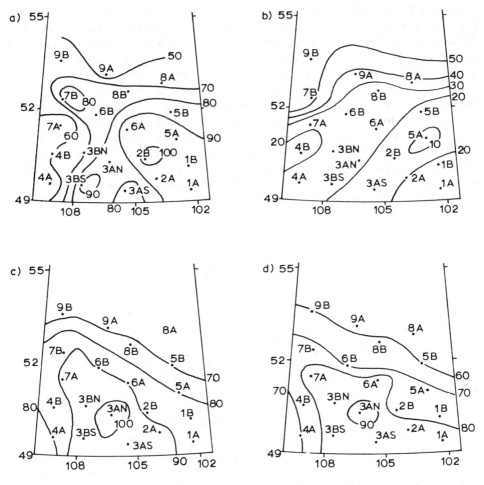

Figure 4.6. Variations in Saskatchewan spring wheat yields (percent of 1961–1979 average): (a) for 1933–1937 (HIST4) scenario, (b) for 1961 (HIST2) scenario, (c) for 2 × CO_2TP (GISS1) scenario, and (d) for 2 × CO_2T (GISS2) scenario.

one's impression that a particular year or set of years are bad or good can be upset terribly by missed facts. The above examples clearly demonstrate this in that, had 1936 and 1937 been looked at in terms of individual years and soil moisture reserves set at normal levels, the results would have been vastly different.

Effects of the HIST2 (1961) scenario

For the HIST2 worst-case individual year scenario [*Figure 4.6(b)*], virtually all of Southern Saskatchewan has computed yields below 50% of normal. Moreover, almost the entire area south of 52°N has estimated yields of less than 30% of normal. When the intensity of the HIST2 scenario impact is compared with those of 1936 and 1937, the large difference is quite evident. In 1936 only four crop districts have estimated yields below 50% of normal. In fact, seven crop districts have estimates near or above normal. For 1937, on the other hand, all the values are at least 30% below normal, with seven crop districts having derived yields at least 70% below normal. The comparison between 1937 and HIST2 shows that the trend in yield isoline patterns is rather similar, with higher values in the northwest and southeast corners and a lower yielding zone in between. However, in the eastern part of the study area, 1937 yields were considerably higher, while in HIST2, the drought zone extends eastwards into Manitoba. The HIST2 drought was generally more severe throughout the wheat-growing region of western Canada, including Alberta and Manitoba, than were 1936, 1937 and other drought years. The impacts of these yield reductions and variability on the economy of Saskatchewan are discussed in Section 5.

Effects of the $2 \times CO_2$ climatic scenarios

With the year-to-year changes exemplified in the HIST2 and HIST4 scenarios discussed above, it is reasonable to assume that no major adjustments, such as shifts to different crops or cropping systems, would take place in response to these conditions. This is because the industry and market could not adapt rapidly to these sorts of unanticipated fluctuations. With shifts in long-term climate as are predicted to accompany CO_2 doubling, however, the situation would be different if the climatic change takes place gradually over several decades. In assessing only a single crop, spring wheat, we cannot show what the likely effects would be on overall productivity. Spring wheat is well adapted to the current climate in Saskatchewan. What we can do is to estimate whether this crop is likely to become less and less productive if the long-term climate changes and, as a consequence, whether adjustment by changing crops is likely to be needed to maintain existing productivity levels. Analyses were undertaken to examine how long-term yields might change, given the predicted climatic changes suggested by GISS1. The results [*Figure 4.6(b)*] show that the overall effect of these changes, as in the historical scenarios, would be to decrease yields. However, the southern area would be generally less affected, with yields remaining within 20% of current levels. The more northerly districts, however, would experience a greater impact with yield decreases of from 25 to 35% projected.

Comparison of the averaged (1933–1937) HIST4 results [*Figure 4.6(a)*] with the GISS1 scenario suggests some similarities are evident. In both cases the largest impact on yields occurs in the southwest and northern parts of the agricultural area. The pattern is consistent in spite of the difference between the HIST4 and GISS1 precipitation. The historical period is noted for its significant decrease in precipitation below normal, whereas the GISS1 scenario projects precipitation increases of approximately 9–14% above the current HIST1 level. This precipitation increase, however, as shown by the below normal yields, would be more than offset by the adverse effects associated with the higher temperatures.

Comparing the HIST4 data to the GISS data, one would think that because of the average 3.5 °C increase in temperature, despite the slight increase in precipitation, the impact on yields would be much more significant. The most striking feature of the GISS scenario results is that the relatively warm dry south is the least affected, while the cooler moister north is more affected. The reason for this is that the northern area is, in actual fact, more affected by the temperature increase than the south. This can be demonstrated, for example, by comparing crop district 9a in the north with 3an in the south in terms of the impact of the GISS warming on spring wheat maturation time in relation to the current HIST2 figures. As shown in *Table 4.3*, the change in maturation requirements is dramatic with a 13-day decrease in crop district 9a in comparison to 5 days in crop district 3an. Maturation requirements under the GISS climate would be equivalent in both districts, averaging approximately 82 days. The reason for this is the change in average temperature during the period of crop growth from planting to maturity. Under the GISS scenario the average temperature is increased in crop district 9a by approximately 2.1 °C above the HIST2 normal as opposed to approximately 0.9 °C in crop district 3an. Consequently, the difference in mean temperature, averaged over the period from planting to maturity, from south to north, decreases from the current 1.8 °C to approximately 0.5 °C under GISS scenario conditions. In this instance the longer day length in crop district 9a makes up for the 0.5 °C difference in temperature in terms of speeding up maturation. These results clearly show that the agroclimate in Saskatchewan under GISS conditions would be much more homogeneous (the difference between the north and south is reduced). They also indicate that in moving toward homogeneity the major adjustment in terms of the impact on spring wheat maturation and production would be in the north.

As mentioned in Section 1, above normal temperatures in Saskatchewan tend to be associated with below normal precipitation. The discussion of the historical scenarios in conjunction with the spring wheat analysis has provided further examples of this tendency. Because of this, it is of particular interest to have some indication of the contribution of the precipitation increase projected by the GISS-GCM results to the estimated impact on spring wheat yields. Estimating the impacts both with the increased precipitation (GISS1), and without it (GISS2), provided such an indication.

The results [*Table 4.7, Figure 4.6(c,d)*] indicate that yields would be reduced overall by 18% for GISS1 and 28% for GISS2. In other words, a further reduction of 10% in spring wheat yields could be expected if the climatic warming predicted by GISS occurred but precipitation remained at the 1951–1980 (HIST1) levels. The spatial pattern for GISS2 [*Figure 4.6(d)*] is quite similar to that for GISS1 [*Figure 4.6(c)*], which reflects the fact that the additional 10% reduction resulting from ignoring the GISS-modeled precipitation increases would probably be spread fairly uniformly throughout Saskatchewan.

The interesting feature shown by the model analysis in converting potential or experimental yields to commercial yields with GISS-model-derived climatic conditions, is that the northern area, in effect, will have commercial yields which are less than the south. The model in actual fact predicts higher yields in the north; however, in converting the model yields to commercial yields, this difference arises. The reason for this is illustrated in *Figure 4.7* which outlines the comparison of commercial yields averaged for the 1961–1979 period to model estimates for the 1951–1980 period (in this instance we assume that the

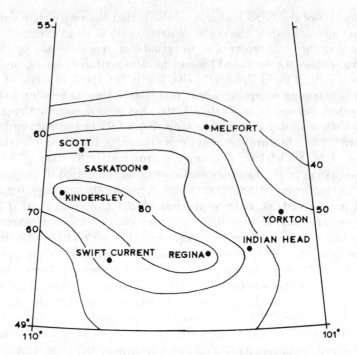

Figure 4.7. Ratio of commerical yields to modeled yields in Saskatchewan.

differences between climate averaged over the periods 1951–1980 and 1961–1979 is minimal). It should be remembered that the model represents the yields obtained at experimental stations under the most optimum of management conditions, and the results do not represent those that would be attained in reality under commercial conditions. The results illustrated in *Figure 4.7* clearly show this fact. Results indicate that the south-central core region of Saskatchewan is producing at about 80% of its potential and that the productivity appears to drop off quite significantly in all directions moving away from this core area. Explanations for this drop in productivity range from the loss in soil fertility since the area was brought into production to the slow adoption of farmers in the region of recommended fertilizer, herbicide and cultivation practices for newer crop varieties (Mack, personal communication, 1985). The model used in this study assumes that soil fertility and management practices by farmers are equivalent throughout the study area. It is obvious from the results shown in *Figure 4.7* that, in terms of commercial yields, this is not a good assumption. Unfortunately, because of the model framework, it is a weakness that has to be accepted.

The data presented in *Figure 4.7* provides a comparison of modeled potential yields with reported commercial yields. To determine if the depicted pattern was real or a modeling problem, yield data obtained from research station variety trials were obtained to see if actual experimental yields showed a similar relationship to commercial yields. Variety trial information obtained from 8

Table 4.8. Ratio of experimental (Exper) and modeled (M) spring wheat yields to commercial (Comm) yields for selected stations in Saskatchewan.

Crop district	Location	Years[a]	Exper[b] (kg/ha)	Comm (kg/ha)	Comm/Exper (%)	C/M[c] (%)
2b	Regina	1975–78	3250	2190	67	80
2b	Indian Head	1975–78	4197	2190	52	64
3bn	Swift Current	1978–79	2400	1735	72	70
5a	Yorkton	1975–78	2710	2266	83	55
6b	Saskatoon	1975–78	3245	1979	64	75
7a	Kindersley	1978–79	2725	2034	75	80
7b	Scott	1978–79	3083	2115	69	70
8b	Melfort	1975–77	3542	2426	68	58

[a] Wheat varieties tested for 1975–1978 include: Benito, Neepawa, Manitou and Sinton, while for 1978–1979, wheat varieties tested included: Neepawa, Sinton, Chester and Columbus.
[b] Experimental yields were obtained from the Central and Western Bread Wheat Cooperative Tests, 1975–1978, and 1978–1979; Seeds Section, Food Production and Inspection Branch, Agriculture Canada, 1980.
[c] Commercial to Model yield ratio, representing the actual 1961–1979 average for commercial yields versus the model calculations for the 1951–1980 climate data.

research locations in Saskatchewan is presented in *Table 4.8*. Experimental yield data obtained for each location within the 5-year period 1975–1979 are compared to published crop district commercial yields averaged over the comparable records for each station. As shown in *Table 4.8*, comparison of experimental yields to commercial yields supports the modeled results illustrated in *Figure 4.7*. A general comparison is made here because the model yields are calculated over a 30-year average and compared to 20 years of actual data. As well, the experimental data represent data from site specific tests while commercial data represent average yields for a large geographic area (crop district) covering a wide range of management conditions. Considering the variation in yields within crop districts is ±20% or more, the results presented in *Table 4.8* are remarkably similar. Despite the discrepancies in the way the data are compared, the comparison of short-term experimental yields with commercial yields clearly supports the yield pattern projected by the modeling results. The fact that commercial yields under GISS-model-derived climatic conditions are predicted to be lower in the north than the south, as mentioned earlier, is based on the assumption that the climatic impact is felt equally by both experimental and commercial yields. It also assumes that the land quality and management practices involving tillage, fertilizer and herbicide application, etc., are equivalent. Given this, the yield potential as projected by the model in the north (crop districts 9a and 9b) should be slightly more than double that of the south (crop districts 4a and 3bs). However, commercially, as shown in *Table 4.6*, the yield difference is currently on the order of 300–500 kg/ha, or about 20–40%. Assuming the current productivity ratio between experimental and commercial yields remains unchanged (i.e., no technology change) the results presented in this study are a good reflection of what yields should be like, given the projected change to GISS scenario climatic conditions.

The above results suggest that the productivity in the dark brown and black soils has fallen dramatically, or that the soils in these zones are underproducing in relation to the agroclimatic resources available for cereal grain production, as a consequence of poor management practices. Results presented here suggest that the fertility loss, or underproduction, ranges from 20–80% of current yields.

In the above, we have estimated the percentages of normal production that could be expected under various changes in climatic conditions as projected by the HIST2, HIST4 and GISS scenarios. The results suggest that the impact on provincial average spring wheat yields would range from 76% below the existing level for HIST2, to between 58% below and 26% above for the HIST4 period (1933–1937, Table 4.7), should a return to these types of climatic condition occur. We have further estimated that yields under a long-term change in climate, as represented by the GISS GCM results for a doubling of atmospheric CO_2 concentration, would range between 18% and 28% below current levels. In Section 5 we explore the economic ramifications of these impacts on the Saskatchewan economy in terms of employment and dollars at the farm and provincial levels.

SECTION 5

The Effects on the Farm and Provincial Economy

5.1. Economic and Employment Impacts

The effects of climatic change on spring wheat dry matter yield and production for several climatic scenarios (*see Table 1.4*) have been estimated in Section 4. In this section the changes in wheat yield are transposed into farm production changes and the direct and indirect economic impacts likely to result from these projected production changes are estimated. In addition to these climatic scenarios, a scenario involving a partial shift to winter wheat is given as an example of how different farm practices could mitigate against potential drier climatic conditions.

The methodology used in determining the impacts of the selected climatic scenarios is as follows:

Base Case		*Scenarios*		*Scenario-induced*
Derived		Derived		Impacts on
financial and	−	financial and	=	economy and
economic		economic		employment
conditions		conditions		
1980				

Economic conditions occurring during a particular scenario year are compared with those existing for the base situation. The base situation in our case can be described as an economic snapshot of Saskatchewan in 1980. It should be emphasized that in interpreting the results the absolute values are not important; the relative differences between the scenarios as compared with the base case are of primary interest. Also, the reader must be cautioned that in reality there are many nonclimatic factors. The scenario effects discussed here are simplistic at best, and many caveats should be attached.

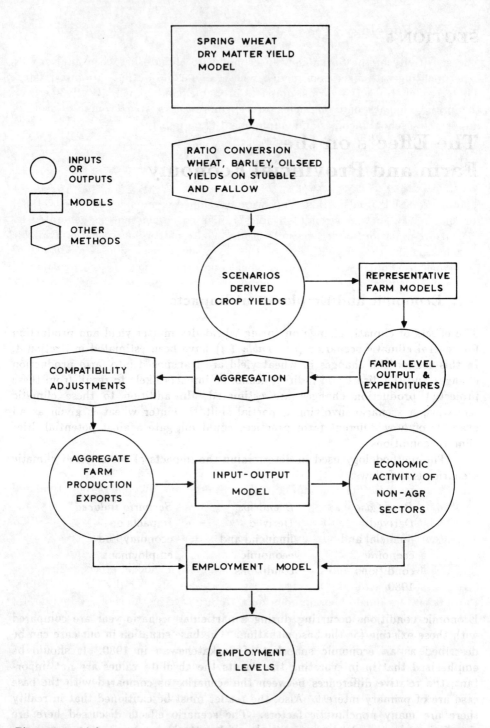

Figure 5.1. Linkages among various models for analysis of economic effects of climatic changes in Saskatchewan.

5.2. Methodology

The sensitivity of the Saskatchewan economy to climate has been analyzed by experimenting with a series of models whose inputs and outputs are interrelated. These models track the various impacts brought about by climatic change from the initial changes of crop yields to the more aggregated economic levels of imports, household incomes and employment (*Figure 5.1*).

5.2.1. Farm level simulation models

The farm models used in this study were developed or modified by Klein (1982) for the "Saskatchewan Drought Studies". The economic analysis generated by the farm simulation models permits study of alternate resource paths available for changes in the organization of the farm enterprise. These models were developed to simulate production of wheat and other grains, oilseeds, cattle and calves, hogs, dairy and poultry. Five models were developed, one for each of five farm types:

(1) Crops.
(2) Beef-Forage-Grain.
(3) Hog-Grain.
(4) Dairy-Forage-Grain.
(5) Poultry.

Model components include:

(1) A main model program and subroutines consisting of a self-contained series of matrices written in FORTRAN IV.
(2) A base data file containing technical data on different production and management strategies.
(3) A farm data file containing inventories, specific production plans, and other items of data that only apply to a specific representative farm.

Though very complex in nature, the simulation models proceed through a series of steps resembling the decision process actually used by farm operators. Farm operators must choose among many alternative methods of production, for instance: which rotation to use, which crops to plant, how many preseed tillage operations to perform, which machine to use, when to plant and what quantities of fertilizers and herbicides to use. Farmers must ensure that they have adequate resources (cash, machines, buildings, labor) to carry out the production plan and they hope the choices deliver suitable rewards. The simulation models mirror this type of decision process.

The models have provisions for accepting a fixed management plan, or the option is available to maximize certain outputs by using an iterative process to select the best management plan. The models incorporate combinations of production and management strategies; they use biological data to differentiate among good and bad growing years; they separate various levels of managerial proficiencies for selecting tasks and they consider integration of enterprises, multiyear planning horizons, and institutional constraints like income tax. In short, the models reflect closely the actual production and management decisions on the farm. The models' output parameters include projected net worth, net income, cash flow and resources required to accomplish selected production plans.

In addition to the management decision inputs, several features were added to the farm models to incorporate specific on-farm adjustments. These included:

(1) Purchases of depreciable assets.
(2) Changes in crop rotation.
(3) Changes in starting date of spring operations.
(4) Changes in preseed tillage operations.
(5) Changes in summer fallow operations.
(6) Different applications of herbicides and fertilizers.
(7) Crop insurance set to maximum.
(8) Let cattle fend for themselves.
(9) Early weaning of calves.
(10) Slower rate of gain.
(11) Haul water, dig dugout or drill well.
(12) Irrigate additional forage.

The Saskatchewan Drought Study (Klein, 1983) conceptualized the aggregation of the farm level results of several representative farms to provincial totals. An appropriate representative group of farms categorized according to type, size and dominant soil zone was selected to represent the more than 65 000 actual farms in Saskatchewan. The following 29 farms, representative of the Saskatchewan farming system, were chosen:

(1) *Mixed farms (4)*: one each on brown and dark brown soils and on black soils with altering mixes of low grain acreages – high cattle numbers (LG-HC) and high grain acreages – low cattle numbers (HG-LC).
(2) *Cow-calf (4)*: one each of small and large farms on brown/dark brown and black soils.
(3) *Feeder farms (4)*: one each of small and large, with finished and unfinished animals.
(4) *Dairy farms (2)*: one small and the other large.
(5) *Hog operations (3)*: one each of small, medium, and large.
(6) *Cereal and Oilseed farms (9)*: one each of small, medium and large farms, in each of the three soil zones: brown, dark brown and black soils.
(7) *Irrigation farm (1)*.
(8) *Poultry farms (2)*: one broiler and the other layer.

Aggregation of these farms to the provincial level totals is accomplished using the distribution of farms shown in *Table 5.1*.

Table 5.1. Types and numbers of farms included in aggregation of the Saskatchewan agriculture sector.

Farm type		Total	Soil type			
			Brown	Brown/ dark brown	Dark brown	Black
Mixed:	Low grain/ High cattle	1 944		508		1 436
	High grain/ Low cattle	1 297		636		661
Cow-calf:	Small	4 153		1 034		3 119
	Large	2 336		1 050		1 286
Feeder:	Unfinished Small	2 319				
	Large	692				
	Finished Small	2 512				
	Large	748				
Dairy:	Small	295				
	Large	332				
Hog:	Small	1 888				
	Medium	1 094				
	Large	300				
Cereal:	Small	18 850	2 834		5 109	10 907
	Medium	20 835	4 704		7 006	9 125
	Large	4 952	1 901		1 785	1 266
Irrigation		81				
Poultry:	Broiler	113				
	Egg	171				
Total		64 831				

5.2.2. Input–output model

Input–output (I–O) analysis is an analytical system that provides a framework for measuring the interdependence among various sectors of a regional economy. An input–output table fulfills two functions: it describes the commodity use relationships between producing and purchasing sectors and between factors of production and output; and given economic assumptions about the nature of production function, it can be used to measure the impact of autonomous disturbances on an economy's output and income.

The input–output model used in this study was developed for the Saskatchewan Drought Studies by Underwood McLellan and Associates (1983). The model contains a transactions table whose entries are based upon the regional income accounting system, except for one difference: in addition to income accounts, the input–output accounts break down the business sector into a number of industries and records the transactions that flow between them.

The commodities purchased by sectors for further production of goods and services are called intermediate demand. The Gross Domestic Product (GDP) of a region is equated to expenditures for consumption by households and other final expenditure components including investment capital used to produce interest income and profits.

Base data for this model are 58 purchasing sectors and 72 commodities, organized in a rectangular input–output table representing the entire Saskatchewan economy. For the purpose of this study, this table has been collapsed into a 12 × 12 square matrix. The 12 purchasing and producing sectors include:

(1) Agriculture.
(2) Nonfuel mines.
(3) Fuel mines.
(4) Forestry and fisheries.
(5) Agricultural processing.
(6) Other manufacturing.
(7) Construction.
(8) Utilities.
(9) Trade.
(10) Transportation, communication and storage.
(11) Finance, insurance and real estate.
(12) Services.

Purchases by other industries (intermediate demand) do not include the total sales of the product. Production is purchased by consumers (households) for consumption, absorbed by governments, exported, or invested for future production. The final users are categorized as final demand. In the end, total output must equal total purchases as outlined in *Table 5.2*. Derived from this matrix is a table of technological coefficients (or budgets). These indicate the value of each commodity required to produce Can$1.00 of output in a given sector.

To determine the economic impact in a given economic sector all that is required is to know what commodities have been changed, and by how much. In this study the agricultural commodities of wheat, barley, oats and canola are allowed to change for each scenario. Changes to wheat were obtained from the crop model results outlined in Section 4. Other grain crop changes were not directly modeled but were calculated by multiplying the current ratio of barley, oats and canola yields to spring wheat yields by the scenario-derived spring wheat yields. In doing this it is assumed that the climatic effects for barley, oats and canola are equivalent to that for wheat.

Since climatic events are local and often isolated occurrences, their impacts should more appropriately be measured for the local economy. To accomplish this, the Saskatchewan economy was divided into four subregional economies.

The ideal means of developing a regional input–output table is to base it upon survey data. Since such an approach is expensive as well as time consuming, the particular nonsurvey method used in this study was the location

Table 5.2. An input-output matrix for Saskatchewan 1980 (millions of Canadian dollars).

Production sectors and primary inputs	Purchasing sectors							
	Agric.	Non-fuel mines	Fuel mines	For. fish.	Agric. proc.	Other manuf.	Const.	Util.
Agriculture	22.94	0.06	0.18	0.10	393.85	1.58	0.66	0.06
Nonfuel mines	204.66	26.18	42.87	0.02	0.30	68.99	130.30	0.71
Fuel mines	1.84	3.09	0.11	0.01	0.03	25.88	2.69	28.84
Forest./fish.	0.21	0.00	0.00	2.56	0.00	22.31	0.18	0.00
Agric. proc.	36.21	0.40	0.02	0.00	62.90	5.58	0.88	0.01
Other manuf.	163.84	12.53	1.72	2.80	7.27	132.02	249.54	2.46
Construction	0.00	18.82	26.03	0.11	2.36	18.65	3.94	81.82
Utilities	0.40	26.77	27.29	0.14	7.33	25.53	2.14	14.66
Trade	1.29	6.89	0.68	0.66	13.67	14.03	53.02	1.31
T/C/S	0.59	19.86	3.91	6.29	45.84	75.18	80.54	23.59
F/I/RE	124.77	57.99	351.70	1.93	2.00	8.96	10.72	5.82
Services	5.88	59.99	21.42	5.84	48.28	103.00	112.00	21.59
Subtotal	562.64	232.56	475.93	20.45	583.82	501.70	646.60	180.87
GDP househ.	998.00	623.00	226.01	13.00	90.00	309.00	500.93	130.00
GDP others	913.00	479.00	207.01	11.00	82.00	283.00	458.94	119.00
Imports	496.70	67.69	26.36	2.26	118.33	484.16	456.09	25.03
CTLS	−6.95	18.77	20.97	1.85	4.66	−54.24	158.40	−6.02
Residual	−7.48	−2.32	−4.68	−0.16	−5.90	−5.82	−7.96	−1.58
Total inputs	2956.00	1318.70	951.60	48.40	872.90	1517.80	2213.00	447.30

Production sectors and primary inputs	Purchasing sectors				Final demand			
	Trade	T/C/S	F/I/RE	Service	House-hold	Other final demand	Residual	Total output
Agriculture	35.45	1.15	2.19	16.63	71.39	2 385.29	24.46	2 956.00
Nonfuel mines	1.06	1.47	3.39	11.15	6.30	811.02	10.29	1 318.70
Fuel mines	1.20	2.53	0.99	0.60	3.20	872.57	8.02	951.60
Forest./fish.	0.01	0.03	0.01	0.75	4.80	17.22	0.31	48.40
Agric. proc.	0.32	0.24	0.03	120.38	75.19	564.26	6.48	872.90
Other manuf.	20.79	44.95	5.99	129.58	184.68	549.77	9.87	1 517.80
Construction	9.89	90.92	275.89	5.66	345.56	1 318.13	15.22	2 213.00
Utilities	25.29	13.98	15.78	15.67	156.48	113.60	2.25	447.30
Trade	9.27	20.06	2.94	150.17	860.50	581.77	6.84	1 723.10
T/C/S	95.29	166.38	56.41	268.59	372.56	288.63	8.17	1 511.80
F/I/RE	36.05	15.04	62.66	34.24	111.79	1 673.01	20.12	2 516.80
Services	186.33	114.20	156.57	480.97	777.21	1 023.62	17.82	3 134.80
Subtotal	420.95	470.94	582.85	1234.37	2569.67	10 198.91	129.94	19 212.20
GDP househ.	593.00	492.00	739.00	600.02	471.95	579.00	0.00	6 264.91
GDP others	543.00	451.00	676.00	549.02	0.00	530.00	0.00	5 301.97
Imports	109.23	78.31	105.16	566.80	2582.61	270.35	17.91	5 407.07
CTLS	60.90	23.39	421.52	197.47	241.77	344.94	8.65	1 436.08
Residual	−3.97	−3.85	−7.73	−12.88	0.00	−92.17	0.00	−156.50
Total inputs	1723.10	1511.80	2516.80	3134.80	6266.00	11 831.03	156.50	37 465.73

T/C/S = Transport/communications/storage
F/I/RE = Finance/insurance/real estate
CTLS = Commodity tax less subsidies

Figure 5.2. Regional boundaries for subprovincial input–output tables.

quotient method. This method assumes that when estimating a regional table from the provincial input–output table, coefficients remain constant whereas self-supply ratios change. The location quotients were based on the share of output of various sectors in various regions. In some cases these weights were based on value of output (agriculture and mining), but for other sectors GDP was the basis for their estimation.

An analysis of economic linkages among the census divisions was carried out involving the following four tests:

(1) Regional distribution of retail sales per capita.
(2) Regional distribution of retail sales per dollar of income.
(3) Interregional traffic flows.
(4) Regional self-sufficiency.

Based on these tests, four regions were selected as shown in *Figure 5.2.*

5.2.3. Employment model

Any changes in yields result in production changes which directly impact the agriculture sector. This impact is translated into a ripple effect on other sectors of the economy, leading to changes for instance in the number of people employed.

When agricultural production declines, as it does, for example, during drought years, it follows that reduced levels of services related to the agricultural sector are required. If farmers reduce fertilizer purchases, or defer new machinery acquisitions, these industries, in the face of the rather bleak economic environment, would lay off workers. The employment model was developed to determine these types of effects.

The employment–production functions in the model are described by linear regression equations. These functions were estimated for the following three categories of workers:

(1) Total or all employees.
(2) Production-related employees.
(3) Administration employees, self-employed individuals, and unpaid family workers.

The production employees for a sector included hired workers directly involved in the actual generation of output of that sector. For agriculture, this category included hired paid workers, while the "total" or "all workers" category included farm operators, unpaid family workers and hired paid workers.

The employment model has been tested and the accuracy found to be within acceptable limits (Kulshreshtha, 1983). It was therefore considered appropriate for use in this study.

5.3. Model Input Data and Simulations

5.3.1. Crop yield input data

Average reported crop yields for the period 1979–1980 are used as the base in the input–output model. These data are given in *Table 5.3* by crop and soil zone for planting on stubble and fallow lands. Two years of yield data are required because in the actual marketing of grain, part of the crop produced the previous year is sold in the current year and part of the current crop is sold the next year. This effect is simulated in the model by assuming that the total production sold in 1980 is composed of 58% of the crop produced in 1979 with the remainder coming from the 1980 crop.

In Section 4 a crop growth model was used to estimate spring wheat yields on stubble under different climatic conditions, expressed in relation to average yields observed over the 1961–1979 period (*see Table 4.7*). These values were in turn modified for use in this section to reflect differences from the input–output

Table 5.3. Crop yields by soil zone, Saskatchewan 1979 and 1980 (kg/ha).

		Brown[a]		Dark brown[b]		Black[c]	
		1979	1980[d]	1979	1980[d]	1979	1980[d]
Wheat:	Stubble[e]	1257	1063	1241	1104	1030	1263
	Summer fallow	1508	1506	1710	1564	1417	1790
Barley:	Stubble[e]	1780	1512	1856	1840	1534	2012
	Summer fallow	1994	1899	2480	2315	2050	2529
Flax:	Stubble[e]	922	785	836	827	616	956
	Summer fallow	1051	906	1180	961	870	1111
Canola:	Stubble[e]	522	583	712	782	736	957
	Summer fallow	632	798	1017	1069	1049	1307

[a] Weighted average of yields in crop districts 3, 4 and 7a. Statistics Canada (1980).
[b] Weighted average of yields in crop districts 1a, 2, 6 and 7b. Statistics Canada (1980).
[c] Weighted average of yields in crop districts 1b, 5, 8 and 9. Statistics Canada (1980).
[d] Crop yields in subsequent years set at 1980 levels for base runs.
[e] Division of average yield into stubble and summer fallow yield ratios estimated on the basis of reported province-wide stubble and summer fallow yields. Statistics Canada (1980).

economic base. The wheat stubble yields for the economic analyses were computed as follows:

$$\text{Wheat stubble yield} = \frac{\text{Scenario model-derived yield}}{\text{1961-1979 average yield}} \times \text{1979-1980 yields} \quad (5.1)$$

As mentioned previously, both stubble and fallow yields are required. The latter were calculated by adjusting the results obtained from equation (5.1) by the reported 1979–1980 ratio of fallow to stubble yields. Yields for barley, oats and canola, as previously mentioned, were then calculated by using the 1979–1980 ratios of yields for these crops to that for spring wheat. This yield information for the various crops and scenarios, expressed in terms of the 1979–1980 base year as explained above (see Table 5.4), was used as input to farm models for determining direct impacts.

5.3.2. Farm level simulation

The farm models were run on a 10-year basis, the GISS1, GISS2, HIST4 average, and HIST2 scenario conditions assumed occurring in year 6 of the 10-year sequence. The first 2 years of data are 1979 and 1980 with the following consequential years at 1980 values or as otherwise stated for the scenario inserts. The half-decade sequence, 1933–1937, was started in year 3; thus the year immediately following (denoted as 1938 in the tables of results) is actually represented by 1980 data. Changes at the farm gate are restricted to only those caused by crop yield reductions and do not consider preharvest input changes. The

Table 5.4. Grain and oilseed yields scenario (kg/ha).

Crop	Seed bed	1961–1979 Mean	1980 base yields	HIST4 1933	HIST4 1934	HIST4 1935	HIST4 1936	HIST4 1937	HIST4 Mean 1933–37	HIST2 1961	GISS1 $2 \times CO_2$ TP	GISS2 $2 \times CO_2$ T	Source
							Brown						
Wheat	ST	1034		424	1082	1069	696	339	735	215	992	877	see Section 4
	ST		1063	436	1112	1126	715	348	755	221	1019	901	
	SMF		1506	617	1575	1595	1013	493	1070	313	1444	1277	Model adjusted in relation to 1961–79 average and 1980 base
Barley	ST		1512	620	1582	1602	1018	496	1075	314	1450	1282	
	SMF		1899	779	1987	2013	1278	623	1350	395	1822	1611	
Flax	ST		785	322	821	832	528	257	558	163	753	665	
	SMF		906	371	948	960	610	297	644	188	869	768	
Canola	ST		583	239	610	618	392	191	414	121	559	494	
	SMF		798	327	835	846	537	262	567	166	766	677	
							Dark Brown						
Wheat	ST	1190	1104	818	1127	1839	1090	490	1084	274	1043	910	see Section 4
	ST		1564	759	1045	1706	1011	455	1006	254	968	844	
	SMF		1564	1075	1481	2417	1432	644	1425	360	1371	1196	Model adjusted in relation to 1961–79 average and 1980 base
Barley	ST		1840	1265	1743	2843	1685	758	1676	424	1613	1407	
	SMF		2315	1591	2192	3577	2120	953	2108	533	2028	1770	
Flax	ST		827	569	783	1278	758	341	753	190	725	632	
	SMF		961	660	910	1485	880	396	875	221	842	735	
Canola	ST		782	538	741	1209	716	322	712	180	686	598	
	SMF		1069	735	1012	1652	979	440	974	246	937	817	
							Black						
Wheat	ST	1323	1263	917	950	1493	983	627	994	346	941	830	see Section 4
	ST		1263	876	908	1426	939	599	950	331	899	793	
	SMF		1790	1241	1286	2021	1331	849	1346	468	1274	1124	Model adjusted in relation to 1961–79 average and 1980 base
Barley	ST		2012	1396	1446	2272	1496	954	1513	527	1432	1263	
	SMF		2529	1754	1817	2856	1880	1199	1901	662	1800	1588	
Flax	ST		956	663	687	1080	711	453	719	250	680	600	
	SMF		1111	770	798	1254	826	527	835	291	791	697	
Canola	ST		957	664	687	1080	711	454	719	250	681	601	
	SMF		1307	907	940	1477	972	620	983	342	931	821	

SMF = Summer fallow. ST = Stubble.

Table 5.5. Farm receipts and outgoings (Canadian dollars) for the base years 1979-1980 and changes to the base under described scenarios, for cereal farms on brown soils.

Scenario	Farm size	Crop receipts	Fuel and oil	Machine repair	Insur. & WGSA[a]	Farm house. income	Total equity	Net farm income	Production in tonnes			
									Wheat	Barley	Flax	Canola
Base 1979-80	S	26 550	1688	1548	531	22 783	140 447	14 343	126	—	1	1
	M	57 329	2961	2755	1147	50 446	296 078	32 227	273	—	2	1
	L	136 879	6212	6062	1500	123 105	714 373	85 234	653	—	4	4
2 × CO_2 Base Precip (GISS 2)	S	−2224	−5	−7	−53	−2159	−2286	−4448	105	—	1	1
	M	−4810	−7	−14	−77	−4712	−4389	−9602	227	—	1	1
	L	−11 091	−15	−34	—	−11 042	−9765	−23 176	543	—	3	3
2 × CO_2 Increase Precip (GISS 1)	S	−705	−2	−2	−10	−691	−764	−1510	119	—	1	1
	M	−1605	−2	−5	−26	−1572	−1471	−3257	257	—	2	1
	L	−3414	−5	−11	—	−3398	−3308	−7856	616	—	4	3
1961 Drought Equivalent (HIST 2)	S	−20 855	−16	−29	−168	−20 642	−17 753	−20 780	27	—	—	—
	M	−45 038	−34	−67	−363	−44 574	−32 786	−44 858	57	—	—	—
	L	−107 014	−75	−162	−43	−106 734	−69 288	−108 305	137	—	1	1
1933-37 Average (HIST 4)	S	−6023	−9	−17	−96	−5901	−8890	−11 967	69	—	—	—
	M	−26 020	−19	−39	−208	−25 754	−13 645	−25 834	149	—	1	1
	L	−61 461	−43	−93	—	−61 325	−27 242	−62 369	356	—	2	2
1933	S	−6119	−11	−294	−40	−5774	−21 711	−14 157	54	—	—	—
	M	−13 354	−64	−134	−268	−12 888	−54 714	−26 637	117	—	1	1
	L	−30 587	+698	+398	—	−31 683	−132 751	−74 028	280	—	2	2

The effects on the farm and provincial economy

Year													
1934	S	−9 621	−18	−207	−192	−9 204	−19 974	−2 138	119	—	—	1	1
	M	−20 522	−78	+22	−411	−20 055	−43 431	−1 106	257	—	—	2	1
	L	−49 853	+664	+975	—	−48 214	−101 948	−12 696	616	—	—	4	3
1935	S	−1 375	−2	−90	−27	−1 256	−17 869	−1 422	121	—	—	1	1
	M	−2 834	−45	+250	−57	−2 982	−33 035	+102	262	—	—	2	1
	L	−6 206	+741	+1 682	—	−8 629	−77 201	−8 979	628	—	—	4	3
1936	S	−4 010	−7	−14	−80	−3 909	−20 603	−8 308	86	—	—	1	—
	M	−8 577	−16	−32	−172	−8 357	−30 471	−17 937	186	—	—	1	1
	L	−20 113	−35	−77	—	−20 001	−63 535	−43 368	447	—	—	3	2
1937	S	−12 428	−23	+37	−249	−12 193	−32 799	−17 812	38	—	—	—	—
	M	−26 931	−45	+91	−539	−26 438	−41 111	−37 271	83	—	—	1	1
	L	−64 074	+209	+750	−4	−65 029	−77 521	−96 502	198	—	—	1	1
1938[b]	S	−11 054	−21	+117	−221	−10 929	−29 452	+998	126	—	—	1	1
	M	−23 984	−39	−95	−480	−23 370	−27 160	+3 044	273	—	—	2	1
	L	−56 882	−100	−166	—	−56 616	−38 563	+889	653	—	—	4	4

[a] WGSA = Contribution to Western Grain Stabilization Act.
[b] 1980 yields to bring the model back to equilibrium following the HIST4 (1933–37) scenario.

Table 5.6. Farm receipts and outgoings (Canadian dollars) for the base years 1979–1980 and changes compared with the base under described scenarios, for cereal farms on dark brown soils.

Scenario	Farm size	Crop receipts	Fuel and oil	Machine repair	Insur. & WGSA[a]	Farm house. income	Total equity	Net farm income	Production in tonnes			
									Wheat	Barley	Flax	Canola
Base 1979–80	S	30854	1859	1834	617	26544	176643	13058	104	51	2	6
	M	75653	3719	4067	1500	66367	453137	39993	260	126	5	15
	L	125387	5619	5865	1500	112403	811403	62047	423	205	9	24
2 × CO$_2$ Base Precip (GISS 2)	S	−3096	−5	−10	−48	−3033	−4284	−6583	81	39	2	5
	M	−6430	−9	−24	−80	−6317	−8483	−16606	201	98	4	11
	L	−12583	−18	−43	—	−12522	−11110	−26687	328	159	7	19
2 × CO$_2$ Increase Precip (GISS 1)	S	−314	—	—	+4	−318	−879	−1683	97	48	2	6
	M	−236	+1	+2	—	−239	−2904	−4336	241	119	5	14
	L	−2491	−1	−4	—	−2486	−2350	−6559	393	194	8	22
1961 Drought Equivalent (HIST 2)	S	−20461	−21	−42	−186	−20212	−20931	−22993	27	13	1	1
	M	−47683	−46	−112	−427	−47098	−43130	−57454	67	32	1	4
	L	−79824	−68	−159	—	−79597	−65917	−94074	109	53	2	6
1933–37 Average (HIST 4)	S	−3196	−2	−5	−24	−3165	−1842	−3619	90	45	2	5
	M	−5385	−3	−8	−20	−5354	−5122	−9214	225	111	5	13
	L	−8799	−2	−5	—	−8792	−5088	−12790	366	193	8	21
1933	S	−4911	−317	−461	−98	−4035	−18517	−12062	67	17	2	4
	M	−7246	−14	−371	−132	−6729	−53882	−13840	186	92	4	11
	L	−13914	+741	+1091	—	−15746	−84854	−25301	375	90	6	17

1934	S	−10094	−330	−406	−202	−9156	−21372	−7102	93	23	2	6
	M	−10680	−21	−110	−201	−10348	−37453	+1937	259	128	5	15
	L	−19510	+712	+1661	—	−22595	−58188	+631	351	125	9	24
1935	S	+525	−308	−289	+11	+1111	−14760	+9227	152	38	4	10
	M	+25252	+51	−179	—	+25380	−2535	+50722	424	210	10	24
	L	+38819	+825	+2118	—	+35876	−3205	+81271	530	205	14	40
1936	S	+2017	0	0	+40	+1977	−19680	−7864	87	22	2	6
	M	+28322	+60	+151	—	+28111	+1467	−4904	243	120	5	14
	L	+43714	+94	+226	—	+43394	+3343	−7576	573	118	8	23
1937	S	−11113	−28	+39	−222	−10902	−33559	−18116	45	11	1	3
	M	−14246	−30	+188	−272	−14132	−1227	−39684	123	61	3	7
	L	−30047	−59	+386	—	−30374	+3821	−64478	318	59	4	11
1938[b]	S	−11163	−26	+133	−223	−11047	−30688	+1340	104	51	2	6
	M	−22161	−45	+260	−430	−21946	+17022	+2009	260	126	5	15
	L	−33490	+63	+1062	—	−34615	+35603	+3602	289	123	9	24

[a,b] See notes to Table 5.5.

Table 5.7. Farm receipts and outgoings (Canadian dollars) for the base years 1979–1980 and changes compared with the base under described scenarios, for cereal farms on black soils.

Scenario	Farm size	Crop receipts	Fuel and oil	Machine repair	Insur. & WGSA[a]	Farm house. income	Total equity	Net farm income	Production in tonnes			
									Wheat	Barley	Flax	Canola
Base 1979–80	S	29 925	1928	1997	599	25 401	118 213	13 688	110	—	3	22
	M	66 439	4431	4995	1329	55 684	299 438	32 428	247	—	7	50
	L	153 496	9337	11 147	1500	131 512	656 013	83 000	568	—	17	117
2 × CO$_2$ Base Precip (GISS 2)	S	−4064	−6	−10	−65	−3983	−5171	−8040	79	—	2	16
	M	−8198	−28	−21	−127	−8022	−7654	−18 274	179	—	5	37
	L	−19 506	−26	−66	—	−19 414	−16 135	−42 040	411	—	12	84
2 × CO$_2$ Increase Precip (GISS 1)	S	−3956	−6	−10	−64	−3876	−4955	−7825	80	—	2	17
	M	−7953	−27	−20	−123	−7783	−7404	−17 790	179	—	5	38
	L	−18 943	−25	−64	—	−18 854	−15 630	−40 918	413	—	12	87
1961 Drought Equivalent (HIST 2)	S	−17 523	−9	−14	−85	−17 415	−18 918	−21 424	31	—	1	6
	M	−50 391	−56	−79	−471	−49 785	−35 457	−48 042	70	—	2	14
	L	−83 021	−77	−193	—	−82 751	−74 968	−111 178	160	—	5	32
1933–37 Average (HIST 4)	S	−9442	−6	−10	−62	−9364	−4781	−7651	80	—	2	17
	M	−10 382	−19	−3	−17	−10 343	−7362	−17 508	181	—	5	38
	L	−36 955	−43	−108	—	−36 804	−15 487	−39 986	416	—	13	87
1933	S	−2845	+60	−29	−57	−2819	−14 005	−9213	77	—	2	16
	M	−5451	+34	−632	−109	−4744	−42 462	−15 025	174	—	5	37
	L	−34 450	−9	−774	—	−33 667	−107 611	−35 127	401	—	12	84

The effects on the farm and provincial economy

Year											
1934	S	−8 592	+50	+61	−172	−8 531	−18 822	−10 340	77	—	2	16
	M	−18 394	+9	−369	−368	−17 666	−38 610	−18 229	174	—	5	37
	L	−31 161	+8	−1 483	—	−29 686	−91 064	−42 787	401	—	12	70
1935	S	−4 707	+57	+229	−95	−4 898	−14 032	+3 965	128	—	4	27
	M	−5 936	+31	−87	−119	−5 761	−20 392	+13 045	288	—	8	60
	L	−24 141	+23	−457	—	−23 707	−45 482	+29 208	662	—	20	138
1936	S	−2 577	−5	−7	−52	−2 513	−17 795	−8 842	79	—	2	17
	M	+699	−16	+4	+14	+697	−20 069	−18 268	178	—	5	37
	L	−8 320	−13	−34	—	−8 273	−41 498	−42 159	410	—	12	86
1937	S	−6 747	−12	+103	−135	−6 703	−27 429	−15 156	54	—	2	11
	M	−23 530	−40	−1 271	−471	−21 748	−26 346	−35 197	122	—	4	26
	L	−55 816	−191	−2 009	—	−53 616	−36 965	−77 752	281	—	8	59
1938[b]	S	−10 122	−17	+196	−203	−10 098	−24 380	+418	110	—	3	22
	M	−19 250	+8	−1 081	−385	−17 792	−14 330	−3 745	247	—	7	50
	L	−24 664	−146	−751	—	−23 767	+1 694	−583	568	—	17	116

[a,b] See notes to Table 5.5

resulting changes of output are, therefore, strictly the result of scenario-derived yield changes for cereal farms. These changes, as compared with the base case and categorized according to soil zone and cereal farm size, are given in *Tables 5.5–5.7*. Absolute changes are then transposed into provincial totals by aggregating according to the number of representative farms in each category (Klein, 1983).

These provincial totals are in turn modified by using the conversion figures established to relate the farm level model and the Province input–output model. *Table 5.8* provides expenditure and income changes in the agriculture sector converted and aggregated to the provincial level for all scenarios. The loss of agricultural employment in terms of person-years derived using the employment model are also presented. As shown, the HIST2 extreme event year (1961) has double the impact of any other scenario at this stage.

5.4. Provincial Economy Impacts

Direct farm impacts brought on by crop yield reductions have repercussions upon other provincial economic sectors. These are commonly known as secondary effects or indirect impacts. These types of impacts are generated by two types of changes:

(1) Reduced agricultural inputs decrease product demand from other sectors, eventually resulting in lower sales of these products and lower production.
(2) Since farmers now have less money available to spend their demand for products for home consumption and farm investments will be lower.

The outcome for both of these changes is an economy that produces less goods and services.

The commodity changes derived from the farm level analysis are used by the input–output model to compute the absolute changes in all sectors. These impacts were derived and are expressed in millions of dollars (Canadian) in *Table 5.9*.

Services, trade, and transportation sectors are affected the most in absolute terms. In relative terms, a significant drop is also felt by agricultural processing, construction, utilities and financial sectors. To illustrate, when primary agricultural output reductions amounting to Can$1800 million (*Table 5.8*) occur in the 1961 (HIST2) scenario, another round of lost values of goods and services by the nonagricultural sectors is triggered amounting to Can$1600 million (*Table 5.9*). This decrease then results in a further reduction in the income levels and other value added in the province. The "Household Income" levels are reduced by Can$2400 million (or a devastating 38% of 1980 base values) by the HIST2 scenario, as well as another Can$547 million which is lost in "Other Value Added". Thus close to Can$3000 million is lost to the provincial economy. Losses are not confined to the borders of Saskatchewan, as provincial industries would be importing Can$1200 million less from Canadian and foreign regions. This would trigger a further chain reaction in these regions similar to the impacts estimated for Saskatchewan.

Table 5.8. Change in aggregated income, expenditures and employment in the agriculture sector of Saskatchewan for the different scenarios, 1980 prices (values in Canadian$ million and person-years).

Commodity	GISS2 $2\times CO_2 T$	GISS1 $2\times CO_2 TP$	HIST4 period 1933–37 average	HIST2 1961
Income changes ($ million)				
Fuel & oil	−0.5	−0.4	−0.5	−1.6
Machine repair	−0.8	−0.4	−0.8	−3.1
Insurance	−3.2	−2.0	−2.2	−11.4
Farm household income	−272.6	−160.7	−595.7	−1795.4
Expenditure changes ($ million)				
Metal products	−0.2	−0.1	−0.2	−0.5
Gas & fuel	−0.8	−0.5	−0.8	−2.4
Petroleum production	−0.1	−0.1	−0.1	−0.3
Insurance	−3.2	−2.0	−2.2	−11.4
Household	−272.6	−160.7	−595.7	−1795.4
Total ($)	−276.9	−163.4	−598.9	−1810.0
Employment changes (person-years)				
Agricultural sector employment	−1224	−722	−2647	−8000

Commodity	1933	1934	1935	1936	1937	1938[a]
Income changes ($ million)						
Fuel & oil	+1.6	−1.6	+4.2	+0.2	−1.1	−1.1
Machine repair	−10.8	−3.1	+11.1	+1.0	−9.2	−4.7
Insurance	−4.4	−10.2	−2.5	−1.3	−12.5	−12.8
Farm household income	−349.7	−663.5	−201.5	+164.6	−801.9	−823.1
Expenditure changes ($ million)						
Metal products	−2.4	−0.7	+2.4	+0.2	−2.0	−1.0
Gas & fuel	+2.4	−2.4	+6.4	+0.3	−1.6	−1.6
Petroleum production	+0.3	−0.3	+0.7	+0.0	+0.2	−0.2
Insurance	−4.4	−10.2	−2.5	−1.3	−12.5	−12.8
Household	−349.7	−663.5	−201.5	+164.6	−801.9	−823.1
Total ($)	−353.8	−677.1	−194.5	+163.8	−817.2	−838.7
Employment changes (person-years)						
Agricultural sector employment	−1564	−2993	−860	+724	−3612	−3707

[a] Represented by 1980 yield data to bring the model back to equilibrium. Denoted as 1938 since it follows the HIST4 (1933–37) scenario.

Table 5.9. Summary of sector level impacts for scenarios (in Canadian $ at 1980 prices and employment in person-years).

Sectors	GISS2	GISS1	HIST2	1933	1934	1935	1936	1937	1938[a]	HIST4
				\multicolumn{7}{c}{HIST4 period}	
				\multicolumn{7}{c}{Provincial economy}	
Nonfuel mines	−3.3	−1.9	−21.8	−4.2	−8.3	−1.9	+2.0	−10.0	−10.0	−7.2
Fuel mines	−1.4	−0.9	−9.3	−1.7	−3.6	−0.7	+0.8	−4.2	−4.4	−3.1
Forest./fish	−0.6	−0.4	−4.2	−0.8	−1.6	−0.3	+0.4	−1.9	−2.0	−1.4
Agric. proc.	−7.7	−4.5	−50.5	−9.8	−18.8	−5.5	+4.6	−22.7	−23.3	−16.7
Other manuf.	−20.6	−12.2	−137.5	−25.9	−53.6	−4.6	+12.7	−64.1	−64.9	−44.7
Construction	−26.8	−15.8	−175.2	−34.4	−65.8	−19.3	+15.7	−79.5	−81.5	−57.8
Utilities	−11.6	−6.9	−76.5	−14.9	−28.5	−8.2	+7.0	−34.4	−35.3	−25.3
T/C/S[b]	−55.3	−32.6	−363.4	−70.9	−135.0	−40.0	+33.1	−163.1	−167.3	−120.4
F/I/RE[c]	−38.4	−22.6	−251.9	−49.2	−94.0	−27.1	+22.9	−113.5	−116.4	−83.3
Services	−13.4	−8.0	−77.9	−17.4	−35.1	−9.6	+4.7	−42.6	−43.7	−24.2
Residual	−69.6	−41.0	−456.8	−89.2	−170.3	−49.6	+41.7	−205.8	−211.1	−151.1
	+0.8	+0.5	+5.5	+1.1	+2.0	+0.5	−0.5	+2.5	+2.6	+1.8
Total	−247.9	−146.3	−1619.5	−317.3	−612.6	−167.3	+144.9	−744.3	−757.3	−537.0
Value added:										
GDP househ.	−365.1	−215.3	−2400.6	−468.3	−891.1	−265.7	+219.0	−1076.8	−1104.9	−795.4
others	−59.5	−35.2	−388.5	−76.2	−146.9	−40.4	+34.7	−177.4	−181.8	−128.0
CTLS[d]	−24.4	−14.4	−158.4	−31.3	−59.9	−18.0	+14.0	−72.4	−74.3	−52.2
Imports	−183.0	−108.0	−1203.4	−234.6	−448.1	−129.5	+109.6	−541.3	−555.1	−398.3

	Employment									
Nonfuel mines	−23	−13	−155	−30	−59	−13	+14	−71	−71	−51
Fuel mines	−2	−1	−13	−2	−5	−1	+1	−6	−6	−4
Forest./fish	−37	−25	−261	−50	−100	−18	+25	−118	−124	−87
Agric. proc.	−20	−12	−132	−25	−49	−14	+12	−59	−61	−44
Other manuf.	−370	−219	−2472	−465	−964	−83	+228	−1152	−1167	−804
Construction	−340	−200	−2227	−437	−836	−245	+200	−1010	−1036	−735
Utilities	−50	−30	−328	−64	−122	−35	+30	−147	−151	−108
Trade	−888	−524	−5840	−1139	−2169	−643	+532	−2621	−2688	−1935
T/C/S	−155	−91	−1015	−198	−379	−109	+92	−457	−469	−336
F/I/RE	−44	−26	−254	−57	−114	−31	+15	−139	−142	−78
Services	−660	−390	−4340	−847	−1617	−470	+394	−1955	−2005	−1435
Residual										
Total	−2589	−1531	−17037	−3314	−6414	−1662	+1543	−7735	−7920	−5617

[a] Represented by 1980 yield data to bring the model back to equilibrium. Denoted as 1938 since it follows the HIST4 (1933–37) scenario.
[b] T/C/S = Transport/communications/storage.
[c] F/I/RE = Finance/insurance/real estate.
[d] CTLS = Commodity tax less subsidies.

Loss in output levels for various economic sectors are next translated into loss in employment. Lost production under the HIST2 scenario will result in an indirect loss of over 17 000 person-years of work (*Table 5.9*) and a direct loss of 8000 person-years of work for the primary producing agriculture sector (*Table 5.8*). This brings the total employment loss to 25 000 person-years or 5.8% of all employed workers in the province. Major losers in sectors other than agriculture are workers in the trade and service sectors. The losses in these sectors will constitute as much as 23% and 17% of the total loss in employment levels, respectively. These results are not unexpected for two reasons. First, both sectors have high labor output coefficients; and second, the output of both sectors is severely affected by lower farmer disposable incomes.

5.5. Effects of Adjustments Made in Crop Rotations in Response to Adverse Climatic Conditions

The following section examines the efficacy of growing winter wheat (rather than spring wheat) to reduce the effects of drought. Winter wheat is most suited to dark brown soils and, therefore, the analysis is restricted to this particular soil zone [i.e., Region I, southeast; Region III, northeast; and Region IV, northwest Saskatchewan (*see Figure 5.2*)]. For the purposes of this scenario, it is assumed that 10% of all the summer fallow land in the dark brown soil zone is planted with winter wheat. On a rotation system of 1 year of summer fallow and 2 years of crop, this is equivalent to approximately 5% of the total cropped acreage. The estimations are for medium-sized farms only.

The following comparisons were made using the above data:

(1) Planting of winter wheat versus no winter wheat under normal conditions.
(2) Planting of winter wheat versus no winter wheat under a "1936-type" drought condition.

Each of these scenarios was examined assuming 1980 prices and economic conditions.

5.5.1. Winter wheat impacts during normal climatic conditions

The results indicate (*Table 5.10*) that with a "no-drought" situation, the sowing of winter wheat had adverse effects upon farm financial performance in the area. An average decrease in net farm worth at year end of Can$5530 per farm, and an average decrease in net farm income of Can$6110 per farm resulted. These decreases were caused mainly by a reduction in crop receipts, which fell Can$12 400. Crop receipts fall because of lower prices for winter wheat versus spring wheat even though winter wheat yields were higher. Chemical costs rose by Can$340 per farm contributing to the decrease in net farm worth and net farm income. The rise in chemical costs is attributed to the cost of spraying in the previous spring. Total crop expenses and costs for seed, fertilizer, fuel and oil, machine repairs and Western Grain Stabilization Act contributions all fell slightly [*Table 5.10(A)*].

These direct expenditures by farmers practicing winter wheat production lead to decreased input use (except for chemicals) and reduced level of household income (operator and hired workers). In fact, this decrease amounts to Can$67.5 million [*Table 5.10(B)*], almost half of which was in Region I, southeast Saskatchewan. Reduced agricultural expenditures lead to reduced nonagricultural sectors output levels and contribution to the value added in the region and the province as a whole. The loss in the Gross Domestic Product of Saskatchewan is estimated at Can$95.5 million, in 1980 dollars [*Table 5.10(C)*]. From these results it appears that the change to winter wheat production would decrease the value added parameters for the provincial economy during a normal period.

5.5.2. Impacts of a "1936-type" drought

The regional and provincial level impact of the drought with conventional cropping is illustrated in *Table 5.11*. The drought results in a loss of Can$216 million in terms of gross domestic product or value-added in the province. Of this total, about Can$79 million is lost in Region I. Further, the rest of Canada is affected to the amount of Can$88 million, in terms of reduced imports into Saskatchewan [*Table 5.11(C)*].

Incorporating 10% winter wheat in the cropping program tended to ameliorate losses caused by drought. Total losses in GDP or value-added income are Can$183 million (compared with Can$216 million) (*Table 5.12*). The average effect of a drought on farms without winter wheat resulted in reduced net worth of Can$10970 per farm [*Table 5.11(A)*], whereas the reduction in net worth for the same farms with the addition of winter wheat was only Can$9700 per farm [*Table 5.12(A)*]. Drought conditions caused reductions in net farm income of Can$11 885 for farms with the conventional crops and of Can$10 110 for those growing winter wheat. Crop receipts fell Can$24 355 per farm for the standard farms, and Can$20 680 per farm for farms sowing winter wheat. Costs for such items as fuel, machine repairs, WGSA contributions and crop expenses were all reduced by drought. Generally, these cost reductions were slightly larger (by one or two dollars/farm/year) for farms using regular planting practices than for those adopting the 10% winter wheat system.

The effect on direct expenditures, of changing from normal weather conditions to drought conditions, for the winter wheat scenario, follows. The on-farm impact of the drought was reduced by diverting some of the crop land to winter wheat. For example, the net income of workers and operators was reduced by only Can$112 million, an improvement of about Can$20 million [*Table 5.12(B)*]. The positive effects of incorporating winter wheat production for drought years is clearly shown in *Table 5.13*. During a 1936-type drought year, if the agricultural sector (as defined in terms of dark brown soil zone farms) converted 10% of cereal crop acreage to winter wheat, it would benefit by Can$19.9 million. This also increases the economic output of nonagricultural sectors by approximately

Table 5.10. Economic impacts of replacing 10% of the spring wheat area with winter wheat on dark brown soils under normal climatic conditions (changes relative to no winter wheat; Canadian dollars).

	Changes per farm (actual $)	Level of change by region in $1000			
		I	III	IV	Province
A. FARM LEVEL IMPACTS					
Number of farms		3 303	1 474	2 224	7 001
Financial indicators:					
Net capital purchases	+750	+2 474	+1 104	+1 666	+5 244
Total assets	−8 230	−27 174	−12 126	−18 297	−57 597
Change in net worth	−5 530	−18 256	−8 147	−12 292	−38 695
Net farm income	−6 110	−20 188	−9 009	−13 593	−42 790
Crop receipts	−12 400	−40 944	−18 272	−27 568	−86 784
Crop expenses	−400	−1 334	−596	−898	−2 828
Seed	−250	−829	−370	−558	−1 757
Fertilizer	−230	−766	−342	−516	−1 624
Chemicals	+340	+1 110	+495	+747	+2 352
Total fuel and diesel	−170	−535	−238	−361	−1 134
Fuel	−140	−456	−203	−307	−966
Lubricants	−30	−79	−35	−54	−168
Machine repairs	−90	−287	−128	−194	−609
WGSA[a] contribution	−70	−231	−103	−156	−490
B. COMMODITY CONVERSION FARM EXPENDITURES					
Wheat seed		−829	−370	−558	−1 757
Machines repairs		−62	−28	−42	−132
Fuel		−813	−363	−547	−1 723
Lubricants		−89	−39	−60	−188
Fertilizer		−966	−431	−651	−2 048
Chemicals		+2 082	+929	+1 402	+4 413
WGSA[a] payments		−231	−103	−156	−490
Household		−30 919	−13 798	−20 818	−65 535
Total expenditures		−31 827	−14 203	−21 430	−67 460
Local purchases					
Wheat seed		−829	−370	−558	−1 757
Machine repairs		−9	−4	−6	−19
Fuel		−512	−228	−345	−1 085
Lubricants		−32	−14	−22	−68
Fertilizer		−29	−13	−20	−62
Chemicals		+63	+28	+42	+133
Insurance		−74	−33	−50	−157
Household		−30 919	−13 798	−20 818	−65 535
Imports		−515	−242	−348	−1 105

Table 5.10. *continued*

	Level of change by region in $1000			
	I	III	IV	Province
C. REGIONAL AND PROVINCIAL IMPACTS				
Agriculture	−706	−315	−475	−1 497
Nonfuel mines	−66	−38	−15	−840
Fuel mines	−33	−2	−9	−350
Forestry and fish	0	−16	−1	−157
Agriculture processing	−149	−136	−10	−1 808
Other manufacturing	−814	−348	−255	−5 418
Construction	−650	−385	−171	−6 243
Utilities	−361	−170	−57	−2 746
Trade	−2 223	−828	−554	−13 453
Transportation	−340	−327	−125	−9 147
Finance, insurance, real estate	−565	−110	−7	−2 547
Services	−2 522	−741	−228	−16 396
GDP: Households	−5 947	−13 783	−22 961	−85 574
Other	−2 125	−738	−378	−4 146
Net indirect tax	−1 834	−681	−990	−5 787
Imports	−5 945	−6 110	−9 672	−43 119

[a] WGSA = Western Grain Stabilization Act.

Can$17.8 million. The GDP for Saskatchewan then increases by Can$32.6 million. Benefits also accrue to the rest of Canada and the world through increased imports into the province. From these results it appears that, as insurance against the impacts of a drought year, putting some land into winter wheat production, as opposed to all under spring wheat, is an advisable farm practice.

5.5.3. Policy implications

The winter wheat scenario as presented here offers but one example of how to evaluate the efficacy of a number of different responses to climatic change. The scenario is simplistic, but by keeping the reference to a single specific cropping change, the impacts are articulated and the complexity of dealing with several mitigative influences simultaneously is avoided. For example, winter wheat in this case is sown on summer fallow but in reality most now is planted on stubble. This latter practice of planting on stubble, however, would enhance the drought mitigative effects and thus the "winter wheat on summer fallow" scenario may be viewed as providing an underestimate of the ultimate benefits that would be likely to result from switching to a cropping system that incorporated winter wheat.

It was also assumed in the scenario that winter wheat yields would be a constant 25% higher than spring wheat yields, as is the case under normal climatic conditions. In cropping trials, however, winter wheat yields were more than 25% above spring wheat yields during drier climatic conditions. In reality then, an appropriate cropping mix might greatly improve farmers' financial

Table 5.11. Economic impacts of 1936-type drought for regular farm practices on dark brown soils (1980 Canadian dollars).

	Changes per farm (actual $)	Level of change by region in $1000			
		I	III	IV	Province
A. FARM LEVEL IMPACTS					
Number of farms		3 303	1 474	2 224	7 001
Financial indicators:					
Net capital purchases	0	0	0	0	0
Total assets	−10 975	−36 247	−16 176	−24 406	−76 829
Change in net worth	−10 970	−36 244	−16 174	−24 404	−76 822
Net farm income	−11 885	−39 253	−17 517	−26 430	−83 200
Crop receipts	−24 355	−80 441	−35 898	−54 163	−170 502
Crop expenses	−295	−971	−433	−654	−2 058
Seed	0	0	0	0	0
Fertilizer	0	0	0	0	0
Chemicals	0	0	0	0	0
Total fuel and diesel	−25	−82	−37	−56	−175
Fuel	−20	−69	−31	−47	−147
Lubricants	−5	−13	−6	−9	−28
Machine repairs	−10	−39	−18	−27	−84
WGSA[a] contribution	−50	−155	−69	−105	−329
B. COMMODITY CONVERSION FARM EXPENDITURES					
Machines repairs		−8	−4	−6	−18
Fuel		−124	−55	−83	−262
Lubricants		−15	−7	−10	−32
WGSA[a] payments		−155	−69	−104	−328
Household income		−62 225	−27 768	−41 898	−131 891
Total expenditures		−62 527	−27 903	−42 101	−132 531
Local purchases					
Machine repairs		−1	−1	−1	−3
Fuel		−78	−35	−52	−165
Lubricants		−5	−2	−4	−11
Insurance		−50	−22	−33	−105
Household		−62 225	−27 769	−41 898	−131 891
Imports		−168	−75	−97	−339
C. REGIONAL AND PROVINCIAL IMPACTS					
Nonfuel mines		−87	−63	−6	−1 584
Fuel mines		−54	0	−11	−678
Forestry and fish		0	−32	−2	−305
Agriculture processing		−295	−312	−19	−3 694
Other manufacturing		−1 118	−536	−178	−9 918
Construction		−1 293	−882	−341	−12 750
Utilities		−714	−389	−113	−5 595
Trade		−4 009	−1 685	−816	−26 673
Transportation		−2 505	−702	−193	−18 431
Finance, insurance, real estate		−1 018	−197	−76	−4 957
Services		−4 112	−1 691	−449	−33 407
GDP: Households		−71 923	−31 845	−45 966	−175 956
Other		−3 920	−1 536	−554	−28 207
Net indirect tax		−3 501	−1 487	−1 872	−11 508
Imports		−31 855	−14 099	−19 326	−88 429

[a] WGSA = Western Grain Stabilization Act.

Table 5.12. Economic impacts of 1936-type drought for 10% converted spring wheat farm practices (1980 Canadian dollars).

	Changes per farm (actual $)	Level of change by region in $1000			
		I	III	IV	Province
A. FARM LEVEL IMPACTS					
Number of farms		3 303	1 474	2 224	7 001
Financial indicators:					
Net capital purchases	0	0	0	0	0
Total assets	−9 660	−31 904	−14 237	−21 482	−67 623
Change in net worth	−9 700	−32 036	−14 296	−21 571	−67 903
Net farm income	−10 110	−33 397	−14 903	−22 487	−70 787
Crop receipts	−20 680	−68 313	−30 485	−45 997	−144 795
Crop expenses	−280	−911	−407	−614	−1 932
Seed	0	0	0	0	0
Fertilizer	0	0	0	0	0
Chemicals	1	0	0	0	0
Total fuel and diesel	−25	−76	−33	−52	−161
Fuel	−20	−66	−29	−45	−140
Lubricants	−5	−10	−4	−7	−21
Machine repairs	−10	−33	−15	−22	−70
WGSA[a] contribution	−50	−152	−68	−102	−322
B. COMMODITY CONVERSION FARM EXPENDITURES					
Machines repairs		−7	−3	−5	−15
Fuel		−118	−53	−79	−250
Lubricants		−11	−5	−7	−23
WGSA[a] payments		−152	−68	−102	−322
Household income		−52 812	−23 568	−35 560	−111 940
Total expenditures		−53 100	−23 697	−35 753	−112 550
Local purchases					
Machine repairs		−1	−1	−1	−3
Fuel		−74	−33	−50	−157
Lubricants		−4	−2	−3	−9
Insurance		−49	−22	−32	−103
Household		−52 812	−23 568	−35 560	−119 940
Imports		−150	−71	−118	−339
C. REGIONAL AND PROVINCIAL IMPACTS					
Nonfuel mines		−74	−53	−5	−1 346
Fuel mines		−46	0	−9	−576
Forestry and fish		0	−27	−2	−259
Agriculture processing		−250	−265	−16	−3 136
Other manufacturing		−954	−457	−155	−8 425
Construction		−1 098	−749	−290	−10 824
Utilities		−606	−331	−96	−4 749
Trade		−3 405	−1 431	−694	−22 644
Transportation		−2 195	−596	−164	−15 647
Finance, insurance, real estate		−870	−170	−69	−4 222
Services		−4 238	−1 436	−382	−28 359
GDP: Households		−61 048	−27 030	−39 014	−149 350
Other		−3 331	−1 305	−473	−23 957
Net indirect tax		−2 974	−1 263	−1 510	−9 773
Imports		−27 047	−11 959	−16 322	−75 114

[a] WGSA = Western Grain Stabilization Act.

Table 5.13. Winter wheat as a mitigative measure in drought years (Canadian $ million).

	1936-type drought impacts with conventional cropping	1936-type drought impacts with 10% acreage switched to winter wheat	Net changes brought about by winter wheat scenario during 1936-type drought
Direct agriculture farm impacts	−132.5	−112.6	+19.9
Regional and provincial secondary impacts	−118.0	−100.2	+17.8
GDP	−215.7	−183.1	+32.6
Imports	−88.4	−75.1	+13.3
Workers (actual numbers)			+280

conditions during a drought. If we consider that a 1936-type drought might be a much more frequent event under the GISS2 scenario conditions, and perhaps even under the GISS1 scenario, the analysis could be altered to assume relatively higher winter wheat yields for these conditions.

It is interesting to view the actual incremental adaptation of a winter wheat crop in recent dry years in Saskatchewan. The amount of winter wheat seeded in Saskatchewan soared to about 141 000 hectares in 1983 from 18 000 hectares in 1982. In the fall of 1984, 8000 farmers signed up for crop insurance for about 405 000 hectares of winter wheat (Saskatchewan Crop Insurance Board, personal communication). After the dry conditions of 1984 (when spring wheat yields were down 25% from normal) farmers were hopeful that the 1984–1985 winter would provide the 10–20 cm snow cover needed to insulate the winter wheat crop and that an early crop would be harvested in 1985. However, the snows did not materialize and a lot of acreage had to be reseeded in the spring. This provides a caveat to the winter wheat scenario in that it only provides an effective drought mitigative measure if conditions favor greater yields than for spring wheat.

5.6. The Role of Economic Models in Impact Studies

While we have attempted to keep the presentation simple, the reader should realize that the economic modeling involved is quite complex. The complexity of economic modeling probably rivals that of general circulation climatic modeling. In impact studies such as this, economic models provide the final link in the chain for translating climatic change hypotheses into socioeconomic terminology. For instance, the abnormalities of Saskatchewan's precipitation and temperature under the HIST2 extreme year scenario are translated here into a provincial employment loss of approximately 25 000 person-years, and an overall loss to the provincial economy of nearly Can$3000 million. The validity of such estimates, of course, depends not only on the economic model, but also on the crop growth and phenological models, and on the procedures used to derive the climate data needed for these models.

In this analysis we considered the most important crop, spring wheat, and one example of a possible response, a partial switch to winter wheat. In doing so we have illustrated the type of analysis that is feasible. In more comprehensive future studies, a number of crops could be tracked through the system for translating climatic differences into economic impacts. Several different types of response and feedbacks could be taken into account. Such analyses could include consideration of the long-term sustainability of the land resource base and could reflect anticipated degradation of this resource base and the resulting yield reductions and economic consequences. Of course, these sorts of analyses are well beyond the scope of this study. The reader is referred to Section 6 for a broader discussion of the results and implications of the economic analyses presented here along with those for all other parts of this study.

SECTION 6

Conclusions and Implications

6.1. The Scope of this Study

The preceding parts of this case study have attempted to answer the question: what would be the impacts on Saskatchewan agriculture if the conditions represented by several climatic scenarios were experienced in the future, assuming the same technology and economic circumstances as in the 1980s? The scenarios considered included those based on an historical extreme year (HIST2), an historical 10- or 5-year period (HIST3 or HIST4), the temperature and precipitation changes implied by one of the recent climatic general circulation models (GCMs) for a doubled CO_2 atmosphere (GISS1), and the temperature changes alone from that model (GISS2). The impacts were assessed by comparing results for these scenarios to results for the 1951–1980 standard climatic normal period (HIST1).

The data on the changes in climate implied in the various scenarios were translated, using several models, into estimates of impacts on the thermal and moisture resources for agriculture, the potential for biomass productivity and for soil erosion by wind, the production levels for spring wheat, and the farm and provincial economy. Estimates were also made of the impacts on drought frequency and severity resulting from the warming due to CO_2 doubling, with or without the predicted precipitation increase (GISS3 and GISS4), in comparison with the drought severity and frequency of a 33-year period in the recent past (HIST5).

This final part of the case study summarizes the results and presents our conclusions and recommendations.

To avoid confusion we should emphasize here some of the things we have *not* done in this case study. We have not attempted to assess the likelihood of any particular climatic change, or to predict future climatic impacts, or to undertake comparative studies of different GCM scenarios, or to consider nonclimatic effects of increased CO_2 on crops, or to make a comprehensive evaluation of all the various models that might reasonably be used to study climatic impacts on

Saskatchewan agriculture. This case study focuses on demonstrating the translation of data on changes in climate into estimates of consequent changes in agriculture. To provide this demonstration it used results from only one GCM, and employed only a few representative impact models.

We should also emphasize again that in general it is impossible to validate objectively models for climatic impacts work. The models may have been validated for other work, but such validation may not necessarily be relevant for the present application. Thus there is usually no basis for quantifying the degree of confidence one should have in impact estimates. Despite these limitations we feel it is important to try to make the best translations of climatic scenarios into estimates of likely effects that can presently be made. Impact models such as those demonstrated in this case study can and should be used for such work.

6.2. Summary of Results

The results suggest that without any fundamental change in climate (*see* HIST scenarios, *Table 6.1*), Saskatchewan agriculture can expect an occasional warmer than normal year with moisture resources so reduced that crop production is only about a quarter of normal, the potential for wind erosion is doubled, and losses to the agricultural economy exceed Can$1800 million and 8000 jobs for cereal crop losses alone. Ripple effects in sectors other than agriculture (*Table 5.9*) translate into a provincial GDP reduction of Can$1600 million and a further reduction in jobs of 17 000. They also imply that an occasional period of 5–10 consecutive years with warmer than normal growing seasons will have moisture resources so subnormal that biomass dry matter production will be reduced by nearly half and spring wheat production by about one-fifth. There would be a loss to agriculture of about 2600 person-years annually, and an annual economic loss of nearly Can$600 million. This loss, resulting from effects on cereal crops alone, translates, for nonagricultural sectors, to a provincial GDP reduction of over Can$500 million and a loss of 5600 more person-years annually.

With respect to the climate inferred for a doubled CO_2 atmosphere, our results suggest that on a long-term average basis there would be a substantial increase in thermal resources, modest increases in moisture resources (according to the precipitation effectiveness index results) and in potential biomass productivity, modest decreases in the wind erosion potential and in spring wheat productivity, and losses to the agricultural industry of about 700 jobs, and about Can$160 million annually (*see* GISS1 scenario, *Table 6.1*). This amounts to a nonagricultural sector reduction of about Can$150 million in provincial GDP and 1500 additional person-years annually. They also suggest more frequent and severe droughts: under the $2 \times CO_2 TP$ scenario the return period for what we would presently call a severe drought would be only about half as long as it is now.

For a climate changed by the warming but not the precipitation increase inferred for a doubled CO_2 atmosphere (GISS2), our results imply, on a long-term average basis, modest reductions in moisture resources and biomass productivity, a moderate increase in wind erosion potential, and a moderate

Table 6.1. Climatic impacts summarized for southern Saskatchewan in relation to normal (HIST1/1951–1980 or HIST5/1950–82).

	HIST2 (1961)	HIST3 (1929–38)	GISS1 ($2 \times CO_2 TP$)	GISS2 ($2 \times CO_2 T$)
Degree-days	+10 to +18%	+3 to +16%	+48 to +53%	+48 to +53%
Precipitation effectiveness	−18 to −53%	−21 to −26%	+1 to +13%	−10 to −12%
Biomass potential	−53 to −100%	−26 to −60%	+1 to +30%	−19 to +3%
Wind erosion potential	+123%	-	−14%	+26%

	HIST2 (1961)	HIST4 (1933–37)	GISS1 ($2 \times CO_2 TP$)	GISS2 ($2 \times CO_2 T$)
Spring wheat production	−76%	−20%	−18%	−28%
Expenditures by agriculture (million$)	−$1810	−$599	−$163	−$277
Employment in agriculture	−8000	−2647	−722	−1224

Palmer Drought Index (PDI)

	Frequencies			Return period (years)	
	HIST5 (1950–82)	GISS3 ($2 \times CO_2 TP$)	GISS4 ($2 \times CO_2 T$)	HIST5 (1950–82)	GISS3 ($2 \times CO_2 TP$)
Severe drought	0.1%	0.9%	10.8%	15 to 35	8.5 to 17.5
Drought	3.0%	9.1%	39.6%	6.5 to 10	4 to 6

See Table 1.3 and text for explanation of scenarios (HIST and GISS); *see* text for details of impacts. For further information on degree-days, precipitation effectiveness, biomass potential and PDI *see Table 2.5*, for wind erosion potential *see Table 3.3*, for spring wheat *see Table 4.7*, and for agriculture expenditures and employment *see Table 5.8*.

reduction in spring wheat productivity. Estimates of losses for the agricultural economy were about 1200 jobs, and about Can$275 million annually, which in turn would give rise to a Can$250 million provincial GDP reduction in sectors other than agriculture and a loss of 2600 more person-years. A severalfold increase in drought frequency could be expected (GISS4).

With both the historically based and the $2 \times CO_2$ scenarios, the results for potential biomass productivity (CA) (*see* Section 2) indicated a more accentuated midsummer dry period than normal (HIST1). This was most noticeable for the extreme decade (HIST3) and for $2 \times CO_2$ warming without the precipitation increase (GISS2). For the extreme year (HIST2) there were so many months without any contribution to CA that the dry midsummer was less marked. For $2 \times CO_2$ (GISS1 and GISS2), the results also indicated that the spring/early summer growth peak period would move back to an earlier part of the year than with HIST1.

The estimated reduction in potential biomass productivity for the extreme year (HIST2) is somewhat greater than for spring wheat production. The HIST3 and HIST4 results suggest that the reduction in biomass potential might be twice as large as that for spring wheat for an extreme 5-year period based on the historical climatic data. Spring wheat is a crop well adapted to the region, and in dry years the grain yield would tend to represent a greater proportion of total biomass than usual (Section 4), so the results are perhaps not too inconsistent.

In contrast to the agreement between biomass and spring wheat results for the historically based extreme year and period, there was a lack of agreement in the case of the $2 \times CO_2$ climatic scenarios. For GISS1 they indicated potential biomass productivity would increase while spring wheat production would decrease; while for CO_2 warming without the increased precipitation (GISS2), it appeared spring wheat production would be reduced considerably more than biomass productivity. The policy implications of these results will be discussed in Subsection 6.4.

The suggestion for GISS1 of a moister climate, as indicated by precipitation effectiveness, accompanied by more frequent and severe droughts, as indicated by PDI, is another apparent contradiction. Its interpretation will be discussed in Subsection 6.3.2.

There were indications, for example in the drought and spring wheat analyses, that the greatest impacts of CO_2-related warming would be in the northern agricultural areas, with the result that the agricultural zone of Saskatchewan would become agroclimatically more homogeneous under a doubled CO_2 atmosphere than it now is.

The spring wheat analysis emphasized the need for distinguishing between individual drought years, and consecutive drought years. The 1937 growing season was rather similar to that of 1936, but the fact that 1937 was a drought year following a drought year contributed to the much greater severity of its impacts.

In the economic analyses it was found that, in drought years, losses were considerably lower if part of the crop land was in winter wheat than if it was all in spring wheat, but the reverse was true in normal years.

6.3. Implications for Climatic Impact Analysis

6.3.1. Simulating GCM-based climatic scenarios

There has been considerable experience, with studies going back more than 70 years, in analyzing historical climatic variations and their impacts in western Canada (reviewed briefly in Section 1), but experience in estimating the impacts for climatic scenarios derived from general circulation models (GCMs) is much more limited. This case study has provided a valuable opportunity to gain some insights regarding such use of GCMs, and we would be remiss if we did not report them here.

The $1 \times CO_2$ equilibrium climate computed by a GCM may seem to correspond fairly well to the observed climate in terms of global or hemispheric annual average values. In any analysis involving agriculture, however, it is necessary to consider quite limited areas on the globe, particular seasons of the year, and the variations from year to year and district to district.

The GISS GCM data for Saskatchewan locations for a $1 \times CO_2$ equilibrium climate (GISSE) represented cool maritime conditions quite different from Saskatchewan's present climate. It would seem desirable that in a regional study the $1 \times CO_2$ equilibrium climate should correspond much more closely to the present climate of the region than was the case here. This correspondence can easily be tested. It is also quite desirable that the change from a $1 \times CO_2$ to a $2 \times CO_2$ climate be realistic, but this would seem impossible to test. When the $1 \times CO_2$ equilibrium climate is quite different from real conditions, then, as Parry and Carter (1984) have remarked, "whether the change between $1 \times CO_2$ and $2 \times CO_2$ equilibrium conditions adequately reflects the real changes that would occur as a result of CO_2 doubling is not known."

In such a case it may still be possible to infer from the GCM results a climatic change that seems "reasonable" or "plausible", to provide a climatic scenario that can be applied in impact analysis, as was done here. However, caution should be exercised in interpreting the computed impacts as estimates of the likely effects of CO_2-related climatic change.

Where the $1 \times CO_2$ equilibrium climate precipitation is much larger than the observed precipitation, it may suggest that the change in precipitation for CO_2. doubling indicated by the GCM results is also likely to be unrealistically high. In view of this it is useful if impact computations are made both with and without the inferred precipitation increase, as was done here, as this provides upper and lower estimates of the impacts.

For inferring the climatic changes from the GISS results, we selected the method of differences for deriving temperature scenario data and the method of ratios for precipitation, as explained in Section 1. This use of ratios for precipitation proved quite controversial. Perhaps this was because in the North Atlantic and in Western Europe the two methods may give almost the same results for precipitation, and Conrad and Pollak (1950) imply that, in that case, it is easier to use the method of differences. In a more continental climate, ratios or percentages have commonly been used in specifying precipitation changes in simulating climates (e.g., Williams, 1970, 1975; National Defense University, 1980; Bootsma et al., 1984).

Experience with the Palmer Drought Index (PDI) analysis of data on a month-by-month basis indicated that using ratios to adjust precipitation to a $2 \times CO_2$ climate (GISS1) led to increased drought frequency, where the use of differences would have led to a much decreased frequency. We suspect that PDI has proved here to be a good indicator of what is, in fact, a major climatic difference. The experience with PDI suggests that using the method of ratios for precipitation in deriving a $2 \times CO_2$ scenario results in a climate that is still quite continental, while using differences for precipitation would lead to a much more maritime climatic scenario. We assumed that it was more appropriate to use methods that would maintain the continental nature of the climate, but this was a subjective judgment.

With the method of differences, the standard deviation of the simulated climate data, i.e., the absolute variability, is the same as in the original historical series. The calculation of PDI thus assumed the absolute variability of temperature to be the same with the simulated climate as with the historical series. The method of computing growing degree-days, which made use of standard deviations of temperature, also made the same assumption. With the method of ratios, the coefficient of variation, i.e., the relative variability, rather than the absolute variability, was assumed to be the same in the simulated climate as in the climate of recent decades. If GCM results could provide more indication of the likely effects of CO_2 doubling on the variability of climatic elements this would be useful for impact analysis.

Experience in this case study suggests several other types of information that would be useful if it were possible to obtain them from GCMs. These include wind speeds, mean daily maximum and minimum temperatures, the likely timing of the summer rainfall peak, and information on expected storm track positions in a $2 \times CO_2$ climate. It would be useful to have a series of results from a GCM representing the climatic transition from a $1 \times CO_2$ to a $2 \times CO_2$ atmosphere, which could then be employed to estimate the resulting transitions in agriculture. There would also seem to be a need for more detailed information on the likely geographic patterns of the anticipated climatic changes within a region such as Saskatchewan. In this study the available information suggested fairly uniform changes across the region, and this may not be realistic.

There is a need for more information to assist users in determining which GCM is the most appropriate for a particular study. In this IIASA/UNEP climate impacts project the GISS model was used as the standard for all the case studies. In a further study, say, of the North American Great Plains, some other GCM might be more suitable, particularly as new research developments are taken into account. There is clearly a need to make dialogue between climatic modelers and impact modelers as easy as possible, and in some cases the GCM selected might be the one for which such dialogue could be most readily promoted.

It would probably be feasible to use GCMs to estimate not only the impacts on climate of CO_2 changes, but also the impacts, for instance, of volcanic eruptions contributing to climatic cooling. Perhaps probabilities could be assigned to these various factors that may affect our future climate, GCMs could be used to assess the likely resulting climate changes, and impact models could be applied to the resulting scenarios to derive a range of estimates, with accompanying probabilities, of likely effects on agriculture.

6.3.2. Implications for estimating impacts

In this subsection we consider the further work that is needed to improve our understanding of:

(1) Climatic aspects relating to soil degradation and soil erosion.
(2) Some aspects of the application of climatic scenarios.
(3) The linkages and interactions between climate models and impact models.

Soil degradation and erosion

Particular emphasis is needed on climatic aspects of soil degradation to make up for past neglect. This subject is of significance to climatology not only because it relates to an important impact of climatic changes and fluctuations, but also because soil degradation affects the capacity of agriculture to respond to changing climates, and soil degradation processes may themselves affect the future climate. In further research the consideration of soil degradation needs to be broadened to include not only wind erosion, which was analyzed in this case study, but also water erosion and salinization, both of which are climate-related processes that are very damaging to agriculture.

In such erosion-prone areas as Saskatchewan, soil erosion losses, and the resulting decreases in agricultural productivity, should be measured routinely, both to assist model development and to monitor the level of the problem. Study of the relation between climate, blowing dust, and wind erosion losses would be useful, as blowing dust may serve as a surrogate for wind erosion losses where the latter are not measured. Improved expressions for the wind erosion potential of climate need to be developed and subjected to frequency analysis. Winter wind erosion, and the relationship of wind erosion to wind speed, need to be studied.

The effects on climate of increasing wind erosion of soil should be assessed. Blowing dust can reduce incoming energy, and soil surfaces that become lighter in color through erosion become more reflective, while dust deposited on snow can decrease surface reflectivity, and all these processes can thus affect the climatic energy balance.

Efforts should be made to develop a wind erosion potential model that could be used operationally on a monthly or weekly basis to provide warnings of imminent wind erosion situations, just as, for example, fire weather indices are computed to warn of climatic conditions conducive to forest fires. Wind erosion potential maps could be used to alert agricultural extension agencies and farmers about the expected times and locations of moderate to high risk conditions.

Applying climatic scenarios

The causes of the differences in the results for the northern agricultural districts in the spring wheat analysis need to be investigated in more detail than was possible here. It may also be possible to improve the modeling to reflect likely differences of stubble and fallow sown crops in how they respond to climate.

In applying the various impact models, scenario data for certain variables were typically used in conjunction with data for some historical period for other variables. Degree-day computations for all scenarios employed standard deviations of monthly mean temperatures from the 1951–1980 period. Calculations of the climatic index of agricultural potential used solar radiation from the 1967–1976 period in all cases. The drought index computations for $2 \times CO_2$ scenarios were performed in such a way that the absolute variability for temperature and the relative variability for precipitation remained the same as those for the 1950–1982 period. The wheat growth model used diurnal temperature ranges for the 1951–1980 period with the $2 \times CO_2$ scenario data, and the economic analyses used economic conditions for the year 1980 in conjunction with all the scenarios. The GISS2 and GISS4 scenarios employed GCM-based temperature scenario data together with historically observed precipitation.

Provided that caution is exercised in interpreting results, the employment of historically observed base-period data for some variables together with scenario data for others is a useful procedure, and there may be justification for applying it more extensively to permit fuller exploitation of scenario data. For example, in the wind erosion potential analysis, historically based wind data for the 1951–1980 period were used in conjunction with the $2 \times CO_2$ scenario temperatures and precipitation. In further work these wind data might also be used with the temperature and precipitation data based on the 1929–1938 decade, a period for which observed wind data were inadequate for this study.

In impact analysis, the use of climatic scenarios has some advantages over analyses that simply examine the sensitivity for a matrix of changing values of input variables such as 2- and 4-degree increases and decreases in temperature and 10% and 20% changes in precipitation. Such sensitivity analyses do not directly answer "What if?" questions about the likely impacts of a future period with climate like that of the 1930s, or like one of the predicted $2 \times CO_2$ climates. Also, such a matrix-type sensitivity analysis typically considers a given change for each variable without seasonal differentiation or regard to changes in other variables. Analyzing just four different temperature changes in conjunction with four different precipitation changes involves a 5×5 matrix, when the no-change conditions are included for each variable, and hence 25 different analyses. Attempting to introduce different changes for different months would produce an unmanageable number of combinations for analysis. The use of scenarios, whether based on historical data or GCMs, helps to overcome this problem.

Linkages and interactions

The climate/agriculture system in Saskatchewan, as in any region, is extremely complex. Many more impact models could be used in future studies than were used here, and it would be desirable to explore many more of the possible linkages and feedbacks than were considered here.

In this case study the link from spring wheat yields to the economy has been developed, but in general the feedbacks such as the effect of economic demand on how much spring wheat or other crops will be grown, and on how much summer fallow and biomass productivity there will be, have not been examined. A few of the many possible linkages that could be explored in future work are shown in *Figure 6.1*.

The biomass potential model (Section 2) could be used with a series of harvest indices for different crops. However, in that case, as in the case of the estimations for barley, oats and canola (Section 5), it is assumed that the different crops respond similarly to climate. Alternatively, the spring wheat model could be replaced by a whole family of crop growth models that would reflect the differing responses. These models, together with feedback regarding economic demand, could be used to simulate crop selection. Significant increases in

Conclusions and implications

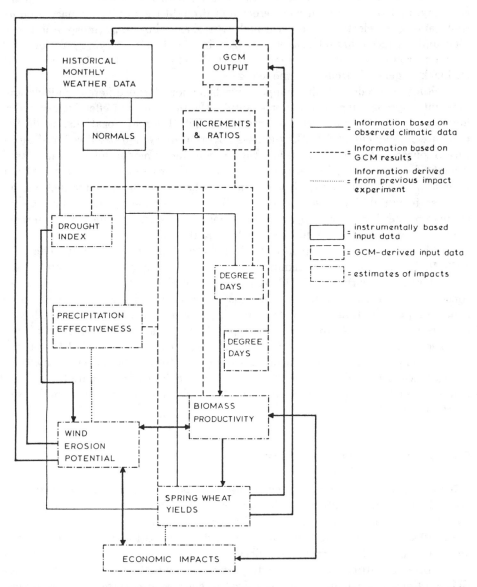

Figure 6.1. Generalized flow chart of the Saskatchewan case study showing some additional linkages that could be explored in the future (*see* arrows).

thermal resources could be expected to lead to more emphasis on crops such as maize which have well-known thermal requirements that could be matched to estimates such as those provided by the degree-day component. A series of economic models might be used to represent different types of agricultural enterprise. Analyses are needed linking year-to-year variability, as examined in the drought study, and economic impacts.

Linkages are needed to reflect the long-term effects of wind erosion on the economy, and the effects on wind erosion of drought frequency, biomass productivity and crop selection. Also, through their effects both on atmospheric turbidity and surface characteristics, changes in the levels of wind erosion and biomass productivity may ultimately affect the climate, so that there would be feedback to general circulation modeling.

Study is needed of the comparative behavior of the different models and different ways of using them. For example, the precipitation effectiveness index indicated improved moisture resources with GISS1 for all locations, while the moisture component (Fs) in the biomass potential model indicated reduced moisture resources at some of them, and the drought index model indicated increased drought frequency at all locations. The increased drought frequency would not necessarily be inconsistent if the climatic change were accompanied by an increase in variability of soil moisture. On the other hand, this apparent contradiction may have arisen because of inconsistencies among the various impact models. More research would be needed to clarify this.

The calculations with respect to drought, and also to the climatic effects of 1933–1937 on spring wheat, demonstrated the value of analyzing data on a year-by-year basis. Mearns *et al.* (1984) have also emphasized the disadvantages of considering only climatic means. Estimates on a year-to-year basis need to be made for precipitation effectiveness and biomass potential and compared with the drought index analysis. The implications regarding variability of the climatic scenario data and the impact estimates need to be subjected to statistical examination. Consideration needs to be given to possible effects of methodological inconsistencies, such as the fact that soil moisture is estimated in at least four different ways in the different impact models, and perhaps in another way again in the GCM.

6.4. Implications for Policy

Practical means exist to answer such questions as, "What are the likely impacts on Saskatchewan agriculture if the climate changes as suggested by some of the recent GCM-based scenarios, or if we have another period with climate like that of the 1930s?" A great deal of work is needed, however, to improve the methods and make the whole process more objective.

In the meantime, this case study provides some tentative answers to "What if?" questions about likely impacts on Saskatchewan agriculture of the particular historical and GCM-based scenarios considered here. It also demonstrates methods that can be used for estimating the impacts associated with any new climatic scenarios that become available. Providing the limitations of the methods are kept in mind and care is taken in interpreting the derived information, these methods can and should be used to assist decision making in planning and formulating policy for agriculture.

The historically based scenario results remind us that if the climate remains basically the same as it has been in recent decades, Saskatchewan can expect to continue to have to cope with occasional years with agroclimatic resources so low that productivity is reduced by 75%, and with some entire decades in which productivity is reduced by 20% or more. Such years or periods can seriously reduce farm and provincial incomes, cause failure to meet export commitments, and make it necessary for governments to provide additional funding for existing agricultural support and assistance programs, and to consider policies to expand such programs. Such climatic anomalies may also significantly increase the rate of soil degradation and accentuate the need for conservation policies.

The temperature and precipitation increases of the $2 \times CO_2$ scenario (GISS1 and GISS3) would change the cultivation, pest control, crop selection and water and snow management practices required to make the best use of the agroclimatic conditions and preserve productive capacity. The results indicated that the biomass potential would increase while wheat production would be reduced (GISS1). We suggest that this is because the long-term climate would have changed to the point where spring wheat would no longer be a well-adapted crop. Crops such as winter wheat, maize and sunflowers, that can better exploit the longer, warmer growing seasons, would be relatively better adapted. A shift to a GISS1 type climate, if accompanied by such crop changes, would probably benefit agricultural production and the economy. However, the results also indicated more frequent and severe droughts, so production could become less stable from year to year.

With CO_2-related warming without the precipitation increase (GISS2, GISS4), the impacts on Saskatchewan agriculture would be largely negative, with substantially reduced productive potential and increased susceptibility to environmental deterioration.

Assuming gradual shifts toward a warmer climate, with or without increased precipitation, the agricultural system might be expected to adjust, in which case a policy to maintain ongoing research, agricultural education and the dissemination of information would help this adjustment. However, more beneficial adjustment could occur if changes could be monitored and coping innovations put in place as the changes occurred, rather than depending on reactive adjustments such as emergency *ad hoc* measures to aid farmers after crop failures or moving people out of severely degraded soil areas. Measures to assist the adaptive process might include the breeding and introduction of new crop varieties suited to the warmer climate and to the shorter daylength at time of peak early season growth, new incentives for conservation, or new policies for publicly funded grain storage facilities.

To obtain more comprehensive answers to questions about likely impacts of CO_2 increases on agriculture, action would be needed to develop and support research that integrated climatic modeling, climatic impacts analysis, and the study of crop response to a CO_2-enriched atmosphere.

Saskatchewan, through its agricultural agents and programs such as "Farmlab", is attempting to make farmers aware of recommended techniques for improving productivity and conservation. "Farmlab" does this through the operation of demonstration farms across the province. The prospects for a changing climate would accentuate the need for policies to support such programs. Related agricultural research is also needed, and much of this is done at the federal level, while the extension workers who deal directly with farmers are employed by the province. There is undoubtedly a need to enhance communication between the different levels of government, and in particular to ensure that the necessary feedback about what research is needed is being communicated to the research organizations.

Soil degradation by wind and water erosion, salinization, and the reduction of soil organic matter is an ongoing concern. This concern has been expressed in reports by the Prairie Farm Rehabilitation Administration (1983), the Standing Committee on Agriculture, Fisheries and Forestry to the Senate of Canada (Sparrow, 1984), the Regional Development Branch of Agriculture Canada (Anderson and Knapik, 1984) and the Lands Directorate of Environment Canada (Bircham and Bruneau, 1985), and in a report on the problems of Canadian agriculture commissioned by the Agricultural Institute of Canada (Fairbairn, 1984). There have also been many related articles in the media. The title of the Senate report (Sparrow, 1984), *Soil at Risk, Canada's Eroding Future*, is an apt statement of the seriousness of the problem, which, as the report noted, could result in Canada losing "a major portion of its agricultural capability." The report includes a comment by the Saskatchewan Institute of Agrologists to the effect that, for the required information and changes in attitude and practice, "it will require a quantum leap in the attention being paid to research and extension activities." The Sparrow report also notes the need for further efforts to demonstrate the benefits, in terms of efficiency and cost effectiveness, of soil conservation techniques for the farmers who apply them. The climate has a quite important influence on soil degradation (World Meteorological Organization, 1983), and any climatic change which resulted in increased drought frequency could be expected both to accelerate soil degradation and to make agriculture more susceptible to reduced soil quality.

The results for GISS1 indicate that the more humid climate, as reflected in the higher precipitation effectiveness levels, would significantly reduce the wind erosion potential. A shift to winter wheat, which leaves the ground covered during the particularly erosion-prone spring period, could further help to reduce wind erosion. However, the accompanying increase in expected drought frequency and severity would increase the risk of serious wind erosion events. The increased precipitation in wet periods could also lead to increased water erosion.

The changes indicated for the GISS1 scenario could have adverse consequences for salinization. This process typically reduces crop productivity of affected soils by 50% or more. In regions of more plentiful rainfall, the downward movement of water keeps salt from becoming a problem. However, in southern Saskatchewan, where rain is not usually plentiful, there is typically considerable upward movement of moisture, which deposits salt at or near the surface as it evaporates. Fairbairn (1984) notes that when years of high rainfall occur, the water goes down far enough to reach saline layers, which can result in the salt being brought up to the surface later. The change to heavier precipitation, particularly during wet periods, combined with more frequent and severe droughts, as indicated for the GISS1 scenario, could thus accentuate soil salinization.

We do not have enough information to be able to comment on whether a change to a warmer climate without increased precipitation (GISS2) would be likely to accentuate salinization or water erosion, but our results suggested that in the case of wind erosion there would be an increase if such a climatic shift occurred. The likely reduction in biomass productivity and increase in drought frequency would accentuate this, although the probable shift from spring to winter wheat could help to limit spring wind erosion.

In interpreting changes implied in the impact model results for the agricultural productivity that may be expected in some future period, the reduced productive capacity due to soil degradation in the meantime should be taken into account. No attempt was made in this case study to incorporate such reductions in capacity. It has been estimated that, at the present rate, the total cost of continued soil erosion losses can be expected to increase by Can$5.66 million per year, and that losses in the Canadian prairies associated with saline soil areas will increase by Can$26 million annually due to expansion of saline areas (Sparrow, 1984).

It would appear from the degree-day, precipitation effectiveness and biomass potential results for Uranium City, that the climatic changes implied in the $2 \times CO_2$ scenarios would lead to significant improvements in the climatic potential for agriculture in northern Saskatchewan and would substantially improve this potential relative to locations in the south. At present (HIST1) Uranium City has the poorest estimated biomass potential, but with GISS1 the biomass potential indicated for Uranium City is higher than the HIST1 values for 9 of the 11 southern stations and higher than the GISS1 values for 6 of those 11 stations. With GISS2 Uranium City ranks higher than 2 of the southern stations. However, this improved climatic potential in northern Saskatchewan would be largely unexploitable, as the land is primarily the rough rockland of the Canadian Shield, which is unsuitable for agriculture (Section 1). Canadian Shield land in the more favorable climate of southern Ontario is similarly unsuited to agriculture, and there is no perceptible trend for this to change. However, forest productivity in northern Saskatchewan might be expected to benefit from warming, particularly if accompanied by increased precipitation (see Part I, Section 5).

6.5. Recommendations

Episodes of climatic adversity have always been part of human experience, and the evolution of civilization may largely be seen as the process of learning to protect ourselves against such vagaries of nature (Lappé and Collins, 1979). If society has, and uses, information on the dynamic nature of climate and its likely impacts, its ability not only to cope with climatic adversity but also to take advantage of potentially beneficial climatic changes will be enhanced. The following recommendations relate either to the climatic impact information itself, or to its application.

6.5.1. Improving climatic impacts information

The major scientific problems in impacts research appear to be the lack of relevant validation of models for impacts work, and the absence of information on whether any given scenario is a realistic representation of a probable future climatic situation. These problems are so considerable, however, that it would probably be a waste of resources to devote greatly increased efforts to them, particularly when there are a number of other problems that could be dealt with fairly readily, thus significantly improving climatic impact analysis:

(1) Highest priority should be given to undertaking work that integrates impacts assessment and climatic analysis. For studies involving scenarios derived from GCMs, the impacts models need to be designed or adapted to fit GCM output; GCMs need to be tailored to provide what impacts analyses require, with compromises and successive approximations on both sides. There needs to be a continual dialogue, otherwise general circulation models will continue to develop into sophisticated systems that provide data that may not be too relevant for impacts work, and impacts modelers are likely to develop models that are totally unrealistic in terms of the data they require from GCMs.

(2) Linkages among different impact models need to be developed (*Figure 6.1*), particularly those relating to major changes in the type of agriculture practiced, and those involving interactions among the economy, cropping practices and long-term resource sustainability.

(3) Impact estimates should be made for a range of different crops and crop varieties which seem likely to be adapted to the climates represented by the scenarios being studied.

(4) The performance and sensitivity of various impact models should be examined and compared critically.

(5) The emphasis on study of soil degradation processes in the context of climatic impacts research should be substantially increased.

(6) Tables showing various climatic change scenarios, impacts and associated probabilities for the region should be assembled for use by agricultural planners and policy makers. These would be analogous to insurance actuarial tables, and they should be based on currently available knowledge and updated routinely to reflect the latest findings from climatic and impacts research.

(7) More detailed consideration of the temporal and spatial variation of impacts than was possible in this case study would be desirable.

(8) In addition to employing models to estimate the impacts of the climatic changes implied by a $2 \times CO_2$ scenario, analogue areas should be sought that have present climates comparable to the study area scenario climate. Study of present-day agriculture and agroclimatic relations in such regions could help give some indication of the probable effects on agriculture if the climate in the study area changed to that indicated by the scenarios.

6.5.2. Recommendations for policy

(1) High priority should be given to supporting research to assess the likely impacts of climatic changes and fluctuations on agriculture.

(2) Agricultural policy formulation and planning should be carried out with a full awareness of the implications for agriculture of a changing climate.

(3) Agencies involved in policy formulation and planning with respect to Saskatchewan agriculture should take the set of impact results for each of the scenarios in this case study and work out the policy package that might be appropriate under the circumstances. In doing so they need to consider such questions as: What policies are needed to deal with the fact that the climate of one of the next three or four decades may be like that of 1929–1938? What policies might be in the best interests of Saskatchewan agriculture if the climate were to become like that depicted by GISS1? This exercise would give such agencies useful experience in translating impact information into policy.

(4) Research into the combined impacts of the direct effects of increasing CO_2 on agriculture, and the climatic effects, should be supported.

(5) Consideration should be given to whether the prospects for a major change in Saskatchewan's climate now seem likely enough to justify intensification of strategies to help agriculture make the best use of the agroclimatic resources. Such enhancement of strategies could include the development of new crops and methods more suited to the changing environment, and the planning of additional economic protection for agriculture if droughts are expected to become more severe and frequent.

(6) Anticipation of climatic change should be taken as an indication for a need for expanded agricultural extension activities; this will help farmers take advantage of beneficial changes and minimize the impacts of detrimental ones.

(7) The results and methods described in this case study should be used to make preliminary assessments of the costs and benefits to Saskatchewan agriculture of preparing to respond effectively to the climatic changes indicated by the scenarios we have analyzed.

Acknowledgments

We thank Dr Martin Parry and the International Institute for Applied Systems Analysis for effectively promoting such work on climatic impacts. We are also grateful to the numerous reviewers who provided constructive criticism and useful comments. We acknowledge the support provided by our respective agencies and extend our appreciation and thanks to the many individuals who graciously contributed to this project.

Appendices to Part II

IIA.1. Climatic Index of Agricultural Potential

In determining the climatic index of agricultural potential (CA), a moisture factor (Fs) is computed for each month using monthly precipitation and temperature (Turc and Lecerf, 1972). Fs is a function that is related to monthly temperature and precipitation through a climatic soil water budgeting procedure, and in general its value is lowered, reflecting depleted moisture conditions, by increases in temperature or decreases in precipitation. Fs ranges from 0 for very dry conditions to 1.0 where moisture is not limiting. The soil moisture capacity used for the water budget calculations was the same as was used for France (Turc and Lecerf, 1972) and Alberta (Williams and Masterton, 1983; Williams, 1985).

A heliothermic factor (HT) is also computed for each month. This is based on a solar factor derived from daylength or global radiation, whichever gives the smallest result, and a temperature factor. The latter is calculated from monthly mean daily temperature, but with modification by mean daily minimum temperatures for values close to freezing, and it is made equal to zero for subfreezing minimum temperatures. Except when working in the range of temperatures below freezing, temperature increases generally increase the value of HT, reflecting improved thermal resources for biomass production. In these impact studies scenario values, i.e., the climate data for the extreme decade and year and the $2 \times CO_2$ simulation climatic data, are used in the computations for Fs and HT in the same way that climatic normals would be used in other types of applications of this model involving mapping spatial patterns of agroclimatic potential.

For each month the heliothermic factor is multiplied by the moisture factor. The index can then be expressed briefly as:

$$\text{CA (for year)} = \text{Sum for 12 months of monthly } HT^*Fs \qquad \text{(IIA.1)}$$

If a negative or zero value is indicated for either Fs or the minimum temperature for any month, the contribution to CA for that month is zero.

IIA.2. Derivation of the Palmer Drought Index

The application of Palmer's methodology, briefly summarized, was as follows:

(1) Potential evapotranspiration was calculated for each month using the model described by Thornthwaite and Mather (1955).
(2) A hydrologic accounting by months was done over a long series of years. Potential soil recharge, potential runoff and potential soil moisture loss were all accounted for month by month.
(3) The average potential runoff, potential recharge and potential soil moisture loss for the 1950–1982 period for each month of the year were then obtained using a data base available from a previous study (Jones, 1984). From these means certain coefficients dependent on climate were derived. They are:

(a) The coefficient of evapotranspiration (α) = $\overline{ET}/\overline{PE}$ (IIA.2)

(b) The coefficient of recharge (β) = $\overline{R}/\overline{PR}$ (IIA.3)

(c) The coefficient of runoff (γ) = $\overline{RO}/\overline{PRO}$ (IIA.4)

(d) The coefficient of loss (δ) = $\overline{L}/\overline{PL}$ (IIA.5)

where $\overline{ET}$ is evapotranspiration, $\overline{PE}$ is potential evapotranspiration, $\overline{R}$ is soil moisture recharge, $\overline{PR}$ is potential soil moisture recharge, $\overline{RO}$ is runoff, $\overline{PRO}$ is potential runoff, $\overline{L}$ is soil moisture loss, $\overline{PL}$ is potential soil moisture loss, and the bar over the symbol indicates that the value would be a 33-year average for that variable for the particular month of the year.

(4) The series of basic data were then reanalyzed using the derived coefficients to determine the "CAFEC" (climatically appropriate for existing conditions) quantities for each month in the 1950–1982 period. The CAFEC quantities were utilized to determine the CAFEC precipitation.

$$\hat{P} = (\alpha)PE + (\beta)PR + (\gamma)PRO - (\delta)L \quad \text{(IIA.6)}$$

(5) The CAFEC precipitation for each month is subtracted from the actual precipitation to obtain a measure of departure, d.
(6) The departure is then weighted to allow the spatial and temporal comparison of stations, using a climatic characteristic factor:

$$K = 17.67 K' / \sum_{i}^{12} \overline{D} K' \quad \text{(IIA.7)}$$

where

$$K' = 1.5 \operatorname{LOG}_{10}\left[\left[\frac{\overline{PE} + \overline{R} + \overline{RO}}{\overline{P} + \overline{L}}\right] / \overline{D}\right] + 0.5$$

and P is monthly precipitation. Rather than recompute K' for Saskatchewan, we used the value computed by Palmer for the conterminous United States to facilitate comparison between Saskatchewan data and that for adjacent parts of the USA (Jones, 1984).

(7) A moisture anomaly index (Z) for each month in the 33 years is calculated, $Z = dK$. The PDI for a particular month is based partly on Z for that month and partly on Z values for earlier months in a way which allows the PDI to take into account the length of the drought in months up to that point.

A more detailed explanation of the methodology outlined above is provided by Palmer (1965). Alley (1984), who notes that this methodology provides what is "perhaps the most widely used regional index of drought", discusses some of the limitations and assumptions of this index.

References

AES Drought Study Group (1986). *An Applied Climatology of Drought in the Prairie Provinces*. Canadian Climate Centre Report No. 86-4, unpublished manuscript.

Agriculture Canada (1966). *Soil Erosion by Wind*. Agriculture Canada Publication 1266 (Reissued 1974), Information Canada, Ottawa, 27 pp.

Agriculture Canada (1976, 1977). *Agroclimatic Atlas, Canada*. Agrometeorology Research and Service, Chemistry and Biology Research Institute, Ottawa, 17 maps.

Agriculture Canada (1980a). Data from Central Bread Wheat Co-operative Tests, 1975 to 1978. In *Description of Variety*. Licence Number 1955, Seeds Section, Food Production and Inspection Branch, Ottawa. 5 pp.

Agriculture Canada (1980b). Data from Central Bread Wheat Co-operative Tests, 1977 to 1979. In *Description of Variety*. Licence Number 2009, Seeds Section, Food Production and Inspection Branch, Ottawa, 4 pp.

Alley, W.M. (1984). The Palmer drought severity index: limitations and assumptions. *J. Clim. Appl. Meteorol.*, **23**, 1100–1109.

Allsopp, T.R. (1977). *Agricultural Weather in the Red River Basin of Southern Manitoba over the Period 1800 to 1975*. Circular CLI-3-77, Fisheries & Environment Canada, Atmospheric Environment Service, Downsview, Ontario. 28 pp.

Anderson, M. and Knapik, L. (1984). *Agricultural Land Degradation in Western Canada: A Physical and Economic Overview*. Regional Development Branch, Agriculture Canada, Ottawa, Ontario.

Bagnold, R.A. (1941). *The Physics of Blown Sand and Desert Dunes*. Methuen & Co., London.

Baier, W., Dyer, J.A. and Sharp, W.R., (1979). *The Versatile Soil Moisture Budget*. Technical Bulletin 87, Land Resource Research Institute, Agriculture Canada, Ottawa. 52 pp.

Ball, T. (1985). A dramatic change in the general circulation on the west coast of Hudson Bay in 1760 A.D.: Synoptic evidence based on historic records. In C.R. Harington (ed.), *Climatic Change in Canada*. Syllogeus **55**, 219–228.

Beltzner, K. (1976). *Living with Climatic Change*. Proceedings of The Toronto Conference Workshop, Nov. 17–22, 1975. Science Council of Canada, Ottawa.

Bennett, J. (1982). Reap the wild wind. *Harrowsmith* **7 (1)**, 60–71, 106–107.

Berry, M.O. and Williams, G.D.V. (1985). Thirties drought on the prairies – how unique was it? In C.R. Harington (ed.), *Climatic Change in Canada*. Syllogeus 55, 63–74.

Bircham, P.D. and Bruneau, H.C. (1985). *Degradation of Canada's Prairie Agricultural Lands: A Guide to the Literature and Annotated Bibliography*. Working Paper No. 37, Lands Directorate, Environment Canada, Ottawa, 137 pp.

Blackwood, O.S., Kelly, W.C. and Bell, R.M. (1967). *General Physics* (3rd ed.). Wiley & Sons Inc., New York.

Bootsma, A., Blackburn, W.J., Stewart, R.B., Muma, R.W. and Dumanski, J. (1984). *Possible Effects of Climatic Change on Estimated Crop Yields in Canada*. Agriculture Canada, Research Branch, Land Resource Research Institute Contribution No. 83-64. Ottawa, 26 pp.

Brady, N.C. (1974). *The Nature and Properties of Soils* (8th ed.). Macmillan Publishing, New York, 639 pp.

Britnell, G.E. (1939). *The Wheat Economy*. University of Toronto Press, Toronto, 260 pp.

Brooks, C.E.P. (1943). Interpolation tables for daily values of meteorological elements. *Q. J. Roy. Met. Soc.*, **69 (300)**, 160–162.

Bryson, R.A. and Murray, T.J. (1977). *Climates of Hunger*. University of Wisconsin Press, Madison, Wisconsin, 171 pp.

Canada Committee on Agrometeorology (1977). *Climatic Variability in Relation to Agricultural Productivity and Practices*. Agriculture Canada, Research Branch, Ottawa.

Canessa, W. (1977). Chemical retardants control fugitive dust problems. *Pollution Eng.*, July, 24–27.

Carder, A.C. (1962). Climatic trends in the Beaverlodge area. *Can. J. Plant Sci.*, **42**, 698–706.

Chakravarti, A.K. (1978). A case of droughts in the Sahara: the effects of removal of natural vegetation on rainfall. *Alternatives, Persp. Soc. and Environ.*, **7**, 55–56.

Changnon, S.A. (1983). Record dust storms in Illinois: causes and implications. *J. Soil Water Conservation*, Jan.–Feb., 58–63.

Chapman, L.J. and Brown, D.M. (1978). *The Climates of Canada for Agriculture*. Revision of Canada Land Inventory Report No. 3, 1966. Environment Canada, Lands Directorate, 24 pp.

Chepil, W.S. and Woodruff, N.P. (1963). The physics of wind erosion and its control. *Adv. Agron.*, **15**, 211–302.

Chepil, W.S., Siddoway, F.H. and Armbrust, D.V. (1962). Climatic factor for estimating wind erodibility of farm fields. *J. Soil and Water Conservation*, **17**, 162–165.

Chepil, W.S., Siddoway, F.H. and Armbrust, D.V. (1963). Climatic index of wind erosion conditions in the Great Plains. *Proc. Soil Sci. Soc. Am.*, **27**, 449–452.

Clayton, J.S., Ehrlich, W.A., Cann, D.B., Day, J.H. and Marshall, I.B. (1977). *Soils of Canada*. Agriculture Canada, Research Branch, Ottawa, Ontario.

Conrad, V. and Pollak, L.W. (1950). *Methods in Climatology* (2nd ed.). Harvard University Press, Cambridge, Massachusetts.

Council for Agricultural Science and Technology (CAST). Beatty, M.T. (Chairman) (1982). *Soil Erosion: Its Agricultural, Environmental and Socioeconomic Implications*. Council for Agricultural Science and Technology Report No. 92, Ames, Iowa.

David, P. (1979). Sand dunes in Canada. *Geos.*, **Spring**, 12–15.
Dominion Bureau of Statistics (1941). The influence of precipitation and temperature on wheat yields in the Prairie Provinces, 1921–1940. *Q. Bull. Agr. Statistics*, **July–Sept.**, 167–187.
Doorenbos, J. and Kassam, A.H. (1979). *Yield Response to Water*. FAO, Rome, 193 pp.
Dumanski, J. and Stewart, R.B. (1981). *Crop Production Potentials for Land Evaluation in Canada*. Technical Bulletin 1983-13E, Research Branch, Agriculture Canada, Ottawa, Canada, 80 pp.
Dyer, J.A., Stewart, R.B. and Muma R.W. (1982). A weather-based early warning system for spring forage in Western Canada. *Can. Farm Econ.*, **17** (4), 9–16.
Edey, S.N. (1977). *Growing Degree-days and Crop Production in Canada*. Agriculture Canada Publication 1635, Ottawa, 63 pp.
Edmonds, T.C. and Anderson, C.H. (1959). Note on climatic trends in the Lower Peace River region of northern Alberta. *Can. J. Plant Sci.*, **40**, 204–206.
Environment Canada (1977). *Manual of Surface Weather Observations*. Atmospheric Environment Service (AES), Downsview, Ontario.
Environment Canada (1982). *Canadian Climate Normals, 1951–1980: Temperature*, Vol. 2; *Precipitation*, Vol. 3; *Degree-Days*, Vol. 4; *Wind*, Vol. 5; *Frost*, Vol. 6. Atmospheric Environment Service, Downsview, Ontario.
Environment Canada (1977–1983). *Monthly Record of Meteorological Observations in Western Canada*. Atmospheric Environment Service (AES), Downsview, Ontario.
FAO (1960). *Soil Erosion by Wind and Measures for its Control on Agricultural Lands*. FAO Agricultural Development Paper No. 71, Agricultural Engineering Branch, Food and Agriculture Organization, Rome. 88 pp.
FAO (1978). *Report on the Agro-ecological Zones Project: Methodology and Results from Africa*. Vol. 1, World Soil Resources Report 48, Food and Agriculture Organization, Rome, 158 pp.
Fairbairn, G.L. (1984). *Will the Bounty End? The Uncertain Future of Canada's Food Supply*. Western Producer Prairie Books, Saskatoon, Saskatchewan, 160 pp.
Glantz, M.H. (1979). *Saskatchewan Spring Wheat Production 1974*. Climatological Studies No. 33, Environment Canada, Atmospheric Environment Service, 27 pp.
Goudie, A. (1982). *The Human Impact, Man's Role in Environmental Change*. MIT Press, Cambridge, Mass.
Gray, J.H. (1978). *Men Against the Desert*. Western Producer, Modern Press, Saskatoon, Saskatchewan, 250 pp.
Guedalia, D., Estournel, C. and Vehil, R. (1984). Effects of Sahel dust layers upon nocturnal cooling of the atmosphere (ECLATS Experiment). *J. Climate Appl. Meteorol.*, **23**, 644–650.
Hansen, J.E., Lacis, A., Rind D., Russell, G., Stone, P., Fung, I., Ruedy, R. and Lerner, J. (1984). Climate sensitivity: analysis of feedback mechanisms. In J. Hansen and T. Takahashi (eds.), *Climate Processes and Climate Sensitivity*. Maurice Ewing Series, 5, Amer. Geophys. Union, Washington, D.C., pp. 130–163.
Heathcote, R.L. (1980). Perception of desertification on the Southern Great Plains: a preliminary enquiry. In R.L. Heathcote (ed.), *Perception of Desertification*. United Nations University, Tokyo, pp. 34–59.
Hinckley, A.D. (1976). *Impact of Climatic Fluctuation on Major North American Food Crops*. The Institute of Ecology, Washington, D.C., 23 pp.

Holmes, R.M. and Robertson, G.W. (1959). *Heat Units and Crop Growth*. Canada Department of Agriculture Publication 1042, Ottawa, 35 pp.

Holy, M. (1980). Erosion and environment (translated by J. Ondrackova), *Environmental Sciences and Applications*, 9, Pergamon Press, Oxford.

Hopkins, J.W. (1935). Weather and wheat yields in western Canada: I. Influence of rainfall and temperature during the growing season on plot yields. *Can. J. Res.*, 12, 306–334.

Jones, K.H. (1984). *An Evaluation of the Palmer Drought Index in Southern Saskatchewan*. Report No. CSS-R84-01, Environment Canada, Atmospheric Environment Service, Scientific Services, Regina, Saskatchewan.

Jong, E. de (1984). Factors of importance in soil erosion. In *Soil Erosion and Land Degradation*. Proceedings of the Second Annual Western Provincial Conference on Rationalization of Water and Soil Research and Management, November 29 to December 1, 1983. Saskatchewan Institute of Pedology, University of Saskatchewan, Saskatoon, Saskatchewan.

Keeling, C.D., Adams Jr., J.A., Ekdahl Jr., C.A. and Guenther, P.R. (1976a). Atmospheric carbon dioxide variations at the South Pole. *Tellus*, 28, 552–564.

Keeling, C.D., Bacastow, R.B., Bainbridge, A.E., Ekdahl Jr., C.A., Guenther, P.R., and Waterman, L.S. (1976b). Atmospheric carbon dioxide variations at Mauna Loa Observatory, Hawaii. *Tellus*, 28, 538–551.

Kendall, G.R. (1959). *Statistical Analysis of Extreme Values*. Proceedings of Symposium No. 1, National Research Council of Canada, Subcommittee on Hydrology, pp. 54–67.

Kimball, B.A. (1983). Carbon dioxide and agricultural yield: an assemblage and analysis of 430 prior observations. *Agron. J.*, 75, 779–788.

Klein Economic Consulting (1982). Regional agricultural structure. *Saskatchewan Drought Studies*. Study Element 5. Saskatchewan Environment, Regina, Saskatchewan, Canada, 94 pp.

Klein Economic Consulting (1983). Framework for regional farm analysis. *Saskatchewan Drought Studies*. Study Element 7. Saskatchewan Environment, Regina, Saskatchewan, Canada, 178 pp.

Kulshreshtha, S.N. (1983). *Relationship Between Employment Levels and Gross Economic Activity in Saskatchewan*. Saskatchewan Drought Studies, Prairie Farm Rehabilitation Administration, Regina, Saskatchewan, Canada, 11 pp.

Lappé, F.M. and Collins, J. (1979). *Food First, Beyond the Myth of Scarcity*. Ballantine Books, New York, 620 pp.

LeDrew, E.F., Dumancic, R., Peivowar, J., Tudin, B. and Dudycka, D. (1983). *Development of a 100 km Grid Square Climatic Data Base*. Report prepared for Agr. Canada, Contract Number DSS 245U.0.1416-2-0754, April (1983), 53 pp.

Lillis, E.J. and Young, D. (1975). EPA looks at "fugitive emissions". *J. Air Pollution Control Assoc.*, 25, 1015–1018.

Lyles, L. (1983). Erosive wind energy distributions and climatic factors for the West. *J. Soil Water Conser.*, **Mar.–April**, 106–109.

McCree, K.J. (1974). Equations for the rate of dark respiration of white clover and grain sorghum as functions of dry weight, photosynthetic rate and temperature. *Crop Sci.*, 14, 509–514.

Mack, A.R. (1982). *Drought Sensitivity Analysis of Crop Yield Estimates for Selected Years 1935 to 1961*. Report prepared for the Prairie Farm Rehabilitation Administration, Dec. (1982), 15 pp.

McKay, D.C. and Morris, R.J. (1985). *Solar Radiation Data Analyses for Canada, 1967-1976, the Prairie Provinces.* Vol. 4, Environment Canada, Atmospheric Environment Service, Downsview, Ontario.
Magill, B. (1980). *Temperature Variations in Southern Saskatchewan, 1921-70.* Thesis, University of Saskatchewan, Saskatoon.
Major, D.J. and Hamman, W.M. (1981). Comparison of sorghum with wheat and barley grown on dryland. *Can. J. Plant Sci.*, **61**, 37-43.
Manabe, S., Wetherald, R.T. and Stouffer, R.J. (1981). Summer dryness due to an increase of atmospheric CO_2 concentration. *Climatic Change*, **3**, 347-386.
Maybank, J. (1985). Climate change is food for you. *Chinook*, **7**, 20-22.
Mearns, L.O., Katz, R.W. and Schneider, S.H. (1984). Changes in the probabilities of extreme high temperature events with changes in global mean temperature. *J. Clim. Appl. Meteorol.*, **23**, 1601-1613.
Mermut, A.R., Acton, D.F. and Eilers, W.D. (1983). Estimation of soil erosion and deposition by a landscape analysis technique on clay soils in Southwestern Saskatchewan. *Can. J. Soil Sci.*, **63**, 727-739.
Moldenhauer, W.C., Langdale, G.W., Frye, W.M., McCool, D.K., Papendick, R.I., Smika, D.E. and Fryrear, D.W. (1983). Conservation tillage for erosion control. *J. Soil and Water Conser.*, **May-June**, 144-151.
National Defense University (1980). *Crop Yields and Climate Change to the Year 2000.* Vol. 1, US Government Printing Office, Washington, D.C.
Newman, J.E. (1980). Climate change impacts on the growing season of the North American corn belt. *Biomet.*, **7**, 128-142.
Novak, M.D. and van Vliet, L.J.P. (1983). Degradation effects of soil erosion by water and wind. In *Soil Degradation in British Columbia.* Proceedings of the 8th Meeting of the British Columbia Soil Science Workshop, B.C. Ministry of Agriculture and Food, Victoria, B.C.
Oehme, F.W. (1978). *The Effects of Dust on Human Health and Well Being.* Proceedings of the 4th International Symposium on Contamination Control, Institute of Environmental Sciences, Mt. Prospect, Illinois.
Palmer, W.C. (1965). *Meteorological Drought.* US Department of Commerce Research Paper No. 45, Washington, D.C., 59 pp.
Parry, M.L. and Carter, T.R. (1984). *Assessing the Impact of Climatic Change in Cold Regions.* Summary Report SR-84-1, International Institute for Applied Systems Analysis, Laxenburg, Austria, 42 pp.
Pasak, V. (1978). The soil erosion by wind. DSc Thesis, Research Institute of Amelioration, Prague (Vétrna eroze pudy, Dokt. diz. prace, Vyskumni ustav meliloraci, Praha). In D. Zachar (ed.), *Soil Erosion.* Developments in Soil Science 10, Elsevier Scientific Publishing, New York.
Penman, H.L. (1948). Natural evaporation from open water, bare soil and grass. *Proc. Roy. Soc. Lond., A.*, **193**, 120-145.
Penman (1963). *Vegetation and Hydrology.* Technical Communication 53, Commonwealth Agricultural Bureau, Farnham Royal, 124 pp.
Prairie Farm Rehabilitation Administration (PFRA) (1983). *Land Degradation and Soil Conservation Issues on the Canadian Prairies.* Soil and Water Conservation Branch, PFRA, Agriculture Canada. Regina, Saskatchewan.
Ramachandra Rao, A. and Padmanabhan, G. (1984). Analysis and modeling of Palmer's drought index series. *J. Hydrol.*, **68**, 211-229.

Rennie, D.A. and Ellis, J.G. (1978). *The Shape of Saskatchewan*. Saskatchewan Institute of Pedology, Publication M41, University of Saskatchewan, Saskatoon. 68 pp.

Richards, J.H. and Fung, K.I. (1969). *Atlas of Saskatchewan*. Modern Press, Saskatoon, 240 pp.

Ritchie, J.T. (1972). Model for predicting evaporation from a row crop with incomplete cover. *Water Res. Res.*, 8, 1204–1213.

Ritter, C. (1935). Bacterial content of the Kansas dust storm on March 20, 1935. *Public Health Reports*, 50, 622-623. Also in F.N. Oehme (ed.) (1978). *The Effects of Dust on Human Health and Well Being*. Proceedings of the 4th International Symposium on Contamination Control.

Robertson, G.W. (1968). A biometeorological time scale for a cereal crop involving day and night temperatures and photoperiod. *Int. J. Biometeor.*, 12, 191–223.

Robertson, G.W. (1974). Wheat yields for 50 years at Swift Current, Saskatchewan in relation to weather. *Can. J. Plant Sci.*, 54, 625–650.

Saskatchewan Drought Studies (1981-1985). Reports prepared for the Interim Subsidiary Agreement on Water Development for Regional Economic Expansion and Drought Proofing. Regina, Saskatchewan.

Schwab, G.O., Frevert, R.K., Edminster, T.W. and Barnes, K.K. (1966). Soil and water conservation engineering. In N. Hudson (ed.), *Soil Conservation*. Cornell University Press, Ithaca, New York.

Shawcroft, R.W., Lemon, E.R., Allen Jr., L.H., Stewart, D.W. and Jensen, S.E. (1974). The soil–plant–atmospheric model and some of its predictions. *Agric. Meteorol.*, 14, 287–307.

Sheppard, M.I. and Williams, G.D.V. (1976). Quantifying the effects of Great Soil Groups on cereal yields in the Prairie Provinces. *Can. J. Soil Sci.*, 56, 511–516.

Shewchuk, S.R. (1984). *An Atmospheric Carbon Dioxide Review and Consideration of the Mean Annual Temperature Trend at Saskatoon, Saskatchewan*. SRC Technical Report No. 160, Saskatchewan Research Council, Saskatoon.

Sly, W.K. (1982). *Agroclimatic Maps for Canada-Derived Data: Soil Water and Thermal Limitations for Spring Wheat and Barley in Selected Regions*. Technical Bulletin 88, Land Resource Research Institute, Agriculture Canada, Ottawa, 25 pp.

Sly, W.K. and Coligado, M.C. (1974). *Agroclimatic Maps for Canada-Derived Data: Moisture and Critical Temperatures Near Freezing*. Technical Bulletin 81, Agrometeorology Research and Service, Research Branch, Agriculture Canada, Ottawa, Canada, 31 pp.

Sparrow, H.O. (Chairman) (1984). *Soil at Risk, Canada's Eroding Future. A Report on Soil Conservation by the Standing Committee on Agriculture, Fisheries, and Forestry to the Senate of Canada*. Senate of Canada, Ottawa, Canada, 129 pp.

Spence, G. (1967). *Survival of a Vision*. Canada Department of Agriculture, Ottawa, 167 pp.

Star Phoenix (1984). Two die in highway accident. Saskatoon, Saskatchewan. June 1.

Statistics Canada (1977). *Canada Year Book 1976–1977*. Ottawa.

Statistics Canada (1980). *Saskatchewan Agriculture, Agricultural Statistics, 1980*. Statistics Canada Catalog No. 22-0020. Ottawa.

Statistics Canada (1981). *Canada Year Book 1980–1981*. Ottawa.

Statistics Canada (1982). *1981 Census of Canada: Agriculture in Saskatchewan*. Statistics Canada Catalog No. 96-909, Ottawa.

Statistics Canada (1983). *Field Crop Reporting Series*, No. 8, Ottawa.

Statistics Canada (1984). *Field Crop Reporting Series*, No. 6, Ottawa.
Stewart, R.B. (1981). *Modelling Methodology for Assessing Crop Production Potentials in Canada*. Technical Bulletin 96, Research Branch, Agriculture Canada, Ottawa, Canada. 29 pp.
Strange, H.G.L. (1954). *A Short History of Prairie Agriculture*. Searle Grain Co. Ltd., Winnipeg, 123 pp.
Street, R.B. and McNichol, D.W. (1983). Historical soil moisture in the Prairie Provinces: a temporal and spatial analysis. In C.R. Harington (ed.), *Climatic Change in Canada*. Syllogeus **49**, 130–143.
Thom, H.C.S. (1954a). The rational relationship between heating degree days and temperature. *Mon. Weather Rev.*, **82**, 1–6.
Thom, H.C.S. (1954b). Normal degree-days below any base. *Mon. Weather Rev.*, **82**, 111–115.
Thomas, M.K. (1961). *June 1961 – A Record Hot Dry Month on the Canadian Prairies*. Canada, Department of Transport, Meteorological Branch, CIR-3539, TEC-372, 6 pp.
Thomas, M.K. (1975). *Recent Climatic Fluctuations in Canada*. Climatological Studies No. 28, Environment Canada, Atmospheric Environment Service, Toronto, 92 pp.
Thornthwaite, C.W. (1931). The climates of North America according to a new classification. *Geog. Rev.*, **21**, 633–655.
Thornthwaite, C.W. and Mather, J.R. (1955). *The Water Balance. Publications in Climatology*. Vol. III No. 1, Drexel Institute of Technology, Centerton, New Jersey, USA.
Treidl, R.A. (1978). Climatic variability and wheat growing in the prairies. In K.D. Hage and E.R. Reinelt (eds.), *Essays on Meteorology and Climatology in Honour of Richmond W. Longley*. Studies in Geography Monograph 3, Department of Geography, University of Alberta, Edmonton, pp. 347–365.
Tung, F.L. (1984). Farm Income Outlook. *Market Commentary*, September 1984. Agriculture Canada, Ottawa.
Turc, L. and Lecerf, H. (1972). Indice climatique de potentialité agricole. *Science du Sol*, **2**, 81–102.
Underwood McLellan and associates (1982). Regional economic structure. *Saskatchewan Drought Studies*. Study Element 3. Saskatchewan Environment, Regina, Saskatchewan, Canada, 104 pp.
Underwood McLellan and associates (1983). Framework for regional economic impact analysis. *Saskatchewan Drought Studies*. Study Element 6. Saskatchewan Environment, Regina, Saskatchewan, Canada, 185 pp.
Unstead, J.F. (1912). The climatic limits of wheat cultivation, with special reference to North America. *Geograph. J.*, **39**, 347–366, 421–446.
US Department of Agriculture, Soil Conservation Service (1982). Conservation tillage – an attractive solution to soil erosion. *Soil and Water Conservation News*, **4** (2), US Department of Agriculture.
Villmow, J.R. (1956). The nature and origin of the Canadian Dry Belt. *Ann. Assoc. Amer. Geog.*, **46**, 211–232.
Wheaton, E.E. (1984). *Climate Change Impacts on Wind Erosion in Saskatchewan, Canada*. SRC Technical Report No. 153, Saskatchewan Research Council, Saskatoon, Saskatchewan.
Whittmore, D.O., Marotz, G.A. and McGregor, K.M. (1982). *Variations in Ground Water Quality with Drought*. Kansas Water Resources Research Institute Report, Sept. 1982, 55 pp.

Williams, G.D.V. (1962). Prairie droughts – the sixties compared with thirties. *Agric. Inst. Rev., Ottawa*, **17 (1)**, 16–18.
Williams, G.D.V. (1969a). Applying estimated temperature normals to the zonation of the Canadian Great Plains for wheat. *Can. J. Soil Sci.*, **49**, 263–276.
Williams, G.D.V. (1969b). Weather and prairie wheat production. *Can. J. Agr. Econ.*, **17 (1)**, 99–109.
Williams, G.D.V. (1970). Effects on weather-based prairie wheat production estimates of increasing precipitation amounts by 10 and 30 percent. In J. Maybank and W. Baier (eds.), *Weather Modification: A Survey of the Present Status with Respect to Agriculture*. Research Branch, Canada Department of Agriculture, Ottawa, pp. 124–133.
Williams, G.D.V. (1971). Wheat phenology in relation to latitude, longitude and elevation on the Canadian Great Plains. *Can. J. Plant Sci.*, **51**, 1–2.
Williams, G.D.V. (1973). Estimates of prairie provincial wheat yields based on precipitation and potential evapotranspiration. *Can. J. Plant Sci.*, **53**, 17–30.
Williams, G.D.V. (1974). A critical evaluation of a biophotothermal time scale for barley. *Int. J. Biometeor.*, **18**, 259–271.
Williams, G.D.V. (1975). An assessment of the impact of some hypothetical climatic changes on cereal production in western Canada. *World Food Supply in Changing Climate*, Proceedings Sterling Forest, N.Y., Conference, Dec. 2–5, 1974.
Williams, G.D.V. (1983). Prairie droughts as indicated by water-based wheat yield estimates. *Abstracts, Canadian Association of Geographers Annual Meeting*, May 30–June 4, 1983. Department of Geography, University of Winnipeg. Winnipeg, Manitoba, pp. 74–75.
Williams, G.D.V. (1985). Estimated bioresource sensitivity to climatic change in Alberta, Canada. *Climatic Change*, **7**, 55–79.
Williams, G.D.V. and Masterton, J.M. (1983). An application of principal component analysis and an agroclimatic resource index to ecological land classification for Alberta. *Climatol. Bull.*, **17**, 3–28.
Williams, G.D.V. and Oakes, W.T. (1978). Climatic resources for maturing barley and wheat in Canada. In K.D. Hage and E.R. Reinelt (eds.), *Essays on Meteorology and Climatology. In Honour of Richmond W. Longley*. Studies in Geography Monograph 3, Dept. of Geography, University of Alberta, Edmonton, pp. 367–385.
Williams, G.D.V. and Robertson, G.W. (1965). Estimating most probable prairie wheat production from precipitation data. *Can. J. Plant Sci.*, **45**, 34–47.
Williams, G.D.V., Joynt, M.I. and McCormick, P.A. (1975). Regression analyses of Canadian prairie crop-district cereal yields, 1961–1972, in relation to weather, soil and trend. *Can. J. Soil Sci.*, **55**, 43–53.
Williams, G.D.V., McKenzie, J.S. and Sheppard, M.I. (1980). Mesoscale agroclimatic resource mapping by computer, an example for the Peace River region of Canada. *Agric. Meteorol.*, **21**, 93–109.
Wilson, C. (1985). Daily weather maps for Canada, summers 1816 to 1818 – a pilot study. In C.R. Harington (ed.), *Climatic Change in Canada*. Syllogeus 55, 191–218.
Wilson, S.J. and Cooke, R.U. (1980). Wind erosion. In M.J. Kirkby and R.P.C. Morgan (eds.), *Soil Erosion*. John Wiley and Sons, Toronto, Ontario.
Wit, C.T. de (1965). *Photosynthesis of Leaf Canopies*. Agricultural Research Report 663, Centre for Agricultural Publications and Documentation, Wageningen. 57 pp.

Woodruff, N.P. (1975). Wind erosion research – past, present and future. In *Proceedings of the 30th Annual Meeting of the Soil Conservation Society of America*, Soil Science Society of America, San Antonio, Texas, pp. 147–152.

Woodruff, N.P. and Armbrust, D.V. (1968). A monthly climatic factor for the wind erosion equation. *J. Soil Water Conserv.*, **23**, 103–104.

Woodruff, N.P. and Siddoway, F.H. (1965). A wind erosion equation. *Proc. Soil Soc. Am.*, **29**, 602–608.

World Meteorological Organization (WMO) (1983). *Meteorological Aspects of Certain Processes Affecting Soil Degradation – Especially Erosion*. WMO Technical Note No. 178, WMO, Geneva, Switzerland.

Zachar, D. (ed.) (1982). *Soil Erosion*. Developments in Soil Science 10, Elsevier Scientific Publishing, New York.

PART III

The Effects of Climatic Variations on Agriculture in Iceland

PART III

The Effects of Climatic Variations on Agriculture in Iceland

Contents, Part III

List of Contributors		385
Abstract		387
1.	Introduction	389
	Páll Bergthórsson	
	1.1. Purpose of the study	389
	1.2. The settlement of Iceland	390
	1.3. Climate impact: a recurring theme	390
	1.4. Land surface characteristics	392
	1.5. Modern agriculture in Iceland	393
	1.6. The climate of Iceland	394
	1.7. Selection of study areas	404
	1.8. Choice of agricultural activities for impact studies	405
	1.9. Impact models and data	406
	1.10. Climatic scenarios	406
	1.11. Climatic variations and shifts in agroecological potential	412
	1.12. Final remarks	413
2.	The effects on agricultural potential	415
	Páll Bergthórsson	
	2.1. Introduction	415
	2.2. The effects on hay yields	416
	2.3. The effects on animal rearing	429
	2.4. The effects on the carrying capacity of improved grassland	431
	2.5. The effects on the carrying capacity of rangelands	435
	2.6. The effects on potential barley cultivation	437
	2.7. The effects on potential for tree growth	439
	2.8. Conclusions	441
3.	The effects on grass yield, and their implications for dairy farming	445
	Hólmgeir Björnsson and Áslaug Helgadóttir	
	3.1. Introduction	445
	3.2. The experimental material	446
	3.3. Yield of pastures cut for hay as a function of temperature	452
	3.4. Mechanisms influencing the effect of temperature on herbage yield	458
	3.5. Implications of temperature changes for Icelandic agriculture	462

	3.6.	Effects of the $2 \times CO_2$ temperature scenario (Scenario V)	471
	3.7.	Conclusions	473
4.	\multicolumn{2}{l	}{The effects on the carrying capacity of rangeland pastures}	475

4. The effects on the carrying capacity of rangeland pastures 475
   Olafur R. Dýrmundsson and Jón Vidar Jónmundsson
   - 4.1. Introduction — 475
   - 4.2. Rangeland grazing by sheep — 477
   - 4.3. Material and methods — 477
   - 4.4. Results — 479
   - 4.5. Interpretation of the results — 482
   - 4.6. Proposed model — 483
   - 4.7. Effects of a warmer climate — 484
   - 4.8. Impact of a colder climate — 485
   - 4.9. Conclusions — 487

5. Implications for agricultural policy — 489
   Bjarni Gudmundsson
   - 5.1. Introduction — 489
   - 5.2. Types of effects and responses — 490
   - 5.3. A summary of the effects on agricultural activity and agricultural output — 493
   - 5.4. Precautions against climatic variations — 498
   - 5.5. Research — 501

References — 505

List of Contributors

BERGTHÓRSSON, Mr. Páll
Vedurstofa Íslands
Bústadavegi 9
150 Reykjavik
Iceland

BJÖRNSSON, Dr. Hólmgeir
Agricultural Research Institute
Keldnaholt
112 Reykjavik
Iceland

DÝRMUNDSSON, Dr. Ólafur R.
The Agricultural Society of Iceland
Baendahöllin
P.O. Box 7080
127 Reykjavik
Iceland

GUDMUNDSSON, Dr. Bjarni
The Ministry of Agriculture
150 Reykjavik
Iceland

HELGADÓTTIR, Dr. Áslaug
Agricultural Research Institute
Keldnaholt
112 Reykjavik
Iceland

JÓNMUNDSSON, Dr. Jón Vidar
The Agricultural Society of Iceland
Baendahöllin
P.O. Box 7080
127 Reykjavik
Iceland

Abstract

Iceland is located at the northernmost limits of technologically developed agriculture and on the edge of the boreal forest zone. Ever since the country was settled over a thousand years ago, climate has always had a significant impact on its agriculture, economy and society. This case study represents an attempt to estimate the impacts of specified climatic changes on Icelandic agriculture and to consider appropriate policies of response.

Considerable attention is devoted to hay production and grazing conditions for animal husbandry, the basis of modern agriculture in Iceland, but other studies are reported on the potential for barley cultivation and for tree growth. Most assessments employ a regression approach to estimate agricultural productivity under specified climatic conditions (represented by sets of meteorological data). Five types of climate (scenarios) have been simulated in the studies:

(1) A baseline climate, representing present-day (1951–1980) climatic conditions.
(2) An anomalously cool decade taken from the historical instrumental record.
(3) An ensemble of the 10 coolest years since 1930, selected from the climatic record.
(4) An ensemble of the 10 warmest years since 1930 (also selected from the climatic record).
(5) The climate derived from the estimates of the Goddard Institute for Space Studies (GISS) general circulation model, for doubled concentrations of atmospheric carbon dioxide.

These scenarios have been based on the long temperature record at Stykkishólmur, which is a representative station for the whole country.

Changes in national hay yields, pasture yields, carcass weight of lambs and carrying capacity of the rangelands have been estimated for each climatic scenario relative to the baseline. The importance of nitrogen fertilizer in determining yield response is investigated, and the implications of this relationship both for stabilizing hay yields and for maintaining dairy production are examined in some depth.

The contrast in effects between a cooler-than-average and a warmer-than-average climate in Iceland is indicated clearly by the results. Under cool conditions analogous to the cool decade 1859–1868, with mean annual temperature at Stykkishólmur 1.3 °C below the baseline, average hay yields could be reduced by 16–19%, the mean carcass weight of lambs decreased by some 4–7%, and the carrying capacity of rangeland reduced by about 20%, relative to the baseline period. The "warm ensemble" scenario, on the other hand, with mean annual temperature 1.1 °C above the baseline, would give percentage *increases* in each of these of a similar magnitude, relative to the baseline. Barley cultivation would be inadvisable anywhere in Iceland during a cool decade, but could be practiced in many lowland areas under warmer-than-average conditions.

The experiments indicate that fertilizers can be used to counteract the variations in growing conditions due to climate. One precondition is that farmers do not apply maximal fertilizers in average or good years. Furthermore, it is demonstrated that these fertilizer application levels can be adjusted on the basis of *predictions* of hay growth, using winter temperatures as an explanatory variable.

A temperature increase of 4.0°C at Stykkishólmur, estimated for the $2 \times CO_2$ scenario, would increase the potential farming area in Iceland and open up new farming options similar to those presently available in Scotland. However, it is clear that farmers should always be prepared for cold years, even if the mean climate is warming, especially if the warmer conditions encourage farmers to cultivate new crops with an equivalent degree of risk as formerly.

Finally, it is recommended that research into climate impacts on Icelandic agriculture should focus on the present marginal areas, in the northern upland regions (where grass production has been uncertain in cool years) and in the more favorable areas (where barley and vegetable cultivation has succeeded in warmer years).

SECTION 1

Introduction

1.1. Purpose of the Study

In the Introduction which follows we shall show that Iceland's climate is extraordinarily marginal with respect to agriculture. Its summers are barely warm enough or long enough for either natural herbage or cultivated grasses to provide fodder for livestock. Its history is one of constant struggle between man and nature at a long-standing frontier of the settled world – a history in which the impact of climate on society is an unbroken theme. Iceland thus offers an attractive laboratory in which to study both the impacts of climatic variations on agriculture and the responses of agriculture to such impacts.

The objectives of this case study are:

(1) To examine the sensitivity of various aspects of Icelandic agriculture to variations of climate.
(2) To attempt to quantify the impacts of specified climatic changes on its agriculture.
(3) To consider what policies of agricultural response might be the most appropriate.

The investigations follow the structure adopted in the four other case studies in cool temperate and cold regions reported in this volume (in Saskatchewan, Finland, Northern European USSR and Japan). Regression equations are developed that can be used to estimate the effects of climatic variations on the primary productivity of grazing and fodder crops, on the carrying capacity of pastures and rangelands, and on the implied livestock production. Experiments are conducted using these equations to evaluate the impacts of several different climatic scenarios representing:

(1) An anomalously cool decade taken from the historical instrumental record.
(2) Individual and groups of anomalously warm and anomalously cool years selected from the recent climatic record.

(3) The climate derived from the estimates of the Goddard Institute for Space Studies (GISS) general circulation model, for doubled concentrations of atmospheric carbon dioxide (Hansen et al., 1983, 1984).

Several adjustment measures for mitigating adverse climatic effects receive a quantitative evaluation, and many more are assessed qualitatively.

This introduction provides background information on the study area and outlines the considerations governing the selection and development of the various impact models and scenarios used in subsequent sections.

1.2. The Settlement of Iceland

Iceland is situated between latitudes 63°23′ N and 66°32′ N and between longitudes 13°30′ W and 24°32′ W. The shortest distance to Greenland to the northwest is 300 km, and to Scotland to the southeast, 800 km. More than one-third of the country is higher than 600 m above sea level (*Figure 1.1*).

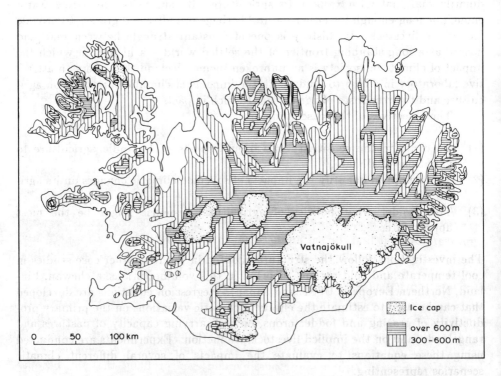

Figure 1.1. The topography of Iceland (for Iceland's location in northwest Europe, *see* *Figure 1.6*).

Apart from Irish hermits who established their cells in a few places about AD 800, the main settlement period in Iceland is usually reckoned to have taken place between 874 and 930 (Jóhannesson, 1974). Most of the settlers were of Norwegian origin. From the earliest times of the settlement, the rearing of livestock, mainly cattle and sheep, was the basis of the domestic economy. The custom was to let the sheep find their pasture all the year round. There were wide areas of birchwood and scrubs providing shelter in the winter. Valleys and lowlands were covered with grass and the uplands with brushwood and heathers. But the arrival of man with his grazing livestock in a virgin country, which before the settlement had no herbivorous animals, greatly disturbed the somewhat unstable equilibrium between the soil-building and soil-eroding processes. Soils in Iceland contain a low percentage of clay, so their structure is weak and they are susceptible to erosion by wind and water, in particular where volcanic activity is most pronounced (volcanic eruptions occur, on average, about once every five years). Through the effects of wood cutting, forest fires and livestock (mainly sheep), and probably to some extent through deterioration of climate, the birch woods were devastated and the result was soil destruction on a catastrophic scale. It is estimated that about half of the country's area was covered with vegetation when settlement began, and about half of that area has now been deprived of its soil cover (Thórarinsson, 1974). Large areas still suffer from overgrazing, in particular in periods of colder climate (*see* Section 4).

1.3. Climate Impact: A Recurring Theme

Throughout the history of Iceland climate has frequently limited the production of food and animal fodder, resulting in hunger and even starvation. It has been suggested that in the early years following the settlement, when 80–90% of the population were involved in agriculture, Iceland could support a maximum of 60 000 inhabitants (Eldjárn, 1981). Two centuries after the settlement, economic conditions began to deteriorate in Iceland. Large numbers of livestock perished of cold and hunger in winters that seem to have increased in severity in the 13th and 14th centuries (Bergthórsson, 1969; Fridriksson, 1973a). Famines became still more frequent in the 17th and 18th centuries. For example, famine spells occurred in the 1600s, 1690s, 1750s and 1780s. After 1800 conditions began to improve, slowly at first, but at an increasing rate by the end of the 19th century. As an example of the fluctuations, the population is estimated to have declined from 48 300 in 1755 to 42 800 in 1759, while it grew from about 38 400 in 1786 to 47 240 in 1801 (Baldursson, 1974).

There is abundant evidence indicating that during the first centuries after the settlement barley was grown in Iceland, mainly in the southern part of the country (Thórarinsson, 1944). This was later abandoned, probably in the 15th or 16th century. In the light of the marginal conditions for growing barley in Iceland (*see* Section 2), it is quite possible and even probable that this occupation was abandoned due to deterioration of the climate.

Historians disagree somewhat as to the role of climatic variations in the economic history of Iceland. Unfavorable trade conditions, failure in fish catch, and volcanic eruptions, especially the great Laki eruption in 1783, undoubtedly had serious consequences. It is, however, fairly evident that the great reductions in livestock numbers preceding many famines were, in most cases, themselves preceded by climatic variations. This can be shown indirectly by the correlation between the temperature and livestock numbers in the 19th century (*see* Section 2).

For several centuries predating instrumental temperature observations, it is possible to utilize the close correspondence between mean annual temperature and the documented prevalence of sea ice off the Icelandic coast to estimate climatic variations (cf. *Figure 1.5*). During the period 1591 to 1846 there is an exponential relationship between the frequency of years with mortality from hunger and the number of months of ice prevalence during the years immediately before the famine (Bergthórsson, 1985). No mortality from hunger occurred during the 26 years when the past ice prevalence had been less than one month/yr, while such famine occurred every other year in those 11 cases when the prevalence had been 4 months/yr or more, according to Bergthórsson's estimates.

1.4. Land Surface Characteristics

Table 1.1 shows the approximate areal extent and elevation of different land surfaces in Iceland. A little less than half of the present vegetated area consists of peat bog although it is generally considered that at the time of the settlement of Iceland in the ninth century the vegetated area was twice as extensive as it is now. Most of the vegetated land at higher elevations is bog because much of the drier soil has been eroded away. At low altitudes a considerable part of the bog has been drained during the last half century. Formerly the unimproved bog was used extensively for hay making, and the peat was a valuable fuel for the farms. In spite of the relatively high vegetative production, grazing is rather limited on the bog because sheep and horses prefer the drier areas. Approximately one-third of the vegetated land consists of naturally drained soils, in many cases characterized by hardy grasses, mosses and dwarf shrubs, particularly at the higher altitudes. In the lowland and on lower mountain slopes, the naturally drained grassland is often rather fertile and useful for grazing. About $1250\,km^2$ are covered by birch trees, which in 80% of the area are shorter than 2 m. Improved grassland, about $1400\,km^2$ in area, where all the hay making now takes place, along with some summer grazing, account for only about 6% of the total vegetated area.

Sandy wastes are generally below the 200 m elevation and they can, to some extent, be improved as grassland, while rocky wastes are the most extensive landscape, mainly at high altitudes. Some of this land at low or moderate elevations was formerly vegetated but has later been eroded. In spite of the very sparse vegetation, mainly along rivers and streamlets, this area is still of some use for summer grazing.

Iceland has a number of active volcanoes and associated lava areas. Much of the lava is unvegetated, and though parts are covered by mosses, these are of very little use for the grazing animals. There are several glaciers in Iceland, the largest, Vatnajökull, covers about $8400\,km^2$ in the southeast of the country. The extent of glaciers has varied with the climate and sometimes they have been

Table 1.1. Land surface, thousands of km² (after Fridriksson, 1973b).

	Vegetated areas	Sandy deserts	Rocky deserts	Lava	Lakes, rivers	Glaciers	Total
Above 600 m	0.8	0.2	22.8	2.5	0.3	11.1	37.7
400–600 m	3.3	0.6	15.9	1.5	0.5	0.4	22.2
200–400 m	6.0	0.6	10.6	0.7	0.2	0.3	18.4
0–200 m	13.7	2.7	4.8	1.6	1.8	0.1	24.7
Total	23.8	4.1	54.1	6.3	2.8	11.9	103.0

detrimental to agriculture, encroaching on productive land during periods of ice advance. In recent decades they have been thinning and retreating. Lakes and rivers are generally owned or used by the adjoining farms. They are in many cases valuable for the fishing of salmon and trout.

1.5. Modern Agriculture in Iceland

Farming has been the main occupation in Iceland during most of the 11 centuries of human settlement, supporting a population of from 40 000 to 80 000 up to the end of the 19th century. Since then the number of farms has only slightly decreased. The agricultural production has, however, increased considerably. *Figure 1.2* shows the approximate extent of the inhabited area today. Interestingly, this bears a close resemblance to the map of mean July temperature (cf. *Figure 1.8*).

During this century, with the development of new industries, the *relative* role of agriculture has declined. In 1980, out of a population of 229 000, only 8% of the population were making a living directly from agriculture compared with 32% of the population of 121 000 in 1940. Before 1920, the total area of improved grassland on all farms was around 200 km² while it is now 1400 km². However, due to changes in management practices, the livestock numbers have not increased at the same rate. *Table 1.2* shows the number of winterfed cattle, sheep and horses kept this century. The winter feeding of cattle usually begins in September and ends in June. Sheep are generally fed from the beginning of November to the middle of May, but the period of winter feeding is dependent upon weather and grazing conditions. The winter feeding of horses is much more variable. In some regions they are even not housed at all and receive only a very limited amount of hay.

Dairy production in 1983 exceeded home demand by 6%, while the corresponding surplus production of meat was 20%. Potato production is sometimes sufficient to meet national demand, but practically all cereals and most fruits and vegetables are imported. *Figure 1.3* shows the composition of agricultural products by value. Altogether, the value of agricultural production in 1984 was equivalent to about 5% of the Gross Domestic Product, and agricultural products, including industrial goods, made up about 7% of total exports – about 3% of the Gross National Product (B. Gudmundsson, personal communication).

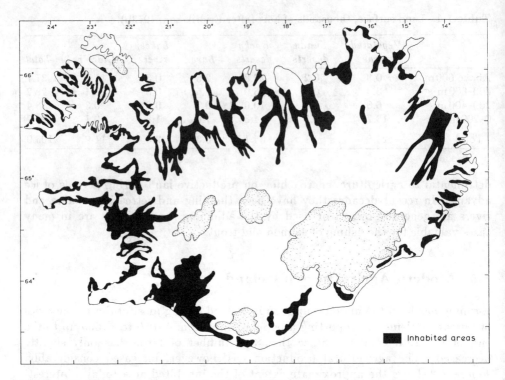

Figure 1.2. Inhabited areas in Iceland 1983 (G.M. Gudbergsson, personal communication).

Table 1.2. Livestock numbers in Iceland (thousands).

Year	Cattle	Sheep	Horses
1900	24	469	42
1920	23	579	51
1940	40	628	56
1960	53	834	31
1984	73	713	52

The industrialization of agriculture has certainly reduced the impact of natural conditions on the human population, but in spite of this, the inhabitants are often painfully reminded that climate still, to a large extent, determines their standard of living.

1.6. The Climate of Iceland

Iceland is situated near the boundary between the relatively warm Atlantic Ocean and the colder waters of Arctic origin. This boundary is not stationary and its fluctuations are clearly reflected by the variable extent of the sea ice in the East Greenland current. *Figure 1.4* illustrates some extreme limits of this ice

Introduction

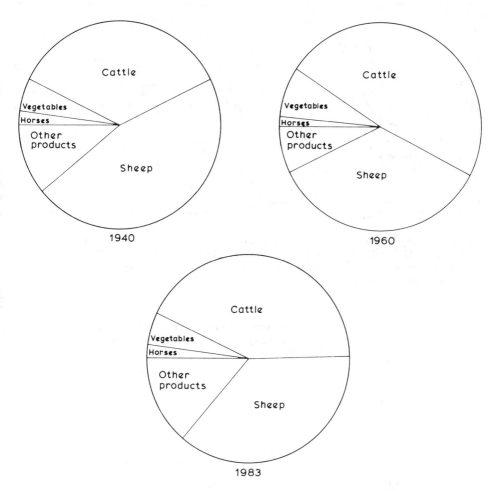

Figure 1.3. Share of commodities in agricultural production in Iceland in the years 1940, 1960 and 1983 (by value).

in autumn and spring. In warmer periods the ice is well distant and Iceland enjoys the relatively warm currents surrounding the country. In the severe ice years, which incidentally tend to come in clusters (e.g., 1859–1868, *see Figure 1.5*), the ocean can be ice-covered half the way from Greenland to Norway in the spring. Then Iceland resembles an icy peninsula extending from the Greenland ice cap. The sea and the ice react slowly to temperature variations in the air due to the great heat capacity of the former and the large quantities of heat required to melt the sea ice once it has been formed (latent heat of fusion). Thus the sea and the ice tend to smooth and delay climatic variations in comparison with continental Europe.

To give an idea of the climatic variations before the advent of quantitative temperature observations, *Figure 1.5* shows the prevalence of sea ice at the

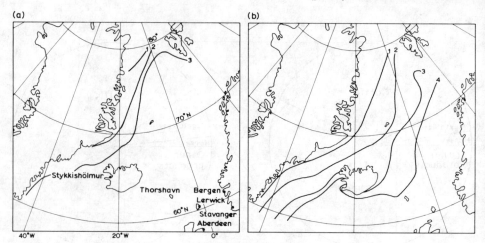

Figure 1.4. (*a*) Recent limits of sea ice in early October: (1) minimum, (2) normal, (3) maximum. (*b*) Limits of sea ice in March–May: (1) recent minimum, (2) recent normal, (3) recent maximum, (4) estimated maximum in historical times. (After Eythórsson and Sigtryggsson, 1971.)

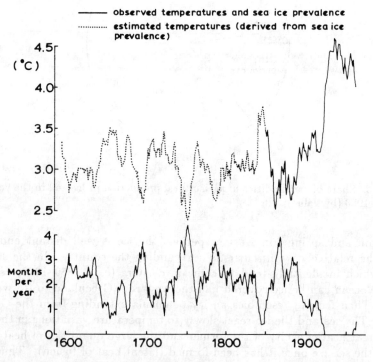

Figure 1.5. Climate variations in Iceland. Below: the 10-year running mean of annual prevalence of sea ice in months per year. Above: the corresponding mean annual temperature in Stykkishólmur, observed after 1845, estimated from ice prevalence before 1846. (Source: Bergthórsson, 1969.)

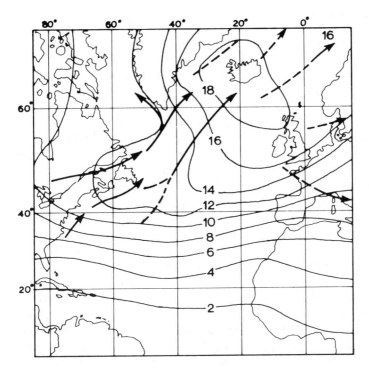

Figure 1.6. Variability of atmospheric pressure and storm tracks in January. Standard deviation of daily pressure (mb) about the monthly mean for January. Solid arrows are primary storm tracks. Dashed arrows are secondary storm tracks. (After Tucker and Barry, 1984.)

Icelandic coasts in months per year, during the period 1591 to the present (Bergthórsson, 1969). The relationship between temperature and sea ice in the observation period after 1945 has been used to construct a proxy record of the temperature before that time. This record has been partly confirmed, for the 17th and 18th centuries, by an independent analysis based only on contemporary written sources (Ogilvie, 1981, 1984).

The country also lies close to the boundary between the mid-latitude westerlies and the polar easterlies. Therefore, cyclones frequently pass close to the country, generally from the southwest, making the climate both unstable and rainy (*Figure 1.6*). In cold periods, which often tend to coincide with extensive sea ice, there is some tendency for the Icelandic low pressure area to be located further south. This seems to be partly a feedback effect, for the displacement of the low pressure lowers the temperature to the north of Iceland by increasing the preponderance of cold polar easterlies, and this cold air tends to raise the surface pressure, thus pushing the low further southwards.

1.6.1. Temperatures

Figure 1.7 shows the distribution of meteorological stations in Iceland that were operating for much or all of the standard reference period, 1951–1980. They are all in the inhabited area, except Hveravellir in central Iceland, most of them at an altitude of less than 100 m above sea level.

Figure 1.8, giving the mean surface air temperature in July in the period 1951–1980, is mainly based on these lowland observations, assuming a lapse rate of temperature of around 1 °C in every 150 m. According to the climatic classification of Köppen, the 10 °C isotherm for the warmest month broadly represents the poleward limit of tree growth (Köppen, 1936). The zone with a mean July temperature greater than 10 °C, which we shall hereafter refer to as the "potential tree zone", comprised about 11% of the country in 1951–1980. As will be discussed later in this section, the extent of the potential tree zone will contract or expand in connection with climatic variations, even if it takes a long time for the vegetation to obtain equilibrium with the new conditions.

Mean monthly temperatures across much of Iceland are close to freezing point for practically the whole winter (*Figure 1.9*), but there are quite large daily fluctuations around these average values. These result in a great variability of snow cover and, through repeated melting and refreezing of snow, can lead to the formation of an ice crust covering the grass. This can have important repercussions for grass growth and hay yields (*see* Section 2). Small variations in the summer temperature can also be of great importance, due to the marginal growing conditions.

The frost-free period is about 120–150 days at coastal stations in southern Iceland, but 80–120 days at other stations in the south (*Figure 1.10*). In the north it is shorter, 100–120 days at the coast and 60–90 days in the interior. A negative deviation of 1 °C in annual temperature will shorten the frost-free period by about 25 days.

The recent climatic variations are illustrated in *Figure 1.11* by the annual mean temperature at Stykkishólmur, during the period 1846–1984. This, the oldest observing station in the country, records values of mean temperature that are fairly representative of the lowlands, both in terms of the mean and the range of the annual temperatures (Sigfúsdóttir, 1969). Variability is high at all time scales, but two features stand out clearly in *Figure 1.11*. First, there is a tendency for clustering of years into sequences of anomalously cool or anomalously warm conditions. Secondly, there are two noticeably abrupt changes in longer-term average temperatures: a sudden warming in the order of 1 °C after 1920, and an equally rapid, but less intense cooling around 1965. Periods such as the 1860s and the 1880s registered mean annual temperatures more than 2 °C lower than those recorded in the warm 1930s and 1940s. Contrasts such as these, with their clear implications for agriculture in Iceland, form the basis for the selection of climatic scenarios (discussed in Section 1.10). Whether the temperature series reflects a longer-term warming trend (perhaps due to the "greenhouse" effect) is a matter for speculation since temperatures over the two most recent decades (1965–1984) have seldom approached the levels observed during the 1930s and 1940s.

Introduction

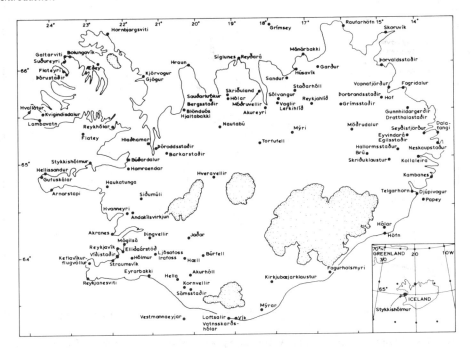

Figure 1.7. Meteorological stations in Iceland 1951–1980. Inset map shows the grid point locations for which GISS-model data were generated.

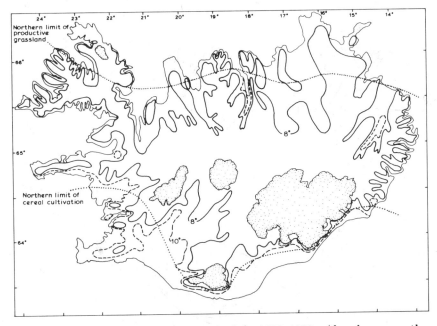

Figure 1.8. Mean surface air temperature in July 1951–1980. Also shown are the conceptualized limits to productive grass growth and to barley cultivation (B. Gudmundsson, personal communication).

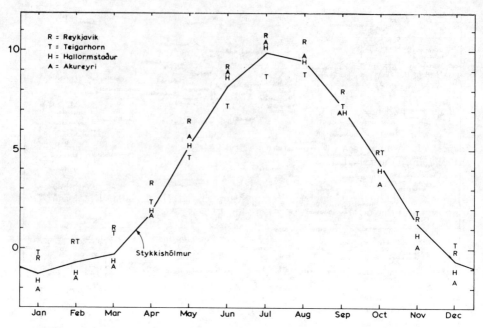

Figure 1.9. Mean monthly temperatures at Stykkishólmur 1951–1980 and at four other stations in Iceland (for location of stations, *see Figure 1.7*).

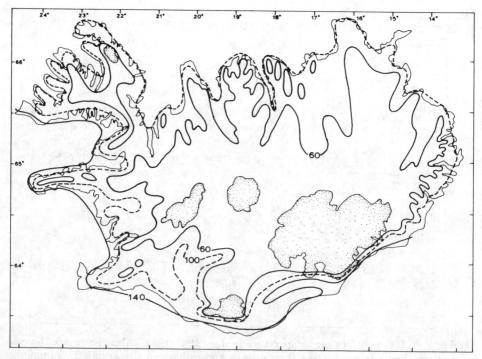

Figure 1.10. Mean duration of the frost-free period 1951–1980 (days).

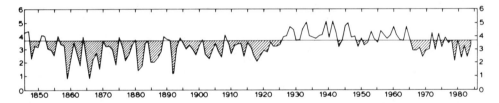

Figure 1.11. Annual mean temperature at Stykkishólmur 1846–1984. Values below the 1951–1980 mean are shaded.

Table 1.3. Characteristics of the temperature climate in Iceland as compared with Edinburgh and Berlin. (Source: Bergthórsson, 1985.)

	Stykkishólmur	Edinburgh	Berlin
Mean annual temperature 1851–1950 (°C)	3.3	8.5	9.2
Standard deviation of *annual* temperature	0.88	0.54	0.76
Standard deviation of *decadal* temperature	0.54	0.17	0.24
Temperature difference: 1901–1950 minus 1851–1900	0.74	0.21	0.14

The high amplitude of temperature variability in Iceland can be readily appreciated when variations on an annual, decadal and 50-year basis at Stykkishólmur (65° N, 23° W) are compared with those at two different locations in Europe, Edinburgh (56° N, 3° W) and Berlin (52° N, 13° E) (*Table 1.3*). Variability at the Icelandic station is the greater at all averaging scales, but the contrast is more marked for the longer averaging periods. This apparent persistence of temperature change may possibly be ascribed to the effects of sea ice discussed above.

Within the country itself, both long- and short-term variability of temperature are greatest in the north. This is illustrated for long-term temperature changes, by comparing mean annual temperatures for the periods 1931–1960 and 1873–1922 [*Figure 1.12*(a) and for interannual variations, using the standard deviation of annual temperatures at single stations during the period 1951–1980, seen in *Figure 1.12*(b)].

1.6.2. Precipitation

Figure 1.13 shows the yearly mean precipitation in Iceland in the period 1931–1960, the data for the reference period 1951–1980 not being available. Precipitation is heaviest in the southerly and southeasterly winds prevailing ahead of cyclones that frequently approach from the southwest. This results in the greatest precipitation on the southeastern slopes of the country, while rain shadows are observed in the north, behind the mountains and glaciers (*Figure 1.13*; and *see Figure 1.1*). Much of the precipitation in northern Iceland, on the other hand, is brought by northerly winds, and under those conditions the rain shadow

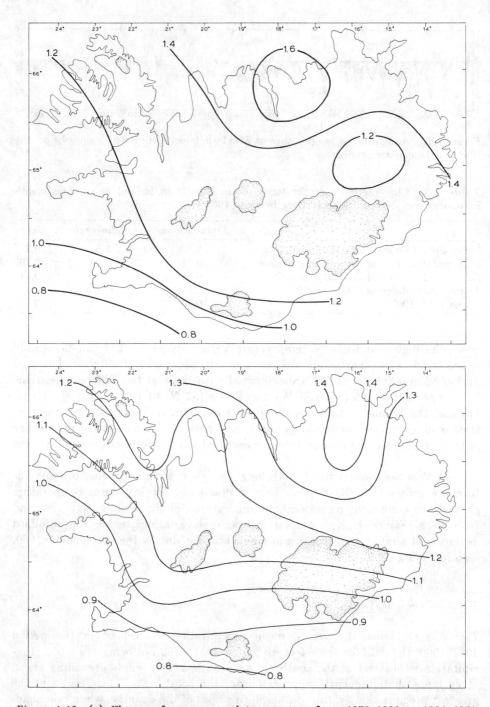

Figure 1.12. (a) Change of mean annual temperature from 1873–1922 to 1931–1960. (Source: Bergthórsson, 1985.) (b) Standard deviation of annual temperature in Iceland during the period 1951–1980 in proportion to the standard deviation at Stykkishólmur.

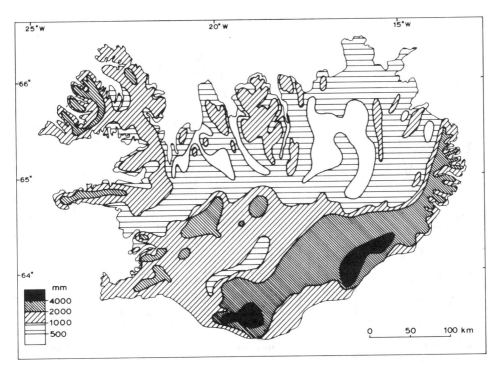

Figure 1.13. Mean precipitation, 1931–1960 (mm).

Table 1.4. Average precipitation 1931–1960 at four stations. Locations in Iceland are indicated in parentheses – *see* also *Figure 1.7*. (Source: Icelandic Meteorological Office.)

	Jan	*Feb*	*Mar*	*Apr*	*May*	*June*	*July*
Reykjavik (SW)	90	65	65	53	42	41	48
Stykkishólmur (W)	83	72	66	47	37	38	36
Akureyri (N)	45	42	42	32	15	22	35
Vík (S)	182	159	164	171	143	167	169
	Aug	*Sep*	*Oct*	*Nov*	*Dec*	*Year*	
Reykjavik (SW)	66	72	97	85	81	805	
Stykkishólmur (W)	50	76	87	89	77	758	
Akureyri (N)	39	46	57	45	54	474	
Vík (S)	188	237	238	212	226	2256	

effect is in the south. *Table 1.4* gives the monthly average precipitation in the period 1931–1960 at four stations in the lowland. This reflects the great geographical variability and the annual distribution, the minimum precipitation occurring in the spring and the maximum in the autumn.

The longest record of precipitation is from Stykkishólmur, dating from 1857. *Figure 1.14* gives an overview of the variation illustrated by 3-year means. A comparison with 3-year-averaged temperatures in the same figure indicates

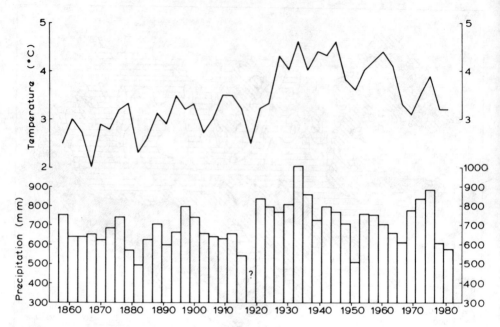

Figure 1.14. Three-year mean temperatures (°C) and precipitation (mm) at Stykkishólmur, 1857–1982.

some positive correlation. A weak positive correlation between temperature and precipitation is also noted in most other regions, both in winter and summer, with the exception of Akureyri, where the correlation is slightly negative, but not significant. Incidentally, a positive relationship between annual precipitation and temperature is consistent with increased values of both variables by the GISS-model, assuming a doubling of the atmospheric carbon dioxide concentration (*see* Part I, Section 3).

1.7. Selection of Study Areas

In contrast to the studies in other high latitude countries reported in this volume, it was decided to carry out the impact assessment in Iceland on a *national scale*. This choice was governed by five major considerations:

(1) While there are quite large regional differences in agro-environmental conditions in Iceland, the main agricultural activity (meat and milk production from herbivores) is common to all regions.
(2) Climatic variations, especially longer-term changes in temperature, tend to affect the whole country at the same time, varying regionally only in terms of their magnitude (*see* Section 1.6).

(3) Long records of agricultural productivity are available at the national level, and representative meteorological data also exist for the same period.
(4) The high sensitivity of Icelandic agriculture to climatic, particularly temperature, variations means that even by modeling impact at an aggregated national scale, the "signal" of the climatic effect on agriculture is still very strong.
(5) The assessments are preliminary, and many of the models and scenarios that are employed have been simplified to provide general estimates or indications of climate impacts for the whole country. Considerable refinement of data, scenarios, models and approaches would be required to conduct local-scale analyses.

Although the general focus throughout this report is on a national-level assessment, one method of accomplishing this has been through the use of local-level information from representative sites in different parts of Iceland (e.g., pasture yields from four experimental stations reported in Section 3, lamb carcass weights from abattoirs in four districts detailed in Section 4, and climatic data from nearby stations to all of these locations). However, the unifying factor in producing results for all Iceland has been the use of the long Stykkishólmur meteorological record, which is a good indicator of variations in climate over the whole country and has been related to most of the agricultural time series employed in the study.

1.8. Choice of Agricultural Activities for Impact Studies

The obvious activity on which to focus the bulk of our attention is livestock production and the provision of grazing and fodder. Traditionally, and at present, this is the major activity in all agricultural regions of Iceland. Furthermore, several of the climatic scenarios employed in the study represent cooler-than-present conditions, and given the already restricted options in agricultural production, while farm practices might have to be adjusted under such conditions (as investigated in the later sections of this report), the *type* of farming would probably remain unaltered.

The awareness that mean temperatures in Iceland could well increase in future decades, particularly in response to the "greenhouse effect" of increased trace gas concentrations (including carbon dioxide) in the atmosphere, has prompted several studies of enterprises that are, at present, close to their northern limits of viability but could be extended under warmer conditions. Barley cultivation would be of value to supplement hay supplies for livestock, and the planting of hardy tree species could help both to protect the soil and to provide shelter for crops and livestock and might even moderate local microclimatic conditions. Afforestation for timber production is also planned in limited areas.

1.9. Impact Models and Data

The direct effects of climate on agriculture have been modeled in Sections 2, 3 and 4 using a regression approach. The dependent variable in each case was the activity of interest (e.g., hay yield, carrying capacity or carcass weight), and the explanatory climatic variables, in all cases, were related to temperature conditions. In addition, other explanatory factors have been included in some models (e.g., fertilizer application on improved pastures) that are of interest in studying possible adjustments to mitigate climate impacts.

The meteorological data requirements as inputs to most models are not demanding: usually mean monthly temperatures over a representative time period and at suitable locations. For the hay-yield models (Sections 2 and 3), and the carrying capacity equations (Section 4), the temperature data are averaged over seasonal or annual periods. For calculations on potential barley cultivation, monthly mean minimum temperatures, monthly mean temperatures and mean monthly precipitation data were required to compute effective temperature sum (ETS) values (Section 2), while monthly mean maximum temperatures and mean monthly temperatures were needed for estimating growth units for tree growth (Section 2).

Data on agricultural productivity are from several sources. At a national level, annual hay-yield data have been collected from survey information and annual production data. Fertilizer applications are estimated from documentary evidence for the early years of this century, but better records exist for the postwar era. Annual values of sheep number and mean carcass weight are available at a national scale for nearly 50 years up to the present, and published estimates of current parameters for dairy production have been used to evaluate climatic effects on that activity.

Information from specific sites has also been used to develop regression models. Experimental stations provided detailed data on herbage yields, fertilizer applications and other management factors (e.g., number of harvests, harvest dates, etc.) for the investigations in Section 3; and slaughterhouse data on carcass weights, along with local information on lambing rates, were used to construct equations in Section 4.

Finally, as mentioned above, wherever impacts have been estimated for the whole country, the agricultural data have been related directly to the Stykkishólmur climatic data set, which is also used as the basis for the selection of climatic scenarios described below.

1.10. Climatic Scenarios

Several climatic scenarios have been adopted in this study that are subsequently used in impact experiments. Temperature is by far the most limiting climatic factor for agriculture in Iceland, and since the temperature conditions in the country are relatively well described by the observations at Stykkishólmur (cf. *Figure 1.9*), this long series (*see Figure 1.11*) is used as a basis for defining the scenarios.

Introduction

Scenarios are of three types:

(1) The reference or baseline scenario, selected from the historical instrumental record at Stykkishólmur.
(2) Instrumental scenarios representing anomalously cool or warm groups of years and including a single extreme event, each of these chosen from the Stykkishólmur record.
(3) A $2 \times CO_2$ scenario, based on the climate simulated by the GISS general circulation model for a doubling of atmospheric carbon dioxide concentration.

1.10.1. The baseline scenario (Scenario I)

In common with other case studies reported in this volume, the standard 30-year period 1951–1980 was selected as the baseline scenario. Mean annual temperature during this period at Stykkishólmur was 3.7°C (horizontal line in *Figure 1.11*), somewhat above the long-term mean (see *Table 1.3*). Mean annual precipitation was 704 mm, 54 mm lower than in the 1931–1960 period (*Table 1.4*). Mean monthly values of temperature and precipitation for this period (Scenario I) are shown in *Table 1.5*.

1.10.2. Instrumental scenarios (Scenarios II, III, IV and VI)

Since many of the most adverse impacts on Icelandic agriculture historically were associated with below-average temperatures, particularly when such conditions occurred in successive years, the coolest decade in the 139-year instrumental record was identified (1859–1868) and adopted as Scenario II. Mean annual temperature during this decade was only 2.4°C (1.3°C below the baseline), while average precipitation was nearly 70 mm below the 1951–1980 value. Interestingly, this period (the 1860s) was also anomalously cold in southern Finland (see Part IV).

Scenarios III and IV are also based on the Stykkishólmur record but employ data only from the period after 1930. Ensembles of the 10 coolest and the 10 warmest years, representing the extreme years over the recent period, were selected to simulate possible future average conditions in a cooler or warmer Iceland, respectively (an analogue approach similar to that described by Lough *et al.*, 1983). The cool period ensemble has a mean temperature 0.8°C below, and mean precipitation 18 mm below the respective baseline values (Scenario III – *Table 1.5*), noticeably smaller anomalies than in the consecutive 10 years of Scenario II. Mean temperature for the ensemble of warm years (Scenario IV) is 1.1°C above the baseline value, and the positive correlation between temperature and precipitation (noted above) is again observable in the high average precipitation value, 82 mm above the 1951–1980 value (*Table 1.5*).

Table 1.5. An overview of the major climatic scenarios selected for Iceland.

	Scenarios				
	I Baseline	II Cold	III Cold	IV Warm	V $2 \times CO_2$
	Stykkishólmur (1951–1980)	1859 to 1868- type	Average of 10 coldest years during 1931 to 1984	Average of 10 warmest years during 1931 to 1984	GISS model- derived
Mean monthly temperature (°C)					
January	−1.3	−2.7	−2.1	−0.2	3.0
February	−0.7	−3.1	−1.9	0.1	3.5
March	−0.3	−4.2	−2.4	1.9	4.0
April	1.8	−0.2	1.0	2.3	6.1
May	5.2	3.9	4.0	6.7	9.2
June	8.2	7.6	7.9	9.0	12.0
July	9.9	9.6	9.5	10.5	13.5
August	9.5	9.0	9.6	10.2	13.1
September	7.2	6.7	6.8	8.3	11.0
October	4.3	3.1	3.0	5.0	8.3
November	1.3	0.0	0.2	3.1	5.4
December	−0.6	−0.8	−1.3	1.2	3.6
Year	3.7	2.4	2.9	4.8	7.7
Mean monthly precipitation (mm)					
January	65	56	61	102	71
February	73	68	61	77	83
March	65	46	62	55	78
April	52	38	49	56	65
May	32	33	31	26	40
June	38	47	46	39	47
July	40	46	42	40	50
August	49	34	68	40	59
September	60	72	63	82	69
October	84	69	72	75	92
November	81	60	57	106	86
December	65	69	74	88	69
Year	704	638	686	786	809

As a variant of Scenario IV, in Section 3 of this report, experiments have been conducted assuming that the magnitude of the positive temperature anomaly in a warm period is equivalent to the negative anomaly in the cool decade (Scenario II). The resulting temperature anomaly (+1.3 °C) is 0.2 °C greater than that for the warm years ensemble, but it offers a sharper, symmetrical contrast between warm and cool periods. We will refer to it as Scenario IV+.

A further scenario, used exclusively in Section 3, is a single extreme event identified from the Stykkishólmur record. As discussed above, marked long-term changes in mean temperature are a characteristic feature of the Iceland climate. Therefore, it is perhaps more realistic for a single year extreme to be assessed in the context of the climate prevailing when it occurs, rather than with respect to

Introduction

an average value representing different conditions in another time period. By comparing temperatures in the four coolest years at Stykkishólmur with the mean temperature for the preceding 10 years, anomalies of between −1.5°C and −2.2°C were obtained. These deviations were similar in magnitude to the value −1.68°C, which is twice the standard deviation for the entire Stykkishólmur series. This value was thus adopted as an example of an event that really could happen, at least in a cold period (Scenario VI).

1.10.3. The GISS 2 × CO_2 scenario (Scenario V)

The climate of Iceland under doubled concentrations of atmospheric carbon dioxide was simulated using estimates from the GISS general circulation model in line with the other high latitude case studies (*see* Part I, Section 3). Data were supplied on a 4° latitude × 5° longitude network of grid points for the Iceland region (cf. inset to *Figure 1.5*). Information on mean monthly temperature (°C) and precipitation rate (mm/day) were required for impact experiments in the Iceland study, and these were obtained in three forms:

(1) GISS model-generated 1 × CO_2 equilibrium values.
(2) GISS model-generated 2 × CO_2 equilibrium values.
(3) GISS 2 × CO_2 − GISS 1 × CO_2 values (change in equilibrium climate for a doubling of CO_2).

As a test of the accuracy of the GISS model in simulating present-day climatic conditions in Iceland, the monthly values of temperature and precipitation generated in the GISS model 1 × CO_2 equilibrium run were examined to see how closely they reproduced the observed conditions over the period 1951–1980 at a representative station, Stykkishólmur. A simple weighted-averaging procedure was used to interpolate the GISS-generated grid point values to the Stykkishólmur location (65°05′ N, 22°44′ W):

Interpolation to Stykkishólmur
$$= 0.42\ GP_{66,25} + 0.35\ GP_{66,20} + 0.13\ GP_{62,25} + 0.10\ GP_{62,20}$$

where GP_{ij} is the grid-point value at latitude i and longitude j.

The comparison of observed values with GISS 1 × CO_2 and GISS 2 × CO_2 values of mean monthly temperature and monthly mean precipitation rate is shown in *Figure 1.15*. The annual range of mean monthly temperatures at Stykkishólmur is 11.2°C, reflecting the maritime nature of the Icelandic climate and contrasting sharply with the annual range in more continental locations at high latitudes reported elsewhere in this volume (e.g., 26.3°C at Oulu in northern Finland − *see* Part IV). The GISS 1 × CO_2 simulation reproduces this maritime effect quite well [*Figure 1.15(a)*]; indeed, it underestimates the summer temperatures at the Stykkishólmur station by over 2°C, giving a temperature range over the annual cycle of only 8.1°C. The GISS model estimates for 2 × CO_2

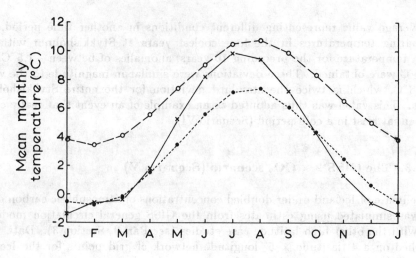

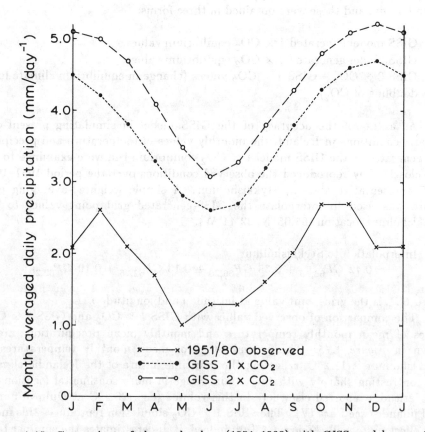

Figure 1.15. Comparison of observed values (1951–1980) with GISS model-generated 1 × CO_2 and 2 × CO_2 equilibrium values at Stykkishólmur of (*a*) mean monthly air temperature (°C), and (*b*) monthly mean precipitation rate (mm/day^{-1}).

Introduction

equilibrium conditions suggest a significantly warmer and even more equable climate than at present (due to slightly greater warming in the winter half year than in the summer).

The GISS 1 × CO_2 estimates of monthly mean precipitation rate, in contrast to the simulated temperatures, grossly overestimate the observed values at Stykkishólmur, often by more than a factor of 2 [*Figure 1.15(b)*]. However, the annual distribution of precipitation is in good agreement with observations, showing a peak in winter and a late spring/early summer minimum. For a doubling of CO_2, the GISS model predicts increased precipitation in all months, the increases being slightly greater in the winter half year than in the summer.

Taken together, the GISS 2 × CO_2 temperature and precipitation estimates for the Stykkishólmur location bear a striking resemblance to the present-day (1931–1960) conditions at Thorshavn, in the Faroe Islands [62°03′ N, 6°45′ W – see *Figure 1.4(a)*], a highly maritime location in the north Atlantic Ocean (Lamb, 1972). However, the poor correspondence of the GISS model 1 × CO_2 estimates to observed conditions (probably due to the inability of the GISS model, with its coarse grid-point network, to account for the modifying effects of the Icelandic land mass on the surrounding oceanic climate), did not inspire confidence that the 2 × CO_2 estimates would be realistic in representing regional conditions for agriculture for the purposes of impact analysis. Rather, in common with the other case studies, it was assumed that the difference between the observed present-day climate and a future 2 × CO_2 climate could be better expressed as the *change* between GISS-generated 1 × CO_2 and 2 × CO_2 equilibrium conditions.

Two different procedures were adopted for adjusting the observed temperature and precipitation values, respectively. For the former, the differences between 1 × CO_2 and 2 × CO_2 estimates were added to the observed mean monthly temperatures for the baseline period – the "differences" technique:

$$T_{2 \times CO_2} = (T_{\text{GISS } 2 \times CO_2} - T_{\text{GISS } 1 \times CO_2}) + T_{\text{REF}}$$

If the differences technique had been used for the latter values of monthly mean precipitation rate, the change to GISS 2 × CO_2 conditions from the highly exaggerated GISS 1 × CO_2 values would have represented a significant increase in precipitation from the much lower observed values at Stykkishólmur. Thus, instead of using differences, the ratios of GISS 2 × CO_2 to GISS 1 × CO_2 values were multiplied by observed values, a procedure recommended by Conrad and Pollack (1962) and also used in the Saskatchewan case study (*see* Part II). This adjustment "by ratios" of the reference data is described by:

$$P_{2 \times CO_2} = (P_{\text{GISS } 2 \times CO_2} / P_{\text{GISS } 1 \times CO_2}) \cdot P_{\text{REF}}$$

The values computed for the 2 × CO_2 scenario (Scenario V) are presented in *Table 1.5*.

Table 1.6. GISS-derived 2 × CO_2 temperature at Stykkishólmur as compared with present climate at some stations abroad [for locations, see Figure 1.4(a)].

	Scenario V at Stykkishólmur	Mean temperature (1931–1960)			
		Aberdeen	Lerwick	Stavanger	Bergen
January	3.0	2.4	3.1	0.7	1.5
February	3.5	2.8	2.9	0.4	1.3
March	4.0	4.5	3.9	2.3	3.1
April	6.1	6.6	5.4	5.5	5.8
May	9.2	9.0	7.8	9.6	10.2
June	12.0	12.0	10.0	12.2	11.7
July	13.5	14.0	12.0	14.7	13.9
August	13.1	13.6	12.1	14.7	13.8
September	11.0	11.7	10.6	12.3	11.6
October	8.3	8.8	8.2	8.5	8.4
November	5.4	5.6	5.9	5.2	5.5
December	3.6	3.7	4.4	2.8	3.3
Year	7.7	7.9	7.2	7.4	7.8

One way of appreciating the extent of these changes, as discussed above for the GISS 2 × CO_2 case, is to pick out weather stations in neighboring countries where the present-day climate is comparable to the derived 2 x CO_2 climate. Table 1.6 shows the mean monthly temperatures at four such stations for comparison with the Stykkishólmur temperatures under Scenario V. It should be noted that the reference scenario 1951–1980 was somewhat colder in Iceland than 1931–1960, but the 1951–1980 values were not available at the foreign stations. The precipitation (not shown here) is generally more abundant at the foreign stations with the exception of Aberdeen, where it is almost the same as under Scenario V in Stykkishólmur.

This comparison indicates that the temperature and precipitation anomalies of Scenario V in Iceland are fairly consistent with an imaginary displacement of Iceland into warmer maritime regions such as northern Scotland or western Norway. These analogies have been pursued further in subsequent sections of this case study.

1.11. Climatic Variations and Shifts in Agroecological Potential

Iceland is located at the northernmost limits of technologically developed agriculture and on the edge of the boreal forest zone (discussed above). Therefore, several margins can be delineated that pass through the country. Two of these are depicted in Figure 1.8 marking the northern limits of viable grassland production and of cereal cultivation, respectively. Relatively small changes in temperature, however, can alter the position of these lines a great deal, i.e., shifting them to the south in cold years and to the north in warm years.

The potential tree zone, defined in terms of the temperature of the warmest month (Köppen, 1936), is also subject to large spatial fluctuations due to climatic variations. The zone is delimited by the 10 °C July isotherm, and covers about 11 000 km^2 for the baseline scenario (*see Figure 1.8*). The 8 °C isotherm delimits the zone as it would appear after a warming of 2 °C. That area is about 41 000 km^2. Since the highest temperature at any station in 1951–1980 is 11.3 °C, a general cooling of 1.3 °C (Scenario II) would approximately eradicate the zone in Iceland.

This information, together with our assumption for the temperature lapse rate (Subsection 1.6.1) and a knowledge of the area between different height intervals in Iceland (*see Table 1.1*), can be used to estimate the extent of the potential tree zone for the different climatic scenarios (*Table 1.7*). As such, the results provide a useful introductory illustration of the potential sensitivity of vegetation to climatic change. In spite of the inevitable inaccuracies of the mapping procedures, it is evident that one of the most important impacts of a warmer climate is the great increase in the area suitable for tree growth and, by implication, for agriculture. It is important to realize, however, that it would take a long time, possibly centuries, for the natural vegetation to obtain an equilibrium with the new climate if man does not intervene to modify this transition.

Table 1.7. Area of potential tree zone in Iceland (km^2).

Scenario	I	II	III	IV	V
Area of tree climate	11 000	8 000	8 000	22 000	64 000

1.12. Final Remarks

In the following sections, more specific points concerning climate impacts on agriculture in Iceland are discussed. Considerable attention is devoted to hay production and grazing conditions, the most important basis of agriculture in the country. In Section 2, regression equations have been used to evaluate the effects of the climatic scenarios on national hay yield and fodder requirements, on the growth of barley, and on the potential for growing birch and Norwegian spruce. The possibility of predicting summer hay yields and of using variable fertilizer applications to stabilize production are also examined. Results of long-term grassland experiments are used, in Section 3, to estimate the effects of scenario temperature changes on pasture yields. The importance of nitrogen fertilizer in determining yield response is investigated, and the implications of this relationship are discussed with particular reference to dairy farming. In Section 4, regression equations are again used to study the relationship between climate and carcass weight of lambs, in an attempt to assess the variation in the carrying capacity of Icelandic rangelands. The impacts of both climatic warming and cooling are examined, and the design of a rangeland model is proposed. In Section 5, the results of the analyses in Sections 1–4 are summarized, and implications for policy making discussed.

SECTION 2

The Effects on Agricultural Potential

2.1. Introduction

This section is concerned with the impact of climate on various aspects of agriculture in Iceland. As discussed in the preceding section, grass growth and hay yield is of fundamental importance in Icelandic farming. Because of this the greater part of this chapter is devoted to the impact of climate on hay yield. A related concern for the farmers is the impact of climate on the winter and summer grazing of animals. The farmers are thus affected by two related problems: poor hay yield after cool seasons and increased need for fodder in severe winters. These two conditions often occur together, and the cumulative effect of a cold winter can be that farmers are faced in the summer with unusually little grass growth, when the fodder consumption in the preceding winter has been especially heavy. To compound matters, such severe years often tend to occur in clusters. An additional limiting factor is the carrying capacity of the summer pastures. In this respect it is therefore advisable to restrict livestock grazing to the carrying capacity of the coldest periods that can be expected in the climatic era. The question of how best to react to climatic impacts in order to avoid fluctuations in livestock numbers is also discussed. Finally, in Subsection 2.7, it is shown that present conditions for growing trees and cereal crops are marginal in Iceland. Their sensitivity to climate and their growth under different climatic conditions are estimated here in terms of the percentage area in Iceland suitable for their growth.

Regression equations have been developed in each of these analyses utilizing historical agricultural data aggregated at the national level and climatic data either from Stykkishólmur alone, or from the station network across Iceland (*see Figure 1.7*). The equations are used to estimate the effects of the five main climatic scenarios described in Subsection 1.10, and to assess some possible adjustment measures to adapt to climatic variations.

2.2. The Effects on Hay Yields

For this study, the effects of temperature and precipitation on annual hay yield for the whole of Iceland during the period 1901–1975 are considered. Precipitation will be considered first.

2.2.1. Effects of precipitation

In the southern parts of the country, the summer (May–September) precipitation ranges from about 250–600 mm (*see* Subsection 1.6). It falls mainly in the relatively warm southerly winds, blowing onshore. As the summer temperature, and hence rates of evapotranspiration, are low, droughts are uncommon in this region. Excessive precipitation, however, may cause some damage to the hay crop. Thus the beneficial growth effects of warm periods will be somewhat reduced when there is a surplus of rain. In the northern part of the country, the summer precipitation is, on average, 150–250 mm. Drought occurs occasionally, mainly with southerly *föhn* winds, blowing from the mountains. This reduces the positive effect of the relatively warm southerlies on plant growth. With cold northerly winds, precipitation in northern Iceland is sufficient but seldom excessive; moreover, the cold air prevents any significant rotting of the hay during the haymaking. Curiously enough, production of silage is very limited in this rainy country. In conclusion, although precipitation is locally important during certain years, it is little related to hay yield for the whole country. Because of this, only temperature and fertilizer application are used as variables in the regression equation expressing the yield shown below.

2.2.2. Effects of temperature

In designing a model of hay yield with respect to temperature, it is necessary to consider the summer (May–September) and winter (October–April) separately. While the use of the summer (growing season) temperature needs no justification, some explanation is required as to why winter temperature is an important parameter (Bergthórsson, 1966). One of the reasons is the winter kill of grass. This may occur as the direct result of severe cold, particularly in relatively snow-free winters (although it should be noted that cold winters are more frequently accompanied by long-lasting snow cover). With the unstable Icelandic climate, another factor which may have an adverse effect on grass is the occurrence of thaws. If these are only light, as is usual in cold winters, the snow does not melt properly but simply becomes wet. When temperatures fall below freezing point again, the partly melted snow refreezes and turns into ice. The result is the formation of a crust of ice on the ground. If this persists for two or three months in late winter (Gudleifsson, 1975), it can be very detrimental to the grass.

The grass may also be badly affected by water, lying on top of frozen soil in the spring. When this occurs, winter cold is an underlying, though indirect cause. Severe winters, which leave the soil frozen and cold in the spring, may anyway delay growth, even in the absence of winter kill. This is particularly true on peat soils, which are very common in the recently cultivated hay fields in Iceland. Indeed, the damage due to severe frost is often difficult to distinguish from other effects. For example, by far the coldest winter in this century occurred in 1918, resulting in the lowest hay yield recorded after 1900. Elderly farmers (E. Gudmundsson, personal communication) have remarked that one of the main reasons for the low yield was the poor development of the grass leaves before the stem extension occurred. In summary, it may be remarked that the effects of cold winters, winter kill being only one of them, all contribute to a lowering of the hay yield in the following summer.

2.2.3. Data requirements

There were several different information sources for the data required in developing regression equations that were used in climate impact experiments. The data were of three types: hay yields, fertilizer applications and temperatures.

Hay yields

A general survey of the improved grassland has been made five times in this century: 1916–1920, 1930, 1936, 1964 and 1970. Furthermore, the number and size of additional cultivated fields have been recorded every year, so it was possible to compile an annual record of the area of improved grassland. Using this, and national production data (recorded annually), the annual hay yield can be computed in kg/ha (85% dry matter).

Before 1960 very little grazing was practiced on these improved fields, all sheep, horses and cows being kept on the unimproved rangelands outside the enclosed fields. Around 1960–1962, a sudden change in the management of the improved grassland occurred when the farmers began grazing dairy cattle there. It has been estimated that this grazing, along with increased grazing of sheep, removed about 700 kg of hay/ha after 1961 (Bergthórsson, 1976), and the hay yield record has been adjusted accordingly (*Table 2.1*).

In addition to the yield from improved grassland, it was also customary in former times to take a considerable part of the hay yield from unimproved rangelands. While the area of these is not known, and thus yields cannot be obtained, the total *production* of this hay has been recorded, which is important because it affects the production of farm manure which, in turn, was used on the improved fields. More detailed information on hay yields is desirable. However, the interannual variations of the yields represented in *Table 2.1* are probably not greatly in error, and it is these variations which are the most important in the correlation with temperature. The relationship between total hay production and amount of farm manure utilized is considered below.

Table 2.1. Seasonal temperatures, fertilizer applications and hay yields, 1901–1983.

Year	Temperature at Stykkishólmur (°C) Oct/Apr	May/Sept	Nitrogen fertilizer (kg/ha)	Hay yield (kg/ha)	Year	Temperature at Stykkishólmur (°C) Oct/Apr	May/Sept	Nitrogen fertilizer (kg/ha)	Hay yield (kg/ha)
1901	0.3	8.8	42	2560	1941	1.2	9.8	66	3850
1902	−1.7	7.6	42	2210	1942	2.6	8.2	68	3730
1903	−0.1	7.1	42	2290	1943	0.5	7.2	72	3290
1904	−0.5	8.3	42	2660	1944	0.9	8.5	72	3660
1905	−0.6	8.0	44	2430	1945	0.6	9.1	79	3800
1906	0.2	7.4	44	2380	1946	2.5	8.5	90	3970
1907	−0.8	6.5	44	1990	1947	1.4	9.1	89	4050
1908	0.6	8.3	42	2510	1948	1.5	7.8	86	3930
1909	1.4	8.4	43	2890	1949	0.0	7.8	104	3710
1910	−1.2	7.5	46	2500	1950	1.0	8.5	102	4060
1911	0.0	7.5	46	2300	1951	−0.8	8.7	103	3410
1912	0.6	8.3	46	2720	1952	0.3	7.3	98	3420
1913	0.4	8.1	47	2650	1953	1.4	8.9	120	4610
1914	−0.8	7.0	47	2620	1954	1.8	7.5	131	4840
1915	−0.4	8.0	47	2420	1955	0.5	8.1	142	4490
1916	−0.5	8.4	47	2590	1956	1.2	7.8	152	4810
1917	0.2	7.6	48	2640	1957	1.8	8.5	163	5260
1918	−3.0	7.5	48	1430	1958	0.6	8.0	171	4670
1919	−0.6	7.9	45	2250	1959	1.5	8.3	175	5160
1920	−1.6	7.6	44	2210	1960	1.7	9.0	160	5190
1921	1.1	6.8	44	2680	1961	1.4	8.7	161	5060
1922	0.7	6.9	46	2520	1962	0.9	7.9	178	5260[a]
1923	1.9	6.9	45	2960	1963	1.5	7.1	182	5170
1924	−0.3	6.9	46	2540	1964	2.5	7.8	181	5220
1925	0.6	8.1	46	3080	1965	1.1	7.8	176	4930
1926	1.1	8.2	52	3210	1966	0.1	8.2	174	4680
1927	0.3	8.4	50	3020	1967	−0.3	7.6	179	4430
1928	1.3	9.2	55	2610	1968	−0.9	8.0	178	4590
1929	3.0	8.0	58	3240	1969	−0.6	7.9	170	3840
1930	0.3	8.8	66	3580	1970	−0.5	7.4	167	3480
1931	−0.5	8.9	68	3010	1971	0.0	8.0	169	4660
1932	1.3	8.9	62	3580	1972	1.3	8.2	168	5170
1933	0.9	10.0	61	4060	1973	0.8	7.6	173	4860
1934	1.7	8.6	63	3990	1974	0.6	8.2	169	4520
1935	0.7	8.9	62	3460	1975	0.1	7.4	168	4570
1936	−0.2	9.6	64	3440	1976	0.4	8.6	176	4770
1937	0.5	8.5	65	2960	1977	0.5	8.6	174	4860
1938	1.6	7.9	65	3180	1978	0.2	8.5	173	4500
1939	1.3	10.3	65	3770	1979	−0.5	6.3	175	3950
1940	1.0	8.2	58	3410	1980	1.1	8.6	173	4630
					1981	−0.9	7.7	163	3860
					1982	−0.3	7.0	164	4080
					1983	−0.4	6.7	156	3900
					1984	0.2	7.9		

[a] After 1961, 700 kg are added to account for summer grazing.

Fertilizer applications

Fertilizers, both natural and artificial, are an important parameter in the hay-yield model. With respect to natural manure, it has been estimated that 100 kg of hay contains about 1.8 kg of nitrogen (Óskarsson and Eggertsson, 1978). The manure resulting from feeding livestock with these 100 kg of hay contains, on the other hand, about 0.8 or 0.9 kg of effective nitrogen fertilizer (if the manure is well preserved). Adding to this the nitrogen in manure resulting from feeding with concentrates, the total content is estimated to be *1 kg of effective nitrogen per 100 kg of the corresponding hay fodder*. The effective phosphorus content of the farm manure is about 36% of the nitrogen, while the corresponding percentage for potassium is 125%.

It is, however, necessary to adjust these figures because during the first decades of this century it was customary to use a proportion of the farm manure as fuel on the farms. From oral information obtained from elderly farmers, it has been estimated that about one-third of the farm manure was burnt in the first part of this century, but the practice had been completely abolished by about 1960. A corresponding correction has been applied in *Table 2.1*, which gives the annual application of nitrogen fertilizer/ha of improved grassland. The estimates of the effective nitrogen fertilizer in farm manure are admittedly very rough since no direct records are available. However, written sources indicate that in the first part of this century a usual application of farm manure was 15 t/ha. If the content of effective nitrogen is 0.3%, this gives about 45 kg/ha, which is in good agreement with the results obtained with the indirect method described above. When the annual applications of farm manure were computed for the whole period 1901–1983, they were found to be remarkably constant over time. Practically all the variations in fertilizer applications are therefore attributable to commercial fertilizers, for which there are accurate records.

In *Table 2.1* only the nitrogen fertilizers are recorded and used as a parameter in the regression equation of the hay yield. The application of phosphorus and potassium can be estimated in a corresponding manner, but these are not included directly in the regression equation. It may be noted that the application of phosphorus can be expressed as an approximate function of the nitrogen:

$$P = 5 + 0.24N \quad (kg/ha) \tag{2.1}$$

The corresponding expression for potassium is:

$$K = 42 + 0.34N \tag{2.2}$$

Thus the constituents P and K comprised a greater proportion of total applied fertilizer when only farm manure was applied, before 1920, than after the introduction of commercial fertilizers.

Temperatures

In addition to the information on hay yield and fertilizers, *Table 2.1* lists annual temperatures at Stykkishólmur for the winter (October–April) and the following summer (May–September). Although it would, perhaps, have been more appropriate to use temperature values representing an average for the whole country, such a measure would have been difficult to devise because the distribution of weather stations has changed considerably over the 83-year record. On the other hand, the weather record from Stykkishólmur is considered to be fairly homogeneous, even from the earliest observations in 1845 (Sigfúsdóttir, 1969). The seasonal temperatures at Stykkishólmur are in close agreement with the mean temperature of all lowland stations, in both summer and winter, and, apart from several exceptions, the correlation of seasonal temperatures between different regions of the country is good. This is confirmed independently in Section 3, where it is noted that Stykkishólmur temperature is almost as effective a predictor of grass growth at experimental stations as are the temperatures at the stations themselves.

Finally, there are some other features of the data in *Table 2.1* that are noteworthy:

(1) A large but gradual rise in the application of fertilizers is evident from about 40–50 kg N/ha after the turn of the century to 150–170 kg in recent decades. Comparing years with similar temperature conditions before and after this increase, it is evident that the additional fertilizer has had a marked effect on hay yields.

(2) Strong year-to-year variations of hay yield are easily detectable as a result of temperature variations. There is also some trend in the temperatures, which are highest between approximately 1925 and 1965. Since this trend only partially coincides with the trend in fertilizer application, regression methods should render a fairly good distinction between the effects of fertilizers and temperature.

2.2.4. The hay-yield regression model

Using the first 75 years of data in *Table 2.1*, an attempt has been made to design a model expressing the mean hay yield from improved grassland as a function of temperature and fertilizer application. By means of successive regression, the following equation was obtained

$$Y = (0.29 + 0.0729\ S + 0.0794\ W)(1820 + 28.1\ N - 0.051\ N^2) \qquad (2.3)$$

where Y is the hay yield from improved grassland (kg/ha), S is the mean summer temperature (May–September) at Stykkishólmur (°C), W is the mean winter temperature (October–April) at Stykkishólmur (°C), and N is the total fertilizer

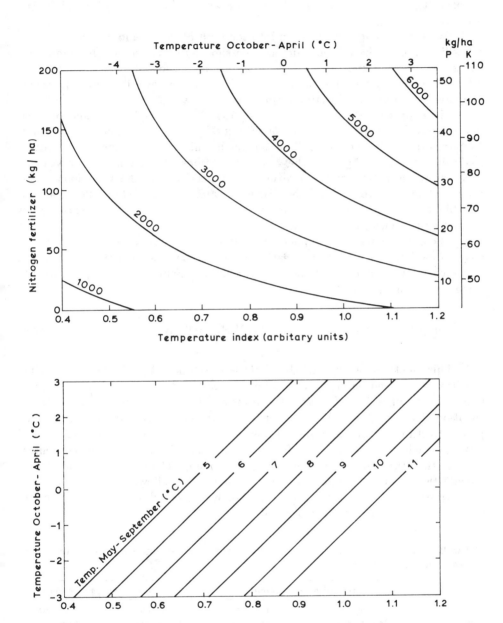

Figure 2.1. Graphical representation of the hay-yield regression model [equation (2.3)], and the hay-yield prediction model [equation (2.5)]. For full explanation, see text. Lower graph: Temperature index is a function of October–April temperature and May–September temperature. Upper graph: Hay yield (kg/ha-85% dry matter) as a function of temperature index, October–April temperature and nitrogen fertilizer application [corresponding conventional applications of phosphorus (P) and potassium (K) according to equations (2.1) and (2.2) are also shown].

nitrogen (kg/ha of improved grassland). A graph of this equation is given in *Figure 2.1*.

By decomposing this equation into the multiple of a temperature index and a fertilizer index, the relationships can be shown to have some agronomic support since it is well known that high fertilizer applications do not improve yields if temperatures are very low, and conversely, grass cannot fully exploit favorable temperature conditions if fertilizer applications are small.

The expression $0.29 + 0.0729\ S + 0.0794\ W$ on the right hand side of equation (2.3) is the temperature index of the equation. The coefficient of the winter temperature is slightly higher than the coefficient of the summer temperature, confirming the importance of the winter conditions. Furthermore, the year-to-year variations in winter temperature are almost double the variations in the summer temperature. As will be discussed later in this section, this makes it possible to use the winter temperature as the only explanatory variable for estimating hay yield, with fair results.

The derivation of this temperature index is somewhat different from that of a corresponding index used in an earlier paper (Bergthórsson, 1985). That index was a curvilinear function of the annual (October–September) temperature (A) using a parabolic expression

$$\text{Temperature index} = 0.169 + 0.281\ A - 0.02\ A^2 \qquad (2.4)$$

This method gave slightly, but not significantly, higher correlation coefficients between observed and estimated hay yields for the observed range of temperatures in Iceland (*Table 2.2*). However, the linear temperature index of equation (2.3) appears to be more realistic than the curvilinear one for the purposes of estimating the hay yield in periods with temperatures not experienced hitherto, such as those predicted for an era of doubled carbon dioxide concentrations in the atmosphere. It is unclear how the slope of the curve would change outside the observed temperature range, but in the following validation of the model (Subsection 2.2.6), evidence from Northern Ireland lends some support to the use of a temperature index of the linear type.

Table 2.2. Correlation between recorded hay yields and estimates based on two alternative temperature indices.

	Correlation coefficients		Root-mean-square error	
Period	Linear temp. index	Curvilinear temp. index	Linear temp. index	Curvilinear temp. index
1901–1940	0.89	0.90	250	240
1941–1975	0.93	0.95	220	200
1901–1975	0.97	0.97	240	220
1976–1983	0.94	0.95	260	240

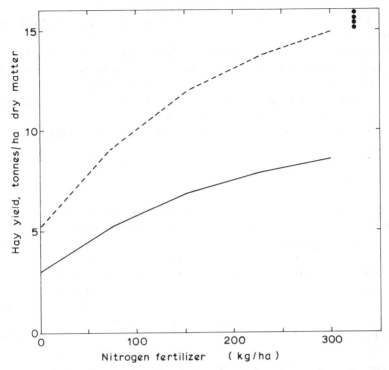

Figure 2.2. Long-term fertilizer experiments at four experimental stations in Iceland mainly in the 1950s. Solid line: Hay yield (t/ha dry matter), as a function of nitrogen fertilizer (kg/ha/yr). Phosphorus: 23% of N. Potassium: 55% of N. Dashed line: Computed yield, assuming the 1931–1960 temperature index of equation (2.3) to be raised from 1.0 at Stykkishólmur to that of Aldergrove in Northern Ireland (= 1.74). Dots: Yield of *Lolium perenne* in Northern Ireland. (Source: Keatinge et al., 1979.)

The function $1820 + 28.1\,N - 0.051\,N^2$ in equation (2.3) is the fertilizer index; it expresses the expected hay yield in kg/ha assuming the temperature index to be unity, (as it turned out to be, incidentally, in the period 1931–1960). The shape of the fertilizer function is reasonable, low rates of application giving the strongest response, which is in good agreement with the results of fertilizer experiments in Iceland (*Figure 2.2*).

2.2.5. A graphical representation of the hay yield model

Figure 2.1 gives an idea of the relationship between the hay yield and the parameters used in equation (2.3). For a given winter temperature (lower left) and summer temperature (sloping lines), the temperature index can be obtained. This in turn gives the yield (curved lines in upper graph), as a function of the applied nitrogen fertilizer (vertical scale). Approximate corresponding values of phosphorus and potassium applications can then be read on the right-hand scale [based on equations (2.1) and (2.2)]. These values of P and K should not necessarily be considered as the most economical rates in relation to the nitrogen application, however.

It is evident from this graph that a given increase in the amount of fertilizer is much more efficient in warm conditions. This possibly explains why, in the cold climate of Iceland, it is not economical to apply as much fertilizer as farmers do in milder climates. While this may be true for *mean* fertilizer applications, however, it may be worthwhile attempting to match the *annual* rates to the annual needs, an idea that will be developed later in this section.

2.2.6. Validation tests

To validate the model, it is possible to compute the correlation coefficients between the computed and the real yield. This computation may be performed for all the 75 years, but a further test is to consider, separately, the first and the second half of the period. Furthermore, the 8 years following the 75-year study period offer an important, independent test of this relationship. Finally, information on hay yield in Northern Ireland is used to test the equation under conditions markedly different from the present Icelandic climate. The correlations and the root-mean-square errors thus obtained are given in *Table 2.2*. For comparison, the results based on the curvilinear temperature index [equation (2.4)] are also shown. All these correlation coefficients are highly significant.

Figure 2.3 shows the relation between real and estimated annual hay yield in the period 1901–1983, together with the regression line based on equation (2.3). The highest observed temperature index in Stykkishólmur, using equation (2.3), is about 1.2. If the annual temperature rises by 4 °C, as might happen with a doubling of atmospheric carbon dioxide (*see* Section 1), the temperature index will be about 1.5. For comparison, the corresponding average temperature index at Aldergrove, near Belfast in Northern Ireland, was 1.74 during the period 1931–1960, while it was 1.0 at Stykkishólmur. If equation (2.3) is valid for the relatively mild climate in Northern Ireland (precipitation, as in Iceland, is usually sufficient for grass growth), one should expect the hay yield in similar experiments to be about 74% higher in Northern Ireland than it is in Iceland. In fact, hay production of 15–16 t dry matter/ha/yr can be achieved without difficulty from swards of perennial ryegrass (*Lolium perenne*) using rates of fertilizer nitrogen of about 325 kg/ha (Keatinge *et al.*, 1979). It is interesting to compare this with a long-term experiment at several experimental stations in Iceland in the relatively mild period before 1965. The mean yield for all these stations is shown as a function of fertilizer nitrogen in *Figure 2.2* (solid line). As is usual in controlled experiments, these yields are much higher than actual farm yields, shown in *Table 2.1* (Bergthórsson, 1976). The dashed line assumes the temperature index (and hence the yield, following the arguments above) to be 74% higher. This indicates about 15 t/ha with a nitrogen fertilization of 300 kg/ha, which is similar to the reported hay yield in Northern Ireland. It should be noted that the curvilinear temperature index [equation (2.4)] for Aldergrove in the period

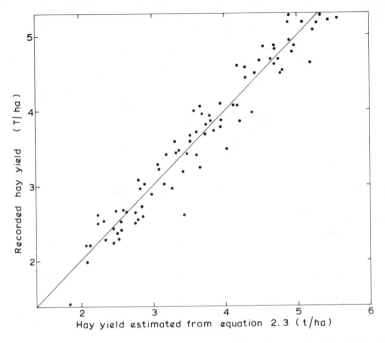

Figure 2.3. Recorded annual hay yield in Iceland (t/ha dry matter) in 1901–1983 plotted against estimated hay yield from equation (2.3). Regression line based on data from 1901 to 1975.

1931–1960 is only 1.08, which would give a completely misleading indication of the growing conditions in Ireland. That index should therefore not be used for extrapolation beyond the temperature interval observed in Iceland.

2.2.7. Estimated yields under different scenarios

Equation (2.3) can now be used to estimate average hay yield in Iceland under the climatic scenarios described in Section 1. Scenario I represents the temperature conditions during 1951–1980. Scenario II is based on the record cold decade of 1859–1868, while Scenario III corresponds to the averaged conditions during the 10 coldest years that occurred after 1930. Scenario IV represents the averaged climate during the 10 warmest years after 1930, while Scenario V is based on estimates from the GISS model, assuming a doubling of atmospheric carbon dioxide concentrations from the present value. *Table 2.3* summarizes winter and summer temperatures under these scenarios, and the estimated hay yield is shown, assuming a constant application of nitrogen fertilizer (160 kg/ha), both in absolute values and in percentages of the reference Scenario (I).

Table 2.3. Estimated hay yield for constant fertilizer application and temperatures under Scenarios I–V.

	Scenarios				
	I Baseline	II Cold	III Cold	IV Warm	V $2 \times CO_2$
	Stykkishólmur		Average of 10 coldest years	Average of 10 warmest years	
	1951 to 1980	1859 to 1868-type	1931 to 1984	1931 to 1984	GISS model-derived
Temperature (October–April)	0.6	−1.1	−0.5	1.9	4.8
Temperature (May–September)	8.0	7.4	7.6	8.9	11.7
Nitrogen application (kg/ha)	160	160	160	160	160
Yield (kg/ha, 85% dry matter)	4620	3723	4035	5467	7646
Yield (% of Scenario I)	100	81	87	118	166

2.2.8. Predicting the hay yield in spring

The preceding discussion has indicated that winter temperatures play an important role in determining the hay yield of the following summer. This is important because it implies that farmers can make predictions of their hay yield as early as the end of April and can adjust their activities accordingly. The effect of a harsh winter can thus be mitigated by applying more fertilizers than usual or, when the outlook is for good yields and the hay stocks are large, farmers can save on fertilizer expenditures.

For this purpose a new regression equation can be computed, expressing the yield as a function of winter temperature only, together with fertilizer applications. For the first 75 years of data in *Table 2.1* this regression equation is

$$Y = (0.876 + 0.0892\ W)(1820 + 28.1\ N - 0.051\ N^2) \tag{2.5}$$

where the symbols are as defined for equation (2.3). In the computation of this equation, by successive regression, the fertilizer index came out so similar to that of equation (2.3) that it was considered unnecessary to alter it. *Figure 2.1* may therefore be used again to compute the *predicted* hay yield according to equation (2.5). For this application, winter temperatures are read off the top horizontal scale of the upper diagram, and fertilizer rates, as before, are taken from the vertical axes.

2.2.9. Validation of the prediction model

The correlation coefficients and the root-mean-square errors of recorded versus predicted hay yield have been computed for different periods and are shown in *Table 2.4*. Even if the correlation is not as good as for the formula considering both the summer and the winter temperatures (cf. *Table 2.2*), it is apparent that more than 80% of the variance in annual yield can be explained by this formula. For the period 1976–1983 which was not included in the 75 years for which the relationship was established, the correlation coefficient is 0.91. In spite of the small sample, this is significant at the 1% level.

Table 2.4. Correlation between recorded and predicted hay yield at the end of April.

Period	Correlation coefficient	RMS error (kg/ha)
1901–1940	0.84	310
1941–1975	0.92	250
1901–1975	0.96	280
1976–1983	0.91	170

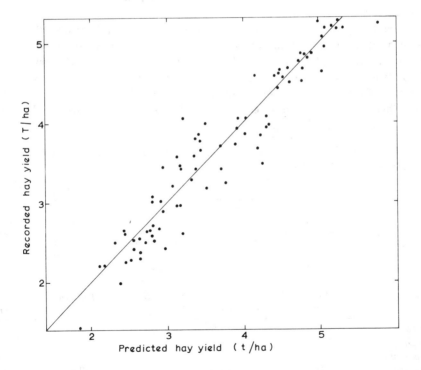

Figure 2.4. Validation of the annual hay-yield prediction model. Recorded yield (t/ha) in 1901–1983 is plotted against yield predicted by equation (2.5). Regression line based on data from 1901 to 1975.

Predicted and recorded hay yields during the whole period, 1901–1983, are indicated in *Figure 2.4*, with the straight line representing the least squares regression equation for the period 1901–1975.

2.2.10. Mitigating the effects of climate on hay yield

By using the horizontal scale of winter temperature and the nitrogen fertilizer scale, *Figure 2.1* can be used to determine the level of applied fertilizer corresponding to a certain desired yield in the coming summer.

Suppose a farmer wishes to obtain a hay yield of about 4000 kg/ha every year. When the winter temperature has been −1.1 °C, as under the coldest Scenario (II), the necessary application of nitrogen fertilizer would be 170 kg/ha, with corresponding levels of P and K of 45 kg/ha and 100 kg/ha, respectively. When the winter temperature has been 1.9 °C, as under Scenario IV, the appropriate values of N, P and K are found to be about 85, 25 and 70 kg/ha, respectively.

Alternatively, it is possible to compute the fertilizer applications required to obtain a certain hay yield for the given winter temperatures from equation (2.5) and the empirical relations between N, P and K [equations (2.1) and (2.2)]. The results are shown in *Table 2.5*.

It has been shown by a farmer in southeast Iceland (Geirsson, 1984) that a variable application of fertilizers can be used effectively to increase the yield when remaining supplies of hay in the spring are limited and the grass is in a poor condition. He applies 125–140 kg/ha of nitrogen in the coldest years, (for example, 1979), while 80–105 kg/ha are sufficient in mild years (such as 1980 and 1984). This is in fairly good agreement with the applications estimated from the

Table 2.5. Fertilizer requirements to maintain a constant annual hay yield (4000 kg/ha) under Scenarios I–V.

	Scenarios				
	I Baseline	II Cold	III Cold	IV Warm	V $2 \times CO_2$
			Average of 10 coldest years	Average of 10 warmest years	
	Stykkishólmur				GISS model-derived
	1951 to 1980	1859 to 1868-type	1931 to 1984	1931 to 1984	
Nitrogen (kg/ha)	110	171	144	84	49
Phosphorus (kg/ha)	31	46	40	25	17
Potassium (kg/ha)	79	100	91	71	59

diagram in *Figure 2.1*, discussed above, considering the winter temperature of these recent years (cf. *Table 2.1*).

In conclusion, it seems that Icelandic farmers could mitigate the effects of climatic variations simply by varying the annual application of fertilizers on the hay fields, thus avoiding the need to reduce their livestock numbers during extremely cold periods lasting for several years.

2.3. The Effects on Animal Rearing

Variations in climate affect not only the hay yield, but also the feeding of hay to livestock during the winter. Some farmers permit their horses and sheep to graze whenever the weather and snow cover permit, even in the middle of the winter, a practice that was particularly common in former times. Cattle are given hay all winter, and in most cases sheep, too, so that in these cases variation in feeding is a function simply of the start and end of the feeding period. In this situation, therefore, the vulnerability to climate is much less, although the fodder requirements are greater than for grazing animals.

Phenological observations from 15–20 weather stations during the period 1941–1949 give valuable information on the relative needs of different animals at that time as a function of winter temperature (Bergthórsson, 1966). In *Figure 2.5* the feeding of lambs, ewes and horses is given as a function of the mean temperature averaged over the corresponding phenological stations during the months October–May. Assuming that the mean temperatures are representative of the whole country and can be substituted by temperatures at Stykkishólmur, these relationships can be expressed by the following regression equations for the whole of Iceland

$$H_L = 67 - 4.9\, t_{OM} \tag{2.6}$$

$$H_E = 60 - 7.0\, t_{OM} \tag{2.7}$$

$$H_H = 48 - 5.3\, t_{OM} \tag{2.8}$$

where H_L is the percentage of full hay feeding of lambs in the period October–May, H_E and H_H are the corresponding percentages for ewes and horses, and t_{OM} is the mean October–May temperature. Equations (2.6) and (2.7) can be combined into a single equation for sheep, assuming the ratio of number of lambs to ewes to be 1:6.

$$H_S = 61 - 6.7\, t_{OM} \tag{2.9}$$

The hay feeding of cattle is not so closely related to midwinter temperatures as for sheep, and variations from year to year are relatively lower. Based on the

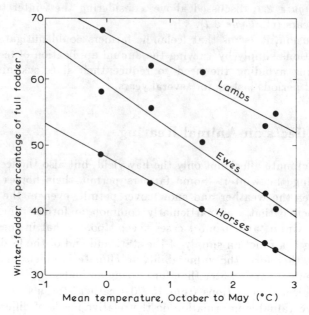

Figure 2.5. Hay feeding of lambs, ewes and horses in the winters 1941–1942, 1942–1943, 1945–1946, 1947–1948 and 1948–1949, as a function of the mean temperature in October–May averaged over all the corresponding 15–20 phenological stations. (Source: Bergthórsson, 1985.)

phenological data and the temperature at Stykkishólmur, the regression equation is

$$H_C = 123 - 1.8\, t_S - 2.5\, t_{DM} \tag{2.10}$$

where t_S refers to the temperature in September and t_{DM} to the temperature in December–May. Temperatures during the months October and November do not appear in this equation because cattle, in contrast to sheep and horses, are kept inside at this time.

As explained above, these regression equations are based on data from the 1940s. While winter feeding of cattle is probably very much the same now as it was then, the management of sheep and horses has changed considerably. One study of the winter feeding of sheep at Hestur, an experimental station near Hvanneyri in western Iceland, indicates that during the twenty years 1963–1982, winter feeding was almost 50% greater than it was in the 1940s in Iceland as a whole and much less variable in relation to the climate (Gudmundsson, personal communication).

However, in the absence of phenological data from recent years, the above regression equations (2.8) to (2.10) have been used to compile *Table 2.6*. The table gives the hay requirement for livestock as a percentage of that of the reference period, 1951–1980, but assumes that management is similar to that of the period 1941–1949.

Table 2.6. Hay requirements for cattle, sheep and horses under Scenarios I–V (percentage of Scenario I).

	Scenarios				
	I Baseline	II Cold	III Cold	IV Warm	V $2 \times CO_2$
	Stykkishólmur		Average of 10 coldest years	Average of 10 warmest years	GISS model- derived
	1951 to 1980	1859 to 1868-type	1931 to 1984	1931 to 1984	
Cattle	100	105	103	95	84
Sheep	100	122	114	84	47
Horses	100	122	114	83	47

According to these computations, the hay requirements of sheep under the $2 \times CO_2$ scenario would be only about one-half of that recorded in the 1940s. This is, of course, an uncertain estimate, but it is by no means unreasonable. In Scotland, where the present climate resembles the $2 \times CO_2$ scenario (*see* Section 1.10), sheep are not housed at all during the winter in some places in the highlands and receive only a very limited amount of hay.

Under the coldest scenario (II), the fodder requirement is high while hay yield is low (*see Table 2.3*). Conversely, under the warm-years scenario (IV), the required fodder is low while the hay yield is high. Thus there are two types of climate impact that have to be considered in combination in order to assess the total effect of climate variations on animal husbandry.

2.4. The Effects on the Carrying Capacity of Improved Grassland

In this section, the two aspects of climate impacts on livestock, discussed above, are combined into a simple model describing the carrying capacity of cultivated grassland.

2.4.1. A simple model of carrying capacity

The numbers of livestock that can be kept on a farm are directly proportional to available hay yield but inversely proportional to the winter feed required. Suppose, for example, that the feed requirement for cattle in relation to the baseline climate is 105% for Scenario II, and the hay yield is 81%. The carrying capacity is therefore $(81/105)100 = 77\%$. *Table 2.7*, which indicates the carrying capacity of improved grassland, can thus be based entirely on *Table 2.3* and *Table 2.6*, assuming that no haymaking occurs outside these areas of improved grassland and that there is no shortage of grassland for summer grazing (the carrying

Table 2.7. Carrying capacity of improved grassland for haymaking.

	Scenarios				
	I Baseline	II Cold	III Cold	IV Warm	V $2 \times CO_2$
	Stykkishólmur		Average of 10 coldest years	Average of 10 warmest years	
	1951 to 1980	1859 to 1868-type	1931 to 1984	1931 to 1984	GISS model- derived
Cattle	100	77	84	124	198
Sheep	100	66	76	140	353
Horses	100	66	76	142	353

capacity of *rangelands*, i.e., non-improved grassland, is considered in Subsection 2.5.) Thus described, carrying capacity is a hyperbolic function of hay yield and winter feeding requirements, increasing rapidly with higher temperatures in a mild climate, as shown by the difference between Scenarios IV and V.

2.4.2. Validation of the model

The relationship between carrying capacity and climate is likely to change over time with changes in management practice, in particular with improvements in feeding. In order to validate the model, two distinct periods are therefore considered, one in the latter half of the 19th century, and the other in the past two decades.

An initial problem concerns the most appropriate method of expressing the total numbers of livestock. The fodder requirement for a dairy cow was 15–20 times that of one sheep in the last century, but this ratio has changed over time. In an attempt to estimate the feeding ratios during the two periods, certain *weights* have been assigned to represent the winter fodder requirements of different animals, partly based on accounts from more than 100 farms for the years 1975–1982.

These weights are expressed as the ratios of dairy cattle:other cattle:sheep:horses in livestock units (one livestock unit is defined as the feeding requirement of one dairy cow) for the period 1846–1900 (1:0.4:0.05:0.125) and the period 1962–1983 (1:0.4:0.08:0.2) (Ólafsson, 1975–1982).

Livestock numbers and temperature 1846–1900

The relation between temperature and total livestock numbers in the period 1846–1900 is shown in *Figure 2.6*. It is recognized that the livestock numbers in a particular year depend considerably on the temperature of the preceding years. Therefore, the temperature is shown as the weighted mean of the recent years, with suitable weights being 1.0 for the immediately preceding year and (5/7),

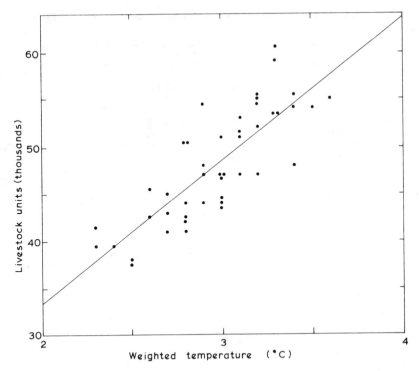

Figure 2.6. Livestock in Iceland (thousands of livestock units) as a function of weighted temperature of the preceding years for the period 1846–1900. (Adapted from Bergthórsson, 1985.)

$(5/7)^2$, $(5/7)^3$, etc. for the progressively earlier years. This is admittedly a rough method, but it is certainly more reasonable than a direct comparison of the livestock and the temperature in the same year (Bergthórsson, 1982, 1985).

From the regression line in *Figure 2.6*, it may be concluded that the relative livestock numbers under the management practices of the 19th century, for the five scenarios discussed above, should be as shown in *Table 2.8*.

For the Scenarios I–IV this result is in good agreement with our estimates of the carrying capacity of the hay fields (*Table 2.7*). The linearly extrapolated livestock numbers for Scenario V should not, however, be expected to correspond with carrying capacity because the latter is a hyperbolic function of hay yield and winter fodder requirements.

Livestock numbers and hay yield 1962–1983

It is probably not surprising that livestock numbers in the 19th century were sensitive to climatic variations, as shown in the discussion above. It is, however, a common belief that this no longer applies because new techniques should enable farmers to counteract the impacts of adverse climate without reducing livestock numbers. It is certainly true that farmers have some new-found flexibility. Imported concentrates are used more extensively when hay is limited;

Table 2.8. Livestock numbers 1846–1900 under Scenarios I–V (percentage of Scenario I).

	Scenarios				
	I Baseline	II Cold	III Cold	IV Warm	V $2 \times CO_2$
	Stykkishólmur		Average of 10 coldest years	Average of 10 warmest years	GISS model-derived
	1951 to 1980	1859 to 1868-type	1931 to 1984	1931 to 1984	
Livestock numbers	100	67	80	128	199

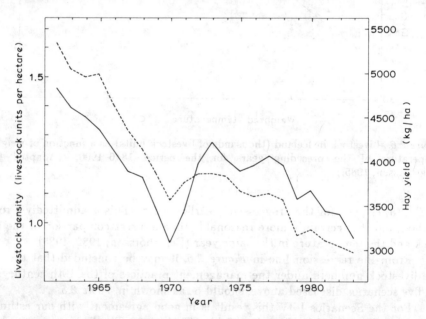

Figure 2.7. Two-year running mean of hay yield (kg/ha) of improved hay fields in kg/ha (solid line), and annual livestock density in livestock units/ha (dashed line).

even winter grazing can be intensified to conserve hay. Reserve supplies of hay, preferably 25–35% of the normal winter fodder, are also an important buffer to save animals in bad years (Jónsson, personal communication; cf. Subsection 5.4.1). Nevertheless, *Figure 2.7* shows that available hay is still crucial in keeping livestock numbers constant from year to year. In order to estimate the appropriate number of livestock to be fed in the coming winter, the hay yield can be expressed as the mean of the yield in the present year and the preceding year.

Around 1972 there was a sudden rise in hay yields after many years of decline. It seems that farmers used this opportunity to improve the feeding of livestock, particularly sheep. From then on, up to the present, the new management practices seem to have remained largely unchanged. This indicates that available hay determines to a high degree the possible numbers of livestock for a given level of management. Since, however, hay yields are strongly dependent on temperature, animal production in Iceland continues to remain sensitive to variations of climate.

2.5. The Effects on the Carrying Capacity of Rangelands

The possibilities of animal husbandry are limited not only by available hay but also by available grazing, which is itself sensitive to climatic variations. In this section we consider the relationship between *optimal* grazing pressure and climate, assuming the carcass weight of livestock (in this application, lambs) to be kept constant under different climatic conditions. To detect and estimate these effects, we shall refer to carcass weight of lambs at the age of 4–5 months, which we shall demonstrate to be a function of climate and grazing pressure on the rangelands in Iceland. The results presented here are used as a basis for further discussions of impacts on rangelands in Section 4. Preliminary research for the study has been reported elsewhere (Bergthórsson, 1985).

The data available (for the period 1941–1984) consist of annual records of the average carcass weight of lambs that are slaughtered in the autumn and the number of sheep as a measure of the grazing pressure. As a climatic parameter the temperature index (T_i) of equation (2.3) has been used

$$T_i = 0.29 + 0.0729\ S + 0.0794\ W$$

While this index is a useful indicator of the grass growth, there has also occurred an overall increase in carcass weight due mainly to the improved feeding of ewes. This increase of the carcass weight over the study period is assumed to have been a linear function of time in the last term of the following regression equation

$$V = 10.79 - 0.0025\ F + 2.3\ T_i + 0.049\ A_n \qquad (2.11)$$

where V is the country average of the carcass weight, F the sheep numbers (in thousands) in the preceding winter, T_i is the temperature index and A_n is the year after 1900 ($A_n = 75$ denotes 1975). The correlation between the computed and the recorded carcass weight is 0.73, and the root-mean-square error is 0.34 kg (significant at the 0.1% level).

Table 2.9. Sheep numbers necessary to maintain a constant carcass weight of sheep under different climatic scenarios.

	Scenarios				
	I Baseline	II Cold	III Cold	IV Warm	V $2 \times CO_2$
	Stykkishólmur		Average of 10 coldest years	Average of 10 warmest years	
	1951 to 1980	1859 to 1868-type	1931 to 1984	1931 to 1984	GISS model-derived
Computed sheep numbers (thousands)	847	683	740	1002	1402
Numbers as a percentage of Scenario I	100	81	87	118	166

The next step is to rearrange equation (2.11) to find the relationship between the temperature index and the number of winter-fed sheep that keeps the carcass weight at a constant level. To do this, it must first be assumed that sheep numbers are representative of the total grazing pressure exerted on the rangelands by all livestock, horses and cattle included. Then, for a given carcass weight (we have chosen 14.71 kg for computational convenience, which is not unreasonable under tolerable climatic conditions) under 1980 management, and substituting into equation (2.11), only the terms F and T_i are left, and the following relationship is derived

$$F = 920\, T_i \tag{2.12}$$

This gives the sheep numbers necessary to keep the carcass weight constant under different climatic conditions, described by the temperature index T_i. Table 2.9 shows the corresponding sheep numbers under the five scenarios. Another row gives the sheep numbers relative to Scenario I. These relative numbers are identical to the percentages of hay yield shown in Table 2.3. It should be stressed that this result has to be taken with many reservations as to the level of management of animals and of land. Nevertheless, it suggests that Icelandic farmers should be aware of the potential effect of climatic variations on grazing conditions in order to avoid long-lasting damage to the rangelands.

In the results presented above, no account has been taken of the considerable changes in the extent of potential grazing land that would occur as a result of longer-term climatic variations, a matter considered in Sections 1 and 3.

2.6. The Effects on Potential Barley Cultivation

Cereal cultivation of even the most hardy crops such as barley is close to its growing limits in Iceland (*see* Section 1, *Figure 1.8*). Thus, relatively small changes in temperature can significantly alter the extent and viability of potential cultivation. To test this for barley, agronomic and weather observations from the experimental station at Sámsstadir in south Iceland (63°44′ N, 20°07′ W), were utilized from the period 1927–1940 (Kristjánsson, 1946; Bergthórsson, 1965). It was found that the sum of the effective daily temperatures above a threshold temperature of 3 °C during the growing period is a good measure of the thermal requirements for the ripening of a fast growing six-row variety of barley (Dönnes). For average May–September precipitation of 375 mm, an effective temperature sum (ETS) of 850 growing degree-days (GDD) was required from sowing to ripening. In drier years the thermal requirements were less: for every millimeter of May–September precipitation below the mean, the ETS requirements are decreased by approximately 0.3 GDD. In Norway, experience has shown that ETS requirements are also lower at higher latitudes (about 30 GDD per 1° latitude), presumably due to longer photoperiods (K. Vik, personal communication). Pending further investigation, it seems reasonable to use this relationship in Iceland also.

Growing conditions for barley were assessed for each of 48 weather stations under the different scenarios using the above information and data on sowing and harvest dates, May–September precipitation, ETS above a base temperature of 3 °C, and station latitude. These were obtained in the following manner.

The earliest suitable sowing date for barley is closely related to spring temperatures. At Sámsstadir sowing can commence on average from April 21, but elsewhere in Iceland the spring tends to be colder, and sowing is somewhat later. The mean temperature on April 21 at Sámsstadir (interpolated from 1931–1960 monthly mean temperatures) was 4.8 °C, and the mean temperature of the 30 preceding days was 3 °C. This latter criterion appears realistic as a requirement for sowing, even in single years, for the snow will then have disappeared and the soil will have melted to such a depth that sowing is possible.

Harvesting at Sámsstadir is often difficult after 20 September, which is close to the mean date of the first autumn air frost. However, as an indicator of harvest dates in individual years, the timing of the first autumn frost is not suitable because of its high variability. Instead, a procedure was adopted that utilizes the 30-day mean temperature (t_{mean}) and the 30-day mean minimum temperature (t_{min}) centered at the day of harvest. Information from several stations was used to derive the following function

$$D_f = 2\, t_{min} - t_{mean} - 1 \qquad (2.13)$$

The first autumn frost occurs, on average, when D_f falls below zero. The difference between t_{mean} and t_{min} assessed for the period 1971–1980 was assumed to be unaffected by climatic variations.

Table 2.10. Percentage of lowland stations with ripened barley under Scenarios I–V.

	Scenarios				
	I Baseline	II Cold	III Cold	IV Warm	V $2 \times CO_2$
	Stykkishólmur		Average of 10 coldest years	Average of 10 warmest years	
	1951 to 1980	1859 to 1868-type	1931 to 1984	1931 to 1984	GISS model- derived
Ripening in 60% of all years	4	0	0	60	100
Ripening in 90% of all years	0	0	0	8	100

Monthly temperature differences between the baseline period (1951–1980) and the scenario climates were evaluated, for individual stations, by adjusting the differences at Stykkishólmur (see Table 1.5, Section 1). It was assumed that both short- and long-term variations in mean monthly temperatures at single stations are proportional to the standard deviation of mean annual temperatures. Thus the variations in mean monthly temperature represented in the scenarios at Stykkishólmur were adjusted by the ratio between the standard deviation of mean annual temperature at the single station and that at Stykkishólmur (1951–1980).

Long-term variations in precipitation at single stations are relatively insignificant in Iceland, and are certainly much less than the difference in precipitation between stations. In terms of the thermal requirements of barley, a 20% deviation in May–September precipitation is equivalent to only 15–20 GDD, which corresponds to a change of little more than 0.1 °C in growing season temperature. Therefore, May–September precipitation in the period 1971–1980 was used, along with the station latitude, to complete the computation of ETS.

The aim of this study was to assess the probabilities of barley ripening at a given station under the climatic scenarios. This depends considerably on the year-to-year variations in ETS during a given scenario period. These can be represented by using the standard deviation of the annual ETS, assuming the distribution to be approximately normal. A study of several stations during recent decades reveals that this standard deviation is between 100 and 150 GDD in Iceland. A value of 125 GDD has been adopted here to express the variation around the mean ETS and to permit the computation of probabilities of annual ETS exceeding the critical value for the ripening of barley (Table 2.10).

It is evident from this study that potential cultivation of barley is extremely sensitive to recorded climatic variations in Iceland. Under the reference scenario (1951–1980) very few localities provide tolerable conditions for cultivation, even with the quick-growing variety considered here. In the warm years scenario (IV), barley could be grown commercially in some areas, while under Scenario V the barley would ripen at all lowland stations almost every year.

2.7. The Effects on Potential for Tree Growth

Climatic conditions in Iceland are marginal for tree growth with low native birch covering only about 1% of the land area. In spite of these marginal conditions there has been a considerable interest in the growth of trees in Iceland, both because of their potential value as timber and as shelter for crops and livestock. We therefore report here some of the different potentials for birch and Norwegian spruce growth corresponding to different climatic scenarios.

These estimates are based on relationships developed by Mork (1968) from experiments in Hirkjølen, a mountain area more than 800 m above sea level, north of Oslo. Mork concluded that the daily elongation of top shoots of Norwegian spruce could be expressed as a function of temperature in terms of growth units, where one growth unit (GU) is defined as the daily growth when the mean temperature of the six warmest hours of the day is 8 °C.

Observations at the timberline of Norwegian spruce and birch during the period 1932–1962 indicate a growth requirement of 255 GU and 222 GU for the respective tree species during the growing period.

Unfortunately, the original relationships cannot easily be applied to the Icelandic conditions (for example, the published climatic data do not include temperature for the six warmest hours of the day), so a simplified method of determining the annual growth units for trees has therefore been developed for Iceland, based on daily maximum and daily mean temperatures (which are available for most Icelandic stations).

Comparison with the well-documented data from Hirkjølen shows that the following temperature index very closely represents the growth units per day

$$GU = (T - 1.6)/(7.4 - 0.215\, T) \qquad (2.14)$$

where T is the average of the daily maximum and daily mean temperatures. This relationship is depicted graphically in *Figure 2.8*.

The monthly sum of growth units is computed by substituting the monthly mean temperature into equation (2.14) and multiplying the resulting GU value by the number of days in the month. By cumulating these monthly values over the whole year, annual growth units can be obtained for each year of the scenario period. The annual growth units obtained with this method are, on average, somewhat higher than the corresponding growth units computed with Mork's method, which considers only growing season conditions. However, comparisons at a number of stations in Iceland indicate that this difference is a relatively constant value of about 45 growth units per year. When this correction is applied, the simplified method described above provides similar results to those obtained by Mork.

Temperature data from 48 Icelandic stations were used to compute the annual growth units for all the five climatic scenarios (*see* Subsection 1.10). In addition, computations for the period 1931–1960 were included, for reasons explained below. For individual stations, the monthly temperature differences between the baseline period (1951–1980) and the scenario climates were evaluated by adjusting the differences at Stykkishólmur (*see Table 1.5*, Section

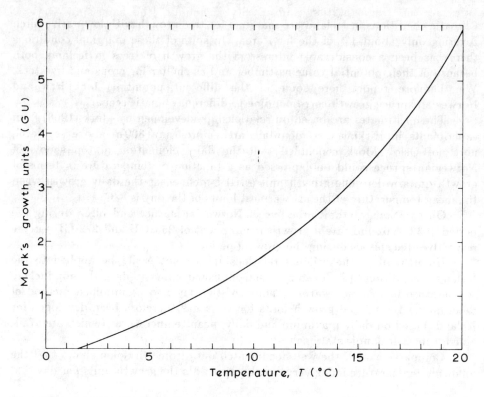

Figure 2.8. Mork's growth units per day (GU) as a function of T, the average of the mean temperature and maximum temperature of the day, according to equation (2.14).

Table 2.11. Percentage of stations permitting cultivation of birch and Norwegian spruce in Iceland under Scenarios I–V and in the period 1931–1960.

	Scenarios					
	I Baseline	II Cold	III Cold	IV Warm	V $2 \times CO_2$	Additional Recent
	Stykkishólmur		Average of 10 coldest years	Average of 10 warmest years	GISS model- derived	Stykkishólmur
	1951 to 1980	1859 to 1868-type	1931 to 1984	1931 to 1984		1931 to 1960
Birch	29	4	10	85	100	60
Norwegian spruce	8	0	0	71	100	27

1), according to the ratio between standard deviations of mean annual temperatures at the single station and at Stykkishólmur (the same procedure as described in the preceding subsection on barley). It was also assumed that the deviation of the monthly mean maximum temperatures [required in equation (2.14)] from the mean monthly temperatures was the same for the scenario periods as in the decade 1971–1980.

The mean annual growth units were compared with the requirements for birch and Norwegian spruce at all 48 stations to evaluate the proportion of stations at which these species could be cultivated under each scenario (*Table 2.11*). It should be noted, however, that the actual timberline of these tree species probably varies very slowly in relation to climatic variations, partly due to the impact of a forest on its own microclimate.

The warm period 1931–1960 has been selected as an additional scenario for Iceland because the growth unit requirements at the treeline in Hirkjølen, reported by Mork, were derived for this approximate period. Warm conditions in this period extended the *potential* for tree growth considerably beyond the actual tree line. If such conditions were typical of Iceland's climate for a century or more, then we might expect birch forest to survive at about 60% of the lowland stations, and Norwegian spruce at 27% of lowland stations (*Table 2.11*). This may support the opinion of many historians that birch woods were relatively extensive during the warm epoch at the time of Iceland's settlement 1100 years ago (*see* Section 1).

2.8. Conclusions

We have examined the likely impacts of five contrasting climatic scenarios on hay yield, livestock fodder requirements and carrying capacity, and the potential cultivation of barley, birch and Norwegian spruce. To conclude, we will summarize the calculated impacts and discuss possible responses to climatic variations.

2.8.1. Summary of results

The estimated effects under Scenarios I–V are listed in *Table 2.12*. For grass growth and livestock farming, the effects of cooler-than-average conditions (Scenarios II and III) are to reduce potential productivity and, simultaneously, to increase fodder requirements. In addition, cool conditions such as those recorded in the 1860s would preclude viable barley cultivation and would be unsuitable for all but the hardiest species of birch in localized areas. In contrast, warmer-than-average conditions (Scenario IV) enhance productivity, permit increased grazing of livestock (thus reducing fodder requirements) and offer the potential for the land to support greater numbers of livestock, for barley cultivation to become viable in many lowland areas and for birch and Norwegian spruce to become established in the lowlands. Under the $2 \times CO_2$ scenario (V), the temperature increase well exceeds the observed variations over recent centuries past. However, it has been demonstrated that there are justifiable grounds for

Table 2.12. Sensitivity of agriculture to temperature in Iceland.

	Scenarios				
	I Baseline	II Cold	III Cold	IV Warm	V $2 \times CO_2$
	Stykkishólmur		Average of 10 coldest years	Average of 10 warmest years	
	1951 to 1980	1859 to 1868-type	1931 to 1984	1931 to 1984	GISS model-derived
Temperature changes					
Dec–May temperature relative to Scenario I (°C)	0.7	−1.2	−0.4	+2.0	+4.9
Oct–Apr temperature relative to Scenario I (°C)	0.6	−1.7	−1.1	+1.3	+4.2
May–Sept temperature relative to Scenario I (°C)	8.0	−0.6	−0.4	+0.9	+3.7
Annual temperature relative to Scenario I (°C)	3.7	−1.3	−0.8	+1.1	+4.0
Effects on agriculture and tree growth					
Area of potential tree zone (% of Scenario I)[a]	100	73	73	200	582
Hay yield from improved grassland (%)	100	81	87	118	166
Necessary hay feeding per animal (%):					
Cattle (management as 1941–1949)	100	105	103	95	84
Sheep (management as 1941–1949)	100	122	114	84	47
Horses (management as 1941–1949)	100	122	114	83	47
Carrying capacity of improved grassland (%):					
Cattle (management as 1941–1949)	100	77	84	124	198
Sheep (management as 1941–1949)	100	66	76	140	353
Horses (management as 1941–1949)	100	66	76	142	353
Carrying capacity of rangelands for sheep (%)[b]	100	81	87	118	166
Percentage of lowland weather stations:					
with barley ripening 60% of years	4	0	0	60	100
with barley ripening 90% of years	0	0	0	8	100
permitting cultivation of birch	29	4	10	85	100
permitting cultivation of Norwegian spruce	8	0	0	71	100

[a] From Section 1.
[b] For constant carcass weight of 14.71 kg.

extrapolating some of the relationships developed for agricultural impacts under present climatic variability to the $2 \times CO_2$ conditions (using information from warmer regions to the south of Iceland). Under this climate, grass growth and hay yields would be dramatically higher than present, the carrying capacity of both improved and unimproved grassland would be greatly enhanced and fodder requirements significantly reduced. As long as soils are suitable, all of the lowland area would become suitable for barley cultivation and for the establishment of birch and Norwegian spruce.

With such a delicate balance between climate and agriculture, farmers and agricultural planners require an array of adjustment measures to respond to climatic variations. Some of these possible responses are introduced below, and a fuller range of adjustments will be discussed in Section 5.

2.8.2. Responses to climatic variations

The types of possible responses to cooler-than-average climatic conditions are somewhat different from those that would be appropriate in a warmer climate. Thus, we will consider these two groups of adjustments separately:

Responses in cool periods

Icelandic farmers have to be in constant readiness for prolonged cold intervals that have occurred repeatedly throughout the past. Experience from 1962–1983 suggests that even today the farmers have not succeeded in adjusting their management to colder weather without reducing their livestock. At best, they have been able to meet the increased demand for winter fodder of the reduced herds. There are several measures that might be of use in improving this situation:

(1) Fertilizers can be used to counteract the variations in growing conditions. Preconditions, however, are that farmers apply only moderate amounts of fertilizer in average or good years, and that hay yield in normal years is sufficient to provide a reserve for use in severe years. An increase in levels of fertilizer application in severe years can then raise yields and thus meet the increased demand for hay that is a characteristic of such conditions. For example, under the cold scenario (II), this increased demand amounts to 5% for cattle and 10–20% for sheep and horses (*Table 2.12*). The levels of fertilizer application appropriate to particular climatic conditions can be estimated on the basis described above (*Table 2.5*).

(2) Increased production of silage from grass might help to reduce losses from hay in wet summers. An increase in the use of concentrates can also help to meet the increased demand for fodder, but if the concentrates are produced from crops grown in Iceland, their production will of course be reduced approximately in proportion to the hay production in unfavorable years. Production and cost of other feeds produced in the country may also be unreliable for climatic or economic reasons, and the import of feeds is not always possible when it is most urgently needed. By far the best

solution, therefore, is that farmers should be as self-sufficient as possible in their fodder production.
(3) Another problem in severe years is the decrease in carrying capacity of rangelands, particularly for sheep and horses. This reduction is at least of the same order of magnitude as the reduction in hay yield in cold periods. In the past this has not been a great problem since the reduction in livestock due to available hay has usually been sufficient to spare the rangelands. If, however, farmers are able to increase their haymaking in severe years, as suggested above, the summer grazing might become a greater problem than hitherto. A possible response is to improve more grassland for grazing in the lowlands. Fertilizers can then be applied on this grassland according to the climatic conditions, and the unimproved rangelands could be spared in cold years, to avoid damage from over-grazing.
(4) In cold periods barley cannot be grown to maturity in Iceland (*Table 2.12*). Thus farmers cannot rely on barley cultivation as a major activity, not even when the climate during individual years is favorable. The possibility remains, however, to convert barley to silage when crops cannot ripen successfully.

It has been noted by some farmers that potatoes usually fail in the same years as barley does (E. Ólafsson, personal communication). In recent years, potato yields have fluctuated widely, the annual production in the country ranging from 3600 tons (in 1983) to 19400 tons in 1984. This seems to suggest that cultivation of potatoes is not advisable as a major occupation on any farm under the present climate, although it may be a profitable enterprise if kept at a small scale, and when restricted to the mildest regions of the country.

Responses in warm periods

The response of Icelandic farmers to warm periods has almost always been to exploit favorable weather by increasing numbers of livestock. But if it is possible to avoid reducing stock in cold years, as discussed above, farmers can well afford to keep their livestock unchanged even in warmer periods. They can thus save on additional expenditures for larger barns, new machinery, etc., and spend less on concentrates and fertilizers. These savings could be redirected towards improvements elsewhere on the farm to prepare farmers for the less favorable periods which may occur at any time.

If farmers can manage to respond to climatic variability as discussed above, it should be a relatively straightforward process to adjust farming gradually in response to more profound climatic change, such as the warming due to an increase in atmospheric carbon dioxide. At the outset it should be remembered that a uniform warming trend is unlikely, and that climatic variability will continue to exist, which may at times disguise any gradual change. Therefore the adjustment to a warm period, unknown in the past, should be implemented with caution. New crops should be introduced in such a way that a temporary cooling will not have too serious consequences, while further adjustments should be possible given a continued amelioration of the climate.

SECTION 3

The Effects on Grass Yield, and Their Implications for Dairy Farming

3.1. Introduction

In this section, we test the temperature–hay yield model developed in Section 2, using the long-term measurements on permanent grasslands from experimental stations in Iceland. Data on national average hay yield, used in Section 2, are only an indirect measure of grass growth, however. Here, and following some preliminary investigations (Björnsson, 1975, 1984), estimates of the effects of climatic variations are refined to consider the yield response to varying fertilizer applications and management practices under different temperature scenarios. Estimates of grass growth are made for five of the six scenarios described in Subsection 1.10:

(1) Scenario I – the baseline period (1951–1980).
(2) Scenario II – the coolest recorded decade (1859–1868).
(3) Scenario IV+ – the ensemble of the 10 warmest years (1931–1984) slightly modified to give a warm anomaly comparable with the cool anomaly in Scenario II.
(4) Scenario V – the GISS-derived 2 × CO_2 scenario.
(5) Scenario VI – the extreme year scenario, defined as a year in which mean annual temperature deviates by at least two standard deviations from the period average (*see* Subsection 3.4.1).

Finally, the implications of these effects for the quality and utilization of herbage are examined with respect to Icelandic dairy production.

3.2. The Experimental Material

3.2.1. The long-term fertilizer experiments

Long-term fertilizer experiments have been carried out at four experimental stations in Iceland, located in each quarter of the country. Basic climatic data for the nearest meteorological stations are given in *Table 3.1*, together with data from the meteorological station at Stykkishólmur. The location of all stations are shown in *Figure 1.7*.

Table 3.1. Climatic characteristics (mean for 1951–1980) of experimental stations in Iceland.

Station	Mean temperature (°C)			Mean annual precipitation (mm)
	January	July	Year	
Akureyri	−2.1	10.4	3.4	474
Reykhólar	−1.4	9.8	3.5	552
Sámsstadir	−0.4	11.3	4.8	1261
Hallormsstadur (near Skriduklaustur)	−1.6	10.4	3.6	679
Stykkishólmur	−1.3	9.9	3.7	704

A number of fertilizer experiments on permanent hayfields were initiated at the experimental stations in about 1950, and many of them have been continued up to the present. These experiments include rates of application of nitrogen (N), phosphorus and potassium, comparisons of N fertilizer types, liming experiments, etc. The experiments were laid out on silty loam and newly reclaimed peaty soils. The vegetation was either a mixture of indigenous grasses, including common bent (*Agrostis capillaris*), tufted hair-grass (*Deschampsia caespitosa*l), red fescue (*Festuca rubra*) and smooth meadow grass (*Poa pratensis*), or a mixture of sown and indigenous species, where meadow fox-tail (*Alopecurus pratensis*) was the most commonly used sown grass. All experiments were managed similarly. They were fertilized in spring, cut for hay once or twice in the summer and protected from grazing animals.

For the present study, experiments were selected in which fertilizer treatments had been constant for a decade or more. Account needed to be taken of variations in harvesting practices in the experiments, both between the experimental stations and between years at individual stations, with respect to number of harvests and harvest dates. At Sámsstadir the experiments have, with just two exceptions, been harvested twice a year. The yields from these two anomalous years were omitted together with the yield in the following year, when residual effect of the deviant cutting treatment can be expected. The experimental station at Akureyri was moved to Möðruvellir, 18 km further north, in 1974. This was followed by a reduction in the number of experiments. The old experiments, located at Akureyri, were only harvested once a year rather than twice, as previously had often been practiced. Other variability in harvesting practices at the four stations is partly due to changes in management

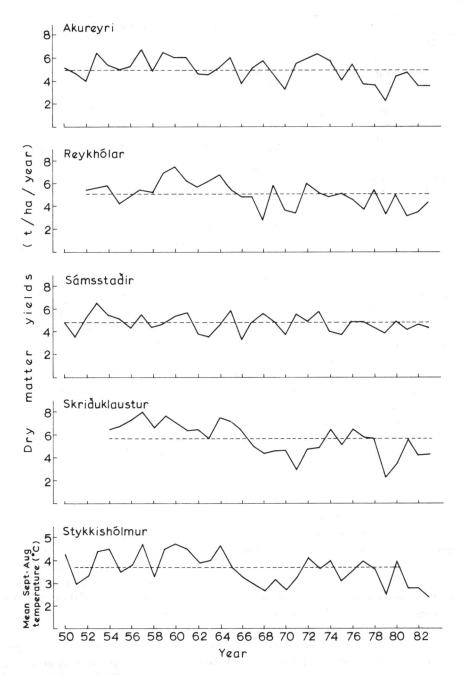

Figure 3.1. Annual mean temperatures (September–August) at Stykkishólmur and annual mean herbage yields in the long-term fertilizer experiments at Akureyri, Reykhólar, Sámsstadir and Skriduklaustur, 1950–1983.

Table 3.2. Number of experimental years in the analysis and the mean nitrogen fertilization.

	Akureyri	Reykhólar	Sámsstadir	Skriduklaustur
(a) The main analysis				
All experimental years	212	214	208	110
Experimental years with two cuts	125	95	208	78
(b) Nitrogen experiments				
All experimental years	57	62	58	56
Experimental years with two cuts	27	30	58	40
(c) Nitrogen application (N kg/ha)				
All experiments	89	102	83	107
Nitrogen plots in the nitrogen experiments	90	102	103	103

practices and partly due to variations in the annual conditions for grass growth. Fertilizer was generally applied in one dose in May or occasionally in early June. In addition, field notes contain information on occasional irregularities in the treatment of the experiments, especially in the fertilizer application. Some of these irregularities are considered to have a carry-over effect so that the yield the following year is also affected. In addition, the first year yields were omitted in two cases. For such reasons altogether 16 single year yields were eliminated. The yields have also been adjusted for incomplete determination of dry matter in the earlier years at two of the stations.

The study period extends from 1950 to 1983. During this period there have been large fluctuations in yield in the experiments. At the same time mean annual temperatures have varied considerably (*Figure 3.1*). *Table 3.2(a)* gives the total number of harvest years for all the selected experiments at the four experimental stations.

Within each experiment, yield measurements from one to three fertilizer treatments were used [*Table 3.2(b)*]. Only those treatments were selected where plant nutrients other than nitrogen were considered nonlimiting for grass growth, though the occurrence of sulphur deficiency cannot be entirely excluded. The nitrogen applications were, in nearly all cases, in the range 67–120 kg N/ha. These, however, are not optimum application rates since significant yield responses, exceeding the cost of fertilization, are obtained for higher N rates. The mean nitrogen application, weighted in the same way as the yield data in the analysis (*see* Subsection 3.2.2), is shown in *Table 3.2(c)*. In addition, in order to study the effect of climate on nitrogen yield response, those nitrogen experiments were used which included a treatment with no nitrogen application. The results from each experiment were used rather than first obtaining yield estimates for each station and year as in the main analysis.

3.2.2. Annual yield level

The method of least squares with years and experiments as factors was used to combine results from experiments of different duration at each station into single yield estimates for each year. The estimation is based on the following linear model

$$Y_{ij} = \mu_i + \gamma_j + \epsilon_{ij}$$

where

Y_{ij} = yield for experiment j in year i,
μ_i = expected yield at the station in year i (the quantity to be estimated),
γ_j = deviation in yield of experiment j from the expected yield,
ϵ_{ij} = interaction between years and experiments + experimental error.

Permanent yield differences between experiments, as a result of different fertility or N fertilization, were thus eliminated. All experiments were given equal weights in spite of the different number of treatments selected. These yield estimates became the dependent variable in the regression analysis of yield on temperature.

3.2.3. Temperature variables

Temperature effects throughout the year were represented in the regression analysis by data on a time basis ranging from monthly to annual averages. Since the experimental plots have usually been harvested by late August or early September, temperatures are calculated for the period beginning September 1. Frequently, however, the experimental plots were cut only once, sometimes in early July, in which case July and August temperatures were excluded from the analysis. When the calculations were restricted to those stations and years when most experiments were cut twice, the temperature for the 12-month period September to August, or subdivisions thereof, was used.

Although soil moisture affects the growth of grasses, a soil moisture index has not been developed for the Icelandic experimental sites. Data on precipitation have not been included in the present regression analysis – selection of an appropriate precipitation variable is difficult, as the growing season is both short and irregular. Untimely or excessive rain may have negative effects on grass growth, so that a linear term for precipitation without some modifying factors may be of little value.

3.2.4. Associate variables

Various associate variables other than temperature were included in the regression analysis. The most important concerned the variability in cutting management, of which two aspects required particular attention. Firstly, the date of

Table 3.3. Number of associate variables in the regression of yield on temperature.

Variables	All experimental years	Experimental years with two cuts only
Date of 1st cut at each station	4	
Days between cuts at each station	4	
Date of 2nd cut at each station the year before	4	4
Change in yield level (at Akureyri and Skriduklaustur)	2	1
Total number	14	5

harvest, which varies from year to year, has a direct influence on the yield, especially through the variable amount of regrowth that remains uncut on the field in the autumn and which affects yield the following year. This is accounted for by adjusting the yield according to average harvest dates for the year of interest and the preceding year (which excludes the yields in the first year of the study period from the analysis). The most significant variation is due to those years and stations when the experiments were cut only once, while the variability of harvest dates is of little importance where there are predominantly two cuts. Furthermore, in the analysis of nitrogen response, the variables representing harvest date were similarly found to have no significance.

A second aspect concerns the effects of weather conditions on the harvest date and the likelihood of a second cut. This not only increases the variability, but also makes the interpretation of the regression analysis difficult. Consequently, the regression analysis has been carried out in two parts. First, all experimental years were included in the analysis, and then the analysis was restricted to those years when all or (occasionally) most experiments were cut twice. The latter procedure almost eliminates the effect of date of harvest, although where years with two cuts follow years with a single cut that was harvested early, an adjustment for harvest date the year before is still required.

Adjustments were also required to account for apparent shifts in the level of yield during the experimental period at two of the stations, Akureyri and Skriduklaustur (*Figure 3.1*). These were made by arbitrarily dividing the total experimental period into two. At Akureyri the changeover occurred between 1976 and 1977, probably due to the movement of the experimental station such that experiments have not been cut twice since 1976. At Skriduklaustur the yield has never reached the earlier, relatively high yield level after the cold period 1965–1970, and the yield decline (assumed between 1970 and 1971) may be the irreversible effect of a cold period, or due to reduced fertility or possibly a result of deteriorating drainage.

The variables adjusted for are, thus, three to four at each station when all experimental years are considered, and one or two when the analysis was restricted to those years with two cuts (*Table 3.3*). The nitrogen response was adjusted only for differences between experiments and for the shifts of yield levels at Akureyri and Skriduklaustur.

Table 3.4. Parameter estimates for the regression of experimental yield on temperature at the experimental stations, accounting for the effects of the date of 1st cut, date of 2nd cut, and the date of 2nd cut the year before, for (a) all experiments, and (b) years when the experiments were generally cut twice.

Experimental station	No. of years	Mean yield (t/ha)	Temperature period	Regression on temperature alone		Adjusted for harvest dates and periods			R^2 due to adjustments alone
				kg/ha/°C	R^2	kg/ha/°C	Residual $MS \times 10^{-2}$	R^2	
(a) All experimental years									
Akureyri	33	5.0	Sept–June	730	0.399	513	3828	0.721	0.584
Reykhólar	32	5.1	Sept–June	987	0.635	1043	4255	0.699	0.476
Sámsstadir	33	4.8	Sept–June	456	0.158	599	5554	0.188	0.056
Skriduklaustur	29	5.7	Sept–June	845	0.308	966	6193	0.747	0.543
All stations	127		Sept–June	786 ± 92	0.374	735 ± 104	5009	0.641	0.474
(b) Years with two cuts									
Akureyri	15	5.7	Sept–Aug	523	0.364	503	2191	0.604	0.275
Reykhólar	17	5.8	Sept–Aug	847	0.406	847	4279	0.407	0.001
Sámsstadir	33	4.8	Sept–Aug	544	0.189	660	4970	0.221	0.002
Skriduklaustur	19	6.3	Sept–Aug	712	0.297	638	5318	0.609	0.406
All stations	84		Sept–Aug	635 ± 115	0.278	644 ± 113	4314	0.429	0.178
All stations	84		Sept–June	554 ± 98	0.281	540 ± 95	4320	0.428	

3.3. Yield of Pastures Cut for Hay as a Function of Temperature

3.3.1. Analysis of yield data

Coefficients for regression of dry matter yield on mean temperature, together with information on the deviations from the regression models, are given in *Table 3.4*. A common estimate for all stations, considering temperature alone, is that the dry matter yield changes by 786 kg DM (dry matter)/ha for a 1 °C change in the mean September–June temperature [*Table 3.4(a)*]. If the calculations are restricted to those years and stations when most experiments were cut twice, the coefficient for regression of yield on temperature in the same 10-month period changes to 544 kg DM/ha/°C, and the regression on mean temperature for the 12-month period September–August is 635 kg DM/ha/°C [*Table 3.4(b)*].

The inclusion of the associate variables has had little effect on the estimates of regression on temperature. The adjustments alone explained 47.4% and 17.8% of the original variability within stations in (*a*) and (*b*), respectively, of *Table 3.4*. With temperature, the respective degree of explanation is 64.1% and 42.9%. Temperature has thus explained 31.7% and 30.5% of the variability remaining after adjustments in (*a*) and (*b*), respectively, of *Table 3.4*. This is an estimate of how far temperature would explain variation in yield if the contribution of cutting management to variability could be eliminated. The effect of harvest dates is somewhat different for the various stations so that different slopes are allowed in the combined analysis. *Table 3.5* gives the coefficients of the associate variables. At Sámsstadir, where the cutting treatment has been relatively constant throughout the period, the adjustments had no effect and the same applies to years with two cuts at Reykhólar (cf. *Table 3.4*).

Table 3.5. The coefficients for harvest dates (associate variables, kg DM/ha/day) in the regression of yield on temperature, estimated for a common regression on mean temperature in (*a*) September–June or (*b*) September–August.

Experimental station	Date of 1st cut	Number of days between 1st and 2nd cut	Date of 2nd cut the year before
(*a*) All experimental years			
Akureyri	17	11	−14
Reykhólar	15	18	2
Sámsstadir	23	−7	−16
Skriduklaustur	43	25	−38
(*b*) Years with two cuts			
Akureyri			−17
Reykhólar			−2
Sámsstadir			−18
Skriduklaustur			−34

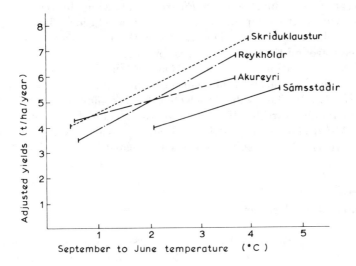

Figure 3.2. Regression lines for the relationship between adjusted yields and September–June temperature at the four experimental stations over the range of temperatures occurring at each station during the experimental period [*Table 3.4(a)*].

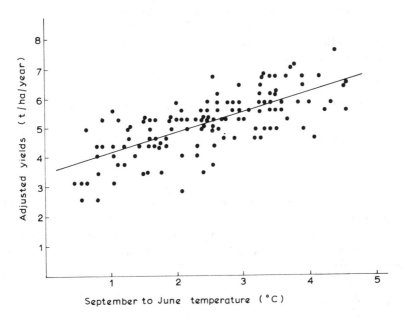

Figure 3.3. The relationship between adjusted yields and temperature, where the dots indicate annual yields at the four stations adjusted towards mean harvest dates at each station and mean yield of all stations. The line of common regression on temperature for all stations is also shown [*Table 3.4(a)*].

The regression is illustrated in *Figure 3.2* showing the regression lines of each station within the temperature range for that station. The differences in slope between stations were not significant and *Figure 3.3* shows the variation of adjusted yields around the common regression line.

For later calculations the value 644 kg DM/ha/°C ± 113 kg DM/ha/°C, which is the adjusted regression of yield on temperature in years with two cuts, is used [*Table 3.4(b)*]. This value is selected because the regression is calculated on the mean temperature for the whole year. It could be argued that results from *Table 3.4(a)* should be used instead, since a wider range of conditions is covered. However, those results have been adjusted more, and hence are less reliable.

3.3.2. Analysis of nitrogen response

The mean yield on plots without nitrogen was 2812 kg DM/ha and with nitrogen 5160 kg DM/ha, giving a mean response to nitrogen of 2348 kg DM/ha. Results of the analysis of the nitrogen experiments in relation to temperature are shown in *Table 3.6*. The results indicate that, in absolute terms, temperature affects yield less where no nitrogen is applied than with nitrogen fertilization. Although the results are rather inconclusive, the value of 171 kg DM/ha/°C for the effect of temperature on the response to 100 kg N/ha is selected for later calculations.

The decline in yield with time, at Akureyri and Skriduklaustur was relatively minor on plots without nitrogen fertilization. The difference between periods is nearly fully expressed in the difference between plots with and without nitrogen fertilization, although it is not necessarily caused by a true reduction in nitrogen response with time.

3.3.3. Models for the effect of temperature on pasture yield

The model for the effect of temperature on yield (kg DM/ha) is

$$Y_i = A + 644\,(T_i - \bar{T}) \tag{3.1}$$

where Y_i is the predicted dry matter yield in year i, A is the mean yield in the reference period (Scenario I), T_i is the mean temperature in year i and $\bar{T}$ is the mean temperature in the reference period. Taking into account the temperature effect on nitrogen response, assuming that it is equally distributed over the whole nitrogen response curve, the model becomes

$$Y_i = A_N + (454 + 171(A_N - A_0)/(A_{100} - A_0))(T_i - \bar{T}) \tag{3.2}$$

where A_N is the yield in the reference period with the nitrogen application of N kg/ha. The coefficient 454 is selected such that the temperature response 644 kg DM/ha/°C is reached at the nitrogen level of 120 kg N/ha for the nitrogen response curve used below (cf. *Table 3.9*).

Table 3.6. Coefficients for regression on temperature of yield with and without nitrogen fertilization, and of nitrogen response. Adjustments in the analysis of experimental yield are as in *Table 3.4*.

		Regression on temperature			
		All experimental years		Years with two cuts	
	Temperature period	kg/ha/°C	Residual $MS \times 10^{-2}$	kg/ha/°C	Residual $MS \times 10^{-2}$
Yield without nitrogen, adjusted	Sept–June	617 ± 99	5593	507	5747
	Sept–Aug			447 ± 81	5506
Yield with nitrogen, adjusted	Sept–June	842 ± 111	6912	627	6328
	Sept–Aug			508 ± 87	6254
Nitrogen response, unadjusted	Sept–June	196 ±60	4178	91	3608
	Sept–Aug			54 ± 64	3625
Nitrogen response, adjusted for periods	Oct–Apr	111 ± 42	3608	88 ± 55	3420
	Sept–June	171 ± 55	3568	125	3419
	Sept–Aug			68 ± 62	3453

Another modification of the model is to assume a multiplicative form of the kind used in Subsection 2.2.4 of this report

$$Y_i = (A_{120} + 644\,(T_i - \bar{T}))A_N/A_{120} \qquad (3.3)$$

Model (3.2) is about midway between (3.1) and (3.3) which can then be regarded as lower and upper limits to the temperature effect on nitrogen response, although in model (3.3) the temperature effect also depends on the yield level at zero nitrogen application.

3.3.4. Validation of the model

Subdivisions of temperature effects over the year

In the analysis of the relationship of grass growth to temperature, some effort was made to differentiate between the effect of different seasons. In general, the seasonal effects were additive. Some results, with the year divided into summer and winter seasons, are given in *Table 3.7*. The regression coefficients are

Table 3.7. Estimation of seasonal contributions to the regression of yield on temperature (kg DM/ha/°C). Winter is September to April and summer May to June or August.

Period	All experimental years			
	Winter	Summer	Total	Residual $MS \times 10^{-2}$
Sept–April	665			4928
Sept–June	656	29	685	4969
Sept–June, without division into seasons			735	5009
Sept–August				
Sept–August, without division into seasons				

Period	Years with two cuts			
	Winter	Summer	Total	Residual $MS \times 10^{-2}$
Sept–April	503			4303
Sept–June	481	41	522	4352
Sept–June, without division into seasons			540	4320
Sept–August	479	103	582	4324
Sept–August, without division into seasons			644	4314

calculated on mean seasonal temperatures. They are thus cumulative and increase with the number of months in the mean.

Winter temperature has a dominating role in the relationship of grass growth to temperature (*Table 3.7*). The coefficients for summer temperature are low and the fit of the regression is not improved by including information on the summer months. A priori, summer temperature might be expected to be of greatest importance. However, the temperature in the summer is much less variable than during other seasons and, moreover, the various irregularities in management of the experiments during the summer and other weather factors such as rainfall will interact with the effect of temperature in the summer. As a result the growing season is of relatively little importance in the overall relationship between grass growth and temperature. This has interesting implications for the possibilities of predicting grass growth in early spring.

A possible explanation for the apparently dominant influence of winter temperature on hay yield is the effect of winter damage of grasses (*see* Subsection 2.2.4). The intensity of winter damage can vary. In extreme cases, extensive areas of a field or even whole fields may be killed. Such winter kill has frequently caused serious failure in hay production in parts of the country. On the other hand, plants may be damaged by low temperatures or injured by frost during winter and early spring without actually being killed. This affects their productivity (e.g., Larcher, 1981) and, hence, will lead to reduced yields in the following growing season. In contrast to winter kill, it is difficult to evaluate this sort of damage directly.

In the experiments analyzed here, observed winter kill is reported to have caused significant damage only at Akureyri (Gudleifsson and Sigvaldason, 1972). The experiments were harvested once in those years. The estimated mean winter kill effect in the 5 years with reported damage, when it is included in the regression model, is 1400 kg DM/ha, adjusted for cutting dates and temperature. The coefficient for the regression on September–June temperature is thus 443 kg DM/ha/°C, compared with 513 kg DM/ha/°C in *Table 3.4(a)*, when the average effect of winter damage is included in the regression model.

These results indicate that observed winter kill has had little effect on the yield-temperature relationship in the long-term experiments and other explanations must be sought.

Common weather variables for all stations

The variation in temperature has been highly correlated between the various parts of the country. Thus, the fit of the regression model is practically unaffected when the temperature at Stykkishólmur is substituted for the local temperatures in the regression analysis (*Table 3.8*). The variability in temperature and yield is, however, not the same in all parts of the country, being generally least in the south, represented by Sámsstadir. The coefficients for regression on temperature become, therefore, different for each experimental station when the same weather station is used for all of them. Previously, Fridriksson (1973b) found that the regression of hay yield on temperature was equal in districts in the northern and southern parts of the country, but that the range in both yield and temperature was greater in the north.

The residual variation from the regression on a common weather variable from Stykkishólmur and with unequal slopes for each station can be divided into the variation between and within years. The residual mean square for the within-years variation is also given in *Table 3.8*. The similarity of these values and those obtained for the regression on the temperature at Stykkishólmur shows that the yield variation common to all stations is well accounted for by the regression on temperature. It indicates that the deviation from the regression is predominantly due to independent variation in local conditions, climatic or other, and experimental and management errors.

Table 3.8. The effect of substituting the temperature at Stykkishólmur for the local temperatures at each experimental station on the residual mean squares in the preferred regression model.

	Residual MS $\times 10^{-2}$	
	All experimental years	*Years with two cuts only*
Local temperature for each experimental station	4916	4438
Common temperature (Stykkishólmur)	5098	4260
Within years	4714	4331

3.3.5. Comparison of the present model with that presented in Section 2

The present model for the effect of temperature on hay yield was obtained by using data from long-term fertilizer experiments on permanent pastures where N application rates were 67–120 kg N/ha. The model reported in Section 2 was, on the other hand, based on measurements of hay yield from farmers' fields all over Iceland since the turn of the century. The regression equation for 120 kg N/ha is 325 kg/ha/°C May–September + 354 kg/ha/°C October–April which is equivalent to 679 kg/ha/°C for the whole year. This value agrees remarkably well with the regression coefficient of 644 kg DM/ha/°C obtained in the present model, although the variables of yield are not the same. The latter coefficient refers to dry matter yield while hay contains at least 15% moisture. This difference is more than balanced out by yield losses on farmers' fields during the haymaking. Considering also the indirect methods involved in estimating hay yield, an exact comparison of the coefficients is not warranted. However, one might anticipate that the regression of yield for the whole country would be steeper than in the present model since winter kill has been much less frequent in the experiments than on farmers' fields in general and the probability of winter kill is related to the coldness of winter (Gudleifsson, 1975).

The effect of temperature on nitrogen response enters more strongly into the model reported in Section 2 than was shown in the present analysis. There are, however, indications in other fertilizer experiments that nitrogen is utilized less by the grass in a cold period on fields where drainage conditions have deteriorated (Björnsson and Óskarsson, 1978). There are several examples of exceptionally poor nitrogen utilization when the fertilizer is applied on wet and cold soil in a late spring like 1970 or 1979 (Björnsson, 1980).

3.4. Mechanisms Influencing the Effect of Temperature on Herbage Yield

In the present analysis a strong relationship between temperature and grass growth has been established and the coefficient of regression is not found to be very sensitive to adjustments for other factors affecting yield. In order to evaluate the possibilities of applying the established relationship to predict the effect of climatic change on farming, it is first necessary to have some understanding of the mechanisms behind this relationship. It is also important to keep in mind that we are finally producing animal products such as meat and milk rather than grass, so that other aspects, such as quality of the feed, must be considered.

3.4.1. Growth rate and the length of the growing season

Major components of the dry matter production on a field over the whole summer are growth rate and the length of the growing season. Growth rate is related to nutrient supply, temperature (Cooper, 1964) and the stage of development of the plant (Parsons and Robson, 1980), but factors such as water stress often counteract the positive effects of optimum temperature. Results of a

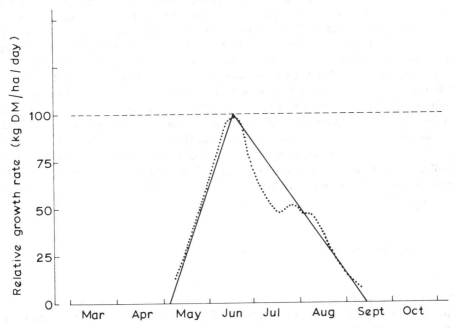

Figure 3.4. The relative growth rate (kg/ha/day) of *Phleum pratense*, Kämpe II. The dotted line indicates a sward grown in Iceland and harvested in 1982 (Corrall, 1984); the solid line indicates a simplified pattern of the relative growth rate over the season.

project organized by the United Nations Food and Agriculture Organization (FAO), and designed to improve knowledge of the potential forage production from cultivated grassland in Europe, show that the growth rate reaches a maximum early in the season and then falls off gradually, although often with a characteristic second peak later in the season (Corrall, 1984). In the majority of cases the maximum growth rate was close to 100 kg DM/ha/day.

The experimental conditions in the FAO project (Corrall, 1984) differ in many ways from those in the Icelandic long-term experiments, especially in more frequent cuts and in the nonlimiting nitrogen supply. Further, the results for the Icelandic site in the FAO series are not typical for the site because the timothy (*Phleum pratense*) variety used, Kämpe II, is poorly adapted to Icelandic conditions. A schematic simplification of the growth rate curve is to assume a linear increase from zero at the beginning of the season to a maximum of 100 and then a linear decrease to zero at the end of the season. This gives an average growth rate of 50 kg DM/ha/day over the growing season (*Figure 3.4*).

The length of the growing season is usually defined as the period with temperatures exceeding a certain base temperature and is, hence, a function of temperature. The growing season has not been mapped for Iceland, but estimates using a base temperature of 4 °C, indicate that the beginning of the growing season may be delayed by 6–13 days for a 1 °C change in temperature, and the effect on the length of the growing season may be twice as much (*see* Section 5). In the east of Scotland near Penicuik, the length of the growing season has been calculated at four different altitudes ranging from 59 to 440 m above sea level, using a base temperature of 5 °C (G. Swift, personal communication). The results correspond to a linear shortening of 16 days/100 m, or 27 days/°C if a temperature lapse rate of 0.6 °C/100 m is assumed and both ends of the growing season are equally affected. With an average growth rate of 50 kg DM/ha/day as suggested above, the effect of altitude or temperature on the length of the growing season would have a yield effect of 1350 kg DM/ha/°C, which is about twice that obtained in the Icelandic experimental yield data. The difference between the two results is not surprising, bearing in mind the schematic nature of *Figure 3.4* and the difference in nutrient supply as well as in the two series of experimental results. In spite of the qualitative nature of the comparison, it serves to illustrate that one of the main reasons for the relationship of yield to temperature may be that it is closely correlated with the length of the growing season.

3.4.2. The importance of nitrogen

In the long-term fertilizer experiments reported here, all nitrogen has been applied in the spring and the first harvest is normally taken late in the period of rapid growth. During this period the fertilizer nitrogen is efficiently taken up by the plant roots. Delayed harvest in a particular year, relative to the normal stage at harvest, will give higher dry matter yields and reduced crude protein content of the herbage. This would, in the above analysis, be adjusted for by the regression on the date of first harvest. As indicated by the lack of response to nitrogen application in the aftermath, shown in *Figure 3.5*, the growth of the aftermath appears to be dependent on natural release of nitrogen in the soil. This release is usually referred to as mineralization, which can be quite substantial in drained peat soils, although rhizospheric nitrogen fixation may often be more important in the long run (Jenkinson, 1976; Witty *et al.*, 1976).

The effect of temperature on nitrogen response is relatively small as the response to temperature is only slightly smaller on plots without nitrogen fertilization than on plots with nitrogen (*Table 3.6*). It is concluded that the response to temperature in the Icelandic experiments has been primarily through nitrogen release in the soil, which depends both on soil conditions and the length of the growing season. Differences in yield between locations are also often easily explained by differences in nitrogen availability in the soil. However, it is not known whether the nitrogen release would continue to increase if the climate were to become warmer. In either case the longer growing season would enable increased nitrogen application, and thus make increased grass growth possible.

3.4.3. Moisture stress

Pasture grasses are known to respond well to soil moisture (Wilson, 1981). In Great Britain, water supply explains to a great extent the variation in herbage yield between sites (Morrison, 1980). In Iceland, water supply can also limit

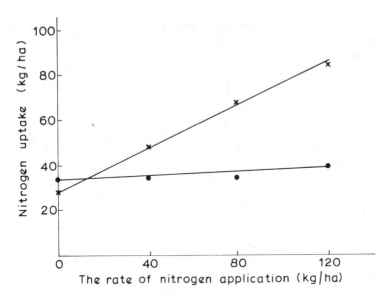

Figure 3.5. Nitrogen uptake for increasing rates of nitrogen application in the first cut (crosses) and the second cut (dots) of an experiment at Skriduklaustur (mean of 14 years).

grass growth. However, the water balance of Icelandic soils has not been studied, and, because of the large temperature effects, it is only expected to explain a small portion of the experimental yield variation. To test this, both temperature and precipitation were included in a multiple regression analysis at Akureyri, where moisture stress is most likely to occur (*see Table 3.1*). No relationship was found between precipitation and yield during the growing season. However, yield was negatively related to precipitation in May (-32 ± 8 kg DM/ha/mm rainfall), indicating a negative effect on yield of rain falling while the ground is still partly frozen and the soil surface is wet. The climatic changes considered in the scenarios of this case study are primarily temperature changes, while precipitation changes little. A warmer climate would lead to moisture stress in parts of the country, thus reducing the otherwise beneficial effect of this climatic change on grass growth on well-drained fields (*see* Subsection 3.6). Cooling of the climate would, on the other hand, in many cases make drained fields wetter and this might reduce the yield response to nitrogen. This occurred on the peat soils at the Hvanneyri Agricultural College in the cold period 1966–1970 (Björnsson and Óskarsson, 1978).

3.4.4. The quality and utilization of the herbage

As previously mentioned, climatic change not only affects animal production through the quantity of fodder produced but also through the quality of the fodder. In addition, the fodder will be used differently. In a warmer climate grazing will replace indoor feeding, while in a colder climate the reverse will occur. The quality of pastures is high in regions with moderate or cool summers, making them favorable for animal husbandry. A review of the literature indicates that there is a reduction in digestibility of 0.4–0.8% of the dry matter for a 1 °C increase in temperature (Wilson, 1982). Deinum et al. (1981) found in a study of timothy (*Phleum pratense*) at six latitudes ranging from 51° N to 69° N that at an equal stage of maturity the digestibility of the grass was higher at higher latitudes but that the rate of decline of digestibility during the growing season was also higher there. In Iceland, where temperatures are relatively low, the digestibility declined by 0.30% per day (Björnsson and Hermannsson, 1983), which is lower than any results published or cited by Deinum et al. (1981) except for some results from Hurley in England at 51° N. For a substantial increase in temperature the quality of grass, both cut for storage as fodder and for grazing, and both on cultivated and natural pastures, will probably be reduced. This would increase the need for concentrate use. Appropriate changes in harvesting procedures might be sufficient to maintain the quality, and moderately increased moisture stress would also, where it occurs, counteract this effect to some extent (Wilson, 1982).

A change in temperature has a large influence on the grazing period of the animals. In earlier times the effects of temperature on grass growth and the grazing period were compounded because grazing was mostly restricted to uncultivated fields (*see* Section 2). However, in modern dairy farming this is not the case since the grazing is limited to cultivated fields. The net utilization of the potential grass growth for fodder, either by grazing or conservation, is not greatly different, though normally the quantity utilized by grazing is slightly less, but the quality is higher. The dairy production per unit area would, therefore, be roughly proportional to the grass growth if herbage quality could be maintained as the climate became warmer.

3.5. Implications of Temperature Changes for Icelandic Agriculture

In the following subsections we shall seek to describe some of the adaptations required of Icelandic agriculture in response to the climatic scenarios adopted in this case study.

3.5.1. Climatic scenarios

Five climatic scenarios are investigated in this study. Four represent standard scenarios used throughout this report (with one minor modification), while the fifth is unique to this section.

The standard scenarios (I, II, IV+ and V)

These are all based on the representative station at Stykkishólmur (*see* Section 1):

(1) Scenario I – *baseline scenario* (1951–1980) – mean annual temperature 3.7 °C.
(2) Scenario II – *cool decade scenario* (1859–1868) – temperature anomaly (relative to Scenario I) −1.3 °C.
(3) Scenario IV+ – *warm ten years scenario* – temperature anomaly (relative to Scenario I) +1.3 °C. This exceeds the anomaly recorded for the ensemble of the ten warmest years (1931–1984), Scenario IV in Section 1, by 0.2 °C. However, by the symmetry between Scenarios II and IV+, the contrast between the cool and warm scenarios is made sharper.
(4) Scenario V – *2 × CO_2 scenario* – temperature anomaly (relative to Scenario I) +4.0 °C. This is far outside the temperature range encountered in Iceland in historical times. Thus an extrapolation of existing relationships into these vastly altered conditions is, at best, hazardous and will not be ventured here. However, different approaches to agriculture will be speculated on, and the potential for cultivation of various arable crops will be given some consideration in Subsection 3.6.

The scenarios are thought of as 10-year averages. Built into them is the year-to-year variation. When the deviation from the average is at maximum, especially if it is superimposed on a climatic trend, it may cause extreme conditions for agriculture and have serious consequences. It is, therefore, of interest to define extreme years and discuss their significance.

The extreme year scenario (VI)

Extreme years can be recognized as arising from two sources of variation, i.e., the year-to-year temperature variation and the prediction error of the yield–temperature relationship. In the following account these components are evaluated. Twice the standard deviation towards lower yield of the two sources of variation combined is then used as an extreme year scenario.

The standard deviation of mean annual temperature at Stykkishólmur for the past 138 years is 0.84 °C. Multiplied by the coefficient for regression of yield on temperature, 644 kg DM/ha/°C [equation (3.1)], it is an estimate of the standard deviation of yield due to temperature. It is an overestimate of the short-term variation as climatic trends enter it. An extreme year, relative to the years that have preceded it, may occur as a combination of the climatic variation around its temporary mean and a trend towards colder climate.

Since the scenarios have been defined as 10-year averages, the extreme years in the recorded history at Stykkishólmur were identified by calculating the deviations of annual mean temperatures from the preceding 10-year averages. The four coldest years deviated by 2.22, 1.62, 1.68 and 1.44 °C respectively, but others by less. The largest positive deviation, on the other hand, is 1.27 °C. The negative deviation of 1.68 °C, twice the standard deviation, can therefore be used as an example of an event that really could happen, at least in a cold period.

From the results reported in Section 2 (*see Table 2.2*) relating to the deviation of actual yield from yield as predicted by temperature, the value 240 kg DM/ha has been adopted as an estimate of the standard deviation of the prediction error. When the two components of variation are added and doubled (for the extreme year scenario) yield will then be 1184 kg DM/ha below the temporary normal, calculated as

$$2\sqrt{[(644)(0.84)]^2 + (240)^2} = 1184$$

Two standard deviation limits on the regression coefficient (418–870 kg DM/ha/°C, see Table 3.4) give the interval 851–1538 kg DM/ha for the yield decline in the extreme year scenario (using the above expression).

3.5.2. Dairy farming in Iceland

An important role of Icelandic agriculture is to supply the population with dairy products. Milk consumption on the home market reached the level of 100 million liters per year around 1960 and is still at this level, although the population has increased by 38% in the past 25 years. The aim of the current agricultural policy is to stabilize the production at this level. In recent years it has meant reduction in the total dairy production in order to reduce surplus production that otherwise is exported at prices below cost. This has led to a slight reduction in the production per unit area of cultivated land.

Although dairy farming is perhaps less climate sensitive than other farming sectors, its well defined role makes it well suited for evaluating climatic scenarios. It was argued in Subsection 3.4.4 that a linear relationship between dairy production and grass growth is a fair assumption if the input of other production factors remains the same. However, if we wish to maintain the production at a nearly constant level, some changes in production input are required to counterbalance a changing potential for grass growth due to climatic change. The short-term response to a colder climate would be to increase the import of concentrates and increase fertilizer use, but, in the long run, the response would be to increase the cultivated area. Warmer climate would have the reverse effect.

In order to evaluate the possible effects of the scenarios on the parameters of milk production, some information on dairy farming is required. The assumptions used by Sigurdsson et al. (1980) in modeling the production on a dairy farm are useful for deriving parameters that serve this purpose. Additional information on, for example, concentrate use, was obtained from the farm extension service (E. Jóhannsson, personal communication). The parameter values derived are shown in Table 3.9. Although partly based on experimental results, they come closer to representing the actual farming practices than to showing the optimal farming practices based on present scientific knowledge. These values could, therefore, easily change in a short period. Such changes, along with intensified research, are actually among the potential methods of adaptation to new conditions.

The calculated effect of the cold and warm scenario, using the models for the yield–temperature–nitrogen relationship put forward in Subsection 3.3.3, is

Table 3.9. Parameter values for dairy farming in Iceland under Scenario I (present climate) (based on Sigurdsson et al., 1980).

Target milk production	100 million l/yr
Nitrogen response curve	$78N^{1/2} + 22N - 0.056N^2$ kg DM/ha[a]
Yield with 0 kg N/ha	2572 kg/ha
Yield with 80 kg N/ha	4671 kg/ha
Yield with 120 kg N/ha	5260 kg/ha
Harvestable yield	78.75% of the dry matter yield
Average forage digestibility	68.7% of the dry matter
Energy value of the forage	0.598 feed units/kg dry matter[b]
Feed requirements, including the raising of replacement heifers	0.927 feed units/l milk[b]
Concentrate use	23.5% of the energy requirements

[a] N = nitrogen, kg/ha; DM = dry matter.
[b] One feed unit for fattening = 1650 NKF (kCal net energy).

Table 3.10. Required input of production factors for maintaining milk production in cold and warm scenarios with yield changing as in model (3.1) (Subsection 3.3.3) using the assumption of *Table 3.9* with 120 kg N/ha. Limits are given for a ± 2 standard deviation interval on the coefficient of regression of yield on temperature.

Scenario:	I	II		IV+	
Character:	Baseline	Cold		Warm	
		(1859 to 1868 -type)		Modified average of 10 warmest years (1931–1984)	
	Stykkishólmur (1951–1980) Mean	Mean	Limits	Mean	Limits
Mean annual temperature (°C)	3.7	2.4		5.0	
Dry matter yield (kg/ha)	5260	4423	4129–4717	6097	6391–5803
Pasture area (ha) without changing fertilizer and concentrate use	28 600	34 000	36 400–31 900	24 700	23 500–26 400
Relative concentrate needs without changing pasture area or fertilizer use (%)	100	152	170–133	48	30–67
Relative effect of changing fertilizer use by 30 kg N/ha (% of predicted yield change due to altered temperature)		37	27–57	50	37–78

Table 3.11. Required input of production factors for maintaining milk production in cold and warm scenarios, at two nitrogen levels, and for models (3.1), (3.2) and (3.3) (Subsection 3.3.3). Results are expressed as in *Table 3.10*

Scenario: Character:	Model	I Baseline Stykkishólmur (1951–1980)	II Cold (1859 to 1868 -type)	IV+ Warm Modified average of 10 warmest years (1931–1984)
Annual mean temperature (°C)		3.7	2.4	5.0
(a) Fertilizer level: 120 kg N/ha				
Dry matter yield (kg/ha)	1,2,3	5260	4423	6097
Pasture area (ha)	1,2,3	28 600	34 000	24 700
Relative concentrate needs(%)	1,2,3	100	152	48
Relative effect of changing	1		37	50
fertilizer use by 30 kgN/ha	2		34	55
(% of predicted yield change	3		31	58
due to altered temperature)[a]				
(b) Fertilizer level: 80 kg N/ha				
	1	4671	3834	5508
Dry matter yield(kg/ha)	2	4671	3888	5454
	3	4671	3928	5414
	1	32 200	39 200	27 300
Pasture area (ha)	2	32 200	38 700	27 600
	3	32 200	38 300	27 800
	1	100	159	42
Relative concentrate needs(%)	2	100	154	46
	3	100	152	48
Relative effect of changing				
fertilizer use by 26.6 kg N/ha	1		50	61
of predicted yield change	2		48	72
due to altered temperature)[a]	3		47	80

[a] The total change in nitrogen use, taking area into account, is the same in parts (a) and (b) of the table.

given in *Table 3.10* [model (3.1) only] and *Table 3.11*. The results are expressed as dry matter yield, area needed for the dairy production, changing concentrate use and changing fertilizer use. *Table 3.10* shows limits to the results corresponding to two standard deviation limits on the regression coefficient in model (3.1), i.e., 418 and 870 kg DM/ha/°C, and *Table 3.11* shows the difference between results by the three competing models at two nitrogen levels. The results are found to be more sensitive to errors in the yield–temperature relationship than to the choice of models. This sensitivity and the sensitivity to the actual yield level in the reference period are also of importance for the impact of an extreme year (*see* Subsection 3.5.5 and *Figure 3.6*). The discussions in the next two subsections, though, are restricted to the mean results.

3.5.3. Effects under Scenario II (cold, 1859–1868-type climate)

The cooling of the climate to a mean annual temperature of 2.4 °C results in a 16% reduction in the available forage feed (*Table 3.10*). As the present concentrate use is low, this reduction can easily be met by import of more concentrates. However, this would increase the production cost considerably. According to results from farm accounts for 59 dairy farms in 1983 (Ólafsson, 1985), concentrate purchase, adjusted to the level of use assumed in *Table 3.9*, would amount to 37% of variable costs, so that a 52% increase in use would increase variable costs by nearly one-fifth.

Increasing the fertilizer rates would, on the other hand, only partially give the forage required if the normal use is around 120 kg N/ha as is presently the practice. Alternatively, it is possible to reduce the fertilizer use per area, but increase the area for forage production, as suggested in Section 2. This is not unrealistic in Icelandic agriculture today, since the production capacity for milk and meat considerably exceeds the market needs. Rather than reduce the cultivated area in order to adjust to the market, the fertilizer use could be reduced. For this reason the effect of changing fertilizer use from the base level of 80 kg N/ha is also considered (*Table 3.11*). As expected, the relative decrease in yield for lower temperatures is greater at this level and it is slightly more costly to meet the reduction by increased concentrate use, except for the multiplicative model (3.3) when the results are equal. However, increased fertilizer use becomes more efficient as a tool for adjustment to the adverse climatic conditions since the response to nitrogen is higher with these low fertilizer rates.

Since the cultivated area is larger, the same total increase in fertilizer use for the whole country is obtained by raising the fertilizer rate from 80 to 106.6 kg N/ha instead of from 120 to 150 kg N/ha as in the earlier example. Using model (3.1), this increase would give 50% of the forage at the lower nitrogen level that is required to compensate for the temperature-induced yield reduction. This compares with 37% at the higher nitrogen level. In addition, the model indicates that all the forage needed could be produced by increasing the fertilizer level from 80 to 143 kg N/ha. According to the other two yield models, increased fertilization is slightly less effective in responding to the decreased yield/ha.

One could anticipate that the first adjustments to a colder climate would be to purchase more concentrates and to increase the fertilizer use. The next response would be to increase the cultivated area, first with annual feed crops that would immediately increase the feed production, and later with perennial grasses. The need for new cultivation of land is greater than that corresponding to the net increase in area requirements since, for several reasons, appreciable areas would be taken out of cultivation. The winter kill of grasses would be aggravated on some fields and the yield would simply become too low on others for continued use as pastures. The cooling of the climate would be more serious in some parts of the country than in others, and it is likely that the effect of lower temperatures would be stronger on some fields than that indicated by the regression obtained from experimental fields. Several farms or even whole communities might well be abandoned.

The extent of the area that would be taken out of cultivation could be derived by two approaches. One is to try to identify the sensitive areas. Around 60% of the new cultivation during the past decades has been on peat soils. Assuming that one-fourth of those would be lost due to adverse effects of the cold climate on the drainage conditions, this would mean 15% of the total area. Another approach is to find the distribution of hay yields in the country and expect fields with yields of, say, 2000–2500 kg DM/ha to go out of cultivation. It must be kept in mind, however, that some parts of the country will be harder hit than others and that in extreme years the yield on such fields would perhaps be around 1000 kg DM/ha or less. Although no estimates are available, it is proposed that, in addition to the 5400 ha of new cultivation required as indicated in *Table 3.10*, as much again would be needed in order to replace abandoned fields. How long this takes would depend on public support given to farmers.

3.5.4. Effects under Scenario IV+ (warm climate – modified mean of 10 warmest years, 1931–1984)

A warmer climate would lead to an entirely different situation. The results given in *Table 3.10* and *Table 3.11* do not realistically show the extent to which warming can be met by reducing the concentrate and fertilizer use. Concentrate use in Icelandic agriculture is already very restricted, reflecting the present agricultural policy in favor of locally produced feed supplies. Reducing the fertilizer use might also create difficulties in maintaining forage quality during the longer growing season. The incentive would rather be to increase the fertilizer use. A considerable area would become free for other use, especially high energy fodder crops such as barley, thus eliminating the need for import of feed grain. The potential for barley cultivation, as far as it depends on temperature, is perhaps best evaluated in terms of the risk involved, i.e., the certainty of barley ripening (*see* Section 2) or, alternatively, the probability of crop failure (Parry and Carter, 1985). Other weather factors are also of importance for the cultivation of grain, such as the frequency of strong winds which can cause great losses during harvest. Soil conditions may also restrict the grain-growing area, although the characteristic acidity of peat soils, the most likely soil constraint, is generally relatively weak in Iceland.

3.5.5. Effects under Scenario VI (extreme years)

Icelandic agriculture often experiences extreme years. In earlier times these were met by restricting the livestock number so that in normal years appreciable hay reserves could be kept in storage for the following years. This was not always possible, however, during "runs" of cold years. More recently, feed production has not been as limiting as in earlier times and new methods of meeting the difficulties have become available as discussed above. Still, extreme years will in the future be of great concern, especially if they mark the beginning of a cold period. Then renewal of the hay reserves that are used up in the extreme years is not possible.

Figure 3.6 shows the effect of an extreme year, defined in Subsection 3.5.1, as a function of mean yield. The central curve depicts the predicted yield reduction (in percent) under an extreme cool year scenario (as described in Subsection 3.5.1), relative to the mean yield estimated for the temperature conditions of the period immediately preceding the extreme year (horizontal scale). The outer dashed curves indicate the uncertainty (± 2 standard deviations) of the estimates, both of the period mean yields (*see* "Limits" columns in *Table 3.10*), and of the yield reductions under the extreme year scenario (vertical scale – *see* Subsection 3.5.1). Thus, for example, if the extreme year occurs during conditions similar to the baseline climate (Scenario I), with a mean yield of 5260 kg DM/ha, the yield reduction (Subsection 3.5.1) would be 1184 kg DM/ha (22.5% – Point A on solid curve), with ± 2 standard deviation limits giving 1538 kg DM/ha (29.2%) and 851 kg DM/ha (16.2%) reductions, respectively, points B and C on the dashed curves.

In contrast, a cool decade, such as under Scenario II, would give a mean yield of 4423 kg DM/ha [based on model (3.1) – *see Table 3.10*] with ± 2 standard deviation limits of 4717 and 4129 kg DM/ha, respectively. The extreme year scenario again gives a reduced yield of 1184 kg DM/ha (26.8% relative to the 4423 kg DM/ha mean yield – Point D), with the uncertainty represented by, for example, the -2 standard deviation limit, giving a 37.2% reduction in yield relative to the lower yield limit, 4129 kg DM/ha (Point E, *Figure 3.6*). Similar procedures can be used to calculate the range of yield reductions for other "eras" of cooler-than- or warmer-than-baseline climatic conditions.

Figure 3.6 indicates that in addition to being more likely in cold periods, the relative effect of an extreme cold year on the total feed production is greater in the cold period, when the feed shortage would be 27% as compared with 19% in the warm period. This result is very sensitive to deviations of the estimated value of the regression coefficient from its true value. On farms that are seriously hit by this extreme year, it would, in many cases, be the final blow to the farming effort.

The awareness of the extreme is at any time very much dependent on recent history. In the relatively warm and climatically stable period from around 1920 to 1965, the largest negative deviation compared to the previous 10-year average was 0.93 °C. At the same time revolutionary changes in hay production and other farming practices took place, making farming less sensitive to weather. Since 1965, however, five years have deviated this much or more. The most extreme deviation was 1.22 °C in 1979. This is the fifth in order of magnitude in recorded history, although the mean temperature was 2.3 °C, almost as high as the cold decade scenario (II). During these recent cold years, some adaptation has already taken place such as contraction of the farming area in parts of the country, and strong emphasis has been placed on the hardiness of pasture species and varieties in agricultural research and farm extension work.

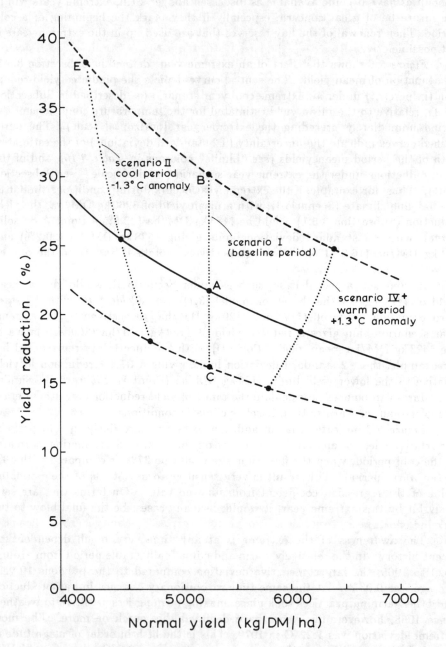

Figure 3.6. Relative yield reduction in a year with grass growth two standard deviations below the normal yield, as a function of period-averaged yield level (extreme year scenario). The central curve is for model (3.1) and the outer curves are for two standard deviation limits on the regression coefficient. The calculated yield levels under Scenario I (baseline period), II (cool decade) and IV+ (warm, 10 years) are indicated by dots and connected. For explanation, *see* text.

3.6. Effects of the 2 × CO_2 Temperature Scenario (Scenario V)

Estimates derived from the Goddard Institute for Space Studies (GISS) general circulation model suggest that a doubling of carbon dioxide concentrations in the atmosphere could cause the mean annual temperature in Iceland to rise by about 4 °C relative to the 1951–1980 baseline climate (*see* Subsection 1.10). That would bring mean temperatures in Iceland to approximately the same level as they are in present-day Scotland (cf. *Table 1.6*). This temperature rise is far outside the range ever encountered in Iceland from the beginning of temperature records, or indeed elsewhere at similar latitudes.

Rather than extrapolate the historical yield-temperature relationship, it is found more enlightening to observe what kind of agriculture can be practiced in regions which at present have climatic conditions similar to those predicted in Scenario V, and to examine the effect in these regions of temperature changes due to elevation. Reference will, therefore, be made to Scottish agriculture, where the maritime nature of the climate resembles that of the predicted climate in Iceland, with similar distribution of rainfall and temperature. Soil conditions, photoperiod and seasonal intensity of solar radiation, though, are factors that will remain quite different.

Icelandic agriculture today is primarily dependent on permanent grasslands for hay production and grazing. This would still be so in a warmer climate as can be seen from the fact that in Scotland about 72% of agricultural land is rough grazing, 18% is enclosed grassland and only 10% is under arable crops (Frame, 1981). However, the productivity and reliability of the grassland could be expected to increase and a substantial lengthening of the potential grazing period would reduce the cost of production. In Section 2 of this report it was predicted that hay yields under Scenario V would increase by 64%, compared with the baseline. From the yield level assumed in this section (*Table 3.10*), this increase would amount to approximately 9 t/ha/y, while the models of Subsection 3.3.3 give 7.7 t/ha/y with the application of 120 kg N/ha. As these results indicate, the extrapolations become sensitive to the choice of models. If, as discussed in Subsection 3.4.2, the historical yield–temperature relationship is primarily due to nitrogen release, through soil temperature and length of the active growing season, yield of grass obtained on fields receiving no nitrogen fertilization could increase to levels that at present require considerable nitrogen use in Scotland. The longer growing season provides for further potential yield increase by increased nitrogen application, distributed over the season, provided that good varieties of high-yielding species such as ryegrass (*Lolium perenne*) can be cultivated. For example, in Penicuik (180 m above sea level), where the mean annual temperature is 7.7 °C, experimental yields are around 14 t/ha/y, obtained in 4 cuts with 400 kg N/ha (G. Swift, personal communication).

In a warmer climate, cultivars of high yielding species would be sought that can utilize the increased growing season. Cultivars that are at present used in warmer areas display either weak or no winter dormancy. It has been shown, for example for ryegrass at various altitudes in Mid-Wales, that growth during the winter is largely controlled by temperature (Thomas and Norris, 1977). Winter growth of Icelandic grasses, on the other hand, is largely controlled by photoperiod, as can be seen from the fact that they become dormant during winter, even when grown in a considerably warmer climate in the south of England (Helgadóttir, 1981). Cultivars adapted to fairly long photoperiods and active growth during the winter might not suit the Icelandic winter even under 2 × CO_2 conditions, for photoperiods are short and light intensity is very low. For example, in December and January the total incoming radiation is approximately equal to one day's radiation in the summer (Einarsson, 1969). The potential for photosynthesis essentially drops to zero for a period of two to three months and, under these conditions nondormant plants could exhaust their nutrient reserves. New varieties, which make use of a long growing season but still are adapted to lack of radiation in winter, would therefore need to be developed for Iceland and other northern regions.

Under Scenario V, present-day marginal areas could be brought into agricultural use. They would be particularly valuable as grazing pastures for a large part of the year. At the same time, the more favorable areas could be taken under other crops such as barley, oats and winter cereals, as well as a range of root crops. In Section 2 it was demonstrated that barley would be expected to ripen in 90% or more of all years at all weather stations in Iceland. In Scotland, arable crops cover about 10% of the agricultural land of which spring barley is the major one. The limited acreage is partly explained by competition from other growing areas. It is, however, dangerous to extrapolate directly to the Icelandic conditions, as both soil and other factors differ between Iceland and Scotland. For example, high wind velocities during harvest time could cause great losses.

In warmer conditions, attack by diseases, at present of little significance, could reduce yield considerably as can be seen from estimates of average cereal losses from disease in Scotland of 10–15% (S.J. Holmes, personal communication). Diseases would also become more important in those crops that are grown now in Iceland. If grasslands were to become dominated by ryegrass (*Lolium perenne*), virus diseases, such as ryegrass mosaic virus, might both reduce the potential yield and have adverse effects on the quality of the crop. Other crops such as potatoes provide better examples of the increased disease problems in a warmer climate. In Scotland, blight (*Phytophthora infestans*) still causes serious losses in wet years, in spite of blight risk warning schemes and effective fungicides (S.J. Holmes, personal communication), but in Iceland no blight epidemics have occurred since 1953.

In the preceding discussion, the effects of an increase in temperature alone have been considered. The otherwise positive effects of the warming of the climate could, however, be seriously restricted by water stress in some parts of the country. An increase in mean annual precipitation of 15% is derived for the 2 × CO_2 scenario (V), with increases ranging from 20 to 25% in the months March–August (*see* Subsection 1.10). Some calculations using Penman's formula, with reference to the work of Einarsson (1972), indicate that the relative increase in potential evapotranspiration could be similar to this in magnitude for the proposed increase in temperature during the growing season. These predictions do not take into account any variations in local conditions. In large parts of the country the potential water balance is near zero or negative during the growing season, especially in the north (Einarsson, 1972; *see also* precipitation for Akureyri in *Table 3.1*). With equal relative increases in both precipitation and potential evapotranspiration, the water deficit would increase in dry areas. The soil would become dry earlier in the season. However, peat soils, where at present there are drainage problems, could become more easily exploited.

3.7. Conclusions

In this section the effect of climatic change on grass growth and its consequences for agriculture have been discussed. Based on data from long-term fertilizer experiments it was found that a change of 1 °C in mean temperature would change the dry matter yield by approximately 644 kg/ha/y. The uncertainty attached to this value, resulting from lack of precision and the nature of the data, is discussed. The effect of the level of nitrogen fertilization on the relationship is also considered, though not determined quantitatively, and it was found that with less nitrogen the absolute temperature effect would be reduced, although the relative change in yield would be greater. However, it is clear that within the wide range of historical values available, the nitrogen effect and the uncertainties involved have little effect on the conclusions drawn.

The yield–temperature model was used to estimate the effect of climatic change on dairy production in Iceland. The main results are that as the climate becomes colder, the cost of supplying milk products for the home market rises and *vice versa*. Similarly, in the next section, an effect of temperature on the carrying capacity of rangelands is also established, expressed as carcass weight of lambs. The effect on carrying capacity is believed to be primarily due to the effect of temperature on grass growth, although confounded by other factors such as the quality of the herbage (also discussed in Subsection 3.4.4).

The model used in the present analysis is based on temperature alone and does not take into account the physiological processes that control grass growth. For such large-scale predictions, as have been attempted here, this may not be a serious limitation. Many factors, that may be important in detailed studies, are either closely related to temperature or are not affected by climatic change and can thus in this connection be viewed as a random disturbance. For example, factors that are known to be important for the seasonal distribution of grass growth may be less important for the total production. However, when temperature changes cause changes in the water regime, leading to (for example) poor soil drainage in a colder climate or to drought in a warmer climate, the temperature–yield relationship no longer holds. Of much greater significance are the structural changes of agriculture that would follow major climatic changes. In a colder climate the use of land would become restricted and winter kill could cause a yield decline exceeding that predicted by the temperature–yield relationship. In a warmer climate the usable land area would increase and a variety of new options for cultivation would open up. In this warmer climate, we may conclude that, in spite of various limitations, the agricultural areas would be more productive and the production could be more profitable than today, though individual farmers may be taking the same degree of risk, only with different crops.

The foregoing attempts to calculate the effect of climatic change on agriculture are to be viewed as indications rather than predictions, because so many other factors will be changing as well. The only thing that we can be sure of is that climate has changed in the past and is likely to change in the future, but we do not know when, or to what extent, or even in which direction. Although a global warming due to increased CO_2 in the atmosphere is likely to occur, it could well be preceded by cooler periods. The challenge is to prepare for these unpredictable changes. Society as a whole can prepare by having some production capacity in reserve for a cold period, and science can prepare by emphasizing research in regions close to the present boundaries of agriculture, for these are the areas in which adjustments may be required first. In the case of boundaries marking the limits of tolerance to low temperatures, the change to a cooler climate might induce shifts of the present-day boundaries into currently important agricultural areas, demanding research into, for example, the cultivation of existing marginal crops, and breeding programs for both grasses and crops in these regions. Similarly, research efforts would be necessary to enable the most suitable and efficient adjustments to be made in exploiting the transition to a warmer climate.

These points are developed further in the concluding section of this report. In the next section, the impact of climatic variations on carrying capacity of rangeland pastures and on lamb carcass weight is examined.

SECTION 4

The Effects on the Carrying Capacity of Rangeland Pastures

4.1. Introduction

In this section, we investigate the relationship between carcass weight of lambs and climate, in an attempt to estimate the variation in the carrying capacity of Icelandic rangelands. The necessary components of a rangeland model are discussed, but lack of suitable data at present prevents the construction of a model. Thus, it has not been possible to predict the effects of the climatic scenarios outlined in Section 1 of this report, but certain general assumptions are made regarding rangeland carrying capacity by considering extreme years and periods.

4.1.1. Rangeland agriculture and climate

The Icelandic rangelands are characterized by native hardy grasses, mosses and dwarf shrub heaths, types of vegetation resulting from the northerly location of the country and its mountainous terrain. In fact, more than half of the total area of Iceland ($103\,000\,\text{km}^2$) is 400 m or more above sea level (Tölfraedihandbók, 1967). The rangelands can be defined as natural pastures, some of which belong to individual farms in coastal areas and adjacent valleys, while the most extensive ranges are the common grazings of the interior. Thus the rangelands stretch from lowland areas up to the most mountainous and rugged parts of the country. Vegetation growth and consequently the carrying capacity of these pastures is strongly influenced by the climate, particularly temperature, and to a lesser extent by precipitation. Furthermore, the adverse effects on plant growth of any decline in temperature are amplified by high altitude. Such relationships have been demonstrated in a recent study where the growth of sown grasses was compared at two contrasting altitudes, 50 and 640 m a.s.l. (*Figure 4.1*). However,

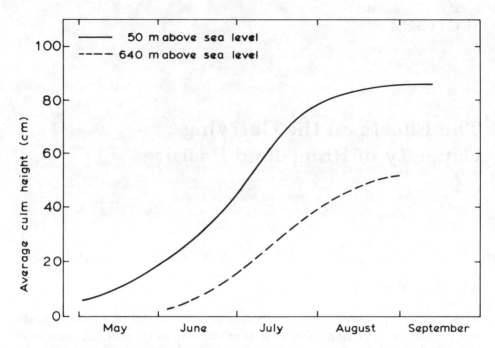

Figure 4.1. Average culm height (cm) of *Deschampsia caespitosa* at two altitudes, 50 meters and 640 meters above sea level. Based on 13 years of data (1969–1981). (From Fridriksson and Sigurdsson, 1983.)

information on variability in the growth of the indigenous range vegetation over a span of several years in relation to climate and altitude is limited. Moreover, many of the ranges, particularly in the highlands, produce very low yields of usable vegetation which cannot be measured accurately (Pálmason, 1982).

There is also a dearth of information on the effects of climate on animal production. In a recent study, however, a positive relationship was established between spring (May and June) temperature and the growth rate of lambs from birth in May to 6–7 weeks of age, presumably the result of differences in grass growth. On the other hand, lamb growth rate was strongly negatively correlated with June rainfall, probably due to the discomfort of exposure to wetness, aggravated by cold (Thorsteinsson *et al.*, 1982). Other evidence also indicates that lamb weight is low in extremely wet years and high in dry years (Ólafsson, 1973). Winter housing of sheep is a general practice in Iceland, and warmer autumns tend to delay the onset of the housing period. Furthermore, winter fodder requirements tend to increase with decreasing mean temperature for the period February–May, reflecting a later onset of the growing season and the need for more extended housing and indoor feeding into the spring (Gudmundsson, 1984). Such findings are clearly of economic importance and are in agreement with general farming experience.

4.2. Rangeland Grazing by Sheep

The sheep industry contributes some 35-40% of the total value of the agricultural sector in the Icelandic economy, and while meat (mainly lamb) is the chief product, offal, wool and skins are valuable by-products. The native Iceland breed, the only breed of sheep in the country, belongs to the northern European short-tailed race of sheep. It is a hardy breed, noted for early maturity, longevity, prolificacy and good milking ability (Dýrmundsson and Hallgrímsson, 1978). Currently the sheep population of Iceland is close to 0.7 million winterfed sheep with an additional 1 million lambs in summer. Lambing takes place at the onset of the growing season in May, and the lambs are killed at 4-5 months of age in September or October, before the onset of winter.

In general terms, the Icelandic system of sheep production aims at maximizing the output of meat (lamb) per winterfed ewe. It is primarily based on intensive (indoor) winter feeding of home produced feeds, mainly hay, and, to a lesser extent, silage. This is in combination with extensive summer grazing on rangelands, often on common land (Dýrmundsson, 1978). Thus any climatic variations affecting vegetation growth and carrying capacity of these pastures may consequently lead to a variation in lamb growth rate and their carcass weight in autumn.

In Iceland carrying capacity is generally expressed in terms of stocking rate or grazing intensity. It has been well established in several grazing trials, both on lowland and upland ranges, that increased stocking rates (i.e., grazing a larger number of sheep per unit area of land), will result in lower live weight gains and declining carcass weights of individual lambs (Gudmundsson and Helgadóttir, 1980; Gudmundsson, 1982). Variation in carcass weight between years is also evident, irrespective of stocking rate. For example, in one of the above trials the mean carcass weight was only 12.7 kg after the exceptionally cold summer of 1979 compared with 15.6, 15.0, 13.9 and 14.2 kg in the four preceding years on the same moderately grazed experimental pasture. Since the profitability of sheep farming is heavily dependent on the carcass weight of the lambs, any reduction will result in lower net income of individual producers. In addition, poorer grading of light carcasses aggravates such a decline in profitability.

It would be an oversimplification to assume a direct relationship between vegetation growth and carrying capacity on the one hand, and lamb growth rate and carcass weight on the other. Variation in herbage quality within each growing season is also of importance, and this in turn is influenced by climate, particularly by the temperature at different times of the year. These and other factors are considered below, in an analysis of several years of data on climate and carcass weight.

4.3. Material and Methods

The study is based on two main sets of data. Firstly, information on carcass weights from four slaughterhouses in northwest, north, east and southeast

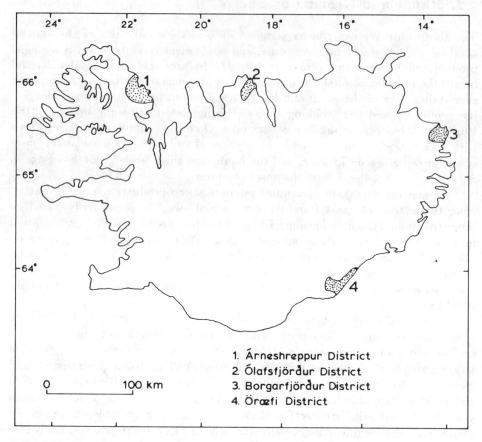

Figure 4.2. Location of the four districts included in the study.

Iceland (*Figure 4.2*), and secondly, on weather records from the Icelandic Meteorological Office representative of these four locations for the period 1965–1983. Earlier livestock data are unreliable because great changes in sheep husbandry practices have taken place in recent years. A study of national averages of carcass weight and temperature was also included. Information on sheep numbers was obtained from official livestock records. Local sheep-recording data were consulted with regard to changes in lambing rate over the period of study, and several points relating to the sheep data were verified by correspondence and discussions with local farmers, advisory officers and slaughterhouse managers.

The carcass weight data were selected on the basis of two main criteria, firstly, that the districts reflected extremes in climate within the country, and secondly, that the available meteorological data reflected as much as possible the climatic pattern of the actual summer grazing area of the sheep in each locality. Furthermore, the choice of carcass weight data was narrowed considerably by the necessity to consider only those localities where all lambs from the district had been slaughtered in the same abattoir throughout the whole period of study.

It should be pointed out that carcass weight is not only influenced by climate, range vegetation growth and grazing pressure (Dýrmundsson, 1979). Some of the variation can be attributed to other factors too. These include the standard of husbandry and winter feeding of the ewes, the ratio between single and twin lambs, the ratio between males and females due to annual differences in ewe replacement rate, whether or not the lambs are grazed on cultivated land for a few weeks before slaughter, and the age of the lambs at slaughter. Some of the data had to be corrected for differences in weighing methods in the abattoirs.

4.4. Results

In *Table 4.1*, data on the means and interannual variability of sheep numbers, carcass weight and four measures of temperature, from the period 1965–1983 are presented, for the four study districts, and for Iceland as a whole. Both sheep numbers and carcass weight differed considerably between districts, and the marked differences in the standard deviation values are of interest in relation to possible causal climatic factors. One of these, temperature, is represented by several measures, all of which show clear differences between districts (highlighted by cooler and more variable temperatures in the north than in the south).

To lend some statistical support to these qualitative comparisons, the results of a regression analysis relating carcass weight and sheep numbers to each of the temperature measures, are summarized in *Table 4.2*. The local variations demonstrate that national averages should be viewed with great care and they may, in fact, be misleading due to the many and complex factors involved. Thus for example, the carcass weight is more strongly associated with temperature variation in the Árneshreppur district than in the other districts studied, carcass weight being reduced by 802g for a decrease of 1 °C in the mean annual temperature (assuming the number of winterfed sheep is held constant, see *Table 4.2*). This relationship is further illustrated in a scatter diagram of the actual values (*Figure 4.3*). It is clear that summer (May–Sept) temperature exerts quite a strong influence on lamb growth in mountainous districts (the Árneshreppur district, Ólafsfjördur district and Borgarfjördur eystri district) in contrast to the Öraefi district where the climate is considerably warmer and most of the sheep graze on lowlands throughout the summer. In this district, correlations are also weak between the the other temperature variables and carcass weight. In the Ólafsfjördur district, where the ground is normally covered by a thick layer of snow in winter, the late winter (January–April) temperature is of little significance for carcass weight, suggesting that the rangeland vegetation is protected by the insulating properties of snow cover during the period when it is most prone to winter kill (*see* Sections 2 and 3).

As mentioned above, lamb growth may be affected by rainfall in summer, particularly when both cold and rain occur together. In the present study, special attention was paid to possible rainfall effects in the Öraefi district, which is a relatively high rainfall area. The analysis did not reveal any significant

Table 4.1. Mean values of sheep number, carcass weight and climatic variables for four districts and for all Iceland, 1965–1983 (standard deviations in brackets).

Variables	Árneshreppur NW-Iceland	Ólafsfjördur N-Iceland	Borgarfjördur eystri E-Iceland	Öraefi SE-Iceland	All Iceland (Stykkishólmur)
Sheep number (winterfed)	3040 (469)	1460 (193)	5095 (558)	4018 (348)	822947 (46227)
Mean carcass weight (kg)	15.81 (1.11)	15.64 (0.86)	14.05 (0.86)	14.11 (0.54)	13.92 (0.43)
Mean temperature (Oct 1–Sept 30, °C)	2.3 (0.68)	2.6 (0.70)	3.2 (0.65)	4.5 (0.53)	3.3[b] (0.58)
Grass growth index (G)[a] based on Oct–Sept temperature (A)	0.71 (0.13)	0.76 (0.12)	0.86 (0.10)	1.02 (0.06)	0.88[b] (0.09)
Mean temperature (Jan 1–Apr 30, °C)	−1.2 (1.14)	−0.7 (1.41)	0.3 (1.25)	1.4 (1.07)	−0.5[b] (1.25)
Mean temperature (May 1–Sept 30, °C)	6.0 (0.73)	6.4 (0.82)	6.2 (0.72)	8.5 (0.51)	7.8[b] (0.64)

[a] Equation (2.4): $G = 0.169 + 0.281A - 0.02A^2$.
[b] Data for Stykkishólmur.

Table 4.2. Regression equations and coefficients of determination (R^2) of carcass weight on the number of winterfed sheep (in hundreds, X_1) and on different climatic variables (X_2) for four districts and for all Iceland 1965–1983 (standard error of the estimate is in parenthesis).

Climatic variable (X_2)	Region	Equations	R^2
Grass growth index[a]	Árneshreppur	$16.91 - 0.144 \ (0.025)X_1 + 4.66 \ (0.91)X_2$	0.84
	Ólafsfjördur	$15.78 - 0.135 \ (0.098)X_1 + 2.40 \ (1.55)X_2$	0.22
	Borgarfjördur eystri	$15.60 - 0.076 \ (0.031)X_1 + 2.73 \ (1.75)X_2$	0.33
	Öraefi	$9.79 + 0.056 \ (0.036)X_1 + 2.01 \ (1.23)X_2$	0.16
	All Iceland	$13.91 - 0.00030 \ (0.00018)X_1 + 2.82 \ (0.96)X_2$	0.41
Mean annual temperature (Oct–Sept, °C)	Árneshreppur	$18.54 - 0.150 \ (0.028)X_1 + 0.802 \ (0.190)X_2$	0.80
	Ólafsfjördur	$16.51 - 0.135 \ (0.098)X_1 + 0.415 \ (0.269)X_2$	0.22
	Borgarfjördur eystri	$16.58 - 0.077 \ (0.031)X_1 + 0.442 \ (0.268)X_2$	0.34
	Öraefi	$11.05 + 0.055 \ (0.036)X_1 + 0.189 \ (0.223)X_2$	0.15
	All Iceland	$14.98 - 0.00030 \ (0.00018)X_1 + 0.416 \ (0.145)X_2$	0.40
Mean temperature (Jan–Apr, °C)	Árneshreppur	$21.20 - 0.167 \ (0.034)X_1 + 0.263 \ (0.113)X_2$	0.68
	Ólafsfjördur	$17.66 - 0.132 \ (0.104)X_1 + 0.107 \ (0.142)X_2$	0.14
	Borgarfjördur eystri	$17.69 - 0.072 \ (0.034)X_1 + 0.067 \ (1.505)X_2$	0.24
	Öraefi	$11.91 + 0.052 \ (0.036)X_1 + 0.077 \ (0.117)X_2$	0.14
	All Iceland	$16.05 - 0.00025 \ (0.00018)X_1 + 0.187 \ (0.068)X_2$	0.39
Mean growing season temperature (May–Sept, °C)	Árneshreppur	$16.20 - 0.146 \ (0.031)X_1 + 0.675 \ (0.201)X_2$	0.75
	Ólafsfjördur	$14.09 - 0.147 \ (0.084)X_1 + 0.579 \ (0.200)X_2$	0.42
	Borgarfjördur eystri	$15.20 - 0.073 \ (0.031)X_1 + 0.412 \ (0.240)X_2$	0.35
	Öraefi	$11.00 + 0.056 \ (0.037)X_1 + 0.102 \ (0.249)X_2$	0.13
	All Iceland	$14.26 - 0.00035 \ (0.00019)X_1 + 0.325 \ (0.140)X_2$	0.32
	Árneshreppur	$21.25 - 0.179 \ (0.038)X_1$	0.57
	Ólafsfjördur	$17.77 - 0.00029 \ (0.0001)X_1$	0.11
	Borgarfjördur eystri	$17.77 - 0.0073 \ (0.0033)X_1$	0.23
	Öraefi	$11.95 + 0.053 \ (0.035)X_1$	0.12
	All Iceland	$16.30 - 0.00029 \ (0.00022)X_1$	0.10

[a] Equation (2.4): $G = 0.169 + 0.281A - 0.02A^2$

relationship between summer rainfall and carcass weight in that district. Therefore it was concluded that the rainfall data available did not warrant further investigation bearing in mind that rainfall is extremely variable within any given district in Iceland, rendering such data much less reliable than temperature records. Moreover, sporadic 2- to 3-day spells of heavy rainfall and low temperature generally considered harmful to ewe and lamb performance, particularly in spring and early summer, are not likely to be properly represented by monthly or seasonal averages.

It is worth noting that the sheep number alone, representing grazing intensity, accounts for 57% of the yearly variation in carcass weight in the Árneshreppur district, carcass weight being reduced by 179 g for an increase of 100 winterfed sheep (*Table 4.2*). Lower values were obtained for other districts indicating that grazing intensity had remained modest or low throughout the period of study.

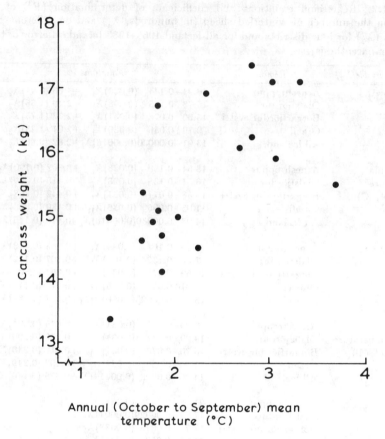

Figure 4.3. Scatter diagram showing the relationship between annual (October–September) mean temperature (°C) and carcass weight (kg) in the Árneshreppur district in northwest Iceland from 1965 to 1983.

4.5. Interpretation of the Results

When interpreting the results obtained, it is important to bear in mind that lamb growth is influenced by several factors, some of which are interrelated. Although considerable information is now available on the relationship between temperature and grass growth on cultivated land in Iceland (*see* Sections 2 and 3), it is not known how strongly the productivity of natural pastures is influenced by variation in temperature *per se*. However, observations indicate very substantial yearly fluctuations, particularly at higher altitudes, and it is generally assumed that carrying capacity is closely associated with vegetation growth, (i.e., both the yield and the duration of the growing season).

The results reported in this section on the impact of temperature on carcass weight certainly support the view that carcass weight of lambs may, to a certain extent, be indicative of the carrying capacity of rangelands. The strong correlation between temperature and carcass weight in the Árneshreppur district in northwest Iceland is in accordance with expectations, and it has become clear that sheep number in a given area, reflecting grazing intensity, aggravates the decline in carcass weight in cold years, particularly in the north. Furthermore, the relationship between temperature and carcass weight is affected by the nutritive value of the grazed vegetation. Thus abundant grazing may in certain cases be associated with poor lamb growth rates in late summer and early autumn due to an early deterioration in nutritive value. Conversely a reduction in vegetation growth and carrying capacity due to a lowering in summer temperature may in some years be associated with satisfactory carcass weight, i.e., when herbage quality is maintained at a relatively high level throughout the grazing season. Low temperatures, providing they are above freezing point, are in fact found to be more favorable to the nutritive quality of pasture than high temperatures (Wilson, 1982).

For the reasons stated above it could be an oversimplification to assume a direct relationship between temperature and carcass weight. An alternative method is to construct a model incorporating the effects of both temperature and precipitation on carcass weight and carrying capacity of rangelands, on a district basis. While the available data were inadequate to permit the construction of a model for the purposes of conducting scenario experiments, we have attempted to identify those factors that are important in developing a model of this kind in Subsection 4.6. In examining this problem special attention has been paid to the modeling approach adopted by Sibbald et al. (1979) working with sheep on hill pastures in Scotland.

4.6. Proposed Model

As a starting point the simple assumption is made that lamb growth is proportional to nutrient supply. From a biological viewpoint this relationship may be linear within a certain range. If the nutrient supply falls below a certain critical level, growth may be permanently stunted by deficiency. It seems unlikely that such extremes would be reached under present Icelandic conditions but values representing nutrient deficiency could be included in the model.

Under rangeland grazing conditions in Iceland it seems likely that stocking rate is sufficiently high in most areas to affect lamb growth, particularly in cold years, as indicated in the results presented above. Thus the relationship between the nutrient supply and the stocking rate has to be allowed for in the model. For this purpose, and to determine the stage at which stocking rate begins to affect growth, the widely documented relationship between gain per animal and stocking rate (Jones and Sandland, 1974) should be considered. Clearly this occurs at a lower stocking rate in cold years due to the adverse climatic effects on pasture yield. Moreover, in some districts horses are in direct competition with sheep for rangeland grazing and their contribution to the stocking load should be accounted for wherever possible.

At this stage it is appropriate to consider the nutrient supply in both quantitative and qualitative terms. While carrying capacity is only a function of pasture yield, lamb growth is also influenced by pasture quality. These factors are considered below.

Pasture yield is clearly affected by climate, mainly by temperature and in some cases by precipitation, and other climatic factors may be of importance too. In fact, such effects have to a certain extent been reflected in the results reported above. The climatic impact could be determined with greater precision by including in the model information representing several rangeland characteristics, such as the area, and distribution of the main plant communities classified according to altitude. It seems realistic to assume that such information will become available from vegetation maps in the not too distant future. Furthermore, deeper understanding of climatic effects is required. Although linear responses may be expected in some cases it is likely that the relationship between climate and pasture yield will be curvilinear when the more extreme climatic conditions are included. If winter temperature is included in the model, it would seem relevant to consider both snow cover and snow depth. Since many of the rangelands are in the interior of the country, at high altitudes, it is important to bear in mind that detailed meteorological records are lacking for most of these areas.

Pasture quality may be limiting to lamb growth irrespective of pasture yield, as indicated above. Abundant vegetation growth in early and midsummer due to favorable climatic conditions may result in nutritive quality of grazed herbage being of inferior quality in late summer giving relatively poor carcass weight. The reverse may apply in colder years provided the stocking rate is not in excess of carrying capacity. Thus the nutritive value of pasture interacts with pasture quantity, stocking rate and climatic factors. It is considered to be of vital importance to include pasture quality parameters in the model, but unfortunately only limited data are available at present and a great deal of work is required in this area, e.g., by the more widespread monitoring of fistulated sheep (animals fitted with special equipment for measuring the composition of ingested feed).

Carry-over effects on pastures of climate and stocking rate from one year to another are also important. It is, for example, reasonable to assume that the vegetation of the rangelands and their carrying capacity was adversely affected not only during the extremely bad year of 1979 but also in subsequent years, particularly in the uplands. General observations certainly suggest such carry-over effects, but it is not clear how they can most efficiently be represented in the framework of a model.

4.7. Effects of a Warmer Climate

As shown in *Table 4.1*, the mean annual temperature for the period 1965–1983 was 3.3 °C at Stykkishólmur, which represents the national average quite well (*see* Section 1). This is 0.4 °C lower than the mean temperature for the baseline period, 1951–1980 (Scenario I), 1.5 °C lower than the mean of the warmest 10 years after 1930 (Scenario IV) and 4.3 °C lower than the climate predicted for 2 × CO_2 conditions (Scenario V) (for details, *see* Section 1.10). The warmest years of the period 1965-1983 were 1972 and 1980, both with an annual mean of 4.2 °C and mean carcass weights of 14.41 and 14.65 kg, respectively, the latter being the highest mean carcass weight recorded for the entire period. On the other hand, the coldest year, 1979, had an annual mean temperature of only 2.3 °C and the lowest mean carcass weight of the period (13.02 kg). A comparison of these extremes, taking into account at the same time the results of the regression analysis in *Table 4.2*, suggests a 0.4–0.8 kg change in mean carcass weight for each 1 °C change in annual mean temperature. The corresponding changes in carrying capacity of rangelands may well be greater, possibly in the range 10–20% per 1 °C, and somewhat greater than the values obtained for cultivated hayfields (*see* Sections 2 and 3). Whether a 4.4 °C increase in temperature relative to this 1965-1983 period (the GISS $2 \times CO_2$ scenario) would (by extrapolating these relationships) result in a 25% increase in carcass weight and a 60% increase in carrying capacity still remains open to speculation.

Although no definite predictions are made, it is reasonable to assume that any warming of the climate would strengthen the rangeland vegetation and increase the carrying capacity, provided other factors such as precipitation would remain favorable. While all rangelands in the country would be affected, it is obvious that climatically marginal regions (i.e., northerly and high altitude areas) could benefit greatly from such changes. A temperature increase of 4.3 °C (Scenario V) would not only increase herbage yield substantially but also increase the length of the annual grazing season and thus reduce the need for winter housing and indoor feeding. This could result in considerable changes in livestock management – for example, earlier lambing and a less seasonal slaughtering of sheep which now only takes place in September and October. However, it would be of limited practical value to consider only carcass weight and carrying capacity in isolation. Gains in herbage quantity may be offset to a certain extent by reduced herbage quality affecting lamb growth as indicated above. A further drawback of a warmer climate would probably be a greater disease risk to grazing animals, particularly parasitism. Nevertheless, the impact of a warmer climate would in most respects be advantageous to rangeland maintenance and utilization. In economic terms grassland-based livestock production in Iceland would certainly become more competitive with livestock industries in neighboring countries.

4.8. Impact of a Colder Climate

In contrast to warming of the climate, cooling could conceivably lead to a marked reduction in carcass weight associated with poorer grading of the meat, and even a relatively greater decline in the carrying capacity of rangelands. Serious problems could arise, their severity depending on the magnitude and duration of the climatic cooling.

The coldest 10 years after 1930 (Scenario III) are all within the period of the carcass weight study, and past experience shows that a noticeable decline in rangeland vegetation growth occurs in such years. It is indeed a matter of great concern to know how Icelandic rangelands would respond to several cold years in succession, perhaps a whole decade similar to 1859–1868 (Scenario II) with a mean annual temperature of only 2.4 °C (only little higher than the value recorded for the extreme year of 1979, and nearly 1 °C lower than the average for the period 1965–1983). To investigate such effects, particular attention should be paid to the northernmost parts and the high altitude regions of Iceland which are most sensitive to climatic cooling (see Section 1). To illustrate, mean carcass weight in the Árneshreppur district in northwest Iceland dropped from 16.22 kg in 1978 (annual mean temperature 2.6 °C) to 13.70 kg in 1979 (1.4 °C) and rose to 15.79 kg in 1980 (3.6 °C). These years were followed by three cold years in succession, leading again to reduced carcass weights. There was a substantial decline in vegetation growth during these years resulting in overgrazing (G. Valgeirsson, personal communication 1984). Thus a few cold years concurrent with increasing sheep numbers apparently caused a drastic reduction in the carrying capacity of rangelands.

From such considerations, a vital question arises concerning how to counteract the greatly reduced carrying capacity of rangelands due to lowered temperature. It can be argued, justifiably, that various developments in agriculture over the last few decades have reduced the vulnerability of, for example, sheep production to impacts of climatic change. These included increased hay production from cultivated land, better indoor feeding and much reduced dependence on grazing in winter, more widespread use of cultivated pastures for grazing in spring and autumn, drainage of bogs and better rangeland management. The Öraefi district in southeast Iceland represents an outstanding example of an area where radical agricultural changes have taken place, mainly due to substantial land reclamation for hay production and grazing. Thus this district has become much less dependent on rangeland grazing which may partly account for the relatively small annual variations in carcass weight compared with the other districts (see Table 4.1). It should be noted, however, that this area enjoys a relatively mild and stable climate by Icelandic standards.

While short-term cooling effects can be met without much difficulty, the implications of Scenario II, with an annual mean temperature of only 2.4 °C compared with the national average of 3.3 °C for the period 1965–1983, would probably call for some major and hitherto unprecedented adjustments of both economic and social importance. At lower altitudes, a decline in carrying capacity could be offset to a certain extent by aerial fertilizer application of rangelands, a costly procedure if needed for several years in succession. However, it seems more likely that a drastic decline in carrying capacity as a result of a shorter grazing season and reduced vegetation yield would have to be met by a substantial reduction in both sheep and horse numbers, perhaps by 30–50%, in order to maintain acceptable animal performance and prevent deterioration in the vegetative and soil resources of the rangelands. Historical evidence certainly indicates that soil erosion in some parts of the country can be substantially enhanced under such low temperature conditions (Sveinsson, 1958). In marginal areas highly dependent on sheep production such changes could prove disastrous to the rural structure unless parallel policy adjustments were made.

Abrupt changes in livestock numbers should be avoided if possible. One means of mitigating such fluctuations in the event of climatic cooling is to utilize the lowland ranges more efficiently, particularly drained bogs that are known to have greater carrying capacity than the upland ranges. Already there is a definite trend in this direction, for example, the grazing of horses has been eliminated on most common upland and mountain ranges where both soil and vegetation are most sensitive to excessive stocking and climatic stress. The widely practiced restrictions on the length of the grazing period for sheep on the commons also serves to reduce grazing intensity. Limits to the number of grazing animals may be enforced by law if necessary.

4.9. Conclusions

The results presented in this section point to a relationship between carcass weight of lambs and temperature, which appears to be a useful indication of climate-related changes in vegetation growth and carrying capacity. The carcass weight–temperature relationship varies with location and is also affected by, for example, the variation in stocking rate and in nutritive quality of grazed herbage. It should therefore be interpreted with care. The design of a rangeland model is proposed and outlined, although lack of data does not allow its construction at this stage.

On the basis of the evidence available, it is postulated that each 1 °C change in annual mean temperature may result in a 0.4–0.8 kg change in mean carcass weight and a 10–20% change in the carrying capacity of the rangelands. Thus any major climatic variations, particularly decreased temperatures, would certainly require substantial adjustments in terms of animal performance and the utilization and preservation of rangeland resources. Special attention should be paid to the possible effects of such changes on ecosystems at the higher altitudes and in northerly districts, for example, in relation to soil erosion.

Against this background it is considered extremely important to implement long-term (10–20 years) grazing experiments representing extremes of rangeland conditions in conjunction with detailed monitoring of vegetation growth and recording of meteorological data. Such policies of preparedness and strategic planning to cope with the effects of climatic change on Icelandic agriculture are explored further in the next, concluding section of this report.

SECTION 5

Implications for Agricultural Policy

5.1. Introduction

There is little diversity in Icelandic agriculture, which is primarily based on grass cultivation, grazing and forage conservation, and the production of meat and milk from ruminants. Around 80% by value of agricultural production is obtained from sheep and cattle farming. The remainder is obtained from horses, pigs and poultry, potatoes and vegetables (including glasshouse products).

An objective of agricultural policy in Iceland is self-sufficiency in those products that can be economically produced locally. To achieve this, it is necessary to control fluctuations in production, such as those caused by climate, so that supply matches demand, thus avoiding shortages and overproduction. A further objective, arising from the isolation of the country, is to depend as little as possible on imported resources. One way to meet these goals is to ensure the retention of fodder reserves and other food products in order to buffer the effects of unavoidable or natural fluctuations.

The influence of climate on agricultural output is largely the result of its effects on plant growth, both on cultivated pastures and on rangelands. As in other industrialized countries, it has been possible to mollify the effect of climate with various technological and economic measures. However, extreme weather conditions can still have considerable impact, as was shown during the late 1960s, when severe winter kill of grasses occurred. The wet summers which from time to time cause great losses to farmers are another example. These losses are caused by unsuccessful fodder conservation, especially during hay-drying, which is the main method of forage conservation in Iceland. In wet summers up to 30-40% of digestible dry matter can be lost, whereas under favorable conditions the loss is only 7-10% (Gudmundsson, 1977).

Figure 5.1 summarizes the effects of climate on sheep farming and shows how its effects on a particular farm can be modified. Success in moderating the effect of climate depends on the ability of the manager, the technological level of

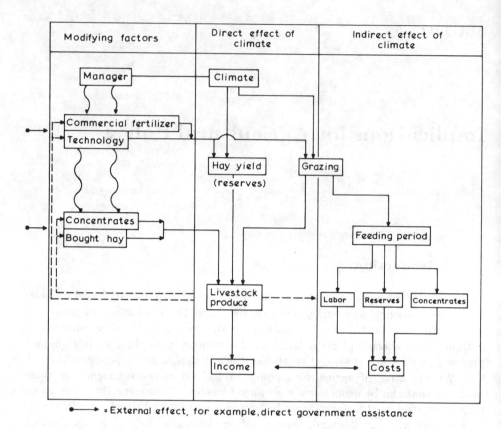

Figure 5.1. Effects of climate on Icelandic agriculture. In the case of specialized crop cultivation, such as cultivation of fodder plants or vegetable production, certain effects do not apply, but otherwise the interactions are the same.

the particular farm activities, the price of various materials and government actions. These points will be discussed later.

5.2. Types of Effects and Responses

With the prevalence of market forces in most industrialized countries, both beneficial and detrimental variations in climate have a considerable effect on market balance and hence on agricultural income. The response must come from government, research and advisory institutions and from farmers themselves.

Responses to climatic fluctuations can be divided into two categories according to the type of fluctuation. On the one hand there are short-term fluctuations lasting only a single season, or one to two years. On the other hand there are longer-term changes that persist more or less continuously over three or more years. The consequences of climatic fluctuations for agriculture and responses to them have been categorized in *Table 5.1*. Beneficial and detrimental variations are considered separately.

Table 5.1. The effects of climatic variations on Icelandic agriculture and responses to them.

Fluctuation	Effects	Responses
	I. Detrimental	
A. SHORT TERM (1 or 2 years)	Crop failure More expensive products Reduced income for farmers Pressure on rangelands	Fodder purchase (home/abroad) Increased fertilizer application Cultivation of green fodder Assistance to farmers Insurance
B. LONG TERM (3 years and more)	Reduction of livestock due to feed shortages and deterioration of rangelands More expensive products Fewer farmers and depopulation of farming areas	Importation of feed and agricultural products Changes in methods of cultivation, harvesting and feeding (including breeding of plants and farm animals) State assistance to discourage rural depopulation
	II. Beneficial	
A. SHORT TERM (1 or 2 years)	High fodder yield Cheaper agricultural production Higher income for farmers Increased production	Accumulation of fodder reserves Accumulation of produce reserves Reduced prices because of increased production
B. LONG TERM (3 years and more)	As for A but also danger of continued overproduction Potential for new fodder and food crops	Search for new markets Farmers encouraged in new production possibilities Afforestation

Farmers respond primarily to short-term fluctuations, whereas the advisory services, research establishments and government must react to longer-term fluctuations. Short-term fluctuations, if they are beneficial, would be expected to assist the running of farms by providing potential fodder reserves for use in poorer, detrimental years, for which every farmer must be prepared in Iceland.

Longer-term fluctuations, on the other hand, require major adaptation to changed conditions. In good years the adaptation would probably occur through the initiative of farmers themselves, scarcely influencing the society as a whole. Conversely, effects of an unfavorable longer-term fluctuation are much more far reaching and serious. These are the changes which have had and will always have the most serious effect on the survival of farmers and which influence society most. For example, when a cold spell has lasted for many years in Iceland, the cumulative effects have sometimes led to a shortage of farm products and the depopulation of farming districts. The cold spell after 1860 is an

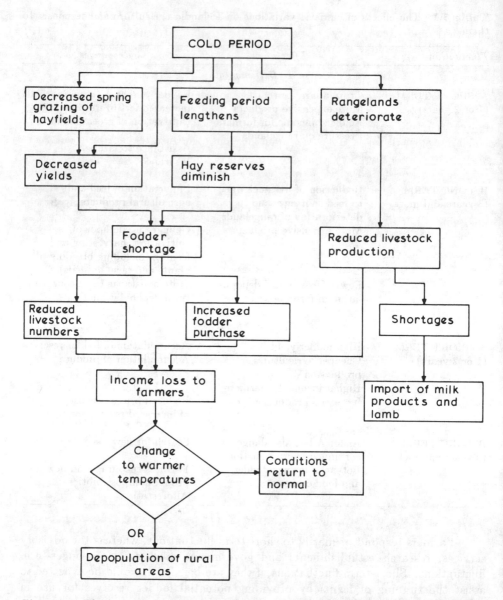

Figure 5.2. The impact of a run of cool years on livestock farming in Iceland. The maximum number of cold years that can be tolerated by a local farm/community before it is abandoned depends upon the economic strength of the farm/community and the ability of the manager to adapt production to changing conditions.

illustration of this, when areas in the northeast became depopulated and people moved to North America. Again, during 1965–1970, agriculture was badly hit in the northeast and also in the western part of the north country due to cold weather. In the worst-hit areas hay yields were up to 50% lower than the mean yield for the period 1900–1968 (Gudleifsson, 1971). *Figure 5.2*, which summarizes the effects of a cold period on sheep and cattle production, indicates that the climatic changes which concern farmers and government most are those associated with colder weather, especially if several cold years occur in succession. However, recent experience shows that a beneficial climate can also lead to serious problems, in this case due to overproduction.

Under certain circumstances precipitation can have just as great an effect as temperature on the quality and, hence, the utilization of home-produced fodder (Gunnarsson, 1983). The most widely acknowledged negative effect in Icelandic agriculture is its effect upon haymaking. In Iceland, rainy summers occur more often and their consequences are more serious in the south and west than in the north and east of the country since the southerly air currents that carry most precipitation lose moisture on their way across the mountains. Excess rainfall can cause reduced quality of the hay and thus makes winter feeding more difficult and more expensive because of the need to buy additional fodder. Poor haymaking conditions have the most serious effects if they follow a cold winter and spring, or if they occur during a long-lasting cold climatic period. Then a limited fodder supply and poor nutritional quality occur together. Experience shows that the impacts from heavy rainfall during haymaking are primarily short-term fluctuations, according to the definition given above.

5.3. A Summary of the Effects on Agricultural Activity and Agricultural Output

5.3.1. The regression models

Some of the more important quantitative relationships reported in detail in the preceding sections are summarized in *Table 5.2*, with some modification to allow comparison between results.

Two estimates, based on different data sets, have been made of the effect of climatic variation on hay yield. Those reported in Section 2 are based on data averaged for the whole country and are collected by the government to estimate hay reserves and to make decisions in the event of hay harvest failure. In actual fact these figures vary considerably between different parts of the country. For example, Fridriksson (1970) has shown that fluctuations in hay yield are considerably larger in the northern part than in the southern part of the country (*Figure 5.3*). Similarly, comparable differences in yield variation were found at the Icelandic experimental stations (*see* Section 3, *Figure 3.2*). The data are from experimental fields (kept for hay production and not grazed), which are generally better managed than farmers' fields and are less sensitive to climatic variations. In spite of these differences, however, the two estimates of effects on hay yield are broadly similar.

5.3.2. The results

The results of the Iceland case study have been analyzed in terms of the five main temperature scenarios described in Subsection 1.10 of this report and are

Table 5.2. Model relationships of some climate-sensitive agricultural sectors to temperature in the Icelandic case study.

Indicator	Study period and area	Temperature measure	Regression on temperature ($R = a + bT$)			Measures of model fit		
			Coefficient b	Units	Standard error	Standard deviation		R^{2d}
National hay yield[a]	1901–1975 all Iceland	Index based on October–April(W) & May–September(S) temperatures[e,f]	628[g]	kg/ha/°C		240 kg/ha[h]		0.94[h]
Dry matter herbage yield[b]	1951–1983 4 experimental stations	October–April temperatures[e,j] 644[m]	368[j]	kg/ha/°C		280 kg/ha[k]		0.92[k]
		September–August[l]	113[m]	kg/ha/°C	113 kg/ha[n]	0.31[n]		
Indoor feed requirements of sheep[a]	1941–1949 15–20 Weather stations	October–May[e]	–6.7[o]	%change/°C				
Mean carcass weight of lambs[c,p]	1965–1983 all Iceland	October–September[e]	0.416	kg/°C	0.145	0.145 kg		0.40

[a] From Section 2.
[b] From Section 3.
[c] From Section 4.
[d] Coefficient of determination – measure of the percentage variability explained by temperature above.
[e] Temperature at Stykkishólmur.
[f] Linear temperature index: $0.29 + 0.0729S + 0.0794W$ [equation (2.3)].
[g] Assuming nitrogen application of 100 kg N/ha and substituting into equation (2.3), for the combined effect of S and W.
[h] From Table 2.2.
[i] Hay-yield prediction model.
[j] Assuming nitrogen application of 100 kg N/ha and substituting into equation (2.5).
[k] From Table 2.4.
[l] Temperature at nearby meteorological stations.
[m] In experiments with nitrogen application around 100 kg N/ha [equation (3.1) and Table 3.4].
[n] Residual variability is dominated by the year vs stations interaction (see Table 3.4 and Subsection 3.2.3).
[o] From equation (2.9).
[p] From Table 4.2.

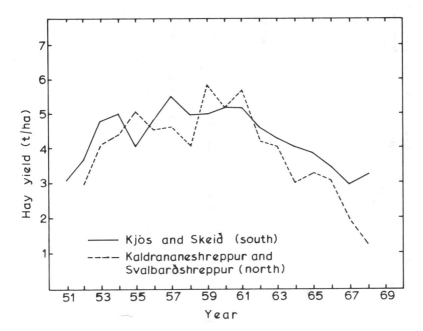

Figure 5.3. Mean yearly hay yields (t/ha) in two districts in the north of Iceland (Kaldrananeshreppur and Svalbardshreppur) and two districts in the south of Iceland (Kjós and Skeid) 1951–1968. (From Fridriksson, 1970.)

summarized in *Table 5.3*. Four points concerning *Table 5.3* should be stressed. Firstly, the results are an average over the whole country. Considerable deviations can be expected from one location to another.

Secondly, the results are given for average temperatures for each scenario not for individual years. It is not certain that the assumption of a linear effect of temperature on grass growth is justified. The averages may thus give a more optimistic estimate of results than would have been obtained if individual years had been considered. The investigations in Section 3, on deviations of individual years from the mean, indicate the importance of this point.

Thirdly, it is assumed that the farming technology used in the reference period (Scenario I) does not differ from that in the other scenarios. The interaction between technology and climate, and the effect of the continual changes in technology with time, are difficult to estimate. However, both will have an important impact upon agricultural output in the various climatic scenarios.

Finally, the estimations (especially those drawn from Section 2) are based on the distribution of agriculture within the country as it has been over the last few decades. This distribution is constantly changing and adapts itself to, among other things, climatic conditions.

Effects of short-term climatic variations (Scenarios II, III and IV)

According to the estimates in *Table 5.3*, under the cold·decade scenario (II), average hay yields for the whole of Iceland could be 16–19% less than in an

Table 5.3. Effects of temperature variations on some aspects of Icelandic agriculture (percentage of baseline unless stated otherwise).

Scenario[a]	No.	Mean annual temperature Stykkishólmur (°C)	Onset of the growing season (mean date)[b]	Hay yield from improved grassland[c,f]	Experimental plot yield[d,g]	Carrying capacity of rangelands[c,e]	Lamb carcass weight[e,h]	Barley ripening (%)[c,i]	Area of potential tree zone (km²)[a]
Baseline (1951–1980)	I	3.7	4 May	100	100	100	100	4	11 000
Cold (1859–1868)	II	2.4	17 May	81	84	81	93–96	0	8000
Cold (10 coldest years 1931–84)	III	2.9	9 May	87	90	87	95–97	0	8000
Warm (10 warmest years 1931–84)	IV	4.8	27 April	118	113	118	104–106	60	22 000
2 × CO$_2$ (GISS model-derived)	V	7.7	17 March	166	149	166	112–120	100	64 000

[a] From Section 1.
[b] Based on Gudmundsson (1974).
[c] From Section 2, for constant carcass weight.
[d] From Section 3.
[e] From Section 4, for constant carcass weight.
[f] Assuming nitrogen application of 160 kg N/ha.
[g] Assuming nitrogen application of 120 kg N/ha.
[h] Assuming 1965–1983 mean sheep numbers.
[i] Percentage of lowland weather stations with barley ripening in 60% of years.

average year under present-day conditions, though in particular areas yield could be reduced by much more. Hay reserves would not go far to meet such a reduction over several consecutive years, but they could make adaptation easier. The first reaction to the yield reduction would be increased use of feed concentrates, the cultivation of annual fodder crops, and increased fertilizer on fields undamaged by winter kill. Technically these measures could make up for production losses caused by a cold period, but farming costs would increase sharply, eroding profits, unless accompanied by a change in pricing policy.

If a 16–19% shortage of feed were to be met by imported feed concentrates, the extra cost to the average farm would equal 29–38% of the farmer's estimated net income (based on a representative farm, as defined in drawing up the Agricultural Price Agreement – Sigthórsson, 1974). This is the estimated *average* loss for a farmer in an average year under Scenario II, which means that individual farmers in particular areas would experience even greater cuts in income in some years. Few farmers would accept such losses for many years. Without public support, the hardship would quickly lead to depopulation of farming districts.

Increased fertilizer could probably compensate for only a part of the fodder shortage, though it would be more economical to keep hay fields large and apply suboptimal rates of nitrogen in average years, in order to obtain the necessary hay reserves for the farm each year (Section 2). Assuming an average rate of 80 kg N/ha, half of the fodder shortage could be met by increasing nitrogen by 26.6 kg/ha (Section 3). Official figures indicate that this could cost up to 5% of net income. The fodder that would still be lacking and would have to be supplemented by the purchase of feed concentrates would equal 15–20% of net income, so that these combined measures would cost something like 20–25% of net income. From this rough estimate it seems more economical to meet cold years by using increased fertilizer and supplementing with feed concentrates, rather than meeting them entirely with the purchase of feed concentrates. There appear to be good reasons, therefore, to study more closely the effects of nitrogen fertilizer in cold years and the possibilities of controlling grass growth by applications of fertilizer that are given according to the mean temperature of the preceding winter.

Cold years do not appear to affect significantly the carcass weight of lambs (*Table 5.3*) but do appear to affect the carrying capacity of rangelands (*see* Section 4). This tends to bear out the experience of the last few years which have been very cold. In order to diminish the grazing pressure on sensitive areas, it is important to protect those habitats which are at high altitudes and concentrate the grazing onto the lowland areas. In contrast, the effects of a warm period (represented by the ensemble of years in Scenario IV) would include increased hay yields (perhaps 13–18% above the baseline), higher carrying capacity and carcass weights, and would provide the opportunity in many lowland areas to cultivate barley and maybe even to plant trees successfully. The further implications are similar to those discussed in relation to longer-term warming discussed below.

Effects of longer-term climatic warming (Scenario V)

The effects on agriculture of a considerable increase in temperature, such as that estimated for $2 \times CO_2$ conditions in Iceland (Scenario V), have been estimated by simple extrapolation of the model relationships (*Table 5.3*) and should therefore be treated with caution. However, the consequences can be further investigated by comparison with countries where comparable climates prevail today (*see* Sections 1, 2 and 3). From the estimates in *Table 5.3* it seems that these conditions would enable cultivation of new fodder crops, and it should thus be possible to cultivate barley for grain every year in the lowland areas of the country. In this way Iceland could become largely self-sufficient in feed barley (*see* Section 2, and Sigurbjörnsson, 1982). Additionally, the area experiencing mean July temperatures above 10 °C would increase sharply (*Table 5.3*), opening up considerably more land for barley cultivation.

It is estimated that the feeding period for sheep, under Scenario V, would be shortened by at least two months, and indoor feed requirements would be reduced by at least one-third (and by a lesser amount for cattle). The competitiveness of livestock farming, both locally and on foreign markets, would probably become considerably stronger.

The most obvious change under a warmer climate would probably be on the rangelands, where more lush vegetation and higher yields could be expected. Relatively small changes in temperature can have a considerable effect on vegetation in the rangelands, especially in areas of high latitude and altitude, although the feeding value of fodder plants would probably be reduced (*see* Section 4). A warmer climate could also be expected to increase the danger of diseases both of plants (e.g., potato blight) and animals (e.g., parasitism), which might well reduce fodder crop yield and the volume of livestock production, and increase production costs. These effects are not considered in *Table 5.3*.

Improved conditions in a warmer climate would not necessarily reduce the risks that farmers take. The climate would still be variable and the risk would move to the new farming options, such as cultivation of the new fodder crops, barley and root crops (*see* Section 3).

5.4. Precautions Against Climatic Variations

Predictions of climatic change are, as yet, unreliable. However, one can be certain that climate will continue to be variable, as it has been up to now. It is essential to keep this variability in mind when planning agricultural production. Furthermore, it is clearly necessary to plan a response to maintain fodder and food production in the colder years. Finally, it should be noted that all long-term fluctuations in the climate are likely to be felt through short-term conditions. Effective reaction to short-term fluctuations can thus make the adaptation of the agricultural production system to longer-term fluctuations a lot easier.

5.4.1. Mitigating the effects of a colder climate

With the above points in mind, as well as the general principles outlined earlier (*see* Subsection 5.2), a policy for cold years can be formulated, as shown below.

Precautions on the farm

It is most important for farmers to produce and maintain sufficient reserves of feed. Experience has shown that a reserve of at least 35% over and above the needs in an average year (1951–1980) should be available. An even higher target (40–50%) is perhaps required in the marginal areas, whereas 25–30% is probably sufficient where growing conditions are more favorable. In connection with this it is important to pay attention to the following points:

(1) Cultivated areas should be sufficiently extensive. Over the last few years there has been a reduction in the number of cattle and sheep, partly due to difficulties of marketing at a national level. Further reduction can be foreseen. This change enables improvement in feed reserves. Rather than allowing fields to fall into disuse, cultivation should be maintained so that a reduction in growth during cold periods can be met by using additional land at a low level of cultivation.

(2) The choice of fodder crops to sow during the renovation of hay fields should be aimed at supplying requirements in cold years. Farmers should, for example, avoid narrow selection and include at least a proportion of hardy fodder crops. The farmers could benefit from the assistance of research bodies and advisory services.

(3) The hay storage facilities should be matched to the reserves which are necessary for the farm in addition to the annual needs.

Planning of agricultural production

Hardships resulting from crop failure caused by severe cold in the north and wet summers in the south affect farming districts to a variable extent in Iceland. It is therefore an essential safety measure to sustain farming production in the various parts of the country. In this way the risk of complete crop failure can be spread, as can the detrimental effects of rain during the haymaking period.

An operations research model has been derived, which allows one to study the most economical ways to meet fodder shortages in certain areas (Adalsteinsson et al., 1978). Depending on the price of hay and concentrates, the economical distances to transport hay and the optimum timing for reducing the number of winterfed livestock can be estimated. In the future, this model could possibly be linked to growth models in order to facilitate an integrated assessment of the response to severe climate.

Harvest insurance

Better organization of harvest insurance is required than exists at present, particularly taking into account short-term fluctuations in climate. A powerful insurance fund should be established which, while bound to cause an increase in agricultural prices, would be a necessary consequence of a policy of self-sufficiency. Direct payments into the fund from the government should therefore be considered. The operation of Bjargrádasjódur (The Emergency Fund) has been based on these principles, but it has in the past been too weak to meet major disasters. For example, financial support for fodder purchase following the cold winter and spring of 1983 in the north and the east was less than a quarter of the estimated direct loss to the farmers (Ministry of Agriculture, unpublished data).

Responsibility of government

It is important that government action should encourage the collection and storage of fodder reserves as well as the maintenance of appropriate reserves of agricultural products. In this context, the following should be mentioned:

(1) Organized government support to maintain the cultivation of hay fields and the building of sufficiently large fodder storage facilities on farms.
(2) Support of grass-pelleting factories in order to enable collection of fodder reserves which can be stored from one year to the next.
(3) Support of farms to encourage grinding and pelleting of hay, both in order to reduce the problems of storage (by reducing bulk) and to improve the utilization of the hay, possibly with the addition of other feedstuffs to the pellets.
(4) Support for the storing of (imported) feed concentrates, so that a 2- to 3-month reserve is always available.
(5) Improved production of fodder from locally produced raw materials, such as waste products from fish processing, which can increase the reliability of local agricultural production.
(6) A target for agricultural production should be a limited level of overproduction, determined so as to enable the nation to provide adequate returns in poorer years. The surplus from better years may be exported, and reserves of agricultural goods would be available to even out short-term fluctuations. Even though prices on the international markets are low, limited exports are economically justified for ensuring a sufficient supply both of lamb and, more importantly, milk in all years.

5.4.2. Taking full advantage of a warmer climate

During a period of anomalously warm conditions, the main problem facing agriculture is to sell the extra production and perhaps also to curtail production. On the other hand, a warmer climate will open up various new possibilities, including:

(1) Increasing the opportunities of cultivating new fodder crops such as barley. This would lead to local production of carbohydrates, the constituent of the fodder which is at present in shortest local supply. In this way the country could become almost self-sufficient in fodder. It is possible to prepare for a warm period with well-organized research into cultivation and utilization of crops that are now marginal such as barley. At present, about 70 000 tons

of concentrates, mostly carbohydrates, are imported. This amounts to 11–13% of all fodder consumption (including grazing), or 1–2% of GNP (by value).

(2) Protecting the land and cultivating more sensitive vegetation in the rangelands as less land is required for hay production and grazing. Afforestation would be of particular interest, both to improve the quality of the land and for timber production. The import of timber makes up around 2% of total Icelandic imports (0.7% of GNP). It is conceivable that these costs could be reduced by 10–15% with local timber production.

(3) Changes in the character of sheep farming, since sheep would probably require shorter periods under shelter, while the growing period of fodder crops could be extended. Winter feeding and the need for hay would decrease. On the other hand, with a longer grazing period, increased grazing pressure might result if the pasture productivity does not increase to the same extent. It is conceivable that beef production would provide increased competition to lamb production. Both beef and sheep farms could extend over greater areas. Taken as a whole, the changes would probably reduce production costs for these meat products.

5.5. Research

Continued research into climatic effects on agriculture is of vital importance for reinforcing our current knowledge and understanding (including the results of this study) and in preparing for likely changes in climate. It would be useful, for example, to examine the more detailed year-to-year effects of weather on agriculture over individual regions as a refinement on the period-averaged impacts at a national scale that have been the focus of this report. Another topic requiring further attention is the effects of climate on the cultivation of food crops such as potatoes and other vegetables, many of which are, or might become, of rapidly increasing importance in Iceland. Preliminary investigations indicate that it may be difficult to find reliable data on the cultivation and productivity of these crops that could be used to establish relationships with climatic factors.

Much of the emphasis of research into the effects of climate on agriculture has hitherto concentrated mainly on temperature. However, other factors are also of importance, such as precipitation (which can have a marked effect on the quality of the hay and losses during harvesting), insolation (determining the growth and development of all productive vegetation), and windspeed and exposure (of importance to both crops and livestock, alike).

It seems likely that the provision of fodder for ruminants will continue to form the basis of Icelandic agriculture in the foreseeable future. Therefore, to accommodate possible future climate changes, research in two main areas is required. Firstly, studies should focus on agriculture in unfavorable conditions, for example, at the northern boundaries of cultivation shown in Section 1 in this report (*see Figure 1.8*). The knowledge and experience of plant selection, breeding, cultivation and harvesting in present marginal areas could be of considerable value, for, should the climate become colder, most of the country might then experience the climate which presently prevails in these marginal areas. It is also necessary, as a complementary study, to estimate the carrying capacity of rangelands under unfavorable conditions and to determine how grazing pressures could be reduced by the cultivation of annual fodder crops and by grazing cultivated land.

Secondly, corresponding research should focus on those parts of the country which currently have the best growing conditions, for example, those areas with a frost-free period greater than 140 days (*Figure 1.10*). One important avenue of research would concentrate on barley cultivation with a view to obtain knowledge, experience and techniques that could be applied to larger areas of the country in the event of a warmer climate. The use of data and experience from other countries with present-day climates analogous to a warmer Iceland can also supplement domestic research (as shown in Sections 2 and 3).

Further research into the use of nitrogen fertilizer, especially in a colder climate, is of fundamental importance both for the biological and the economic success of fodder production in Iceland (as demonstrated in Section 3). In addition, valuable insights into the effect of climate on carrying capacity of rangeland vegetation could be obtained through the development of a production model of the type proposed in Section 4. To achieve this, it may be necessary to initiate specialized long-term grazing experiments, particularly in marginal areas.

An important research and development project for stabilizing Icelandic agricultural production would be to find a method to reduce yearly fluctuations due to weather in order to save the expense and inconvenience of shortages and over-production. One method of stabilizing hay yield from one year to another would be to use varied levels of fertilizer application (*see* Section 2), and this is currently being tested in an experiment where fertilizer use each year is determined with respect to mean October–April temperature preceding the harvest (*see* Subsection 2.2.10). Other methods should also be sought.

The following research projects, which are all concerned with improving growing conditions, are of special importance:

(1) The heating of soil with geothermal heat and/or electricity to assist cultivation of vegetables is already in a research and development phase. Geothermal energy, along with the use of plastic covers, can increase the yield of vegetables, even in the colder parts of the country or under a considerably colder climate, and could help the country to become self-sufficient in vegetables.

(2) Shelterbelts can be used to permit the cultivation of more sensitive crops (barley, potatoes and other vegetables) as well as providing shelter for grazing animals and buildings. The cultivation of shelterbelts is in an initial stage in Iceland and is being encouraged by cooperation between The Agricultural Society and the government, with direct financial support to farmers already available from the government, on request.

(3) It is necessary to encourage research into forestry both for land improvement and for timber production. Forestry will be increasingly important, especially if the climate becomes warmer and the land requirements for sheep farming decrease. Moreover, forestry initiated in a warm period

could conceivably moderate the effects of a subsequent short-term and even longer-term cooling of the climate, especially at the microscale.

It would also be interesting to perform some economic investigations as indicated below:

(1) In order to investigate the reaction of individual communities to a cooler climate, it would be necessary to map accurately the influence of cold years such as 1965–1970 on agriculture and population in those parts of the country that suffered the most (e.g., Árneshreppur in the northwest, and districts in the northeast). In this way it would be possible to obtain valuable recent information on the actual reaction of communities to climatic change. These findings could then be applied to the wider context of the nation as a whole.
(2) The research that is discussed here is concerned primarily with the agricultural sector alone. A next step would be to investigate the repercussions elsewhere in the Icelandic economy and society that would accompany the changes in agricultural capacity and production due to climatic changes.

History shows that individuals and communities are quick to adapt to changed conditions, especially to improved conditions. But history also indicates that they are often quick to forget previous experience. This is an important point both for research and advisory agencies. A few good years should not cause us to forget the bad ones. These agencies must have readily available the resources and the know-how to help farmers if the climate were to become cooler. Agricultural colleges must provide continued education on climatic fluctuations, their effects on agriculture, and ways to counteract these effects.

Of one thing we may be certain: variations in climate will continue to occur in the future, as they have in the past.

References

Adalsteinsson, S., Jensson, P., and Stefánsson, G. (1978). Skipulagning á jöfnun heyforda (The organization of hay redistribution). *Rannsóknaráð ríkisins*, **Rr. 4 '78**.
Baldursson, G. (1974). Population. In J. Nordal and V. Kristinsson (eds.), *Iceland 874-1974*, Central Bank of Iceland, Reykjavík, pp. 22-32.
Bergthórsson, P. (1965). Throskalíkur byggs á Íslandi (Probability of barley ripening in Iceland). *Veðrið*, **10**, 48–56.
Bergthórsson, P. (1966). Hitafar og búsaeld á Íslandi (Temperature and agriculture in Iceland). *Veðrið*, **11**, 15–20.
Bergthórsson, P. (1969). An estimate of drift ice and temperature in Iceland in 1000 years. *Jökull*, **19**, 93–101.
Bergthórsson, P. (1976). Töðufall á Íslandi frá aldamótum (Hay yield of improved grassland in Iceland). *Freyr*, **72**, 250–262.
Bergthórsson, P. (1982). Áhrif loftslags á búfjárfjölda og thjóðarhag (The impact of climate on livestock and economy in Iceland). *Eldur er í norðri*, papers in honor of Professor Sigurdur Thórarinsson, Sögufélag, Reykjavík, pp. 283–294.
Bergthórsson, P. (1985). Sensitivity of Icelandic agriculture to climatic variations, *Climatic Change*, **7**(1), 111–127, Special Issue.
Björnsson, H. (1975). Köfnunarefni og grasspretta (Nitrogen and grass growth). *Freyr*, **71**, 330–337.
Björnsson, H. (1980). Áburdartap (Fertilizer losses). *Freyr*, **76**, 462–470.
Björnsson, H. (1984). Er unnt ad jafna árferdismunmed breytilegri áburdargjöf? (On the possibilities of reducing annual yield variation by varying the amount of fertilizer). *Freyr*, **80**, 93–96.
Björnsson, H. and Hermannsson, J. (1983). Samanburdur á meltanleika nokkurra túngrasa [Comparisons of the *in vitro* (T.T.) digestibility of some pasture grasses]. *Ráðunautafundur 1983*, 145–160.
Björnsson, H. and Óskarsson, M. (1978). Comparison of nitrogen fertilizers on permanent pastures. I. Yield of hay and chemical composition of grass on peat soil at Hvanneyri (English summary). *Íslenskar landbúnaðarrannsóknir*, **10**, 34–71.
Conrad, V. and Pollack, L.W. (1962). *Methods in Climatology*, Oxford University Press.
Cooper, J.P. (1964). Climatic variation in forage grasses. *Journal of Applied Ecology*, **1**, 45–62.

Corrall, A.J. (1984). Grass growth and seasonal pattern of production under varying climatic conditions. In H. Riley and A.O. Skjelvåg (eds.), *The Impact of Climate on Grass Production and Quality.* Proceedings of the 10th General Meeting of the European Grassland Federation, As, pp. 36–45.
Deinum, B., Beyer, J. de, Nordfeldt, P.H., Kornher, A., Ostgard, O. and Bogaert, G. van (1981). Quality of herbage at different latitudes. *Netherlands Journal of Agricultural Science,* 29, 141–150.
Dýrmundsson, Ó.R. (1978). Utilization of Icelandic rangelands by sheep. *European Association for Animal Production,* 29th Annual Meeting, Stockholm, 1978.
Dýrmundsson, Ó.R. (1979). Beitarthol og fallthungi dilka (Carrying capacity of rangelands and carcass weight of lambs). *Árbók landbúnadarins,* pp. 78–86
Dýrmundsson, Ó.R. and Hallgrímsson, S. (1978). Reproductive efficiency of Iceland sheep. *Livestock Production Science,* 5, 231–234.
Einarsson, M.Á. (1969). *Global Radiation in Iceland,* Vedurstofa Islands, Reykjavík.
Einarsson, M.Á. (1972). *Evaporation and Potential Evapotranspiration in Iceland,* Vedurstofa Islands, Reykjavík.
Eldjárn, K. (1981). Hugleidingar um land og thod i timanna rás (Reflections on a country and nation through the ages). *Freyr,* 77, 9–15.
Eythórsson, J. and Sigtryggsson, H. (1971). The climate and weather of Iceland. *Zoology of Iceland,* 1, Part 3, Ejnar Munksgaard, Copenhagen and Reykjavík.
Frame, J. (1981). *Unlocking the Potential of Grass,* Technical Note Nr. 131, The West of Scotland Agricultural College, Auchincruive, Ayr.
Fridriksson, S. (1970). Kal og búskaparhaettir (Winterkill and farming). *Freyr,* 66, 220–232, 131.
Fridriksson, S. (1973a). *Líf og land,* Vardi, Reykjavík.
Fridriksson, S. (1973b). Crop production in Iceland. *International Journal of Biometeorology,* 17, 359–362.
Fridriksson, S. and Sigurdsson, F.H. (1983). Áhrif lofthita á grassprettu (The effect of air temperature on grass growth). *Journal of Agricultural Research in Iceland,* 15, 41–54.
Geirsson, Th. (1984). Thegar sandarnir voru raektadir (When the sands were improved – an interview). *Freyr,* 80, 690–695.
Gudleifsson, B.E. (1971). Extent and Causes of Winter-damages in Icelandic Grasslands, Ph.D. thesis, NLH, As, Norway.
Gudleifsson, B.E. (1975). Um kal og kalskemmdir IV. Samband vedurfars og kalskemmda (On winter kill and damage due to winter kill IV. The relationship between weather and winter kill damage). *Ársrit Raektunarfélags Nordurlands,* 72, 45–64.
Gudleifsson, B.E. and Sigvaldason, J. (1972). Um kal og kalskemmdir II. Kalskemmdir í tilraunareitum Tilraunastödvarinnar á Akureyri (On winter kill and damage due to winter kill II. Winter kill damage on experimental plots at the experimental station in Akureyri). *Ársrit Raektunarfélags Nordurlands,* 69, 84–101.
Gudmundsson, B. (1974). Spring temperature and growth conditions with particular reference to grassland. *Íslenskar landbúnadarrannsóknir,* 6, 23–36.
Gudmundsson, B. (1977). Studies of *Phleum pratense* (Engmo) II. – Losses of dry matter and reduction of digestibility of hay during drying in the field (English summary). *Íslenskar landbúnadarrannsóknir,* 9, 3–15.
Gudmundsson, B. (1984). *Áhrif tídarfars á gjafartíma saudfjár* (The effect of climate on the duration of housing of sheep), unpublished.

Gudmundsson, Ó. (1982). Effect of stocking rate and fertilization on production from Icelandic lambs grazing mountain pasture. *European Association for Animal Production*, 33rd Annual Meeting, Leningrad, 1982.
Gudmundsson, Ó. and Helgadóttir, S. (1980). Mixed grazing on lowland bogs in Iceland, *Proceedings, Mixed Grazing Workshop, Galway, 1980*, pp. 20–31.
Gunnarsson, G. (1983). Grasspretta, nýting og heyfengur 1630–1900 samkvaemt sögulegum heimildum (Grass growth, utilization and hay yield, 1630–1900, according to historical accounts). *Freyr*, 79, 250–255.
Hansen, J., Russell, G., Rind, D., Stone, P., Lacis, A., Lebedeff, S., Ruedy, R. and Travis, L. (1983). Efficient three-dimensional global models for climate studies: Models I and II. *Monthly Weather Review*, 111, 609–662.
Hansen, J., Lacis, A., Rind, D., Russell, G., Stone, P., Fung, I., Ruedy, R. and Lerner, J. (1984). Climate sensitivity: analysis of feedback mechanisms. In J. Hansen and T. Takahashi (eds.), *Climate Processes and Climate Sensitivity*, Maurice Ewing Series, Vol. 5, American Geophysics Union, Washington D.C., pp. 130–163.
Helgadóttir, Á. (1981). The Genecology of Icelandic and British Populations of *Poa pratensis* L. and *Agrostis tenuis* Sibth, Ph.D. thesis, University of Reading, UK.
Jenkinson, D.S. (1976). The Nitrogen Economy of the Broadbalk Experiments. I. Nitrogen Balance in the Experiments. *Rothamsted Experimental Station, Report of 1976*, Part 2, pp. 103–109.
Jóhannesson, Th. (1974). An outline history. In J. Nordal and V. Kristinsson (eds.), *Iceland 874–1974*, Central Bank of Iceland, Reykjavík, pp. 33–57.
Jones, R.J. and Sandland, R.L. (1974). The relation between animal gain and stocking rate. Derivation of the relation from the results of grazing trials. *Journal of Agricultural Science*, 83, 335–342.
Keatinge, J.D.H., Stewart, R.H. and Garrett, M.K. (1979). The influence of temperature and soil water potential on the leaf extension rate of perennial ryegrass in Northern Ireland. *Journal of Agricultural Science*, 92, 175–183.
Köppen, W. (1936). Der geographische System der Klimate. In W. Köppen and R. Geiger (eds.), *Handbuch der Klimatologie*, Vol. 1, Part C, Bornträger, Berlin.
Kristjánsson, K.K. (1946). Kornraektartilraunir á Sámsstödum og vídar (Grain-growing experiments at Sámsstadir and other places). *Rit landbúnadardeildar, B-flokkur, nr. 1*, Atvinnudeild Háskólans, Reykjavík.
Lamb, H.H. (1972). *Climate: Present, Past and Future*. Vol. 1. *Fundamentals and Climate Now*. Methuen, London.
Larcher, W. (1981). Effects of low temperature stress and frost injury on plant productivity. In C.B. Johnson (ed.), *Physiological Processes Limiting Plant Productivity*, Butterworths, London, pp. 253–269.
Lough, J.M., Wigley, T.M.L. and Palutikof, J.P. (1983). Climate and climate impact scenarios for Europe in a warmer world. *Journal of Climate and Applied Meteorology*, 22, 1673–1684.
Mork, E. (1968). Økologiske undersøkelser i fjellskogen i Hirkjølen forsoksområde (Ecological investigations in the mountain forest at Hirkjølen experimental area – in Norwegian with English summary). *Meddelelser fra Det Norske Skogforsøksvesen*, 93, XXV, 461–614.
Morrison, J. (1980). The influence of climate and soil on the yield of grass and its response to fertilizer nitrogen. In W.H. Prins and G.H. Arnold (eds.), *The Role of Nitrogen in Intensive Grassland Production*. Proceedings of an International Symposium of the European Grassland Federation, Wageningen, pp. 51–57.

Ogilvie, A.E.J. (1981). Climate and Society in Iceland from the Medieval Period to the Late Eighteenth Century, PhD. thesis, University of East Anglia., Norwich, UK
Ogilvie, A.E.J. (1984). The past climate and sea-ice record from Iceland, Part 1: Data to A.D. 1780. *Climatic Change*, **6**, 131–152.
Ólafsson, G. (1973). Nutritional studies of range plants in Iceland. *Journal of Agricultural Research in Iceland*, **5(1–2)**, 3–63.
Ólafsson, J. (1975–1982). Annual reports of the Farm Accounting Office, Búnadarfélag Íslands, Reykjavík.
Ólafsson, J. (1985). Ársskýrsla Búreikningastofu landbúnadarins 1983 (Annual report of the Farm Accounting Office 1983 – English summary), *Búreikningastofa landbúnadarins*, nr. 47.
Óskarsson, M. and Eggertsson, M. (1978). *Áburdarfraedi* (A textbook on fertilizers). Búnadarfélag Íslands, Reykjavík.
Pálmason, F. (1982). Beitarálag metid med maelingum á gródri (Estimation of grazing pressure by use of dry matter yield and crude protein content of herbage – English summary). *Journal of Agricultural Research in Iceland*, **14**, 47–54.
Parry, M.L. and Carter, T.R. (1985). The effect of climatic variations on agricultural risk. *Climatic Change*, **7**, 95–110.
Parsons, A.J. and Robson, M.J. (1980). Seasonal changes in the physiology of S24 perennial ryegrass (*Lolium perenne L.*). 1. Response of leaf extension to temperature during the transition from vegetative to reproductive growth. *Annals of Botany N.S.*, **46**, 435–444.
Sibbald, A.R., Maxwell, T.J. and Eadie, J. (1979). A conceptual approach to the modelling of herbage intake by hill sheep. *Agricultural Systems*, **4**, 119–134.
Sigfúsdóttir, A.B. (1969). Hitabreytingar á Íslandi 1846–1968 (Temperature variations in Iceland 1846–1968). *Hafísinn*, Almenna bókafélagid, Reykjavík, pp. 70–79.
Sigthórsson, G. (1974). Agriculture. In J. Nordal and V. Kristinsson (eds.), *Iceland 874–1974*, Central Bank of Iceland, Reykjavík.
Sigurbjörnsson, B. (1982). Endurvakning kornraektar á Íslandi (Resumption of barley growing in Iceland). *Rádunautafundur*, **1982**, 59–62.
Sigurdsson, G., Sigvaldason, H., Björnsson, H., Hannesson, K.A., Jensson, P. and Ólafsson, S. (1980). Reiknilíkan af mjólkurframleidslu kúabúa (A computer model of the milk production on dairy farms). *Fjölrit RALA* (RALA Report) No. 76.
Sveinsson, R. (1958). Greinargerd frá 1953 um gródureydingu, sandfok, uppblástur, sjálfgraedslu og raektun sanda og foksvaeda á Íslandi, vegna beidni um taeknilega adstod frá FAO (Memorandum from 1953 on soil erosion and reclamation problems in Iceland in support of an application for technical assistance from the FAO). In A. Sigurjónsson (ed.), *Sandgraedslan*, Búnadarfélag Íslands og Sandgraedsla ríkisins.
Thomas, H. and Norris, I.B. (1977). The growth responses of *Lolium perenne* to the weather during winter and spring at various altitudes in Mid-Wales. *Journal of Applied Ecology*, **14**, 949–964.
Thórarinsson, S. (1944). *Tefrokronologiska Studier på Island. Thjórsárdalur och dess förödelse*, Ejnar Munksgaard, Copenhagen.
Thórarinsson, S. (1974). Geology and physical geography. In J. Nordal and V. Kristinsson (eds.), *Iceland 874–1974*, Central Bank of Iceland, Reykjavík, pp. 1–11.
Thorsteinsson, S.Sch., Pálsson, H. and Thorgeirsson, S. (1982). Vöxtur lamba og vedurfar (Growth rate of lambs in relation to climate). *Rádunautafundur*, **1**, 76–81.
Tölfraedihandbók Hagstofu Íslands (1967). Statistical Abstract of Iceland, Statistical Bureau, Iceland.

Tucker, G.B. and Barry, R.G. (1984). Climate of the North Atlantic Ocean. In H. van Loon (ed.), *Climates of the Oceans* (World Survey of Climatology, Vol. 15), Elsevier, Amsterdam, pp. 193–262.

Wilson, D. (1981). The role of physiology in breeding herbage cultivars adapted to their environment. In C.E. Wright (ed.), *Plant Physiology and Herbage Production*, British Grassland Society, pp. 95–108.

Wilson, J.R. (1982). Environmental and nutritional factors affecting herbage quality. In J.B. Hacker (ed.), *Nutritional Limits to Animal Production from Pastures*, Commonwealth Agricultural Bureaux, Farnham Royal, UK, pp. 111–131,

Witty, J.F., Day, J.M. and Dart, P.J. (1976). The Nitrogen Economy of the Broadbalk Experiments. II. Biological Nitrogen Fixation. *Rothamsted Experimental Station, Report for 1976*, Part 2, pp. 111–118.

PART IV

The Effects of Climatic Variations on Agriculture in Finland

Contents, Part IV

List of Contributors		515
Abstract		517
1.	Introduction	519
	Veli Pohjonen, Lauri Kettunen and Uuno Varjo†	
	1.1. The purpose of the study	519
	1.2. Geography, soil and farm structures	520
	1.3. Agroclimatic background	520
	1.4. Trends in Finnish agriculture	526
	1.5. Finnish agricultural policy	531
	1.6. Choice of areas for impact studies	532
	1.7. Choice of crops for impact studies	533
	1.8. Impact models and data	533
	1.9. Choice of scenarios for impact studies	534
	1.10. Policy considerations – the mapping of future field crop zones	542
2.	The Effects on Barley Yields	547
	Jaakko Mukula	
	2.1. Introduction	547
	2.2. Description of the study areas	549
	2.3. Weather and yield data	553
	2.4. Description of the models	554
	2.5. Model validation	557
	2.6. The effect of climatic variations on barley yields	558
	2.7. Discussion	562
3.	The Effects on Spring Wheat Yields	565
	Olli Rantanen	
	3.1. Introduction	565
	3.2. Weather and yield data	567
	3.3. Development of the model	568
	3.4. Description of the model	569
	3.5. Model validation	569
	3.6. The effect of climatic variations on wheat yields	569
	3.7. Discussion	576

4. The Effects on the Profitability of Crop Husbandry 579
   Uuno Varjo†
   4.1. Introduction 579
   4.2. The variability of gross margins 579
   4.3. Climatic variability and marketable yield 584
   4.4. Model sensitivity analyses 586
   4.5. Comparison of model results with observations 586
   4.6. Effects of climatic variations on the profitability of crop husbandry 589
   4.7. Conclusions 596

5. Conclusions and Implications for Policy 599
   Lauri Kettunen
   5.1. Summary of results 599
   5.2. Implications of CO_2-induced warming in Finland 604
   5.3. Effects on production, income and regional policy 605

Acknowledgments 609
References 611

List of Contributors

KETTUNEN, Professor Lauri
Agricultural Economics Research Institute
Luutnantintie 13
SF-00410 Helsinki
Finland

MUKULA, Professor Jaakko
Institute of Plant Husbandry
Finnish Agricultural Research Center
SF-31600 Jokioinen
Finland

POHJONEN, Dr Veli
UNDP – Field Office
P.O. Box 5580
Addis Ababa
Ethiopia

RANTANEN, Mr Olli
Institute of Plant Husbandry
Finnish Agricultural Research Center
SF-31600 Jokioinen
Finland

†VARJO, Professor Uuno
Department of Geography
University of Oulu
Linnanmaa
SF-90570 Oulu
Finland

---

†We were saddened to learn that Professor Varjo passed away shortly after the completion of this case study (*the Editors*).

Abstract

Cereal farming in Finland is very sensitive to year-to-year variations in climate. Any changes in climate over the longer term are therefore likely to affect crop yields and consequently cereal production and farm incomes. This case study attempts to quantify the effect of climate on Finnish agriculture and, in particular, to study how variations in climate influence yields, production and the farm economy.

Three studies of yield responses to climatic variations are reported. Each utilizes regression models to estimate yields for a particular climate (represented by a set of meteorological data). Four types of climate (scenarios) have been simulated in each investigation:

(1) The reference or baseline climate, representing present-day climatic conditions.
(2) A warm-period scenario (selected from the historical instrumental climatic record.
(3) A cool-period scenario (also taken from historical data).
(4) The climate estimated for doubled concentrations of atmospheric carbon dioxide, based on predictions by the Goddard Institute for Space Studies (GISS) general circulation model.

Changes in the yields of barley, spring wheat and oats have been estimated for each climatic scenario relative to the baseline. The assessments have been conducted for sites both in northern and in southern Finland to illustrate the geographical contrasts in present-day climate, changes in climate and changes in yield. The variability of crop yields is also examined, to evaluate the likely reliability of crop yields (crop certainty) under changed climate.

To assess the farm level impact of climate-induced changes in crop yields, the profitability of cultivating cereal crops is also investigated. The gross returns per unit of production are balanced against the variable expenditure and farmer's wages, to determine net returns under each climatic scenario, assuming 1980 levels of prices and costs. An assessment is also made at national level, based on 1984 prices.

Results indicate that changes in temperature and precipitation implied in the climatic scenarios can have a marked effect on crop yields. A cooler climate generally leads to below-average yields both in northern and in southern Finland, but this effect may be compensated, to some extent, in southern Finland by increases in precipitation that would remove the risk of summer drought. In the north of Finland precipitation excess, rather than deficit, depresses yields so that for climatic warming to be beneficial for yields in this region, precipitation totals should not be too high. In general, however, warmer conditions would raise crop yields (particularly those implied in the $2 \times CO_2$

scenario), and increase farm incomes if prices and costs do not change. This assumption is, however, questionable: Finland is already producing more agricultural products than the domestic market can absorb. An increase in yields would worsen the over-supply problem and the government might be forced to take further actions to cut production. Nonetheless, for farmers the effects of an increase in temperature would be more tangible. Cultivation boundaries of certain crops will move northwards, crop selection will be less difficult and new, later developing varieties could be introduced into southern regions, with implications for plant breeding and agriculture extension services in Finland.

SECTION 1

Introduction

1.1. The Purpose of the Study

Yields of crops vary in Finland from year to year owing to the variations in climate, particularly in temperature. Since the growing season is relatively short early sowing is important and this is sometimes delayed owing to late springs. Preparations of the seedbed cannot begin before the snow has melted and the soil has both thawed and dried sufficiently. Because of this, annual crops generally cannot benefit from the first 1-3 weeks of the growing period.

Drought in the early summer and frosts also lower the yields. Rains in the harvesting period are particularly damaging and have caused, perhaps, the greatest losses in yields. On the other hand, winter cereals and grasses also suffer from the length of the winter and sometimes from unpredictable thawing and refreezing of the soil in the early spring. Especially in areas with much snow in the east and the north of the country "low temperature" parasitic fungi are a problem. They are able to grow under the snow and to kill the overwintering plants.

Boundaries of cultivation have shifted from time to time. For example, above-average summer temperatures in the 1930s coincided with wheat and rye cultivation in the northern parts of Finland. But climate only provides a partial explanation for subsequent retreat of these crops. Economic considerations are also important (*see* Subsection 1.4.1).

While the general effect of climate on yields is well known there have, however, been few quantitative studies of the effects of climatic variations on Finnish agriculture (Keränen, 1931; Hustich, 1952). This case study estimates this effect by examining how variations in climate influence yields and by considering the economic consequences of these effects. Yield functions are estimated and are then applied to meteorological data from several warm and cool periods and, as a special case, to data describing the climate predicted for a doubled concentration of atmospheric carbon dioxide by the Goddard Institute for Space Studies (GISS) general circulation model (Hansen *et al.*, 1984).

1.2. Geography, Soil and Farm Structures

Finland is situated between the northern latitudes of 60 and 70 degrees. Biologically and geographically the country is in the boreal zone. This biome is sparsely populated (the average density 16 inhabitants/km^2), and is dominated by coniferous taiga forests. The Gulf Stream, which skirts around the Scandinavian peninsula, has a warming effect on the Finnish climate. As a result, nowhere else is agriculture practiced extensively so far north (*Figure 1.1*). Climatic conditions at these same latitudes in Sweden and Norway are suitable for agriculture, but the topography and soils are not as favorable as in Finland.

Peat soils are characteristic of many areas (*Figure 1.2*), especially in the center and north (Ilvessalo, 1960). Tills are predominant in the central region, and most alluvial soils occur in the coastal areas of the south and southwest. Some degree of areal differentiation in the production capacity of the soils may be found in southern Finland between the coastal clays and the inland till areas, whereas a relatively high spatial homogeneity is typical of soil conditions in the north (Varjo, 1977).

Nearly two-thirds of the country is under forest, while less than a tenth is cultivated. Owing to rapid post-war urbanization and structural change the share of agriculture of the gross domestic product has decreased (from 16% in 1950 to 4.6% in 1982). Over the same period, partly as a result of mechanization, the population engaged in agriculture has decreased from 46% to 13%. In 1981 the typical farm averaged 11.2 ha of arable land and 35.1 ha of forest (*Farm Register*, 1984).

The country is divided into 18 agricultural districts, each served by a regional center (*Figure 1.3*). Farming in Finland is based essentially on three elements: field cultivation, animal husbandry and forestry. Animal production (especially milk production) has typically been the backbone of the farm economy and has succeeded in providing reasonably stable incomes which do not fluctuate greatly from year to year in response to climatic variations. Despite the shortness of the grazing period, dairy cows are raised even in Lapland, in some cases being kept indoors throughout the year. The short grazing period is unfavorable for beef cattle and the number of these is small. Sheep breeding has almost totally disappeared despite many attempts to promote it.

1.3. Agroclimatic Background

Finland is the northern-most country where agriculture is practiced comprehensively, although only a half of the year (in the north even less) is suitable for plant production. Climatic and natural conditions vary greatly from the south to the north. Permanent snow cover lasts from 100 days in the south to over 200 days in the north [*Figure 1.4(a)*]. It protects winter crops from frosts but causes damage in some years. Winter crops cover actually only about 3–5% of the total arable area, although milder winters would favor the expansion of their area.

The length of growing season (the period of time with long-term mean daily temperatures above 5 °C) ranges from more than 175 days in the south to less

Introduction

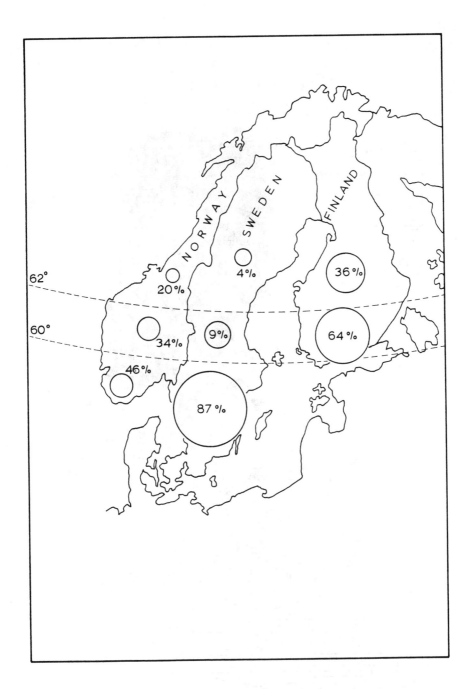

Figure 1.1. The acreage of seed crops (cereals, pulses, oilseed rape) between different latitudes in the Scandinavian countries. The size of each circle corresponds to the cropped area (Kivi, 1983).

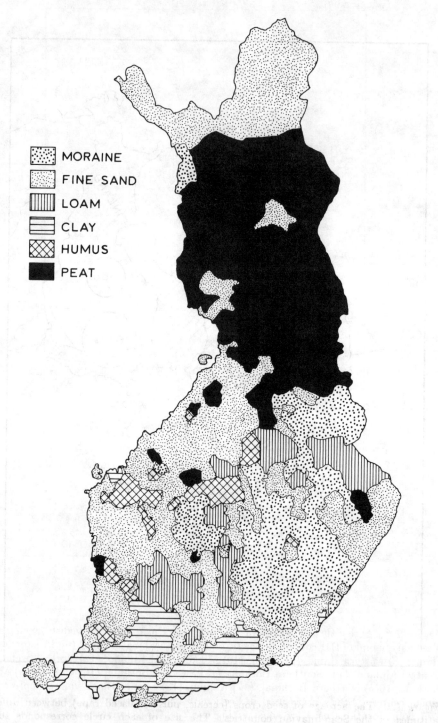

Figure 1.2. Dominant soil types in Finland (Kurki, 1972).

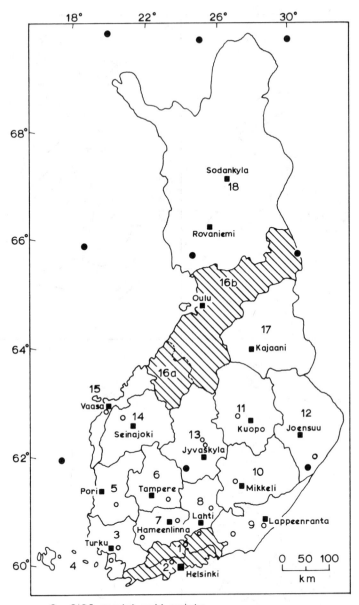

Figure 1.3. Agricultural districts of Finland: (1) Uusimaa, (2) Swedish Uusimaa, (3) Varsinais-Suomi, (4) Economic Society of Finland, (5) Satakunta, (6) Pirkanmaa, (7) Häme, (8) Eastern Häme, (9) Kymi, (10) Mikkeli, (11) Kuopio, (12) North Karelia, (13) Central Finland, (14) Southern Bothnia, (15) Swedish Bothnia, (16) Oulu [16(*a*) Middle Bothnia, 16(*b*) Northern Bothnia], (17) Kainuu, (18) Lappi. Black squares represent regional centers. Study areas used in Sections 2 and 4 are shaded. Also shown are the grid point locations used for constructing the $2 \times CO_2$ climate scenario (solid circles) and the meteorological stations referred to in Section 3 (open circles).

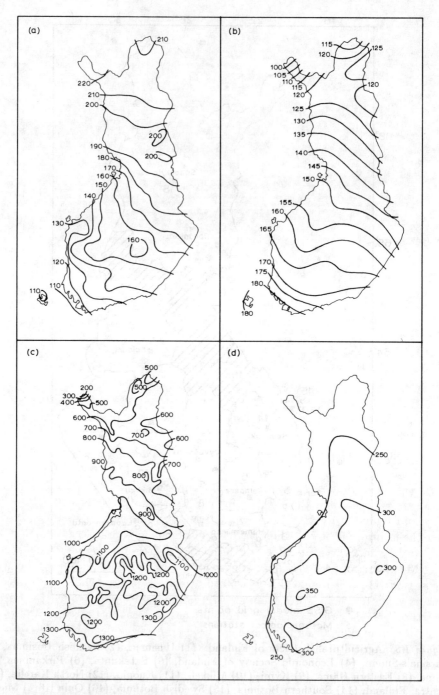

Figure 1.4. The agroclimate of Finland for the period 1931–60: (*a*) Days of permanent snow cover (Kolkki, 1969). (*b*) Length of growing season (days with mean temperature above 5 °C) (Solantie, 1975). (*c*) Effective temperature sum (degree-days above 5 °C) (Solantie, 1975). (*d*) May–September precipitation (mm) (Helimäki, 1967).

Introduction

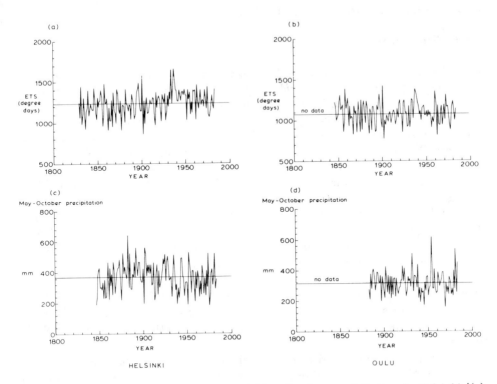

Figure 1.5. Effective temperature sum and May–October precipitation in Helsinki [(a) and (c)] and Oulu [(b) and (d)], 1829–1983. (Source: Finnish Meteorological Institute.)

than 130 in the north [*Figure 1.4(b)*]. The total receipt of solar radiation is, however, about the same over the whole country in the middle of the summer owing to the longer days in the north, and this compensates, in part, for the shortness of the growing season. The effective temperature sum (ETS), calculated over the threshold temperature of 5 °C varies, on average, from 800 to 1350 degree-days over the latitudes 60–67°N where arable crops farming is practiced. North of this area, in Lapland, the ETS value falls below 500 degree-days [*Figure 1.4(c)*].

Mean annual precipitation varies from 400 to 650 mm and during the growing season (May–September) from 250 to 370 mm [*Figure 1.4(d)*]. Autumns are normally wetter than springs and early summer drought frequently impedes growth of cereals and grasses, especially in coastal areas in the south and in the west. Autumn rains often hamper harvesting and a sizeable amount of crops (up to 51 000 ha or 4% of the area of grains in some years) have had to be left unharvested.

Long-term time series of both ETS and May–October precipitation are illustrated in *Figure 1.5* for southern Finland (Helsinki) and for the north-central region (Oulu). These climatic series form the basis for the selection of climatic scenarios in the case study (*see* Subsection 1.9). Sections 2, 3 and 4 include further information on the time series of recorded temperature and precipitation.

1.4. Trends in Finnish Agriculture

1.4.1. Field crops

There has been a marked increase in crop yields in Finland throughout much of this century (*Figure 1.6* and *Table 1.1*). The greatest increases have been in spring wheat yields, especially in the west-central region (Southern Bothnia, 4.8% p.a. over 1951–80) and in the extreme south-west (Varsinais-Suomi, 3.5% p.a.). Increases were most marked during the 1960s, especially in Southern and Northern Bothnia, the trend being less pronounced from 1970 onwards.

It should also be observed here that, although some similarity can be discerned between interannual fluctuations in yields and the temperature conditions during the growing season (as reflected by ETS), the significant increase in yields over the longer term (*Figure 1.6*) is not dependent in any way on the ETS, but is due largely to the general progress made in crop husbandry methods. Nearly half of the rise in yield levels has been attributed to the development of domestic plant varieties, the other half to increased fertilizing (Elonen, 1983).

Recent agricultural development has seen a shift from cattle farming to crop production, particularly in south and southwest Finland. This has led to a reduction of grass and an increase in cereals and oilseed (*Figure 1.7*). The share of barley has increased most, particularly in the south and southwest of the country.

Figures for cropped area, average yields and total production in 1982 are given in *Table 1.2*. These values, however, can be regarded only as an approximate indicator of present land use because large year-to-year fluctuations in production have been a characteristic of recent years (Kettunen, 1985a). Feed grains (barley and oats) account for the largest part (48.8%) of arable land kept in production; 33.7% is used for cultivation of grasses (leys), while 7.8% is under the bread grains – wheat and rye. In 1982 an additional 8% of total agricultural land was in pasture, 3% in fallow and 8% uncultivated. The area mentioned last consists of soils with poor quality. It could be recultivated if the Finnish production policy were to be changed.

Grasses are commonly grown throughout the country, barley up to latitude 66°N (sometimes up to the latitude 67°N or even 68°N), oats up to 65°N, spring wheat and rye up to 63°N and other crops only in the southwestern part of the country. Finland is, in fact, the only country in the world where widespread grain crop production extends north of the Arctic Circle. There is evidence for rye and barley being commonly grown in Lapland about 25 km north of the Arctic Circle during the early eighteenth century (Outhier, 1744). Small-scale barley production is practiced in that area today, but the limit for rye now lies much further south.

Shifts of the limits of crop cultivation since 1930 are shown in *Figure 1.8*. Having extended its area of cultivation up to 1930, barley appears to have altered its position very little recently (Valle and Aario, 1960; Varjo, 1977, 1980). However, much more extensive changes have taken place in the cultivation of rye and spring wheat. While rye was grown in the middle of Lapland in the 1930s, its limit had shifted southward by 1975 to the level of Vaasa in the west and

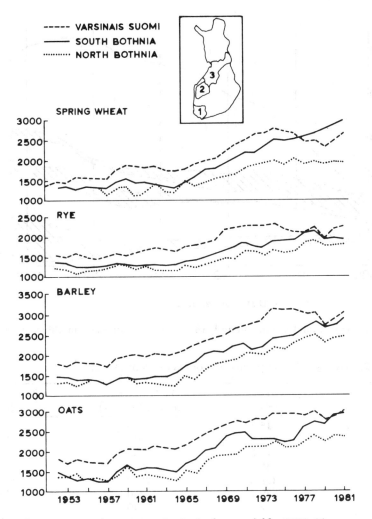

Figure 1.6. Three-year moving means of principal crop yields, 1951–81.

Table 1.1. Average annual changes in yields (%) obtained from the major grain crops in Varsinais-Suomi, Southern Bothnia, Northern Bothnia and Lapland, 1951–80.

Grain	Varsinais-Suomi	Southern Bothnia	Northern Bothnia	Lapland
Spring wheat	3.5	4.8	1.8	–
Rye	1.4	1.4	0.7	–
Barley	2.4	3.2	2.7	1.5
Oats	2.1	3.1	3.2	2.8

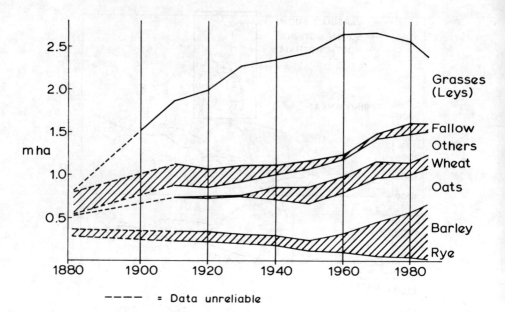

Figure 1.7. Arable land use in Finland, 1880–1983 (J. Mukula, unpublished data, 1985).

Table 1.2. Area, yields and production of major crops (1982).

	Area (000 ha)	Yield (kg/ha) (rounded values)	Total production (000 t)
Winter wheat	15.7	3090	48.5
Spring wheat	127.2	3040	386.9
Rye	16.3	2150	35.0
Barley	540.4	2960	1598.5
Oats	459.3	2870	1319.9
Potatoes	39.1	15370	601.1
Sugar beet	32.4	23340	756.1
Hay	445.3	3790	1689.4
Silage	244.4	17670	4319.2
Oil seed	63.7	1510	96.3
Other crops	64.3		
Total	2048.1	2526[a]	5094.4[b]
Pasture	205.4		
Fallow	74.2		
Uncultivated	188.9		
Total area	2516.6		

[a] Feed units per hectare without straw (1 feed unit is equivalent to the energy content of 1 kg barley).
[b] Million feed units without straw.
Source: *Monthly Review of Agricultural Statistics*, 1982.

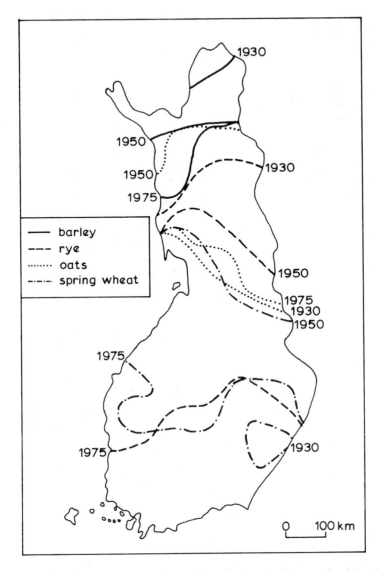

Figure 1.8. Geographical changes in the northern limits of widespread cultivation (more than 1% of field area) of certain crops, 1930–1975 (Varjo, 1977).

Joensuu in the east. The limit for spring wheat had similarly been pushed as far north as southern Lapland by 1950, but retreated to around the central parts of the Lake Region and the vicinity of Vaasa in the west by 1975. More or less the same trend may also be noted for oats, a crop that was still common in northern Lapland in 1950, but has since retreated to Southern Lapland and Kainuu (*Figure 1.8*).

These shifts seem to have less to do with changes of climate than changes in the relative profitabilities of the crops. For instance, in terms of the annual

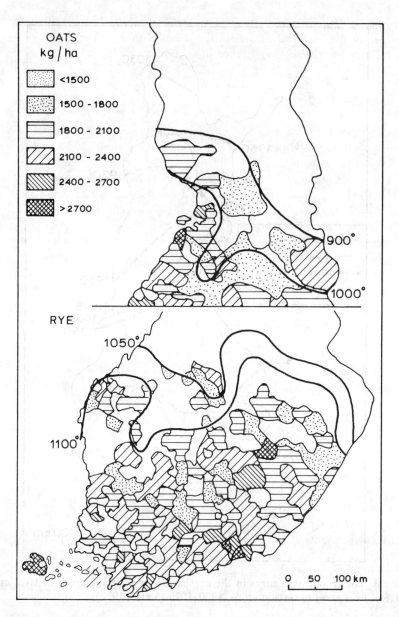

Figure 1.9. Yields of oats and rye in 1974–76 (Varjo, 1980) and isopleths for effective temperature sums of 900, 1000, 1050 and 1100 growing degree-days in 1931–60 (Kolkki, 1969).

ETS at which harvests were obtained, the limits within Finland in 1974–1976 would correspond to approximately 1000 degree-days for oats and 1100 degree-days for rye (*Figure 1.9*). Cultivation of oats and rye is apparently restricted in Finland to those places where harvests usually exceed 1500 kg/ha. However, the past cultivation limits bear little relation either to shifts in the isolines of ETS or to changes in yields. More important may be other climatic and/or economic factors. These are considered further in Section 4.

1.4.2. Animal husbandry

Livestock account for about 70% of the total value of agricultural production (Kettunen, 1985a), milk and beef alone contributing about 50%. The number of cattle has declined from a peak of 2.2 m in 1963 to 1.7 m in 1984 (Talman, 1978; *Monthly Review of Agricultural Statistics*, 1985). The highest densities, over 45 cows per 100 ha of arable land, are found over extensive areas of eastern and southeastern Finland and the inland lake region, in contrast to the more arable areas of the south and southwest (Varjo, 1983). Southern parts of the country where farms are larger have recently seen a shift to crop production, while smaller farms in the north have maintained an emphasis on livestock, particularly milk production (which is labor intensive and well suited for small farms). Levels of animal production for the years 1980–1984 are shown in *Table 1.3*.

Table 1.3. Animal production in 1980–1984.

Product (millions)	1980	1981	1982	1983	1984
Milk (l)	3174	3082	3086	3136	3129
Beef (kg)	114	122	117	118	127
Pork (kg)	169	179	181	177	169
Eggs (kg)	79	80	82	83	88
Poultry (kg)	15	17	17	18	19
Other meat (kg)	2	2	2	2	2

Source: Kettunen, 1985a.

1.5. Finnish Agricultural Policy

Agricultural productivity in Finland is influenced to a great extent by national policies of agricultural support. The effect of climate on productivity would therefore be determined in part by adjustments in these policies. The main goals of Finnish agricultural policy are to safeguard the supply of food, to secure an acceptable standard of living for farmers while keeping consumer prices at a reasonable level, and to maintain the rural population (Kettunen, 1981). These partly conflicting goals have led both to surpluses in animal products (*Table 1.4*) and also to high producer prices, well above world market prices (*Table 1.5*). Exports of agricultural products therefore have to be subsidized in order to support farmers incomes and these subsidies place a heavy burden on the state budget. Consequently, an aim of agricultural policy has been to restrain the growth of production and ultimately to reduce output. There are two types of restrictions on production. The first involves collective quotas for milk, meat, eggs and feed grains. If these collective quotas are exceeded, farmers get only

Table 1.4. Self-sufficiency of the main agricultural products (production expressed as a percentage of requirements; 100% indicates self-sufficiency).

Product	1980	1981	1982	1983
Bread grains	70	40	56	93
Milk	129	125	122	133
Beef	102	113	110	115
Pork	120	128	126	117
Eggs	140	153	159	165

Source: Kettunen, 1985b.

Table 1.5. Producer and world market prices for main products in January, 1985, Finnish Marks (FIM)/kg[a].

Product	Domestic price	World market price
Wheat flour	1.90	0.82
Butter	37.05	7.59
Pork	14.89	7.40
Beef	23.29	3.66
Eggs	12.82	5.00

[a] 1 US Dollar = 6.64 FIM (January, 1985).
Source: Anon., 1985a.

the world market price for the excess production. The farmers' share of extra costs is collected by a complex system of marketing fees and taxes (Kettunen 1985b). From the beginning of 1985 a second type of restriction was applied: individual farm quotas for milk. For excess production above quotas farmers receive only the world market price, this "penalty" being calculated for each individual milk farm.

These restrictions on production mean, in fact, that there is no room for any increase of agricultural production. If the productivity increases (i.e., production per hectare improves), arable land has to be reduced accordingly. We shall see later that increases in productivity under scenarios of a warmer climate indicate a need for substantial adjustment of agricultural policy and land use to avoid increased overproduction.

1.6. Choice of Areas for Impact Studies

Finland is marginal for the cultivation of all types of cereals, particularly in the extreme north. Since reliable data are not available for the north the present study is restricted largely to southern and middle parts of the country. Study areas have been chosen to illustrate, in particular, the regional contrast between north and south, although comparison of crop yields between east and west is also undertaken in Section 3. Some of these regional differences are exhibited in *Table 1.6*.

Introduction

Table 1.6. Regional differences in agriculture and forestry.

	Yield of barley (kg/ha)	Average area of:		Average net farm income of agriculture 1981
		Arable land (ha/farm)	Forest land (ha/farm)	
South and west	3415	14.7	26.8	25 358
East	2732	9.9	40.6	19 443
North	2520	9.6	56.4	16 731

Source: Central Statistical Office of Finland, 1983.

1.7. Choice of Crops for Impact Studies

Section 2 focuses on the effects of possible climatic changes on barley, the most extensive annual crop in Finland. Section 4 considers effects on regional yields of both barley and oats to examine the impact on profitability of raising these crops.

Spring wheat cultivation in Finland has been increasing recently and a possible future amelioration of the climate might be expected to accelerate this trend. Section 3 considers the effect of temperature and precipitation change on wheat yield and crop certainty zones for different climatic scenarios.

1.8. Impact Models and Data

In Sections 2, 3 and 4, regression models are used to estimate the effect of climatic variations on yields of spring wheat, barley and oats. The dependent variable was crop yield, while several explanatory climatic variables were included in the regression equations. In Sections 2 and 4, to investigate the contrast in conditions between northern and southern Finland, long-term time series of climatic and yield data were analyzed for Oulu (65°N, 25°E), in the subdistrict of Northern Bothnia, and Helsinki (60°N, 25°E), in Uusimaa district (*see Figure 1.3*). Monthly data for both ETS and precipitation exist for the periods 1846–1983 at Helsinki, and 1883–1983 at Oulu (*see Figure 1.5*). Information on barley yields was also available for these areas and time periods, enabling a simple regression model to be developed in Section 2. However, the temporal resolution of the climatic data (monthly) was not sufficient to reflect those weather variations that are important for explaining barley yield, so this model was discarded in favor of another, based on daily data, for the period 1959–83. In this second model, the growing season is divided into several periods according to the growth stages or "phases" of the crop, and detailed daily meteorological data are allocated to each phase. A selective regression estimation method was applied on the climatic and annual yield (deviations from the trend) data to determine the statistically significant climatic variables.

In Section 3, a similar procedure was used for estimating spring wheat yields. Climatic and yield data from nine crop districts in southern Finland, for the period 1959–83, were used to develop the model. In both Sections 2 and 3, the influence of non-climatic factors (such as increased use of fertilizers and chemicals, and technological improvements) on yields are eliminated by concentrating on residuals from the yield trends. The approach can thus be characterized as a residual analysis.

In Section 4, regression models of a different nature are presented for both barley and oats. These have been developed on the basis not only of temporal variations in climate and yields (the analysis uses data for the period 1979–81), but also on spatial variations between regions in Finland during these years. The long-term climatic records for Helsinki and Oulu are used in the construction of climatic scenarios in this application, and experiments have been conducted to simulate impacts in those two regions.

1.9. Choice of Scenarios for Impact Studies

In common with the other case studies in this volume (in Saskatchewan, Canada; Iceland; northern European USSR; and Japan), climatic scenarios were selected to reflect issues of particular concern in Finland, simulating typical short-term climatic anomalies and their relation to possible longer term, CO_2-induced climatic change. The scenarios chosen in this study were of two broad types: (i) an extreme weather-decade or period of weather-years taken from the historical instrumental record, and (ii) the climate simulated by the Goddard Institute for Space Studies (GISS) general circulation model for doubled concentrations of atmospheric CO_2 (Hansen et al., 1984).

1.9.1. Instrumental scenarios

Instrumental meteorological data are used for two purposes in the Finnish case study. Firstly, data were required to construct the climatic scenarios. Long-term time series of monthly temperature and precipitation at Helsinki and Oulu have been taken to represent the climatological "histories" of southern and northern Finland, respectively. These are the records from which historical "cool" and "warm" periods have been selected. They are also required in constructing the 2 × CO_2 scenario, and provide reference data against which to assess all the climatic information used in the impact experiments. Furthermore, they are the representative sites for which several of the climate impact experiments have been conducted.

Secondly, data were needed both to develop and to run the impact models. The specific requirements of certain models necessarily limited the length of period for which data were available (see Subsection 1.8), and therefore restricted the selection of the instrumental climatic scenarios. Nonetheless, scenarios and reference periods have, wherever possible, been used consistently throughout the case study.

The major climatic constraint on cropping, particularly in the more northerly regions of Finland, is temperature, through its effects on the duration and intensity of the growing season. Historically, several of the more notable

Introduction

episodes in Finnish agriculture such as famines and production surpluses have coincided with unusually cool or warm sequences of years. Thus, temperature has been used as the governing criterion for selecting the instrumental scenarios. In order to reflect better the growing season temperature conditions for field crops, mean daily temperatures have been converted to values of ETS, and these were used to choose the scenarios.

Long-term records of annual ETS for the period 1946–1983 at Helsinki and Oulu are presented in *Figure 1.10*. Temperature data for the Helsinki station (Kaisaniemi, in the city center) are corrected for the warming effect of urbanization. Since this report was prepared, the data have been further modified (Mukula, personal communication), but the changes are minor and do not affect the results or conclusions of the investigations (Finnish Meteorological Institute, unpublished data). The Oulu station site is sufficiently rural in nature for any similar effects to be ignored. On average, ETS values at Helsinki are about 200–300 degree-days higher than at Oulu, but the fluctuations depicted on the two plots correspond quite closely. A number of periods of anomalous temperatures can be identified at both stations, in particular, sequences of cool years from the late 1850s to the 1900s, in the 1920s and during the late 1950s and early 1960s, and warmer-than-average periods, especially in the 1930s and 1940s.

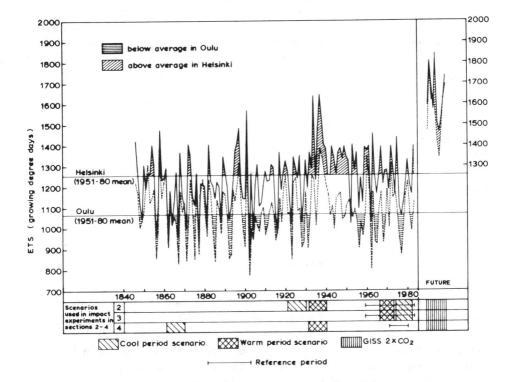

Figure 1.10. Annual effective temperature sums, 1846–1983, at Helsinki (upper plot) and Oulu (lower plot) and the position of the climatic scenarios used in impact experiments. ETS values for the GISS 2 x CO_2 scenario are as described in Section 4.

Table 1.7. Mean annual effective temperature sum (growing degree-days) and May–October precipitation (mm) for the period 1951–80, and percentage deviations from these for the reference, warm period, cool period and $2 \times CO_2$ scenarios used in different sections of the Finland study.

Climatic scenarios	Section of report	Mean annual ETS	Mean May–Oct precipitation
		Helsinki	
	Reference periods		
1951–80	1	1255 GDD	344 mm
1959–83	2,3	+1%	−1%
1971–80	4	−2%	−1%
	Warm periods		
1931–40	2,4	+13%	−2%
1966–73	2,3	+2%	−3%
	Cool periods		
1861–70	4	−10%	+1%
1921–30	2	−3%	+25%
1974–82	2,3	−4%	+6%
$2 \times CO_2$ scenario	2,3,4	+31%[a]	+49%
		Oulu	
	Reference periods		
1951–80	1	1068 GDD	313 mm
1959–83	2,3	+1%	0
1971–80	4	+2%	−8%
	Warm periods		
1931–40	2,4	+11%	+11%
1966–73	2,3	+4%	−9%
	Cool periods		
1861–70	4	0	–
1921–30	2	−0%	+6%
1974–82	2,3	−1%	+2%
$2 \times CO_2$ scenario	2,3,4	+46%[a]	+58%

[a] ETS calculated according to the method described in Section 4.

The 1860s was a period of famines in Finland, and the decade 1861–70 was chosen in Section 4 as a cool period scenario. Decadal mean values of ETS are well below the 1951–80 average in southern Finland although around average in the north. May–October precipitation was slightly below average in the south while data are not available for Oulu, in the north (*Table 1.7*). Unfortunately, the relevant daily meteorological data for this period, required as inputs to the impact models in Sections 2 and 3, were not available. Instead, alternative cool periods were selected: 1974–82 and 1921–30 (*Table 1.7*).

The obvious choice for a warm period scenario was the decade of the 1930s (1931–40). This represents a very large positive temperature anomaly in both the south and the north (ETS values 13% and 11% above the 1951–80 mean, respectively), while precipitation was a little below average at Helsinki, but considerably above average at Oulu (*Table 1.7*). This scenario was used in Sections 2 and 4, but an alternative warm period (1966–73) was also employed in Sections 2 and 3 because of constraints on data for the 1930s period.

Limitations of data also help to explain the choice of reference periods. The common baseline period 1951–80 (used in the other case studies in this volume, and for comparison in this section – see *Table 1.7* and *Figure 1.10*) could not be applied to model experiments in Sections 2 and 3 because appropriate daily meteorological data were available only for the period 1959–83. (Data for the period 1920–58 were subsequently made available for further scenario experiments in Section 2, but arrived too late to allow for a complete reworking of the analysis.) In Section 4, it was found convenient to conduct model experiments for 10-year sequences rather than for longer periods, thus explaining the selection of 1971–80 as the baseline period. However, the differences between the ETS and precipitation values for these three periods (1951–80, 1959–83, and 1971–80) are slight, with the exception of Oulu precipitation in 1971–80, which is considerably lower than for the other two periods (*Table 1.7*).

1.9.2. The GISS $2 \times CO_2$ scenario

Outputs from the GISS general circulation model, in common with the other case studies in this volume, were used to simulate a $2 \times CO_2$ climate in Finland. (Details of the GISS general circulation model are described in Part I, Section 3 of this volume.) Data were provided on a 4° latitude × 5° longitude network of grid points for an area encompassing the whole country (*Figures 1.3* and *4.6*). Values of mean monthly temperature (°C) and precipitation rate (mm/day) were required for the Finnish study, and information on these was obtained in three forms:

(1) GISS model-generated $1 \times CO_2$ equilibrium values.
(2) GISS model-generated $2 \times CO_2$ equilibrium values.
(3) (GISS $2 \times CO_2$) − (GISS $1 \times CO_2$) values (change in equilibrium climate for a doubling of CO_2).

Before using the GISS $2 \times CO_2$ data for developing the $2 \times CO_2$ scenario, it was first necessary to evaluate the performance of the GISS model in simulating present climatic conditions in Finland. For this purpose, monthly values of temperature and precipitation generated in the GISS model $1 \times CO_2$ equilibrium run were examined to see how closely they reproduced the observed conditions over the period 1951–80 at Helsinki and Oulu. The GISS $1 \times CO_2$ grid point values were interpolated to the two station locations using a simple averaging procedure. Since both stations lie very close to the 25°E longitude line, the following weighting factors were used:

(1) Interpolation to Oulu (65° N, 25° E) = $(3GP_{66,25} + GP_{62,25})/4$
(2) Interpolation to Helsinki (60° N, 25° E) = $(GP_{62,25} + GP_{58,25})/2$

where GP_{ij} is the grid point value at latitude i and longitude j.

Observed values are compared with GISS 1 × CO_2 and GISS 2 × CO_2 values of mean monthly temperature in *Figure 1.11*, and of monthly mean precipitation rate in *Figure 1.12*. The observed winter temperatures at both stations are quite well reproduced by the GISS model (*Figure 1.11*), but the amplitude of the annual cycle is badly underestimated, giving much lower summer temperatures than are observed [e.g., estimated mean July temperature at Oulu is almost 7°C below the observed value – *Figure 1.11(a)*]. In fact, the observed summer temperatures exceed even the GISS 2 × CO_2 estimates, although in winter the GISS estimates are considerably higher than the observed (one result of the much greater warming in winter than in summer predicted by the GISS model).

The pronounced summer maximum of precipitation observed at Oulu, and the autumnal maximum at Helsinki are not well matched by the GISS 1 × CO_2 values (*see Figure 1.12*). Annual rainfall estimates are both greater in quantity, and more evenly distributed throughout the year. At Oulu, the winter and spring precipitation estimates are as much as double the observed values [*Figure 1.12(a)*], and the late-spring values are grossly exaggerated at Helsinki [*Figure 1.12(b)*]. The GISS 2 × CO_2 precipitation estimates are, in all cases, greater than the 1 × CO_2 values, with the most pronounced increases estimated for the summer months (in contrast to the temperature increases).

The poor correspondence of GISS model 1 × CO_2 estimates to observed conditions in Finland implied that the 2 × CO_2 estimates, if used directly in impact experiments, were likely to be similarly unrealistic in representing local agroclimatic conditions under doubled CO_2. For this reason, as in the other case studies, it was assumed that the *change* between GISS-generated 1 × CO_2 and 2 × CO_2 equilibrium conditions could be taken to represent the difference between the observed present-day climate and a future 2 × CO_2 climate. Thus, in each of the subsequent sections (2, 3 and 4) the 2 × CO_2 scenario has been constructed by adding the difference between 1 × CO_2 and 2 × CO_2 temperature and precipitation values to the mean values for the appropriate reference period – the "differences" technique:

$$T_{2 \times CO_2} = \left(T_{\text{GISS } 2 \times CO_2} - T_{\text{GISS } 1 \times CO_2} \right) + T_{\text{REF}}$$

$$P_{2 \times CO_2} = \left(P_{\text{GISS } 2 \times CO_2} - P_{\text{GISS } 1 \times CO_2} \right) + P_{\text{REF}}$$

Note that the adjustment of precipitation values differs from the "ratios" method used in the Canadian and Icelandic case studies.

Introduction

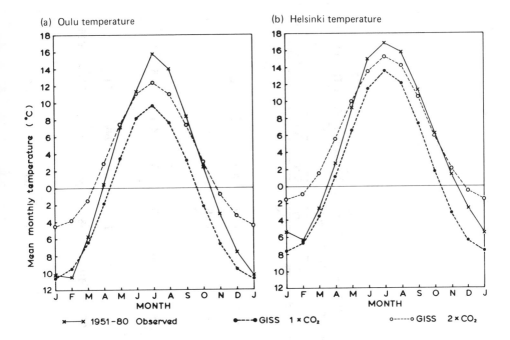

Figure 1.11. Comparison of observed mean monthly temperatures (1951–80) with GISS model-generated $1 \times CO_2$ and $2 \times CO_2$ equilibrium temperatures at (*a*) Oulu, and (*b*) Helsinki. Observations at Helsinki are not corrected for urbanization effects. (Source: Finnish Meteorological Institute, unpublished data.)

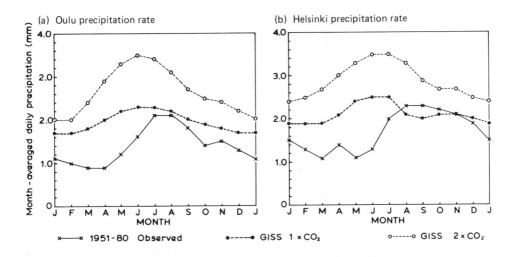

Figure 1.12. Comparison of observed monthly mean precipitation rate (1951–80) with GISS model-generated $1 \times CO_2$ and $2 \times CO_2$ equilibrium precipitation rates at (*a*) Oulu, and (*b*) Helsinki. (Source: Finnish Meteorological Institute, unpublished data.)

Table 1.8. Mean monthly temperature (°C) and precipitation (mm) observed (1951–80) and for the $2 \times CO_2$ scenario at Helsinki and Oulu, and for two stations with a present-day (1931–60) climate analogous to the $2 \times CO_2$ conditions.

Climatic Scenario	Jan	Feb	Mar	Apr	May	June	July	Aug	Sep	Oct	Nov	Dec	Year
HELSINKI[a]													
Temperature (°C)													
1951–80 Observed	−5.3	−6.2	−2.7	2.8	9.2	15.0	16.9	15.9	11.3	6.2	1.4	−2.5	5.2
$2 \times CO_2$	0.9	−0.5	2.4	7.2	12.5	17.1	18.5	18.0	14.5	10.5	6.6	3.4	9.3
Analogue: Friedrichshafen (W. Germany)	−1.0	0.2	4.1	8.6	13.2	16.7	18.4	17.6	14.3	8.9	4.2	0.5	8.8
Precipitation (mm)													
1951–80 Observed	48	37	33	41	33	40	62	73	72	67	66	59	629
$2 \times CO_2$	63	54	57	70	62	68	96	108	98	86	81	74	915
Analogue: Friedrichshafen (W. Germany)	63	56	53	60	95	112	137	113	93	65	59	54	960

OULU

Temperature (°C)

	J	F	M	A	M	J	J	A	S	O	N	D	Year
1951–80 Observed	−10.2	−10.5	−5.8	0.4	7.1	13.3	15.8	14.0	8.4	2.4	−3.1	−7.6	2.0
2 × CO_2	−3.8	−4.9	−0.8	5.2	11.2	16.3	18.4	17.2	12.7	7.6	2.8	−1.2	6.7
Analogue: Oslo (Norway)	−4.7	−4.0	−0.5	4.8	10.7	14.7	17.3	15.9	11.3	5.9	1.1	−2.0	5.9

Precipitation (mm)

	J	F	M	A	M	J	J	A	S	O	N	D	Year
1951–80 Observed	35	27	27	27	37	48	65	65	55	45	44	41	514
2 × CO_2	46	35	44	54	73	84	101	94	77	65	61	56	790
Analogue: Oslo (Norway)	49	35	26	44	44	71	84	96	83	76	69	63	740

[a]Temperature values are uncorrected for urbanization effects – station Kaisaniemi. (Sources: Finnish Meteorological Institute, unpublished data; Müller, 1982.)

In Sections 2 and 3, period-mean daily temperatures (1959–83) for specific locations have been adjusted using interpolated monthly differences and then converted to daily ETS values. Only growing season (April–October) precipitation totals were required in impact modeling. A different method of adjusting ETS values is reported in Section 4, where differences in mean annual temperature have been mapped for the whole of Finland, and values for Helsinki and Oulu interpolated by eye. A simple conversion factor of temperature change (°C) to change in annual ETS is employed to adjust the annual ETS observations during the reference period 1971–80 (and also to adjust the 1951–80 mean values in *Table 1.7*). In addition, May–September precipitation values are required as inputs to the impact models.

The $2 \times CO_2$ scenario values for Helsinki and Oulu are compared with observed monthly values for the 1951–80 period in *Table 1.8*. Note how the amplitude of $2 \times CO_2$ temperatures at Oulu is much greater than that at Helsinki, and annual precipitation is approximately the same at both locations, reflecting the greater changes in conditions implied by the GISS model for northern Finland relative to the south.

Finally, it is interesting to compare the climatic conditions implied by the $2 \times CO_2$ scenario with climates found in other regions of Europe at the present-day. Examination of climatological statistics for the 1931–60 period suggest that the $2 \times CO_2$ climate of Oulu might resemble the present climate in southern Norway (Oslo, 60°N, 11°E), and that of Helsinki, the climate in southern Germany (Friedrichshafen, 48°N, 10°E; see *Table 1.8*). These "analogue regions" might offer some clues to assist in assessing the potential agricultural activities and productivity in Finland, under a $2 \times CO_2$ climate.

1.10. Policy Considerations – The Mapping of Future Field Crop Zones

In a marginal agricultural area such as Finland, changes in the climate can affect revenue from all branches of farming. Arable cropping is a valuable component of Finnish agriculture, therefore the impact of climatic change on arable crops merits the most careful study.

At present, useful guidance on the choice of different crops and varieties for different areas in Finland can be obtained on the basis of field cropping zones (*Figure 1.13*; Mukula, 1984). These zones are essentially geographical indicators of ripening limits, beyond which a certain variety of cereal crops will not reach maturity during an average growing season. The zoning is based mainly on the known close dependence of crop development rates on temperature (ETS), but also incorporates information on precipitation, soil type, altitude, effects of lakes and the sea, etc. The system is widely utilized in Finnish agriculture and, together with a comprehensive series of long-term field experiments with different cereal crops, forms the basis for practical extension work at the farm level.

The field cropping zonation has been developed using long-term average climatic data. As such, the zone boundaries are hypothetical, for climatic

Introduction

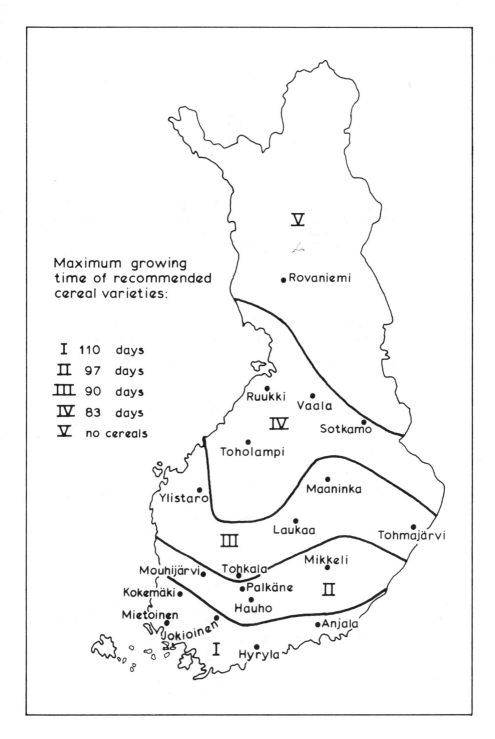

Figure 1.13. Zones for cultivars of field crops in Finland (Mukula, 1984).

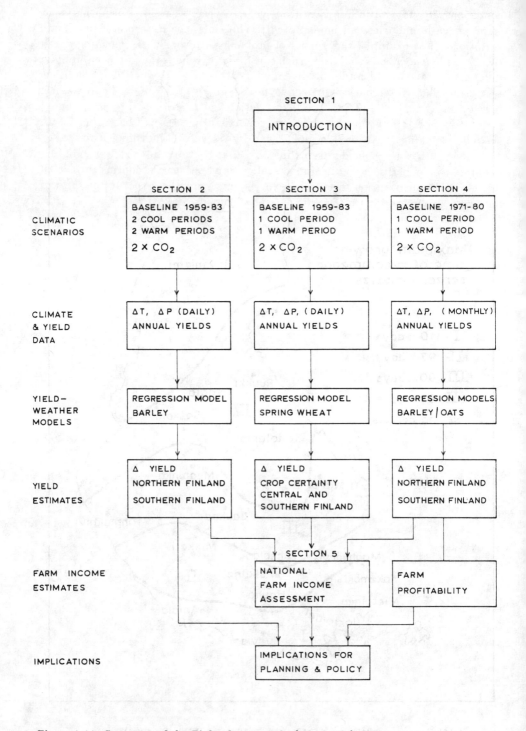

Figure 1.14. Structure of the Finland case study (schematic).

conditions are not static but vary from year to year, as well as over the longer term. Zone boundaries can be expected to shift in space in response to climatic variations. For example, the impact of probable future climatic warming due to the "greenhouse effect" could be crucial in Finland, where the northern limits of different cereals would undoubtedly move northwards. Under these conditions new crops like winter barley, maize, soybean, sunflower, etc. may be introduced in Finnish agriculture. Therefore investigations such as those presented here – connecting climatic, biological and also economic aspects – will have their use in the policy making at national, regional, and farm level in Finnish agriculture.

It is with these considerations in mind that Sections 2–4 examine, through a number of experiments, the impacts of various climatic scenarios on certain aspects of Finnish agriculture. The experiments, and the structure of the Finland case study, are illustrated schematically in *Figure 1.14*.

SECTION 2

The Effects on Barley Yields

2.1. Introduction

In Section 1 several scenarios based on instrumentally observed historical climatic data, and on results from a general circulation model (GISS), were selected. In this section we use models to estimate the impacts associated with these scenarios on barley yields in Finland. Two methods were used, one based on the longest available temperature and precipitation records expressed as monthly averages and the other one on more detailed daily meteorological records for the years 1959–83. Two study areas were selected: Uusimaa province, covering the area around Helsinki at latitudes 60–61°N, to represent southern Finland; and the Talousseura region, centered on Oulu in the agricultural subdistrict of northern Bothnia, at latitudes 63°50′–66°N, to represent the northern limit of the present barley growing area (*Figure 1.3*).

Barley is the oldest cereal crop cultivated in Finland. Cultivation spread from south to north with the settlement of the country. At the beginning of the eighteenth century, the Finnish settlers brought barley cultivation up to the coast of the Arctic Ocean in Alta Fjord at latitude 70°N in Finnmark, part of Norwegian Lapland (Sundbärg, 1895). In Finnish and Swedish Lapland the northern limit of occasional barley cultivation reached latitude 69°50′N on some individual farms in the middle of the eighteenth century (Rein, 1867; Hellstenius, 1871; Grotenfelt, 1897; Hellström, 1917).

At present, barley is commonly grown in Finland between latitudes 60° and 66°N, less commonly up to 67°N and occasionally up to 68°N [*Figure 2.1(a)*]. It is usually cultivated in rotation with other field crops, in the Helsinki region mostly in clay soils, in the Oulu region mostly in sandy or peat soils (cf. *Figure 1.2*).

Until the 1950s, the area annually under barley was 0.1–0.2 m ha. In the 1970s this area increased to 0.4–0.5 m ha and in the 1980s to 0.55 m ha,

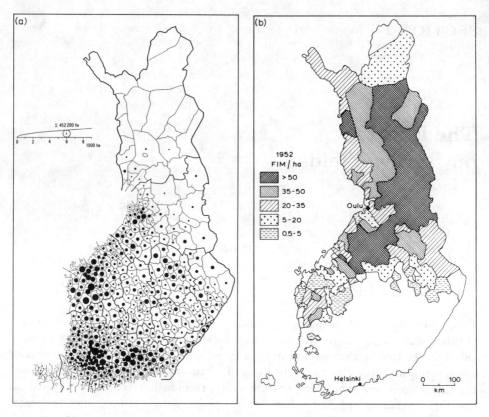

Figure 2.1. (a) Distribution of barley cultivation by municipalities (*Atlas of Finland*, 1982). (b) Regional distribution of compensation paid to farmers for crop losses caused by night frosts in 1952 (Valmari, 1966).

corresponding to 20–25% of the country's arable field area. In early times barley was used mostly for human consumption. Nowadays, the major part of the crop is used for animal feed on farms, 10% for malting and 7–8% for seed; only 1.5% is hulled for food. Present production meets domestic demand.

The average barley yield was less than 1000 kg/ha until the 1920s, exceeded 1500 kg/ha in the 1930s and reached an all-time record of 3206 kg/ha in 1983. This remarkable improvement has been a result of intensive plant breeding, improved cultivation techniques, increased use of fertilizers, etc. This trend shows no signs of abating. The yearly deviations from the average barley yields have been as much as 15–25%, except in the Helsinki area during the period 1870–1930, where they were less than 10% (cf. *Figure 2.3*).

Owing to restrictions imposed by the climate, only spring barley can be grown in Finland. Winter barley is not hardy enough to survive the long Finnish winter. The great majority of the barley cultivars currently grown in Finland are domestic varieties, adapted to long days and a short, cool growing season. The first Finnish cultivars were selected from a local six-row strain in the

1920s. The earliest of them, *Olli*, needed only 77–78 days or an ETS (base temperature 5 °C) of 700 degree-days to ripen. *Olli* was also cultivated in Chipiny, USSR, at latitude 67°44′N during the 1930s (Valle, 1935) and is still cultivated in Alberta (Valle, 1959; Valle and Carder, 1962) and in other Canadian provinces between the latitudes 53° and 57°N (Williams, 1974a).

The present Finnish barley cultivars need 80–95 days to ripen. The required ETS is 750–800 degree-days for the earliest cultivars and about 1000 degree-days for the latest ones (Lallukka et al., 1978). In a cool climate the Finnish barley needs more time and a higher ETS to ripen than in a warmer climate. On the other hand, long growing days speed up ripening and reduce the ETS requirement, especially when long days are combined with high ETS to increase the photothermal resources available (Nuttonson, 1957; Valle and Carder, 1962).

Late summer night frosts often damage barley in northern and central Finland before the grains have ripened (Valmari, 1966; Kolkki et al., 1970). This seldom occurs in southern Finland as indicated by the lack of compensation paid to farmers [*Figure 2.1(b)*] but early summer droughts in the south frequently result in losses. In addition, a wet harvesting season is a potential risk for barley production throughout the country.

2.2. Description of the Study Areas

2.2.1. Helsinki region

Temperatures in the Helsinki region are favorable for the cultivation of barley, much more so than elsewhere in the country. The long-term (1846–1983) mean ETS for this area is 1242 degree-days and is assumed to increase up to 1730 degree-days under the $2 \times CO_2$ GISS scenario (using methods described in Subsection 1.9.2). The average monthly precipitation of the present growing season (from April to October) varies from 33 to 73 mm. During the early summer months (May–June) the present precipitation does not usually meet the optimum requirements for barley. According to the proposed $2 \times CO_2$ scenario, the monthly precipitation of the growing season would, however, increase up to a range of 62–108 mm.

The soil in the Helsinki region is silty clay, and is therefore particularly sensitive to drought and excess moisture. It is less suitable for barley than for some other grain crops because barley has poorly developed roots that make it particularly vulnerable to early summer drought. High temperatures in the early summer accentuate the moisture stress, while excess moisture in clay soils often prevents oxygen from reaching the plant roots.

The area under barley in the Helsinki region was only between 10 000 and 12 000 ha until the 1960s, after which it started to increase rapidly, rising to 49 200 ha, or 22.6% of the arable field area of this region by 1983 (*Figure 2.2*). The old Finnish land-race barleys were replaced during the 1930s and 1940s in the Helsinki region by the Danish two-row cultivars, *Binder* and *Balder*, which had stronger straw and a considerably higher yield. *Binder* and *Balder* were in

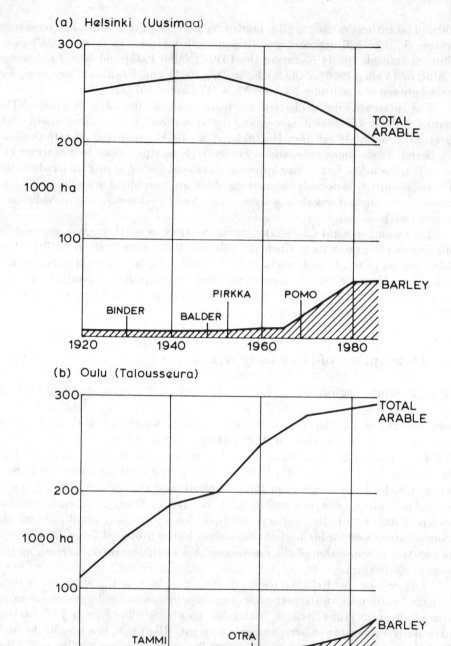

Figure 2.2. The area of barley in the regions of (*a*) Helsinki and (*b*) Oulu over the period 1920–1983. The time of introduction and names of the most important new cultivars are given.

turn replaced during the 1950s and 1960s by several six-row cultivars developed in Finland. These new six-row varieties had even stronger straw and higher yield. The most important of them were *Pirkka* and *Pomo*. Since then two-row varieties have been cultivated in the Helsinki region only in small quantities for malting purposes.

The rising trend in barley yields started as early as the 1920s and, with the exception of the wartime slump, continued at a growing pace up to the end of this study period in 1983 [*Figure 2.3(a)*]. One important reason for the exceptionally early upswing in the Helsinki region was the stabilization of agricultural and political conditions and the improvement in the social status of tenant farmers, who acquired possession of their smallholdings after the First World War. Even more important was the breakthrough in animal husbandry in this area, providing more manure for the fields. The favorable trend in the Helsinki area was strengthened by the warm period of the 1930s, by increasing use of chemical fertilizers and by the adoption of high-yield cultivars.

2.2.2. Oulu region

In the Oulu region, the climate is unfavorable for agriculture because the growing season is short, the average temperatures for May and August are low, and night frosts are frequent (Valmari, 1966; Kolkki et al., 1970) [*Figure 1.4(b)*, *Table 1.8*]. The present long-term mean ETS (1883–1983) in the Oulu region is less than 1100 degree-days, but is 1455 degree-days under the GISS $2 \times CO_2$ scenario. The monthly precipitation of the present growing season (May–September) in the Oulu region varies from 37 to 65 mm, which is sufficient for barley production in the conditions in this area. Under the GISS $2 \times CO_2$ scenario the monthly precipitation of the now longer growing season is between 54 and 101 mm per month (*see* Subsection 1.9.2).

Until the end of World War II, only about 8% of land in the Oulu region was under barley. After the war, much of the agricultural population of the ceded territories of Finland was resettled in the region, and between 1945 and 1960 45 000 ha, mostly in peat soils, were added to the arable area. Land reclamation for barley proceeded steadily after 1960, and in 1983 it covered 20.3% of the total farmed area of the region.

The rising trend in barley yield began later in the Oulu region than in the south [*Figure 2.3(b)*]. Initially, this was because the early ripening barley cultivars were not available. In addition, the old Finnish land-race barleys had weak straw, as did the bred-cultivar *Olli*. In the 1930s, however, the *Tammi* cultivar, which ripened sufficiently early in the Oulu region, was developed from *Olli*. Its straw was slightly stronger and it produced a 20% higher yield than *Olli*. The next improvement was *Otra*, developed from *Tammi* in the 1950s. It ripened even earlier and raised the yield by a further 10%. Thus in 30 years plant breeding resulted in a 30% rise in barley yields in the Oulu region (Kivi, 1960).

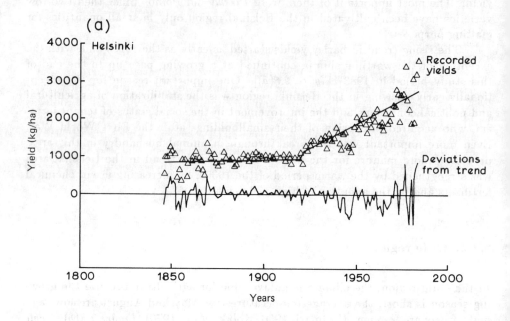

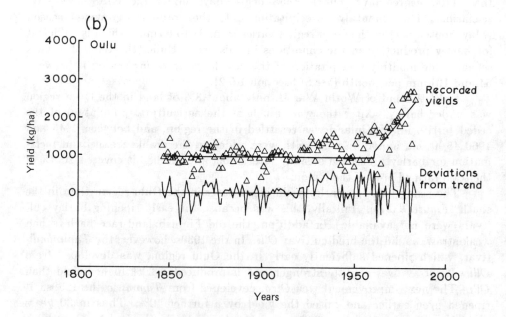

Figure 2.3. Barley yields in (*a*) Helsinki and (*b*) Oulu regions, 1846–1983.

2.3. Weather and Yield Data

2.3.1. Meteorological data

Data on daily mean temperatures and precipitation were obtained for the period 1846–1983 for the station at Helsinki, and for the period 1883–1983 at Oulu. These were the longest periods for which both weather variables were available together with the yield records. For part of this study, daily temperatures were converted into monthly ETS values, and precipitation was expressed as monthly totals (*Figure 1.5, Tables 1.7* and *1.8*). More detailed daily meteorological records are available since 1921 for Helsinki (Kaisaniemi) and since 1959 from several other meteorological stations in Finland (Finnish Meteorological Institute; Heino and Hellsten, 1983). These are used in further experiments reported below.

2.3.2. Yield data

Records of grain yields, expressed in kg/ha, have been published annually in the *Official Statistics of Finland* since 1920 (Central Statistical Office of Finland, 1983). Yields expressed as "grain indices" were published in the yearbook from 1861 to 1919. Older, less accurate, estimates from 1811 are available in the Finnish State Archives (*Annual Reports of Provincial Governors*, 1846–1860).

The grain index indicates the number of harvested grains obtained from one seed grain or the number of grain barrels obtained from a barrel of seed grain. The grain indices were converted into kg/ha by using historical records of sowing densities and volume weights of barley. This method has previously been used by Keränen (1931), Hustich (1952) and Soininen (1975). For the present study, the original records were carefully revised since some systematic errors were found in the figures for sowing densities and volume weights for the barley from northern Finland. To improve accuracy when converting the grain indices into kg/ha, some corrections were made by using records from neighboring agricultural districts or provinces and from the respective latitudes in Sweden (cf. Sundbärg, 1895). *Table 2.1* shows the adjusted sowing densities and volume weights used in this study. The sowing densities for Oulu are lower than those recorded by Böcker (1834–1835) and used by Soininen; the volume weights are also lower than those suggested by Soininen.

The published records of yields from 1920 onward are more accurate and reliable than the grain indices, with the exception of the years around World War II. Yield data for these years are evidently inadequate and too low. Some systematic errors in the published yields may also exist because the sampling methods were adjusted every 10 years up to 1972. Furthermore, during the first decade after the war yields were often collected as hl/ha and converted into kg/ha without detailed surveys on yearly or regional variations in the volume weights.

Table 2.1. Adjusted sowing densities and volume weights of barley.

Variable	Region	1846	1864	1910	1920
Sowing density (kg/ha)	Helsinki	3.1	3.0	2.8	3.1
	Oulu	3.5	3.3	3.1	3.0
Volume weight (kg/hl)	Helsinki	57	58	60	60
	Oulu	52	54	56	58

2.4. Description of the Models

In the first stage of the analysis of the yield records, linear yield trends were calculated in order to eliminate the variation caused by long term factors as distinct from seasonal weather conditions. Deviations from these trends were then examined. For optimum consistency, the diagrams depicting the yield (*Figure 2.3*) were divided into two periods according to the transition in the yield trends in the respective areas, as follows:

Helsinki: 1846–1919 and 1920–1983 [*Figure 2.3(a)*].
Oulu: 1883–1958 and 1959–1983 [*Figure 2.3(b)*].

This method of fitting long term yield trends by a pair of straight lines has previously been used, e.g., by Waggoner (1979) and Dennett (1980). As stated previously, the trend in the yield of barley rose substantially in the Helsinki region already in the 1920s, whereas in the Oulu region the change was delayed until the late 1950s.

Two statistical approaches (Methods 1 and 2) were developed and tested.

2.4.1. Method 1

Deviations from the trend (dY) were evaluated with selective multiple regression analysis, where effective temperature sum in month i (ETSi), monthly precipitation (Pi), monthly P/ETS ratio (Ii), and total annual ETS (ΣETS) were the independent variables.

This method produced the following two equations where the numbers following each letter symbol denote the month:

$$\text{Helsinki: } dY = -433.4 - 1.2\,\text{ETS7} \tag{2.1}$$
$$r^2 = 0.058$$

$$\text{Oulu: } dY = -23.2\,I5 + 2.63\,\text{ETS8} + 0.52\,\Sigma\text{ETS} \tag{2.2}$$
$$r^2 = 0.300$$

The equation for Helsinki explained only 5.8% of the variation, although statistically its significance was satisfactory (F = 7.11, Prob. < 0.01). Only a high July temperature seemed to slightly reduce the yield, but this effect was less than would have been expected in the light of previous studies (Mukula et al., 1977b). It seems unrealistic to draw any specific conclusions from this equation.

The explanatory value of the equation for the Oulu area was considerably better (30.0%), with a good degree of significance (F = 10.09, Prob. < 0.01). Low temperature combined with high precipitation in May ($I5$) reduced yields in Oulu. Evidently crop sowing was delayed under such conditions. A large temperature sum for the entire growing season (ΣETS), and particularly in August (ETS8), increased yields, apparently by ripening the crop in time.

In conclusion, as far as the applicability of the Method 1 is concerned, in southern Finland the calculations based only on monthly ETS and average monthly precipitation data do not seem to explain consistently the effects of climatic variations on barley yields. One of the difficulties is that between the beginning and the end of June the thermal and moisture requirements of barley change quite radically. At the beginning of June (the tillering stage of barley), cool and rainy weather is favorable. At the end (the shooting stage, when photosynthesis and development of the leaf area is at a maximum), warm weather is most advantageous (Mukula et al., 1977b; Lallukka et al., 1978). These effects can be hidden if monthly ETS and averages of precipitation for June are used. It proved necessary to try another method in which more detailed daily meteorological data would be used and the number of variables increased.

2.4.2. Method 2

In this method more detailed daily meteorological records from 1959 to 1983 were used. The growing season was divided, not into months, but into successive "phases", based on the phenological development of the barley crop, consisting of steps of 100 degree-days and 88 degree-days for the areas of Helsinki and Oulu, respectively (represented schematically, for cereal crops in general, in *Figure 2.4*). These ETS values were selected to fit the transition from one phase to another at a specific phenological growth stage of the crop. The average rate of growth of cultivars per unit temperature in the Oulu region is faster than that of cultivars in the Helsinki area. Hence, the length of the phases had to be shorter (88 degree-days) in the Oulu area than in the Helsinki area (100 degree-days). This method is a simplification of another method previously used, e.g., by Williams (1974a). Several additional variables, such as the mean maximum daily temperature of each phase, the lowest minimum and the average daily minimum, etc., were also included in the model.

This revised approach highlighted a new problem. While early summer drought is the most important crop-limiting factor in the south, other factors (especially low temperature) become increasingly important farther north. Since the Oulu area is almost at the northern-most limit of barley cultivation, even a small decrease in the temperature sum causes crop losses; and since there is a strong negative correlation between precipitation and temperature at Oulu, greater than normal rainfall in early summer is usually accompanied by a reduced temperature sum. Another climatic difference is that night frosts are common in the Oulu region but rare in the Helsinki area [see *Figure 2.1(b)*].

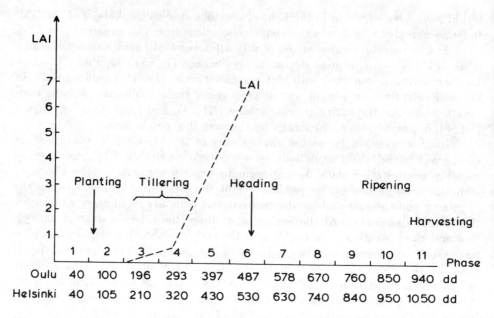

Figure 2.4. Schematic illustration of the ETS (degree-days) at Helsinki and Oulu applied to fit with the phenological growth stages of a cereal crop. LAI is Leaf Area Index.

However, the long-term meteorological records used in Method 1 for the Oulu region were obtained from the neighborhood of the city of Oulu, which is close to the Gulf of Bothnia, and is not affected by night frosts, typical of the inland area of the Oulu region. (The location of the station has varied during the long period represented by the Oulu record, but all values are corrected to be consistent with the recent station data from Oulu airport.) To improve the accuracy of Method 2, it was considered necessary to include in the calculations meteorological records from another station, Ähtäri, located in the night frost area southwest of the Oulu region. No suitable station for this purpose was available in the Oulu region itself.

Similarly, for the Helsinki region, the long-term meteorological records from the Kaisaniemi station in the city of Helsinki were replaced by records from two neighboring stations, Vantaa and Tikkurila, which better correspond to the average climatic conditions of the Uusimaa district as a whole. The use of station data from other locations also avoids possible trends in the long-term climatic records at both Oulu and Helsinki due to urbanization effects.

This method produced the following estimated function for the Helsinki area:

$$dY = -1772.8 + 359.6 P_3 - 68.2 S_4 + 72.4 Tx_6$$
$$- 140.1 C_7 - 112.6 C_9 + 116.1 Tn_{10} + 130.2 S_{11} \quad (2.3)$$
$$r^2 = 0.72$$

and for the Oulu area:

$$dY = -2528.0 + 107.2\,Tx_1 + 42.5\,Tn_3 - 6.0\,SP_4$$
$$+ 74.2\,S_5 + 35.2\,Tx_7 \qquad (2.4)$$
$$r^2 = 0.56$$

where:

dY = the deviation of the barley yield from the trend
P_i = precipitation in phase i
C_i = average daily temperature in phase i
Tx_i = maximum daily temperature in phase i
Tn_i = minimum daily temperature in phase i
S_i = $Tx - Tn$ in phase i
SP_i = sum of precipitation over the first i phases $(P_1 + P_2 + ... + P_i)$

The equation for the Helsinki area explained 71.8% of the variation, and its significance was good (F = 13.84, Prob. < 0.001). As was expected, high precipitation in early summer (P_3) increased the yield in this area. Similarly, high maximum temperatures during the ripening stage (Tx_6) seemed to have a favorable effect on the yield. In the harvesting season, calm, clear weather, well represented by the difference between Tx and Tn (S_{11}) (see Section 3.3), is of importance for avoiding harvest losses due to lodging.

The explanatory value of the equation for Oulu area was also satisfactory (55.8%), and its degree of significance was good (F = 11.09, Prob. < 0.001). Excess rainfall during the early summer (SP_4) seemed to reduce the yield significantly in this area, while calm, clear weather before the ear formation of the crop increased the yield (S_5). In this respect the results are in agreement with the findings of Elomaa and Pulli (1985); at the time of intensive photosynthesis and the most active growth of the leaf area, the light requirements of plants are at their maximum (*Figure 2.4*).

2.5. Model Validation

The validity of the model used for the Helsinki area in Method 2 was tested by using independent data from the Agricultural District of Satakunta, located on the western coast of Finland, between latitudes 61°N and 62°N (*Figure 1.3*, District 5). Climatic conditions in Satakunta differ slightly from those in the Helsinki area and the soil is more sandy.

The annual yield variations estimated by the Helsinki model were in good agreement with the observed yearly variation in Satakunta (*Figure 2.5*). The differences from observed values were greatest in 1974 and 1981, which is probably explained by exceptional flooding in Satakunta during these two years.

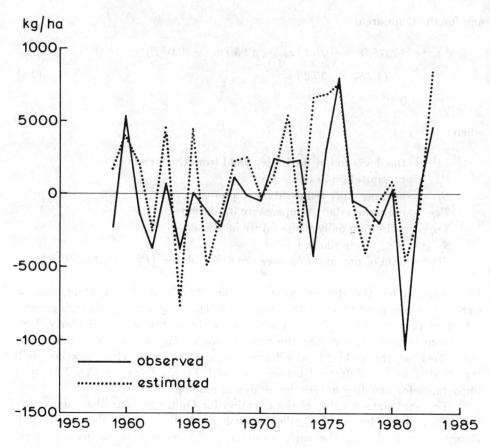

Figure 2.5. Validation of Method 2 (Helsinki equation): model yield estimates and observed yields at Satakunta.

2.6. The Effect of Climatic Variations on Barley Yields

2.6.1. Base case yields

Equations (2.3) and (2.4) from Method 2 were used to estimate barley yields under changed climate. As a point of departure against which to compare each of the climatic scenario results, mean values of the climatic variables for the period 1959–83 were substituted into equations (2.3) and (2.4) and, by adding the trend term, yields were projected to 1983. The average "base case" yields were therefore calculated as:

Helsinki area (1959–83 mean climate): baseline yield = 3075 kg/ha
Oulu area (1959–83 mean climate): baseline yield = 2560 kg/ha

In addition, because daily meteorological data for the Oulu region were not available for years prior to 1959, equation (2.2) from Method 1 was used to compute yields for the warm and cool decade scenarios 1931–40 and 1921–30, respectively. The "base case" yield projected to 1983 in equation (2.2) was:

Oulu area (1883–1983 mean climate): baseline yield = 2570 kg/ha

Note that this is 10 kg/ha greater than for the 1959–83 baseline climate calculation using Method 2.

2.6.2. Effects of the GISS $2 \times CO_2$ scenario on barley yields

The procedures described in Subsection 1.9.2 were used to simulate the change in climate between the present base case and the GISS model $2 \times CO_2$ estimates. Differences in mean monthly temperatures between the GISS model estimates for $1 \times CO_2$ and $2 \times CO_2$ equilibrium climates were added to the daily maximum, minimum and mean temperatures over the period 1959–83 for each location. It was assumed that the diurnal temperature range would not change under $2 \times CO_2$ conditions. A similar procedure was employed for precipitation: differences between mean daily precipitation rates for each month in the growing season under $1 \times CO_2$ and $2 \times CO_2$ GISS-modeled climates were added to daily precipitation totals over the period 1959–83. The adjusted daily mean temperatures were transformed into daily ETS values and these were used to divide the growing period into phenological phases (*Figure 2.4*). The adjusted climatic data were allocated to the appropriate phases in a form suitable for application in equations (2.3) and (2.4). Outputs from Method 2 are in the form of deviations from trend yields (d Y). Adding the trend yields gives the estimated yields in a $2 \times CO_2$ climate projected to 1983.

The results of applying Method 2 in the Helsinki and Oulu regions are shown on *Figure 2.6* and in *Table 2.2* (where standard deviations of annual yields and coefficients of variation are also given). In the Helsinki area an increase in barley yield under the GISS $2 \times CO_2$ scenario of 270 kg/ha is estimated (9% above the baseline yields). In the Oulu region an increase in yields of 554 kg/ha (14%) is estimated. Since calculations have been conducted for yearly data, it is possible, in addition, to compare the estimated interannual variability of yields under $2 \times CO_2$ conditions with present-day variability. The coefficient of variation (standard deviation expressed as a percentage of the mean) of barley yields at present in the Helsinki region (14%) would be reduced to about 8%, whereas in the Oulu region the calculations imply an increase in variability from 17% to 20% (*see Table 2.2*).

2.6.3. The effects of warm and cool decades on barley yields

To enable comparison with the results for spring wheat (Section 3), the same years, 1966–1973 and 1974–1982 were selected to represent a warm and a cool decade, respectively (*see also* Subsection 1.9.1 and *Figure 1.10*). Deviations from

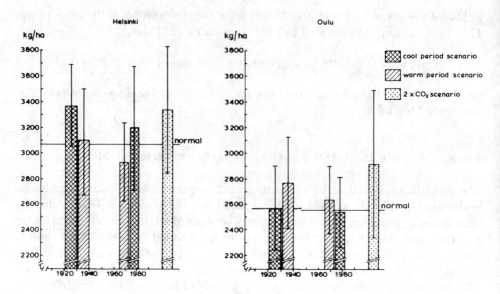

Figure 2.6. Yield estimates for barley at (a) Helsinki and (b) Oulu under different climatic scenarios. The "normal" yields projected to 1983 are indicated by a horizontal line; the vertical bars are standard deviations. Estimates are by Method 2 except for the 1921-30 and 1931-40 scenarios at Oulu, calculated using Method 1.

the reference climate (1959–83) of both the mean ETS and the mean precipitation during the growing season for both periods can be computed from *Table 1.7*. The warm period (1966–73) is characterized by above-normal ETS values in both Oulu and Helsinki regions (3% and 2% higher, respectively) and below-average precipitation (−9% and −3%). ETS values in the cool period (1974–82) are below-average in both areas (2% and 5% lower, respectively) and precipitation in both regions is higher than normal (2% and 6% higher).

Subsequently, daily meteorological data for the period 1921–59 were made available for the Helsinki region, and this permitted experiments to be conducted using Method 2 for the same warm decade (1931–40) as is employed in Section 4, and for an adjacent cool decade (1921–30). The same scenario periods were also selected for experiments in the Oulu region, employing Method 1 rather than Method 2 in the absence of daily input data for these decades. The period 1931–40 was one of the warmest in this century in both Helsinki and the Oulu regions (ETS values 11% and 13% above the 1959–83 baseline, respectively), while precipitation was lower than average in the Helsinki region (−2%), but considerably above average (11% higher) in the Oulu region (*Table 1.7*).

The meteorological records from the extreme decade periods were entered into the regression equations for each period in turn, providing estimated deviations from the trend. The yields were then projected to the 1983 level. Note that for the 1920s and 1930s decades, due to data limitations, daily meteorological values from Helsinki Kaisaniemi station were used to represent the Helsinki region in equation (2.3). Results of these analyses are given in *Figure 2.6*.

Table 2.2. Estimated means, standard deviations and coefficients of variation of barley yields under different climatic scenarios in the Helsinki and Oulu areas, using Method 2 (unless otherwise stated).

			Helsinki area		
				Barley yields	
Climatic scenarios	ETS (degree-days)	May–Oct precipitation (mm)	Mean (kg/ha)	SD	CV (%)
Baseline 1959–1983	1263	342	3075	428	14
Cool periods 1921–1930	1218	431	3369	321	10
1974–1982	1202	364	3200	500	16
Warm periods 1931–1940	1417	337	3092	416	13
1966–1973	1285	333	2840	310	11
$2 \times CO_2$	1730	512	3345	270	8
			Oulu area		
				Barley yields	
Climatic scenarios	ETS (degree-days)	May–Oct precipitation (mm)	Mean (kg/ha)	SD	CV (%)
Baseline 1959–1983	1080	313	2560	429	17
Cool periods 1921–1930	1067	331	2577[a]	335	13
1974–1982	1060	320	2540	260	10
Warm periods 1931–1940	1188	346	2775[a]	361	13
1966–1973	1111	286	2630	260	10
$2 \times CO_2$	1455	496	2930	586	20

[a] Estimates using Method 1.

When comparing the yields obtained by the "warm-decade" scenarios with those obtained by the $2 \times CO_2$ GISS scenario in the Helsinki area, marked differences are found. The "warm-decade" scenario (1966–73) resulted in reduced yield while the still warmer decade (1931–40) and the $2 \times CO_2$ scenario increased yield. These differences are apparently due to a combination of differences in mean precipitation and contrasts in interannual weather variability.

Estimated yields increased under the "cool-decade" scenarios, especially for the 1921–30 conditions where the above-average precipitation contributed to a mean yield estimate higher than that calculated for the $2 \times CO_2$ scenario.

In the Oulu area the effects of warm and cool decades were reversed, i.e., the warm periods increased the yields relative to the baseline, while the cool periods produced little change in yields. Under the $2 \times CO_2$ scenario estimated yield was markedly higher than the baseline value.

2.7. Discussion

2.7.1. Summary of impacts on barley yields

In southern Finland, warmer than average periods lead to slightly decreased or slightly increased yields, depending on annual precipitation conditions, while in the Oulu area in northern Finland warm periods increase yields. Cool periods have a reverse effect in both areas.

Under the considerably warmer and also more moist climate estimated in the GISS model-derived $2 \times CO_2$ scenario, the present (1983) trend yield of 3075 kg/ha increases by 9% in the Helsinki area. The present trend yield of 2560 kg/ha in the Oulu area increases by 14%, according to the best-fitting method used in this study. Neither of these predictions allow, of course, for future improvements in plant breeding and cultivation techniques or for the possible direct effect of increased atmospheric carbon dioxide concentrations on the photosynthesis and water use efficiency of crops. The $2 \times CO_2$ scenario produces a decrease in the yearly variation coefficient for barley yields in the Helsinki region, but a slight increase in the Oulu area.

2.7.2. Implications of climatic warming in southern Finland

Although it was difficult, with a single model, to estimate yield levels that are representative of the wide range of barley cultivars that are grown in southern Finland, some interesting observations emerge from this analysis. In southern Finland low temperature alone seems not to be a growth-limiting factor for barley. More harmful in this area is the lack of moisture during the early summer, and the less suitable soil type for barley.

Therefore, barley would not benefit greatly from a warmer $2 \times CO_2$ climate. However, other crops with higher temperature requirements, such as spring wheat, would probably benefit more. This is discussed in Section 3.

2.7.3. Implications of climatic warming in northern Finland

In northern Finland, as represented by the Oulu area, present-day temperature is already a growth-limiting factor. Here barley production would inevitably benefit from a warming of the climate. However, if warming were accompanied by increased precipitation as suggested by the GISS model, harvest losses might increase. Nevertheless, longer growing seasons would allow earlier sowing and consequently earlier harvesting and also more flexibility in the proper timing of both sowing and harvesting of crops. Even the higher yielding cultivars, which need higher temperatures and longer growing periods, may become suitable for this area.

A significant warming of the climate would also allow the extension of widespread cultivation of barley to northern Lapland, i.e., up to latitudes 67-69° N. The soil and topography are suitable for this purpose along the river valleys and on the swamp areas.

2.7.4. Adaptation

A slow change of climate, such as might be expected given a continued gradual rate of accumulation of atmospheric CO_2, may allow sufficient time for plant breeders to develop crop varieties that are suited to altered climate. But we know little about the possible effects of long-term climatic changes on the incidence of pests and diseases. These and other implications are discussed in Section 5.

SECTION 3

The Effects on Spring Wheat Yields

3.1. Introduction

In this section we consider the effects of the same three climatic scenarios (a warm period, a cool period, and a "$2 \times CO_2$" climate) on spring wheat yields.

Winter wheat can be grown only in a small area in southwestern Finland, while spring wheat extends over much of the south and center of the country, covering 0.10–0.15 m ha (*Figure 3.1*). Average yield rose from 1000–1500 kg/ha to 2500–3000 kg/ha between 1920 and 1983, largely a result of plant breeding, better cultivation techniques and increased use of fertilizers.

At present, average spring wheat yields of marketable quality are highest (2400–2500 kg/ha) in southwestern and southeastern Finland (*Figure 3.2*). In the best years the yield rises to 3400–3500 kg/ha in southern Finland and 2800 kg/ha in the north. In poor years, the yield falls to about 1700 kg/ha even in the most productive regions. High yields are found on the west coast as far north as 63° N, crops being grown there on the best arable land (*Figure 3.2*).

Most of the spring wheat cultivars grown in Finland are of domestic or Scandinavian origin, and are suited to short, cool summers. Their growing period is about 110 days. The earliest varieties require an ETS of about 900 degree-days and the latest varieties 1110 degree-days (Mukula *et al.*, 1977a; Kontturi, 1979). In general, varieties with the highest yields are those requiring the longest growing time but only, of course, when the growing season is sufficient for the variety to ripen. For example, while late cultivars (which have a higher yield than early ones) are preferred in the south of Finland because of the region's longer growing season, the yield of these cultivars in a cool year drops more than that of early-maturing cultivars.

The quality of spring wheat depends to a large extent on weather conditions during the harvesting season. Since the major component of wheat endosperm is starch, at least one of the enzymes of the grain may be expected to be

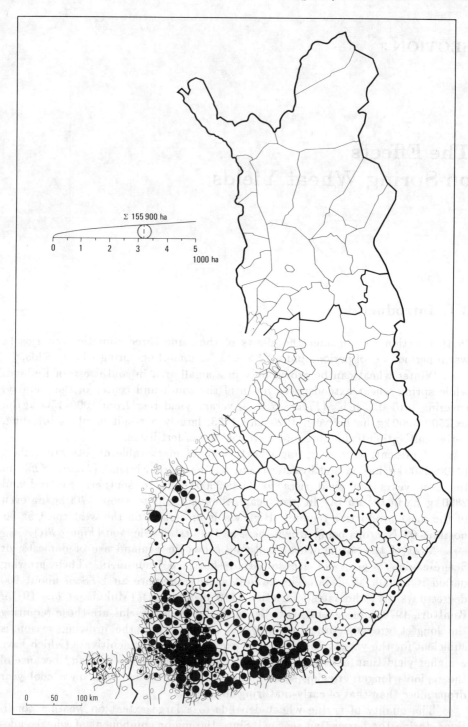

Figure 3.1. The distribution of spring wheat cultivation in Finland by municipality (*Atlas of Finland*, 1982).

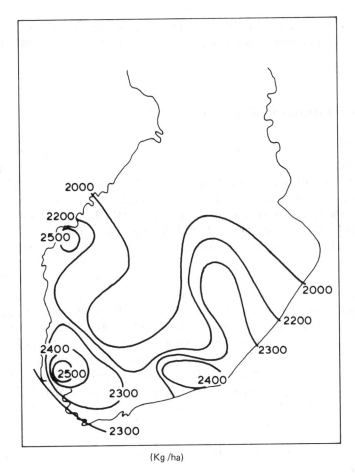

Figure 3.2. Mean yields (kg/ha) of spring wheat estimated for 1959–83 climate and adjusted to 1983 technology.

suitable for the digestion of this substance. Two forms of starch-hydrolyzing enzyme are known to be present, α-amylase and β-amylase. The former is highly destructive of starch and sound wheat should contain very little; it is abundant in wheat which is beginning to germinate. Rain tends to increase α-amylase activity during the ripening stage or even results in the sprouting of grains on ears. Under normal weather conditions the grains seldom dry out sufficiently in the field and usually need artificial drying after harvesting.

3.2. Weather and Yield Data

The yield data, which refer only to yield of marketable quality [*Official Statistics of Finland*, 1950–1983 (Central Statistical Office of Finland, 1983)], were collected from 15 agricultural districts (districts 1–15, inclusive – *see Figure 1.3*). Daily weather data were obtained from 19 observation stations, providing averages for each of the 15 districts. They comprised daily mean, minimum and maximum temperature and daily precipitation for the period 1959–83 (see Section 1).

3.3. Development of the Model

A regression analysis was performed in order to assess the degree of variation between deviations from the yield trends and climatic variables. Regional linear trends in the yields were first determined for the baseline period 1959–83.

The rate of development of a crop is closely related to the sum of effective temperatures (ETS) over the growing season (Davidson and Campbell, 1983; Bauer et al., 1984; Russelle et al., 1984). For the purposes of this study, the growing period is assumed to extend from April 1 to October 31, and the development of wheat is described by dividing the growing period into phenological stages on the basis of ETS. The procedure is the same as that illustrated in Figure 2.4. This allows for synchronization of weather conditions with the phenological stages of wheat development, and for estimation of the impact of weather conditions on the yield. The length of the first phase (I) was defined as 40 degree-days on the basis of the sowing dates for 1959–1983 (Official Statistics of Finland). This phase represents conditions before sowing. The following phase (II) amounts to 60 degree-days, and comprises sowing and subsequent emergence. The length of the following phases (III–XII) is determined by region on the basis of the ETS for the entire growing season, so that at least 10 phases are obtained for each year, in addition to the pre-sowing and pre-emergence phases (I, II).

The length of the phases III–XII together varies from 100 degree-days for late cultivars to 89 degree-days for early cultivars. The late cultivars are grown in the south (districts 1, 2, 3, and 4 in Figure 1.3), where the length of a phase was defined as 100–98 degree-days. Districts 5–9 at latitudes 61–62°N have a phase length of 97–94 degree-days, and districts 10–15 at latitudes 62–64°N, where early cultivars are grown, have a phase length of from 93 degree-days to 89 degree-days, depending on the variety. Very little spring wheat is cultivated in the northern-most districts of the country.

The amount of precipitation for the phases was calculated, as were the means for the following variables: mean, maximum (Tx) and minimum (Tn) daily temperatures. Since no radiation data were available, the variable $S = Tx - Tn$ was used as a substitute index for radiation. This is because when the daily variation in temperature is high, cloudiness is assumed to be low – i.e., there is a high incoming radiation by day, and large long-wave radiation losses by night. Logarithmic modifications had to be made to the precipitation figures because of the skewness of the distributions.

Multivariate selective regression analysis was applied to explain deviations in yield by meteorological variables. The analysis was applied to the meteorological versus yield characteristics in each of districts 1–9, in which most of Finland's spring wheat ($>97\%$) is grown.

3.4. Description of the Model

The regression analysis resulted in a model which explained 81.5% of the variation in wheat yields (F = 95.0; Prob. < 0.001). The model equation is:

$$dY = -3812 + 0.99 \cdot Trend - 264.0 P_2 - 65.6 S_2$$
$$+ 30.1 Tn_3 - 50.5 C_4 + 2.2 SP_4 + 72.7 Tn_5$$
$$- 37.4 Tx_7 + 79.9 Tx_{10} + 55.1 Tn_{10}$$
$$+ 55.7 S_{11} + 122.9 S_{12} - 29.6 Tx_{12}$$

(3.1)

where:

dY	= the deviation of spring wheat yield from the trend
$Trend$	= technology trend indexed to years (e.g., a value of 1983 represents 1983 technology)
C_i	= average daily temperature in phase i
Tx_i	= maximum daily temperature in phase i
Tn_i	= minimum daily temperature in phase i
S_i	= $(Tx_i - Tn_i)$ in phase i
P_i	= total precipitation in phase i (logarithmic modification)
SP_i	= sum of log (precipitation) over the first i phases $(P_1 + P_2 + ... + P_i)$

We can interpret the importance of each climatic variable shown in equation (3.1) in terms of the requirements of a wheat crop (*Table 3.1*).

3.5. Model Validation

The model was tested in North Karelia (district 12, latitude 63°N), a region sufficiently distant from the area in which the model was formed to be regarded as suitable for independent verification (*Figure 1.3*). The differences between the observed yield and the model estimates are small (*Figure 3.3*). They are largest in years of early night frosts (such as 1964, 1974 and 1978, indicated by arrows in *Figure 3.3*) when the observed yield was higher than that estimated by the model, probably as a result of the local ameliorating effect on minimum temperatures of many lakes in the region.

3.6. The Effect of Climatic Variations on Wheat Yields

3.6.1. Scenario inputs

As described in Sections 1 and 2, the 2 × CO_2 scenario involved the addition of changes in mean monthly temperature and precipitation, as simulated by the

Table 3.1. Requirements of a spring wheat crop at different phases of growth.

Phase	Important climatic variable(s)	Influence on yield	Phase characteristics
II	S	−	Sowing and germination occurs during this phase. Precipitation must be low so that crops can be sown under favorable conditions. Excessively high temperature can form a crust, particularly in loam.
	P	−	
III	Tn	+	Floral initiation. Low temperatures are harmful.
IV	C	−	Growth development should not be too rapid at this phase. As temperature is decreased, duration of the vegetative and spikelet phases of apex development is prolonged, resulting in an increase in total spikelets formed and kernel number. Low minimum temperatures, however, are harmful. Sufficient moisture in the early summer guarantees safe sprouting conditions. Kernel number is fixed and the roots become strong enough to withstand drought.
	SP	+	
V	Tn	+	Stem elongation and booting. Minimum temperatures should not be too low.
VII	Tx	−	Anthesis. Excessive maximum temperatures are harmful.
X	Tn	+	Ripening time. The plants are very sensitive to frost. High temperatures accelerate ripening and improve grain quality.
	Tx	+	
XI	S	+	Weather conditions at harvest time determine the final yield volume and quality.
XII	S	+	A cool northerly air current is best for harvesting in Finland. Then the weather is clear and bright, with little precipitation.
	Tx	−	

GISS general circulation model, to the daily normals for 1959–83 (i.e., the baseline scenario). Adjustments were made by interpolation of the GISS model-derived changes to the 19 station locations in districts 1–15 (*see Figure 1.3*). The new data thus obtained were divided into phases as for the baseline climate, the sole difference being that the length of each phase, beginning with phase II, was extended by 10 degree-days. The growth phases were lengthened on the assumption that later ripening cultivars – requiring a higher temperature sum – would be sown as the growth period lengthened.

The first phase in the scenario began as soon as the temperature reached the effective level (40 degree-days), i.e., 2–3 weeks earlier than under the baseline climate. Because of the short growing period in Finland, sowing must take place as early as possible. Under the GISS $2 \times CO_2$ scenario winter temperatures would rise by 5–6 °C, shortening the winter season to the period from January until March or April. Temperatures in April would be some 4 °C higher than at present (*see Table 1.8*) contributing to a reduction in the duration and severity of soil frosts.

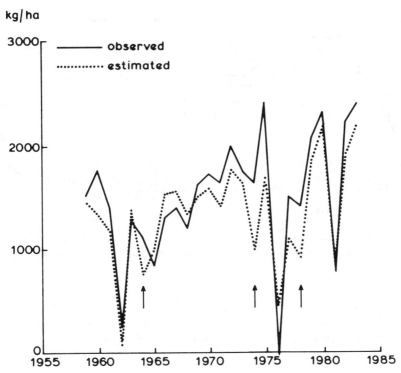

Figure 3.3. Model validation: model yield estimates, and observed spring wheat yields in North Karelia.

The other two (instrumentally derived) scenarios were for two periods described in Sections 1 and 2 and shown in *Figure 1.10* and *Table 1.7*: a "warm" period (1966-73) and a "cool" period (1974-82). Climatic data for these scenarios were applied to the original phenological phases described above (Subsection 3.3).

The meteorological data are presented by growth phases in *Figure 3.4*. It is important to note that these data are only comparable with respect to the modeled growth phases and are not coincident in calendar time. So, for example, $2 \times CO_2$ temperatures in Phase I of crop development probably represent temperatures from March or early April, whereas "cool-period" temperatures may well reflect a late start to the growing season in May. Comparison of "cool", "warm" and $2 \times CO_2$ scenarios showed that early growing season temperatures were practically identical. Early growing season precipitation under the $2 \times CO_2$ scenario was almost double that of the warm-period scenario. In the middle of the growing season, average temperatures were 1-2°C lower under the cool-period scenario than under the warm-period scenario: at harvest time the differences were even greater (from 2 to 3°C). Temperatures were higher still under the $2 \times CO_2$ scenario: relative to the warm-period scenario, the difference from ear formation to ripening was around 1°C in Uusimaa [*Figure 3.4(a)*] and between 1 and 2°C in North Karelia [*Figure 3.4(b)*]. At harvest time (phases

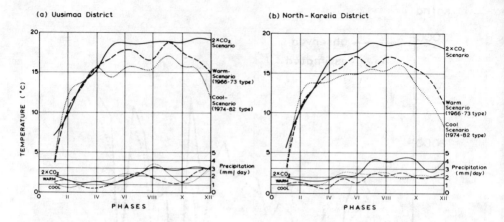

Figure 3.4. Mean daily temperature and precipitation at different phenological growth stages for the three scenarios at (*a*) Uusimaa (Helsinki) and (*b*) North Karelia (for locations, *see Figure 1.3*).

XI–XII), the differences were much greater, from 4 to 7 °C. The highest precipitation under the $2 \times CO_2$ scenario was in the second half of the growing season, being as much as twice that during the warm-period scenario.

3.6.2. The effects of the baseline climate on wheat yield

Meteorological data from each year of the baseline period (1959–83) were input to equation (3.1) to obtain annual yield estimates for each meteorological station location in the study region. These location-specific yields were then ranked from highest to lowest to obtain the 5 and 95 percentile yields. Period-mean yields were also computed for each location. Values of each of these three measures of baseline yield could then be located on a map and isolines constructed [*Figures 3.5(a), 3.6(a)* and *3.7(a)*].

3.6.3. The effects of a "$2 \times CO_2$" climate on wheat yield

Under the $2 \times CO_2$ scenario the yields of spring wheat increase by an average of 6–20% [*Figure 3.5(b)*]. The yields rise about 10% (to 2750 kg/ha) in the coastal areas of southwestern Finland, but less in the east because assumed increases in precipitation indicate losses due to heavy rains at harvest time. The yields for the "best" years (5 percentile) rise the most (by 15%, or 450 kg/ha) in the northern parts of the present cultivation area [*Figure 3.6(b)*]. Yields in southern Finland increase by no more than 2–5%. The yields for the "poorest" years (95 percentile) rise by 25–40% in the south and by 10–15% in the north. In general, the greatest yield increases under the $2 \times CO_2$ scenario occur in the regions that have the lowest yield at present.

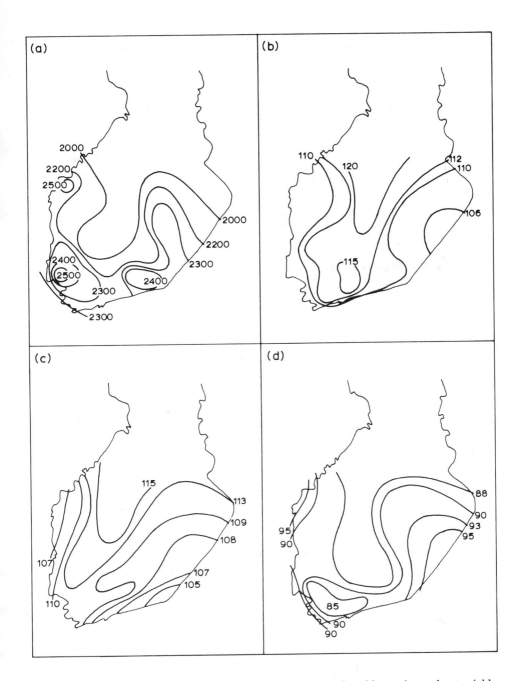

Figure 3.5. Effects of climatic scenarios on mean marketable spring wheat yields (kg/ha): (a) average climate, 1959–83; (b) GISS $2 \times CO_2$ scenario with variety having thermal requirement 120 GDD greater than present varieties; (c) warm scenario (1966–1973 type) with present-day varieties; (d) cool scenario (1974–1982 type) with present-day varieties.

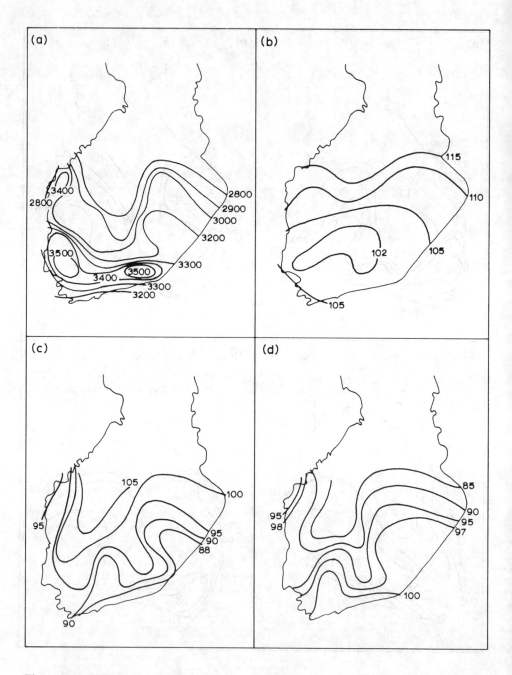

Figure 3.6. Effects of climatic scenarios on highest (5 percentile) marketable spring wheat yields (kg/ha): (*a*) average climate, 1959–83; (*b*) GISS $2 \times CO_2$ scenario with variety having thermal requirement 120 GDD greater than present varieties; (*c*) warm scenario (1966–1973 type) with present-day varieties; (*d*) cool scenario (1974–1982 type) with present-day varieties.

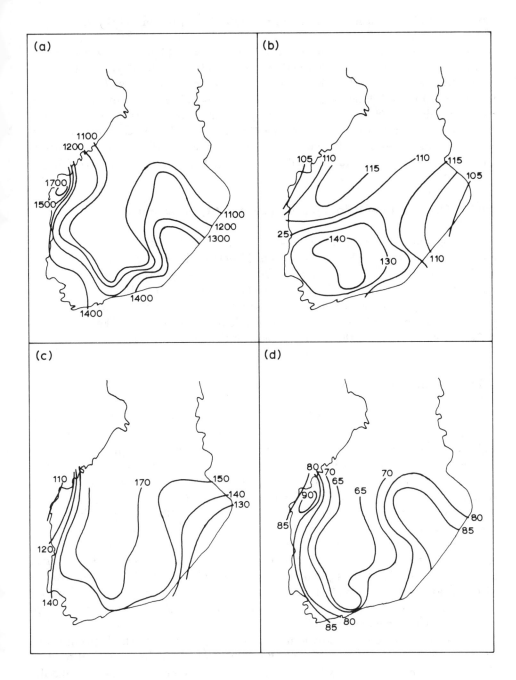

Figure 3.7. Effects of climatic scenarios on lowest (95 percentile) marketable spring wheat yields (kg/ha): (*a*) average climate, 1959–83; (*b*) GISS 2 × CO_2 scenario with variety having thermal requirement 120 GDD greater than present varieties; (*c*) warm scenario (1966–1973 type) with present-day varieties; (*d*) cool scenario (1974–1982 type) with present-day varieties.

3.6.4. The effects of a warm period (1966–73 type scenario)

Under the warm-period scenario wheat yields are highest (2825 kg/ha) in southwestern Finland. In such a period the yields for the "best" years would rise most – by 5%, or 130 kg/ha – in the northern parts of the country [*Figure 3.6(c)*]. In southern Finland, however, early summer drought during warm summers can reduce yields by 10%. Yields in these years in southeastern Finland do not exceed 3000 kg/ha in the south. During the "poorest" (95 percentile) years of a warm period, however, yields increase by as much as 50–70%. The difference is greatest inland, where yields are 800 kg/ha higher than "normal" [*Figure 3.7(c)*].

3.6.5. The effects of a cool period (1974–82 type scenario)

Under the cool-period scenario average yields are 115–350 kg/ha, or 5–15% below normal [*Figure 3.5(d)*]. In the best years (5 percentile) of the cool period "normal" yield levels are achieved in southern Finland, but yields are reduced toward the north in line with the fall in mean yields [*Figure 3.6(d)*]. In the worst years (95 percentile) of the cool period the possibility of total crop failure is evident. Even in southern Finland the yields can fall by 35% to 800–900 kg/ha [*Figure 3.7(d)*]. Departure from the normal is least in the western coastal area.

3.7. Discussion

On the basis of the experiments for the GISS $2 \times CO_2$ scenario, changes in temperature and precipitation caused by an increase in carbon dioxide can be expected to have a marked effect on spring wheat yields and on plant production in general. With the rise in temperature, winters in particular would become milder. The snow cover would not last as long as it does now, and soil frost would decrease. The cultivation of winter crops could be increased and their cultivation area could be extended to central Finland. Cultivation of fodder crops may be transferred to the central and northern parts of the country.

Higher temperatures would allow sowing 2–3 weeks earlier than at present, thus enabling the cultivation of spring crops with a longer period of vegetative growth. The brevity of this period is at present one of the obstacles to increasing yields in Finland. Moreover, under the warmer climate, both spring and winter crops could make better use of early spring radiation, which is barely exploited at present owing to snow and soil frost.

According to the model experiments, the vegetative growth of spring wheat under the $2 \times CO_2$ scenario takes place in temperatures almost identical to baseline values (*see Figure 3.4*), while from the shooting stage on, temperatures differ from the "warm period" temperatures by only 1 °C. This observation assumes that wheat varieties with higher heat requirements are substituted for present-day varieties in southern Finland. In the north of the country, in contrast, the thermal conditions would resemble those at present found in the south. Soils are not generally limiting to cereal cultivation in the north, so it is possible that present-day quick-ripening spring wheat varieties could be cultivated there. However, over the whole country under the GISS 2 × CO_2 scenario high rates of precipitation in early summer could delay growing and in the late summer could increase the risk of losses due to wet weather during harvest. Higher rates of evaporation might be a countervailing factor, but excess moisture could remain a long-term problem.

In this section we have not considered the "direct effects" of increased concentrations of atmospheric carbon dioxide on wheat yields through increases in crop photosynthesis (Lemon, 1983; Sionit et al., 1981). However, we can conclude from the model experiments that the indirect effect of increased CO_2 on the climate will improve crop cultivation conditions in Finland.

Under the warm-period scenario yields vary considerably less about the mean ($\pm$ 10-15%) than during the cool period ($\pm$ 27-32%). In cold years, ripening often does not occur until September, which increases the risk of night frost before harvest. Higher relative humidity greatly hampers threshing and increases harvesting costs in autumn. Excessive moisture in autumn also obstructs the operation of heavy harvesting machines. Cool summers normally result in good yields of spring wheat in southern Finland (60° N, districts 1-4). Between latitudes 61° and 62° N (districts 5-9) the yield level may be high in cool years, although the probability of very low yields increases also. Further north, in districts 10-15, the low ETS prevents the attainment of good yields in most years.

During a warm period, crop yield "certainty" increases. In the best years yields of spring wheat may be fairly similar throughout the country. However, in the worst years yields are reduced much less than they are in a cool period.

Cool periods tend to result in underproduction in Finland, and wheat has to be imported. Warm periods result in overproduction. The implications of this for national agricultural policy are discussed in Section 5.

SECTION 4

The Effects on the Profitability of Crop Husbandry

4.1. Introduction

Sections 2 and 3 of this case study have considered the effects of climate on crop yields. There is, however, no simple relationship between yield and profitability – and yet it is profitability that ultimately determines which crops are grown, and where. In this section we consider the effect of changes in climate on yields of barley and oats in two regions of Finland, and go on to examine likely changes in the profitability of cultivating these crops.

4.2. The Variability of Gross Margins

We may define "profitability" as the income that remains for the farmer once his variable expenditure, i.e., the costs of seed and fertilizers, use or hire of machinery, wages paid to outside workers, etc., have been deducted (Varjo, 1974). We shall term this gross margin in order to distinguish it from net return (*Figure 4.1*).

The data required here to determine the gross margin per hectare for a given crop are the marketable quality yield per hectare, its market price and the variable expenditure in its cultivation, which are published by the Association of Agricultural Centers (MKL; Varjo, 1977). In addition to the marketable yield, the farmer will, at least in some years, be left with a certain amount of grain of inferior quality owing to adverse weather conditions which can be fed to livestock on the farm. Because of this, grain cultivation can often prove profitable even though the gross margin may not be sufficient in itself to assure a farmer of a reasonable wage for his work excluding state agricultural subsidies.

GROSS RETURN 2970 mk							
VARIABLE COSTS 1247 mk				GROSS MARGIN 1723 mk			
Cost of seed	Cost of fertilizers	Cost of machine hire	Other costs	Wages for farmer 360 mk	NET RETURN 1362 mk.		
					Interest on capital invested		Net profit

Figure 4.1. Gross return, gross margin and principal cost factors per hectare for the cultivation of barley in 1980. The numbers refer to average national costs per farm in Finland in 1980. 1 US$ = 3.6 FIM in 1980 (MKL, 1980).

The manner in which the marketable yield, the cost and the selling price mentioned above affect the profitability of arable farming is illustrated in *Figure 4.2.* Comparison of the marketable barley yield (Official Statistics of Finland – SVT III), variable expenditure and gross margins for 1965, 1975 and 1980 in Lapland and Varsinais-Suomi shows harvest yields to have increased by about 1200 kg/ha in both areas over the period 1965–1980, most of this increase occurring between 1975 and 1980 in Lapland but between 1965 and 1975 in Varsinais-Suomi. With the variable expenditure increasing and the market price decreasing in real terms in a roughly similar manner in both areas (*Table 4.1*), the outcome was a pronounced decline in the profitability of barley cultivation in Lapland, which had in any case been low. For the sake of simplicity, the values employed here are those for central Finland, which deviate from those for Varsinais-Suomi and Lapland by less than 3%.

Thus the gross margin, which had been negative in Lapland in 1965, at −230 FIM/ha, reached figures of −900 FIM/ha or worse by 1975. Later, however, the rapid rise in yields, together with a reduction in costs, meant that profitability improved markedly, so that by 1980 the gross margin in Lapland was just over 400 FIM/ha. By comparison, the trend in Varsinais-Suomi is seen to have been more much more favorable throughout. The gross margin in 1965 was 1975 FIM/ha, of which about 750 FIM represented a direct net return. Again profitability declined up to 1975 on account of the lower market price and increased costs, being 1720 FIM/ha in that year. Nevertheless, since the proportion of this gross return accounted for by wages decreased more rapidly than did the market price of barley, partly as a result of increased mechanization, the net return obtained increased to almost 1400 FIM/ha, approximately the same figure as that recorded in 1980. Thus the difference in the gross margin for barley cultivation between Lapland and Varsinais-Suomi diminished from some 2200 FIM/ha in 1965 to around 1300 FIM/ha in 1980. This still means, however, that the figure for Varsinais-Suomi was more than four times that for Lapland, suggesting that although the great difference in temperatures between southern and northern Finland naturally does contribute to the differences in harvest yields, this effect is overshadowed as far as the profitability of crop cultivation is concerned by the very considerable influence of economic factors and agricultural policy measures. This also explains why the substantial increase in yields

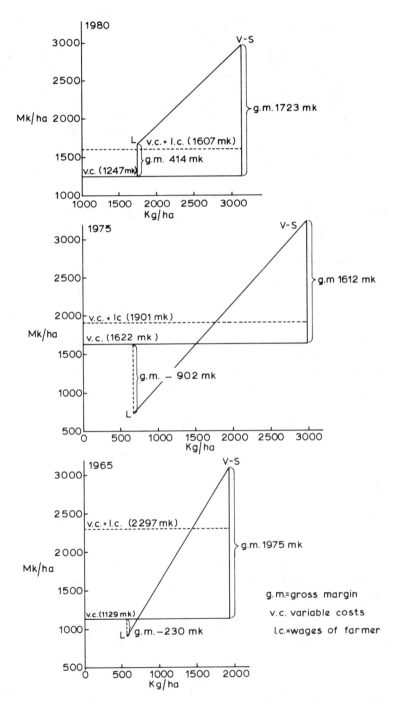

Figure 4.2. Profitability of the cultivation of barley and principal factors in its calculation in Varsinais-Suomi (V-S) and Lapland (L) in 1965, 1975 and 1980 (MKL, 1980). For location of the two regions, see Figure 1.3.

Table 4.1. Trends in producer prices, variable expenditure and producer's wages between 1965, 1975 and 1980 adjusted to 1980 prices.

Crop	Producer price (FIM/kg)[a]			Variable expenditure (FIM/ha)			Wage for farmer (FIM/ha)		
	1965	1975	1980	1965	1975	1980	1965	1975	1980
Spring wheat	2.31	1.27	1.41	1596	1975	1405	300	280	360
Rye	2.32	1.46	1.56	1129	1911	1606	360	310	400
Barley	1.57	1.07	0.96	1129	1622	1247	300	280	360
Oats	1.47	0.96	0.88	1129	1607	1606	300	280	360

[a] 1 US dollar = 3.6 FIM in 1980.
Source: MKL (1965, 1979, 1980).

achieved all over Finland in recent years has not led to any extension of cultivation in northern Finland, nor even to an improvement in profitability in every case (see Figures 1.6 and 1.8).

The considerable differences in harvest yields from one year to another serve to emphasize the difference in profitability of crop husbandry between northern and southern Finland. This is evident in *Figure 4.3* which shows the marketable harvest yields for spring wheat, rye, barley and oats, the gross margins for these crops and their labor costs expressed in kilograms of grain. The figures given apply to four regions over the period 1953–1983. The yields obtained in Varsinais-Suomi are seen here to have been sufficient to cover the variable expenditure in all years except for some rare cases such as oats in 1953–1956, barley in 1957, spring wheat in the early 1950s and in 1977 and 1978. The gross margin also served to cover the labor costs in almost all cases, generally leaving a substantial net return. Thus the profitability of crop husbandry in this area may in general be regarded as good.

However, profitability decreases progressively further north, so that even by the latitude of South Bothnia (see *Figure 1.3*) the cultivation of spring wheat has become a much more uncertain undertaking on account of the frequent, and sometimes consecutive, years of crop failure with only occasional intervening years in which an actual net return is achieved. In North Bothnia, where spring wheat is no longer grown on any large scale, years of crop failure occurred regularly during the period examined, with only very exceptional years in which the gross margin succeeded in covering the farmer's wage requirement. Lapland is the least favored of all the regions, wheat being cultivated only in very exceptional cases.

The yields of rye are also much lower in South Bothnia than in Varsinais-Suomi, frequently failing to cover the farmer's wage requirement and sometimes not even meeting the variable expenditure. This latter state is more or less the rule in North Bothnia except in a few of the most recent years examined when rye has even yielded a small net return. The trend in Lapland is unusual because, while rye cultivation was just as successful here as in South Bothnia over *ca.* 1953–65, it has frequently failed in recent years.

A similar pattern may also be discerned in the fodder grains, barley and oats, in that while their cultivation proved economically profitable almost

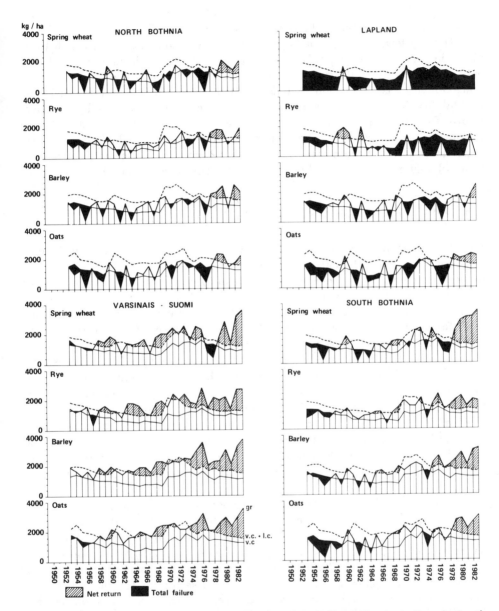

Figure 4.3. Profitability of crop husbandry in 1953–1983 in Varsinais-Suomi, South Bothnia, North Bothnia and Lapland (MKL, 1980). For district locations, *see Figure 1.3.*

without exception in Varsinais-Suomi, and has been extremely profitable since about 1970, their profitability became increasing uncertain from South Bothnia northwards owing to the progressively greater frequency of crop failure and the smallness of the gross margin. In this case, too, the most recent years proved more favorable, to the extent that these fodder crops even gave a small net return. This is exceptional, however, it being common for their profitability to be extremely low, especially in Lapland, but also to some extent in North Bothnia. The fact that, despite this, they continue to be grown in Lapland (although oats admittedly only in the south of the region) can be explained by their importance for use as fodder on the farms themselves, in which case their low profitability is of little consequence, as it can be compensated for by improved returns from animal husbandry. It does mean, however, that oats and barley are seldom grown in excess of the farm's own requirements, which leads to the importation of much more cattle fodder into Lapland from elsewhere than would be absolutely necessary were the region's full production capacity to be realized.

4.3. Climatic Variability and Marketable Yield

Variations of climate are readily reflected in agricultural performance in the form of fluctuations in yields, particularly in the north of Finland. Data on climate-induced changes in crop yields are available in Finland from the early nineteenth century onward (Keränen, 1931; Hustich, 1945, 1950, 1952; Soininen, 1974; and see Section 2). The value of these figures for statistical analysis is nevertheless doubtful since the variables are few in number and since a considerable increase in yields has been observed recently which cannot be attributed to any climatic change. It is thus necessary to distinguish between changes in yields resulting from climate effects and those brought about by the development of more productive strains of cereals, more intensive use of fertilizers, etc. At the same time the uncertainties of the statistics, especially those for marketable yield in the nineteenth century, and to some extent also covering more recent times, mean that the figures for individual years would in any case be inadequate for proper statistical analysis. The difficulties encountered in attempting to relate barley yields to climate over such a long period have already been discussed in Section 2.

Another method is available, however, which avoids this problem by using commune-specific forecasts for marketable yields and relating these to effective temperature sums and May–September rainfall figures. By restricting the examination to a short period of just a few years, the effects of advances in agricultural technology are reduced. The yield forecasts are inexact, being based on visual inspection of the fields by specially appointed assessors in the individual communes. However, the assessors have many years of experience in crop forecasting behind them and the data can be taken as at least a good indication of ultimate yields (Varjo, 1980).

As described in Section 1 in this study, ETS and rainfall data for the individual communes in the calculations were obtained from maps by interpolation between the corresponding isopleths. The analysis is applied to data from 1979, 1980 and 1981, years with quite different summer weather (average, warm and dry, and cool and wet conditions, respectively). This gives a total (over 3 years and for more than 370 communes, nationwide) of 1123 observations for barley and 1117 for oats.

The influence of ETS on marketable yields of barley and oats was examined first, with the climatic and yield data divided into three parts, representing for each year the communes receiving a precipitation of less than 250 mm, 250–300 mm and over 300 mm. Correlations between yields per hectare (Y) and ETS were then calculated for each group and each crop separately, using the GIMMS cartographical program (Waugh and McGalden, 1983). The results were as follows:

Barley

< 250 mm	$N = 73$	Equation not significant	$R^2 = 0.01$
250–300 mm	$N = 506$	$Y = 1.6 \cdot \text{ETS} + 530.2$	$R^2 = 0.33$ (4.1)
> 300 mm	$N = 544$	$Y = 2.5 \cdot \text{ETS} - 1044.7$	$R^2 = 0.28$ (4.2)

Oats

< 250 mm	$N = 74$	Equation not significant	$R^2 = 0.01$
250–300 mm	$N = 504$	$Y = 0.9 \cdot \text{ETS} + 1328.4$	$R^2 = 0.13$ (4.3)
> 300 mm	$N = 539$	$Y = 1.7 \cdot \text{ETS} + 164.0$	$R^2 = 0.39$ (4.4)

As may be seen from the above figures, changes in temperature sum do not seem to affect the marketable yields of the crops concerned when rainfall during the growing season remains below 250 mm, whereas with a rainfall of over 250 mm they show a clear positive correlation with the ETS.

The effect of rainfall during the growing season on harvest yields was then examined in a similar fashion, by dividing the data sets into three groups with reference to the ETS values and determining the correlations between precipitation ($P\text{-}t$) and yields per hectare in these groups. The following results were obtained:

Barley

< 1000 GDD	$N = 360$	$Y = -6.1 \cdot P\text{-}t + 3271.0$	$R^2 = 0.11$ (4.5)
1000–1300 GDD	$N = 530$	$Y = -6.1 \cdot P\text{-}t + 3271.0$	$R^2 = 0.10$ (4.6)
> 1300 GDD	$N = 228$	Equation not significant	$R^2 = 0.00$

Oats

< 1000 GDD	$N = 363$	$Y = -4.1 \cdot P\text{-}t + 3154.6$	$R^2 = 0.25$ (4.7)
1000–1300 GDD	$N = 530$	$Y = -1.2 \cdot P\text{-}t + 2679.3$	$R^2 = 0.10$ (4.8)
> 1300 GDD	$N = 230$	Equation not significant	$R^2 = 0.04$

These figures suggest that at temperature sums below 1300 growing degree-days (GDD), increased precipitation led to a marked decline in marketable yields of oats and barley per hectare; whereas in warm summers, with effective temperature sums of over 1300 GDD, increased precipitation had no effect on harvests.

According to the above both ETS and precipitation during the growing season appear to affect the marketable yields of barley and oats independently of the other. In order to ascertain their combined influence, a statistical multivariant analysis program was employed (Brown et al., 1981). The calculation for this gave the following equations for estimating the marketable yields of barley and oats given in the above variables:

$$Y_B = 3755.7 - 13.12\,P\text{-}t + 79.9 \left(\frac{P\text{-}t}{100}\right)\left(\frac{\text{ETS}}{100}\right) - 3.4 \left(\frac{\text{ETS}}{100}\right)^2 \tag{4.9}$$

$R^2 = 0.56$ (standard error of est. 552.4)

$$Y_O = 3219.4 - 8.4\,P\text{-}t + 56.17 \left(\frac{P\text{-}t}{100}\right)\left(\frac{\text{ETS}}{100}\right) - 2.7 \left(\frac{\text{ETS}}{100}\right)^2 \tag{4.10}$$

$R^2 = 0.34$ (standard error of est. 537.4)

in which:

Y_B = Marketable yield of barley per hectare
Y_O = Marketable yield of oats per hectare
$P\text{-}t$ = May–September precipitation (mm)
ETS = Effective temperature sum (GDD)

It is important to note that the formulation of equations (4.9) and (4.10) involves the pooling of two types of weather variability: variability over space (differences between regions) and variability over time (differences between the three weather-years). It is therefore assumed implicitly that the response of yields to given weather conditions is the same at all locations, regardless of regional differences in management practices, technology and soils.

4.4. Model Sensitivity Analyses

By systematically entering a range of arbitrary values of growing season precipitation and ETS into equations (4.9) and (4.10), an indication of the modeled sensitivity of yields to each of these climatic variables can be assessed. The results of this analysis are shown in *Figure 4.4*. The positive effect of high ETS values is evident for yields both of barley and oats, with high precipitation exerting a negative influence on yields under low temperature conditions, but proving beneficial to yields during the warmest summers.

4.5. Comparison of Model Results with Observations

Bearing in mind the reservations, expressed above, concerning the pooling of spatial and temporal climatic information, the regression equations can be used to depict trends with time at a given place or areal differences occurring at a given point in time. As an example of the latter we have calculated the areal pattern of marketable barley yields in 1979 by means of equation (4.9), using commune-specific ETS values and May–September rainfall figures (*Figure 4.5*). In this year we estimate that the highest yields (over 2400 kg/ha) occur along the

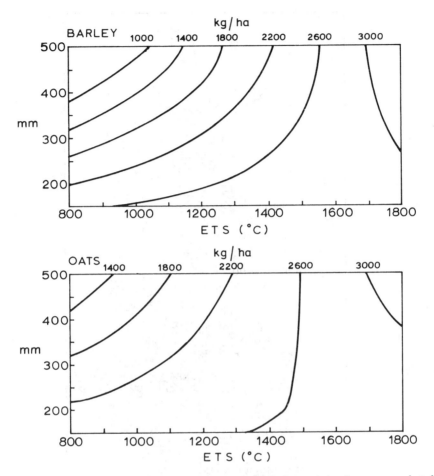

Figure 4.4. Influence of effective temperature sum and precipitation on marketable yields (kg/ha) of (*a*) barley and (*b*) oats.

south coast and in a few communes of Northern Karelia (district 12 in *Figure 1.3*). Even in the north, yields of over 1800 kg/ha are estimated for almost all of Kainuu (district 17) and some communes in central Lapland.

The accuracy of the estimates for 1979 can be checked by reference to the observed yield and cultivated area data for that year (SVT III, 1979). These data are available only for agricultural center districts and not for communes, so it is necessary to sum the estimated total yields and field areas under barley for the communes in each district to calculate mean yields for the districts. Moreover, since information for 1979 was used to construct the original model equation and does not constitute an independent temporal data set, this method of model verification is not ideal. It does, however, provide some indication of the correspondence of model estimates to real conditions in Finland.

Table 4.2 shows the predicted national yield to be quite similar to the observed figure. Discrepancies in the values for the individual agricultural

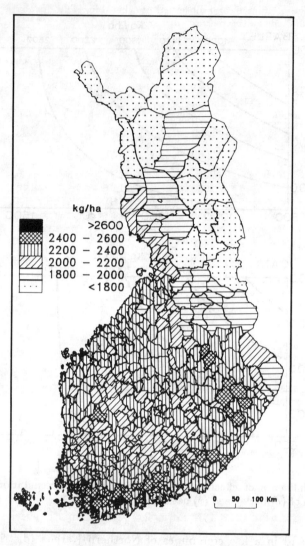

Figure 4.5. Estimated marketable yield of barley by commune in 1979 (kg/ha).

centers are in general small, the only exceptions being the Kuopio, Kainuu and Lapland Agricultural Districts. In the first two cases, the explanation may lie in the custom of concentrating the cultivation of barley on the tops and slopes of hills on account of their more favorable microclimate (cf. Kivinen, 1941), so that conditions would be more advantageous and yields greater than one would expect from the ETS values and rainfall figures. In Lapland the low summer temperatures in 1979 may well have caused the occasional frost which would have reduced the yields more than would be predicted from the ETS and rainfall data (cf. Huovila, 1974).

Table 4.2. The predicted and observed marketable yield for barley, discrepancies between these, and the cultivated area of barley in 1979.

Agricultural district	Marketable yield (kg/ha)		Percentage differences from the official yield	Area of barley (000 ha)
	Observed	Predicted		
Uusimaa	2480	2410	−3	43.1
Swedish Uusimaa	2570	2360	−8	19.5
Varsinais-Suomi	2320	2280	−2	91.0
Economic Society of Finland	2560	2520	−2	5.6
Satakunta	2290	2160	−6	56.1
Pirkanmaa	1950	2140	+10	26.9
Häme	2140	2210	+4	58.4
East-Häme	1950	2140	+10	21.0
Kymi	2220	2320	+4	38.9
Mikkeli	2170	2240	+4	18.2
Kuopio	2620	2230	−15	35.6
North Karelia	2440	2290	−6	19.8
Central Finland	2020	2210	+10	21.6
South Bothnia	2330	2210	−5	80.2
Swedish South Bothnia	2310	2340	+2	35.4
Oulu	2110	2200	+5	54.3
Kainuu	2320	1810	−22	5.5
Lapland	1650	1940	+18	2.0
Finland	2270	2250	−1	633.1

Source of observed data: SVT III (1979).

4.6. Effects of Climatic Variations on the Profitability of Crop Husbandry

The effects of three scenarios of climatic change were considered: a cool period (characterized by the observed meteorological record for the period 1861-70), a warm period (data for 1931-40) and a "$2 \times CO_2$" climate derived from the GISS general circulation model simulations (see Subsection 1.9). Impact experiments were conducted utilizing the long-term temperature and precipitation records for Oulu and Helsinki (see Figure 1.5). Unfortunately, due to a deficiency in the source material, precipitation values were not available for the "cool" scenario period (1861-70) at Oulu.

As described in Subsection 1.9.2, the $2 \times CO_2$ scenario was constructed by first mapping the changes in mean annual temperature (°C) and May-September precipitation (mm) between GISS $1 \times CO_2$ and $2 \times CO_2$ equilibrium conditions (Figure 4.6). As may be seen, the increase in mean annual temperature is 4.0-4.5°C in southern Finland and 5.0-5.5°C in northern Finland. In order to calculate the change in effective temperature sum (ETS) implied by these temperature changes, an approximate conversion factor was adopted of 100 degree-days per 1°C change in mean annual temperature. This provides only a rough estimate, but if we bear in mind that the GISS model-generated data include

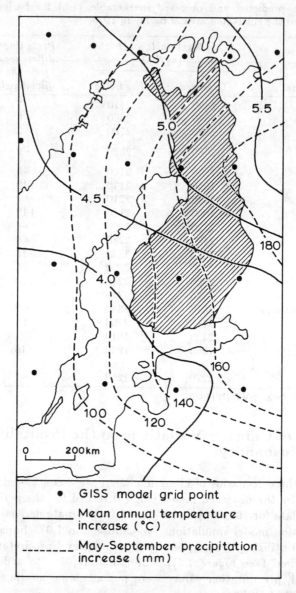

Figure 4.6. Increase in mean annual air temperature (°C – solid isolines) and May–September precipitation (mm – dashed isolines) over eastern Fennoscandinavia generated by the GISS model for a $2 \times CO_2$ climate relative to the $1 \times CO_2$ control climate.

only four grid points within, and about 12 others adjacent to, the national boundary of Finland, from which to construct isopleths of temperature change, these figures may be regarded as an adequate regional basis for the development of the $2 \times CO_2$ scenario for the purposes of this investigation.

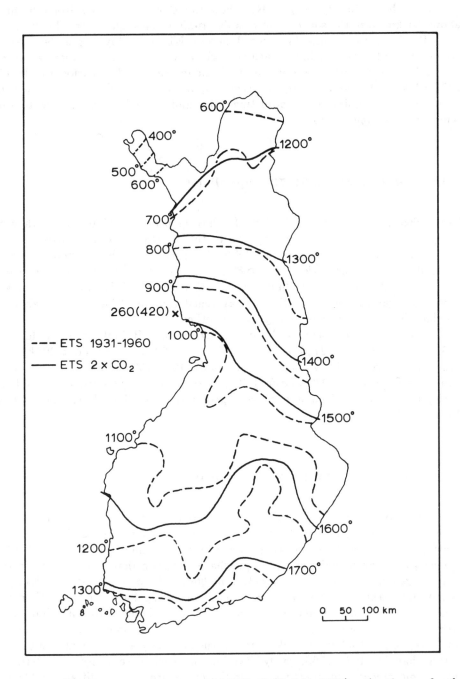

Figure 4.7. Effective temperature sums in 1931–60 (Kolkki, 1969) and estimates for the $2 \times CO_2$ climatic scenario (degree-days).

Transformation of the ETS values observed for the period 1931–60 [*Figure 1.4(c)*] in accordance with the method described above, gives mean annual ETS values as shown in *Figure 4.7*. Note how the 1300 degree-days isopleth shifts about 800 km from the south coast to a line running through Central Lapland.

The impact experiments required data on ETS and May–September precipitation for Oulu and Helsinki. Each scenario was designed to represent a decade of weather, so for ease of comparison with the instrumental scenarios, the decade 1971–80 was selected as a reference scenario, and the $2 \times CO_2$ scenario was constructed correspondingly by adjusting ETS and precipitation relative to this reference (*see Figure 1.10*).

4.6.1. Effects of an "1861–70" type climate

The 1860s are renowned in Finland for having been a cold period, with very many years of crop failure, usually due to the occurrence of night frosts before the grain had ripened. The mean ETS for the period was about 100 degree-days lower than during the period 1971–1980 in southern Finland, although there was very little difference in the Oulu area (cf. *Figure 1.10, Table 4.3*). The May–September rainfall figure in Helsinki averaged around 290 mm, although with considerable variations from one year to another.

The estimated mean yield of barley in the Helsinki area under the 1861–1870 scenario is 2120 kg/ha and of oats is 2270 kg/ha, about 200 kg and 220 kg lower, respectively, than the means for 1971–1980. Employing the price and costs figures for 1980 (*Table 4.1*), it may be shown (*Figure 4.8*) that both barley and oats would have given very poor returns in Helsinki except in 1861, 1865 and 1868.

4.6.2. Effects of a "1931–40" type climate

This period differs radically from that considered above in its temperature conditions, with an exceptionally high mean ETS especially in the south of Finland, the figure for Helsinki being about 190 degree-days higher than in 1971–1980 (*Table 4.3*). The period was also somewhat drier than that of 1971–1980 in the Helsinki region, but very much wetter in Oulu. As a result estimated barley yields are 260 kg higher in Helsinki but about 120 kg lower in Oulu, than the mean for 1971–1980. Similarly oat yields are 150 kg higher in Helsinki, but about 50 kg lower in Oulu than for the baseline period. Thus oats give a high return in all years, except 1935 in Helsinki; and returns on barley are excellent in all years, better than for 1971–1980, easily covering the variable expenses and farmer's wage requirement and allowing an ample net return on its cultivation (*Figure 4.8*).

The effects on the profitability of crop husbandry

Table 4.3. Mean values and standard deviations of ETS and May–September precipitation at Helsinki and Oulu, and estimated marketable yields of barley and oats for the scenarios 1971–80 (baseline), 1861–70 (cool decade), 1931–40 (warm decade), and 2 × CO_2 (GISS model-derived).

Scenario	ETS (degree-days)		May–September Precipitation (mm)		Estimated yields barley (kg/ha)		Estimated yields oats (kg/ha)	
	Mean	S D	Mean	S D	Mean	S D	Mean	S D
Helsinki								
1971–1980 (baseline)	1231	126.3	275	78.1	2315	328.7	2386	173.8
1861–1870	1130	151.1	287	74.4	2119	414.4	2265	235.7
1931–1940	1417	155.5	252	68.7	2584	284.8	2540	146.2
2 × CO_2	1638	126.3	427	78.1	2808	293.8	2821	194.8
Oulu								
1971–1980 (baseline)	1094	133.9	226	40.1	2359	194.9	2386	109.9
1861–1870	1068	128.4	–	–	–	–	–	–
1931–1940	1188	150.1	275	74.5	2238	449.9	2335	258.2
2 × CO_2	1562	133.9	389	40.1	2679	270.5	2707	184.4

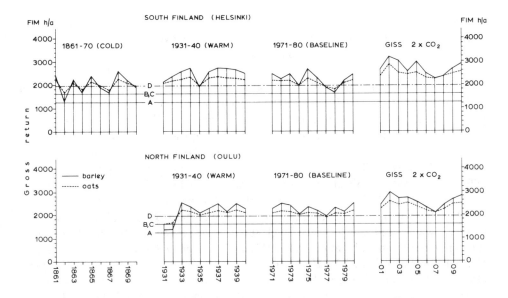

Figure 4.8. Estimated effects of climatic variations on the profitability of barley and oats cultivation at Helsinki and Oulu (FIM/ha). A: Variable costs (barley). B: Variable costs and wages (barley). C: Variable costs (oats). D: Variable costs and wages (oats).

4.6.3. Effects of the GISS $2 \times CO_2$ scenario

Under the GISS model-derived $2 \times CO_2$ climate described in Subsection 1.9.2 and above, the mean ETS at Helsinki is estimated to be about 410 degree-days higher, and at Oulu nearly 470 degree-days higher than in the reference period (1971–1980). May–September precipitation is also greatly enhanced at both stations. Because of the greater sensitivity of barley to high precipitation, yields of oats are slightly higher than those of barley under the warmer, wetter $2 \times CO_2$ climate in the two areas (*Table 4.3*), yields of both crops being considerably higher than for the baseline period. Barley yields are about 510 kg higher in Helsinki and 320 kg higher in Oulu; oats yields, respectively, 430 kg and 320 kg higher. However, the higher producer price for barley than for oats, combined with lower variable expenditure on barley (*Table 4.1*) given 1980 prices and costs, provides farmers with a greater profit margin for barley than for oats in most years (*Figure 4.8*).

In interpreting these results, it is important to note that the climatic conditions simulated by the $2 \times CO_2$ scenario for any single location in Finland lie well outside the range of conditions recorded historically at that location. Thus, although the regression models (4.9) and (4.10) have been developed on the basis of climatological data from many different parts of Finland, their applicability for estimating crop yields in those regions (particularly in southern Finland) where the $2 \times CO_2$ climate is analogous to the present climate in regions further south in Europe (*see Table 1.8*), is somewhat uncertain, being dependent on extrapolation of the derived equations.

4.6.4. Mapping zones of productivity and profitability under the GISS $2 \times CO_2$ scenario

On the basis of equation (4.9), it is possible to map the national distribution of yields estimated for the GISS $2 \times CO_2$ scenario according to the approximate pattern of ETS and precipitation conditions (*Figure 4.9*). The analysis ignores the significant differences in May–September precipitation averages between east and west Finland [*Figure 1.4(d)*], which are enhanced under the $2 \times CO_2$ scenario (*Figure 4.6*). A marketable yield of barley averaging approximately 3000 kg/ha would be obtained near the coast of southern Finland, approximately 2480 kg/ha around Oulu and approximately 1850 kg/ha in Inari (northern Lapland). These yields would be sufficient to provide a gross margin, covering the farmer's wages and the net return, of about 1620 FIM/ha in the south, decreasing northwards to only about 530 FIM/ha in Inari, a figure some 28% higher than the mean obtained in 1980 (*see Figure 4.2*). In the case of oats the yield close to the coast of southern Finland would also be nearly 3000 kg/ha, corresponding to a gross margin of about 1000 FIM/ha, or 620 FIM less than for barley. At the latitude of Oulu in central Finland the marketable yield would be about 2620 kg/ha and the gross margin approximately 700 FIM/ha, about 430 FIM/ha less than for barley, and in Inari the yield would be about

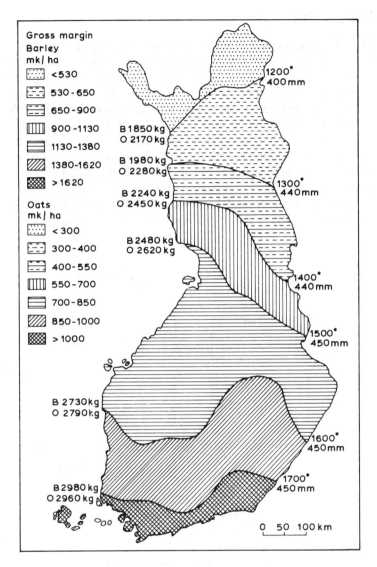

Figure 4.9. Harvest yield per hectare and gross margin for barley and oats in certain effective temperature sum regions for the GISS $2 \times CO_2$ scenario.

2170 kg/ha, about 320 kg/ha higher than for barley, and the gross margin around 300 FIM/ha, some 230 FIM/ha lower. Thus barley would appear to be a more profitable crop to cultivate across the whole of Finland under the conditions of the $2 \times CO_2$ scenario (and assuming 1980 costs and prices). However, if expenditure on oats cultivation were to fall or prices to rise relative to barley, then oats (the higher yielding crop in northern Finland) would become the more profitable. Equally, any further increase in precipitation would probably be more detrimental to barley than to oats yields, and might favor oats cultivation.

4.7. Conclusions

The results of the foregoing analysis are shown in *Figure 4.10* where the ETS values for the various climatic scenarios are depicted on the horizontal axis and the net returns from the cultivation of oats and barley according to 1980 price and cost levels on the vertical axis. The limits of variation in each case are expressed as the means for the three best and three poorest years. These details are presented for the Helsinki area in the lower diagram, which shows the difference in mean ETS between the coldest three years in the period 1861–1870 and the warmest three years in 1931–1940 to have been nearly 650 degree-days, while the mean ETS for these periods differed by about 290 degree-days. Values for the period 1971–80 fall in between the 1931–40 values and 1961–70 values, both the extreme and mean ETS values being about 100–150 degree-days warmer than in the 1861–70 period, and 150–200 degree-days cooler than in 1931–40.

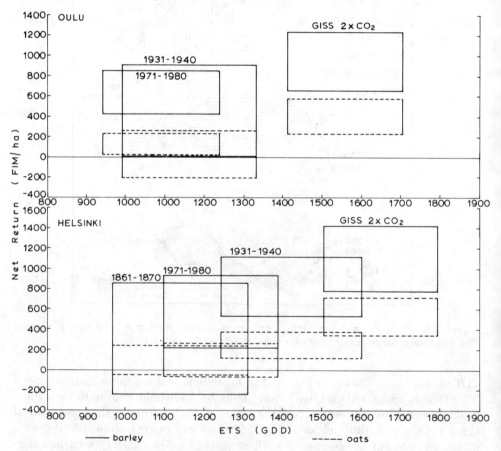

Figure 4.10. Trends in extreme values (averages of the 3 highest and of the 3 lowest estimates) for the profitability of barley and oats cultivation in the Oulu and Helsinki areas for various climatic scenarios. Net returns are based on 1980 prices and costs. For explanation, see text.

The Oulu area (upper diagram) differs markedly from the Helsinki area in that temperatures in the periods 1931–40 and 1971–80 were fairly similar (*Figure 4.10*). In terms of ETS, the 1930s did not constitute the same kind of "warm period" in the Oulu area as was the case around Helsinki, but instead featured conditions in many ways comparable with those that prevailed in Helsinki in the period 1861–1870.

The period 1971–80 was very much cooler in the Oulu region than in Helsinki, the difference in mean ETS being more than 140 degree-days. However, with low rainfall figures during the growing season, the marketable yields of barley and oats are similar in both areas (*Table 4.3*). Annual fluctuations in harvests are small in Oulu, and similar net returns are achieved as in the Helsinki area; better returns in the cooler years. Indeed, the baseline scenario is more favorable for both crops in the Oulu area than the warm-period scenario, especially with respect to estimated returns in the coolest years, although slightly higher returns are achieved in the warmer years of the 1930s than in 1971–80. The returns estimated for the warm-period scenario in the Helsinki region, in contrast, are higher and less variable than under the baseline conditions.

A warming of the climate to the extent envisaged under the $2 \times CO_2$ scenario increases yields and thus net returns of both crops. The mean net returns in the cooler years at Helsinki are only a little greater than those obtained in the poor years within the period 1931–40. This is mostly a consequence of the high rainfall figures predicted by the GISS model, which serves to reduce marketable yields considerably in cool summers.

It can be assumed that effects on other grain crops are similar to those described for barley and oats, and it seems likely that greater diversity would be introduced into the crop combinations growing on farms in Finland as the more demanding species came to be grown further north. Thus the rise in ETS under the $2 \times CO_2$ scenario could bring about an extension of winter wheat cultivation in northern Finland as far north as Rovaniemi (latitude 66.5°N – *Figure 1.3*) and would enable spring wheat and rye to be grown commercially at still higher latitudes. Temperatures now experienced at the south coast of Finland would then prevail in southern Lapland and Kainuu, allowing widespread cultivation of grains at present restricted to the south.

SECTION 5

Conclusions and Implications for Policy

5.1. Summary of Results

5.1.1. The experiments

The purpose of this study was to estimate the effect of selected climatic variations on crop yield and farm profitability in Finland. In Section 2, yield-weather functions for barley were established for two regions (Helsinki and Oulu) using data for 1846–1983. At first, monthly climate data were employed, but since the results were not satisfactory, new yield functions were developed using daily data for 1959–1983. The same method was used for spring wheat in Section 3, but for a network of locations in southern and central Finland. Daily values of ETS and precipitation, as well as minimum and maximum daily temperatures were allocated to 12 growth phases of the crop and used as explanatory variables in these studies. In Section 4, yield functions for both barley and oats were developed using cross-sectional data from different locations over much of Finland for the period 1979–1981. In this case the explanatory variables were only effective temperature sum and growing season precipitation.

To illustrate the impact of climate on yields, the estimated yield functions were applied to simulate yield levels under different climatic conditions at present technological levels. The effect of cooler- or warmer-than-normal conditions was studied by applying the estimated functions to meteorological data which characterized warm and cool periods recorded in the recent past. In addition, outputs from the Goddard Institute for Space Studies (GISS) general circulation model were used to derive climatological information on the possible conditions in Finland under doubled concentrations of atmospheric carbon dioxide. These data were also employed in the yield simulations.

Table 5.1. Summary of yield estimates for barley, spring wheat and oats in southern and northern Finland under different climatic scenarios. Baseline values are absolute; scenario estimates are percentage changes relative to the baseline.

Scenario type	Scenario period	Section of study	Climate[c]		Estimated crop yields (kg/ha) (coefficients of variation in parentheses)		
			ETS (degree-days)	May–October precipitation (mm)	Barley	Spring wheat	Oats
				Southern Finland[a]			
Baseline	1959–83	2,3	1263	342	3075 (14%)[d]	2300 (20%)[e]	–
	1971–80	4	1231	342[f]	2315 (14%)[g]	–	2386 (7%)[g]
Cool periods	1861–70	4	–8%	+2%	–8% (+38%)	–	–5% (+43%)
	1921–30	2	–4%	+26%	+10% (–32%)	–	–
	1974–82	2,3	–5%	+6%	+4% (+12%)	–15% (+40%)	–
Warm periods	1931–40	2	+12%	–1%	+1% (–3%)	–	–
		4	+15%	–1%	+12% (–22%)	–	+6% (–21%)
	1966–73	2,3	+2%	–3%	–8% (–22%)	+10% (–40%)	–
2 × CO_2	Future	2,3	+37%[d]	+50%[d]	+9% (–42%)	+10% (–20%)[h]	–
	Future	4	+33%	+50%	+21% (–26%)	–	+18% (–5%)

			Northern Finland[b]				
Baseline	1959–83	2,3	1080	313	2560 (17%)[d]	2000 (20%)[e]	—
	1971–80	4	1094	288[f]	2359 (8%)[g]	—	2386 (5%)[g]
Cool periods	1861–70	4	−2%	—	—	—	—
	1921–30	2	−1%	+6%	+1% (−22%)	—	—
	1974–82	2,3	−2%	+2%	−1% (−39%)	−12% (+15%)	—
Warm periods	1931–40	2	+10%	+11%	+8% (−22%)	—	—
		4	+9%	+20%	−5% (+143%)	+15% (−60%)	−2% (+140%)
	1966–73	2,3	+3%	−9%	+3% (−41%)	—	—
$2 \times CO_2$	Future	2,3	+35%[d]	+58%[d]	+14% (+19%)	+20% (+5%)[h]	—
	Future	4	+43%	+64%	+14% (+22%)	—	+13% (+48%)

[a] Represented by sites in the Helsinki area (Section 2), by mapped information for the Helsinki area (Section 3), and by Helsinki (Kaisaniemi) meteorological station (Section 4).
[b] Represented by Ähtäri meteorological station, southwest of Oulu (Section 2), by mapped information for central Finland at approximate latitude 64°N (Section 3), and by Oulu meteorological station (Section 4).
[c] Information from Table 1.7.
[d] Information from Table 2.2.
[e] Approximate values derived from mapped yields in Figures 3.6, 3.7 and 3.8.
[f] May–September precipitation (not shown here) is used in the Section 4 model experiments.
[g] Information from Table 4.3.
[h] Estimated yields for an adapted spring wheat variety (Section 3).

5.1.2. Results: Effects of climatic variations on cereal yields

The results of the model experiments conducted in Sections 2–4 are summarized in *Table 5.1*. Although different baseline periods and study areas have been used in separate experiments, the differences are slight and do not affect the general conclusions that can be drawn from a comparison of the results in each region. However, an apparent discrepancy between estimates of barley yields for the baseline periods used in Sections 2 and 4 does need to be accounted for. In Section 2, estimated yields are significantly higher in southern than in northern Finland, whereas in Section 4 the estimated yields are slightly higher in the north. The climatic data show a greater contrast in temperatures between south (warmer) and north (cooler) for the Section 2 baseline (1959–83) than for the Section 4 baseline (1971–80), but a smaller contrast in growing season precipitation (drier in the north than the south). Since barley yields are particularly sensitive to precipitation, low temperatures and wetter conditions in the north combine to give low yields for the Section 2 baseline climate. The higher Section 2 yield estimates for southern Finland are mainly a result of the trend yields for the regression model being calculated relative to 1983, while in Section 4 yields were assessed at 1979–81 levels.

These inconsistencies apart, the impacts of the climatic scenarios may still be compared by expressing each as a change in yield relative to the appropriate baseline. The results for "cool period" scenarios indicate that spring wheat and oats yields are reduced under cooler and wetter-than-average conditions, probably reflecting a dominant effect of temperatures on yields. In contrast, estimates of the response of barley yields, while partly contradictory, suggest a greater sensitivity to precipitation anomalies especially in the drier south of Finland. Cooler average conditions are often associated with high interannual variability of climate in Finland, and this helps to explain the increased variability in many of the yield estimates (expressed as percentage changes in the coefficient of variation relative to the baseline). Only during the 1921–30 and 1974–82 periods at Oulu and the 1921–30 period at Helsinki are yields (in each case of barley) less variable, but these are relative to a highly variable baseline period.

Warm periods are characterized, in general, by higher-than-normal yields, with reduced interannual variability. The exceptions to this rule (*see Table 5.1*) are attributable to precipitation anomalies, particularly with respect to barley yields. In southern Finland, where precipitation is below average (as in 1966–73), the effects of drought may outweigh the positive influences of higher temperatures on crop yield. Conversely, in northern Finland, it is precipitation excess that often tends to reduce yields despite considerably warmer temperatures (as demonstrated for both barley and oats in Section 4). In addition, the highly variable climate of the 1930s in northern Finland (*see Table 4.3*) leads to estimates of high yield variability, the extent of which is somewhat obscured in *Table 5.1* by the contrasting variability in the different baseline periods.

The considerable climatic changes simulated for the $2 \times CO_2$ scenario give rise to increased yields of all the three cereals studied both in southern and in northern Finland. Interestingly, the impact experiments indicate that yields would become more stable in southern Finland but variability would increase in the north. This suggests that, even if increases in mean yields are greater in northern than in southern Finland under $2 \times CO_2$ conditions (an arguable point from these results – Table 5.1), the comparative advantage for cultivating cereals on the basis of climate alone would still lie in southern Finland, where year-to-year variations in yield would be significantly lower than in the north.

5.1.3. Results: Effects of yield changes on farm profitability

Assuming that other factors remain constant, increases (or decreases) in yields due to a change in the climate will improve (or worsen) the profitability of farming. The increase of the net income in plant production is simply the quantity multiplied by the producer price of the crop. No extra costs need to be taken into account. Thus, the increase in the net income is much greater than in cases where the yield increase is a result of increasing the use of inputs (e.g., fertilizers, pesticides, etc.).

Some examples of this are given in Section 4 and summarized here in Table 5.2. Net returns are calculated simply as gross return (i.e., producer price)

Table 5.2. Summary of profitability estimates for barley and oats production (average net return and variability) under different climatic scenarios in Helsinki and Oulu (see Section 4).

			Barley			Oats		
Climatic scenario	Mean ETS^a	May–Sept precip. (mm)	Mean yield (kg/ha)	Net return[b] Mean	S D	Mean yield (kg/ha)	Net return[b] Mean	S D
Helsinki								
1971–80 (baseline)	1231	275	2315	616	316	2386	134	153
1861–70 (cool)	1130	287	2119	427	398	2265	27	207
1931–40 (warm)	1417	252	2584	873	273	2540	269	129
$2 \times CO_2$ (future)	1638	427	2808	1089	282	2821	517	171
Oulu								
1971–80 (baseline)	1094	226	2359	657	187	2386	134	97
1861–70 (cool)	1068	–	–	–	–	–	–	–
1931–40 (warm)	1188	275	2238	542	432	2335	89	227
$2 \times CO_2$ (future)	1562	389	2679	964	260	2707	416	162

[a] Degree-days.
[b] FIM/ha.

minus the sum of variable expenditure and farmer's wages (all fixed at 1980 levels – see Subsection 4.2).

In southern Finland, profitability of both barley and oats increases for the warm period scenario (1931–40) and the $2 \times CO_2$ scenario, but decreases, relative to the baseline, during the cool decade (1861–70). Oats apparently provide more stable income than barley, both on an interannual basis and for the different scenario periods, but the mean levels of net income are considerably higher for barley, mainly as a result of lower production costs.

In northern Finland, estimated incomes from both crops are similar to those in the south for the baseline climate, in line with yields that are enhanced by dry growing season conditions. The wetter conditions of the 1931–40 warm period lead to reduced and highly variable net incomes for both crops, while the $2 \times CO_2$ conditions would improve farm incomes to slightly higher levels than those estimated for the warm-period scenario in the south. Again, it should be noted that estimated yields of oats in the north are higher than those of barley, and a narrowing of the difference in production prices and costs between the two crops relative to the 1980 values might well be an inducement to plant oats rather than barley, especially since the former gives more stable yields.

Some other implications of climatic change for farm incomes in Finland are examined in Subsection 5.3.

5.2. Implications of CO_2-Induced Warming in Finland

Under the GISS $2 \times CO_2$ scenario the average annual temperature in Finland is 4–5 °C higher than at present (see Figure 4.6). This can be expected to result in an increase of crop yields. Drought may, however, slightly lower the yields in the southern parts of Finland. It is difficult to estimate the average rise of the yields in the whole country, since all crops were not studied and since the structure of cultivation may change. Grassland for hay and silage now covers about 34% of the total cultivated land, barley about 26% and oats about 22% (see Table 1.1). The area under grain production may increase thus lowering the grassland share, and resulting in a rise in the average yield level. Winter crops may also give higher yields than spring crops.

The variation in crop yields has increased in recent years. Since the fit of the estimated functions is not very good, the standard error of the estimate of the estimated functions is relatively large. Thus the statistical confidence limits of the results from the scenarios are also large, which means that the range of variation in likely yields is substantial. Finnish farmers thus have to be prepared for greater uncertainty, similar to that they experienced recently when (in 1981) the average cereal yields fell by about 23% from the long-term trend.

As a result of the warmer climate the boundary of viable cultivation would move slightly northwards. This would have some effect in the case of wheat, rye, oilseeds and sugar beet, the present growing area of which is limited to south Finland and which compete for land in the south. Thus the warming-up of the climate could be expected to improve self-sufficiency in bread grains, which has been under 100% in recent years (see Table 1.4; Kettunen, 1985b).

The cultivation of winter crops is likely to increase owing to milder winters. New varieties can be bred and introduced because of the longer growing season. The quality of grain crops would improve since temperatures in the harvesting period will be higher, and since there will be more winter crops which are harvested earlier than spring crops.

The increase in yields would improve regional farm income distribution and thus affect the conduct of regional policy. However, the problem of oversupply of foodstuffs would be aggravated. These issues are discussed in Subsection 5.4.

5.3. Effects on Production, Income and Regional Policy

We may estimate the economic effect of climatic changes on farm income by generalizing from the results of yield effects reported in Sections 2, 3 and 4. The rise of yields has a direct effect on the growth of income in agriculture. When the growth of plant production takes place without any increase in inputs, the incremental income can be calculated from the increase in quantity multiplied by the producer price. To be exact, there are some extra costs due to the additional production, i.e., those caused by harvesting, drying and transporting of the additional output. Since most of the plant production is used as an intermediate input in animal production and since these intermediate inputs (such as hay or silage) usually do not have any market price, the assessment of the effect of higher yields in money terms is difficult. A rough estimate can be obtained by calculating the increase of production in feed units and using some estimate such as the price of a feed unit. In Finland the feed unit (f.u.) refers to the energy content of each product, the basic unit being the energy content of a kilogram of barley.

The average ("norm") yield for 1984 is estimated to be 2580 f.u./ha (Kettunen, 1985a). Since the arable land area is 2.26 million ha, the total production is 5830 million f.u. One percentage point of this is 58.3 million f.u. The producer price of barley may be used as a common measure for the value of each feed unit (minus additional cost for drying and harvesting, estimated to be about 0.05 FIM/kg). The present (1986) target price is 1.59 FIM/kg. Thus, the 1% increase in yield makes a 89.8 million FIM increase in farm income. Total farm income was 6764 million FIM in 1984. The final result of this rough calculation is that a 1% increase in yield raises the farm income by about 1.3%. This figure can be applied to the results given by the scenarios above.

The estimates above refer to an average across the whole country. Under the GISS $2 \times CO_2$ scenario, the increase in yields is, in some cases, greater in north Finland than in south Finland. This would tend to assist in the achievement of a particular goal of regional policy which is to reduce existing differences in farm incomes between north and south (Anon., 1985b). At present, prices of inputs are subsidized in the north (Hanhilahti, 1980). Assuming that this regional policy is not altered, northern farmers would benefit if yields rise more in the north than in the south. It should be noted, however, that these regional differences in yield changes are far from certain (see Sections 2 and 4).

Thus far it has been assumed that the increase in yields will not have any effect on prices. This assumption is based on the present price policy of the Farm Income Act which subsidizes rises of costs due to the rise of input prices (Kettunen, 1981). In addition, the Farmers' Union and the state negotiate on the increase of farm income (which compensates the farming family's own labor and capital inputs). There are no rules for raising the farm income but a general practice has been to look upon the increase in wages and salaries in other sectors of the economy and to apply the same to agriculture (Kettunen, 1981). Although it would benefit farm incomes, the rise of productivity in agriculture is not taken into account in these negotiations. Increases of yields would therefore mean a sizeable increase in farm income.

These increases in income may, however, be less than is at first apparent because Finnish farmers obtain only the world market price for an increase in production, and world market prices have been less than 50% of the Finnish producer prices (see *Table 1.5*), because Finnish agriculture already produces more food than is needed by the domestic population. In 1983, the self-sufficiency ratio (production divided by consumption) for milk was 133%, for pork 117%, for beef 115% and for eggs 165% (see *Table 1.4*; Kettunen, 1985b). There was also sufficient production of bread grains, and a part of these was exported from the harvest of 1983. Finnish agricultural policy is struggling with the problem of excess supply, because the state must subsidize exports and this is seen as an unwanted burden on the taxpayer. Attempts to curtail production have met with poor results.

The budget expenditures caused by agricultural overproduction have been regulated by setting so-called production ceilings or export ceilings for milk, meat, eggs and grains (Kettunen, 1981). If production ceilings are exceeded, agriculture must export the excess amount at its own cost, i.e., farmers get only the world market price for that production, and it is much lower than the domestic producer price. Production costs are higher in Finland than the price received at the world market, so extra production gives no income increases in normal conditions. However, if the increase in production is obtained as a "gift from heaven" (with no extra costs) it also gives extra incomes, but only calculated to world market prices. If, however, production has been adjusted to consumption, or no marketing fees are collected from farmers for export costs, the income effect can be calculated to the domestic market prices.

Although the consumption of agricultural products is not expected to grow from the present level (Rouhiainen, 1979), the structure of consumption is likely to change a little. Meat consumption should increase slightly, but consumption of milk products (except that of cheese) will probably decrease. Consumption of plant products is expected to be static. Thus the present excess supply can be abolished only by lowering production. In this sense, the higher yields will make the conduct of agricultural policy more difficult. Without a warming of the climate yields are expected to grow by about 1% per year as a result of developments in technology. The effects of a warmer climate on production suggest that control of supply would become an even more important objective of state agricultural policy. However, the detection of such trends is also hampered by the large variation of yields from one year to another, making it difficult to act at the right time and in the right manner.

In general a warming of the climate will mean an extra increase in yields. This increase may be hidden by a larger upward trend in yields due to technology. In order to keep excess supply at a manageable level the cultivated area will have to be further reduced or export subsidies raised correspondingly. Large variations in yields can also be expected, but food stocks are already sufficient to cover these effects.

Finally, if the state retains semi-official production targets (they have never been officially accepted by the parliament), plant production potential has to be decreased – for example, by fallowing or by other set-aside programs. That will, in turn, hamper structural policy making since less land will be available for increasing the size of farms, which is itself a major means of increasing farm income.

In conclusion, we have seen that while post-war Finnish agriculture has developed to a position of self-sufficiency and even excess supply of many food products, crop productivity and farm incomes remain sensitive to short- and medium-term variations in climate. Moreover our results suggest that under a future warmer climate mean crop yields would increase substantially, while interannual variability of yields would decrease. Cultivation boundaries of certain crops would move northwards and additional varieties could be introduced into southern regions. A real challenge is presented: for plant breeders to develop new crop varieties, for the extension services to supply timely information and advice to farmers, for the farmers themselves to respond quickly to exploit the changing conditions, and for the government to manage the transition in an economical and regionally equitable manner.

Acknowledgments

The authors express their sincere thanks for the generous assistance provided by the following persons and institutions in completing this report. First of all they wish to thank Martin Parry, Tim Carter and Nico Konijn for all their encouragement, assistance and support in preparing the Finnish part of the project. It has been most instructive and delightful to work with them. P. Kauppi assisted in the finalizing of Section 1. The Finnish State Archives and the Provincial Archives of Oulu provided data required in Section 2, and the Finnish Meteorological Institute and the Statistical Centre of Finland were valuable sources of data for Sections 2 and 3. Special thanks are also due to Mr. M. Saarinen for data processing and PhD. Ilkka Havukkala for comments on the manuscripts in Sections 2 and 3. For Section 4, Mrs. Anja Kaunisoja and Mrs. Marjaterttu Roiha prepared the maps, Markku Kemppainen M.A. compiled the statistical data, Mr. Alfred Colpaert M.A. carried out the computer work and Mr. Malcolm Hicks M.A. translated the manuscripts, all generously supported by grants from IIASA, Laxenburg and the University of Oulu. Finally, John Sumelius carefully read and corrected the manuscripts in Sections 1 and 5. The authors, of course, take full responsibility for any possible errors in the report.

References

Annual Reports of Provincial Governors. In *The Letter Diary KD of the Senate: 1846-1860*, Finnish State Archives, Helsinki (in Swedish).
Anon. (1985a). Maatalouden ja elintarvikkeiden tukia selvittiäneen työryhmän muistio (Support of agriculture and food). *Työryhmämuistio 1985*: VM 26, Helsinki.
Anon. (1985b). *Viljelijäväestön ja palkansaajien tuloeroja selvittävä tutkimus* (A study of income differences between farmers and wage earners). Research Reports of the Agricultural Economics Research Institute, Finland, No. 116, Helsinki, 168 pp.
Atlas of Finland (1982). Helsinki.
Bauer, A., Frank, A.B. and Black, A.L. (1984). Estimation of spring wheat leaf growth rates and anthesis from air temperature. *Agron. J.*, 76, 629-835.
Brown, M.B., Engelman, L., Frane, T.W., et al. (1981). *BMDP Statistical Software 1981*. University of California Press, Berkeley.
Böcker, C.Ch. (1834-1835). Statistical survey of Finnish agriculture in the 1830s. *A Private Collection*, Finnish State Archives (in Swedish).
Central Statistical Office of Finland (1983). *Farm Economy, 1981*. Official Statistics of Finland 39, 11, Helsinki.
Davidson, H.R. and Campbell, C.A. (1983). The effect of temperature, moisture and nitrogen on the rate of development of spring wheat measured by degree-days. *Can. J. Plant Sci.* 63, 833-846.
Dennett, M.D. (1980). Variability of annual wheat yields in England and Wales. *Agric. Meteorol.* 22, 109-112.
Elomaa, E. and Pulli, S. (1985). Variation of global radiation, effective temperature sum, precipitation, potential evapotranspiration and precipitation deficit in relation to the growth of field crops in southern Finland. *NJF Seminarium* 7, 19-27 (in Swedish).
Elonen, P. (1983). Plant production research in Finland. *Ann. Agric. Fenn.* 22, 258-263.
Farm Register 1981 (1984). National Board of Agriculture, Helsinki.
Finnish Meteorological Institute (unpublished). Meteorological observations in Helsinki (1828-1983) and Oulu (1846-1983), Helsinki.
Grotenfelt, G. (1897). *Agriculture in Finland*, Helsinki (in Swedish, Finnish, French and Russian).

Hanhilahti, H. (1980). Subsidies in Finnish agricultural policy. *Publ. Agric. Econ. Res. Inst. Finland* **43**, 19-31.

Hansen, J., Lacis, A., Rind, D., Russell, G., Stone, P., Fung, I., Ruedy, R. and Lerner, J. (1984). Climate sensitivity: analysis of feedback mechanisms. In J. Hansen and T. Takahashi (eds.), *Climate Processes and Climate Sensitivity*, Maurice Ewing Series, Vol. 5, American Geophysics Union, Washington, DC, pp. 130-163.

Heino, R. and Hellsten, E. (1983). Climatological statistics in Finland 1961-1980. *Meteorol. Yearb. Finland* **80(1a)**, 1-500.

Helimäki, M. (1967). Tables and Maps on Precipitation in Finland, 1931-1960 Suppl. *Meteorol. Yearb. Finland 1966* **66(2)**, 1-22, Helsinki.

Hellstenius, J. (1871). Crop yields in Sweden and their impacts. *Sveriges Officiella Statistik* **I**, 77-127 (in Swedish).

Hellström, P. (1917). Agriculture in Norrland. *Norrländskt Handbibliotek* **VI**, 1-685.

Huovila, S. (1974). Lämpöolojen vaihtelurajat Pohjois-Suomessa (Summary: The temperature range in Northern Finland). *Acta Lapponica Fenniae* **6**.

Hustich, I. (1945). Om det nordfinska jordbrukets utveckling och årliga produktionsvariationer. *Fennia* **69(2)**, 1-102.

Hustich, I. (1950). Recent yields of cereals in Finland and the recent climatic fluctuation. *Fennia* **73(3)**, 1-32.

Hustich, I. (1952). Agricultural production in Finland and the recent climatic fluctuation. *Fennia* **75**, 97-105.

Ilvessalo, Y. (1960). Peatlands as a percentage of the total land area. *Atlas of Finland 1960: Map 11/15*, Helsinki.

Keränen, J. (1931). Über die Abhängigkeit der Ernteerträge von den Temperaturen und Regenmengen während der Vegetationszeit in Finnland (in Finnish with German Summary). *Acta Agr. Fennica*, **23(1)** 1-31.

Kettunen, L. (1981). Objectives and means in Finnish agricultural policy. *J. Sci. Agri. Soc. Finland* **53**, 285-293.

Kettunen, L. (1985a). *Finnish Agriculture in 1984*, Research Reports of the Agricultural Economics Research Institute, Finland, No. 112a, Helsinki.

Kettunen, L. (1985b). *Maatalouden omavaraisuus Suomessa vuosina 1970-1983* (Summary: Self-sufficiency of agriculture in Finland in 1970-1983), Publication of the Agricultural Economics Research Institute, Finland, No. 49, Helsinki (in press).

Kivi, E. (1960). Otra barley. *Siemenjulkaisu* **1960**, 194-198 (in Finnish).

Kivi, E. (1983). The goals of plant breeding. *Hankkijan Saroilta 1983* **11-12**, 4-5 (in Finnish).

Kivinen, E. (1941). Mistä vaara-asutus johtuu [The reasons for the settlement of "vaaras" (large hills in northern Finland)]. *Terra* **53(4)**, 255-262.

Kolkki, O. (1966). Tables and maps of temperature in Finland during 1931-1960. *Suppl. Meteorol. Yearb. Finland.* **65** 1a, 1-42.

Kolkki, O. (1969). Katsaus Suomen ilmastoon. *Ilmatieteen Laitoksen Tiedonnantoja* **18**, 1-18, Helsinki.

Kolkki, O., Huovila, S. and Valmari, A. (1970). The regional boundaries of crops production. *Nordia* **3**, 1-19 (in Finnish).

Kontturi, M. (1979). The effect of weather on yield and development of spring wheat in Finland. *Ann. Agric. Fenniae* **18**, 263-273.

Kurki, M. (1972). *Über die Fruchtbarkeit des finnischen Agrarbodens auf Grund der in den Jahren 1955-1970 durchgeführten Boden-Fruchtbarkeitsuntersuchungen* (in Finnish, with German Summary), Helsinki.

Lallukka, U., Rantanen, O. and Mukula, J. (1978). The temperature sum requirement of barley varieties in Finland. *Ann. Agric. Fenn.* **17**, 185–191.
Lemon, E.R. (Ed.) (1983). *CO_2 and Plants*, American Association for the Advancement of Science, Selected Symposium 84, Westview Press, Colorado.
Major, D.J. (1980). Photoperiod response characteristics controlling flowering of nine crop species. *Can. J. Plant Sci.* **60**, 777–784.
MKL (various dates). Katetuottomenetelmän mukaisia mallilaskelmia, *Maatalouskeskusten Liito, Ser.*, Association of Agricultural Centers, Helsinki.
Monthly Review of Agricultural Statistics, No 11 (1982), No 6 (1985), National Board of Agriculture, Helsinki.
Mukula, J. (1984). Zones for cultivars of field crops. *Maataloushallinnon Aikakauskirja* **1984**(4), 1–7, Helsinki.
Mukula, J., Rantanen, O. and Lallukka, U. (1977a). Crop yield certainty of spring wheat in Finland 1950–1976. *Kasvinvilj. lait. Tiedote* **8**, 1–72.
Mukula, J., Rantanen, O. and Lallukka, U. (1977b). Crop yield certainty of barley in Finland 1950–1976. *Kasvinvilj. lait. Tiedote* **9**, 1–83.
Müller, M.J. (1982). *Selected Climatic Data for a Global Set of Standard Stations for Vegetation Science*, Junk Publishers, The Hague, The Netherlands, 306 pp.
Nuttonson, M.Y. (1957). *Barley-Climate Relations and the Use of Phenology in Ascertaining the Thermal and Photo-thermal Requirements of Barley*. American Institute of Crop Ecology, Washington.
Outhier, R. (1744). *Journal d'un Voyage au Nord en 1736 et 1737*, Paris.
Rein, G. (1867). Material for statistics of Finland II. Oulu Province. *Bidrag till Kännedom of Finlands Natur och Folk* **10**, 1–371 (in Swedish).
Rouhiainen, J. (1979). *Changes in Demand for Food Items in Finland 1950–1977 with Consumption Forecasts for 1980, 1985 and 1990*. Publication of the Agricultural Economics Research Institute, No. 40.
Russelle, M.P., Wilhelm, W.W., Olson, R.A. and Power, J.F. (1984). Growth analysis based on degree-days. *Crop Sci.* **24**, 28–32.
Sionit, N., Strain, B.R. and Helmer, H. (1981). Effects of different concentrations of atmospheric CO_2 on growth and yield components of wheat. *J. Agric. Sci.* **97**, 335–339.
Soininen, A.M. (1974). Old traditional agriculture in Finland in the 18th and 19th century. *J. Sci. Agri. Soc. Finland* **46**, 1–459 (in Finnish, with English Summary). (Published also in *Historiallisia Tutkimuksia*, **96**.)
Solantie, R. (1975). The influence of the lakes in Finland on air temperatures. *Vannet i Norden* **4**, 9–20.
Sundbärg, G. (1895). Production and consumption of the main crops. *Kongl. Lantbr. Akad. Handl. och Tidskr.* **1895**, 284–382 (in Swedish).
Suomela, H. (1976). Ilmastolliset edellytykset. In J. Paatela et al. (eds.), *Tuottava maa. Osa 1: maataloustuotanto*, Kirjayhtymä, Helsinki, pp. 139–151.
SVT III (various dates). *Suomen virallinen tilasto III* (Official Statistics of Finland III), Annual Statistics of Agriculture, National Board of Agriculture, Helsinki.
Talman, P. (1978). Dairy farming in Finland: geographical aspects of the development, typology and economics of Finnish dairy farming. *Acta Universitatis Ouluensis A* **62**, *Geographica* **5**, 1–139.
Valle, O. (1935). Über die russische Pflanzenzüchtungsarbeit besonders der subarktischen Pflanzenbaustation in Chibiny. *J. Sci. Agric. Soc. Finland* **7**, 135–148 (in Finnish, with German Summary).

Valle, O. (1959). Success in cultivation of Finnish crop varieties in North America. *Terra* **1959(2)**, 98–108.

Valle, O. and Aario, L. (1960). Changes in northern limits of general cultivation of winter wheat, spring wheat, barley, rye, oats, leguminous plants, winter turnip and rape. *Atlas of Finland 1960*, maps 21/1–2, Otava, Helsinki.

Valle, O. and Carder, A.C. (1962). Comparison of the climate at Tikkurila, Finland, and at Beaverlodge, Alberta, Canada with particular reference to the growth and development of cereal crops. *J. Sci. Agri. Finland* **100(1)**, 1–31.

Valmari, A. (1966). On night frosts research in Finland. *Acta Agr. Fennica* **107**, 191–214.

Varjo, U. (1972). Über die Productivität der Acker- und Waldböden Finnlands. *Fennia* **113**, 1–50.

Varjo, U. (1974). Agriculture in Northern Lapland, Finland: profitableness and trends since World War II. *Fennia* **132**, 1–72.

Varjo, U. (1977). Finnish farming: typology and economics. In G. Enyedi *Geography of World Agriculture*, **6**, Akademiai Kiadó, Budapest, 1–146.

Varjo, U. (1980). Crop cultivation in Finland: Recent temporal and regional trends. *Nordia* **14(2)**, 1–30.

Varjo, U. (1983). Farming in Finland in the pressure of change. *Nordia* **17(1)**, 153–164.

Waggoner, P.E. (1979). Variability of annual wheat yields since 1909 and among nations. *Agric. Meteorol.* **20**, 41–45.

Waugh, T.C. and McGalden, J. (1983). *GIMMS Reference Manual*, Edinburgh.

Williams, G.D.V. (1974a). Deriving a biophotothermal time scale for barley. *Int. J. Biometeor.* **18**, 57–69.

Williams, G.D.V. (1974b). A critical evaluation of biophotothermal time scale for barley. *Int. J. Biometeor.* **18**, 259–271.

PART V

The Effects of Climatic Variations on Agriculture in the Subarctic Zone of the USSR

Contents, Part V

List of Contributors		619
Abstract		621

1. Introduction — 623
   S.E. Pitovranov, V.I. Kiselev, V.N. Iakimets and O.D. Sirotenko
   1.1. Aims and background — 623
   1.2. Agricultural planning and the problem of weather-induced crop yield fluctuations — 626
   1.3. Agroclimatic conditions in the subarctic farming zone of the USSR — 629
   1.4. Land uses — 631
   1.5. Policy considerations for increasing agricultural productivity — 633
   1.6. Assessing the effects of climatic changes — 635

2. The effects on agriculture and environment in the Leningrad region — 639
   V.N. Iakimets and S.E. Pitovranov
   2.1. Introduction — 639
   2.2. Agroclimatic resources — 640
   2.3. Variability of winter rye yields — 642
   2.4. Winter rye cultivation and climate — 643
   2.5. Generating scenarios for regional climatic change — 644
   2.6. Selection and validation of impact models — 649
   2.7. The sensitivity of winter rye yields and of some environmental parameters to climatic change — 655
   2.8. Mitigating the impacts of climatic change — 657
   2.9. Conclusions — 659

3. The effects on spring wheat yields in the Cherdyn region — 663
   O.D. Sirotenko
   3.1. Introduction: methods of estimating crop yield responses to climate — 663
   3.2. A dynamic crop–weather model for estimating spring wheat yields — 664
   3.3. Model sensitivity experiments for the baseline climate — 668
   3.4. Scenario results — 671
   3.5. Conclusions — 675

4. Planned responses in agricultural management under a changing climate 677
    V. I. Kiselev
    4.1. Introduction 677
    4.2. A scheme for model-based, long-term agricultural planning 678
    4.3. The regional agricultural planning model 682
    4.4. Use of the regional model in planning responses to climatic change 684
    4.5. Centralized planning at the state level 692
    4.6. Conclusions 693

5. Results and policy implications 695
    S.E. Pitovranov, V.N. Iakimets, V.I. Kiselev and O.D. Sirotenko
    5.1. Introduction 695
    5.2. Summary of results 696
    5.3. The implications of climatic changes for policy 701

Acknowledgments 707
Appendices 709
References 719

List of Contributors

PITOVRANOV, Dr. Sergei
International Institute for
  Applied Systems Analysis
Schloss Laxenburg
Schlossplatz 1
A-2361 Laxenburg
Austria

IAKIMETS, Dr. Vladimir
International Institute for
  Applied Systems Analysis
Schloss Laxenburg
Schlossplatz 1
A-2361 Laxenburg
Austria

KISELEV, Prof. Victor
Laboratory of Optimal Planning
  in the Agroindustries
Central Institute for Economics
  and Mathematics
Moscow
USSR

SIROTENKO, Dr. Oleg
All-Union Hydro-Meteorological
  Research Institute
12, Pavlik Morozov Str.
123376 Moscow
USSR

Abstract

Despite considerable increases in Soviet agricultural output during the post-war period, crop production has remained quite sensitive to interannual variations in weather conditions. Furthermore, in regions toward the northern limit of widespread crop cultivation, termed the "subarctic farming zone", limited thermal resources exert a major constraint on crop productivity. However, measures of the bioclimatic potential of this zone indicate that it should be possible, through appropriate land management and crop selection, to raise productivity considerably above the present level. Moreover, model estimates of future climatic change due to enhanced concentrations of "greenhouse" gases imply substantial warming in the northern USSR, which would have profound impacts on agriculture in the subarctic zone.

This case study seeks to evaluate the present-day sensitivity of agriculture, particularly crop yields, to climatic variations in the northern European USSR, and to assess the likely impacts of future climatic change. Studies that consider three different aspects of this issue are reported, each conducted in a different region: effects of climate on spring wheat yields in the Cherdyn region; on winter rye yields in relation to the wider natural environment in the Leningrad region; and on regional production and crop allocation for a range of crops in the Central region (around Moscow). Each utilizes mathematical models to convert climatic anomalies to estimates of impact. Four types of model are employed: a regression model and two dynamic crop–weather models to estimate crop yields, a global ecological model for evaluating environmental impacts, and a regional agricultural planning model for assessing the costs of adjusting crop allocation strategies to optimize production.

Several approaches have been used for simulating possible future climate, with a major emphasis directed towards CO_2-induced climatic change. Each climatic scenario bears comparison with a reference or baseline scenario representing present-day climatic conditions.

Two methods were utilized for simulating the transition between baseline climate and an altered climate. Some experiments considered this transition as an instantaneous "step-like" change, others simulated the time dependent "transient" nature of change over a prescribed period into the future. The scenarios include:

(1) An empirically derived scenario developed by Vinnikov and Groisman linking recorded global temperature changes to regional changes in temperature and precipitation.
(2) Arbitrary changes in temperature and precipitation, used mainly for testing model sensitivity.

(3) The climate estimated for doubled concentrations of CO_2, based on predictions by the Goddard Institute for Space Studies (GISS) general circulation model.
(4) A transient change scenario combining elements from (1) and (3).

Two of the study areas are in the subarctic zone, and were chosen to illustrate the contrast in agroclimatic conditions between the wetter western part (Leningrad) and the relatively dry eastern part, near to the Ural mountains (Cherdyn). The third area is to the south of this zone (Central region), and was selected to demonstrate the use of the regional planning model.

Results show that under present-day climate, above-average temperatures during the growing season are favorable for yields both of winter rye in the Leningrad region and of spring wheat in the Cherdyn region. Fluctuations in precipitation, however, completely alter this assessment. A surplus of precipitation during the growing period has a strong negative impact on winter rye yields in the Leningrad region, but a positive effect on spring wheat yields at Cherdyn.

A more extreme, impulsive "step-like" change in climate, simulated using the 2 × CO_2 GISS model-derived estimates, would cause a slight decrease in spring wheat yields in Cherdyn, the benefits of increased temperature and precipitation being outweighed by a significant reduction in the vegetation period. Winter rye yields would also be reduced under this scenario in the Leningrad region, predominantly because of greatly increased precipitation.

Further results indicate that climatic warming up to the end of the century (assembled from different scenario experiments) would be beneficial to yields of both grain crops in the subarctic zone as well as the more heat-requiring crops, like corn or winter wheat, in the Central region, but unfavorable in this latter region for crops adapted to moister, cooler conditions. Over the longer term (1980–2035) in the Leningrad region, even given a significant increase in fertilizer applications, simulated soil quality is degraded in response to gradual increases in temperature and precipitation. Yields of current winter rye varieties would increase up to the year 2010, but then unfavorable soil conditions combined with the direct effects of the more extreme climatic changes would cause reductions in yields. Higher precipitation would also lead to rising groundwater levels and increased surface water pollution from agricultural watersheds.

Finally, several possible adjustment measures are tested for effectiveness in mitigating the adverse impacts of climatic change. These include substitution of new crop varieties, development of improved drainage systems, increased fertilizer applications, and increased allocation of crops performing well at the expense of those recording reduced yields. The implementation of these, and other alternative measures, is discussed within a framework of centralized decision-making at the regional and state level in the USSR.

SECTION 1

Introduction

1.1. Aims and Background

The purpose of this case study is to quantify the effects of specified climatic changes on crop yields in selected parts of the USSR and to consider the adjustments in land use and land management that represent the most appropriate responses to these effects. In common with other case studies of the IIASA/UNEP project a number of scenarios of both long-term and short-term climatic changes have been adopted. Different types of models have been used to examine both the potential impacts of these scenarios and some possible responses. The methods, results and implications are considered with respect to agricultural productivity in three regions of the northern European USSR.

During the post-war period, considerable attention has been paid in the Soviet Union to the intensification of agricultural production. This has led to increases in the amounts of mineral fertilizer applied, the numbers of tractors, combine harvesters and other machinery in use, and expenditures on energy. Due to these efforts, average agricultural output has been considerably increased (*Table 1.1*).

However, crop production has remained quite sensitive to variations in weather conditions, which is reflected by the variability in grain output from one year to another (*Table 1.2*).

There are at least three contrasting grain producing regions in the Soviet Union. Most grain is produced in the chernozem soil belt which includes the traditional grain growing areas (the Ukraine, Moldavia and part of Belorussia), and in the "new lands" developed during the 1950s in the cooler, drier eastern areas of Kazakhstan and Western Siberia. The third region, in the non-chernozem zone of the cool and moist northwest, has become increasingly important in recent years. The significance of these regional contrasts for total production is that poor conditions in one area are usually compensated by more favorable conditions in another – the so-called Regional Compensation Effect (Tarrant, 1985). This is considered in further detail below.

Table 1.1. Five-year averaged Soviet agricultural output, 1961–1980.

Index	1961–1965	1966–1970	1971–1975	1976–1980
1. Gross farm output (in comparable 1973 prices – 000 million roubles)	82.8	100.4	113.7	123.9
2. Grain (m tons)	130.3	167.6	181.6	205.0
3. Raw cotton (m tons)	5.0	6.1	7.7	8.9
4. Sugar beet (m tons)	59.2	81.1	76.0	88.7
5. Vegetables (m tons)	16.9	19.5	23.0	26.3
6. Meat slaughter weight (m tons)	9.3	11.6	14.0	14.8
7. Milk (m tons)	64.7	80.6	87.4	92.7
8. Eggs (000 m)	28.7	35.8	51.4	63.1

Source: *The USSR Economy for the Period 1922–1982*.

Table 1.2. Variability of Soviet grain production by five-year period, 1961–1980.

	Grain output (m tons)			Variability (maximum–minimum)	
Period	Annual average	Maximum	Minimum	Million tons	Percentage of average
1961–1965	130.3	152.1	107.5	44.6	34.2
1966–1970	167.6	186.8	147.9	38.9	23.2
1971–1975	181.6	222.5	140.1	82.4	45.4
1976–1980	205.0	237.4	179.3	58.1	28.3

Source: Yakunin (1982).

The most damaging effect of climate on total grain production in the USSR comes from drought during the growing season. The period 1951–80 was marked by severe droughts in 1954, 1955, 1972, 1975 and 1979 which affected large areas of the grain producing regions (Ulanova, 1984). In contrast, the most favorable agroclimatic conditions, in terms of total grain production, occurred in 1956, 1958, 1966, 1968, 1970, 1971, 1973, 1976 and 1978, when average grain yield was between 11% and 25% above the trend (Ulanova, 1984). These averages, however, mask considerable regional variations which can be identified by examining production data for individual oblasts (Tarrant, 1985). For example, in 1975 positive anomalies of spring and summer temperatures combined with low precipitation (*Figure 1.1*) led to severe drought over most of the grain growing regions, and Soviet grain production was only about 140 million metric tons, the lowest total for over ten years (*Table 1.2*). However, the positive anomalies of temperature contributed to above-average production in the northwestern region of the subarctic zone (*Figure 1.2*).

The reverse effect was evident in 1966 and 1968, when anomalously wet and cool summers led to above-average production in the southern and eastern grain growing areas, while in the western part of the subarctic zone (from the Leningrad region to the Ural mountains) grain production was well below trend [see *Figures 1.2(b)* and *1.2(c)* for grain production by oblast in 1966 and 1968, respectively, and *Figure 2.2* for winter rye yields in the Leningrad region].

Introduction

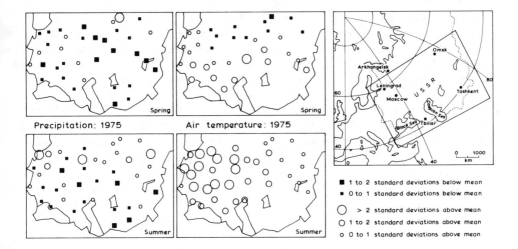

Figure 1.1. Spring (March–May) and summer (June–August) anomalies of precipitation and air temperature in the Soviet grain producing regions in 1975. Anomalies are calculated (where data are available) relative to the average values from the period 1951–83 (from Tarrant, 1985).

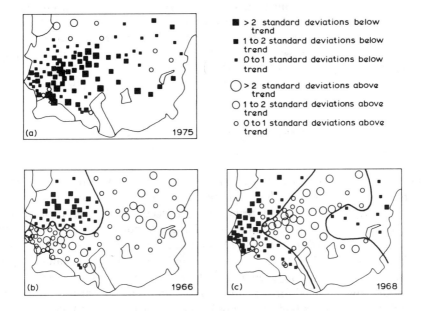

Figure 1.2. Grain production in the Soviet Union by oblast: deviations from trend in selected years; (a) 1975, (b) 1966, (c) 1968 (from Tarrant, 1985).

Following the mode of analysis adopted in other case study areas (Saskatchewan, Finland, Iceland and northern Japan), this case study considers possible impacts of climatic changes on some features of high latitude agricultural activity. The study focuses on crop production in the subarctic farming zone of the European USSR, where severe winters and a short growing season impose limitations on the range of crops cultivated and their productivity. There is also a noteworthy tendency for levels of grain production in this zone to run counter to those in the other major production regions of the Soviet Union. Thus, any change in climate, particularly in average temperatures, might have significant impacts on crop yields in these regions, as well as economic implications extending well beyond the boundaries of the regions themselves.

Recent research into the possible effects of increased concentrations of atmospheric carbon dioxide suggests that high latitude regions may undergo considerable warming during future decades (Bolin et al., 1986). It is this CO_2-induced warming effect that forms the basis of the climatic scenarios developed in this study. The scenarios are used as inputs to impact models that indicate likely responses of crop yields in the study regions. In addition, adaptive measures to cope with climatic change are investigated by simulating possible responses such as the introduction of new crop varieties, and the adjustment of land allocation strategies.

The remainder of this section provides background material on the influence of climate on Soviet agriculture, and describes the selection of study areas and choice of crops in each area, the development of climatic scenarios, and the selection of impact models. Sections 2 and 3 examine the effects of climatic change and variability on winter rye yields in the Leningrad region, and spring wheat yields in the Cherdyn region, respectively. In Section 4, the effects of climatic warming on the yields of a number of crops are investigated, and the optimum land allocation is calculated that fulfills planned objectives under the changed climatic conditions. Finally, the results are summarized in Section 5, where they are interpreted in terms of their implications for policy and planning.

1.2. Agricultural Planning and the Problem of Weather-Induced Crop Yield Fluctuations

Economic activity in Soviet agriculture is carried out on the basis of planned assignments which involve the state procurement of crops and livestock products, delivery of material and technical resources, centralized investment into industrial and non-industrial sectors and state involvement in other schemes such as land reclamation. Details of the plan are refined throughout its duration. The agricultural units (i.e., collective and state farms) examine the proposals in the plan and submit their demands for material and technical resources, being guided by the need for maximum profit and operational production capacity. The state planning bodies, taking into account the budgeting and allocation of resources throughout the regional economy, calculate the preliminary targets for all items for each producing unit and inform the regional authority of these targets. Any discrepancies that appear between units' demands and authorities' proposals, due to competing interests or disagreement over the necessary criteria governing their development, are dealt with in an iterative process involving compromises arising from the detailed scrutiny of each proposal.

Market levers have been little utilized until recently. [At present cost-accounting in agriculture is being strengthened and the role of economic levers is being increased in order to promote a rise in production efficiency (M. Gorbachev, 1982).] For example, procurement prices are revised every few years following significant changes in the profitability of agricultural products. As a result, there is weakening of cost-accounting incentives, but a strengthening of administrative methods of management. Because there exist both highly profitable crops (e.g., grain, sunflower, cotton, grapes, etc.) alongside marginally profitable and sometimes unprofitable crops (e.g., until 1982, meat, milk, potatoes, sugar beet, etc.), collective and state farms are impelled to understate the required output of some products and to overstate the output of others. The influence of the previous years' weather is also used as one of the arguments in the dialogue between producing units and the local authority. One of the ways to achieve a compromise between planning bodies and production units is to establish a combination of profitable and unprofitable products in the plan that are suitable for both sides. Other ways of compensating for losses from unprofitable products include additional delivery of resources to production units, or an increase in state investment, or writing off a part of any debts, etc.

The main measures for stabilizing crop output during individual seasons include irrigation, breeding of new crops resistant to stresses such as drought or low temperatures, selection of crops with different physiological characteristics in a single region to minimize the risk of simultaneous impact from the same adverse event, and dispersion of crops over several regions of the country to decrease the risk of low output because of unfavorable weather in single "core" regions for the given crop. These methods are used widely at present although their scientific grounds can still be improved substantially.

The determination of optimal capacities for harvesting, transportation, storage and processing of crops (activities termed the "harvest procurement system") can play an important role in saving high crop yields for storage in favorable years. The existing planning practice is to prepare for each harvest using a set of indicators (namely, the demand for material and technical resources, for harvest machinery, for transport facilities, and for storage, and an estimate of the processing capacities for agricultural products) that are based on the average trend of production (*Figure 1.3*, line A).

However, in the past even this level of development of the system had not been achieved. For example, when an average or somewhat above-average harvest occurs, harvest laborers can usually be transported to the regions in sufficient numbers to complete the harvest without large losses (*Figure 1.3*, line B). With harvests that promise substantially above-average productivity, the losses are increased. The optimal level of development of the harvest procurement system (*Figure 1.3*, line C) has to be determined in the planning process taking into account two factors:

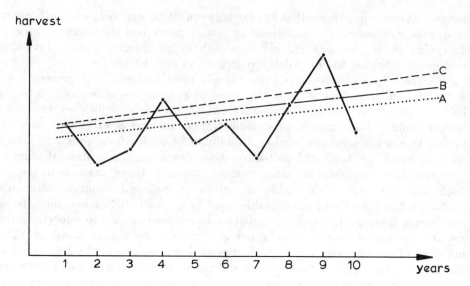

Figure 1.3. Harvesting capacities in relation to annual agricultural production (solid curve): A = base capacity related to trends in average production; B = maximum capacity related to deployment of additional resources during surplus years; C = optimum capacity to accommodate high production levels in most years (schematic).

(1) By developing the capacity of the system estimated on the basis of average production, there are likely to be large losses, and the average procured harvest over the long term will be lower than the average biological production.

(2) By tailoring the capacity to accommodate high production levels, large-scale investment would be necessary, in the full knowledge that this capacity will not be used completely during most years of the planning period.

As a consequence of the second strategy, there is an increase in the cost of each additional harvest unit saved.

To find the optimal level of development of this system it is necessary to examine the relative importance of several indicators that are difficult to measure, such as the comparative importance of the different investment strategies in the national economy, social effects arising from failure to satisfy fully the public demand for food products in particular years, and the deficit of foreign currency to compensate for bad harvests at the expense of foreign trade. Under existing planning practices, these problems are not formalized and remain the subject of heuristic decisions. Thus, at the national level, the capacity of the harvest procurement system depends on the investment which the state can allocate for this purpose, but in the regions these capacities depend on the comparative efficiency of regional agricultural production.

In view of these considerations there are two main stages of planning which merit special investigation: planning at the state and at the regional levels. These are considered within a modeling framework in Section 4 of this study.

1.3. Agroclimatic Conditions in the Subarctic Farming Zone of the USSR

The northern boundary of intensive farming in the European territory of the USSR coincides with the 1400 degree-days isoline of effective accumulated temperatures during the warm period (above 10 °C). [Throughout this case study, two measures of thermal resources are employed. The first, termed *effective accumulated temperatures*, refers to the annual accumulated sum of air temperatures above 0 °C during those days when daily mean air temperature is greater than a specified baseline (usually 10 °C or 5 °C). This measure is also used in the Japan case study (Part VI, this volume). The second, referred to as *effective temperature sum* (ETS), describes the annual accumulated sum of air temperatures above a specified threshold temperature that represents a minimum requirement for the commencement of active plant growth (commonly 5 °C). The measure is employed in all other case studies in this volume.] To the north of that line, farming activity is insignificant, mainly comprising localized cultivation of oats and vegetables in the valleys of the larger rivers. The southern boundary of the subarctic farming zone is normally drawn along the 1800 degree-days isoline of effective accumulated temperatures. The zone stretches for 2500 km from the coast of the Gulf of Finland eastward to the Urals. It covers most of the Leningrad and Novgorod regions, the Vologda and Perm regions, the southern part of the Karelian ASSR (Autonomous Soviet Socialist Republic) and Arkhangelsk region, and the northern part of the Kirov region and the Udmurt ASSR [*Figure 1.4(a)* and *(b)*].

Woodlands, essentially coniferous, occupy 60–70% of the area. Soils are podzols and turfy podzols, interspersed with many lakes, rivers, and peat bogs. The abundant rainfall (500–800 mm annually), combined with low evaporation and relatively flat topography, leads to problems with waterlogging [*Figure 1.4(c)*].

Although the zone is relatively uniform with regard to heat supply, it can be divided into two subzones, the western and eastern, with respect to water supply and winter conditions. The boundary between these subzones runs from north to south, crossing the Vologda region at the longitude of the White Lake [*Figure 1.4(a)*]. The western subzone has conditions of excessive moisture but a relatively mild winter, making it possible for many fruit crops to survive the winter. The eastern subzone is characterized by warmer and drier summers but longer and more severe winters. The duration of the frost-free period is about 95 days in the east and 130 days in the west. Soil frosts end around 20 May to 10 June (in the southern Leningrad and Pskov regions from 15 to 20 May) and begin again on around 10 to 17 September (22 to 30 September in the southwest) [Ulanova, 1978; *see Figure 1.4(d)*].

Generally the subarctic zone is characterized by adequate or even excessive moisture. During the warmer period of the year (from April to October), 350 mm of rainfall can be expected in the east and 450 mm in the west, exceeding evaporation by 70–80 mm. At the beginning of the vegetative period, in early spring, the available moisture is over 200 mm, normally about 20 mm in excess of the capacity of the soil. By the time grain crops have reached maturity, water

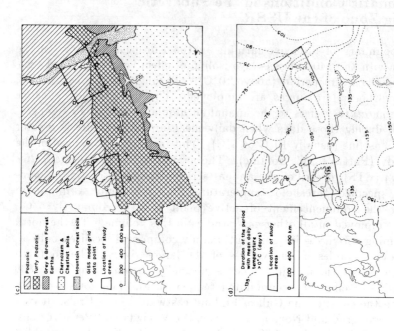
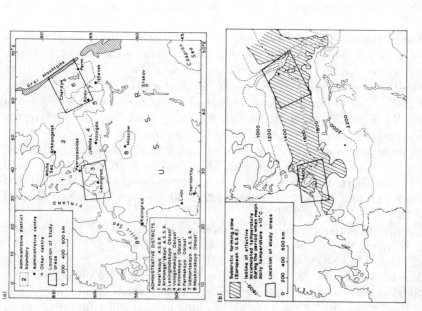

Figure 1.4. European territory of the USSR. (a) Location of study areas and administrative districts (source: D Mil Survey, 1974). (b) Thermal resources and location of the subarctic zone (source: World Atlas of Physical Geography, 1964). (c) Dominant soils and location of GISS model grid data points (source: World Atlas of Physical Geography, 1964). (d) Duration of the frost-free period (source: World Atlas of Physical Geography, 1964).

Table 1.3. Thermal regime in the subarctic farming zone of the European USSR.

Region	Effective Temperature Sum (ETS) above $5°C$	Dates of transition across $0°C$ (mean daily air temperature)		Duration of frost-free period (days)	Average July air temperature (°C)
		Spring	Autumn		
Leningrad	850–1250	1–10/4	30/10–15/11	90–150	17.7
Vologda	950–1000	1–7/4	23–29/10	105–120	17.1
Kirov	1000–1400	20–28/4	29/9–8/10	103–131	18.0
Perm	680–1330	20/4–10/5	25/9–8/10	70–116	18.0

	Frost dates ($Tmin \leq 0°C$)			
	Spring (last)		Autumn (first)	
Region	Air	Soil	Air	Soil
Leningrad	22/5–4/6	5–15/6	9–24/9	28/8–13/9
Vologda	26–31/5	6–11/6	10–13/9	31/8–3/9
Kirov	14–29/5	–	7–23/9	–
Perm	22/5–3/6	9–20/6	7–15/9	26/8–4/9

Sources: Sirotenko, unpublished data; Müller (1982).

storage has decreased to 120–150 mm. In some years, high soil water content and high air humidity may lead to adverse conditions for the ripening of the grain and complicate harvesting (Ulanova, 1978).

The average duration of the autumn growing season for winter crops varies from 50 days in the west of the zone to 45 days in the east. During that period, the heat available permits successful cultivation of winter rye, provided it is sown at the right time. Despite severe frosts across the zone (the average absolute annual temperature minima being $-40°C$ to $-45°C$ in the east, $-35°C$ to $-40°C$ in the central part, and $-30°C$ to $-35°C$ in the west), the snow cover (approximately 40–44 cm in the east and 30–45 cm in the west) protects the soil against freezing to great depths. Soil temperature is normally maintained at a level of $-8°C$ to $-10°C$ at the depth of the tillering node of winter crops. A more detailed picture of the heat and moisture regimes of the zone is given in Tables 1.3 and 1.4.

1.4. Land Uses

In the west of the zone (the Leningrad region), forage crops prevail, being mainly red clover mixed with grasses and swedes. Grain crops do not account for more than 18% of the arable land in the region. Among the spring grain crops, oats, which tolerate waterlogged soil and higher soil acidity better than other cereals, occupy the largest areas in the Leningrad region. Further eastwards in the Vologda region, there is an increase in the proportion of arable lands under grain crops, among which spring barley, oats, and winter rye predominate (Table 1.5).

Table 1.4. Moisture regime in the subarctic farming zone of the European USSR.

Region	Total rainfall for period when mean daily temperature is above 10°C (mm)	Water availability (HTC)[a]	Available water (mm) in the top 1 m layer of soil by the date of transition across 10°C in the spring		Probability of droughts and dry winds (%)
			Winter crops	Spring crops	
Leningrad	225–300	1.4–1.8	130–275	140–220	4
Vologda	250–290	1.6–1.7	150–250	–	8
Kirov	225–250	1.0–1.6	174–188	165–202	20
Perm	200–250	1.2–2.0	135–195	125–205	24

[a] The hydrothermal coefficient (HTC) is equal to ten times the ratio of the total rainfall (mm) to the effective accumulated temperature (degree-days) for the period when mean daily temperatures are above 10°C.
Source: Sirotenko, unpublished data.

Table 1.5. Land use in the subarctic farming zone of the European USSR.

Region	Total arable land in 1974 (000 ha)[a]	Percentage of total arable land[b]			
		Grain crops	Forage crops	Tilled crops	Other crops
Leningrad	386	18.0	60.0	20.0	2.0
Vologda	868	41.2	45.3	5.0	8.5
Kirov	2,705	72.0	19.5	4.0	4.5
Perm	2,113	54.0	34.0	5.0	7.0

Region	Percentage of grain crop area[b]				
	Winter wheat	Winter rye	Barley	Spring wheat	Oats and others
Leningrad	5.8	21.3	28.7	1.6	42.6
Vologda	5.4	25.6	31.0	6.9	31.1
Kirov	0.5	47.6	14.1	4.5	33.3
Perm	–	25.1	12.9	33.3	28.7

[a] *Statistical Yearbook of the RSFSR* (1974).
[b] 1983 (Sirotenko, unpublished data).

The percentage of arable land under grain is highest in the Kirov region, where climatic conditions are most favorable for growing winter rye, to which almost half the cereal area is allocated. In the Kirov region, winter crops hardly ever suffer from freezing, since the deep snow cover reliably protects the crops against the low temperatures. The superiority of winter rye over the spring crops is basically due to its earlier ripening (27 to 30 June as against 8 to 13 August for spring crops) so that it can take advantage of more favorable moisture and harvesting conditions. However, in about one year in five (*see Table 1.4*) droughts and dry winds occur in the Kirov region, thus reducing the yield.

In the Perm region, the percentage of arable land under winter rye is only about half that in the adjacent Kirov region, essentially due to the more severe winter conditions. The leading grain crop of the Perm region is spring wheat, which accounts for one-third of the whole cereal area. If sown on optimal dates, early and semi-late cultivars of spring wheat over the greater part of the region are adequately supplied with heat in practically every year (probability > 95%), and late cultivars in eight or nine years of every ten. Barley and oats, less heat-loving crops than spring wheat, experience almost no problems in obtaining the necessary heat supply. Water deficiency is a limiting factor on crop production in the Perm region, causing excessive drying of the arable layer of soil in about one year in four. Meanwhile, in the northern parts of the region and the foothills of the Urals yield reduction due to excessive soil moisture occurs on average once or twice in ten years, while in certain years grain may be injured by early autumn frosts.

1.5. Policy Considerations for Increasing Agricultural Productivity

The bioclimatic potential of a region can be measured using the bioclimatic index of plant productivity (Kashtanov, 1983):

$$BCP = K_p \frac{\Sigma T_{10}}{1000}$$

where K_p is the coefficient of water availability to plants (proportional to the ratio between the annual level of precipitation and the sum of the mean daily values of the saturation deficit); ΣT_{10} is the effective accumulated temperature for the period above 10 °C; 1000 is the annual effective accumulated temperature near the northern boundary of conventional agriculture (degree-days). For convenience, BCP values are expressed as index values (B_c), based on a value of 100 for the average productivity of the USSR grain belt. In the Leningrad, Vologda, Kirov, and Perm regions, B_c values fall in the ranges 81–99, 80–102, 80–103, and 80–95, respectively. The lower limit of each of these estimates normally coincides with the northern boundary and the higher with the southern boundary of the region concerned.

A comparison of the bioclimatic potential and the actual yields in the subarctic farming zone indicates that it should be feasible to raise the productivity of agriculture in this area to 2–3 times the present level, through appropriate increases in fertilizer application rates, soil reclamation (liming, drainage), and the introduction of high-yielding cultivars adapted to the local conditions. As an example of what is possible, winter wheat yields in the Leningrad region increased from 0.75 t/ha in 1951–1955 to 2.3–2.5 t/ha in 1971–1980 and winter rye yields from 0.66 t/ha to 2.0 t/ha, over the corresponding period, due to the application of these measures. In the northeastern parts of the subarctic farming zone, which experience the most severe climatic conditions, two- or three-fold yield increases are still possible. For example, spring wheat yields at the

Cherdyn experimental station (in the northeast of the Perm region) have been 3.5–4.0 t/ha during recent years, three or four times higher than the average regional yield.

It is evident that increasing agricultural productivity in zones with soddy podzolic soils and low natural fertility requires the input of labor and material resources. For example, considerable investments are needed for liming (some 4–5 t/ha of $CaCO_3$ are recommended for application on soils in the Leningrad region, approximately once every five years: *Handbook of Agronomy*, 1983), and drainage activities [costs estimated at about 300–400 roubles (1 rouble is approximately equivalent to 1 US dollar) to construct 100 m of open drainage channel (Agricultural Academy, personal communication)]. The effects of improved drainage are explored in Section 2.

Future measures for increasing yield without decreasing natural soil fertility are, in some senses, the most important elements of agricultural policy for the subarctic farming zone. Until recently, the normal measure employed for increasing yield levels has been to apply greater quantities of mineral fertilizer to the land. However, these practices may have to be reconsidered in the light of some damaging side-effects that have been observed. Firstly, over-fertilization of the soil can occur if there is insufficient moisture in the soil (particularly in eastern parts of the subarctic zone during dry years). Secondly, as reported by Valpasvuo-Jaatinen (1983) for Finnish conditions and important too in the western part of the subarctic zone, a protracted annual period of freezing-over, combined with the shallowness of lakes and a slow turnover of water in waterways, all contribute to the danger of severe water pollution. Pollution by nitrates is considered further in Section 2.

One of the measures proposed for avoiding such problems aims to increase the top soil through the application of organic fertilizers. Normally, the quantities of organic fertilizers available are limited by the scale of livestock production in the area, but the regions studied here have huge reserves of a valuable organic fertilizer, namely peat. One disadvantage is that applying peat as fertilizer leads to a sharp decrease in the downward flow of heat from soil heated during the day and hence the risk of plants being killed by night frosts is increased.

The eastern part of the subarctic farming zone is characterized by high frequencies of years with insufficient moisture for agriculture. Agroecological systems in this area are already sensitive to drought because the root systems of plants are located in relatively thin fertile layers of soddy podzolic soils. Thus, even short periods without rainfall (10–20 days) lead to overdrying of the soil and decreased yields (*see* Section 3). Therefore, stable agricultural production in these regions requires additional capital investment for irrigation.

On the other hand, the western part of the subarctic farming zone often suffers from excessive moisture levels. Crop losses can occur during the vegetative and ripening stages, as well as during the harvesting period, due to untimely rain and snowfall (Section 2). Once again, additional investments in drainage, grain drying and mechanical equipment are required to facilitate harvesting under unfavorable conditions.

Capital investments are also made throughout the whole subarctic farming zone to minimize harvest losses due to late spring and early autumn frosts.

Introduction

Plant breeders are developing new crop varieties that are both high yielding and resilient to climatic variations. Furthermore, as responses to possible future climatic change, the introduction of new varieties (*see* Section 3) and the reallocation of existing crops (Section 4) would be important mechanisms of adjustment.

In summary, the subarctic farming zone of the European USSR has significant potential for the intensification of agriculture and the production of high and stable yields unaffected by droughts. However, at present the environment in general, and agriculture in particular, are highly sensitive to variations and changes in climatic conditions, which must therefore be considered carefully when determining strategies for the economic development of the region.

1.6. Assessing the Effects of Climatic Changes

A schema of the approach followed in this study is given in *Figure 1.5*. Some details are discussed below.

1.6.1. Choice of study areas

Because of the great contrast in agroclimatic conditions within the subarctic farming zone, climate impact experiments have been conducted in two subzones, one in the wetter western region around Leningrad, and one in the drier eastern subzone of Perm [*Figure 1.4(a)*]. Section 2 considers the impacts of climatic

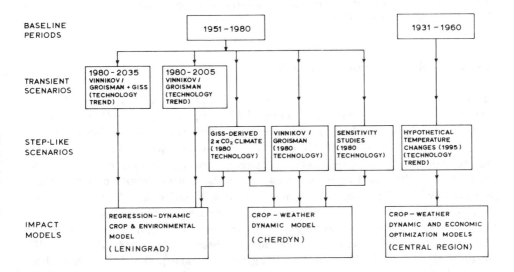

Figure 1.5. Schematic representation of the climatic scenarios and impact models employed in the USSR case study.

variations on crop yield, and the associated environmental impacts of climate and agricultural activity in the Leningrad administrative region. Section 3 focuses on the district of Cherdyn, which lies close to the Ural mountains in the Perm administrative region. Section 4 examines the centralized management options most appropriate as a response to changing climatic conditions. The principles of the approach are demonstrated with worked examples from the central zone of the European part of the USSR (Central region).

1.6.2. Choice of crops

The major crop cultivated in the Perm region is spring wheat, accounting for 33.3% of planted grain crop area (*Table 1.5*). Section 3 analyzes the sensitivity of spring wheat yields to present-day climatic variability and to possible longer-term carbon dioxide-induced climatic change in the Cherdyn district.

The choice of crop in the Leningrad region was not so straightforward. The region is not a major crop producer relative to the country as a whole, and grain production is limited to the provision of local foodstuffs. Historically, winter rye was one of the most important crops in the region, representing more than 40% of the planted grain crop area up until the early 1950s. This proportion is now 21.3% (*Table 1.5*), which is less than the respective areas of barley and oats. However, recent increases in winter rye yields up to levels of around 2 t/ha has permitted the export of the surplus to other regions. Recent plans are for expansion of the area sown to winter rye over the next 5–10 years to increase this export potential. For this reason winter rye has been chosen for climate impact investigations in Section 2.

The third study, in the Central region of the European USSR, is concerned with the optimal mix of a wide range of crops needed to attain specific production goals under different environmental conditions.

1.6.3. Choice of climatic scenarios

In common with the other case studies in this volume, Sections 2 and 3 report results of experiments analyzing crop yield responses to the GISS (Goddard Institute for Space Studies)-modeled regional estimates of CO_2-induced climatic change (Part I, Section 3). Estimates were provided of changes in mean monthly air temperature and precipitation between model-simulated climates under current ($1 \times CO_2$) and under doubled ($2 \times CO_2$) concentrations of atmospheric carbon dioxide. The data were generated for a 4° latitude × 5° longitude network of grid points. The location of these is indicated in *Figure 1.4(c)*.

In order to test the validity of the GISS model for simulating present-day climate in the study areas, a comparison has been made of the GISS $1 \times CO_2$ model estimates with observed data (long-term averages). The results have been graphed for the Leningrad region (*see Figure 2.3*) and for the Kirov/Perm regions (*see Figure 3.5*). It can be noted here that while the GISS model reproduces the observed values of mean monthly air temperature reasonably well, the fit for mean monthly precipitation data is rather poor. The deviations between estimated and observed precipitation may, in some months (particularly in winter and spring) exceed 100%. However, while it is evident that the estimates of CO_2-induced climatic change are subject to large uncertainties, they should be regarded in this study not as predictions, but merely as possible scenarios of future climate. The main objective of the investigations has been to demonstrate how general circulation model outputs can be used in climate impact experiments.

The GISS data have been utilized in two ways in the Soviet study. In Section 3 the climatic changes are introduced as a discrete, step-like perturbation between the present climate and the doubled CO_2 climate, while in Section 2 the GISS estimates are combined with the empirical approach developed by Vinnikov and Groisman (1979b) and Groisman (1981), to construct a scenario of transient climatic change. Assuming a doubling of atmospheric CO_2 concentration to occur in the year 2050, a smooth linear change in climate is then assumed between the present-day and the year 2005 according to the Vinnikov and Groisman estimates, followed by a second linear trend from 2005 to 2050, to reach the levels of change estimated by the GISS model.

The standard climatological period, 1951–80, was adopted in Sections 2 and 3, and the period 1931–60 in Section 4, to serve both as convenient periods in which to validate the agroclimatic models, and as references against which to compare scenario results. Section 3 includes, in addition, several sensitivity experiments for this period to estimate the influence of soil water storage on crop-emergence dates, and to assess the effects of altering systematically the levels of mean growing season temperature and precipitation.

In Section 4, synthetic scenarios are adopted to simulate three situations of hypothetical regional warming projected to the year 1995: increases in growing season temperatures relative to present of 0.5 °C, 1.0 °C and 1.5 °C, respectively.

1.6.4. Impact models

The estimation of winter rye yields (in Section 2) has been conducted using a combined regression–dynamical modeling approach. The regression component incorporates 11 meteorological input variables and is used to estimate interannual deviations from "average" of winter rye yield in response to short-term climatic fluctuations (Obukhov, 1949). The assessment of average yield is carried out using the so-called VNIISI model (Pegov et al., 1983). This model is made up of a set of ordinary differential equations that describe the dynamic relationships between important constituents of the ecological environment in the region (e.g., soils, climate, vegetation, water, etc.). For these experiments, this allows investigations to be conducted of longer-term changes in soil quality in response to climatic change and to variable applications of fertilizer. In addition the model simulates other agro-environmental effects of the climatic changes, such as changes in ground water level and water pollution by nitrogen.

Simulations of the impact of climate on spring wheat yields have been conducted using a dynamic crop–weather model (Section 3). The model comprises a closed set of ordinary differential equations for calculating the dynamics of crop phytomass and water content of the root zone of the soil. Input climatic data are required on a 1-day time step for the whole growing season, and procedures have been developed for converting mean monthly values supplied for the GISS 2 × CO_2 scenario to the necessary daily resolution for model input.

In Section 4, a crop production model (Konijn, 1984a) has been used to compute approximate percentage changes in crop yields for each of the three hypothetical climatic scenarios outlined above, and for each of the crops cultivated in the Central region. An optimization model is then used to calculate the reallocation of crop land that would be required, given the changes in crop yield estimated by the Konijn model, to minimize total expenditure while maintaining production at present levels.

SECTION 2

The Effects on Agriculture and Environment in the Leningrad Region

2.1. Introduction

In this section we investigate the impacts on agriculture and the environment in the Leningrad region of possible climatic changes that might be induced by an increase of carbon dioxide and other "greenhouse" gas concentrations in the atmosphere.

While agriculture is one of the most sensitive areas of human activity to climatic changes, such changes could also have wider environmental implications in the region affecting, for example, water flow, groundwater level, soils, etc. Therefore it is extremely important to analyze, in an integrated fashion, the complex dynamic behavior of climate, agriculture and environment. The aim of this section is to investigate impacts of several scenarios of climatic change on crops at present cultivated in the region and on other aspects of the local environment.

In Subsection 2.2, the main climatic characteristics of the region are given along with some specific agroclimatic characteristics relating to winter rye cultivation. The scenarios of climatic changes, which have been used for analyzing impact are described in Subsection 2.3. The scenarios are built around the GISS general circulation model-derived $2 \times CO_2$ scenario and are constructed to simulate transient climatic changes. In Subsection 2.4, a short description is presented of the combined regression–dynamic modeling approach that has been used in the study. The validity of the model has been tested using climatic and winter rye yield data from the period 1951–1980.

In Subsection 2.5, some results are presented demonstrating the sensitivity of winter rye yields, ground water level, soil quality and surface water pollution to the specific climatic scenarios. Finally, possible adjustment measures for mitigating the impacts of climatic change are considered in Subsection 2.6, such as the increasing of drainage activity.

2.2. Agroclimatic Resources

The region can be divided into five main agroclimatic districts and one subdistrict bordering the Gulf of Finland (*Figure 2.1*). This classification is based on the thermal regime calculated as the effective temperature sum (ETS). The annual sum of effective air temperature above 5 °C varies, on average, from about 850 growing degree-days (GDD) in the east to 1250 GDD in the southwest (*Table 2.1*). The latter value resembles closely the values reported for the southern part of Finland (Part IV, this volume) and the northern part of Saskatchewan in Canada (Part II, this volume).

The annual accumulated sums of above-zero temperatures (effective accumulated temperatures) for the warm period (when mean daily temperature is

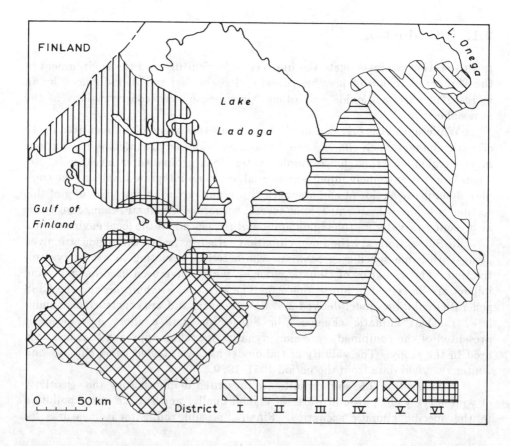

Figure 2.1. Agroclimatic districts of the Leningrad region. Climatic characteristics of each district are shown in *Tables 2.1* and *2.2*. (Source: Agroclimatic Resources, 1971.)

Table 2.1. Thermal features of each district for the vegetative period. Values relate to the period 1891–1967.

District	I	II	III	IV	V	VI
Effective accumulated temperatures during period when mean daily temperature exceeds 10 °C	1400–1600	1600–1800	1500–1700	1500–1700	1700–1900	1700–1800
Effective Temperature Sum (ETS) above 5 °C (GDD)	850–1000	950–1150	900–1100	1000–1100	1050–1250	1050–1150
Duration of the period without frosts (days)	90–105	105–115	115–130	115–120	120–130	135–150

Source: Agroclimatic Resources (1971).

Table 2.2. Estimates of the probable annual duration (days) of the warm period with average temperatures above 10 °C.

District	Percentage of years in which warm period is of a given length or longer					
	100	90	75	50	25	10
I	80	94	102	110	119	126
II	87	101	109	117	126	133
III	85	99	107	115	124	131
IV	82	96	104	112	121	128
V	90	104	112	120	129	136
VI	88	102	110	118	127	134

Source: Agroclimatic Resources (1971).

higher than 10 °C) are also shown in *Table 2.1*. An estimate of the probable occurrence of various warm-period durations (measured in days with temperatures above 10 °C) for different districts is given in *Table 2.2*. For example, for District I a warm period lasting for more than 110 days is observed in 5 years out of 10 (50%), while a period longer than 126 days is observed in only 1 year in 10 (10%).

Total precipitation during the period of active vegetation for agricultural crops averages 225–300 mm and has fluctuated between 66 and 476 mm from year to year. Excessive rain during the harvesting period is observed in 4 out of 10 years in August, and in 6 out of 10 years in September.

The average duration of the periods between the last spring and first autumn frosts ranges from 90–105 days for District I in the east, to 135–150 days for the coast of the Gulf of Finland (*see Table 2.1*). The average incoming solar radiation in the region is approximately $356\,\mathrm{kJ\,cm^{-2}}$ per year. Solar radiation is not a limiting factor for cultivated crops in the region.

Soils in the region are mostly heavy, with a low natural humus content. The main soils are podzolic, peaty-podzolic, and boggy-podzolic (peaty and peaty-podzolic gley), but there are also areas of sandy-podzolic and soddy-gley soils. In general, these soils are suitable for growing vegetables (particularly potatoes) and winter cereals. For the same management and agrotechnology, winter rye is higher yielding than winter wheat on these soils, being more resilient to the unfavorable winter conditions. That is why winter rye is recommended for allocation in the northwestern and central regions of the European USSR (*Handbook of Agronomy*, 1983).

2.3. Variability of Winter Rye Yields

Winter rye is one of the major grain crops in the Leningrad region, being cultivated for both grain and feed. The two main varieties in use are Vyatka and Vyatka-2, both of which are widely distributed. The total area of winter rye in 1983 was about 12.5×10^3 ha (21.3% of the regional area for all grain crops).

The variations in annual winter rye yields measured during the period 1947–1980 in the Leningrad region are shown in *Figure 2.2*. The yield trend $[y(t)$, in units of 0.1 tonnes/ha] obtained by the method of smoothing may be described as follows:

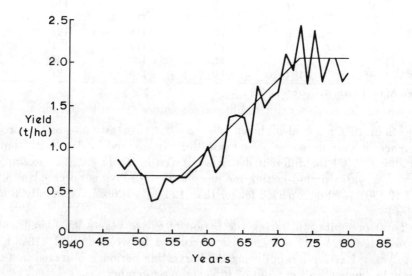

Figure 2.2. Observed yield of winter rye (t/ha) in the Leningrad region (1947–80). Yield trends are also marked. (Source: Maximov, personal communication, 1984.)

$$y(t) = \begin{cases} 6.6 & 1947 \leq t \leq 1956 \\ 11.6 + 0.8975\,(t - 1962.5) & 1957 \leq t \leq 1972 \\ 20.3 & 1973 \leq t \leq 1980 \end{cases} \qquad (2.1)$$

This is also indicated in *Figure 2.2*, which demonstrates that before the year 1956 and after 1973, increasing or decreasing trends in winter rye yields in the region are absent. The increasing trend in yields between the mid 1950s and the early 1970s can be largely explained by improvements in agricultural technology and management (e.g., increasing use of fertilizers and pesticides, improved machinery, etc.).

The interannual variations in winter rye yields over the period 1951–1980 have been summarized by dividing the period into its three component decades. As well as the decadal mean yields, the absolute variability of yields (standard deviation) and the variability of yields relative to the decadal mean (coefficient of variation) are given (*Table 2.3*).

Table 2.3. Statistical analysis of winter rye yields in the Leningrad region by decades for the period 1951–1980.

	1951–60	*1961–70*	*1971–80*
Mean yield (t/ha)	0.72	1.83	2.02
Standard deviation	0.12	0.21	0.23
Coefficient of variation	0.17	0.12	0.11

It should be noted that while the absolute variability of yield has increased over time, the increase in mean yields has been proportionally greater, so that the coefficient of variation has decreased over time.

2.4. Winter Rye Cultivation and Climate

We now consider some specific conditions for winter rye cultivation in the region. The yield of winter rye depends on the state of the crop when growth ceases at the beginning of winter, on the winter conditions themselves, and on the weather conditions during the period of resumed growth up until ripening.

During the period from sowing to germination, productive moisture reserves of less than 20 mm are observed once or twice in 20 years. The duration of this period is limited by temperature conditions. In favorable years (e.g., 1948, 1960, 1963), it averages 6–8 days, but in years with a moisture deficiency (less than 20 mm in the topsoil layer) or with lower air temperatures (below 14 °C), the period between sowing and germination increases (e.g., to 16–20 days in 1959). Just before winter the average productive moisture reserves are usually excessive. By the time the hard frosts begin, sufficient snow cover (above 30 cm) has accumulated thus protecting the seeds from the direct influence of the low temperatures. Only in the most severe winters does the soil temperature fall low enough to cause damage, to around -11 to $-13\,°C$ at the depth of the rooting zone (e.g., in 1955–1956 and 1959–1960). On the other hand, snowy and milder conditions can lead to crop-rotting under the snow cover. More than 10% of the crops sown were lost because of rotting in the winters of 1960–1961, 1961–1962 and 1965–1966.

During the reproductive period (at the end of May), the moisture reserves in the soil are generally adequate. Only in 1 year in 10 are these reserves insufficient (less than 50 mm), i.e., unfavorable for promoting development of the ear and flower of the grain. Waterlogging of fields during the period of grain ripening occurs on average once every 3–5 years, when moisture reserves are above 125 mm. The waxy stage of grain ripeness generally occurs at the end of July, or by the middle of August in cooler summers. Taken together, the average climatic conditions in the region today are quite favorable for winter rye cultivation.

Having presented some information about the present-day conditions for winter rye cultivation in the region, we will now move on to consider the implications in this region of possible global climatic changes. The examination of this problem includes three main elements:

(1) The construction of a regional climatic scenario consistent with global changes in climate expected with a doubling of atmospheric CO_2.
(2) Development of a model for investigating the behavior of selected crop and environmental parameters.
(3) The simulation of adjustments in agricultural management which would enable sustainable development of agriculture and environment in the region.

2.5. Generating Scenarios for Regional Climatic Change

Two approaches can be identified for generating scenarios of changing regional climatic conditions. The first utilizes specific models such as the general circulation models for the atmosphere (GCMs), for example, the GISS model which has been adopted in other case studies in the IIASA/UNEP climate impacts project (this volume). The second approach employs empirical climatic data to construct a simple heuristic model for forecasting climatic changes in different regions. In this study we shall follow both approaches.

2.5.1. GISS scenario of doubling CO_2

Comparison of interpolated GISS gridpoint data (generated for an equilibrium climate under present-day, $1 \times CO_2$ conditions) with the present climatic normals for the period 1951–1980 for the Leningrad region (observation station Pulkovo, 30 km northeast of Leningrad; it should be noted that the city of Leningrad itself would probably show an urban warming effect) revealed a fair correspondence between observed and simulated temperatures, but a poor fit for

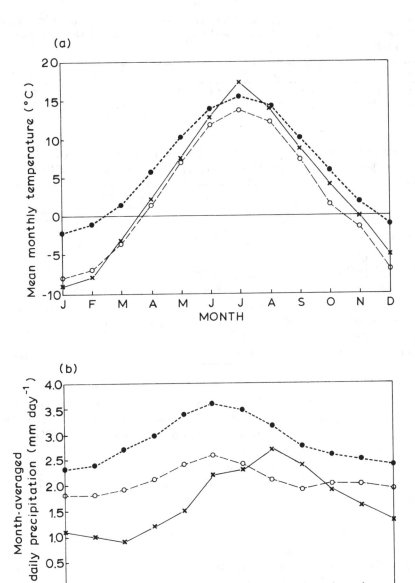

Figure 2.3. Validation of the GISS model estimates for: (a) monthly mean temperature and (b) precipitation rate at Leningrad (gridded GISS data are averaged for the location 60 °N, 30 °E).

Table 2.4. Changes in temperature (ΔT) and precipitation (Δr) for the Leningrad region under the GISS scenario of a doubled CO_2 concentration by the year 2050 (values are rounded).

Month	J	F	M	A	M	J	J	A	S	O	N	D	Year
$\Delta T(°C)$	6.4	6.0	5.2	4.4	3.4	2.1	1.5	2.0	3.2	4.5	5.4	6.2	4.2
$\Delta r(mm)$	16	20	24	29	32	34	37	34	26	20	17	15	301

Table 2.5. Comparison of projected $2 \times CO_2$ climatic data for Leningrad with observed data for that station and three stations in the southern European USSR.

Climatic factor (mean annual value)	Leningrad (60°N; 30°E)	
	Observed (1951–80)	Projected (GISS)
Temperature (°C)	4.0	8.2
Precipitation (mm)	570	871

Climatic factor (mean annual value)	Other stations (observed, 1951–80)		
	L'vov (50°N; 23.5°E)	Kaliningrad (54.4°N; 20.4°E)	Chernovtsey (48.1°N; 25.6°E)
Temperature (°C)	7.1	5.9	7.8
Precipitation (mm)	708	740	660

precipitation (*Figure 2.3*). This is a common problem when attempting to use GCMs to give regional-scale estimates of climatic change (*see* Part I, Section 3). However, assuming that the estimated differences between $1 \times CO_2$ and $2 \times CO_2$ equilibrium climates are credible estimates of possible changes in future climate, we have used the following procedure to construct the climatic scenario for doubled CO_2 concentration using the GISS data to adjust observed (1951–1980) monthly normals:

$2 \times CO_2$ mean monthly temperature = (GISS $2 \times CO_2$ – GISS $1 \times CO_2$) + observed mean

$2 \times CO_2$ mean monthly precipitation rate = (GISS $2 \times CO_2$ – GISS $1 \times CO_2$) + observed mean.

In the Saskatchewan case study (Part II, this volume) an alternative technique is reported for applying GISS data to precipitation normals. However, in the Leningrad region, with an annual precipitation approximately 1.5 times greater than in Saskatchewan, the difference between the two techniques is not very great and does not affect the analysis here. The GISS estimates for the region (average of four grid point values) are given in *Table 2.4*.

In order to illustrate the scenario *Table 2.5* presents observed and projected data of mean annual temperature and precipitation for the Leningrad region, alongside the 1951–1980 observed data for three more southerly locations in the European part of the USSR [see Figure 1.4(a)].

Comparing the historical records for L'vov, Chernovtsey and Kaliningrad with the GISS scenario estimates demonstrates that the projected climate in Leningrad would be warmer and wetter than any of these regions are today.

2.5.2. Transient climatic change scenario

The GISS data provide a scenario of climatic change in terms of a step function (i.e., an abrupt change from the present-day climate to a doubled CO_2-induced climate). However, present projections indicate that a doubling of atmospheric CO_2 concentrations will not occur for several decades, and we should expect any associated climatic change to exhibit a gradual transition over time. Hence, in order to investigate the dynamic behavior of agriculture and the environment we need to consider the *transient* change of climate in our scenarios. A simple interpolation between the present-day climatic conditions and the GISS projection may on its own be too crude for representing the future 20–25 years. That is why we have used an alternative approach for simulating the first 25 years of changing climate, based upon an empirical climatic data set according to the regression scheme developed by Vinnikov and Groisman (1979a, b).

In common with other researchers, they selected the annual mean surface air temperature as the major physical parameter of global climate. This choice of parameters has a straightforward physical background, since hemispheric air temperature and the associated temperature difference between equator and pole are the key parameters defining baroclinic stability and the intensity of atmospheric processes. Note that the determination of statistical relationships between global temperature and a set of climatic parameters such as zonal and regional temperature, precipitation, etc., calls for the analysis and treatment of a huge body of empirical data.

Table 2.6. Estimates of the CO_2 concentration in the atmosphere.

Year	1980	1995	2005
CO_2 concentration (ppm)	336	378	416

Source: Sov–Amer (1982).

We shall construct the scenario based upon some simple estimates. The increase of atmospheric CO_2 concentration and the consequent "greenhouse effect" are thought likely to be the major contributors to global climatic change in future decades. *Table 2.6* contains estimates of the probable increase in atmospheric CO_2 concentration up to the year 2005. These estimates were produced on the basis of analyses of energy-development forecasts and computer calculations of CO_2 cycles (Sov–Amer, 1982).

Let T_c be a function of the global equilibrium temperature corresponding to a given CO_2 concentration in the atmosphere. Simulations show that T_c is approximately proportional to the logarithm of the CO_2 concentration over a considerable range (Sov–Amer, 1982). This can be written as follows:

$$T_c(t) - T_0 = \Delta T_c \cdot \log_2\left[n(t)/n_0\right] \qquad (2.2)$$

Here $n(t)$ is the forecast CO_2 concentration at time t, n_0 is the concentration corresponding to the year 1980 (336 ppm), and ΔT_c is the increase in equilibrium temperature corresponding to a doubling in the value of the CO_2 concentration relative to the 1980 level. The value of ΔT_c is estimated as 3 °C. This is the upper limit of a range of estimates reported from the Sov–Amer (1982) study and bears the closest resemblance to the GISS estimates.

The increasing concentration of trace gases of anthropogenic origin in the atmosphere (ozone in the troposphere, methane, nitrous oxide, halocarbons, and some others) also contributes to a strengthening of the greenhouse effect. Estimates suggest that the increase in equilibrium temperature due to trace gases may be as large as 50% of the contribution due to the build-up of CO_2 (Bolin et al., 1986).

The thermal inertia of the oceans will cause a delay in global warming. In terms of the response of average annual surface air temperature, this will amount to approximately 10 years. This means that the increase in equilibrium temperature $T_c(t)$ for year t will be approximately 20% less than if the delay were not taken into account (Butner, 1983). In short, the average annual surface air temperature of the globe is expected to increase by about 0.6 °C by 1995 and 1.1 °C by 2005, relative to the 1980 temperature.

Vinnikov and Groisman (1979a) suggested a relatively simple statistical model for relating the values of regional temperature and precipitation to northern hemisphere temperature:

$$z_i^j(t) = \alpha_i^j T(t) + \beta_i^j + \epsilon_i^j(t) \qquad (2.3)$$

Here z_i^j is the mean local (for the ith region) annual or seasonal temperature ($j = 1$) as well as mean local annual or seasonal precipitation ($j = 2$), $T(t)$ is the hemispheric mean annual temperature (for the extra-equatorial part of the Hemisphere), α_i^j is a regression coefficient, β_i^j is the regular error, and $\epsilon_i^j(t)$ is a random error connected with errors of measurement and other factors. Vinnikov and Groisman calculated the parameters of model (2.3) for the Northern Hemisphere (see also Pitovranov et al., 1984). The estimates of regression coefficients for the Leningrad region can be seen in Table 2.7 where α^1 are coefficients of linear relationships between hemispheric temperature changes and regional seasonal and annual temperature response (Vinnikov and Groisman, 1982), and α^2 are coefficients for regional summer and annual precipitation response (region N2 in Groisman, 1981).

Thus, given a scenario of *global* temperature increase up to 2005, we obtain the scenario of temperature and precipitation changes in the Leningrad region given in *Table 2.8*. According to this empirical assessment the most significant changes of regional temperature would be expected in the winter months. It

Table 2.7. Estimates of regression coefficients between northern hemisphere temperature changes and temperature and precipitation changes in the Leningrad region (the standard deviations of estimates are given in parenthesis).

Regression coefficient	December – February	March – May	June – August	September – November	Year
Temperature α^1	1.7 (0.9)	0.8 (0.6)	0.9 (0.4)	0.9 (0.5)	1.2 (0.4)
Precipitation α^2 (%/0.1 °C)	–	–	0.0	–	0.7

Sources: Vinnikov and Groisman (1982); Groisman (1981).

should be noted that the empirical approach suggests a similar tendency of increasing annual precipitation in the region as the GISS data, but indicates no increase in precipitation during the summer months. In contrast, according to the GISS data, maximum changes of precipitation could be expected in the summer months of the year.

The empirical approach is based upon the data for the past century, a period which was characterized by an amplitude of global mean annual temperature changes of about 1 °C (Vinnikov and Groisman, 1982). That is why the forecast has been restricted only to the period up to the year 2005, at which time estimates of likely CO_2-induced temperature increases are of a similar magnitude (Sov–Amer, 1982).

The GISS scenario of a 4 °C global warming for doubled atmospheric CO_2 might be expected to occur somewhat later, towards the middle of the next century (according to a variety of projections of CO_2 emissions over the future decades – Nordhaus and Yohe, 1983; Bolin et al., 1986). Here we assume that date to be 2050. On the basis of the empirical forecast for the years until 2005 and assuming a linear change in the temperature and precipitation from 2005 until the doubling date of 2050, we obtain a scenario of transient climatic change in the Leningrad region (*Table 2.9*).

It is clear that this scenario is only a very crude forecast of possible climatic change in the Leningrad region, but it does give us the opportunity to test the methodology for investigating broad-scale environmental changes in the region.

2.6. Selection and Validation of Impact Models

2.6.1. The Obukhov model

In order to assess the sensitivity of winter rye yields to weather variations in the Leningrad region, we have chosen to use the classical regression model developed by Obukhov (1949). This model was constructed from meteorological and winter rye yield data for the period 1883–1916. It should be noted that meteorological and yield data for this period are reliable for the region (Obukhov, 1949). The regression equation incorporates 11 meteorological variables which, according to Obukhov, are the major climatic factors influencing winter rye yield. A description of the Obukhov model is given in *Appendix 2.1*.

Table 2.8. Changes of temperature and precipitation in the Leningrad region up to the year 2005 (relative to the period 1951–1980).

Years	Global mean annual[a]	Changes of temperature (°C) Leningrad region					Changes of precipitation Leningrad region	
		Year	December–February	March–May	June–August	September–November	Year (mm/year)	June–August (mm/3 months)
1995	+0.6	+0.7	+1.0	+0.5	+0.55	+0.55	+34	0.0
2005	+1.1	+1.3	+1.9	+0.9	+1.0	+1.0	+62	0.0

[a] Calculated by linear interpolation from the Sov–Amer (1982) estimates of global climate for doubled CO_2 (assumed to occur in the year 2050).

Table 2.9. The scenario changes (per year) of temperature and precipitation in the Leningrad region.

Period		J	F	M	A	M	J	J	A	S	O	N	D	Year
1980–1995	ΔT (°C)	0.07	0.07	0.03	0.03	0.03	0.04	0.04	0.04	0.04	0.04	0.04	0.07	0.045
	Δr (mm)	0.25	0.25	0.25	0.25	0.25	0.0	0.0	0.0	0.25	0.25	0.25	0.25	2.25
1996–2005	ΔT (°C)	0.09	0.09	0.04	0.04	0.04	0.045	0.045	0.045	0.045	0.045	0.045	0.09	0.06
	Δr (mm)	0.3	0.3	0.3	0.3	0.3	0.0	0.0	0.0	0.3	0.3	0.3	0.3	2.7
2006–2050	ΔT (°C)	0.1	0.09	0.095	0.08	0.055	0.025	0.01	0.02	0.05	0.08	0.1	0.115	0.065
	Δr (mm)	0.2	0.3	0.4	0.5	0.55	0.75	0.8	0.75	0.4	0.3	0.2	0.2	5.35

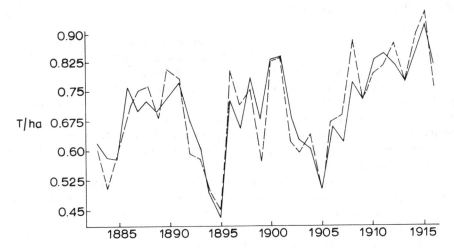

Figure 2.4. Winter rye yields in the Leningrad region (1883–1916). (Tonnes/hectare; converted from Poods/Dessiatina, where 1 Pood = 16.38 kg and 1 Dessiatina = 1.09 ha). Dashed lines represent observed yields; solid lines represent model estimates. (Source: Obukhov, 1949.)

The model allows us to calculate the deviation of yield (y_z) from an average value. The winter rye yields calculated by Obukhov (1949) are given in *Figure 2.4*. Analysis of this figure reveals quite a good fit between calculated and observed yields (correlation coefficient, $r = 0.89$).

However, our confidence in the ability of the Obukhov model to simulate short-term weather-induced fluctuations in crop yield cannot be extended to the examination of climate impacts over the longer term for two important reasons. Firstly, Obukhov's analysis did not include a consideration of the influence of technological and management factors on yield levels, probably because any trends that might have been attributable to these factors during the early period for which the model was developed were too slight to merit inclusion. However, the marked improvements in crop yields in the decades following World War II undoubtedly are strongly related to trends in technology and management (*Figure 2.2*). These are likely to continue into the future, making it imperative that they should be incorporated into any scheme for simulating crop yields.

Secondly, over the long-term and even in the absence of anthropogenic influences, soil quality is itself related closely to the climate. For example, temperature and precipitation influence the processes of biomass decomposition and humus creation in topsoil. The amount of precipitation also affects rates of erosion, soil salinity, waterlogging, etc. None of these effects are built into the Obukhov model.

To supplement these shortcomings, it was decided to link the Obukhov model to another model that is capable of accommodating long-term effects of technology and climate.

2.6.2. The VNIISI model

The so-called "VNIISI" model, which incorporates the long-term effects discussed above, was chosen for assessing the impact of projected climatic changes on average yields of winter rye as well as impacts on some other environmental parameters. This model was developed in Moscow at the All-Union Research Institute for Systems Studies (VNIISI) by Pegov and others (Kroutko *et al.*, 1982a,b; Pegov, 1984; Pegov *et al.*, 1983, forthcoming). The model takes into consideration the interaction between regional environmental parameters and thus provides an averaged, but integrated representation of the ecological development of the region.

The VNIISI model is termed an "index" model, since it contains indexes of four principal environmental factors: soil fertility, vegetation, water and climate efficiency. The model itself consists of a *control block*, and an *ecological processes block* comprised of four submodels (sectors): the biotic sector, the water sector, the pollution sector, and the health sector (*Figure 2.5*). This study considers estimates from all sectors, with the exception of the health submodel. Ordinary differential equations are used to describe the dynamic process contained in each of the submodels.

The dialogue and system management are operated in a control block, which also incorporates a large data bank of ecological information that is required to initialize the system variables. The information includes characteristics of each type of regional landscape found globally, e.g., climate, permafrost, soil moisture, soil erosion, soil salinity, fire prevalence, pollutant load, etc.

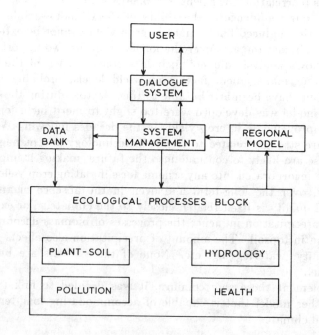

Figure 2.5. Structure of the VNIISI model (adapted from Pegov *et al.*, forthcoming).

In the processes block, the *biotic (plant–soil) submodel* contains the following components: vegetation, organic matter and mineral soil. The components of the *hydrological (water) submodel* include: soil moisture, evapotranspiration, surface water flow, groundwater flow, deep groundwater flow, natural and artificial surface dissection, soil salinity, and aggregate flow parameters. Finally, the *environmental pollution submodel* incorporates components that include: air pollution, and transformation of pollutants in each of vegetation, soil, groundwater, deep groundwater and surface water. A description of the equation systems for each of the above-mentioned blocks and a sensitivity study of the model are reported in Pegov *et al.* (1983, forthcoming).

The VNIISI model can be used to simulate the response of a range of different agricultural crops to climatic change. Its sensitivity has been evaluated for a wheat crop growing in an environment representative of the US Great Plains (Watts 1984). The results (which are broadly applicable also to winter rye) demonstrate that it is more appropriate to use the model to simulate *long-term responses* to climatic change (mainly through the effects of climate on soil properties) than to assess the impacts of short-term (interannual) variations in climate (simulated in some versions of the VNIISI model as stochastic variations). Thus, the VNIISI model is utilized in this section purely for its sensitivity to long-term climate/soil/fertilizer relationships, and is coupled to the Obukhov model which accounts for the interannual variability of yields about these longer-term trends.

2.6.3. Combining the Obukhov and VNIISI models

In order to estimate winter rye yield (y) with the help of both models the following equation is used:

$$Y = \hat{Y} + y_z \tag{2.4}$$

where $\hat{Y}$ is the yield estimate from the VNIISI model, and y_z is the deviation of the expected yield from its average value, produced by the Obukhov model. We will refer to this combined model as the VNIISI–Obukhov model.

For validation of the VNIISI–Obukhov model, regionally-averaged observed station data of ten-daily temperature and precipitation were used. The data were obtained from the All-Union Research Institute for Hydrometeorology – World Data Center (VNIIGMI–MCD). Data on fertilizer applications for winter rye (from national and regional statistics) were also required and these are given in *Table 2.10*.

The observed and calculated yields of winter rye are given in *Figure 2.6*. The computations using the VNIISI model have been conducted for the fertilizer application rates given in *Table 2.10*, and climatic data of mean annual temperature and precipitation for the period 1951–80. For comparison, the average yield at experimental sites is currently around 2.55 t/ha.

The yield trend obtained using this model can be seen in *Figure 2.6* (curve VNIISI). *Figure 2.6* also shows the computed winter rye yields simulated using

Table 2.10. Average fertilizer applications for winter rye in the Leningrad region.

Period	1947–51	52–55	56–60	61–70	71–75	76–80	Experimental sites[a]
N (kg/ha)	8	14	24	25	28	30	55
P_2O_5 (kg/ha)	2	4	11	15	19	22	45
K (kg/ha)	3	6	17	20	24	28	45
Organic (t/ha)	8.7	9.2	9.9	10.8	11.0	10.8	15.0

[a] Recommended fertilizer applications for winter rye on typical soil types in the Leningrad region, based on experimental site data.
Source: Maximov, personal communication (1984).

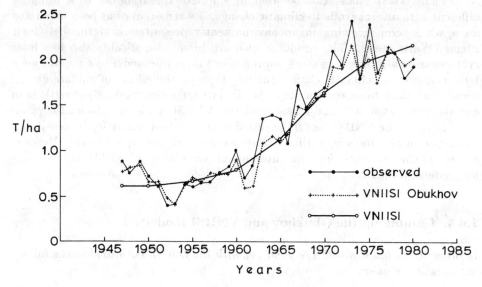

Figure 2.6. Observed and calculated yields of winter rye (1947–80).

the combined model (curve VNIISI–Obukhov). The comparison shows that the results obtained using the VNIISI and VNIISI–Obukhov model approximate quite closely both the observed trend, and the observed interannual fluctuations of yield. The correlation coefficient of deviations from trend of observed yields against deviations of estimated yields (VNIISI–Obukhov model) was calculated as r = 0.74, indicating fairly good agreement.

It is important to stress that while such a close correspondence of model estimates with observed yields indicates the model's suitability for applications over the observed range of climatic conditions, confidence in the model relationships is necessarily reduced when climatic changes are simulated that occur *outside the range* of conditions for which the model was developed. In these cases it is necessary to extrapolate model relationships outside their natural range of applicability, unless suitable modifications can be introduced to the model functions. The reader is cautioned, therefore, that the results of the following experiments rely, in some instances, on the extrapolation of model functions and are thus attended by considerable uncertainty. However, these are preliminary experiments, designed as much to test the approach as to generate results, which themselves are subject to future refinement.

2.7. The Sensitivity of Winter Rye Yields and of Some Environmental Parameters to Climatic Change

2.7.1. GISS scenario impact on winter rye yields

The response of winter rye yields to the GISS "step-like" scenario of climatic change has been analyzed using the Obukhov model. A uniform distribution of GISS monthly precipitation changes for each 10-day period within a month is assumed. The calculations show that the deviation y_z of winter rye yields from average is as follows:

$$y_z = -0.27 \text{ t/ha}$$

This represents a 13% decrease from the average yield of 2.03 t/ha which was observed over the period 1973–1980 [see equation (2.1)].

2.7.2. Transient scenario impact on winter rye yield and some environmental parameters

To introduce more realism into the scenario experiments, the response of winter rye yield and some environmental parameters to the transient climatic scenario changes was investigated, using the combined VNIISI–Obukhov model. This response was compared with the results of the reference scenario computer run.

The Reference Scenario (Variant 1)

The climatic conditions for the reference scenario were defined according to the data corresponding to the period 1951–1980 (10 day, monthly and annual temperature and precipitation). The initial conditions of mineral and organic fertilizer application were fixed at the levels observed in the period 1976–1980 (see *Table 2.10*) and these were increased linearly up to the levels currently applied at "experimental sites" (*Table 2.10*) over a simulation period of 55 years from 1980 to 2035. During this period, the modeled system variables were able to achieve a statistically stable state.

The output data generated for this reference scenario by the VNIISI–Obukhov model include: winter rye yield (g/l), a soil fertility index (dimensionless), surface water pollution with nitrogen (t/ha) and depth of the first ground water table (meters).

The results of computations are given in *Table 2.11*. The soil index increased over the 55-year period by 33% and winter rye yield rose to 2.7 t/ha. As noted above, the average yield at experimental sites is now around 2.55 t/ha. Surface water pollution also increased (by about 13%).

Table 2.11. Simulations of absolute and relative impacts of Scenario Variants 1 and 2 on winter rye yields, on the soil fertility index, on the depth of the first groundwater table, and on surface water pollution due to nitrogen.

	1980	1990	2000	2010	2020	2035
Change in mean May–October climate under Variant 2						
Temperature (°C)	0	+0.4	+0.8	+1.2	+1.6	+2.2
Precipitation (mm)	0	+8	+16	+38	+74	+127
Winter rye yields						
Variant 1 (t/ha)	1.99	2.3	2.5	< 2.6	> 2.6	2.7
Variant 2 (t/ha)	1.99	2.4	2.6	2.7	2.5	2.1
Impact of Variant 2 relative to Variant 1 (%)	0	+4	+5	+5	−4	−23
Soil fertility index						
Variant 1 (arbitrary units)	10.7	12.3	13.1	13.6	13.9	14.2
Variant 2 (arbitrary units)	10.7	12.2	12.6	12.1	10.8	8.3
Impact of Variant 2 relative to Variant 1 (%)	0	−1	−4	−11	−22	−42
Depth of the first groundwater table						
Variant 1 (m)	−5.02	−5.1	−5.1	−5.2	−5.2	−5.3
Variant 2 (m)	−5.02	−5.0	−4.9	−4.7	−4.5	−3.3
Impact of Variant 2 relative to Variant 1 (%)	0	+2	+4	+9	+14	+38
Surface water pollution due to nitrogen						
Variant 2 (g/l)	0.152	0.20	0.24	0.33	0.41	0.63
Impact of Variant 2 relative to Variant 1 (%)	0	+18	+30	+75	+114	+225

The Transient Climatic Change Scenario (Variant 2)

The changes in climate were generated according to *Table 2.9*. All other parameters were the same as in the reference scenario. The results of the calculations are presented in *Table 2.11*. Note that the model simulations were not conducted for the entire period from 1980 until the GISS doubling date of 2050. Rather, it was decided to restrict the experiment to climatic changes occurring up to 2035, beyond which time both the climatic and the technology projections were considered to be highly speculative. Thus, the climatic scenario considered here does not reflect the full magnitude of changes estimated by the GISS model for a doubled CO_2 situation; merely a transitional state that approaches these changes.

Analysis of the data in *Table 2.11* shows that for the climatic scenario (Variant 2), the winter rye yields would increase up to 2010 and then would fall by 23% up to 2035, compared with those projected in Variant 1. *Table 2.11* also contains corresponding data related to the soil fertility index, depth of the first ground water table, and surface water pollution. These data show that the soil fertility index in Variant 2 would decrease continuously over time to a value 42% of that in Variant 1 by 2035. The increase of winter rye yield in Variant 2 compared with Variant 1 up to the year 2010 is explained by beneficial increases of temperature during the growing period combined with unchanging summer precipitation. The reduction in yields during the period 2010–2035 is the result of large increases in precipitation in the summer months. Though offset to a certain degree by increased evapotranspiration in the warmer climate, the high precipitation would lead to enhanced soil degradation (noted above). The decreasing of the soil fertility index is the result of increases in both soil moisture and water erosion, reflecting the response of the VNIISI model to increasing annual precipitation.

From *Table 2.11* we can see that the depth of the first ground water table is decreased, which may accelerate waterlogging processes in the region, while perhaps the most drastic changes involve surface water pollution due to nitrogen. These data show that increasing precipitation is likely to increase markedly the leaching of nitrates into surface water, which would result in degradation of water quality, with associated impacts on population and the environment in the region. The calculation of nitrogen concentration in the runoff requires specific information about watershed hydrology, topography, etc. A crude assessment of the impact may be made by comparing computed nitrate leaching with calculations of leaching in Finland (Valpasvuo-Jaatinen, 1983). The average fertilizer application in the country as a whole is 69 kg nitrogen, 59 kg phosphorus and 52 kg potassium per hectare. According to this assessment, the total leaching of phosphorus in southern Finland is about 22 kg/km^2 ($100 \text{ kg/km}^2 = 1 \text{ kg/ha}$) and that of nitrogen 400 kg/km^2. One result is that the river Durajoki, supplying drinking water for the city of Turku, cannot be totally purified of these nitrates and phosphates using the purification technology of present-day water treatment.

2.8. Mitigating the Impacts of Climatic Change

According to results for Variant 2, climatic changes in the region are accompanied by decreases in soil quality. The soil quality may be improved, however, by applying greater quantities of fertilizer. Such a situation is built into a third scenario, Variant 3.

2.8.1. Scenario of increasing fertilizer application (Variant 3)

While the climatic conditions in this scenario were the same as in Variant 2, it is supposed that mineral and organic fertilizer application would increase by 50% compared to the Variant 1 scenario. The computation of winter rye yield under this scenario can be seen in *Figure 2.7*. The results demonstrate considerable increases in yields. For example, yields in 2010 are estimated to be more than

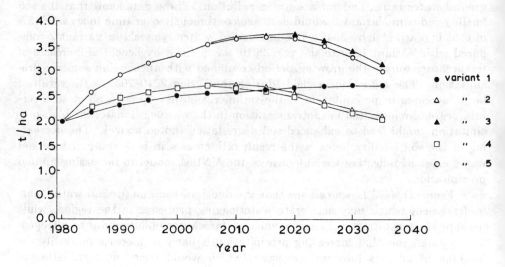

Figure 2.7. Simulated changes in winter rye yield for five scenario variants, 1980–2035 (tonnes per hectare).

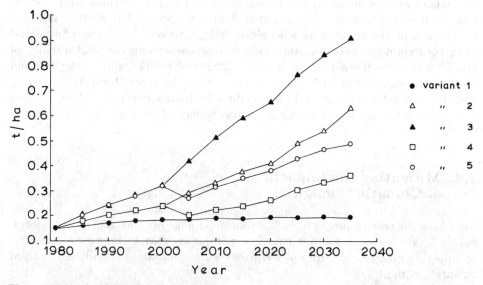

Figure 2.8. Simulated changes in surface water pollution with nitrogen for the five scenario variants, 1980–2035 (tonnes per hectare).

40% greater than under Variant 1. It should be noted that this assessment is likely to be an overestimate because the effectiveness of fertilizer application has been found to decrease under conditions of overmoistening, and this, along with other factors that may also be important, is not built into the current version of the VNIISI–Obukhov model.

According to the simulations, increases of fertilizer application lead to an increase of surface water pollution. The computation of changes in water pollution can be seen in *Figure 2.8*. After the year 2010 this pollution is over 50% more severe than under Variant 2.

The amount of water surface flow depends partly on the dissection of the region by drainage systems. The VNIISI model includes an option to specify a drainage system in the model. Variant 4 uses this option, assuming an enhancement of drainage activity in the region.

2.8.2. Scenario of improved drainage (Variant 4)

In this scenario we used the same climatic estimates and data on mineral fertilizer application as for Variant 2, but for the period 2001–2003 the following additional drainage activity was foreseen: increases in 2001 and 2002 of $0.5\,km/km^2$, and an increase of $1\,km/km^2$ in 2003. All drainage ditches were assumed to be of 1.5 m depth.

The winter rye yields are slightly decreased compared with Variant 2 yields (*Figure 2.7*). This can be explained through increases in leaching of soil nutrients. Considerable changes in the local surface water pollution in the region are also estimated (*Figure 2.8*), as most of the pollutants are carried away in the improved drainage network.

These improvements in soil quality and increases in the drainage density have been combined into a single scenario (Variant 5).

2.8.3. Scenario of increasing fertilizer application and improved drainage (Variant 5)

In this scenario the fertilizer application was as in Variant 3, and drainage activity as in Variant 4. It can be seen from *Figures 2.7 and 2.8* that Variant 5 is the preferred scenario overall. Winter rye yield increases by approximately the same amount as for Variant 3, but there are considerably lower values of surface water pollution due to nitrogen, approximately the same values of the soil fertility index as in Variant 3, and improvements in the depth of the first ground water table (*Figures 2.9 and 2.10*).

2.9. Conclusions

Analysis of the results presented in this paper allows us to draw some conclusions about the impacts of climatic change on agriculture and the environment in the Leningrad region. The global warming projected by the GISS model for a doubling of atmospheric CO_2 would be accompanied by significant increases of precipitation in the region. Even accounting for increases in evapotranspiration associated with higher temperatures, these changes could lead to increases in soil moisture and water erosion of soils. The ground water level in the region would

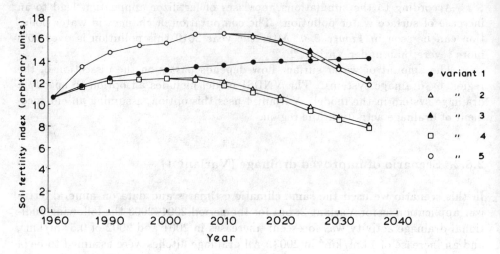

Figure 2.9. Simulated changes in soil fertility index for the five scenario variants (1980–2035).

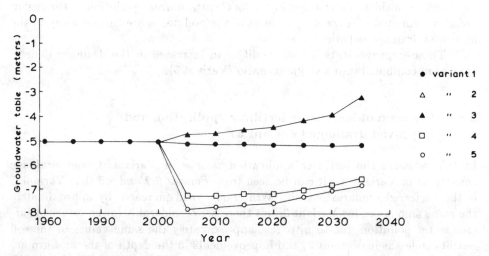

Figure 2.10. Simulated changes in depth of the first groundwater table for the five scenario variants (1980–2035).

probably rise thus enhancing waterlogging processes in the region. Degradation of soil quality would occur as a result of these processes.

The impact of climatic changes on yields of present-day cultivated winter rye may occur in two phases in this region. In the first phase, for the period up to the beginning of the next century, precipitation increases may be relatively moderate and temperature conditions slightly more favorable for winter rye. In the next phase of warming when precipitation (especially in summer) would increase considerably, the winter rye yields would decrease.

Such climatic changes will influence other aspects of the environment of the region. For example, the water pollution from agricultural watersheds has a demonstrably high sensitivity to climate. It was shown that the water pollution by nitrogen increases considerably given a wetter climate in the region. These processes may result in the degradation of water quality in the region and acceleration of eutrophication in fertile water.

Similar negative impacts on environment are possible in some other case study regions. Those especially sensitive to climate impacts include Finland, where the numerous water courses are highly susceptible to pollution (Valpasvuo-Jaatinen 1983).

One method for improving agricultural and environmental conditions in regions with an excess of water is to develop efficient drainage systems. The computations presented here for scenarios of enhanced drainage activity show that these measures could serve to mitigate some of the negative impacts of climatic changes and increased fertilizer use in the Leningrad region. Changes in crop type and the optimal land allocations for each crop under altered climates are considered in Section 4.

SECTION 3

The Effects on Spring Wheat Yields in the Cherdyn Region

3.1. Introduction: Methods of Estimating Crop Yield Responses to Climate

In this section we investigate the effects of both short-term climatic variations and long-term climatic changes on spring wheat yields in the Cherdyn region, just west of the Ural Mountains in the northern USSR [cf. *Figure 1.4(a)*]. There are three approaches which might be adopted to make this assessment: expert judgment, regression analysis and dynamic modeling.

3.1.1. Expert assessment

As an example of the first approach, the US National Defense University (NDU 1978, 1980) used an expert method to estimate the yield response of major food crops to five predetermined scenarios of future climatic conditions. Crop-weather models were presented in the form of arrays developed by experts to represent the sensitivity of the average yield per unit area for each country studied to specified variations in temperature and rainfall. These estimates were made for the United States, Canada, Argentina, India, China, Australia, and the Soviet Union. But such an approach is unlikely to be sufficiently reliable or to embody the most recent state of our knowledge on the yield effects of weather conditions. It may, however, be useful for treating certain qualitative aspects of the problem that cannot yet be analyzed in terms of physics or mathematics.

3.1.2. Regression models

The second approach is based on the use of regression analysis, a widely applicable tool for yield forecasting (*see* Part I, Section 2, this volume). However, the climatic variables of interest are rarely incorporated into forecasting schemes since correlation analysis of the initial sets of factors potentially affecting crop yields often shows a bias toward biometric, edaphic, or agronomic parameters. Analysis of the intensive work in the field of agrometeorological forecasting during the last 30 years indicates that it is seldom possible to discover universal and reliable statistical relationships between meteorological parameters and crop yields. However, in those cases where such correlations have been established their application has proved to be of some value.

3.1.3. Dynamic models

The third and most promising approach is based on the quantitative theory of agroecosystem productivity currently under development. This method has already resulted in the construction of several dynamic crop production models (e.g., Bihele *et al.*, 1980; Sirotenko, 1981), but these rather sophisticated models have not yet been used to analyze the impact of climatic changes and variability, hitherto being associated more with complex problems of yield programming and forecasting. Some positive results have been achieved by studying intermediate outputs from dynamic models, such as the analysis of carbon dioxide exchange in agrocenoses, in relation to climatological problems (Terjung *et al.*, 1976; Menzhulin and Savvateev, 1980). But clearly, crop photosynthesis and final grain yield are far from identical characteristics. Therefore, although the use of such models seems attractive on the basis of the relatively simple calculations involved, they look less promising now that dynamic models are being developed for specific crops.

3.2. A Dynamic Crop–Weather Model for Estimating Spring Wheat Yields

For a number of years, we have been developing and improving a dynamic crop–weather model to provide operational agrometeorological support to agriculture (Sirotenko *et al.*, 1982). This model comprises a closed set of differential equations for calculating the dynamics of crop phytomass and water content of the root zone of the soil:

$$\frac{dm_p}{dt} = \alpha_p(1 - R_G)(F + Q) - D_p - q_p - P_p \tag{3.1}$$

$$\frac{dW_i}{dt} = q_{i-1} - q_i - TR_i - \delta_i E \tag{3.2}$$

where m_p is the phytomass of leaves $(p = 1)$, stems $(p = 2)$, roots $(p = 3)$, spike envelopes $(p = 4)$, and grains $(p = 5)$ per unit area of the crop canopy; α_p is the set of growth functions $(\alpha_p \geq 0, \sum_{p=1}^{5} \alpha_p = 1)$; R_G is the growth respiration coefficient; F is the gross photosynthesis of the canopy; Q is the total decay of the structural matter $(Q = \sum_{p=1}^{5} q_p)$; D_p is the maintenance respiration; P_p is the rate of phytomass shedding; W_i is the water content of the ith layer of the soil; $q_i - 1$, q_i are the rates of water flow through the upper and lower boundaries of the ith layer of soil, respectively (q_p is the precipitation absorbed by the soil, and $i = \overline{1,15}$); TR_i is the water loss through transpiration from the ith layer; E is the evaporation; δ_i is a logical variable ($\delta_i = 1$ if $i = 1$; and zero for the rest of the layers); and t is time.

Equation (3.1) is integrated numerically with a one-day time step throughout the growing season up to maturation. Initial values for biomass m_p ($t = 0$) and water storage are determined by the date of emergence. *Figure 3.1* presents a diagram of the relationships defined in equation (3.1) between the processes of plant photosynthesis, respiration, growth, development, and water transfer within the integrated soil-plant-atmosphere system. The inputs to the model are the mean daily air temperature (T), vapor pressure deficit (d), hours of sunshine (S), and daily rainfall (R). The outputs are the dynamics of plant phytomass by organs (m_p), the final yield ($\hat{Y}$), and the dynamics of the components of soil water budget (W_i) to a depth of 1.5 m by 10 cm increments throughout the growing season.

Model parameters have been determined for a number of crops and, in particular, for spring wheat. *Figures 3.2* and *3.3* illustrate the results of model calculations; these indicate that the model performs satisfactorily when using data from experimental sites located in contrasting areas [Ershov in the dry steppe zone of the Soviet Union and Cherdyn in the humid forest zone – the northern limit of the spring wheat belt; cf. *Figure 1.4(a)*] and also for data on average regional yields over the Volga Basin area as a whole. Moreover, the calculations correctly accounted for the response of the spring wheat crop to extreme conditions such as the severe droughts of 1972 and 1975 in the steppe zone (Ershov), when crops actually failed, as well as the conditions of 1978, when a record harvest was produced. The large discrepancy between estimated and actual yields at Cherdyn, in 1975, is explained by the development of disease following lodging of the crop near the end of the growth period. Meteorological conditions in 1975 otherwise favored record yields.

Dynamic crop production models may be applied in two main ways to climatological studies:

(1) To estimate the effects on crop production, of current short-term (monthly or seasonal), medium-term (up to 2 or 3 years), and longer-term climatic variations (i.e., as observed in the instrumental climatic record).
(2) To forecast the response of yields to possible climatic changes.

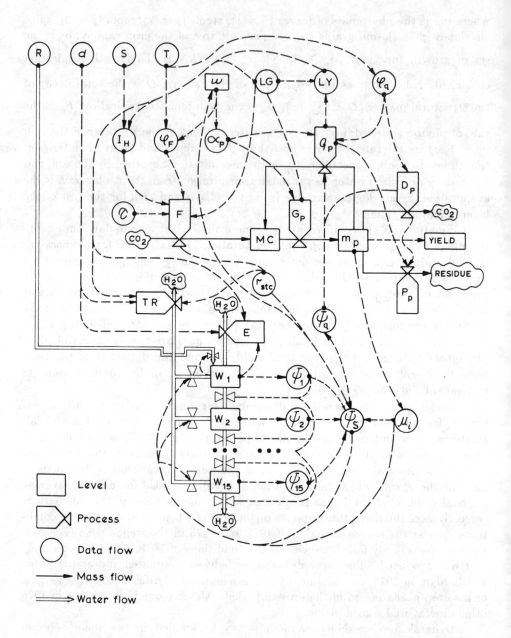

Figure 3.1. Flow diagram of the crop-weather model. Variables are defined as follows: ω, accumulated temperature; LG, green leaf area; LY, spike area; I_H, average intercepted photosynthetically active radiation (PAR) in the canopy; Ψ_q and φ_q, moisture and temperature coefficients of respiration; φ_F, thermal coefficient of photosynthesis; G_p, growth; MC, mass of carbohydrate pools; r_{stc} stomatal/cuticular resistance; Ψ_i, water potential of the ith layer of soil; $\bar{\Psi}_s$ mean weighted water potential; τ, day length; μ_i is the root density in layer i. Other symbols are explained in the text.

Spring wheat yields in the Cherdyn region

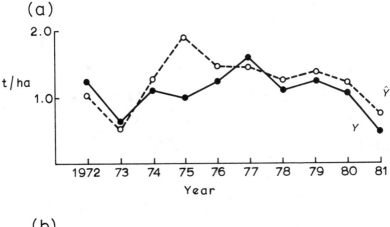

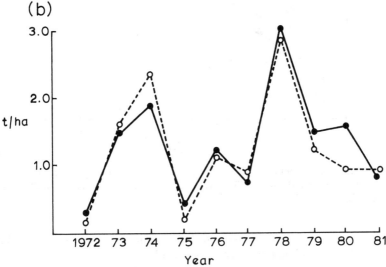

Figure 3.2. Dynamics of actual (Y) and calculated ($\hat{Y}$) spring wheat yields during the period 1972–1981, based on data from the Cherdyn (*a*) and Ershov (*b*) agrometeorological stations (latitudes 60.4°N and 51.3°N, respectively).

In order to solve both these problems, it is necessary to:

(1) Specify the estimated (X) and standard (X_0) climatic scenarios for the time interval to be studied.
(2) Calculate the yield levels corresponding to X and X_0, i.e., to derive $Y(X)$ and $Y(X_0)$.
(3) Estimate the difference between $Y(X)$ and $Y(X_0)$.

In fact, the problem of estimating the effects on yield of climatic variations differs from that of estimating yield responses to climatic change only in the way in which the climatic scenarios are developed. In the first case, the vector

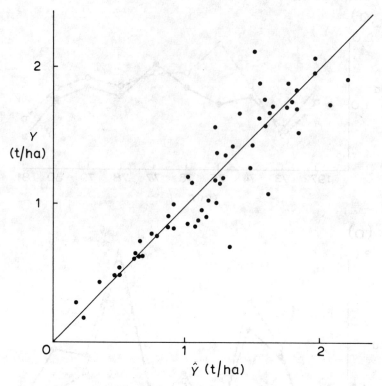

Figure 3.3. Relationship between actual (Y) and calculated ($\hat{Y}$) mean regional spring wheat yields in the Volga Basin (1975–1982) (correlation coefficient, r = 0.92; significant at the 1% level).

function X is constructed from observed data, while in the second case a forecast is used instead. Let us begin with the first problem.

3.3. Model Sensitivity Experiments for the Baseline Climate

Table 3.1 provides an example of the estimation of climate-induced yield variations during the period 1977–1981. Expected yields were calculated from the observed initial water-storage and meteorological data from the beginning of the water-storage period (taken as the autumn of the preceding year) up to 20 June, 20 July, and 20 August in each year. The experiments utilized data collected from climatological manuals. The average actual yields proved to be close to the calculated, $Y(X_0)$. The data of *Table 3.1* indicate rather high within-seasonal variability in the estimates of yield for observed climatic conditions in the forest zone of Cherdyn.

Thus, for example in 1977 estimates of final spring wheat yield based on early season conditions were nearly 50% higher than estimated mean long-term yields. Thereafter a gradual deterioration of moisture and thermal conditions

Table 3.1. Estimates of final spring wheat yields $Y(X)$ at Cherdyn based on observed climatic conditions over three different estimation periods during the years 1977–81 [expressed as percentages of mean estimates based on the long-term climate $Y(X_0)$].

Year	End dates of the estimation periods		
	20 June	20 July	20 August
1977	149	138	118
1978	82	69	74
1979	129	130	113
1980	129	135	121
1981	80	86	80

occurred, which in two months resulted in a 31% decrease in the estimate. In this example the expected yield may vary by between 6% and 31% from the expected mean due to fluctuations in climatic conditions.

Figure 3.4(a) illustrates the dynamics of similar estimates of crop responses to climatic conditions during the period 1951–1981. It can be seen that climate-related yield variations are quite significant. Crop-yield estimates for the area around Cherdyn range from 38% (1973) to 171% (1967) of the long-term mean (1.15 tonnes per ha: calculated as the average of estimated yields for each weather-year in the 1951–81 period. Note that the yield calculated for climatic values averaged during this 31 year period was 1.08 tonnes per ha.) Conditions for crop production were exceptionally favorable in 1967; the average temperature, total rainfall and average sunshine duration during the growing season were, respectively, 1 °C, 15% and 8% above the long-term mean levels.

These crop production models can also be used to evaluate the significance of fluctuations in specific elements of the climate. For example, *Figure 3.4* also shows estimates of the extent to which yields are affected by temperature and rainfall during the growing season as well as by the hydrothermal conditions during the preceding autumn, winter, and spring. These estimates were obtained by substituting observed values for the respective long-term averages in the "normal" scenario.

As might be expected, temperature-related yield variations can be fairly significant in the moderately cool forest zone. In particular years, yields varied by 25–35% from the long-term mean due to variations in the growing-season air temperatures [*Figure 3.4(d)*]. As a rule, given positive temperature anomalies, yield estimates were above the long-term mean, indicating the limited nature of thermal resources in the area. On the other hand, when an ample water supply was available in spring, as in Cherdyn, yield estimates based on the initial water storage [see *Figure 3.4(b)*] hardly differed at all from the long-term mean over the long period from 1951 to 1974. Nevertheless, it can be seen that cases where an initial water-storage deficiency had a significant effect on yields became more frequent in the late 1970s.

The approach described makes it possible to examine in greater detail the influence of specific components of the climate. Thus, for example, favorable

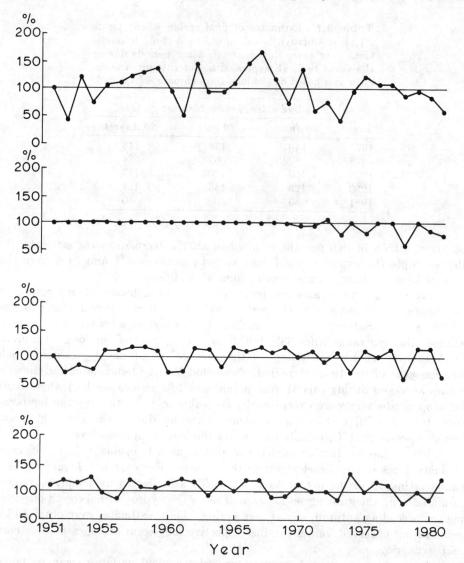

Figure 3.4. Dynamics of estimates of effects of climatic conditions on production of spring wheat in the forest zone (Cherdyn agrometeorological station): (a) combined influence of all modeled climatic variables; (b) influence of soil water storage at emergence date; (c) influence of rainfall during the growing season; (d) influence of mean air temperature during the growing season.

temperature conditions, which were estimated at 134% and 119% of the long-term mean, respectively, occurred in 1974 and 1981. In the same years, however, the moisture regimes were rather unfavorable, with estimates of the effects of rainfall on crop yields of only 68% and 46% of the mean, respectively. As a result, despite the favorable influence of temperature the overall estimates for the agricultural year were slightly below average in 1974, and significantly below average in 1981 (at 53% of mean levels); the year 1981, in particular, was one of the most unfavorable years on record in that area in terms of agrometeorology (*Figure 3.4*).

The examples presented above by no means exhaust the great diversity of possible approaches to this type of research using dynamic models. We will now turn our attention to the use of dynamic models for estimating the response of yield to possible climatic changes.

3.4. Scenario Results

The estimation of yield responses to longer-term climatic change is more complicated than the estimations for short-term climatic variations, since neither now nor in the foreseeable future will climate forecasts be detailed enough to serve, without any preliminary refinement, as an input data base for calculating crop yields. The most detailed available scenarios of climate forecasts involve, at best, mean monthly values of temperature, rainfall, and cloudiness (*see* Part I Section 3), and there is naturally considerable uncertainty associated with transforming monthly mean values into daily averages. Nevertheless, in calculating the daily values of climatic parameters, the basic mean monthly values were approximated by trigonometric polynomials to account for the annual cycle. However, because the crop-weather model is essentially non-linear, the resulting "smooth" input data led to significantly overestimated yield values. Therefore, the air temperature dynamics during the growing season, were expressed as:

$$T = \bar{T} + \sigma\delta(t) \tag{3.3}$$

where $\bar{T}$ is the smoothed mean daily temperature, σ is the standard deviation, and δ is a function of time (t) representing a random normal process with zero mathematical expectation and unit variance. A similar procedure was developed to "reconstruct" the seasonal pattern of daily rainfall amounts from the mean monthly values and available data on the number of days with rainfall of varying intensity.

To calculate the sunshine hours and vapor pressure deficit data that are usually absent from climate forecast scenarios, regression equations were developed relating these values to air temperature, respectively:

$$S = aT + b \tag{3.4}$$

and

$$d = cT + e \tag{3.5}$$

Where S is hours of sunshine (a proxy for solar radiation), d is the vapor pressure deficit, T is temperature and a, b, c and e are empirical, local coefficients. In addition, statistical relationships were derived that allowed changes in soil water storage to be calculated from the precipitation increment during the preceding cold season.

3.4.1. Experiments with synthetic scenarios of climatic change

These procedures were combined to give a technique for estimating the response of crop yield to any changes in climatic conditions within the model's range of validity. *Table 3.2* provides an example of normalized yield values calculated for nine different combinations of temperature and rainfall during the growing season in Cherdyn. All the calculations were made using long-term average values for the initial soil water storage. Under humid forest conditions, an increase in temperature can contribute to an increase in crop yields, and this is particularly noticeable when it occurs in conjunction with increased rainfall.

Table 3.2. Response of spring wheat yield (as percentages of the long-term mean) to variations in air temperature (ΔT) and rainfall (ΔR) during the growing season (Cherdyn, forest zone).

ΔR (mm)	ΔT (°C)		
	-1.0	0	$+1.0$
-20	93	97	99
0	95	100	103
$+20$	97	101	107

It should be noted that all the climatic scenarios used in calculating *Table 3.2* were developed using linear shifts in the "long-term average" scenario, X_0. Many other ways of modifying the X_0 scenario could of course be devised. It seems advisable to proceed in the future from a unitary long-term average scenario to an ensemble of annual representations $x_1, x_2, ..., x_n$ for the climatic variables during the period studied, with a unit disturbance L being similarly replaced by an ensemble of possible disturbances $l_1, l_2, ..., l_m$. Thus, for each element in *Table 3.2* we would obtain an array of calculated yield values $y(x_i, l_j)$ ($i = 1, 2, ..., n; j = 1, 2, ..., m$) that could then be averaged. Such an approach would certainly be more reliable, but its full-scale elaboration is associated with a number of technical problems, such as the lack of "daily" meteorological time series stored on machine-processable media.

3.4.2. Experiments with empirically derived scenarios of climatic change

Nevertheless, we have elaborated a simplified version of this technique for estimating possible changes in spring wheat yields over the Soviet Union under the assumption of a 0.5 °C global warming. We used a spatial and temporal expansion of the temperature and rainfall scenario suggested by Vinnikov and Groisman (1979) and Groisman (1981) and already used in Section 2 of this case study. Long-term average scenarios of climatic conditions, X_0, modified in a linear manner according to Vinnikov and Groisman's forecast, were developed from data contained in the climatic and agroclimatic reference books of 300 observation stations more or less uniformly covering the area of the Soviet grain belt. Without presenting a detailed account of these largely preliminary estimates (Sirotenko et al., 1984), we merely note that, under the scenario in question, a process of global warming would be accompanied by a substantial decrease of rainfall in the steppe and forest-steppe zones. Together with the increasing temperatures, this might aggravate the sort of drought phenomena that are typical of the spring wheat belt.

3.4.3. Experiments with a GCM-derived scenario of climatic change

Estimates of the expected changes in spring wheat productivity resulting from the GISS model-derived doubled CO_2 climatic scenario have been calculated for the region near Cherdyn. The performance of the GISS model in simulating present day, $1 \times CO_2$ climate has been assessed for this region in *Figure 3.5*. The GISS-modeled $1 \times CO_2$ equilibrium conditions for gridpoint 58°N, 50°E are compared with the climatic normals for the period 1931-60 at Kirov meteorological station, to the southwest of Cherdyn [cf. *Figure 1.4(a)*]. Observation data for the whole year were not available for the Cherdyn station.

As was noted in the Leningrad region (Subsection 2.5.1), estimates of mean monthly temperatures are quite close to observed values, although the GISS-derived values tend to overestimate the magnitude of the annual cycle [*Figure 3.5(a)*]. Estimates of monthly precipitation rate show a pronounced late summer/early autumn peak, in contrast to the spring/early summer peak actually observed [*Figure 3.5(b)*]. Furthermore, the simulated total annual precipitation is greater than the observed value at Kirov. The poor correspondence of the precipitation values is, again, akin to that reported for Leningrad.

Despite our reservations about the accuracy of the GISS model simulation of present-day climate, we have proceeded on the assumption that the estimated differences between $1 \times CO_2$ and $2 \times CO_2$ equilibrium conditions can be taken to represent the change between the observed present-day climate and a future $2 \times CO_2$ climate. Using the same procedure to construct the $2 \times CO_2$ scenario as that employed for the Leningrad region (Subsection 2.5.1), the *differences* between GISS-generated $1 \times CO_2$ and $2 \times CO_2$ temperature and precipitation values (*Figure 3.5*) have been added to the mean values for the growing season (May–September) at Cherdyn for the reference period 1951-80. The scenario implies an increase of 2.7°C in the average air temperature for the growing season, and a 50% increase in rainfall.

According to calculations with the crop-weather model, under these changes in the hydrothermal régime and at the current level of farming efficiency, a slight (3%) decrease in yields could be expected. The combined effect of these changes in hydrometeorological conditions and the direct effects on the crops of a doubling of the CO_2 concentration in the atmosphere (through enhanced intensity of photosynthesis – modeled using the semi-empirical "Rabinowitch–Chartier" approach: Rabinowitch, 1951; Chartier, 1970) would

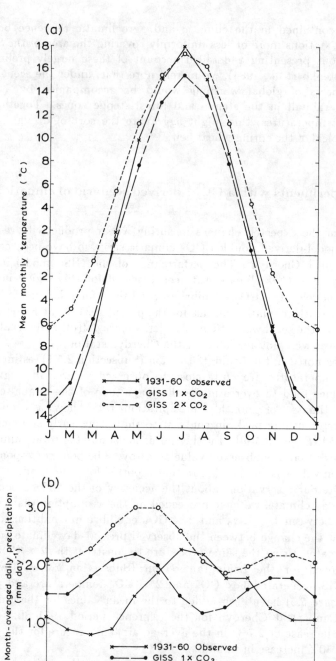

Figure 3.5. Comparison of observed values (1931–60) with GISS model-generated $1 \times CO_2$ and $2 \times CO_2$ equilibrium values at Kirov (58°39′N, 49°37′E) of: (*a*) mean monthly air temperature (°C); and (*b*) monthly mean precipitation rate (mm per day).

lead to a 17% increase in the calculated yields of the currently grown spring wheat cultivar.

3.4.4. Adaptation to climatic change – experiments with new varieties

It is clear from the above results that it is essential to consider the adaptation of agricultural production methods to changes in the climate. For example, if climatic warming led us to replace the existing wheat cultivar by two more heat-loving cultivars suited to accumulated temperatures respectively 50 degree-days and 100 degree-days higher than the original over the whole growing season, then the warming would directly enhance yields by 16% and 26%, or by 39% and 49%, respectively, if CO_2 effects were taken into account.

3.5. Conclusions

The theory of agroecosystem productivity promises further useful insights, in view of its successful application to the problem of predicting changes in agricultural production caused by the dynamics of environmental factors. But unless the dynamic nature of agriculture and its adaptive potential are taken into account it is impossible properly to judge the probable effects that would follow a climatic change of the type described here.

The results described above for the Cherdyn region indicate that yields of current spring wheat varieties exhibit a positive response both to above-average growing season temperatures and to above-average precipitation, when these anomalies lie within the present range of observed climatic fluctuations. If, however, the magnitude of climatic change is markedly greater than the present variability, as in the case of the GISS-derived $2 \times CO_2$ scenario, then the effects on yield of large increases in both temperature and precipitation is no longer multiplicative, or positive. The main reason for the resulting negative yield response is the accelerated development and early ripening of the crop under the much higher temperatures, although the high precipitation probably compensates for the negative effects of enhanced water loss through evapotranspiration.

Selection of a slower-maturing variety of wheat is an obvious farming response to such conditions, as indicated from the results for more heat-loving varieties. By prolonging the period of growth, the crop is able to exploit more fully the thermal, radiative and moisture régimes under a changed climate.

The adaptation of cultivars is not the only possible response to climatic change, however. Certain crops might be replaced by others, the system of agricultural management might be altered, or a wide range of land reclamation measures applied. In Section 4, some of these responses are considered further, by examining how information on climate-induced changes in crop productivity can be used in adapting agricultural planning systems to cope with changing climate.

SECTION 4

Planned Responses in Agricultural Management under a Changing Climate

4.1. Introduction

The purpose of this section is to study how agrometeorological information can be used in centralized planning for improving the adaptation of agriculture to variations in climate. This problem is particularly important in the USSR where, in spite of existing measures for yield stabilization, the fluctuations in annual crop production remain considerable. As a consequence there is, for example, an increasing tendency toward wastage of agricultural crops in favorable production years, incomplete utilization of the industrial capacities for processing agricultural products in bad years, and irregular employment of the agricultural population.

Two types of agrometeorological information can be identified that may be of use in the management process. Firstly, seasonal weather forecast data for the agricultural year would be of great practical use in determining the most appropriate strategies to be employed at particular locations (e.g., deciding on crop mixes, fertilizer applications, pest and disease control strategies, timing of activities, etc.). Such information is of special importance in deciding the most appropriate measures for stabilizing crop yields. Secondly, information on likely medium- to long-term changes in climate would indicate possible future conditions in agriculture that would affect the determination of optimal strategies for agricultural development. This information can be efficiently used at the state level, where large-scale measures are implemented for crop yield stabilization and economic adaptation to yield fluctuations. In the following, we shall consider only the second type of information, examining possible planning responses to a changing climate on the basis of model experiments. The models utilize agrometeorological information to calculate the likely productivity of crops under a changed climate. This information can then be used to assess the economic viability of alternative agricultural strategies, e.g., alterations in regional specialization in agriculture, adjustments to the structure of interregional exchanges of production, and optimizing the use of industrial processing capacity to minimize waste and maintain low production costs.

4.2. A Scheme for Model-Based, Long-Term Agricultural Planning

A recent initiative to assist the process of long-term agricultural planning in the Soviet Union has been the wider and more efficient application of economic-mathematical models (Kiselev, 1979). Used as part of an automated system of plan estimations, models should help to reduce the lead-time and labor intensity involved in developing a plan, while also improving its final efficiency. One such system of models is examined in this chapter, its function being to provide an improved understanding of the intersectoral and territorial structure of the agricultural production environment (the "agro-industrial complex" – AIC), the imperfections of which are currently detrimental to the national economy.

4.2.1. A scenario approach to agricultural planning

For a given situation where the climate is changing over the longer term, the model system aims to provide plan estimations that ensure the maximum volume of final production over the period of the climate change, as well as to determine the most efficient ways of adapting to the changed conditions. In the course of an estimation it is necessary to calculate the optimal allocation of limited investments in determining the mix of crops, the extent of irrigation, the level of supply of material and technical resources, the level of harvesting capability and the delivery of agricultural products to the consumer.

A stochastic model of the Dantzig–Madansky type (Dantzig and Madansky, 1961) might be an effective tool for solving such a problem: identifying the maximum expected value of the final product under circumstances of random production capacity and taking into account the cost of overcoming weather anomalies with a given amount of capital investment. However, there are considerable computational difficulties in realizing such large-scale tasks. Moreover, even experiments directed to smaller-scale tasks indicate that when using such a model it is difficult for planners to analyze and evaluate the reliability of the simulated decision; the process of decision management becomes complicated because an expert does not always have a clear understanding of the extent to which the initial conditions (e.g., the probabilistic characteristics of production capacity, the possible damage to crops, the expenditures required to avoid damage, etc.) influence the decision made.

In view of these difficulties, it is considered appropriate to use a *scenario approach*, where the optimal strategy for agricultural development is calculated assuming some realistic variant of weather conditions in a planned period. The applicability of the plan estimations to other likely conditions is then scrutinized and the decision can be modified informally. Thus, one may speak of an inexplicit approximate solution of a stochastic problem, where the loss of accuracy of the simulated decision is to a certain extent compensated by the wide use of expert assessments.

4.2.2. Defining the scales of analysis

In the course of national economic planning, a comparison is made between possible capital investment in agriculture and the likely increase in final production that could result. On this basis, the allocation of limited material and technical resources and the necessary investment of capital are decided. In each region the investments will be spent on improving the biological yields, increasing their stability in time, and ensuring that the crop product reaches the consumer. The increment of final product per unit of capital investment is a useful indicator of investment efficiency. The higher this index is, the more economically justified is the concentration of material, technical and financial resources in the region. Other social factors must, of course, also be taken into account requiring, for instance, the allocation of capital investment to regions with a low level of return on investment for the purposes of retaining rural population and increasing the regional agricultural production. However, these can be assessed on the basis of informal decisions without the use of models.

Two contiguous levels are therefore of importance in the system of long-term planning:

(1) The *regional level*, at which the conditions for agriculture (weather conditions included) are examined, a system of measures for increasing the level of yields and their stabilization is determined, and thus the function of return on increased capital investment can be evaluated; and
(2) The *state level*, at which investment efficiency in various regions is compared, possible expenditures on social development are evaluated, the proportions allocated between sectors are set up, and foreign trade relations are taken into account. On this basis, the long-term strategy of development of the whole agro-industrial complex is determined, as well as the distribution of capital investment and material and technical resources among branches and regions.

A general schema for estimating the long-term plan is presented in *Figure 4.1*. The main tools in these plan estimations are the regional and national models of the AIC. In order to take maximum account of yield fluctuations in plan estimations, and to reflect their influence on the formation of the intersectoral structure of the AIC, a dynamical formulation of the planning models is necessary.

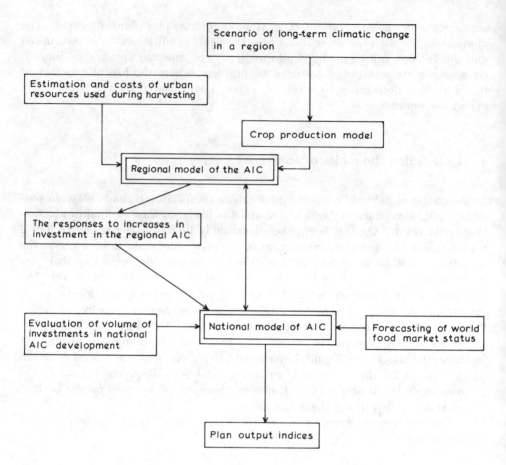

Figure 4.1. A system of models for long-term agricultural planning. AIC = agro-industrial complex.

4.2.3. Planning at the regional level

An important component at the regional scale is the estimation of future agricultural crop productivity. This can be conducted on the basis of extrapolation of yield trends (attributable to improved agrotechnology and management), taking into account any cyclical fluctuations. Under conditions of a changing climate, dynamic crop-weather models can be used that are capable of simulating levels of crop yield for a given set of environmental input variables that include the climate. A single model can usually only provide accurate yield estimates at the regional level, however, where it is possible to assume a more or less homogeneous climate. In this application (*Figure 4.1*), altered climatic conditions are input to a dynamic crop production model developed by Konijn (1984a). By superimposing crop responses to climatic change onto the underlying trends (described above), this model can provide estimates of regional crop yields over the simulation period.

One of the basic indices determined in the course of estimations is the level of development of the "harvest procurement system" (cf. Subsection 1.2). At harvest time, the demand for labor resources and transport facilities intensifies, and it becomes necessary to seek these additional resources from nearby urban areas. These are the reserves that can be mobilized to cope, to a certain extent, with peaks in crop production during favorable years. The amount of labor that might be required for the harvesting period, the cost of their involvement, their productivity, and the damage to the national economy incurred by relocating them from their normal places of employment all need to be evaluated. A similar estimation should be made for the provision of transport facilities. These data are used in a regional model when defining the quantity of harvest which can be brought in without waste and when estimating additional expenditures on harvesting.

Regional model estimations can be carried out as a series of variants representing an increasing volume of capital investments. For each variant, estimates of the volume of the final product and its cost per unit are obtained. In accordance with the considerations described above, we can expect each additional unit of production to be more costly. Thus, by using calculations for the series of variants, the form of a function can be obtained that describes production costs per rouble of final product over a range of different regional production volumes. A simpler relationship can be defined to estimate the value of regional production according to the limits on allocated investments. These functions, together with the estimates of regional production in physical terms for each variant, comprise the input information for the national AIC model.

4.2.4. Planning at the national level

In the national AIC model, individual regions can be presented at varying phases of development. Each is described by an input–output vector incorporating expenditures on investment into the particular phase of development. The creation of stocks of unperishable agricultural products and possible purchases of foodstuffs on the world market reduce somewhat the requirements for an obligatory minimum of production in the country as a whole. To ensure such a minimum production level, even in years of bad harvest, would require too great an investment to be realistic. Accordingly, a model block is required to estimate the food resources available on the world market, and to forecast the market prices.

From the joint national economic development plan, the limit on capital investment can also be introduced into the AIC model (*Figure 4.1*). This can be allocated to the AIC development in the plan period, permitting the coordination of development plans for individual regions, allowing a determination of the volume of interregional exchange, and ensuring the balancing of the whole system at the prescribed limit of capital investment.

4.2.5. The scope of this study

The experiments that are reported in this section are designed to show how this model system can be used in assessing the implications of climatic change for agricultural planning. As explained above, the effects of changes in climate over a territory as large as the USSR must, necessarily, be treated at a regional rather than a national scale. To illustrate this process, experiments were conducted for a single region, using the regional-level models that are shown in the top half of *Figure 4.1*. Descriptions of the models, the experiments and the results are given in the following subsections. Similar exercises could also be conducted for each of the other regions in the USSR, the results together providing the appropriate inputs for the national model (depicted in the bottom part of *Figure 4.1*). Such a comprehensive multi-regional scale analysis was outside the scope of this study, however, so no results were generated using the national model. Bearing in mind, though, that most of the major decisions concerning national agricultural development are made at the state level, the national model is an essential component in the overall assessment process. Therefore, Subsection 4.5 contains a description of the national model and shows how it can be linked to the regional models to provide useful estimations for centralized planning at the state level.

4.3. The Regional Agricultural Planning Model

The following model description is intended as a brief exposition of the most important model relationships. Numbers in the text refer to the formal equations that are included, along with a glossary of terms, in *Appendix 4.1*.

The model should not be considered either as the major or as the sole instrument available for long-term (more than 5 years) regional planning. In formulating the model, the objective was not to provide a balanced accounting of all factors affecting the rates and proportions of agricultural development, but only to ensure a planned balance of the elements dependent on weather fluctuations, and an appropriate allocation of capital investments. In particular, the model is used to identify, for a given level of biological production, the capacities for harvesting, transportation, storage and processing of agricultural crops that minimize losses in good production years and, at the same time, reduce the idle time of these capacities in poor years. Simultaneously, the economic efficiency of two measures for reducing losses can be compared: measures to stabilize the biological production and measures for provision of additional expenditures (relative to the average level) for harvesting and storage of surplus production in favorable years.

The problem under investigation involves identifying the maximum final production that can be achieved, bearing in mind the purchasing prices of the products and the limits on capital investment into improved production (relationships A4.1 and A4.18). A double restriction is placed on the types and quantities of products required in a region (relation A4.2). Firstly, while attempting to satisfy the regional demand for semi-perishable products (e.g. dairy and meat products), in order to avoid monoculture in a region it is important to maintain the production of less profitable products, for which an appropriate production infrastructure is required. Secondly, an upper limit to potential production is imposed by the level of agrotechnology in a region.

For determining the optimal level of trend yield capacity, a production function is used (in this case linear; in a general case it could be arbitrarily selected). It is assumed that four production factors determine the yield capacity: the capital stocks necessary for cultivating crops, mineral and organic fertilizers, capital investments directed towards improvements in soil fertility and efforts at reclamation to expand the cropped area, as well as the period within which scientific and technological progress can be integrated (e.g., breeding of new varieties, application of new means for plant protection, etc.). The actual yield capacity is determined by applying a correction factor ξ_i^t arbitrarily selected or determined in a particular scenario (relation A4.3). In the linear form of the production function, upper boundaries of production factors are set approximately at the limits of each (relation A4.5).

The annual increment of durable resources (i.e., capital stocks such as harvesting machinery, transport facilities, storage capacities and processing capacities) is determined taking into consideration the annual depreciation of funds at the beginning of the plan period in terms of a set coefficient, as well as the required funds in the current year (relations A4.4, A4.8, A4.10 and A4.12).

The extent of the cropped area is quite variable. However, in order to avoid abrupt changes in area from year to year due to fluctuations in yield in previous years, restrictions are set on the annual changes of area under the same crop (a maximum permitted change of approximately 5% – relation A4.7). The relation describing gross production (A4.6) is non-linear, but can be linearized by predetermining discrete values of the production capacity according to appropriate levels of resource supply. Relations A4.9, A4.11 and A4.13 determine the value of production that can be actually harvested as a function of the level of development of the production system calculated by the model. (As a simplification, urban resources usually involved in harvesting are not accounted for in the model.) Relation A4.15 defines the volume production that cannot be harvested in good years. The shares of processed and of fresh produce required for consumers are shown in A4.14.

Production from cattle-breeding is related only to the available feed as feed input per unit of production. Concentrated feeds can be imported from other regions, although limits on the import value can also be set, or penalty functions imposed for overdelivery of feed. At the same time limits are set on the composition of feed in order to avoid the unnecessary reliance on a single crop (relations A4.16 and A4.17).

Scenario analyses are carried out by changing the following control parameters:

(1) The limit on capital investments (K), which will determine the sequence of balanced measures for increasing the volume of the final product.
(2) Lower and upper limits on production in year t of crop products (i) and cattle products (j) in the region $(\underline{P}_i^t, \bar{P}_i^t, \underline{P}_j^t, \bar{P}_j^t)$, which would partially

formalize some of the socially motivated decisions in planning agrotechnical factors, such as the level of domestic consumption.

(3) Purchase prices for the products (C_i^t, C_j^t), reflecting the social use of products, can determine which products a region should specialize in, allowing surplus products to be taken out of the region.

Consideration of all possible combinations of control parameters would take too much time, but it is particularly important to select those parameters that characterize the likely future socio-economic development of a region. In this case the estimations are limited to 4 to 6 variants, broadly reflecting the perspectives of agricultural development.

It is possible to generalize the results of alternative estimations using the regional model in order to make them applicable at the national level. We can use a scheme in which the results of estimations on each variant are described by a vector B_s (where s is the index number of the variant), the components of which cover all the products (except domestic consumption). A plus ("+") symbol indicates the situation where regional production *exceeds* domestic consumption in accordance with the concept of regional development (a_i^t), and a minus ("−") symbol indicates the reverse situation of *deficit* $(-a_i^t)$ requiring an increment in resource allocation (ΔZ_k^t); to achieve the required volume of production and level of capital investment (Z_k^t, K^t). The results of the estimations by the regional model can thus be represented by a set of vectors:

$$B_s = \left\{ + a_i^t, + a_j^t, - a_i^t, - a_j^t, \Delta Z_k^t, Z_k^t, K^t \right\}_s \quad (4.1)$$

which are submitted at the state level of planning for final decision making.

4.4. Use of the Regional Model in Planning Responses to Climatic Change

The purpose of this section is not to forecast future changes in the structure of agricultural production in response to climatic change, but rather to show, using a specific example, how the model system can be practically applied in the forecasting and planning of such changes. For illustrative purposes, we present calculations showing the possible changes in production structure resulting from climatic change in the Central Region of the European part of the Soviet Union (Moscow region).

4.4.1. Characteristics of the study region

The Central Region of the European USSR, centred on Moscow, is located immediately to the south of the subarctic zone [see Figure 1.4(a)], but it exhibits many characteristics that are similar to the subarctic regions. The region has a moderate continental climate with cold winters (1931–60 mean January temperature $-10.3\,°C$) and warm summers (mean July temperature $17.8\,°C$). Despite a much greater absolute range of temperatures (between $-42\,°C$ and $+37\,°C$), the frost-free period of 130–140 days [see *Figure 1.4(d)*] and a mean annual precipitation of 575 mm generally permit a wide range of agricultural activities. Soils of a sward-podzolic type predominate in the region [*Figure 1.4(c)*].

The following crops are cultivated in the area: winter wheat, barley, oats, corn (maize) for silage, hay, potatoes and other vegetables. Spring wheat, winter rye, buckwheat, and flax are also grown, but their share in the total is fairly small. The region is industrially developed and its population density is relatively high. Agriculture is therefore accorded a correspondingly high level of resource supplies. Yields here are better, but product costs are higher, than in neighboring regions.

Table 4.1. Characteristics of the seven main crops grown in the Central region of the European USSR.

Crop	Yield trend (quintals/ha)[a,b]	Maximum yield fluctuation (quintals/ha)[c]	Cost based on trend data (roubles/quintal)	Share of total cropped area (%)
Winter wheat	$y_1 = 20.3 + 0.067t$	16.2	15.0	14.2
Barley	$y_2 = 17 + 0.053t$	16.0	13.0	12.3
Oats	$y_3 = 15.5 + 0.063t$	13.7	13.0	4.7
Corn (silage)	$y_4 = 240 + 0.13t$	52.0	2.8	12.3
Hay	$y_5 = 44 + 0.21t$	25.8	4.5	42.6
Potatoes	$y_6 = 132 + 0.53t$	84.0	15.0	11.3
Vegetables	$y_7 = 328 + 0.54t$	127.0	9.2	2.3

[a] 1 quintal = 0.1 metric tons.
[b] t is time (in years after 1980).
[c] Difference between the highest and lowest recorded yields in the 10-year period 1971–80.

Table 4.1 presents some basic characteristics of the main crops. Because of the high population density the region is not self-sufficient in agricultural products, and considerable quantities of food products and concentrated cattle feed are imported from other regions. It is essentially impossible to expand the area under cultivation so that agricultural production can only be increased through more intensive farming. Achieving the latter will depend not so much on a growth in resource supplies as on strengthened economic incentives, enhanced production efficiency, and the utilization of high-yielding crops.

4.4.2. Climate-warming scenarios and simulated changes in crop yield

Three synthetic scenarios of climatic change were used in this study. The scenarios assume, respectively, that there will be a rise in the average annual temperature in this region of 0.5, 1.0 and $1.5\,°C$ over the period up to 1995. This

Table 4.2. Projected yields of seven major crops, based on trend extrapolation with present climate, and on climatic scenarios 2 and 3.

Crop	Trend with present climate (q/ha)	Scenario 2, $\Delta T = 1°C$			Scenario 3, $\Delta T = 1.5°C$		
		AV^a (q/ha)	Changes from trend		AV^a (q/ha)	Changes from trend	
			(q/ha)	(%)		(q/ha)	(%)
Winter wheat	21.0	26.9	5.9	28.14	26.4	5.4	25.5
Barley	17.7	16.9	−0.8	−4.90	16.4	−1.3	−7.8
Oats	16.4	15.6	−0.8	−4.90	15.2	−1.2	−7.8
Corn for silage	242.0	256.9	14.9	6.15	261.2	19.8	7.9
Hay	47.0	45.1	−1.9	−3.93	44.0	−3.0	−6.4
Potatoes	140.0	136.2	−3.8	−2.74	133.7	−6.3	−4.5
Vegetables	336.0	325.5	−10.5	−3.13	319.2	−16.8	−5.0

[a] AV = absolute value.

information was used to adjust the climatological input data required to run the crop production model (CPM). A description of the CPM is included in *Appendix 4.2*.

For this study, climatological data for Moscow were assumed to be broadly representative of conditions in the Central Region as a whole. In order to run the model, data for mean air temperatures, relative humidity, windspeed, global radiation and precipitation are required at 10-day intervals throughout the year. Since only mean monthly data were available (1931–60 means, extracted from Müller, 1982), these were processed to obtain ten-day averages. Model runs were made assuming no nutrient constraints, and although water was allowed to be limiting, in practice the water limitation was of only minor importance, due to the use of average data, and because of the relatively high precipitation totals in the region.

The model was first run for each of the seven crops using the original climatic data to establish "baseline" yields. Subsequently, the 10-day mean temperature data for the whole year were perturbed according to the three scenarios (i.e., increased by +0.5, +1.0 and +1.5 °C, respectively). Model runs were then conducted for each of the scenarios, producing yield estimates that could be expressed as percentage changes relative to the baseline. These changes in "potential" yield were assumed to be representative of changes in actual yields that could be expected for the same climatic scenarios.

The climatic changes are simulated to occur by the year 1995, but in the absence of climatic change, actual yields for that year have been estimated by extrapolating the linear trends shown in *Table 4.1*. These "baseline" trend yields are presented in the left-hand column of *Table 4.2*. The effects on yields of Scenarios 2 and 3 are shown as estimates relative to the trend yields in the remaining columns of *Table 4.2*. The results of Scenario 1 (a 0.5 °C temperature increase) were intermediate between those of the baseline and Scenario 2, but are not included in *Table 4.2*.

For most crops a reduction in yield is estimated, though not for all. The increase in temperatures led to an increased evaporative demand by the atmosphere and consequently increased the water stress towards the end of the growing period. Winter wheat and maize show increased yields, the positive effect of enhanced temperatures outweighing the negative water stress effect. Winter wheat, for example, would begin growing earlier under a warmer climate and therefore ripens early enough to escape the dry period. The other crops all showed yields lower than the baseline, although if the sowing times had been adjusted, the estimated effects might have been less negative.

4.4.3. Changes in crop allocation and expenditures for the climatic scenarios

In this subsection, simulated changes in crop yields for the climatic scenarios are input to the regional model to estimate appropriate planned changes in crop allocation and expenditure in the region. The region studied is an importer of all types of agricultural products including concentrated feedstuffs. The existing production structure has been developed so as to meet the overriding demand of the urban population for milk and the demand by cattle-breeders for bulky feedstuffs that are expensive to transport from elsewhere.

Variable soil quality, the widespread need for irrigation, and the risk of low yields in climatically unfavorable years have been important factors in determining the present-day mix of crops, as well as influencing the particular skills developed by the local population and the structure of capital stock. This is a major explanation of why both profitable and less-profitable crops are grown in the region.

In the course of calculations using the regional model, two main problems have been solved:

(1) Estimation of changes in the optimal crop mix for each climatic scenario.
(2) Determination of appropriate functions describing the expenditures required to increase production in the region in favorable years by improving harvesting capacity.

Some of the information required for carrying out model forecasts was lacking in this application. For instance, in the climatic change scenarios there are no realistic assumptions of possible precipitation changes in the region under consideration, all retrospective statistical data refer only to actual harvested yields but not to biological yields, there are no statistical data on the quantities of machinery and labor involved in harvesting from urban areas. To fill in these gaps, the calculations were carried out using expert evaluations of the missing data and some simplifying assumptions. For example, the capacity of the "harvest procurement system" at the beginning of the plan period was determined by the value of maximum harvested yield in the pre-plan period. The increase in capacity of this system presumes some increment in the provision of harvesting machinery, transport facilities and storage within the existing structure of agriculture (seasonal factories for processing of the perishable agricultural raw produce are actually missing in the zone). The cost of capital stocks in the system, necessary for the harvesting of one tonne of a crop, is determined by their actual cost divided by the value of maximum harvested yield.

Four main assumptions were made in preparing the forecast calculations. First, production of goods for domestic consumption will rise in proportion to population growth and will be the same in all scenario variants. Second, changes in crop mix can only take place on land and soil of the same type (for example, land under irrigation where there is "competition" for the soil between corn for silage, potatoes and vegetables) or within groups of similar crops (for example, the cereal crops). Third, the area of perennial grassland for hay production remains constant; and fourth, if the production within a region is not sufficient to meet the demands of the population, then imports can be increased, but for considerably higher prices.

The calculations were conducted for 10 years. The task is then formulated as minimizing the expenditure required to satisfy a fixed share of the demand for agricultural products. Two-sided restrictions, designed to exclude monocultural farming, are imposed on the area of individual crops (*see* Subsection 4.3, above).

In order to present the results from the crop production model in a form suitable for input to the regional allocation model, the following procedure was adopted. Values of biological yield were available for each crop over a sequence of ten years, representing the baseline, pre-plan period (for example, winter wheat yields are shown in *Table 4.3*). Harvested yields are slightly lower than biological yields, so a correction was necessary to replicate the actual fluctuations in yields (required as inputs to the allocation model). The average harvested yield capacity for the plan (i.e., scenario) period was forecasted by trend (*Table 4.1*). Scenario estimates from the crop production model were of percentage changes in ten-year averaged yields, and it was assumed that these average changes could be applied to each of the ten annual baseline yields. Thus, for Scenario 2 ($\Delta T = +1\,°C$), the actual harvested yield capacity of winter wheat relative to trend (Column 1) for each of the ten years was increased by 28% (*Table 4.2*).

Table 4.3. Biological yield fluctuations of winter wheat for the baseline scenario without climate change (quintals/ha).

Years	1	2	3	4	5	6	7	8	9	10
Biological yield	33	25	17	22	34	22	39	30	21	19

Runs with the regional allocation model were made for four scenario variants comprising yield estimates for the baseline scenario, and for the three climatic change scenarios. The results of calculations for the baseline (trend) yields are shown in *Table 4.4*. This variant is the benchmark with which subsequent results are compared.

The costs were calculated from the expenditure per hectare under crops together with the resource supply, which in turn was defined on the basis of trends. The mix of crops was chosen primarily to meet the human demand for

Table 4.4. The optimal model solution with trend-based crop yields (baseline scenario).

Crop	Yield (q/ha)	Area (10^3 ha)	Volume of production (10^3 quintals)			Unit cost of imported feed (roubles/q)
			Total	Feed-stuffs	Human consumption & processing	
Winter wheat	21.0	150	3150	1700	1450	15.0
Barley	17.7	130	2301	1200	1100	13.0
Oats	16.4	50	820	500	320	13.0
Corn for silage	242.0	130	31460	5033	–	2.8
Hay	47.0	450	21150	7825	–	4.5
Potatoes	140.0	120	16800	224	15400	15.0
Vegetables	336.0	25	8400	–	8400	9.2

Table 4.5. The development costs of fulfilling increased grain production targets (baseline scenario).

Realizable grain production (m quintals)	Expenditures to fulfill production target (m roubles)	Production increment (m quintals)	Expenditures on additional 1 quintal of grain (roubles/q)	Average cost of the whole grain crop (roubles/q)
Up to 64.3	–	–	–	13.9
64.3–66.3	30	2.0	15	14
66.3–67.4	30	1.1	27.3	14.4
67.4–68.1	30	0.75	40	14.7

vegetable products. All the remaining agricultural land was then allocated to feedstuffs. Thus, the feedstuff volume and the cost of the product – two indicators that change rather frequently – may be regarded as measures of the efficiency of a given run or variant. In the baseline variant, a total of 16482 thousand quintals of feedstuffs are produced, of which winter wheat, barley, and oats together account for 3400 thousand quintals. Some other feedstuff resources taken into account in the models have fixed values in all the variants: they include imports of concentrated feed from other regions, pasture, straw, nutritive wastes, etc. The price of 1 quintal of imported feed in the present variant is set at 15 roubles.

Since the expenditures on basic production are already accounted for, additional expenditures on improved production for the whole period (capital investments plus operational costs related to harvesting of crops) will determine the costs of obtaining additional produce. The costs of obtaining three levels of increased production are presented in *Table 4.5*.

Evidently, the expenditures on improving grain production through more complete harvesting of the biological yield *increase rapidly*, representing one possible functional form of capital investment efficiency. This function will be used for choosing optimal allocations of capital investments.

A second variant studied corresponds to the second climatic change scenario (a 1 °C rise in mean annual temperature by 1995). For each type of crop an obligatory production volume was specified, sufficient to meet the demands of the population. Substitution was allowed only within separate feed groups (concentrates and succulents). From the point of view of feed value, wheat, barley, and oats are essentially interchangeable. On the other hand, if only the *yield* of feedstuffs is considered then barley and oats would be completely replaced by wheat. However, since wheat is less resistant to unfavorable weather impacts, demands greater expenditure of labor, and has less nutritional value for cattle, some agricultural units prefer to grow barley and oats in spite of their lower yield.

Under the new climatic conditions, the relative "efficiencies" of various crop mixes will change. But because of management inertia, crop mixes will in fact not change to a great extent, which is why we introduce lower limits for the areas of the "less-efficient" crops (barley, oats), which we assume to be equal to 50% of the areas formerly given over to these crops. The optimal crop mixes calculated for these conditions are presented in *Table 4.6*. We can see that the areas under the crops with the highest yields have not increased up to the maximum possible. Feed volume has grown to 17745 thousand quintals, of which the concentrated feeds – winter wheat, barley, and oats – now account for 4605 thousand quintals. The cost of 1 quintal of feed unit has now fallen to 11.7 roubles. Clearly, therefore, the 1 °C rise in temperature has positive consequences for agriculture.

However, the existing level of development of the yield procurement system (for the same types of fluctuation in yields as shown in *Table 4.3*, but a higher level of average yields) permits the procurement of only 84% of the total biological yield over ten years. The effects of developing the harvest procurement system to higher levels of efficiency but for the same expenditures (at the expense of investments into other yield-improving activities) are shown in *Table 4.7*.

It can be seen that the returns on capital investment into developing the system of yield procurement are favorable for all levels of development considered. This is explained by the fact that, along with the increase in harvesting

Table 4.6. The optimal model solution calculated on the basis of Scenario 2 ($\Delta T = +1$ °C).

| Crop | Yield (q/ha) | Area (10^3 ha) | Volume of production (10^3 quintals) | | Human consumption & processing | Unit cost of imported feed (roubles/q) |
			Total	Feedstuffs		
Winter wheat	26.9	193.5	5205	3755	1450	11.70
Barley	16.9	100.0	1700	600	1100	13.60
Oats	15.6	36.5	570	250	320	13.66
Corn for silage	256.9	136.2	35 195	5631	–	2.64
Hay	45.1	450.0	20 295	7509	–	4.70
Potato	136.2	113.0	15 400	–	15 400	15.40
Vegetables	325.5	25.8	8400	–	8400	9.50

Table 4.7. The development costs of fulfilling increased grain production targets (Scenario 2).

Realizable grain production (m quintals)	Expenditures to fulfill production target (m roubles)	Production increment (m quintals)	Expenditures on additional 1 quintal of grain (roubles/q)	Average cost of the whole grain crop (roubles/q)
Up to 73.4	–	–	–	12.2
73.4–79.8	30	6.4	4.6	11.6
79.8–82	30	2.2	13.6	11.63
82–83.2	30	1.2	25.0	11.82

Table 4.8. The optimal model solution calculated on the basis of Scenario 3 ($\Delta T = +1.5\,°C$)

Crop	Yield (q/ha)	Area (10^3 ha)	Volume of Production (10^3 quintals)		
			Total	Feedstuffs	Human consumption and processing
Winter wheat	26.4	188.8	4984	3534	1450
Barley	16.4	103.7	1700	600	1100
Oats	15.2	37.5	570	250	320
Corn for silage	261.2	133.5	34870	5579	–
Hay	44.0	450.0	19800	7326	–
Potatoes	133.7	115.2	15400	–	15400
Vegetables	319.2	26.3	8400	–	8400

capacity, the number of years when the system's capacity is fully utilized is also increasing (due to the higher production under a warmer climate).

Calculations using the third climatic change scenario were carried out using similar premises. In this case the projected rise in temperature is 1.5 °C, and leads to a drop in the yields, relative to the previous scenario, of all crops except corn. The results for this scenario are given in *Table 4.8*.

It can be seen that, after all other demands have been met, the optimal area under feed crops has fallen compared to Scenario 2. Feed production has been reduced from 17745 to 17289 thousand quintals of feed units, and the average cost is now 13.56 (relative to 11.7) roubles per quintal (not shown in *Table 4.8*). Both these indicators are improvements on the baseline scenario, however.

On the whole, therefore, the hypothetical changes brought about by each of the "warming" scenarios would appear to be favorable for agriculture. The introduction of new crops that favor warmer weather would ensure more complete utilization of the changed climatic potential, but we could not realistically include them in the model due to the lack of even approximate data. In the following section we show how the results from regional allocation models can be used in models of national-level agricultural planning.

4.5. Centralized Planning at the State Level

Calculations made using separate regional models provide useful information, but decisions on the rates of agricultural development in each region, the magnitude of the measures to be taken, and the funding arrangements for them can only be made at the state level after a thorough comparison of the merits of the different measures proposed for each region. It is also necessary to take into account measures such as the creation of reserves of agricultural products and the potential of international trade. Therefore a national-level model must include the possibility of increasing agricultural reserves in good years when there is a surplus and drawing on these reserves in bad years.

Foreign agricultural trade needs to be balanced so that the value of imports must equal that of exports. In particularly bad years, currency subsidies may be used. These factors have been incorporated in several national-scale models, for example, in the Soviet Model of Agricultural Production (SOVAM), developed at the International Institute for Applied Systems Analysis (Iakimets and Kiselev, 1985; Iakimets and Lebedev, 1985).

The yield and cost of crops in particular regions are determined using regional models. The results can be represented as regional development variants, described either with the input–output vector B_s [(equation (4.1)] or with a function defining the inputs for an increase of final product in the region: $Y_{ri}^v (K_r^2)$. Further calculations using a special long-term planning model at the state level can then be performed.

The economic formulation of the problem may be described as follows. It is necessary to determine a structure of agricultural products in each region, volumes of interregional product flows, the allocation of investments, volumes of reserves and a foreign trade structure, so as to maximize the supplies of food while minimizing production and transportation costs. This is of course a very general formulation, reflecting the targets of state socioeconomic policy, but more disaggregated versions can be developed where necessary.

It is assumed that all production reserves and adaptation measures are investigated at the regional level and that they are adequately reflected in cost functions and also functions connecting the possible production levels with the capital investment required. If other types of constraints also appear in the regional models, these can also be included in the national model. A description of a formalized version of a national-level model is included in *Appendix 4.3*.

Neither the regional nor national models described in this section reflect the entire process of long-term planning; but both illustrate particular, formalized aspects of the process. Runs may be conducted according to different development variants or different scenarios. Results obtained at the two different levels can be integrated and then used together with expert estimations in developing the final variant of the plan.

4.6. Conclusions

If we consider a region as an open economic system, then any decisions concerning the structure of agricultural production under new climatic conditions need to take into account possible changes in interregional product flows. One possible means of adaptation is to reduce the domestic production of some crops and increase imports from other regions. In the numerical examples described above it was not possible to draw any final conclusions concerning production structure in the regions. The calculated changes may be regarded as one possible solution, permitting the achievement of the same targets under the new, more favorable climatic conditions, as under the old ones. But it may well be that this approach will need modification because the climatic changes in neighboring regions may prove to be more favorable agriculturally so that it becomes more profitable, for example, to import barley and oats and to increase the domestic production of winter wheat. In this way, the planning and the investigation of the problem may well turn out to be extremely complex.

Note that our estimates of the impacts on yields of hypothetical warming or cooling, whichever model we use (whether it be the crop production model or another such as that used in Section 3 of this case study), assumes *stable* weather conditions. But any long-term rise (or fall) in temperature will be accompanied by shorter-term weather anomalies, which will continue even after the average climate becomes stable. These anomalies certainly influence the yield, but they are not taken into account in the present application of the crop production model. Forecasts made using such models are likely to be optimistic because any enhancement of the amplitude of fluctuations would probably lead to a lowering of average yield. That is why in the regional model a small decrease of the area under barley and oats is specified under the new conditions. All of these factors indicate a need for models to include a facility for estimating yields stochastically (involving estimates of probabilities and risks of likely crop damage, and of likely compensations for crop damage, etc.). Only this integrated approach can allow us to evaluate fully the rational inputs for maintaining or increasing production in a region and to determine their optimal allocation under conditions of a changing climate.

SECTION 5

Results and Policy Implications

5.1. Introduction

This section summarizes the results of three investigations conducted within the framework of the Soviet case study for analyzing the impacts on and necessary adjustments to agriculture in the Leningrad, Cherdyn and Central region of the European territory of the USSR, in response to possible climatic changes. In these investigations there were two central problems:

(1) The analysis of impacts of climatic changes on the yield of agricultural crops and the improvement of the methods of such analysis.
(2) The elaboration of a method to aid decision makers in planning the future development of agriculture, involving the use of economic models that can incorporate environmental factors including information on the likely impacts of climatic changes.

The first problem was considered in two sections that describe experiments conducted on two different crops in contrasting areas within the subarctic farming zone of the European USSR. In the Leningrad region changes of winter rye yield were estimated in response to different climatic change scenarios (Section 2). In the drier Cherdyn region in the eastern part of the zone, estimates were made of the responses of spring wheat yield to climatic variability (Section 3).

The second problem was considered in an illustration of the use of a regional optimization model for the Central region of the European USSR, to demonstrate how changing climate can be accommodated in planning strategies to stabilize regional production or to minimize expenditures (Section 4).

5.2. Summary of Results

Each experiment in this case study has been conducted according to specific circumstances and assumptions (including, for example, model type, location, crop type, climatic scenarios, technology, soils, etc.). As such, it is not possible to make direct comparisons between the results from different investigations. However, some idea of the broad tendencies of responses can be presented. Before attempting to summarize the impacts, however, it is important to review the climatic scenarios and to show how they differ.

5.2.1. Review of the scenarios

The scenarios of temperature and precipitation change are depicted in *Figure 5.1(a)* and *5.1(b)*, respectively. A common "baseline" climate was adopted in the Leningrad and Cherdyn regions, representing the 1951–80 period. For the Central region of the European USSR, the baseline period is shorter (represented by climatic data from the 1960s and 1970s). The vertical bars on the diagrams indicate that the GISS general circulation model $2 \times CO_2$ scenario was analyzed in the Leningrad and Cherdyn regions as a step function "shock" scenario, assuming no gradual transition between the present equilibrium climate and the $2 \times CO_2$ climate. Sensitivity analyses conducted in the Cherdyn region, and the synthetic scenarios of changes in temperature alone of 0.5 °C, 1.0 °C and 1.5 °C above the baseline in the Central region, were of the same type, though in the latter case a technology trend was assumed between 1980 and 1995.

The remainder of the experiments were of a slow-change, transient nature (Section 2). They include scenarios of temperature and precipitation changes up to the year 1995, based on the Vinnikov and Groisman empirical approach for the Leningrad region (sloping lines in *Figure 5.1*), an extension of the Vinnikov and Groisman results to the year 2005, and then a projection of the scenario through to the year 2050 by interpolating between these estimates for 2005 and the GISS $2 \times CO_2$ climate (assumed to occur in 2050). However, the impacts of this scenario have been analyzed only up to the year 2035. All of the experiments with scenarios of transient climatic changes were conducted assuming a concurrent change in agrotechnology (based on recent or projected trends).

Once developed, the scenarios provided the inputs for the impact models, as illustrated in *Figure 1.6* in the introductory section to this report.

5.2.2. Short-term sensitivity of crop yields to climate

Investigations of the impacts on crop yield of climatic variations during the baseline periods have shown that above-average temperatures during the growing season are favorable both for winter rye in the Leningrad region and for spring wheat in the Cherdyn region. Fluctuations in precipitation, however, completely alter this assessment. For example, in the drier Cherdyn area, even given beneficial temperature conditions such as occurred in the years 1974 and 1981,

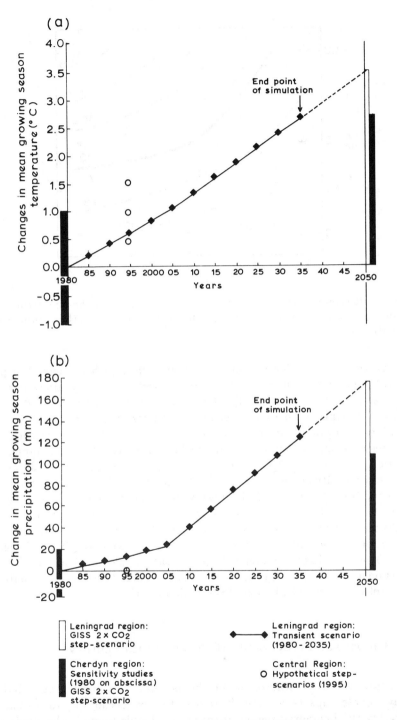

Figure 5.1. Scenarios of (*a*) temperature and (*b*) precipitation changes employed in the USSR case study.

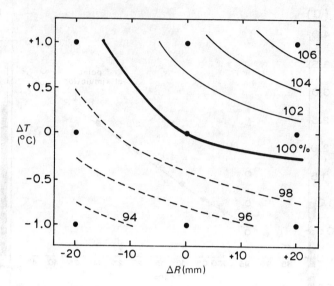

Figure 5.2. Response of spring wheat yield (as percentages of the long-term mean) to variations in air temperature (ΔT) and precipitation (ΔR) during the growing season in the Cherdyn region.

below-average precipitation can seriously reduce spring wheat yields (for precipitation 68% and 46% of the mean, respectively, yields in both years were approximately halved relative to normal). The dual role of temperature and precipitation has been explored further in sensitivity studies (*Figure 5.2*). In contrast, in the Leningrad region a *surplus* of precipitation during the growing period has a strong negative impact on winter rye yields (as reflected, for example, by the correlation coefficient of -0.5 between final yield and precipitation in June).

The sensitivity of these two crops has been tested in a more extreme experiment, using the GISS model-derived estimates to simulate the likely impacts of a sudden climatic "shock" on present-day cultivars. For spring wheat, a slight (3%) decrease in yields is estimated, the benefits of increased temperatures and precipitation being outweighed by a significant reduction in the vegetative period. The considerable increase in precipitation calculated for the Leningrad region under the GISS scenario is the main explanation for a simulated 13% decrease of winter rye yields there (Subsection 2.7.1).

5.2.3. Climate and crop yields at the end of the century

Results from the three studies enable the assembly of some tentative indicators of likely climate impacts on agriculture over the two decades up to the end of the century. In the Leningrad region, it seems reasonable to apply the Vinnikov and Groisman transient scenario based on historical climatic data over the past century (and assuming a linear increase in fertilizer applications) for a projection to

Table 5.1. Summary estimates of climate impacts on crop yields towards the end of the century (based on Sections 2, 3 and 4 of this report).

	Leningrad		Cherdyn		Central region
Technology:	Fertilizer trend (interpolated)		Fixed (1980)		Trend (extrapolated)
Year:	1990	2000	Arbitrary Scenarios	Arbitrary	1995
Temperature change[a]	+0.4 °C	+0.8 °C	+1 °C	+1 °C	+1 °C
Precipitation change[a]	+8 mm	+17 mm	0	+20 mm	0
			Changes in crop yield		
Spring wheat	–	–	+3%	+7%	–
Winter rye	+4%	+5%	–	–	–
Winter wheat	–	–	–	–	+28%
Barley	–	–	–	–	–5%
Corn for silage	–	–	–	–	+6%
Potatoes	–	–	–	–	–3%

[a] Temperature and precipitation changes represent growing season (Leningrad and Cherdyn) and annual (Central region) values relative to the baseline climate.

the end of the century (*see Table 5.1*). Small increases in winter rye yield, relative to trend, are anticipated for this scenario.

The increases in temperature and precipitation that might be expected to occur over a similar period in the Cherdyn region would probably fall within the range of values depicted in *Figure 5.2* (hypothetical changes of +1 °C temperature with both normal and +20 mm precipitation are also included in *Table 5.1* for comparative purposes). Both scenarios are favorable for present-day spring wheat cultivars, even *without* changes in agrotechnology.

In the Central region, the effects of warming alone up to the year 1995 would be beneficial for heat-requiring crops such as corn or winter wheat, but unfavorable for those crops adapted to moister, cooler conditions, such as barley and potatoes. Results for a 1 °C mean annual temperature change up to 1995 (yield changes relative to trend) are shown in *Table 5.1*.

Of course, given that different crops respond in different ways to climatic change, the comparative advantage of cultivating one crop as against another is also likely to change. An optimization model was used in Section 4 to show how regional agricultural planners could alter the allocation of cropped land in order to exploit the changed climatic conditions. *Table 5.2* summarizes the results for the 1 °C temperature change scenario, illustrating how cropped area of winter wheat and corn (which we have seen respond favorably to temperature increases in this region), might be expected to increase at the expense of some of the crops recording reduced yields. This is just one example of possible strategies for mitigating the impacts of climatic change. These are pursued further in a later subsection.

Table 5.2. Impact of a 1 °C temperature increase on crop yields in the Central region European USSR and an optimized response through changes in crop allocation.

Crop	Reference scenario		$+1\,°C$ temperature change scenario	
	Yield (quintal/ha)[a]	Present-day Area (10^3 ha)	Yield (quintal/ha)[a]	Optimized Area (10^3 ha)
Winter wheat	21.0	150	26.9	193.5
Barley	17.7	130	16.9	100.0
Oats	16.4	50	15.6	36.5
Corn for silage	242.0	130	256.9	136.2
Hay	47.0	450	45.1	450.0
Potatoes	140.0	120	136.2	113.0
Vegetables	336.0	25	325.5	25.8

[a] 1 quintal = 100 kg = 0.1 tonnes.

Table 5.3. Changes relative to trend of winter rye yield (ΔY), soil fertility index (ΔSI), ground water level (ΔGW) and surface water pollution by nitrogen from agricultural watersheds (ΔPOL) under the transient climatic change scenario, 1980–2035, in the Leningrad region.

Year	$\Delta T(°C)$	ΔR(mm)	$\Delta Y(\%)$	$\Delta SI(\%)$	$\Delta GW(\%)$	$\Delta POL(\%)$
1990	+0.4	+8	+ 4	− 1	+ 2	+ 18
2000	+0.8	+16	+ 5	− 4	+ 4	+ 30
2010	+1.2	+38	+ 5	− 11	+ 9	+ 75
2020	+1.6	+74	− 4	− 22	+ 14	+ 114
2035	+2.2	+127	− 23	− 42	+ 38	+ 225

5.2.4. Agroenvironmental impacts of longer-term climatic change

The implications of longer-term climatic change for agricultural productivity and environmental conditions of soil quality, ground water level and surface water pollution, have been investigated in the Leningrad region for the period 1980–2035. The results are summarized in *Table 5.3*. Even given a linear increase in fertilizer application over this period, simulated soil quality is degraded due primarily to a 6–12% increase in precipitation in the region. Yields of current winter rye varieties are expected to increase up to the year 2010, but then the unfavorable soil conditions combined with more extreme climatic changes would cause reductions in yields. Higher precipitation could also lead to rising groundwater levels and, with the increased leaching of nutrients, to increased surface water pollution from agricultural watersheds.

In Section 2 some experiments were also conducted to investigate how these negative effects might be mitigated. It was shown that implementation of more intensive drainage activity would lead to a lowering of the groundwater level and would reduce local surface pollution by transporting pollutants more efficiently away from their sources.

5.3. The Implications of Climatic Changes for Policy

The policy implications of climatic change can be considered at two temporal scales. The first concerns efforts to stabilize agricultural production in the face of short-term climatic variations. The second involves possible adjustments of agriculture to accommodate long-term climatic changes. In practice, because these two types of adjustment are by no means mutually exclusive, we have not sought to separate them in the following. Indeed, it has been argued that any future impacts from long-term changes in mean climate are embedded in present-day impacts from short-term climatic variability, the short-term impacts being the medium through which any long-term change is felt (Parry and Carter, 1985). Thus, adjustment activities that are designed to mitigate the impacts of short-term climatic anomalies may, under a changing climate, also assist in coping with an altered magnitude and/or frequency of these anomalies (i.e., changes in risk of impact). If the changes in weather conditions are large, however, systematic centralized measures will be necessary to adapt agriculture to these conditions.

For adjustment activities to be applied successfully in agricultural planning, two problems have to be addressed:

(1) Defining the adjustment measures and the likely results of their implementation.
(2) Assessing the location and scale of each of these measures, and the material and technical resources required to implement and maintain them.

We consider each of these problems below, using both the results of this study and other published information.

5.3.1. Types of adjustment to climatic change

Possible measures for the adaptation of Soviet agriculture to changing climatic conditions are listed below. Some of these are illustrated in *Figure 5.3*, and most are not restricted solely to cold marginal areas, but are widely applicable.

Altered Crop Varieties

One method of adaptation to changing climate is to replace existing crop cultivars by other cultivars that are better suited to the prevailing conditions (for example, through an altered crop vegetative period or improved resistance to lodging). In Section 3 it was demonstrated that if climate warming led us to replace the existing spring wheat cultivars in Cherdyn by two more heat-loving cultivars suited to effective temperature sums, respectively, 50 and 100 degree-days higher than the original over the whole growing season, then, under the GISS-derived 2 × CO_2 scenario, warming would directly enhance yields by 16% and 26%, respectively, relative to the baseline.

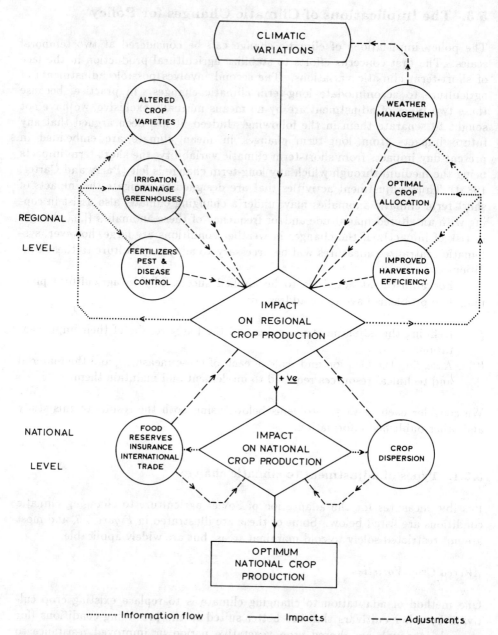

Figure 5.3. Some mitigating responses to climatic variations at regional and national levels in the USSR.

Alternatively, certain crops might be replaced by others. For instance, the substitution of spring crops by winter crops in eastern regions of the subarctic zone would seem to be expedient given significant climate warming (mirroring the strategy suggested for Saskatchewan of substituting winter wheat for spring wheat – cf. Part II, this volume). In the southern regions of the zone, further cultivation of feed crops may be the best management strategy for utilizing the additional thermal resources.

Altered Land Management

Several types of management practice could be employed in response to climatic change. Irrigation can be improved and developed allowing a greater land area to be artificially irrigated. The results of sensitivity studies of crop response to fluctuations in rainfall and soil moisture, such as those reported for spring wheat in Cherdyn (Section 3), can provide useful information on the water requirements and potential benefits from irrigation schemes. Further development of techniques for retaining moisture in soil and for preventing erosion would also be required in this context.

The post-war decades have been characterized by increased applications of artificial fertilizers. This trend is likely to continue in the foreseeable future, and is accounted for in several of the experiments in this case study (Sections 2 and 4). Fertilizers can be used both to enhance crop yields and to stabilize annual yield variations (cf. experiments in the Iceland case study, Part III, this volume). However, their increased usage has other environmental implications that may run counter to their positive effects in adjusting crop yields (as indicated in the experiments with increased fertilizer application levels in the Leningrad region in Subsection 2.8).

Improvements in soil drainage may be necessary in some regions, given the magnitude of precipitation increases suggested under CO_2-doubling, to prevent waterlogging of crops and accelerated soil erosion. Again, as shown for the Leningrad region (Subsection 2.8), the effects may not all be beneficial, for increased drainage activity can also lead to increased leaching of soil nutrients and reduced crop yields.

Expansion of the area under greenhouses is a topic that has received considerable attention in the Soviet Union in recent years because of development of the Siberian and Soviet Far East resource areas and the urgent need for local provision of fresh food in the harsh northern environment (Wein, 1984). The utilization of greenhouses is limited by the high costs of energy, although modern techniques make it possible to cover relatively large areas with plastic film. Further drawbacks are the high cultivation costs and the poor quality of crops, particularly fruit and vegetables, cultivated on "covered" soil. However, the latter problem can be overcome with the help of better cultivar selections and improved agrotechnology. Future climatic warming would probably obviate many of the extra costs, possibly the need for greenhouses at all in some regions, and could be a boon to development in the pioneering regions of the USSR.

Altered Land Allocation and Optimized Regional Production

The optimal allocation and combination of different crops and the development of an efficient "harvest procurement system" are important elements in adaptive response to climatic change at the regional level, and one of the main components of planning activity. Here, potential adaptive responses are based on the fact that different crops react in different ways to changes in weather, as illustrated for the Central region in Section 4. Those experiments, however, did not consider the very important problem of maintaining reserves of crops or agricultural products as a buffer against production shortfalls. It is very difficult to examine this problem separately from considerations of transportation costs, the demands of food processing industries, storage costs, etc. Nevertheless, it is reasonable to take note of the assessments made in the Icelandic case study (Part III, this volume) where it is estimated that an average farm in the more marginal areas should maintain reserves equivalent to about 40–50% of the normal annual production, as insurance against unfavorable weather.

The reliability of weather forecast information also plays an important role in determining crop allocation strategies. As a rule, the greater the potential yield of a crop, the higher is its sensitivity to weather conditions and to the state of agrotechnology. The better the accuracy of weather prediction, the greater the guarantee of selecting crops with the highest productivity under actual weather conditions. Conversely, if accurate weather predictions cannot be assured, then it is important to use a wider variety of crops that are appropriate to a range of weather anomalies, even if their intrinsic yield is not very high.

Weather Management

A further possibility rests with the development of methods to prevent or to limit the extent of damaging weather events. These may take the form either of protective measures for crops (e.g., shelter belts to reduce wind damage, mulching to conserve moisture during drought periods, etc.) or direct modification of the weather itself (e.g., cloud seeding to enhance precipitation or to limit the extent of damaging hailstorms). Both sets of measures have, however, up to now achieved little or no success (Riebsame, 1985).

Most of the foregoing discussion has focused on adjustments that are applicable at the farm and regional levels. However, in a centralized economy such as the USSR, there are other measures that need to be decided at the state level. These are depicted in the bottom part of *Figure 5.3*. Two types have been identified.

Crop Dispersion

If the territory concerned is large (as is the case in the Soviet Union) there is also the possibility of dispersing crops over the whole of the territory. In this way a bad harvest in one region can be compensated by a good one in another region, and specialization of regional agriculture becomes less important (*see* Subsection 1.1). Under stable weather conditions this strategy may lead to a fall in the overall output, but under unstable weather conditions the average harvest can be increased.

Trade, Insurance and Reserves

Finally, the development of international agricultural trade, the creation of an international system of insurance in case of bad yields in particular countries, and the creation of agricultural reserves all have extremely significant roles to play in stabilizing national production. These points are particularly important for small countries, where the entire territory used for agriculture may suffer from unfavorable weather at the same time.

5.3.2. Implementation of adjustments

The adjustment measures that have been described require time and resources to be implemented successfully. In this concluding subsection, we consider, briefly, some examples of adjustment strategies, discuss the importance of climatic information in the planning process, and close with some general remarks on adjustment to climate impacts in the subarctic zone.

Strategies of Adjustment

If significant changes in climate occur over the longer term, it is necessary to prepare agriculture for them. In particular, attention should be directed towards the breeding and substitution of new kinds of crops (*see* above) where this is economically expedient. Such a transition would call for the training of agricultural workers and employees, partial changes in the fixed stocks, and probably some modification to elements of the agricultural infrastructure. New agrotechnologies need to meet the requirements of the future, and this may mean that initially they are less efficient than the older ones. At present, the selection of new varieties of crops takes at least 5–7 years. Some 3–4 years more are then spent in testing these varieties in real-world conditions, and in reproducing the seeds. The whole cycle in the creation of a new agrotechnology can thus take at least 10 years. This means that current decisions must be made concerning new agrotechnologies that will only come into widespread use in 10–15 years' time.

The more extreme the climatic changes, the sooner agrotechnologies will become obsolete. Some of the sharpest climatic changes appear to occur in marginal areas. Given a climatic change in such areas, there is a choice between waiting for the climate to stabilize, leading to a decrease of production, or implementing adaptation measures more frequently, resulting in increased expenditure. It is also possible to choose an intermediate strategy, by balancing the expenditures for adaptation, the productivity lost because of imperfect functioning of various agrotechnologies in the new climatic conditions, and the probable shortfall in agricultural products.

Climatic Information in Planning

In order to implement any adjustment measures, there is a need for appropriate meteorological data to describe the weather variations that cause impact. Weather anomalies can be divided into two categories:

(1) Those that can be overcome (e.g., by limiting the consequences of droughts with the help of artificial irrigation).

(2) Those that are inevitable and essentially insurmountable.

For responding to these, there needs to be further information concerning:

(i) Agroclimatic descriptions of average weather situations.
(ii) Probabilities of weather anomalies that can be overcome.
(iii) Probabilities of isolated but insurmountable weather anomalies.
(iv) Probabilities of consecutive years of insurmountable weather anomalies.

Such probabilistic scenarios are convenient for economic planning, since they are simple and flexible to apply in terms of the data needed: only the general long-term regularities in the climate need to be known. For planning, the most important function of the climatic scenario is to provide the necessary information for evaluating the responses of agricultural crops under the expected or possible meteorological conditions. For example, the models used in Section 4 consider both existing and proposed new crops, providing projections of:

(1) Mean annual productivities of the more common crops in a particular region and fluctuations from these means due to weather anomalies.
(2) Expenditure required to obtain prescribed average yields and additional foreseeable expenses that may be needed to counteract unfavorable weather conditions.

Adjustment in the Subarctic Zone

Summarizing, one can say that identifying economically and ecologically appropriate ways for adjusting agriculture to possible climatic changes allows us not only to mitigate impacts but also to meet the future requirements of agriculture in the cold-marginal zone of the USSR. Different analytical approaches have been adopted in this study. The results of these analyses for different regions suggest that a number of economic and planning decisions may be necessary in the future, such as increasing production expenditures to compensate for decreasing yields of certain crops, expansion of crop areas of those crops for which yield increases could be expected, introduction of new varieties which are better suited to the changed climatic conditions, and so on. As a result, the agricultural specialization of regions, the structure of fixed capital assets and interregional flows of agricultural commodities may well change markedly. It is clear that such policies of adjustment to changing climate will require corresponding changes in investment and organization.

Of course, the examples of adjustment presented above serve merely as illustrative, preliminary experiments. However, with the help of further model development, with refinement of data on possible regional climatic changes and with the implementation of multivariant simulations, we have a real opportunity to improve our understanding of the potential problems and to devise more appropriate methods for their solution.

Acknowledgments

The authors would like to express their appreciation to E.V. Abashima and V.N. Pavlova for their valuable contributions to Section 3, to Nicolaas Konijn for conducting simulation experiments and preparing a description of his model in Section 4, and to Tim Carter for his valuable comments and advice which helped us to improve the papers in our case study and to provide linkages to the other case studies.

Appendices

Appendix 2.1: Coefficients and Variables in the Obukhov Model

Let us denote the average temperature and precipitation of the jth month as T^j and ρ^j, respectively, and the precipitation of the ith decade in the jth month as ρ_i^j. Obukhov's (1949) statistical analysis showed that, in the Leningrad region, winter rye yield, y, is mostly dependent on the following meteorological factors, which he grouped into four categories:

(1) *During the period before the beginning of plowing*: High average monthly temperatures for July of the previous year, $_{-1}T^7$, show a positive correlation with yield, $(r_{y,-1}T^7 = 0.4)$, while surplus precipitation during the fallow period, ρ_f, is negatively correlated with yield $(r_y,\rho_f = -0.46)$. This is because high July temperatures increase the degree of grain maturity, i.e., the quality of the grain viewed as seed for the following year's crop; conversely, surplus precipitation, which frequently occurs in this region during the fallow period, leads to undesirable waterlogging of the soil, thus depressing the yield.

(2) *During autumn, from plowing until November*: Surplus precipitation has a strongly negative effect on winter rye yield in the Leningrad region when it occurs during the following periods: during the final plowing ρ_w $(r_y,\rho_w = -0.5)$, between sowing and emergence of the crop ρ_e $(r_y,\rho_e = -0.35)$, and between emergence and bushing-out ρ_s $(r_y,\rho_s = -0.23)$. This is because the surplus moisture during these periods causes a reduction in soil aeration. Conversely, precipitation following the bushing-out stage, ρ_y and high temperatures during September and October, $_{-1}T^{9,10}$, both have a positive correlation with the yield $(r_y, \rho_y = 0.3; r_y, {_{-1}T^{9,10}} = 0.3)$.

(3) *During the spring and summer growing period*: There is a negative correlation between yield and surplus precipitation in May ρ^5 $(r_y,\rho^5 = -0.28)$, June ρ^6 $(r_y,\rho^6 = -0.5)$, and July ρ^7 $(r_y,\rho^7 = -0.36)$. On the other hand, high average monthly temperatures for July, T^7, show a positive correlation to yield $(r_y, T^7 = 0.26)$.

(4) *During the ripening and harvesting period*: There is a positive correlation between yield and mean August temperature, T^8 $(r_y, T^8 = 0.34)$, and a negative correlation between yield and August surplus precipitation, ρ^8 $(r_y,\rho^8 = -0.33)$.

The general regression equation allows us to calculate the deviation of expected yield from an average value, which itself depends on the level of technology. This equation includes eleven meteorological parameters and has the form:

$$y_z = 0.06y_1 + 0.12y_2 + 0.07y_3 + 0.05y_4 \tag{A2.1}$$

where y_1, y_2, y_3, y_4 are defined by the following formulae:

$$y_1 = -0.1 - 0.07(\rho_f - 145) + 1.42(_{-1}T^7 - 17.6) \tag{A2.2}$$

where

$$\rho_f = \rho^5 + \rho^6 + \rho_1^7 + \rho_2^7$$

$$y_2 = 0.2 + 0.93\tilde{y} + 0.018(\rho_3^9 - 17) + 0.4(_{-1}T^{9,10} - 8.9) \tag{A2.3}$$

where

$$\tilde{y} = 3.0 - 0.4(\tilde{\rho} - 50) - 0.02(\tilde{\rho} - 50)^2 + 0.0004(\tilde{\rho} - 50)^3$$

$$\tilde{\rho} = 0.47(0.5\rho_2^8 + \rho_3^8 + 0.5\rho_1^8) + 0.36(0.5\rho_3^7 + \rho_1^8 + 0.5\rho_2^8)$$
$$+ 0.17(0.5\rho_3^8 + 0.75\rho_1^9 + 0.75\rho_2^9 + 0.25\rho_3^9)$$

and

$$_{-1}T^{9,10} = 0.76\,_{-1}T^9 + 0.224\,_{-1}T^{10}$$

$$y_3 = 0.1 - 0.04(\rho^5 - 44) - 0.15(\rho^6 - 57) - 0.08(\rho^7 - 70) \tag{A2.4}$$
$$+ 0.58(T^7 - 17.6)$$

$$y_4 = 0.03 - 0.07(\rho^8 - 78) + 1.465(T^8 - 15.1) \tag{A2.5}$$

Appendix 4.1: Model Relationships in the Regional Agricultural Planning Model

Indices

i	– Agricultural crop types
j	– Cattle-breeding product types
t	– Years of the plan period
k	– The main production factors
$k = 1$	– Capital stocks
$k = 2$	– Capital investments into land reclamation and increased fertility
$k = 3$	– Mineral and organic fertilizers
$k = 4$	– Time over which technological improvements such as breeding of new varieties, etc., become significant
$k = 5$	– Harvesting machinery

Appendices

$k = 6$ – Transport facilities
m – Types of storage
n – Types of crop processing capacities

Sets

J_1 – Crops, harvested at the most resource-demanding period
J_m – Crops, stocked in storage of mth type
J_n – Crops, processed at factories of the nth type
F – Feed crops

Variables

$\tilde{V}_i^t$ – Volume of the final product from crop i in year t
U_j^t – Volume of cattle-breeding product i in year t
Z_{kit} – Volume of production factors of the kth type needed for cultivating crop i in year t
ΔZ_{kit} – Increment of production factor k used on crop i in year t
Y_i^t – Yield of crop i in year t
X_i^t – Area of crop i in year t
V_i^t – Gross production of crop i in year t
ΔM_{mt} – Increment of storage capacity m in year t
P_i^t – Volume of crop i processed in year t
R_i^t – Volume of crop i unprocessed in year t
ΔN_{nt} – Annual increment of the processing capacity n of agricultural products in year t
W_i^t – Possible production losses of crop i in year t caused by a lack of capacity of the "yield procurement system"
I_i^t – Import of feed concentrates of type i from other regions in year t

Coefficients

C_i^t, C_j^t – Purchase prices for crop and cattle-breeding products
$\underline{P}_i^t, \underline{P}_j^t$ – Minimum production of crop and cattle products in the region required to satisfy the demand for semi-perishable products, and to ensure continued crop rotation
$\bar{P}_i^t, \bar{P}_j^t$ – Maximum possible levels of production in the region
β_{ci}, β_{ki} – Coefficients of regression equations describing the dependence of the crop production capacity on the most important production factors
ξ_{ti} – Coefficient of the deviation of crop production capacity from its estimated value for a scenario of altered climate

$\varphi_{ki}, \varphi_m, \varphi_n$ — Coefficients of annual depreciation of the capacities of capital stocks
Z_{ki}^0, M_m^0, N_n^0 — Availability of capital stock by types at the beginning of the plan period
D_{ki} — Maximum possible influence of production factors determined by the agrotechnical limit
W_i — Coefficient of the maximum change of crop area permitted under a certain crop in one year
δ_{ki} — Average capacity of harvesting machinery and transport facilities during harvesting
σ_1, σ_2 — Average annual proportion of total production sent for industrial processing (1) and marketed to the consumer as unprocessed, fresh produce (2)
α_i — The quality of feed in feed units
γ_j — The quantity of feed units necessary for the production of a unit of cattle-breeding products
δ_i — Permitted share of certain feed crops in the total mix of feeds
K_{kit}, K_m, K_n — Specific capital investments for the increment of the basic capital funds

Model Relationships

$$\max \left\{ \sum_{t,i} \tilde{V}_i^t C_i^t + \sum_{t,j} U_j^t C_j^t - \sum_{t,i} I_i^t C_i^t \right\} \tag{A4.1}$$

— maximum of the final production in the region for the plan period.

$$\underline{P}_i^t \leq \tilde{V}_i^t \bar{P}_i^t \; ; \quad \leq \underline{P}_j^t \leq U_j^t \leq \bar{P}_j^t \tag{A4.2}$$

— minimum volume of production required in the region.

Relations (A4.3)–(A4.5) describe the estimation of the biological yield capacity:

$$Y_i^t = \left[\beta_{0i} + \sum_{k=1}^{4} Z_{kit} \cdot \beta_{ki} \right] \cdot \xi_i^t \tag{A4.3}$$

— the dependence of biological yield capacity on production factors and weather conditions, the impact of which are described by the coefficient ξ_i^t.

$$Z_{kit} = Z_{kit-1} \cdot \varphi_{ki} + \Delta Z_{kit} \; ; \quad Z_{kit=0} = Z_{ki}^0 \quad (k=1) \tag{A4.4}$$

— the relation between capital funds for crop cultivation to their availability in the preceeding year (i.e., depreciation or growth of funds during the current year); the initial conditions for the beginning of the plan period are put down separately.

$$Z_{kit} \leq D_{ki} \quad (k=1,2,3) \tag{A4.5}$$

— the value of a production factor for all years cannot exceed the prescribed agrotechnical limit.

$$X_i^t \cdot Y_i^t = V_i^t \tag{A4.6}$$

– gross production of crops.

$$X_i^{t-1}\left(1 - W_i\right) \leq X_i^t \leq X_i^{t-1}\left(1 + W_i\right) \tag{A4.7}$$

– the condition of smooth change of crops areas at the rate not more than $W_i = 5\%$ per year.

Relations (A4.8)–(A4.15) describe the estimation of the gross harvest, a function of the level of development of the "harvest procurement system":

$$Z_{kit} = Z_{kit-1} \cdot \varphi_{ki} + \Delta Z_{kit} \ ; \quad Z_{kit=0} = Z_{ki}^0 \quad (k = 5, 6) \tag{A4.8}$$

– availability of harvesting machinery and transport facilities in year t regarding physical wear and the purchase of new machinery.

$$Z_{kit} \cdot \delta_{ki} \geq \sum_{i \in J_i} \tilde{V}_i^t \quad (k = 5, 6) \tag{A4.9}$$

– the average realizable capacity of harvest machinery in year t (expressed in terms of harvested volume).

$$M_{mt} = M_{mt-1} \cdot \varphi_m + \Delta M_{mt} \ ; \quad M_{mt=0} = M_{m^0} \tag{A4.10}$$

– availability of storage capacities of type m in year t regarding physical wear and new construction.

$$M_{mt} \geq \sum_{i \in J_m} \left(\tilde{V}_i^t - P_i^t - R_i^t\right) \tag{A4.11}$$

– necessary storage capacity for storing the remainder of the harvested crop not provided to the consumer in the form processed or fresh produce.

$$N_{nt} = N_{nt-1} \cdot \varphi_n + \Delta N_{nt} \ ; \quad N_{nt=0} = N_n^0 \tag{A4.12}$$

– availability of crop processing capacity in year t, regarding physical wear and new construction.

$$N_{nt} \geq \sum_{i \in J_n} P_i^t \tag{A4.13}$$

– capacity required for crop processing.

$$P_i^t = \sigma_1 \tilde{V}_i^t \ ; \quad R_i^t = \sigma_2 \tilde{V}_i^t \tag{A4.14}$$

– the proportion of harvested crop used in processing and in direct marketing to the consumers.

$$V_i^t - \tilde{V}_i^t = W_i^t \tag{A4.15}$$

– volume of crop product that cannot be harvested.

$$\sum_{i \in F} \tilde{V}_i^t \cdot \alpha_i + \sum_i I_i^t \cdot \alpha_i = \sum U_j^t \gamma_j \tag{A4.16}$$

- the availability of feed (in feed units and digested protein) balanced against the requirement for feed per volume of cattle-breeding product.

$$\tilde{V}_i^t \cdot \alpha_i \overset{\leq}{>} \delta_i \sum_{i \in F} \tilde{V}_i^t \cdot \alpha_i \tag{A4.17}$$

- restriction on the composition of feed.

$$\sum_{t,k=1,5,6} \Delta Z_{kit} \cdot K_{kit} + \sum_{t,k=2} Z_{kit} + \sum_{t,k=3} \left(Z_{kit} - Z_{kit-1} \right) \cdot K_{kit}$$

$$+ \sum_{t,m} \Delta M_{mt} K_m + \sum_{t,n} \Delta N_{nt} K_n \leq K \tag{A4.18}$$

- restriction on the volume of capital investments (for an increase in capital stocks in agriculture including harvesting machinery, transport facilities and capital investments in land reclamation, for increased fertilizer supply, and for increased storage and processing capacities).

Appendix 4.2: Description of the Dynamic Crop Production Model (CPM)

Background

Instead of using statistical information and expert judgment to determine crop yields for different crops under altered climates, a more quantitative approach using a dynamic crop production model has been employed for the Central region. This model, developed at IIASA, has been used in other regions with satisfactory results (see Konijn 1984b; Akong'a et al., 1988), though while some preliminary attempts at model verification have been presented for the semi-arid climate of Kenya (Akong'a et al., 1988), so far no exhaustive validation has been possible because of the lack of a full data set. Confidence in the model estimates is strongly dependent on how we judge the validity of the relationships employed. Most of the relationships described below were derived from published investigations and have proved to present fairly satisfactory descriptions of reality.

The Crop Production Model

The crop production model is used to generate crop yield data for seven crop types: winter wheat, barley, oats, corn for silage, hay, potatoes and vegetables. The functional relationships in the model are organized hierarchically and estimations are carried out on a 10-day interval basis (for full details see Konijn, 1984a). Therefore, the dynamic input characteristics such as the climatic data need to be supplied on that time basis.

Three levels of calculations can be distinguished:

(1) Estimation of the photosynthetic dry matter production.
(2) Dry matter production subject to constraints of water availability.
(3) Dry matter production as affected by nutrient constraints.

In the estimation of photosynthetic dry matter production, a distinction has been made between two types of plant. Under the present atmospheric composition the biochemical pathway used by the so-called C-3 type of plant operates most efficiently under cooler climates, while C-4 plants grow relatively better in the warmer, lower latitudes. Crops like barley and wheat are C-3 plants, while corn (maize) and sorghum are examples of the C-4 plant type.

The determination of the photosynthetic dry matter production is carried out following the procedure proposed by de Wit (1965), and amended by Goudriaan and van Laar (1978). For a given radiative and temperature regime, the production value obtained represents the maximum possible total production (i.e., for a closed canopy and omitting the growth and maintenance respiration).

The partitioning of the total dry matter produced, together with the effects of maintenance and growth respiration, are determined after possible water constraints have been quantified. Water supplied to the rooting zone can originate from different sources: precipitation, irrigation, and/or capillary rise from ground water. Their relative importance depends on the local circumstances. In periods when evapotranspiration exceeds the water supply to the rooting zone, the capacity of the soil to store water determines how much a plant will suffer from drought stress. Changes in soil water tension are determined at 10-day intervals and dictate, together with the evaporative demand of the atmosphere, to what extent the plant will transfer water from the soil to the atmosphere. Should the plant need to close its stomata to prevent wilting, no carbon dioxide assimilation will then be possible, leading to reduced production.

Although lack of nutrients often limits the production, in these model runs nutrients are considered to be non-limiting. Similarly, the effectiveness of pest, disease and weed control is assumed to be optimal.

Input Requirements

Together with the climatic data that are required as inputs to the model, it is important to have information about certain crop characteristics, for the better the varieties are adapted to the production environment the less will be the risk of reduced production. These characteristics describe the length of the growing season, the length of different growing stages, the partitioning of dry matter over the various plant organs, the type of carbon dioxide assimilation pathway, the rates of growth respiration and maintenance respiration, and the response to water stress. They all are crop-specific, and differ from location to location depending on local developments in crop selection and plant breeding.

Appendix 4.3: The National-Level Optimization Model

Model Indexes

i	– Crop products
j	– Cattle-breeding products
t, ϑ	– Years of the period investigated
r, s	– Regions
S^*	– Regions through which foreign trade passes

Model Variables

$Z_{ri}^t, \bar{Z}_{rj}^t$	– Final consumption of crop and cattle-breeding products in region r
U_{ri}^t	– Volume of crop production in region r

X_{sri}^t, X_{rsi}^t – Volume of product transported from region s to region r and vice versa
FS_{ri}^t – Withdrawals from stocks
IS_{ri}^t – Replenishment of stocks
S_i^t – Current value of stocks
I_{s^*ri} – Products imported to region r
$E_{rs^*i}^t$ – Products exported from region r
I_i^t, E_i^t – Volumes of imports and exports
$K_r^\vartheta, \bar{K}_r^\vartheta$ – Capital investment for the development of crop cultivation and cattle breeding

Model Parameters

B_i, B_j – Penalty for failure to satisfy consumer demand for a product
$R_{ri}^t, \bar{R}_{rj}^t$ – Planned consumption levels for crop and cattle-breeding products in region r
$C_{ri}^t(U_{ri}^t)$ – Expenditure per unit of production as a function of production volume
C_{rsi}^t – Expenditure for transporting unit product from region r to region s
f_{rij}^t – Feed expenditure for one unit of cattle-breeding production
P_i^t – World market prices for product i
VD^t – Currency subsidy in year t
K^ϑ – Capital investment allocated for the development of agriculture in year ϑ.

Description of the Model

The basic model equation represents a mixed optimization problem, simultaneously maximizing production and minimizing total national expenditure on production and transportation. It takes the following form. Minimize:

$$F = \sum_{r,i,t} B_i \left(R_{ri}^t - Z_{ri}^t \right)^2 + \sum_{r,j,t} B_j \left(R_{rj}^t - \bar{Z}_{rj}^t \right)^2 + \sum_{r,i,t} C_{ri}^t \left(U_{ri}^t \right) \cdot U_{ri}^t$$
$$+ \sum_{r,s,i,t} C_{rsi}^t X_{rsi}^t \quad (A4.19)$$

subject to:

$$U_{ri}^t + \sum_s X_{sri}^t + FS_{ri}^t + \sum_{s^*} I_{s^*ri} = Z_{ri} + \bar{Z}_{rj}^t \cdot f_{rij}^t \quad (A4.20)$$
$$+ \sum_s X_{rsi}^t + IS_{ri}^t + \sum_{s^*} E_{rs^*i}^t$$

which balances production and allocation for each region. The left-hand side of equation (A4.20) specifies the produce available (domestic production, imports from other regions, withdrawals from stocks, imports from abroad), while the right-hand side deals with its allocation (consumption by the population, consumption in cattle-breeding, exports to other regions, replenishment of stocks, exports abroad).

Restrictions on the amounts of produce that may be taken from the stocks are given by:

$$FS_{ri}^t \leq S_i^t \; ; \quad S_i^{t=1} = 0 \tag{A4.21}$$

while the balance for products rotating through the stocks is specified as:

$$IS_i^{t-1} + S_i^{t-1} - FS_i^{t-1} = S_i^t \tag{A4.22}$$

Equation (A4.23) correlates imports (exports) of products with their allocation to the various regions:

$$\sum_{s^*r} I_{s^*ri}^t = I_i^t \; ; \quad \sum_{rs^*} E_{rs^*i} = E_i^t \tag{A4.23}$$

and the currency balance on the foreign trade account, including import subsidies, is represented by:

$$\sum_i I_i^t P_i^t \leq \sum_i \sum_{\vartheta=1}^t E_i^\vartheta P_i^\vartheta - \sum_i \sum_{\vartheta=1}^{t-1} I_i^\vartheta P_i^\vartheta + VD^t \tag{A4.24}$$

Functions determining the capital investment needed to achieve the planned production levels in crop cultivation and cattle-breeding are specified as follows:

$$U_{ri}^t = \sum_{\vartheta=1}^{t-1} \varphi_{ri}^\vartheta \left(K_r^\vartheta\right) \; ; \quad \bar{Z}_{rj}^t = \sum_{\vartheta=1}^{t-1} \varphi_{rj}^\vartheta \left(\bar{K}_r^\vartheta\right) \tag{A4.25}$$

Finally, general restrictions on capital investments for individual years are represented as:

$$\sum_r K_r^\vartheta + \sum_r \bar{K}_r^\vartheta \leq K^\vartheta \tag{A4.26}$$

In order to simplify the presentation here we have not given explicitly the balance equations for cattle-breeding, which are similar in structure to equations (A4.19)–(A4.24) but with the subscript j rather than i. The transportation and consumption of processed products are evaluated in fresh-product units.

References

Agroclimatic Resources (1971). *Agroclimatic Resources of the Leningrad Region.* Gidrometeoizdat, Leningrad (in Russian).
Akong'a, J., Downing, T.E., Konijn, N.T., Mungai, D.N., Muturi, H.R. and Potter, H.L. (1988). The effects of climatic variations on agriculture in central and eastern Kenya. In M.L. Parry, T.R. Carter and N.T. Konijn (eds.) *The Impact of Climatic Variations on Agriculture. Volume 2. Assessment in Semi-Arid Regions.* Reidel, Dordrecht, The Netherlands.
Bihele, Z.N., Moldau, H.A. and Ross, J.K. (1980). *Mathematical Modeling of Plant Transpiration and Photosynthesis under Soil Water Stress.* Gidrometeoizdat, Leningrad (in Russian).
Bolin, B., Döös, B.R., Jäger, J. and Warrick, R.A. (eds.) (1986). *The Greenhouse Effect, Climatic Change and Ecosystems.* Wiley, Chichester, 542 pp.
Butner, E.K. (1983). The ocean response to atmospheric greenhouse effect changes. *Izvestia Academii Nauk SSSR, Ser. Phys. Atmosph. i. Ocean,* pp. 892–895 (in Russian).
Chartier, P. (1970). A model of CO_2 assimilation in the leaf. In I. Setlik (ed.) *Prediction and Measurement of Photosynthetic Productivity.* PUDOC, Wageningen, The Netherlands, p. 307.
D. Mil Survey (1974). *USSR and Adjacent Areas: Administrative Map 1:8 000 000.* Ministry of Defence, United Kingdom.
Dantzig, G. and Madansky, A. (1961). On the solution of two stage linear programs under uncertainty, *Proceedings of the Fourth Berkeley Symposium on Mathematical Statistics and Probability,* **Vol. 1.** University of California Press, Berkeley.
Frere, M.H. (1978). *Model for Predicting Water Pollution from Agricultural Watersheds.* Proceedings of the IFIP Conference on Modeling and Simulation of Land, Air and Water Resources Systems, North Holland Publishing Co., Amsterdam, pp. 501–508.
Gorbachev, M. (1982). Prodovolstvennaja programma i zadachi jejo realizatsii. *Kommunist,* **58**(10), (The Food Program and the Tasks of its Realization).
Goudriaan, J. and van Laar, H.H. (1978). Calculation of daily totals of the gross CO_2 assimilation of leaf canopies. *Neth. J. Agric. Sci.,* **26**, 373–382.

Groisman, P.Ya. (1981). Empirical estimates of the reaction of precipitation in the USSR to global warming and cooling processes. *Izvestia Academii Nauk SSSR, Ser. Geograph.*, 5, 86–95 (in Russian).

Handbook of Agronomy (1983). Kolos, Moscow, 320 pp. (in Russian).

Hansen, J., Lacis, A., Rind, D., Russell, G., Stone, P., Fung, I., Ruedy, R. and Lerner, J. (1984). Climate sensitivity: Analysis of feedback mechanisms. In J. Hansen and T. Takahashi (eds.) *Climate Processes and Climate Sensitivity*. Maurice Ewing Series, Vol. 5, American Geophysics Union, Washington, D.C., pp. 130–163.

Hansen, J., Russell, G., Rind, D., Stone, P., Lacis, A., Lebedeff, S., Ruedy, R. and Travis, L. (1983). Efficient three-dimensional global models for climate studies: Models I and II. *Monthly Weather Review*, 3, 609.

Iakimets, V. and Kiselev, V. (1985). *First version of SOVAM*. Working Paper WP-85-60. International Institute for Applied Systems Analysis, Laxenburg, Austria, 12 pp.

Iakimets, V. and Lebedev, V. (1985). *Revised Version of the SOVAM*. Working Paper WP-85-66. International Institute for Applied Systems Analysis, Laxenburg, Austria, 14 pp.

Kashtanov, A.N. (ed.) (1983). *Natural and Agricultural Zoning and Land Use in the USSR*. Kolos, Moscow (in Russian).

Kiselev, V.I. (1979). *Organization and Planning of an Agroindustrial Complex*. Nauka, Moscow (in Russian).

Kiselev, V.I. (1983). Synthesis of planning and economic management methods in an agroindustrial complex. *Economics and Mathematical Methods*, XIX (I).

Konijn, N. (1984a). *A Crop Production and Environment Model for Long-Term Consequences of Agricultural Production*. Working Paper WP-84-51. International Institute for Applied Systems Analysis, Laxenburg, Austria, 37 pp.

Konijn, N. (1984b). *Modeling the Impacts of Climatic Variation on Agricultural Production in Dry Regions. An Application of Data from Stavropol*. Working Paper WP-84-52. International Institute for Applied Systems Analysis, Laxenburg, Austria, 15 pp.

Kroutko, V.N., Pegov, S.A., Khomyakov, D.M. and Khomyakov, P.M. (1982a). *Model of the Dynamics of Environmental Farming Factors*. VNIISI, Moscow (in Russian).

Kroutko, V.N., Pegov, S.A., Khomyakov, D.M. and Khomyakov, P.M. (1982b). *Evaluation of Environmental Quality*. VNIISI, Moscow (in Russian).

Menzhulin, G.V. and Savvateev, S.P. (1980). Present climate change influence on crop yields. In *Problems of Atmospheric Carbon Dioxide*. Gidrometeoizdat, Leningrad (in Russian).

Mischenko, Z.A. (1984). *Agroclimatic Mapping of the Continents*. CAgM Report No. 23, World Meteorological Organization, Geneva, 110 pp.

Müller, M.J. (1982). *Selected Climatic Data for a Global Set of Standard Stations for Vegetation Science*. Junk Publ., The Hague, The Netherlands, 306 pp.

NDU (1978). *Climate Change to the Year 2000*. National Defense University, Fort McNair, Washington, D.C.

NDU (1980). *Crop Yields and Climate Change to the Year 2000*. National Defense University, Fort McNair, Washington, D.C.

Nordhaus, W.D. and Yohe, G.W. (1983). Future paths of energy and carbon dioxide emissions. In Carbon Dioxide Assessment Committee, *Changing Climate*. National Academy of Sciences, Washington, D.C., pp. 87–153.

Obukhov, V.M. (1949). Estimation of expected yield (Moscow and Leningrad regions). *Yield and Meteorological Factors*. Gosplanizdat, Moscow, pp. 244–298 (in Russian).
Parry, M.L. and Carter, T.R. (1985). The effect of climatic variations on agricultural risk. *Climatic Change*, **7**, 95–110.
Pegov, S.A. (1984). Simulation of ecological development processes. *Environmental Management in the USSR*, **3**, 51–64, Commission of the USSR for UNEP, Moscow.
Pegov, S.A., Khomyakov, P.M. and Kroutko, V.N. (1983). *Forecasting and Estimating Environmental Changes on a Regional Basis*. Working Paper WP-83-83. International Institute for Applied Systems Analysis, Laxenburg, Austria.
Pegov, S.A., Khomyakov, P. and Kroutko, V. (forthcoming). *A modeling system of regional environment state dynamics*. Working Paper. International Institute for Applied Systems Analysis, Laxenburg, Austria.
Pitovranov, S., Pegov, S.A. and Khomyakov, P. (1984). Modeling the Impact of Climatic Change on Regional Ecosystems. *Environmental Management in the USSR*, **3**, 65–83, Commission of the USSR for UNEP, Moscow. Also available as Collaborative Paper CP-84-7. International Institute for Applied Systems Analysis, Laxenburg, Austria, 27 pp.
Rabinowitch, E. (1951). *Photosynthesis and Related Processes. Vol. II, Part 1*. Interscience, New York.
Riebsame, W.E. (1985). Research in climate society interaction. Ch. 3 in R.W. Kates, J.H. Ausubel and M. Berberian (eds.) *Climate Impact Assessment. Studies of the Interaction of Climate and Society*. SCOPE 27, Wiley, Chichester, pp. 69–84.
Sirotenko, O.D. (1981). *Mathematical Modeling of Hydrothermal Regimes and the Productivity of Agroecosystems*. Gidrometeoizdat, Leningrad (in Russian).
Sirotenko, O.D., Abashina, E.V. and Pavlova, V.N. (1984). An estimate of the influence of possible climate variations and changes on the productivity of agriculture. *Izvestia Akademii Nauk SSSR, ser. Phys. Atmosph. i Ocean.*, **20**(11), 1104–1110 (in Russian).
Sirotenko, O.D., Abashina, E.V., Pavlova, V.N. and Dolgiy-Trach, V.A. (1982). *A Technique for Quantitative Estimates of the Influence of Dry Conditions on Spring Wheat Yields*. Gidrometeoizdat, Moscow (in Russian).
Sov–Amer (1982). *The Impact of Increasing Atmospheric Carbon Dioxide on Climate*. Proceedings of the Soviet-American Workshop on Atmospheric Carbon Dioxide Increasing Study, Leningrad, 15–20 June 1981. Girometeoizdat, Leningrad.
Statistical Yearbook of the RSFSR (Russian Soviet Federal Socialist Republic) (1974). Finansy i Statistika, Moscow (in Russian).
Tarrant, J.R. (1985). An analysis of variability in Soviet grain production. In P. Hazell (ed.) *Sources of Increased Variability in Cereal Yields*. Unpublished Proceedings IFPRI/DSE Workshop, Feldafing, FRG, November, 1985.
Terjung, W.H., Louie, S.S.-F. and O'Rourke, P.A. (1976). Toward an energy budget model of photosynthesis predicting world productivity. *Vegetatio*, **32**(1).
The USSR Economy for the Period 1922–1982 (1982). Finansy i Statistika, Moscow (in Russian).
Ulanova, E.S. (ed.) (1978). *Agrometeorological Conditions and Productivity of Agriculture in the Non-Chernozem Zone of the RSFSR*. Leningrad, Gidrometeoizdat (in Russian).
Ulanova, E.S. (1984). Agrometeorological conditions and the productivity of grain crops. *Meteorologia i Gidrologia*, **5**, 95–100 (in Russian).

Valpasvuo-Jaatinen, P. (1983). Water-related environmental problems of agriculture in Finland. In G. Golubev (ed.), *Environmental Management of Agricultural Watersheds*. Collaborative Proceedings Series CP-83-S1. International Institute for Applied Systems Analysis, Laxenburg, Austria, pp. 121-129.

Vinnikov, K.Ya. and Groisman, P. Ya. (1979a). An empirical model of present-day climatic changes. *Meteorologia i Gidrologiya*, **3**, 25-28 (in Russian). English translation available in *Soviet Meteorol. Hydrol.* **3**.

Vinnikov, K.Ya. and Groisman, P. Ya. (1979b). An empirical study of climate sensitivity. *Izvestia Academii Nauk SSSR, Ser. Phys. Atmosph. i. Ocean.*, **6**, 5-20 (in Russian).

Vinnikov, K.Ya. and Groisman, P.Ya. (1982). The empirical study of climate sensitivity. *Atmos. Oceanic Phys.* **18**, 1159-1169.

Watts, R. (1984). *Predicting Changes in Crop Yield due to CO_2-Induced Climatic Change – Some Cautionary Comments*. Working Paper WP-84-15. International Institute for Applied Systems Analysis, Laxenburg, Austria, 11 pp.

Wein, N. (1984). Agriculture in the pioneering regions of Siberia and the Far East: Present status, problems and prospects. *Soviet Geography: Review and Translation*, **25(8)**, 592-620.

Wit, C.T. de (1965). Photosynthesis of leaf canopies. *Agric. Res. Rep.*, **663**, PUDOC, Wageningen, The Netherlands.

World Atlas of Physical Geography (1964). USSR Academy of Sciences, Main Office of Geodesy and Cartography, Moscow, 294 pp.

WWR (1974, 1980). *World Weather Records, Vol. 2, Europe.* U.S. Department of Commerce, Washington, D.C.

Yakunin, I. (1982). Increasing grain resources – the key task of the agroindustrial complex. *Planovoe Khoziaystvo*, **11**, 62-70 (in Russian).

PART VI

The Effects of Climatic Variations on Agriculture in Japan

PART VI

The Effects of Climatic Variations on Agriculture in Japan

Contents, Part VI

List of Contributors		727
Abstract		729
1.	Introduction: The policy and planning issues *Masatoshi Yoshino*	731
	1.1. Background and purpose	731
	1.2. The environment and agriculture in Japan	732
	1.3. Climatic variations and rice production	740
	1.4. Techniques and policies for mitigating climate impacts	744
	1.5. Structure of this case study	747
2.	Development of the climatic scenarios *Masatoshi Yoshino*	751
	2.1. Purpose	751
	2.2. Climatic changes in the historical period	751
	2.3. Recent climatic variations	753
	2.4. Instrumentally derived scenarios	754
	2.5. The GISS $2 \times CO_2$ scenario	757
	2.6. Relating regional scenarios to synoptic processes	763
3.	The effects on latitudinal shift of plant growth potential *Zenbei Uchijima and Hiroshi Seino*	773
	3.1. Introduction	773
	3.2. Statistical characteristics of thermal resources	774
	3.3. Yield variability of rice in relation to temperature variations	779
	3.4. Climatic variations and plant growth potential	781
	3.5. Conclusions	793
4.	The effects on altitudinal shift of rice yield and cultivable area in northern Japan *Tatsuro Uchijima*	797
	4.1. Introduction	797
	4.2. Critical temperatures in rice production	797
	4.3. Altitudinal shifts of the cultivable area	797

	4.4.	Changes in rice yield and cultivable area under different climatic scenarios	804
	4.5.	Conclusion	806
5.	The effects on rice yields in Hokkaido *Takeshi Horie*		809
	5.1.	Introduction	809
	5.2.	The model	809
	5.3.	Crop parameters and climatic scenarios	815
	5.4.	Yield simulations, technology adjustment and sensitivity analysis	817
	5.5.	Estimated rice yield for stable and unstable decades, and for anomalously cool years	820
	5.6.	Estimated rice yields for the $2 \times CO_2$ scenario under present and adjusted technology	822
6.	The effects on the Japanese rice market *Hiroshi Tsujii*		827
	6.1.	Introduction	827
	6.2.	The rice production system	828
	6.3.	A model of the Japanese rice market	832
	6.4.	Scenario experiments and results	839
	6.5.	Policy implications	847
7.	The implications for agricultural policies and planning *Masatoshi Yoshino, Zenbei Uchijima and Hiroshi Tsujii*		853
	7.1.	Summary of results	853
	7.2.	Discussion	858
	7.3.	Implications for policy and planning	860

Acknowledgment   864
References   865

List of Contributors

YOSHINO, Professor Masatoshi
Institute of Geoscience
University of Tsukuba
1-1-1 Tennodai
Sakuramura
Niihari-gun
Ibaraki Prefecture
305 Japan

HORIE, Professor Takeshi
Institute of Crop Sciences
Faculty of Agriculture
Kyoto University
Kitashirakawaoiwakecho
Sakyo-ku
Kyoto
606 Japan

SEINO, Dr. Hiroshi
Department of the Environment
Kyushu National Experimental Station
496 Izumi
Chikugo-shi
833 Japan

TSUJII, Dr. Hiroshi
Farm Accounting Institute
Faculty of Agriculture
Kyoto University
Kitashirakawaoiwakecho
Sakyo-ku
Kyoto
606 Japan

UCHIJIMA, Dr. Tatsuro
Experimental Farm Division
National Institute of Agro-Environmental
 Science
3-1-1 Kannon-dai
Yatabemachi
Ibaraki Prefecture
305 Japan

UCHIJIMA, Dr. Zenbei
Agrometeorology Division
National Institute of Agro-Environmental
 Science
3-1-1 Kannon-dai
Yatabemachi
Ibaraki Prefecture
305 Japan

Abstract

Over the last few decades in Japan, significantly improved agricultural practices and technology have been introduced in order to increase and stabilize food production. However, despite these advances, the influence of climatic variations on agricultural production remains considerable. In particular, the effects of climate on rice production are of special importance in Japan, where rice is the staple food, and represents a vital political as well as economic commodity. This case study evaluates the effects of specified climatic changes on plant productivity, on rice production and on the national rice market in Japan, and considers possible technological or policy responses to these changes.

The focus of the study is on the coolest regions of Japan (in the northernmost districts of Hokkaido and Tohoku, and at higher elevations) which are the areas where rice crops are most susceptible to cool temperature damage (particularly in the summer months). Three types of "impact model" have been employed to assess the effects of climate on production:

(1) Empirical-statistical models that relate plant production or crop yields to climate using regression equations.
(2) A process-based rice yield simulation model.
(3) An integrated regional econometric model that simulates the relationships between supply, demand, government policy and trade in the Japanese rice market.

In line with the four other case studies in cool temperate and cold regions considered in the IIASA/UNEP project and reported in this volume (in Saskatchewan, Iceland, Finland and the northern European USSR), experiments have been conducted to estimate the impacts of several different climatic scenarios. Four types of scenario are represented, usually defined on the basis of temperature (the major limiting factor for rice cultivation):

(1) The baseline climate, indicative of present-day (1951–80) climatic conditions.
(2) Single anomalous years, either unusually warmer or cooler than the baseline, that have been recorded in the historical instrumental climatic record.
(3) Anomalous decades, also identified in the instrumental climatic record, that have been selected either on the basis of mean temperature deviations from the baseline, or as a reflection of unusually stable or unstable agroclimatic conditions (indicated by historical time series of annual crop yields as well as of temperatures).

(4) The climate estimated for doubled concentrations of atmospheric carbon dioxide, based on the predictions of the Goddard Institute for Space Studies (GISS) general circulation model.

The results demonstrate the clear contrasts between the impacts of future cooling or warming on Japanese agriculture. Cool conditions are inevitably associated with cool summer damage to rice in northern Japan, and a cool year such as 1980 (with mean July–August temperatures 1.6°C below the baseline in Hokkaido) could lead to a 7% decrease in relative total net production of vegetation (RTNP – over the whole of Japan), but a 40% drop in rice yields relative to long-term values (or 17% relative to recent values for higher yielding cultivars) in Hokkaido. In fact, the shift in cultivable limits for rice under a 1980-type climate effectively demarcates nearly all of the island of Hokkaido as unsuitable for rice cultivation. Thus consecutive impacts of this type of event could have dramatic consequences for rice production in Japan. These are illustrated using the econometric model and assuming 10 consecutive occurrences of a 1980-type year. Considerable reductions in regional rice supply are estimated and the rapid decline in stocks that would result if national policies were not changed would lead to a rice shortage situation that could jeopardize national food security.

Anomalously warm conditions, in comparison, are generally beneficial to agricultural productivity under present conditions. For example, a warm year such as 1978 (with mean July–August temperatures 2.5°C above the baseline value in Hokkaido) could lead to a great expansion of cultivable rice area, an increase in rice yield (assuming that crops are, as at present, fully irrigated) of 15% relative to the long-term mean (or 6% relative to values for modern cultivars), and an increase in RTNP of 3%. This temperature anomaly represents an *extreme event*, but under the GISS model-derived 2 × CO_2 scenario the *average* temperature conditions would be higher still (about 3.5°C above the baseline values in Hokkaido).

Such a warming would be a mixed blessing for rice producers in northern Japan, however. Although a warmer mean climate (assuming interannual temperature variability remained at its baseline levels) would automatically lead to a decreased frequency of cool summer events, the higher temperatures in average and warm years would probably have a detrimental effect on rice yields, because the crop would ripen too quickly, before the potential for grain assimilation has been fully realized. This is shown by a predicted rise of only 5% in rice yields under the 2 × CO_2 climate (compared with 15% under 1978-type conditions), relative to the long-term baseline. This is clearly a poor response for a well-managed commercial crop since the modeled impact of the 2 × CO_2 scenario on net production of natural vegetation is greater (+9% relative to the baseline; this, moreover, for the whole of Japan including areas where the estimated warming is less than for Hokkaido). Nevertheless, the impact of the scenario on rice supply and on national rice stocks would be considerable, leading to an acute rice surplus that would require drastic government measures for its disposal. This situation would only be exacerbated if farmers were (as is probable) to adapt their cultivation methods to the new climate, for example by introducing more heat-loving rice varieties with earlier planting dates. Such adjustments might result in increases of rice yields in Hokkaido of up to 25% relative to the recent levels, and further increases in rice stocks.

The implications of the modeled impacts on rice point to several adjustments in government pricing and subsidization policies that would be necessary to stabilize the domestic rice balance, including a longer-term movement towards minimized government intervention and a liberalized rice market. Finally, the results of this assessment indicate that research into methods of coping with climatic change should be a high priority in government planning during the coming years.

SECTION 1

Introduction: The Policy and Planning Issues

1.1. Background and Purpose

The influence of climate on agriculture in Japan is quite striking. Its influence on rice production is especially important because rice is the main staple in Japan and, being a tropical plant, is grown close to its limits of tolerance to low temperatures in the northernmost regions of the country.

Almost all the rice produced in Japan is cultivated in paddy fields flooded with water that is supplied through well developed waterways. Thus, unlike most other countries in the world, water hardly plays a role as a yield-reducing factor in Japanese rice production. The major climatic factors that significantly influence rice production are temperature (especially in the north), strong winds and heavy rainfall associated with typhoons, and solar radiation. The location and timing of the typhoon damage is unpredictable, but as a whole Japan loses a certain amount of rice production every year through their effect. Typhoon damage is considered to be unavoidable, and typhoons themselves bring much precipitation that is necessary for production of rice.

Some of the greatest losses to rice crops result from cool summer conditions that occur quite frequently in northern Japan (particularly in Tohoku and Hokkaido districts, where mean July–August temperature ranges from 18 to 20 °C). Low summer temperature can result in delay of growth, reduced plant height, or sterility of the grains of rice plants. If the temperature of irrigation water is lower than 18 °C this may also obstruct growth and result in standing g.een plants remaining in autumn that are unsuitable for harvesting. It has been estimated that, for example, cool summer damage occurs somewhere in Tohoku district about one year in four, and in Hokkaido the frequency is higher still.

During the last 20 years, the position of agriculture in the national economy of Japan has been rapidly declining. For example, the proportion of those engaged in agriculture to the total active population was 26.8% in 1960 but

11.2% in 1975. During the same period, the contribution of agriculture to the gross domestic product decreased from 10.2% to 5.0%. The basic relationship between climate and rice production has, however, altered very little. Year-to-year variations in rice production due to climate remain significant.

The purpose of this case study is to quantify the effect of both short- and long-term climatic changes on Japanese agriculture and, in particular, to study how variations in climate influence rice yields and how these variations in yields can affect national agricultural production. In common with other case studies in the IIASA/UNEP project these estimations (a) involve experiments with a hierarchy of models – climatic, crop-yield and economic – and (b) are derived for different climates described by meteorological data from recent warm and cool periods and, as a special case, for the climatic conditions estimated for a doubled concentration of atmospheric carbon dioxide by the Goddard Institute for Space Studies (GISS) general circulation model (Hansen et al., 1983, 1984).

This introduction provides background information on the physical environment and agriculture in Japan, and outlines considerations behind the selection of the crops, regions and models used in this study.

1.2. The Environment and Agriculture in Japan

1.2.1. Topography and ocean currents

The Japanese Islands extend in a southwest-northeast direction through 22 degrees of latitude, from 24 to 46°N over a distance of about 2200 km; i.e., from the subtropical to alpine zones. The total land area is 369 883 km^2, of which the four main islands, Honshu (227 414 km^2), Hokkaido (78 073 km^2), Shikoku (18 256 km^2) and Kyushu (36 554 km^2), make up the major proportion (*Figure 1.1*). All the islands are mountainous or hilly and the flat lowland plains tend to be very narrow features found mostly along the coast and banks of larger rivers. The rivers on the islands are commonly very short and even the longest, the Shinano River, is less than 400 km. The area under cultivation comprises only 17% of the total land area.

The most important water mass near Japan is the Pacific Ocean, because of its physical influence on Japanese climate. Between the Japanese Islands and continental Asia lies the Sea of Japan, enclosed by the central arc of the Japanese Islands to the south and east and by the Korean Peninsula to the west. The Sea of Okhotsk, to the north of Japan, is bounded to the East by the island chain of the Kurils and the Kamchatka Peninsula. It is the source region of the cold and moist Okhotsk air mass, which exerts a considerable effect on the weather and climate of northern Japan. In contrast to the two above-mentioned seas, the East China Sea is merely a marginal stretch of shallow water lying on the continental shelf (*Figure 1.2*).

Three principal oceanic currents wash the shores of the Japanese Islands: the warm Kuro-shio, cold Oya-shio, and warm Tsushima Currents, and each has a strong influence on the climate. The Kuro-shio or Japan Current is the most important, carrying a total water volume of some $3-5 \times 10^7 \, \text{m}^3/\text{s}$, together with

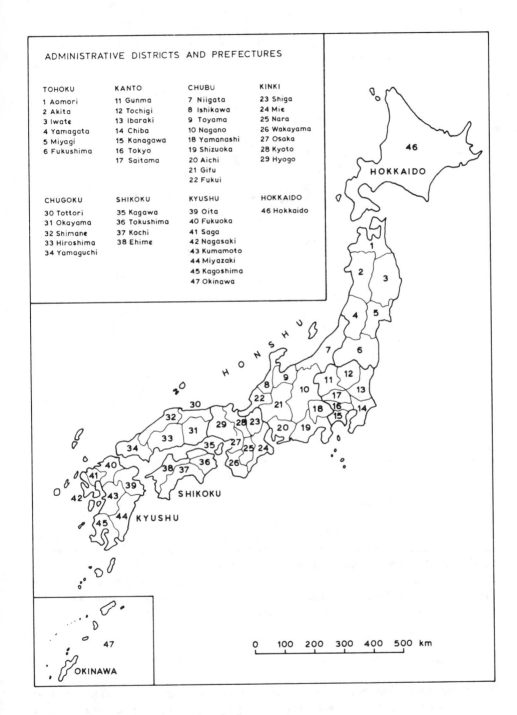

Figure 1.1. Administrative regions in Japan.

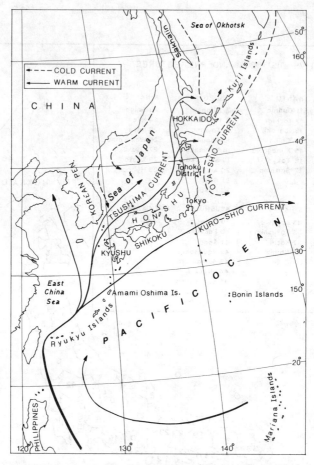

Figure 1.2. Location of the Japanese Islands, seas and ocean currents.

an enormous amount of heat and salt, and for this reason is often compared with the Gulf Stream that washes the southeastern coast of the USA.

The Oya-shio or Kuril Current originates in the cold water on the eastern side of Sakhalin, together with the southward extensions of the cold currents that lie off the eastern coast of Kamchatka. It flows southwards via the southern coast of the Kuril Islands, to the east of Hokkaido, and then extends along the eastern coast of Tohoku district (or Sanriku), finally meeting the Kuro-shio current to form a marked convergence zone (*Figure 1.2*).

1.2.2. Climate

The three major features of Japan's climate are the winter monsoon, the Bai-u (rainy) season and the typhoon (tropical storm). In winter the development of a low pressure area (the "Aleutian Low") and its position relative to the "Siberian

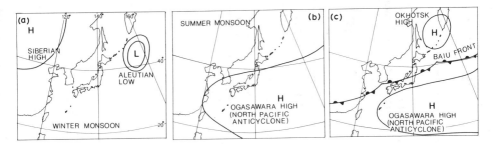

Figure 1.3. Position of the dominant air masses during (a) the winter monsoon, (b) the summer monsoon, and (c) the Bai-u season.

High" pressure [Figure 1.3(a)] gives rise to northwesterly winds (the winter monsoon). This results in a distinctive weather pattern due to orographic controls: abundant snowfall and cloudy weather in areas facing the Sea of Japan and dry, clear weather in the areas facing the Pacific Ocean. In summer, southwestern Japan is usually dominated by the Ogasawasa high pressure area [Figure 1.3(b)] giving southeasterly winds (the summer monsoon) and fine weather throughout Japan in the high summer season. The Bai-u season (the early-summer rainy period) and the Shurin season (autumn rainy period) are characterized by the stagnation of the polar front as it moves northwards (Bai-u) and southwards (Shurin), respectively [Figure 1.3(c)].

As a consequence of the influence of these contrasting air masses, Japan's climate is rather continental in character with a large annual range of temperature. For example, the annual range of 1931–60 mean temperature in Tokyo is 22.6 °C which compares with 12.0 °C in Gibraltar and 9.6 °C in Los Angeles, at similar latitudes in Western Europe and on the Pacific coast of the United States, respectively (Table 1.1). Thus Japan can be seen to exhibit a low mean winter temperature relative to other locations at the same latitude, and higher summer temperatures, more synonymous with those of tropical regions. This, together with the high amount of insolation during summer months, permits the cultivation of rice, which was originally a product of the tropics.

In addition to the two maxima in the seasonal march of precipitation in Japan (the Bai-u and the Shurin), another precipitation maximum commonly occurs during the winter monsoon, bringing large snowfalls in coastal areas adjacent to the Sea of Japan. The airstream absorbs a large amount of water vapor during its passage across the Sea of Japan and winds then blow up the slopes of the coastal mountains, depositing snow in abundance on their windward slopes (Figure 1.4). Hokuriku district (on the Sea of Japan side of central Honshu) holds the world record for maximum observed snow depth.

In addition to these rainy seasons typhoons bring occasional torrential rain in autumn, so that Japan has an overall precipitation receipt that is remarkably high for a temperate region. In Hokkaido, mean annual precipitation at all locations is more than 1000 mm, and totals may exceed 3000 mm in the Pacific slope areas of south-western Japan and on the coast of the Sea of Japan.

Table 1.1. A comparison of the climatic characteristics at Tokyo with those at Gibraltar and Los Angeles (at approximately the same latitude) and at stations in the other case study regions reported in this volume. (Sources: Müller, 1982; Lamb, 1972; G.D.V. Williams, personal communication.)

Country	City	Latitude	Longitude	Height a.s.l. (m)	No. of years of record[a]	Mean air temperature (°C)				Annual precipitation (mm)
						Jan	July	Annual	Annual range	
Japan	Tokyo	35°41′N	139°46′E	4 m	30	3.7	25.1[b]	14.7	22.7	1563
Gibraltar	Gibraltar Town	36°06′N	05°21′W	27 m	30	12.5[c]	23.8[b]	17.8	12.0	863
USA	Los Angeles	34°03′N	118°14′W	103 m	30	13.2	22.8	18.0	9.4	373
Canada	Saskatoon	52°08′N	106°38′W	157 m	29	−18.1	19.0	1.7	37.1	359
Iceland	Reykjavik	64°08′N	21°56′W	18 m	30	−0.3	11.4	5.1	11.7	779
Finland	Oulu	65°01′N	25°29′E	17 m	30	−9.8[c]	16.5	2.3	26.5	514
USSR	Leningrad	59°58′N	30°18′E	4 m	30	−7.6[c]	18.4	4.6	26.3	559

[a] Period 1931–60.
[b] Maximum in August.
[c] Minimum in February.

Introduction

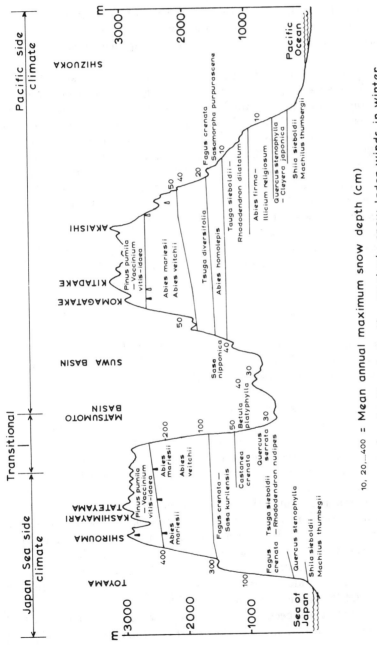

Figure 1.4. Cross section from the Sea of Japan coast to the Pacific coast showing altitudinal vegetation zones and bioclimatic conditions (after Yoshino, 1978a).

1.2.3. Vegetation

The vegetation in Japan exhibits a great diversity and high productivity. Zones of natural vegetation can be identified that are closely associated, both latitudinally and altitudinally, with the three major forest types. These are:

(1) Evergreen broad-leaved or laurel forest, represented by evergreen oaks.
(2) Deciduous broad-leaved forest, represented by beech trees.
(3) Evergreen coniferous forest, represented by fir and spruce trees.

The corresponding annual mean temperatures of these zones are 13–21 °C, 6–13 °C, and <6 °C, respectively. The first forest zone in Japan occupies the southern half of the main island chain and the northern half of the Okinawa Islands. Its northern limit reaches 37.5°N on the Pacific coast in Fukusima Prefecture. The second forest zone occupies the northern half of Honshu and the lowlands of Hokkaido south of 43.5°N. The third forest zone occupies the remainder of Hokkaido. The altitudinal upper limit of the evergreen broad-leaved forests of Kyushu and Shikoku occurs at a height of 700–1000 m a.s.l., with the upper levels containing mixed forests of broad-leaved and coniferous trees. A cross-section of the altitudinal zonation of vegetation in central Japan is given in *Figure 1.4*.

The potential productivity of natural vegetation in Japan under different climatic conditions is explored further in Section 3 of this report.

1.2.4. Agricultural land use

Japan's staple agricultural product is rice, which accounts for more than 40% of the nation's cultivated land (*Tables 1.2* and *1.3*). Its cultivation is strongly promoted by national price-support policies which guarantee reasonable incomes to producers.

In many mountain areas, rice became a more popular crop after World War II, and today it is cultivated up to an altitude of 1370 m, at the foot of the Norikura volcano (36°10′N, 137°30′E) in the so-called Japanese Alps. Expansion of rice cultivation into the northern regions has also been a gradual process. The area of rice fields in Tohoku District, northeast Japan, increased by 60% (compared with 30% over the whole country) during the period 1890–1970. In Hokkaido rice was not important until the late nineteenth century, American advisors suggesting that barley or potatoes were more suitable. However, the Hokkaido Government recommended rice cropping in 1901 and the Ishikari Plain and Kamikawa Basin were subsequently developed into paddy, the area under rice in Hokkaido increasing from 3000 ha in 1880 to 290000 ha in 1970 (Tabayashi, 1983). The gradual northward shift of rice cultivation areas in Japan since about 300 BC is illustrated in *Figure 1.5*.

This northward shift has been enabled by the development of cold-resistant rice cultivars and the development of heated rice nurseries. The effective

Table 1.2. Area of cultivated land in Japan, 1981. (Source: Crop Statistics, Ministry of Agriculture, Forestry and Fisheries.)

Type of cultivated land	Area (1000 ha)	Proportion of total (%)
Paddy field	3031	55.7
Upland field	1241	22.8
Temporary meadow	589	10.8
Other permanent crop field	581	10.7
Total	5442	100.0

Table 1.3. Planted area in Japan, 1981. (Source: Crop Statistics, Ministry of Agriculture, Forestry and Fisheries.)

Type of crop	Area (1000 ha)	Proportion of total (%)
Rice	2278	40.2
Wheat, barley, oats and rye	352	6.2
Sweet potatoes	65	1.1
Potatoes	120	2.1
Miscellaneous cereals	26	0.5
Pulses	265	4.7
Vegetables	647	11.4
Fruits and nuts	404	7.1
Industrial crops	269	4.8
Tea	61	1.1
Forage and manure crops	1057	18.7
Mulberries	117	2.1
Total	5661	100.0

accumulated temperature (ΣT_{10}) required for ripening of most rice varieties is normally 3500–4500 degree-days, but cold-resistant varieties require only 2500–3000 degree-days. [Effective accumulated temperature refers to the annual accumulated sum of air temperatures above 0 °C during those days when daily mean air temperature is greater than 10 °C. This measure is also used in the USSR case study (*see* Part V, this volume).] In Hokkaido rice is mainly grown in the regions in which the effective accumulated temperature exceeds 2500 degree-days (*Figure 1.6*). Its chief producing districts are the Ishikari Plain and the Kamikawa Basin, where average temperatures during the warmest month (August) are above 20 °C. On the Pacific Coast of the eastern and northeastern part of Hokkaido, the mean August temperature is lower than 20 °C and summers are particularly cool when persistent easterly winds blow from the Sea of Okhotsk, often leading to cool weather damage of the rice crop. Central Japan also suffers occasional cool weather damage in the highland areas. These effects are discussed later.

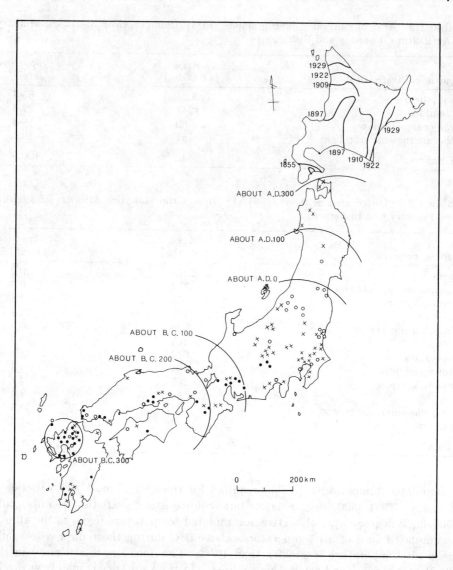

Figure 1.5. Expansion of paddy cultivation in Japan since about 300 BC (isolines). Symbols denote sites of early cultivation: black circles – early Yayoi Period (until about 100 BC); white circles – middle Yayoi Period (until about 100 AD); crosses – later Yayoi Period (until about 300 AD) (after Tabayashi, 1983).

1.3. Climatic Variations and Rice Production

The cultivated area, yield and total production of rice in Japan over the period 1883–1984 are shown in *Figure 1.7*. Years of poor harvests (e.g., 1883, 1889, 1897, 1902, 1934, 1941, 1945, 1953, 1971, 1976 and 1980) were mainly caused by cool summer conditions in northern regions and summer drought in the southwest. On the northern island of Hokkaido, where mean crop yields have

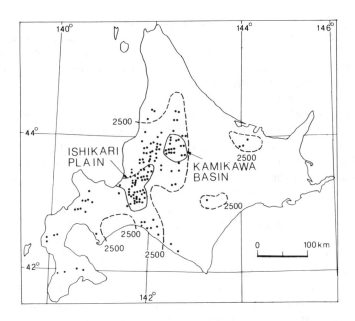

Figure 1.6. Areas of rice cultivation in Hokkaido in 1978. One dot represents 1000 ha and the broken lines are isopleths of 2500 degree-days (after Ichikawa *et al.*, 1984).

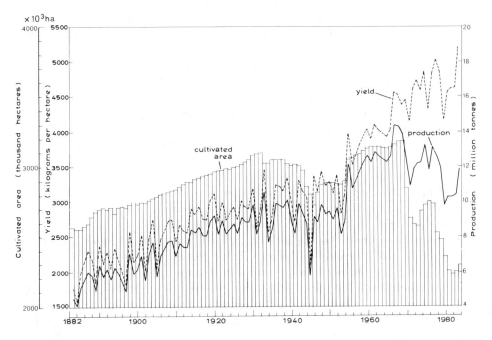

Figure 1.7. Secular change of paddy area, yields and rice production in Japan, 1883–1984 (Japanese Meteorological Agency, 1985).

increased considerably since the turn of the century due to improved cultivation methods and new varieties, the sensitivity of rice to low summer temperatures is still marked today. This is demonstrated in *Figure 1.8*, where mean July–August air temperatures have been plotted against rice yield for five periods between 1900 and 1973. If the air temperature drops below 20 °C yield drops sharply in each period, regardless of the level of mean yield (Tani, 1978). Hokkaido and Tohoku suffer more frequent damage by cool summer weather than elsewhere in Japan (*Figure 1.9*). During the period 1965–1983, for example, the years 1971, 1976, 1980, 1982 and 1983 were all characterized by cool summer damage. Farmers in these areas, however, continue to cultivate rice because it produces four times as much income as millet or sorghum, per unit area.

Estimates of the extent of damage to Japanese rice production by abnormal weather in the period 1969–1983 are given in *Table 1.4*. The annual production data are as shown in *Figure 1.7* and shortfalls attributable to weather, disease and pests are assessed relative to the potential (calculated on the basis of production trends). The total area affected by abnormal weather is obtained by a summation of the damaged field area.

During the 15 years from 1969 to 1983, the best harvest (lowest percentage losses relative to potential production) of paddy field rice was obtained in 1978, and the worst in 1980. Yields both in 1971 and in 1980 were very low, but the production loss attributable to anomalous weather was considerably greater in 1980 (1603 thousand tonnes) than in 1971 (1090 thousand tonnes).

In the years 1981–1983, while yields were close to the 1969–1983 average, cultivated area was much smaller than in the preceding years and thus production losses relative to the potential were greater than previously. However, the proportion of these production losses attributable to abnormal weather was markedly higher than before, and the area affected by cool summer damage was large in each year. Cool summers prevailed in each of the years 1980–83, although the effects were partially compensated in 1983 by favorable conditions for rice production in southwest Japan (*Table 1.4*).

For comparison, data on combined wheat and barley production in Japan over the same period (1969–1983) are given in *Table 1.5*. Grain crops occupy less than one-sixth of the area under rice (*Table 1.3*) and yields are considerably less. The proportion of annual damage to cultivated wheat and barley is significantly greater than for rice, although total production losses are less. Interestingly, there appears to be little correspondence between the characteristic weather events that cause damage to wheat and barley and the events that are detrimental to rice production (cf. *Tables 1.4* and *1.5*).

These interannual fluctuations in national production disguise significant differences between regions in Japan. *Table 1.6* relates rice yields to summer (July–August) temperatures in five districts representative of the full range of Japanese conditions. Temperatures are expressed as anomalies relative to the 1951–1980 mean, and examples are given of a "good" harvest year (1978), a "bad" harvest year (1980) and two "moderate" harvest years (1982 and 1983).

Introduction

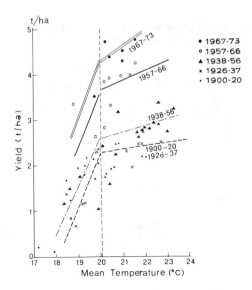

Figure 1.8. Relationship between yield of paddy rice and mean July–August air temperature during five periods in Hokkaido (Tani, 1978).

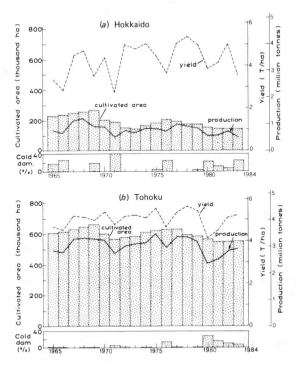

Figure 1.9. Secular change of paddy area, yields and rice production, and the area affected by cool summer damage, 1965–1983, in (*a*) Hokkaido, and (*b*) Tohoku (Japanese Meteorological Agency, 1985).

Table 1.4. Paddy field rice yield, production losses and damage attributable to weather, 1969–1983 (after Yoshino and Yasunari, 1986).

Year	Yield (kg/ha)	Ratio of damage[a] (%)	Damage by abnormal weather		Notes
			Area (th.ha)	Production loss (thousand tonnes)	
1969	4350	10.2	1201.0	709.3	
1970	4420	9.2	1173.0	375.9	
1971	4110	15.2	2032.0	1090.0	Bad harvest (cool summer) in N Japan
1972	4560	7.3	722.9	358.5	
1973	4700	6.6	544.2	275.3	Good harvest (warm summer)
1974	4550	9.3	736.6	308.6	
1975	4810	6.4	898.5	298.0	Good harvest (warm late-summer)
1976	4270	16.6	1963.0	1321.0	Bad harvest in N and Central Japan
1977	4780	5.9	517.2	213.9	
1978	4990	5.6	664.2	294.6	Best harvest (warm summer in N Japan)
1979	4820	7.1	976.8	420.4	
1980	4120	22.0	1992.0	1603.0	Bad harvest (cool summer in N Japan)
1981	4530	13.0	1504.0	956.3	Cool summer
1982	4580	13.5	1792.0	896.9	Cool summer
1983	4590	12.3	1646.0	711.5	Cool summer in N Japan, but warm summer in SW Japan
Average	4590	10.7	1224.2	655.5	

[a] Percentage reduction in production relative to potential production, due to abnormal weather, diseases and pests.

1.4. Techniques and Policies for Mitigating Climate Impacts

Countermeasures for damage caused by weather can be divided into *permanent* and *temporary* measures, according to the time scale of their execution and the continuity of their effect. The permanent measures are usually implemented at scales ranging from regional to state level, while the smaller-scale farm or local responses tend to be more temporary in character (Yoshino, 1983; Parry and Carter, 1984; Clark, 1985).

In general, three types of countermeasures can be identified (Tani, 1978):

(1) Mitigation of climatological conditions themselves (the direct cause of impacts).

Table 1.5. Wheat and barley yield, production losses and damage attributable to weather, 1969–1983 (after Yoshino and Yasunari, 1986).

			Damage by abnormal weather		
Year	Yield (kg/ha)	Ratio of damage[a] (%)	Area (th. ha)	Production loss (thousand tonnes)	Notes
1969	2650	15.8	185.8	102.3	Warm winter
1970	2070	38.8	194.8	183.1	
1971	2650	18.1	100.5	64.2	
1972	2500	23.1	93.0	60.0	
1973	2700	18.7	47.5	32.5	
1974	2800	16.2	49.5	33.5	Cold, snowy weather
1975	2690	18.3	54.8	34.6	
1976	2500	27.6	68.7	58.3	
1977	2750	20.2	53.2	42.4	
1978	3270	7.7	40.0	21.0	
1979	3630	5.3	37.5	18.6	Abnormally warm winter
1980	3050	14.5	120.0	72.7	
1981	2620	32.8	174.1	215.8	Heavy snow, cold winter
1982	3260	16.5	151.4	106.8	
1983	3030	24.8	161.9	160.0	Cold winter

[a] Percentage reduction in production relative to potential production, due to abnormal weather, diseases and pests.

Table 1.6. July–August air temperature anomalies relative to the period 1951–1980 and rice yield in four districts of Japan, in 1978, 1980, 1982 and 1983. (Source: Ministry of Agriculture, Forestry and Fisheries, various dates; Japanese Meteorological Agency, 1981.)

	1978		1980	
Districts	Temperature anomaly (°C)[a]	Yield (kg/ha)	Temperature anomaly (°C)	Yield (kg/ha)
Hokkaido	+2.6	5360	−1.7	3850
Tohoku	+2.2	6640	−2.5	3990
Chubu	+1.9	9360	−1.5	4420
Kyushu	+1.1	4670	−1.3	3980
Okinawa[b]	−3.4	2760	−2.6	2550

	1982		1983	
Districts	Temperature anomaly (°C)	Yield (kg/ha)	Temperature anomaly (°C)	Yield (kg/ha)
Hokkaido	+0.8	5010	−0.5	3550
Tohoku	−2.5	5080	−1.0	5220
Chubu	−1.0	4540	0.0	4610
Kyushu	−1.0	4380	+0.7	4440
Okinawa[b]	−3.2	2700	−2.9	2700

[a] Different stations used for averaging than in *Table 2.1*.
[b] Part of Kyushu district.

(2) Development of more resistant crops.
(3) Changing the nature, timing and location of crop cultivation in order to avoid encountering adverse climatic conditions.

1.4.1. Modifying the direct effects of climate

One method of moderating the most harmful effects of strong wind on crops has been to use wind breaks or shelter belts. These have been effective as protection against the cool northeasterly *yamase* winds, originating from the Okhotsk Anticyclone [cf. *Figure 1.3(c)*]. Rice plants generally show improved growth and yield for distances on the leeward side of shelters (normally trees) of up to ten times their height (Yoshino, 1975).

1.4.2. Breeding and propagation of cold-resistant varieties

The breeding of high yielding, early ripening varieties of rice has gone some way towards combating the problems of cold damage, while markedly increasing average yields. For example, yields have increased in Hokkaido from about 2.5 t/ha in normal years around 1925, to about 6.0 t/ha around 1975 (with yields during cool summer years exceeding 3 or 4 t/ha).

In Tohoku District an early maturing and high yielding variety called "Fujisaka-5", having a tolerance to injury through sterility, was developed in 1949. This was a significant advance for rice cultivation in northern Japan: in two subsequent cool summer years, 1953 and 1954, it was estimated that the benefit from planting "Fujisaka-5" compared with older varieties was an

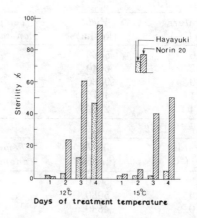

Figure 1.10. Varietal differences of sterility caused by coolness at the booting stage (after Satake, 1978).

additional 12 000 tonnes and 50 000 tonnes, respectively (Toriyama, 1978). The variation in tolerance to cool conditions between different varieties can be further demonstrated in experimental conditions. For example, cooling treatment at 15 °C for 4 days induced 51% sterility in a cool-sensitive variety "Norin 20", while only 5% sterility in a highly tolerant variety "Hayayuki" (*Figure 1.10*).

There are few examples of agricultural policies directed specifically toward mitigating the effects of climatic variations. However, developments of the techniques, experiments and researches mentioned above have been encouraged and supported at the national level and agricultural insurance schemes have been developed to reduce impacts on farm incomes. In general policies to mitigate climatic impacts *have* emerged in an indirect fashion. More direct approaches that could be developed in the future are discussed in Section 7.

1.5. Structure of This Case Study

The approach adopted in this study has involved estimating the effects of a range of possible future climatic conditions (specified by a number of climatic scenarios) on the potential productivity of vegetation, on crop cultivation and yields, and on the national economy in Japan (described by a set of "impact models"). The implications of these effects are then examined and possible adjustment strategies are considered, both descriptively and quantitatively, for adapting to the changed conditions.

A broad outline of the case study structure is presented in *Figure 1.11*, illustrating the hierarchy of effects that are investigated. We will now explain briefly the choice of crops, study areas and impact models that are depicted in *Figure 1.11*.

1.5.1. Choice of crops for impact studies

The main focus of the study is on rice cultivation, since this is the staple food in Japan, and is grown in most parts of the country. As discussed above, the yield and production of rice are extremely weather sensitive, especially in the north of Japan, where low summer temperatures may often be damaging to the crop. However, the Japanese government is committed to the avoidance of rice shortages because the *Japonica* rice variety favored by the population is not widely grown outside Japan, and imports would thus be difficult to obtain.

In fact, the greater problem in Japan in recent years has been to prevent a rice *surplus*, rather than a shortage, and the government has been forced into large expenditures to subsidize prices and land transfers in order to maintain a balance between domestic supply and demand. Climate-induced changes in rice production and supply would disrupt this precarious balance and could have serious economic implications for Japan, an ample justification for the selection of rice in this study.

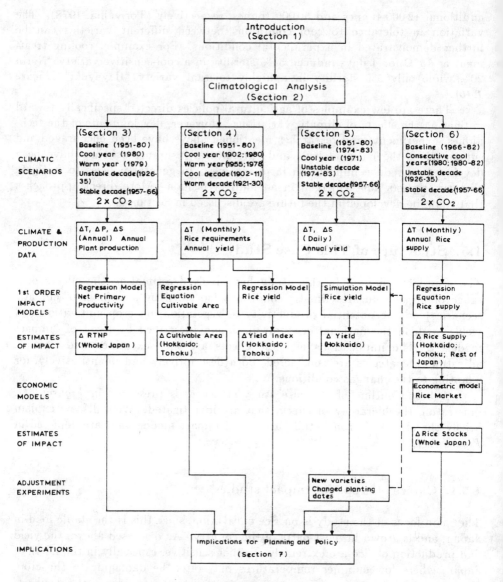

Figure 1.11. Structure of the Japan case study (schematic). Δ = change in T (temperature), P (precipitation), S (sunshine hours), $RTNP$ (relative total net production).

In addition, a national-scale assessment of the potential net primary productivity of natural vegetation in Japan is undertaken in the report. This provides a useful overview of the general resources for plant growth under changed climatic conditions, and presents a background for the crop-specific studies of rice.

1.5.2. Choice of areas for impact studies

The emphasis in impact experiments conducted for this study is on evaluating the effects of climatic variations on the productivity of rice, and on the spatial limits of its cultivation. The investigations have therefore been concentrated in those regions where the sensitivity of rice production to interannual weather variability is the greatest, namely the coolest areas in Japan (in the north and at higher elevations). The two northernmost districts, Hokkaido and Tohoku, thus provide the focus of study in each section of the report, while national assessments have been conducted for a few specific conditions.

District-level crop yield data are available, both for Hokkaido and for Tohoku, back to the end of the last century, and representative meteorological stations have been selected in each region to provide the spatially averaged climatic record for corresponding time periods. In the economic assessment (Section 6 of this report), rice supply is estimated both for Hokkaido and for Tohoku independently of the rest of Japan in order to evaluate the regional impacts on rice of cool summer damage in the north of Japan. The regional totals are then aggregated for the whole of Japan to permit a national assessment of impacts on the Japanese rice market (*see Figure 1.11*).

1.5.3. Impact models and data

Three types of model are employed in evaluating impacts of climate on agriculture and vegetation in Japan. The first, most common approach (used in Sections 3 and 4) for assessing impacts on net primary productivity (NPP) of natural vegetation and on rice yields, utilizes regression equations of the relationship between climate (the explanatory variable) and plant productivity (the dependent variable). For estimating rice yield (Section 4), only one explanatory climatic variable, temperature, is required while the NPP calculations (Section 3) involve combining a set of climatic variables into a dryness index and then relating this to productivity data (*see Figure 1.11*).

The second type of modeling approach is more process-oriented, and is represented by the rice yield simulation model presented in Section 5 of the report. Here, the development and yield of a rice crop are simulated from planting until harvest, as a function of the solar radiation incident on the crop and of the temperature conditions during its growth. The model operates on a daily time step, and is thus more demanding of meteorological data than the simple monthly- or annual-resolution regression models.

Thirdly, the economic assessment in Section 6 involves the use of an integrated regional econometric model for Japan which relates aggregated regional rice supply (defined as a function of technology, producer price and weather in Hokkaido, Tohoku and the rest of Japan) to national consumption, trade and stocks. As in Sections 3 and 4, the model equations are based on a regression approach, using relationships between variables developed for time series of data over a 17-year simulation period (1966–1982).

Each of the models used in the study has specific requirements for meteorological data. The least demanding are the rice yield regression models and the econometric model that only require mean July–August temperatures, for which data are available at many stations over long time periods back into the nineteenth century. More variables are required (though only on a mean annual basis) for the NPP calculations in Section 3 and these are not all available over long periods or at so many geographical locations. The daily weather data required in the simulation model (Section 5) limit the model analysis to the post-World War II period because data are extremely restricted before this time.

Data on plant production, rice yields, supply, consumption and trade have been obtained from several sources. National statistics are published annually that provide information on rice production by prefecture and at a national scale in relation to the domestic rice market and international trade. Plant production data are those compiled for the International Biological Programme (*see* Section 3).

Finally, where appropriate, the climatic data have been spatially averaged to be as representative as possible of the areas for which impacts have been measured or are estimated. The next step is to choose a set of climatic scenarios for Japan to provide the input data for modeling the likely impacts of future climatic conditions. The selection of climatic scenarios is described in the next section.

SECTION 2

Development of the Climatic Scenarios

2.1. Purpose

This section describes the range of climatic variations observed in Japan in recent history, and the scale of the changes that may accompany future increases in atmospheric carbon dioxide. The description provides a basis for the characterization of three types of climatic scenario that will subsequently be used to estimate the potential impact of climatic variations on rice yield and rice production in Japan:

(1) The weather in extreme individual years selected from historical records of agroclimatic conditions.
(2) The climate during anomalous sequences of years, also determined from historical records.
(3) A simulated climate under doubled concentrations of atmospheric carbon dioxide based on estimations from the Goddard Institute for Space Studies (GISS) general circulation model.

Finally in this section, the patterns of climatic change implied in the scenarios for the Japan region are examined in relation to the synoptic climatological features that might explain them.

2.2. Climatic Changes in the Historical Period

Evidence for climatic variations in the period before instrumental meteorological observations have been assembled for two regions in eastern Asia, labeled broadly as China and Japan (*Figure 2.1*). Documentary information and proxy data were used to derive a temperature classification characterizing conditions in these two countries during the last two millenia. A composite chronology for East Asia is presented on the right of the diagram. The most striking features

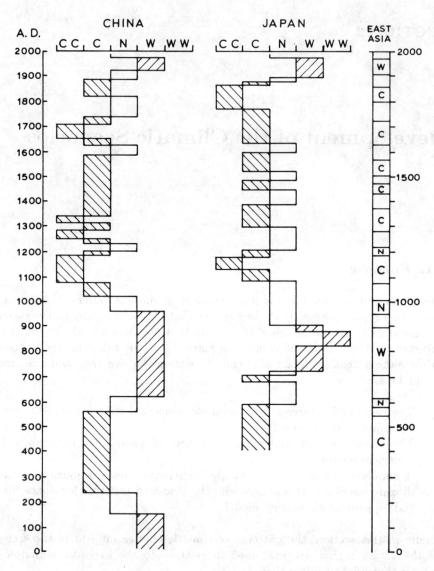

Figure 2.1. Temperatures in China and Japan during the historical period. CC, very cold; C, cold; N, normal; W, warm and WW, very warm. The temperature chronology for East Asia (right-hand column) is a composite of the other two graphs (after Yoshino, 1978b).

are the warm period from about AD 700–900 (with similar conditions to the climatic optimum or hypsithermal of 6000–5000 years BP), and the repeated cold periods between about 1070 and 1870, with a particularly cold spell in Japan in the early nineteenth century when the northern part of Honshu suffered frequently from severe famine due to crop failures in cool summers.

2.3. Recent Climatic Variations

Three meteorological variables are of particular importance for describing the thermal, moisture and light conditions that determine crop growth: air temperature, precipitation and solar radiation. Data on the last of these are only available from a few stations for recent years, but temperature and precipitation records are of greater longevity and denser spatial coverage in Japan. The emphasis in this study is on the effects of changes in temperature on rice production since, in general, rice is a fully irrigated crop. The locations of meteorological stations supplying data used in the study are shown on *Figure 2.2*.

Figure 2.2. Location of meteorological stations used in the Japan case study.

2.3.1. Air temperature

Homogeneous records of observed temperatures are available for the period since about 1900 in Japan. *Figure 2.3* shows the secular changes of annual, summer

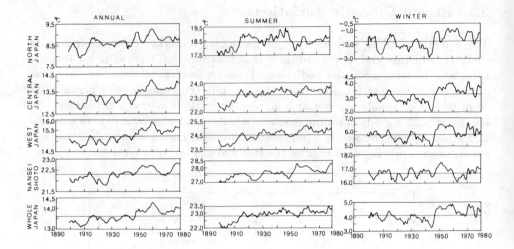

Figure 2.3. Secular change of annual, summer (June, July and August) and winter (December, January and February) air temperature (5-year running mean) in 4 regions and in the whole of Japan, 1900–1980 (Japanese Meteorological Agency, 1984).

and winter air temperature in 4 regions of Japan and in the country as a whole (Japanese Meteorological Agency, 1984). The curves for northern Japan (average of 6 stations), central Japan (6 stations) and western Japan (7 stations) correspond quite closely in the coincidence of minima and maxima, but the curve for the Nansei Islands (2 stations) is somewhat different. However, all curves of annual temperatures display minima around 1908–1910 and a maximum around 1960, and both annual and seasonal curves exhibit an upward tendency, upon which is superimposed wave-like fluctuations. The range of these fluctuations is greater in the north than in the south, resembling the tendency elsewhere in the higher latitudes (*see* Part I, Section 4 of this volume).

2.3.2. Precipitation

Long-term instrumental data on seasonal and annual precipitation are presented in *Figure 2.4*. Annual precipitation in all regions increased between 1940 and 1955, this wet period coinciding with the warming period mentioned above. Since 1955, precipitation has generally decreased and the late 1970s were characterized by drier than average conditions with lower temperatures following a peak.

2.4. Instrumentally Derived Scenarios

The instrumental meteorological observations described above provide the data base with which to characterize recent climatic variations in Japan. By studying these records alongside those of agricultural productivity for the same period, it

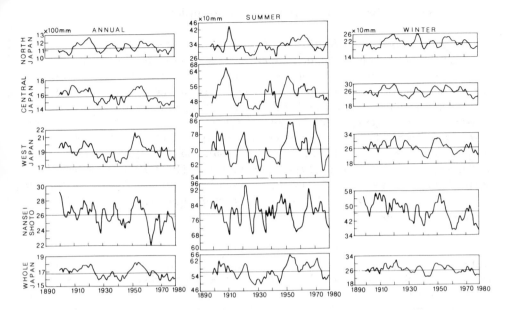

Figure 2.4. Secular change of annual, summer (June, July and August) and winter (December, January and February) precipitation (5-year running mean) in 4 regions and in the whole of Japan, 1900–1980 (Japanese Meteorological Agency, 1984).

is possible to identify anomalous climatic events that coincide with particular levels of agricultural productivity. This was the method employed in the Japanese case study to select instrumental climatic scenarios.

The emphasis of this report on rice productivity (an irrigated crop), and its focus in northern Japan, mean that air temperature is the dominant meteorological variable influencing crop yields. It has thus formed the basis of the scenario descriptions. The instrumental scenarios are of three types:

(1) The reference or "baseline" climate.
(2) Individual years or periods selected with respect to temperature anomalies alone.
(3) Single years or periods chosen on the basis of observed rice yield anomalies.

2.4.1. The baseline climate

In order to provide a reference against which to compare the effects of the different climatic scenarios, the standard 30-year period 1951–1980 was chosen as the "baseline" scenario. To be consistent with investigations reported for the other cool temperate and cold regions in this volume, temperature and precipitation anomalies, including those derived for doubled concentrations of atmospheric CO_2 by the GISS model, were each evaluated relative to this baseline (except where appropriate data were unavailable).

2.4.2. Scenarios based on temperature anomalies

The approach adopted in Section 4 for selecting instrumental climatic scenarios involved the inspection of the air temperature records for specified study areas in Japan, and the identification of anomalously warm or cold years during the period of observations. Temperature conditions in the months of July and August are probably the most critical for rice yields, and so most scenarios were constructed from averaged July–August temperatures. In Section 4 six temperature anomalies relative to the baseline climate were identified for Hokkaido and Tohoku districts: warmest (1978), warm (1955), cool (1980) and coolest (1902) individual years, and warm (1921–1930) and cool (1902–1911) decades (*Table 2.1*). Another scenario, in Section 5, representing a climatic "worst case", was the individual year (1971), selected on the basis of large negative anomalies during the growing season both of temperature and of sunshine hours.

Table 2.1. Mean July–August air temperatures (°C) for the period 1951–80 in Hokkaido and Tohoku districts, and deviations from these for the climatic scenarios used in different sections of the Japan study.

Climatic scenarios	Section of this study	Mean July–August temperature (°C)[a]	
		Hokkaido[b]	Tohoku[c]
Reference periods:			
1951–80	3,4,5	20.6	22.8
1966–82	6	20.6	22.7
"Bad" years			
1902	4	−3.3	−3.1
1971	5	−1.0	+0.2
1980	3,4,6[d]	−1.6	−2.6
"Good" years			
1955	4	+2.0	+1.9
1978	4	+2.5	+2.3
1979	3	−0.2	−0.3
"Bad" periods			
1902–11	4	−1.1	−1.0
1926–35	3,6	−0.3	−0.1
1974–83	5	0.0	−0.5
1980–82	6[e]	−0.3	−1.0
"Good" periods			
1921–30	4	+0.4	+0.7
1957–66	3,5,6	−0.1	0.0
2 × CO_2 scenario	3,4,5,6[f]	+3.5	+3.2

[a] As specified in Section 4.
[b] Averaged values for 2 stations: Sapporo and Asahigawa (*see Figure 2.2*).
[c] Averaged values for 6 stations: Aomori, Miyako, Ishinomaki, Akita, Yamagata and Fukushima (*see Figure 2.2*).
[d] Scenario assumes ten consecutive occurrences of these conditions.
[e] Scenario assumes three repetitions of these conditions.
[f] Includes a scenario of transient change from present to 2 × CO_2 conditions.

2.4.3. Scenarios based on rice yield anomalies

Using the 102-year record of national rice yields (1883–1984), years with unusually high or low yields (relative to trend) were identified in Section 3. Scenarios for individual years were selected on the basis of significant positive ("good" year) or negative ("bad" year) yield anomalies. Sequences of years were chosen according to the stability of yields during a period, the unstable periods characterized by a high frequency of summer damage years. Meteorological information for each of the chosen years and periods was subsequently obtained from the instrumental observation records to quantify the climatic scenarios.

The scenarios selected were: good (1979) and bad (1980) individual years (for which May–October temperature anomalies are mapped for all Japan in *Figure 3.12*), and stable (1957–1966) and unstable (1926–1935) decades. Some other variants were adopted elsewhere in the report. In Section 5, because suitable data were not available prior to 1945, the period 1974–1983 was selected as an unstable decade, and in Section 6, where climatic data have been used as inputs to a national economic model, two further "synthetic" scenarios are introduced that are based on the instrumental record. The first embraces ten consecutive occurrences of the "bad" year, 1980, and the second involves three repetitions of the consecutive "cold damage" years 1980–82 assumed to occur over a 9-year period (*Table 2.1*).

2.5. The GISS $2 \times CO_2$ Scenario

In common with the other case studies in this volume, outputs from the GISS general circulation model were used to simulate the climate of Japan under doubled concentrations of atmospheric carbon dioxide (*see* Part I, Section 3 of this volume). Values of mean monthly air temperature (°C), precipitation rate (mm/day) and cloud cover (tenths) were provided on a 4° latitude × 5° longitude grid point network for the Japan region [cf. *Figure 2.6(a)*]. These were obtained in three forms:

(1) GISS model-generated $1 \times CO_2$ equilibrium values.
(2) GISS model-generated $2 \times CO_2$ equilibrium values.
(3) GISS $2 \times CO_2$ – GISS $1 \times CO_2$ values (change in equilibrium climate for a doubling of CO_2).

Since the cloud cover values were not actually required for impact experiments, these were noted but no further computations were conducted. Temperature values are required as inputs to all the impact models in Sections 3–6, and for estimating net primary productivity in Section 3, precipitation data have also been used.

2.5.1. GISS model verification

In order to test the performance of the GISS model in simulating present climatic conditions in Japan, comparisons were made between the monthly values of temperature and precipitation generated in the GISS model $1 \times CO_2$ equilibrium run and instrumentally observed station data for the period 1951–1980. Three stations (Sapporo, Sendai and Fukuoka) were selected to represent northern (Hokkaido district), north-central (Tohoku) and southwestern (Kyushu) Japan, respectively, and GISS model data from nearby grid points were extracted for comparison:

Hokkaido: Sapporo (43°03′N, 141°20′E); GISS model grid point (42°N, 140°E).
Tohoku: Sendai (38°16′N, 140°54′E); GISS model grid point (38°N, 140°E).
Kyushu: Fukuoka (33°35′N, 130°23′E); GISS model grid point (34°N, 130°E).

Observed values are plotted on the same axes as GISS $1 \times CO_2$ and GISS $2 \times CO_2$ values of mean monthly temperature and monthly mean precipitation rate in *Figure 2.5*. At all locations, the GISS model successfully reproduces the annual march of temperatures, with an August maximum, but the amplitude of the annual cycle is underestimated, with too low summer temperatures at Fukuoka [*Figure 2.5(c)*], highly exaggerated winter temperatures (by nearly 10°C) at Sendai [*Figure 2.5(b)*] and a combination of these at Sapporo [*Figure 2.5(a)*]. The discrepancies are probably attributable to the proximity of Japan to the moderating influence of the Pacific Ocean (warmer than the land in winter, cooler in summer), and the inability of the GISS model, with its coarse grid point network, to account for the modifying effect of the Japanese land mass on this oceanic climate. The GISS $2 \times CO_2$ temperature estimates imply a warming, relative to $1 \times CO_2$ conditions, in every month with the greater increases occurring in the first than in the second half of the year at all stations (*see* below).

While the GISS $1 \times CO_2$ temperature estimates are inaccurate in representing the present conditions, the estimates of monthly mean precipitation rate are grossly in error and extremely misleading (*Figure 2.5*). The GISS model generates a fairly even distribution of monthly precipitation throughout the year at all stations, with higher values in the north than the south, and with a peak in late summer or early winter. Although the observation data show a late summer peak at Sapporo, values in every month are lower than predicted by the GISS model. The pronounced summer peak in observed precipitation rate at both Sendai and Fukuoka (the Bai-u rains – *see* Section 1) is not simulated by the GISS model at all, values are overestimated for Sendai [*Figure 2.5(b)*] and underestimated (by as much as a factor of 3) for Fukuoka [*Figure 2.5(c)*]. GISS model estimates for $2 \times CO_2$ precipitation are similar in pattern to the $1 \times CO_2$ estimates but, in contrast to the estimates for the other high latitude regions reported in this volume, the $2 \times CO_2$ values for some months are *lower* than the $1 \times CO_2$ values (particularly at Fukuoka, in southwestern Japan – *see* discussion below).

Development of the climatic scenarios

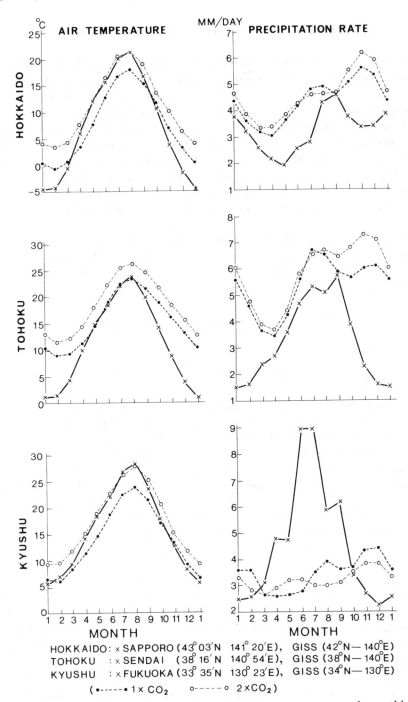

Figure 2.5. Comparison of observed mean monthly temperatures and monthly mean precipitation rate (1951–80) with GISS model-generated $1 \times CO_2$ and $2 \times CO_2$ equilibrium temperatures at (a) Sapporo, (b) Sendai, and (c) Fukuoka. (Source: Japanese Meteorological Agency, 1981.)

2.5.2. Constructing the $2 \times CO_2$ scenario

Given the poor correspondence between the GISS model $1 \times CO_2$ estimates and observed climate in these three regions of Japan, it was considered likely to be similarly unrealistic to use the $2 \times CO_2$ estimates directly to represent local agroclimatic conditions under doubled CO_2. Instead, and in line with the other case studies, it was thought more appropriate to use the *change* between the GISS-generated $1 \times CO_2$ and $2 \times CO_2$ equilibrium conditions to represent the difference between the observed present-day climate and a future $2 \times CO_2$ climate. Although this assumption is of somewhat doubtful validity for constructing the $2 \times CO_2$ precipitation scenario, that scenario is employed in only one application (in Section 3) and is not used for estimating impacts on rice production.

In each of the subsequent sections, therefore, the $2 \times CO_2$ scenario has been constructed by adding the difference between $1 \times CO_2$ and $2 \times CO_2$ temperature and, where applicable, precipitation values to the mean values for the baseline period – the "differences" methods:

$$T_{2 \times CO_2} = [T_{\text{GISS } 2 \times CO_2} - T_{\text{GISS } 1 \times CO_2}] + T_{\text{REF}}$$

and

$$P_{2 \times CO_2} = [P_{\text{GISS } 2 \times CO_2} - P_{\text{GISS } 1 \times CO_2}] + P_{\text{REF}}$$

It should be noted that the adjustment of precipitation values differs from the "ratios" method used in the Canadian and Icelandic case studies, there being no particular advantage in using the latter technique in Japan since baseline precipitation totals are quite high (especially during the growing season) relative to the changes estimated by the GISS model. Further discussion of the "differences" and "ratios" methods is given in Part II, Section 1 of this volume.

The geographical patterns of changes in air temperature and precipitation rate are shown in *Figures 2.6(a)* and *2.6(c)* for the warmer half year (April–September), and in *Figures 2.6(b)* and *2.6(d)* for the colder half year (October–March), respectively. Some specific features of the distributions of air temperature include:

(1) Temperature increases of about 3–3.5°C in the warmer half year, compared with 2–4°C in the colder half year (in contrast to greater warming in the winter than summer months in the other high latitude case study regions).
(2) The smallest temperature increases in both years are in southwestern Japan.
(3) There is a zone of maximum warming running from the Korean Peninsula to Hokkaido during the warmer half year, while Hokkaido experiences the greatest warming in Japan in the winter half year [*Figure 2.6(a)* and *(b)*].

The pattern of changes in precipitation rate shows a marked contrast between southwestern Japan and the central and northern regions, the former

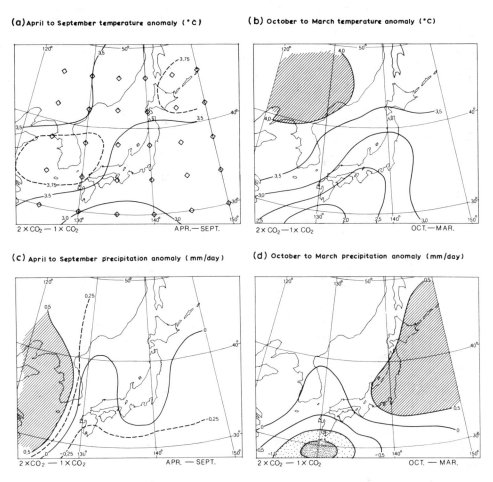

Figure 2.6. Geographical distribution of GISS model-generated changes between 1 × CO_2 and 2 × CO_2 climate: (a) mean April–September air temperature (°C), (b) mean October–March air temperature, (c) mean April–September precipitation rate (mm/day), and (d) mean October–March precipitation rate.

area receiving decreased precipitation in both half years (particularly so in the winter), while Hokkaido and Tohoku districts show small positive anomalies [*Figure 2.6(c)* and (d)].

In the following sections (3–6) of this study, climate impact experiments have been conducted for specific regions in Japan, with each regional 2 × CO_2 scenario constructed relative to different baseline station data and for different months or seasons. *Table 2.1* compares temperatures under the 2 × CO_2 scenario with those for the various instrumental scenarios in Hokkaido and Tohoku districts. For comparative purposes, all scenarios are expressed in terms of mean July–August temperatures, although this averaging period is not employed in all studies.

Table 2.2. Mean monthly temperature (°C) and precipitation (mm) observed (1931–60) and for the $2 \times CO_2$ scenario at Sapporo and Sendai, and for two stations with a present-day (1931–60) climate analogous to the $2 \times CO_2$ conditions. (Sources: Müller, 1982; Arakawa and Taga, 1969.)

Scenario	Jan	Feb	Mar	Apr	May	Jun	Jul	Aug	Sep	Oct	Nov	Dec	Year
							SAPPORO						
							Temperature (°C)						
1931–60 Observed	−5.5	−4.7	−1.0	5.7	11.3	15.5	20.0	21.7	16.8	10.4	3.6	−2.6	7.6
$2 \times CO_2$	−1.5	−0.8	3.0	9.8	15.3	19.1	23.2	24.8	19.9	13.4	6.5	0.7	11.1
Analogue Yamagata	−1.6	−1.1	2.1	8.7	14.7	19.1	23.2	24.4	19.4	12.7	6.7	1.4	10.8
							Precipitation (mm)						
1931–60 Observed	111	83	67	66	59	67	100	107	145	113	112	104	1134
$2 \times CO_2$	120	86	73	75	66	67	91	98	145	122	130	123	1196
Analogue Yamagata	101	78	75	78	66	99	165	136	127	107	82	119	1233
							SENDAI						
							Temperature (°C)						
1931–60 Observed	0.1	0.6	3.5	9.0	13.9	17.8	22.0	23.8	19.8	13.8	8.2	2.9	11.3
$2 \times CO_2$	2.6	3.4	6.5	12.2	17.3	21.0	24.9	26.7	22.8	16.6	10.7	5.2	14.2
Analogue Nagoya	2.9	3.6	7.1	12.7	17.5	21.5	25.7	26.6	22.7	16.5	10.9	5.6	14.4
							Precipitation (mm)						
1931–60 Observed	37	44	62	95	100	155	167	136	191	133	61	50	1231
$2 \times CO_2$	49	44	62	98	103	152	164	139	209	177	100	81	1378
Analogue Nagoya	49	64	100	137	145	204	178	155	212	160	86	57	1547

For a more detailed comparison between the observed climate and the $2 \times CO_2$ scenario climate, *Table 2.2* shows the respective mean monthly air temperature and precipitation data for Sapporo (Hokkaido district, altitude 17 m) and Sendai (coastal Tohoku, altitude 38 m) for the period 1931–60. In addition, to give an impression of the types of geographical shifts of the climatic regime implied in the $2 \times CO_2$ scenario, meteorological stations have been identified (by examining climatological data for the same 1931–60 period from elsewhere in Japan) that might serve as present-day analogues of the estimated scenario conditions. Thus, the $2 \times CO_2$ climate of Sapporo might resemble the present climate of Yamagata (38°15′N, 140°21′E, altitude 151 m – inland from Sendai in central Tohoku) or, further south, that of Nagano (36°40′N, 138°12′E, altitude 418 m – an upland site in northeast Chuba district), and the $2 \times CO_2$ Sendai climate would be analogous to that of Nagoya (35°10′N, 136°58′E, 51 m – at the extreme southwest of Chuba district (*Table 2.2* – for station locations, *see Figure 2.2*).

Finally, mention should be made of the methods with which the $2 \times CO_2$ scenario has been applied to data in the impact experiments. In most cases, the experiments are conducted for climatic conditions averaged over a period of years or for individual years, and the $2 \times CO_2$ scenario is assumed to represent the equilibrium climatic condition at a hypothetical date in the future. However, in Section 6, an attempt has been made to simulate the transition between the present climate and the $2 \times CO_2$ conditions. The impact model used in this application operates over 16 years; therefore this time period has been selected as the hypothetical transition phase over which the climatic change occurs (it is assumed to be linear over time). Of course, this represents a very rapid, probably unrealistic, change in climate, but it does provide a temporal component to the scenario that is lacking in the other experiments.

2.6. Relating Regional Scenarios to Synoptic Processes

2.6.1. Composite maps of anomalously warm months

We now attempt to evaluate the realism of the $2 \times CO_2$ scenarios by seeking analogues in Japan from the instrumental record.

Four anomalously warm months were selected to represent each of the warmer and cooler half years over Japan. These were April and May 1963, May 1964 and May 1967, representing the warmer half year; and February 1967, December 1968, February 1972, and December 1972, the colder half year. The use of specific warm months to represent a half year period was considered to be an acceptable procedure, since under $2 \times CO_2$ conditions, all months in a particular half year might, on average, be expected to exhibit similar (semi-permanent) mid-tropospheric circulation patterns. That assumption is not tested here, but it should be noted before attempting to interpret the results. The 4-month composite surface air temperature anomalies relative to the period (1951–80) are mapped for the Japan region in *Figures 2.7(a)* and *2.7(b)* (for the summer and winter half years, respectively).

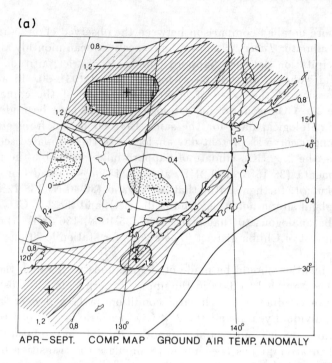

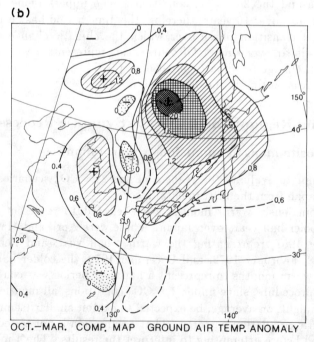

Figure 2.7. Composite maps of surface air temperature anomaly (°C) relative to the period 1951–80 for 4 months representing (*a*) the summer half year (April–September) and (*b*) the winter half year (October–March).

In the summer half year [*Figure 2.7(a)*], a warm anomaly extends from southwestern to central Japan, and a cool anomaly runs from the Yellow Sea via the Korean Peninsula to Tohoku District. Another warmer than average zone extends from northeast China in an east-northeast direction. Comparing *Figure 2.7(a)* with *Figure 2.6(a)*, which shows the temperature anomalies estimated by the GISS model for $2 \times CO_2$, it is evident that the anomalies are of much greater magnitude under the latter conditions, and the anomaly patterns are also different. This suggests that different mechanisms in the synoptic climatology of the region may be of importance under conditions of a 0–1 °C warming than under the postulated 3.0–3.8 °C change given $2 \times CO_2$ conditions.

The composite anomaly map for the winter half year [*Figure 2.7(b)*] is characterized by large positive anomalies over the northern part of the Sea of Japan extending to northern Japan, and a small negative anomaly to the southwest of Japan. Comparing *Figure 2.7(b)* with *Figure 2.6(b)*, ($2 \times CO_2$, winter half year), it is clear once again that the patterns are somewhat different. However, the two patterns do both show an increase in the temperature anomaly towards northern Japan.

The vertical distance in the atmosphere between air at 1000 mb pressure (near to the ground at sea level) and air at 500 mb pressure (at about 5–6 km height) varies according to the temperature of the air. Warm air expands, and the 1000–500 mb distance (geopotential height) increases, while the reverse occurs with cold air. Composite maps of the 1000–500 mb geopotential height anomaly were constructed for the same months as described above [*Figures 2.8(a)* and *2.8(b)*]. This measure was used because it shows the characteristic situation of the troposphere better than surface temperature or surface pressure patterns alone. Note that these maps cover a much larger area of eastern Asia and the north Pacific than the maps of surface air temperature anomalies. During the summer half year, a positive height anomaly covers the zonal region extending from east China to the Aleutian Islands. This zone is still present but less pronounced in the winter half year. On the other hand, the negative anomaly region in Siberia is more striking in the winter than in the summer half year.

Combining the information in *Figures 2.7(a)* and *Figure 2.8(a)*, the following synoptic climatological processes can be interpreted for the summer half year. The zonal region with positive air temperature anomalies over southwest Japan is a result of weaker activity of the polar frontal zone, which can be observed as a positive geopotential height anomaly. On the other hand, the positive anomaly of air temperature from northeast China to Sakhalin may be caused by a westward shift of the low pressure trough that extends from the Arctic Sea to higher latitudes in Siberia. The eastern part of this trough, which under the anomalous condition covers the area from northeast China to Sakhalin [*Figure 2.8(a)*], brings warmer than normal conditions to northern Japan because of the warmer southwesterly winds that circulate around the trough. Furthermore, comparing *Figure 2.8(a)* with *Figure 2.6(a)*, it can be inferred that the maximum warming zone from the Korean Peninsula to northeast Japan might be related to a positive geopotential height anomaly zone under $2 \times CO_2$ conditions. In addition, the weakening of the polar frontal zone in southwest Japan could explain the reduced precipitation inferred for the region [*Figure 2.6(c)*].

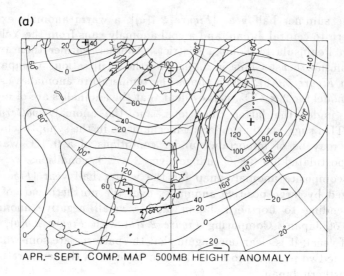

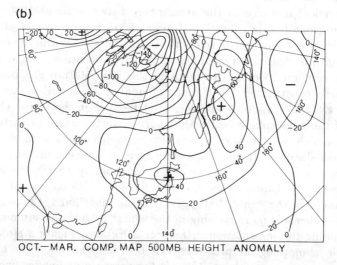

Figure 2.8. Composite maps of the geopotential height anomaly (gpm) at the 500 mb level relative to the period 1951–80 for 4 months representing the (*a*) summer half year and (*b*) winter half year.

Comparing *Figure 2.7(b)* with *Figure 2.8(b)*, for the winter half year, a positive height anomaly extends across the whole of Japan indicating that, at least to the south of 55°N, the trough in East Asia is weaker than normal, resulting in a reduced northwesterly winter monsoon and positive temperature anomalies in northeast China and in and around Japan [*Figure 2.7(b)*].

The striking positive anomalies in northeast China and northern Japan under $2 \times CO_2$ conditions [*Figure 2.6(b)*] are therefore thought to be real features. Also, the marked negative anomalies of precipitation south of Kyushu might be caused (as in the summer half year) by the weakened polar frontal activities [*Figure 2.6(d)*].

The polar frontal zone at the surface, and the trough in the mid-troposphere, discussed above, appear as semi-permanent features on seasonal pressure maps, and play an important role in the climate of East Asia (Yoshino, 1978a). Results such as these can therefore help to clarify some aspects of the synoptic climatology both of past and of future (including $2 \times CO_2$) climates in the region.

2.6.2. Composite maps of "good" and "bad" harvest years

In the same way that composite maps were used to examine the synoptic features of the $2 \times CO_2$ scenario, the instrumental scenarios of "good" and "bad" harvest years (described above in Subsection 2.4.3) can also be characterized in terms of regional-scale spatial climatic patterns. During the period 1945–1983, a number of good harvest years (1952, 1955, 1962, 1967, 1975 and 1979) and bad harvest years (1953, 1954, 1956, 1971, 1976, 1980, 1981, 1982 and 1983) occurred that have been identified in Section 3 of this report. Some of these have been selected as extreme year scenarios (*see Table 2.1*), so we produce here composite maps of the surface climatic conditions for all of the respective good and bad harvest years, in an attempt to identify characteristic features of these events. Anomaly maps (relative to 1951–80 averages) have been constructed for mean growing season (May–October) air temperature and precipitation during six good and six bad harvest years, respectively. Due to limitations on data availability for certain stations in China and North Korea, the years 1981, 1982 and 1983 were omitted from the calculations.

Air Temperature Anomalies

Figures 2.9(a) and *2.9(b)* show the distribution of May–October temperature anomalies (relative to 1951–80) for good and bad harvest years, respectively. In *Figure 2.9(a)*, the regions of greatest positive anomalies are located on the Pacific side of Hokkaido and northern Honshu, where the cool northeasterly Yamase wind normally prevails, and also over northeast China. This suggests that during the good harvest years, air temperature is frequently higher than the normal years in the higher latitude regions, probably because of the weakened influence of the northeasterlies from the Okhotsk anticyclone. In *Figure 2.9(b)* (bad years), the negative anomaly regions cover a broad area encompassing nearly all of the four major Japanese islands. In contrast to this negative anomaly region, a positive anomaly appears in the Nansei Islands, southwest Japan. This sharp gradient along the southern coast of Japan may be attributable to a marked polar frontal zone located along the coast. The magnitude of the temperature differences between the composite good and bad harvest years are about 1 °C in north Japan, 0.8°C in Central Japan and 0.1–0.5 °C in Kyushu and Shikoku, southwest Japan.

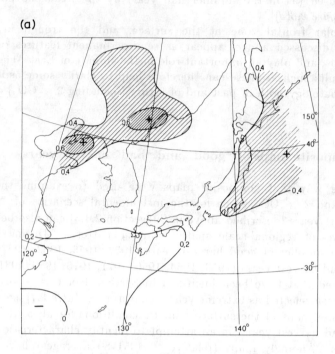
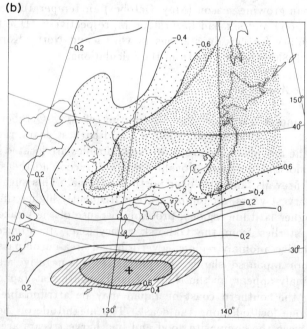

Figure 2.9. Distribution of May–October temperature anomalies (relative to 1951–80) for (a) good and (b) bad harvest years.

Precipitation Anomalies

In order to examine the relative magnitude of precipitation anomalies in "good" and "bad" harvest years, absolute values of precipitation from the composite sets of good and bad harvest years have been mapped in *Figures 2.10(a)* and *2.10(b)*. The maps reveal a zone of heavy rainfall (>300 mm) running from the Nansei Islands in an east-northeast direction, coinciding with the Pacific polar frontal zone that controls the Bai-u and Shurin rainy seasons. Three interesting features can be observed:

(1) In the good harvest years, the axis of the rainfall zone is located latitudinally about 1–3° to the south of the southern coast of Japan, whereas in the bad harvest years, the zone covers the southern coasts of Kyushu and Shikoku.
(2) The zones of greatest precipitation amount are more extensive in the bad than in the good harvest years.
(3) The combination of the above two effects results in Kyushu and Shikoku receiving 70–100 mm more rainfall in the composite bad harvest years than in the good harvest years. In Central and Northern Honshu, the differences are 20–40 mm.

The distribution of precipitation anomalies for the good and bad harvest years are shown in *Figures 2.11(a)* and *2.11(b)*, respectively. Consistent with the observations mentioned above, negative anomalies of about −20 to −40 mm are found in Kyushu and Shikoku during good harvest years, while positive anomalies of 60–80 mm are common over much of southwest Japan with only minor negative anomalies in Hokkaido and the Nansei Islands in the bad harvest years. Interestingly, the pattern of precipitation anomalies in the good harvest years is similar to that implied for the warmer half year under the $2 \times CO_2$ scenario [see *Figure 2.6(c)*].

2.6.3. Interpreting the $2 \times CO_2$ scenario

The foregoing analysis of anomaly patterns over the Japan region suggests the following interpretation of the synoptic mechanisms operating under a $2 \times CO_2$ climate. In the summer half years, the large positive temperature anomalies over Japan [*Figure 2.6(a)*] could be explained by the enhanced development of the Ogasawara high (subtropical anticyclone) bringing warm air to the higher latitude regions in the study area. A marked belt of positive pressure anomalies would restrict the intensive development of the Bai-u and Shurin fronts, resulting in scarce rainfall over southwest Japan and the southern seas [as in *Figure 2.6(c)*].

For the colder half year, the striking positive air temperature anomaly regions in northeast China (about 4 °C), and smaller positive anomalies (about 2 °C) along with decreased precipitation in the lower latitudes of the study region may be related to the appearance of a positive pressure anomaly in the middle

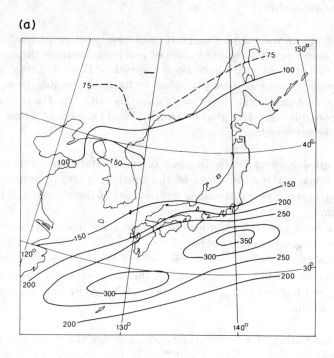

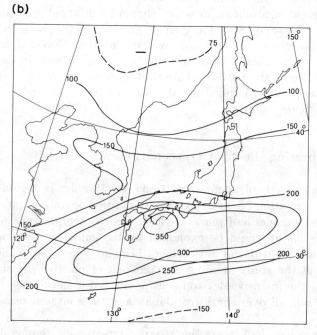

Figure 2.10. Absolute values of precipitation (mm) from composite sets of (a) good and (b) bad harvest years.

Development of the climatic scenarios

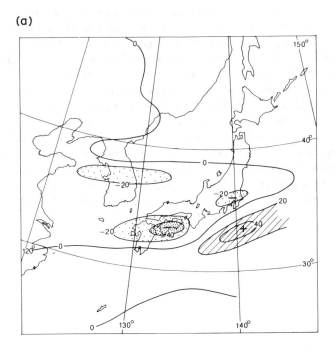

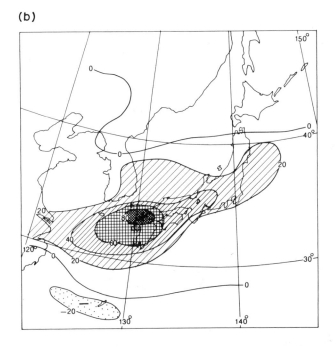

Figure 2.11. Distribution of precipitation anomalies (mm, relative to 1951–80) for (a) good and (b) bad harvest years.

troposphere in the latitude of Hokkaido. This implies a weakening of the trough that usually extends along 140°E longitude, and a weakened winter monsoonal circulation – leading to warmer conditions – over the whole country, with less precipitation in southern regions due to decreased frontal activity [*Figure 2.6(d)*].

Whatever the controlling mechanisms behind these climatic changes, it is likely that further changes *will* occur in the future, so the remainder of this study investigates what the probable impacts of a range of climatic variations (as specified by the scenarios) may be. In Section 3, the sensitivity of national rice yields to climatic variability is discussed, and changes in net primary productivity of vegetation are evaluated for the different scenarios. In Section 4, the limits on cultivable area for rice in Hokkaido and Tohoku districts are computed for a range of scenarios, while in Section 5 levels of rice yield in Hokkaido are estimated using a simulation model. Section 6 examines the effects of climate-induced productivity changes on the supply and stocks of rice at national level using an integrated econometric model.

SECTION 3

The Effects on Latitudinal Shift of Plant Growth Potential

3.1. Introduction

The main purposes of this section are:

(1) To study the latitudinal distribution of thermal resources important for the cultivation of rice.
(2) To estimate the extent of latitudinal shift due to climatic variations.
(3) To assess variations in plant growth potential which might occur under the different climatic scenarios described in Section 2.

Rice plants, being indigenous to tropical or subtropical humid regions, need enough thermal and water resources throughout their growing season to yield a good crop. Since rice cultivation has been introduced into Japan, much effort has been concentrated on the development of agricultural techniques for increasing its yield. The breeding of rice varieties more tolerant to cool summer conditions and the development of irrigation systems for water supply have succeeded in expanding the boundary of rice cultivation as far north as Hokkaido (located between 41°N and 45°N) and at as high an altitude as the foot of Mt. Yatsugatake (at about 1000 m, and at latitude 36°N). With the development of rice cultivation techniques, the main producing area of rice in Japan has moved northwards from Kyushu district in the Meiji era (1868–1912), to Hokkaido, Tohoku and Hokuriku at the present.

The climate in Japan varies considerably from south to north Japan, because the four main islands – Hokkaido, Honshu, Shikoku and Kyushu – stretch in an arc 2200 km long between 24°N and 46°N. Temperature conditions in the northern districts are not favorable for rice production and rice cultivation here is very sensitive to variations of climate, particularly to cool summer damage. It is this sensitivity of rice production in marginal areas of Japan that is the focus of this section.

3.2. Statistical Characteristics of Thermal Resources

3.2.1. Latitudinal distribution of thermal resources

Meteorological data from about 150 stations in Japan and about 229 stations with an altitude of less than 150 m elsewhere in the northern hemisphere, have been analyzed to study the latitudinal change in thermal resources important to plant productivity (Uchijima, 1976a, 1976b, 1976c; Uchijima and Horibe, 1977). Four agroclimatic indexes – effective accumulated temperature (ΣT_{10}, degree-days), effective period (D_{10}, days), warmth index (WI, degree-months) and coldness index (CI, degree-months) – were calculated from the following relations:

$$\Sigma T_{10} = \sum_{i=1}^{k} T_{d,i} \qquad \text{for } T_d \geq 10\,°\text{C} \qquad (3.1)$$

$$D_{10} = \sum_{i=1}^{m} i \qquad \text{for } T \geq 10\,°\text{C} \qquad (3.2)$$

$$WI = \sum_{j=1}^{m} (T_{m,j} - 5) \qquad \text{for } T_m \geq 5\,°\text{C} \qquad (3.3)$$

$$CI = \sum_{j=1}^{n} (5 - T_{n,j}) \qquad \text{for } T_n \leq 5\,°\text{C} \qquad (3.4)$$

where $T_{d,i}$ is mean temperature on day i, $T_{m,j}$ is mean temperature in month j, k is the number of days on which daily mean temperature was equal to or above 10 °C, m is the number of months in which monthly mean temperature was equal to or above 5 °C and n is the number of months in which monthly mean temperature was equal to or below 5 °C. The effective period (D_{10}) characterizing the duration of period in which summer crops and plants can grow vigorously is the length of the consecutive period with T_d equal to or above 10 °C, in days.

The computed data for the respective indexes were grouped over latitudinal bands, each with a width of 5 degrees, to illustrate their latitudinal distribution. They are plotted along with mean annual temperature over the northern hemisphere, in *Figure 3.1*. The points in *Figure 3.1(a)* and *3.1(c)* denote the zonal average of the climatic indexes. The number of stations representing each latitude band are shown together with vertical bars on the points that indicate the standard deviation of the indexes in the respective latitudinal bands. As can be seen in these plots, $\overline{\Sigma T_{10}}$, $\overline{D_{10}}$ and $\overline{WI}$ are nearly constant in a latitudinal range lower than about 20°N, and then decrease sharply in the middle latitudes, becoming approximately zero at a latitude of about 75°N.

On the other hand, the mean coldness index ($\overline{CI}$) is zero at latitudes lower than 30°N, and increases considerably with increasing latitude. The change in

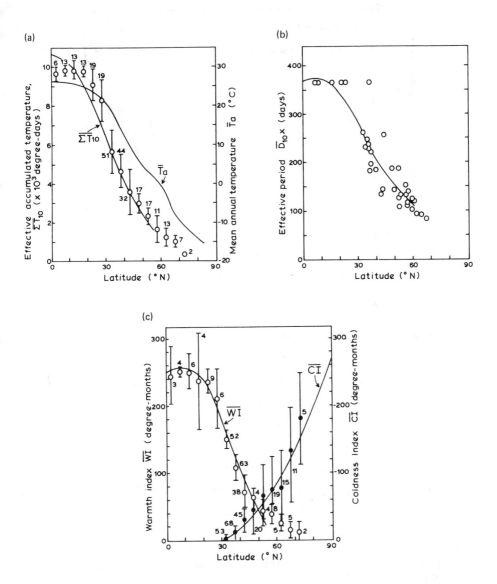

Figure 3.1. Latitudinal distribution of (a) effective accumulated temperature, $\overline{\Sigma T}_{10}$, and annual mean temperature, $\overline{T}_a$ (Uchijima, 1976a); (b) effective period, $\overline{D}_{10}$ (adapted from Uchijima, 1976b); and (c) warmth index, $\overline{WI}$, and coldness index, $\overline{CI}$ (Uchijima and Horibe, 1977). For explanation, *see* text.

$\overline{\Sigma T}_{10}$, $\overline{D}_{10}$, $\overline{WI}$ and $\overline{CI}$ with latitude, φ, over the northern hemisphere can be well approximated by:

$$\overline{\Sigma T}_{10} = 10663 + 41.1\varphi - 9.0\varphi^2 + 0.096\varphi^3 \qquad \text{for } \varphi \leq 60°\text{N} \qquad (3.5)$$

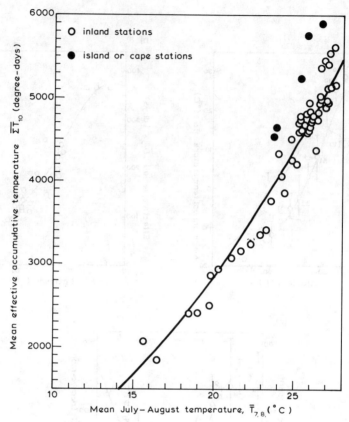

Figure 3.2. The dependence of $\overline{\Sigma T}_{10}$ on mean July–August temperature, $\bar{T}_{7.8}$.

$$\bar{D}_{10} = 363.1 + 3.54\,\varphi - 0.3\,\varphi^2 + 0.0029\,\varphi^3 \qquad \text{for } \varphi \leq 60°\,\text{N} \qquad (3.6)$$

$$\overline{WI} = 244.3 + 3.94\,\varphi - 0.29\,\varphi^2 + 0.0026\,\varphi^3 \qquad \text{for } \varphi \leq 50°\,\text{N} \qquad (3.7)$$

$$\overline{CI} = 0.676\,(\varphi - 30)^{1.47} \qquad \text{for } \varphi \geq 30°\,\text{N} \qquad (3.8)$$

where the upper bars denote the period average values of each quantity. The curves calculated from the above empirical relations are shown by solid lines in *Figure 3.1*.

Analyses of field experimental data of rice growth and yield in relation to weather conditions have revealed that the yield index of rice [described in equation (3.11), below] correlates closely to the mean temperature for the warmest period (July and August) (NIAS, 1976; Uchijima, 1981). This is mainly because the young ears of rice plants, which are vulnerable to temperature, are formed and developed in this period. Hence, it is reasonable to assume that the mean air temperature for the period July to August ($\bar{T}_{7.8}$) is also an important climatic index characterizing the thermal resources for rice cultivation. Using meteorological data from 150 stations over Japan, the relationship between $\overline{\Sigma T}_{10}$ and $\bar{T}_{7.8}$ was studied and is presented in *Figure 3.2*. $\bar{T}_{7.8}$ was found to range from 19°C in the northern part of Hokkaido to 27°C in the southern part of Kyushu. The dependence of $\overline{\Sigma T}_{10}$ on $\overline{T}_{7.8}$ in Japan can be approximated by:

$$\overline{\Sigma T}_{10} = 9.08\, \overline{T}_{7.8}^{1.92} \tag{3.9}$$

This empirical relation can be used to convert $\overline{T}_{7.8}$ into $\overline{\Sigma T}_{10}$, and vice versa.

3.2.2. Latitudinal shift of isopleths of ΣT_{10}

Using climatic data from the European plain of the USSR, Sapozhnikova et al. (1957) studied the northward or southward shift of ΣT_{10} due to recorded variations in the climate. As already reported in a previous paper (Uchijima, 1981), the variability of thermal resources increases gradually with decreases in the annual mean temperature, and these variations should lead to a shift of the isopleths from their average position, influencing the growth and yield of crops. Using the assumption that values of $d\varphi/d\overline{\Sigma T}_{10}$ obtained from equation (3.5) can be applied to estimate the latitudinal shift of ΣT_{10}-isopleths from their normal positions, the following relation was obtained (Uchijima, 1976a):

$$\Delta\varphi = A(1 - Z)\, \overline{\Sigma T}_{10} \tag{3.10}$$

where $\Delta\varphi$ is the latitudinal shift of the ΣT_{10}-isopleth (in degrees), $\overline{\Sigma T}_{10}$ is the period average of ΣT_{10}, and A and Z are respectively given by:

$$A = d\varphi/d\Sigma T_{10}$$
$$= -1.96 \times 10^{-2} + 5.16 \times 10^{-6}\, \overline{\Sigma T}_{10} - 3.96 \times 10^{-10}\, \left(\overline{\Sigma T}_{10}\right)^2,$$

and

$$Z = 1 \pm \left[\tau_0 / \left(370 + 40.8\, \tau_0\right)\right] \cdot CV_{\Sigma T}$$

τ_0 is the return period in years and $CV_{\Sigma T}$ is the coefficient of variation of ΣT_{10} fluctuations. Equation (3.10) indicates that when Z is larger than unity (implying that climate becomes warmer than the normal), the ΣT_{10}-isopleths shift northwards, while values of Z smaller than unity lead to a southward shift of the isopleths, indicating that the climate has become cooler than normal.

The results obtained from equation (3.10) are presented as the curve in *Figure 3.3(a)*. This is compared, on the same diagram, with the results obtained from the *World Atlas of Agroclimatic Resources* (Golts'berg, 1972). The good agreement of results from equation (3.10) and from the *World Atlas* gives strong support to equation (3.10), which has thus been applied to climatic data from Japan, for estimating the shift of ΣT_{10}-isopleths over the latitude range of Japan due to climatic fluctuations [expressed in terms of their return period – *Figure 3.3(b)*]. The conclusion to be drawn from this figure is that the shift of ΣT_{10}-isopleths is larger in the northern districts than in the southern districts of

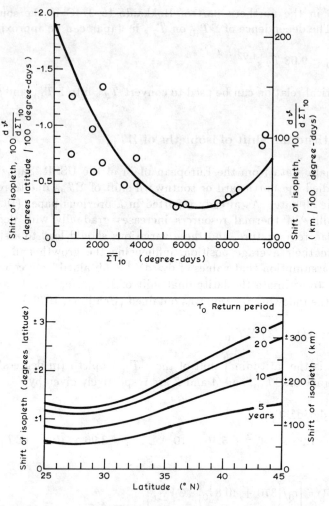

Figure 3.3. (*a*) Dependence of shifting rate of ΣT_{10}-isopleth on period-averaged effective accumulated temperatures, ΣT_{10}. Open circles denote the shifting rate of isopleths evaluated on an agroclimatic map (Golts'berg, 1972). (*b*) Estimated shift of ΣT_{10}-isopleths that could be expected if the temperature regime changed as indicated (after Uchijima, 1976c).

Japan. This implies that the fluctuation in thermal resources is larger in the northern districts, giving climatic conditions that are less favorable to rice cultivation, than in southern districts where thermal resources for rice cultivation are plentiful. A southward shift of ΣT_{10}-isopleths indicates an anomalous lowering of temperature and a poor crop of rice, while their northward shift is associated with anomalously high temperatures and the likelihood of a good rice crop.

3.3. Yield Variability of Rice in Relation to Temperature Variations

Although improved agricultural techniques have been successful in increasing and stabilizing rice production in Japan, climate (particularly temperature) is still a most important constraint (*see* Sections 4 and 5 of this report). The poor rice crop in Japan for the 4 consecutive years from 1980 to 1983 indicates clearly the close correlation between rice yield and temperature conditions.

The following yield index (IY) was calculated in order to distinguish the effects of the development of agricultural techniques and of fluctuating climate on rice yield:

$$IY(t) = Y(t)/Y_T(t) \qquad (3.11)$$

where $Y(t)$ and $Y_T(t)$ are, respectively, the actual rice yield and the trend value of rice yield (t/ha) in year t. The trend yield of rice $[Y_T(t)]$ was estimated approximately by fitting a polynomial equation to the time series of annual rice yields using a simple computer package (Gen and Ida, 1983). *Figure 3.4* shows the variations in the yield index of rice in representative prefectures of Japan.

In Hokkaido and Iwate where the rice production has been frequently affected by cool summers, the variations in $IY(t)$ are of greater magnitude than those in southern prefectures with sufficient heat supply (as indicated by the standard deviation values shown in *Figure 3.4*). As a result, the occurrence probability of a yield index lower than 0.7 is about 0.18 in Hokkaido and about 0.07 in Iwate prefecture. In regions further south, the magnitude of the coefficient of variation of the IY time series decreases considerably, reaching a minimum level in the Kansai district (central Honshu, prefectures 23 and 25-29, see *Figure 1.1*), before increasing again in the southernmost districts of Japan, probably because of the higher frequency of typhoons in those areas.

The yield variability (CV_{IY}) of rice has been plotted against the mean July-August temperature ($\bar{T}_{7.8}$) for all Japanese prefectures to investigate quantitatively the dependence of yield variability of rice on thermal resources [*Figure 3.5(a)*]. The magnitude of CV_{IY} in the IY time series increased approximately linearly with decreasing $\bar{T}_{7.8}$ for the range of $\bar{T}_{7.8}$ below 24 °C, reaching a level of about 30% at $\bar{T}_{7.8}$ of 19 °C (in Hokkaido). Although there is somewhat large scatter of points in the range of $\bar{T}_{7.8}$ above 24 °C, it appears that the magnitude of CV_{IY} is lowest at about 10–12% between 25 and 26 °C (in the Kansai district), followed by a gradual increase in CV_{IY} with higher mean temperatures. *Figure 3.5(b)* depicts the relationship between yearly variations of the yield index and of $\bar{T}_{7.8}$. The figure indicates how the standard deviation of the IY time series (σ_{IY}) increases nearly linearly with increments of the standard deviation of the $\bar{T}_{7.8}$ time series, for the range of σ_T over 1.0 °C. The dependence of the yearly variability of rice yield (as characterized by CV_{IY} and σ_{IY}) on the thermal resources ($\bar{T}_{7.8}$) and the yearly variability of thermal resources (σ_T) can be well approximated:

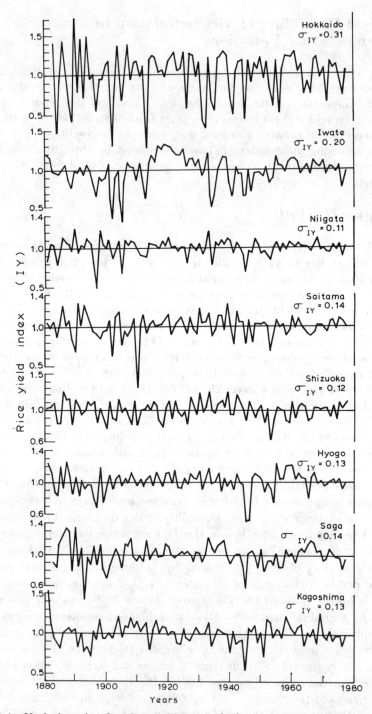

Figure 3.4. Variations in the rice yield index (IY) in representative prefectures of Japan. Also shown are standard deviations of yield variation, σ_{IY}. For location of prefectures, see *Figure 1.1* (after Uchijima, 1981).

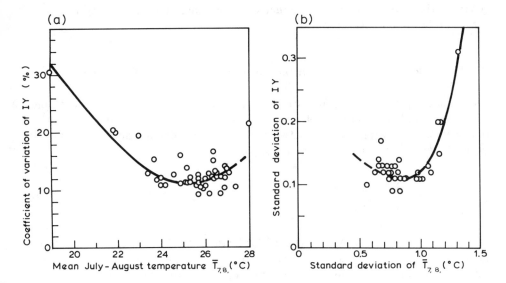

Figure 3.5. (a) Dependence of variability of rice yield (CV_{IY}) on mean July–August temperature $(\bar{T}_{7.8})$. (b) Dependence of variability of rice yield σ_{IY} on temperature variability, σ_T (after Uchijima, 1981).

$$CV_{IY} = 275.19 - 19.933\,\bar{T}_{7.8} + 0.0376\left(\bar{T}_{7.8}\right)^2 \qquad (3.12)$$

$$\sigma_{IY} = -1.068 + 4.582\,\sigma_T - 5.737\,\sigma_T^2 + 2.332\,\sigma_T^3, \qquad (3.13)$$

$$0.65 \leq \sigma_T \leq 1.3$$

where $\bar{T}_{7.8}$ is the period average of $T_{7.8}$ (Uchijima, 1981). Relationships similar to equations (3.12) and (3.13) were also obtained by Hanyu and Ishiguro (1972) between the effective accumulated temperature during the safe cultivation season for rice plants, and the crop situation index of rice. They used weather and rice yield data from Hokkaido and Tohoku districts. Because these districts have somewhat unfavorable weather conditions for rice cultivation, the relationship between rice yield and thermal resources was more clearly defined than in the results presented in *Figure 3.5*.

3.4. Climatic Variations and Plant Growth Potential

Different climatic factors tend to fluctuate simultaneously with some correlation between them, and their combined effect can influence the growth of crops and natural vegetation. Therefore, it is usually necessary to consider the simultaneous effects of climatic factors on plants. In the following section an agroclimatic model for evaluating the net primary productivity (*NPP*) of natural vegetation is described. Impacts of climatic variations on plant growth potential (characterized by net primary productivity) are then quantified using the model for a number of climatic scenarios.

3.4.1. Chikugo model for evaluating net primary productivity

By considering the vertical fluxes of carbon dioxide and water vapor due to photosynthesis and transpiration of a plant community, Uchijima and Seino (1985a) obtained the following relation between NPP [tonnes dry weight per hectare per year, tDW/(ha/yr)] and climatic factors:

$$NPP = \frac{A_0 \cdot R_n}{d(1 + \beta)} \quad (3.14)$$

where R_n is the annual net radiation (kcal/cm^2), d is the water vapor deficit of air (mm Hg), β is the Bowen ratio characterizing the partitioning of solar energy into sensible and latent heat fluxes, and A_0 is a numerical constant related to stomatal and canopy resistances, carbon dioxide concentration in the air, and stomatal cavities of leaves. Equation (3.14) indicates that NPP should increase linearly with increasing annual net radiation and is inversely proportional to the product of d and $(1 + \beta)$ (related to the dryness of the climate – see Budyko, 1971). Equation (3.14) is the theoretical basis for the Chikugo model to be described below. Data on phytomass production, obtained through the International Biological Programme (IBP) during the period 1964 to 1972 (Cannell, 1982) and climatic data from Japan and the rest of the world (Müller, 1982; Golts'berg, 1972; Japanese Meteorological Agency, 1982) were used to verify the above theoretical expression and to construct the Chikugo model.

Figure 3.6 illustrates the dependence of NPP on the annual net radiation for respective climatic zones that are characterized on the basis of the radiative dryness index ($RDI = R_n/lr$, where l is the latent heat of evaporation, cal/g H$_2$O, and r is the annual precipitation, cm). The circles on *Figure 3.6* denote the average of NPP values over R_n bands with a 5 kcal/cm^2 interval width. Inspection of *Figure 3.6* indicates that NPP values in each RDI zone are approximately proportional to annual net radiation and that the proportionality constant (α) between NPP and R_n decreases considerably as the climate becomes drier. This suggests that the efficiency of plant photosynthesis decreases rapidly as climate becomes more dry, agreeing well with results obtained in crop experiments (for example, see Gifford, 1979).

The relationship between RDI, R_n and NPP can thus be expressed as follows:

$$NPP = \alpha \cdot R_n \quad (3.15)$$

$$\alpha = f(RDI) \quad (3.16)$$

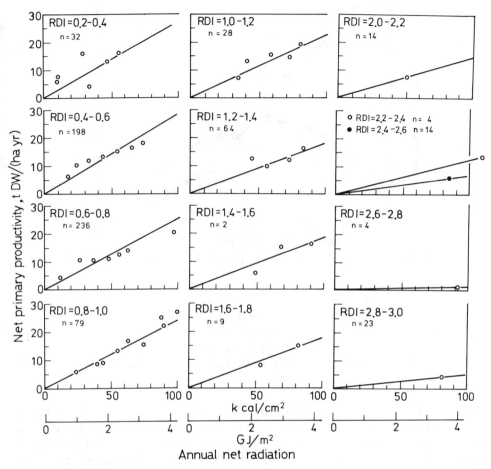

Figure 3.6. Dependence of *NPP* of natural vegetation on annual net radiation (R_n) for respective *RDI* bands. The numeral (n) for each zone denotes the number of data on net primary productivity (after Uchijima and Seino, 1985a).

The proportionality constant (α), determined from the data in *Figure 3.6*, is presented in *Figure 3.7* as a function of the radiative dryness index (*RDI*). As indicated in this figure, the value of α decreased drastically with increasing *RDI* from about 0.29 in a *RDI* range below 0.2, to a level of 0.01 in a *RDI* range above 3.0. This relation can be approximated by:

$$\alpha = 0.29 \exp\left[-0.216(RDI)^2\right] \tag{3.17}$$

By combining Budyko's and Hare's proposals about the relationship between climatic dryness indexes and vegetation types (Budyko, 1956; Hare, 1983), we can suggest the following correlation between *RDI* and vegetation types:

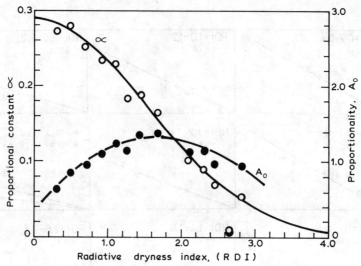

Figure 3.7. Dependence of α and A_0 on *RDI* (after Uchijima and Seino, 1985a).

RDI Range	Vegetation type
> 10	True desert
7–10	Desert margin
2–7	Semi-arid zone
1–2	Subhumid, steppe or savanna
< 1	Humid forest zone

Figures 3.6 and *3.7* and the above list, indicate that *RDI* is a useful index for characterizing the efficiency of phytomass production by natural vegetation. A dry climate, by causing a shortage of available moisture in the root zone of plants, thus reduces the energy efficiency of plant phytomass production (expressed as the percentage calorific value of dry matter relative to the intercepted solar radiation energy). This agrees well with the conclusion obtained from equation (3.14).

Substituting equation (3.17) into equation (3.15) yields:

$$NPP = 0.29 \left[\exp\left\{ -0.216(RDI)^2 \right\} \right] \cdot R_n \tag{3.18}$$

This is the Chikugo model for evaluating net primary productivity of natural vegetation from weather data (Uchijima and Seino, 1985a).

3.4.2. Verification of the Chikugo model

Evaluation of net primary production

The total net production (*TNP* – tonnes dry weight per year, tDW/yr) in each of the Japanese prefectures was evaluated using the Chikugo model and compared with results reported by Iwaki (1984), in order to test the model's validity. Total net production in the ith prefecture was estimated as:

$$TNP_i = \sum_{j=1}^{4} A_{ij} E_j \overline{NPP_i} \qquad (3.19)$$

where A_{ij} is the area (ha) of the jth land class in the ith prefecture, E_j is the production efficiency of the jth land class, and $\overline{NPP_i}$ is the mean net primary productivity of natural vegetation in the ith prefecture. In our calculation, the mean NPP of individual prefectures was determined using a geographical distribution map of NPP (Uchijima and Seino, 1985b).

In an independent analysis, Iwaki (1984) evaluated the values of $\overline{NPP_i}$ using data on net primary productivity for 30 typical vegetation types, derived mainly from production data obtained experimentally in the Japan IBP studies (Kira, 1976; Tadaki and Hachiya, 1968). The land in each prefecture was divided into four classes – forest, orchard, cultivated land and grassland – having production efficiencies of 1.0, 0.8, 0.81 and 0.625, respectively.

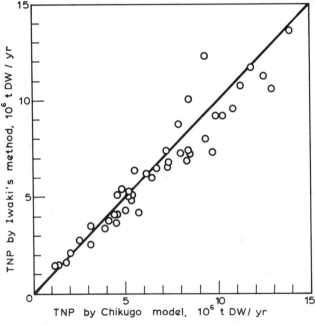

Figure 3.8. Comparison of total net production values, by prefecture, obtained using the Chikugo model and Iwaki's method (after Seino and Uchijima, 1985).

Figure 3.8 shows the comparison of TNP_i obtained by the Chikugo model with the results evaluated by Iwaki (1984) using plant ecological data. As can be seen in this figure, the points are well distributed near the line of perfect agreement, indicating that the Chikugo model is satisfactorily valid, and can be applied to assess the influence of climate variations on net production of natural vegetation.

Applicability for estimating crop production

The following relation derived from equations (3.14) and (3.15) was used to estimate the magnitude of the numerical constant A_0:

$$A_0 = \alpha \cdot d \cdot (1 + \beta) \tag{3.20}$$

Values of A_0 estimated from equation (3.20) using empirically derived values of α and weather data are also presented as a function of *RDI* in *Figure 3.7*. The magnitude of A_0 increases curvilinearly with increased *RDI*, up to an *RDI* value of between 1.4 and 1.8, and decreases gradually for higher values of *RDI*. The value of A_0 is strongly affected by physiological characteristics of the plants and weather conditions at each location. Experimental data show that the net photosynthesis of crop leaves exceeds that of forest vegetation by a factor of about 1.5–2.5 (for example, *see* Kira, 1973). Therefore, we can conclude that the value of A_0 for crops grown in a well cultivated field are likely to be larger than that of natural vegetation. Further studies are needed to verify this assumption and to extend the use of the Chikugo model to crop production.

3.4.3. Climatic scenarios

Instrumental scenarios

As already outlined in Subsection 2.4.3, in order to construct instrumental climatic scenarios for evaluating impacts of climatic variations on plant growth potential, the secular change in the national average of rice yield in Japan was studied. Data on yield, extending over 102 years (1883–1984), were obtained from the *Crop Statistics* published by the Ministry of Agriculture, Forestry and Fisheries of Japan [*Figure 3.9(a)*]. Also shown, on the same time axis, are the secular variations in decaded variability of yields, represented as 10-year moving averages of the coefficient of variation of the yield index, CV_{IY} [*Figure 3.9(b)*]. Rice cultivation in Japan was affected 28 times by unusually cool summer conditions during the past 102 years [black dots on *Figure 3.9(a)*], implying that cool summer damage of rice cultivation in Japan occurs, on average, once in every 4 years. However, the variations in CV_{IY} indicate that highly unstable subperiods alternate with relatively stable subperiods at an interval of about 20–30 years [*Figure 3.9(b)*]. Considering the characteristics of these fluctuations in rice yield, a number of individual years and periods were adopted as climate scenarios to study the impact of climate variation on *NPP*-distribution over Japan [*see Figure 3.9(a)*].

In selecting "good" and "bad" decades, we considered the number of cool summer years in each period. The 10-year period, 1957 to 1966, contained only

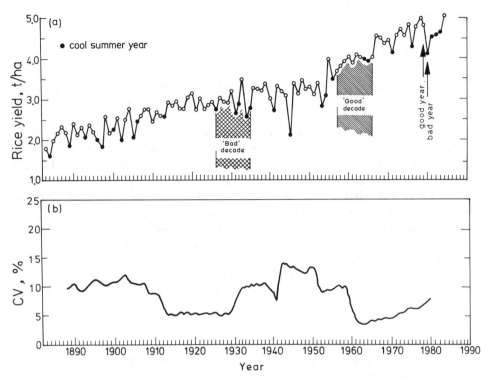

Figure 3.9. Secular changes, 1883–1984, in (a) average national rice yields (t/ha), and (b) coefficient of variation (10-year running mean) of detrended rice yields (yield index), CV_{IY} (%). Also marked are the positions of the instrumental climatic scenarios.

Table 3.1. Index of observed rice yields under the instrumental climatic scenarios used in this section.

Scenario	Year(s)	Yield index (%)
Good year	1979	103
Good period	1957–1966	103
Baseline	1951–1980	100
Bad period	1926–1935	100
Bad year	1980	87

two weak cool summer years (good decade), while the bad period, 1926 to 1935, included five cool summer years [*Figure 3.9(a)*]. Because of the difference in the number of cool summer years between the two periods, the fluctuation of rice yield was larger in the bad period than in the good period, as indicated in *Figure 3.9(b)*. However, there was little difference in the mean yield index between these periods (*see Table 3.1*).

Data on air temperature and precipitation for each of the selected instrumental scenarios were extracted from meteorological statistics (Japanese

Meteorological Agency, 1981), and were the basic input data for the impact experiments.

$2 \times CO_2$ scenario

The change in climate from the baseline climate (1951–1980) to that due to a doubling of carbon dioxide concentration in the atmosphere was also studied as a climatic scenario. As described in Subsection 2.5, changes in mean monthly air temperature, precipitation and cloud cover between $1 \times CO_2$ and $2 \times CO_2$ equilibrium climates were generated by the Goddard Institute for Space Studies (GISS) general circulation model (*see* Part I, Section 3 of this volume). For the purposes of these experiments, distribution maps of the equilibrium changes in each variable were shown using the grid point data. The distribution maps were then superimposed onto the distribution map of meteorological stations in Japan to determine the changes at each station. Finally, these values were added to the baseline (1951–80) values of the respective variables at the stations, in order to obtain scenario values for a $2 \times CO_2$ climate. Since the changes in cloud amount were found to be only slight, these were subsequently ignored in the calculation of global solar radiation and net radiation required for the Chikugo model.

3.4.4. Variations in plant growth potential under different climatic scenarios

Using the climatic scenarios shown in *Table 3.1* together with the $2 \times CO_2$ scenario for meteorological station locations in Japan, estimates were made with the Chikugo model to evaluate the impacts of changes in climate on *NPP* distribution and total net production of Japan (TNP_0, tDW/yr). *Figure 3.10* compares the *NPP* distributions for the respective climatic scenarios. Several important conclusions can be drawn:

(1) When the climate is as warm as in 1979 (relative to normal), the isopleths of *NPP* shift northwards from their average positions, mainly due to the prolongation of the plant growing period. It was found that the area with *NPP* values below 10 tDW/(ha/yr) narrows considerably [shaded area in *Figure 3.10(e)*], a result of the upward altitudinal shift and northward latitudinal shift of *NPP*-isopleths.

(2) When the climate resembles the cool conditions of 1980, the isopleths of *NPP* shift southwards and towards lower elevations relative to their normal positions. These shifts of *NPP*-isopleths lead to an enlargement of the land area with *NPP* values below 10 tDW/(ha/yr) [*Figure 3.10(c)*].

(3) Under the GISS $2 \times CO_2$ scenario mean annual air temperatures are about 2.5 °C above the 1951–80 baseline in southern districts (Kyushu and Okinawa) and about 3.5 °C above in northern districts [Hokkaido and Tohoku – see Section 2, *Figure 2.6(a)* and *2.6(b)*]. In response to this

warming of climate, the NPP-isopleths shift a considerable distance northwards [Figure 3.10(f)].
(4) The latitudinal and altitudinal shifts of NPP-isopleths are greater in Tohoku and Hokkaido districts than in the southwestern districts of Japan. This result agrees well with the conclusion obtained from equation (3.10) and presented in Figure 3.3(a).

The distribution maps of NPP over Japan shown in Figure 3.10 were used to assess the change in area of zones with various NPP values due to variations in climate (Figure 3.11). The thin line shows the distribution curve of percentage area for the baseline climate (1951–1980). As can be seen in this figure, the curve shifts towards a higher NPP range or lower NPP range, reflecting the change in climate. Comparison of the distribution curves between the good year (1979) and the bad year (1980) shows that the favorable climate observed in 1979 diminishes the contribution from the NPP range below 10 tDW/(ha/yr) and increases the contribution from the intermediate NPP range between 10 and 15 tDW/(ha/yr). On the other hand, the unfavorable climate in 1980 decreases the contribution from the NPP range above 15 tDW/(ha/yr). No great difference in the distribution curve is observed between the good period and the bad period. This is assumed to be due to the smoothing effect of averaging weather conditions over 10 years. Warming of climate under the $2 \times CO_2$ scenario has a considerable effect on the distribution curve of the percentage area, shifting the curve towards a higher NPP range.

To quantify the effect of climatic change on the total net production (TNP_0) of Japan, the relative total net production ($RTNP$) was calculated from:

$$RTNP = \frac{TNP_{0,s}}{TNP_{0,b}} \tag{3.21}$$

where $TNP_{0,s}$ is the total net production for a specified climatic scenario, and $TNP_{0,b}$ is the total net production for the baseline climate (380×10^6 tDW/yr). The value of TNP_0 was determined from:

$$TNP_0 = \sum_{i=1}^{47} TNP_i$$

where TNP_i is the total net production for the ith prefecture evaluated from equation (3.20). As indicated in Figure 3.11, the values of $RTNP$ range from 0.93 in the bad year (1980) to 1.09 for the $2 \times CO_2$ climate. Moreover, an environment with higher atmospheric CO_2 concentrations would increase the photosynthetic rate in plants, with a concomitant increase in dry matter production (for example, see Goudriaan and de Ruiter, 1983). The value of $RTNP$ under a $2 \times CO_2$ climate might thus be larger than 1.09, if we consider the combined effects on plant production of enhanced photosynthesis and a warmer climate.

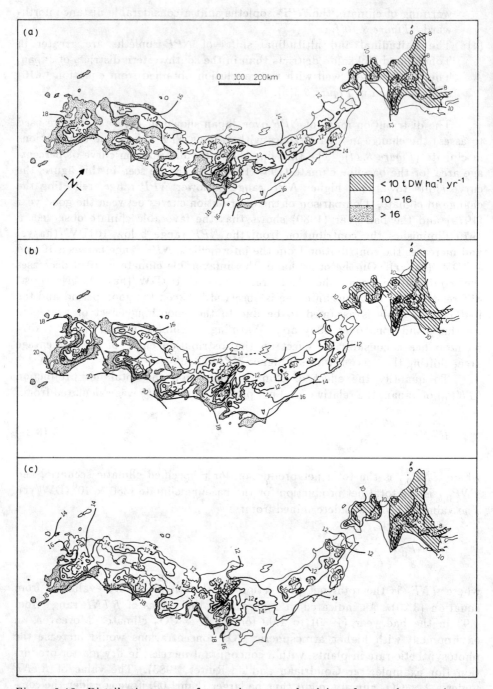

Figure 3.10. Distribution maps of net primary productivity of natural vegetation estimated using the Chikugo model for specified climatic scenarios: (*a*) baseline period (1951–80), (*b*) bad decade (1926–35), (*c*) bad year (1980), (*d*) good decade (1957–66), (*e*) good year (1979), (*f*) GISS $2 \times CO_2$ experiment.

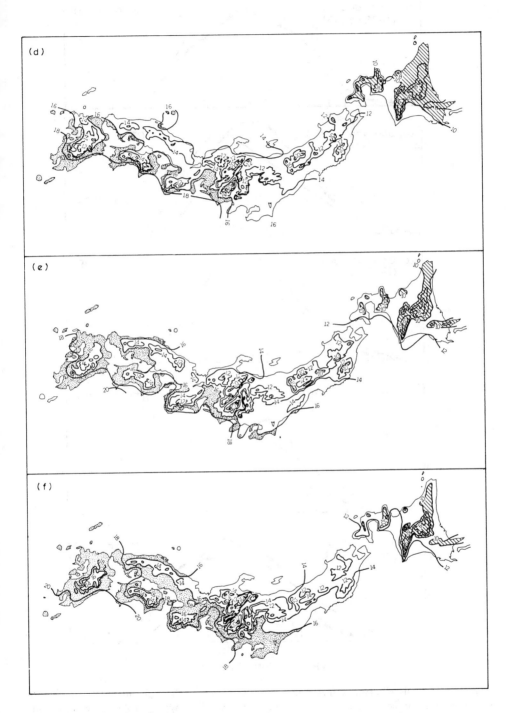

Figure 3.10. Continued.

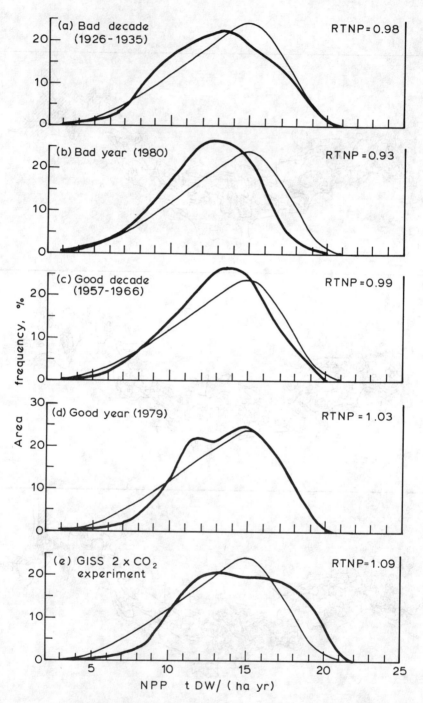

Figure 3.11. Distribution curves of percentage area with different NPP values for specified climatic scenarios. Class intervals are $2\,\text{tDW}/(\text{ha/yr})$; broken line represents distribution for the baseline climate (1951–1980).

3.4.5. Effect of the extreme year scenarios on rice yield

As described above, rice cultivation in Japan is very sensitive to climatic conditions, particularly to variations in thermal resources. To investigate the impact of climatic variations on rice yield, the geographical distribution of the yield index (IY), by prefecture, was compared between the good year and the bad year. The results are presented in *Figure 3.12* together with the geographical distribution of average (May–October) air temperature anomalies relative to the baseline (1951–80). As shown in this figure, there are large differences in IY values between 1979 and 1980, particularly on the Pacific Ocean side of Tohoku and Hokkaido, where east or northeasterly cool winds ("Yamashe-kaze") dominated during the rice growing season in 1980. Marked differences in IY values between 1979 and 1980 were also observed in southwestern parts of Japan, and these can be attributed to the active Bai-u front being located in that area for a prolonged period during July and August of 1980, reducing global solar radiation and decreasing air temperatures. The delayed growth of rice plants that resulted in that area caused the decreases in rice yield recorded in 1980.

Given such large contrasts in yields between 1979 and 1980, the average temperature anomalies for the same years appear modest by comparison. The May–October temperature anomalies in 1979 and 1980 ranged from 0.0 to 2.0 °C, and from 0.2 to −1.0 °C, respectively. As indicated in *Table 3.1*, the IY values in 1979 and 1980 were 103% and 87% respectively, while corresponding $RTNP$ values were 1.03 and 0.93 (*Figure 3.11*). These results indicate that the effects of climatic variations on rice yield in Japan are somewhat greater than the effects on the net production of natural vegetation. This is probably because rice plants are more sensitive to temperature conditions than the natural vegetation found in Japan.

3.5. Conclusions

The results indicate that the effective accumulated temperature ($\overline{\Sigma T}_{10}$) is highly correlated with July–August air temperatures ($\overline{T}_{7,8}$) that are important for rice cultivation. Isopleths of ΣT_{10} move northward or southward in response to changes in the general climate. Such changes in ΣT_{10} distribution are more substantial in northern Japan than in the southern region, indicating that climatic change has a more important influence on rice cultivation in the northern districts. This is confirmed by the fact that variability of rice yields increases approximately linearly with decreasing $\overline{T}_{7,8}$ over the range of $\overline{T}_{7,8}$ values below 24 °C, the coefficient of variation reaching 30% in Hokkaido district where temperature resources are frequently insufficient for rice cultivation.

The Chikugo model was used to study the impact of climatic changes (including that due to a doubling of atmospheric CO_2 content) on plant productivity potential. The results show that the potential, defined as net primary productivity (NPP), is significantly affected by changes in climate. The impact of climatic change (represented by specified climatic scenarios) on NPP was more considerable in the northern districts of Japan than further south, as observed

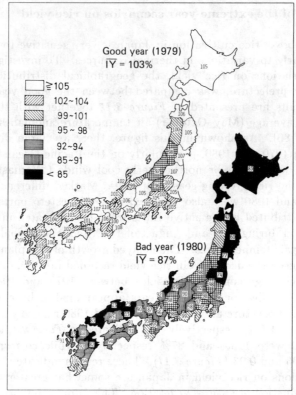
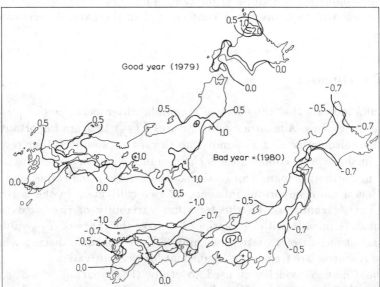

Figure 3.12. (a) Regional distribution of rice yield index (*IY*) in 1979 and 1980. (b) Distribution map of average May–October air temperature anomalies in 1979 and 1980 relative to the baseline (1951–1980).

for the relationship between rice yield and climate. The relative total net production ($RTNP$), a measure of the ratio of total net production over Japan in scenario weather years to that in the baseline (1951–80) average weather year, ranged from 0.93 under the "bad" year scenario (1980) with a cool summer, to 1.09 under the GISS model-derived $2 \times CO_2$ climatic scenario. This latter result indicates the potentially important role of climatic changes due to increasing CO_2 for plant productivity potential and crop production in Japan.

In the next section, an empirical statistical approach is employed to examine the sensitivity of rice yields to temperature variations, and the altitudinal limits on potential for rice cultivation are also investigated for a number of climatic scenarios.

SECTION 4

The Effects on Altitudinal Shift of Rice Yield and Cultivable Area in Northern Japan

4.1. Introduction

The Hokkaido and Tohoku districts in northern Japan (cf. *Figure 1.1*) are major rice producing areas, accounting for about one-third of total Japanese rice production. However, these districts exhibit large interannual fluctuations of rice yield, because they are situated near the northern limit of production. The coefficient of variation of the yield index (representing annual yields) is higher here than elsewhere, amounting to 30.7% in Hokkaido and 13.9% in Tohoku (Uchijima, 1981; *see also* Subsection 3.3 above).

Rice growth is particularly sensitive to temperature conditions during the heading phase in the summer season, and cool weather damage due to low temperature is a recurrent phenomenon in the Hokkaido and Tohoku districts. The frequency of occurrence is about one year in six in northern Tohoku, and about one year in three or four in Hokkaido, the most northerly island in Japan (Uchijima, 1983). In this section we explore the effect of variations of summer temperature on rice yields and the cultivable area for rice in northern Japan.

4.2. Critical Temperatures in Rice Production

The usual cropping season for rice in Japan is from May to October. In order to complete the normal growth cycle, effective accumulated temperatures (ΣT_{10}) of at least 3200 degree-days in Tohoku and at least 2600 degree-days in Hokkaido (where early-ripening rice varieties are widespread) are required (Uchijima, 1983). Note that the estimate for Hokkaido is a refinement of that presented in Section 1 (*see Figure 1.6*).

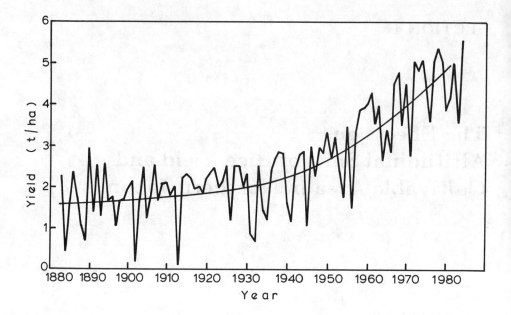

Figure 4.1. Variations in annual rice yield (tonnes/hectare) in Hokkaido, 1888–1983. Curved line indicates polynomial trend fitted to annual yields.

Rice cultivation in Tohoku and Hokkaido is commonly restricted to those areas with suitable average ΣT_{10} values. However, rice yields in any one year are closely related to the temperature in mid-summer, which includes the heading period. *Figure 4.1* shows the secular change of rice yield in Hokkaido over the period 1888–1983. The effect of cool summer damage is reflected in significantly reduced yields in some years. Since about 1950 average yield has increased considerably due to the development of agricultural technology.

The fluctuation of yield is closely related to yearly variations in temperature. To separate the effects of the development of agricultural technology and fluctuating weather upon rice yield, a yield index was calculated as:

$$IY = Y(t) / Y_T(t) \tag{4.1}$$

where IY is the yield index, and $Y(t)$ and $Y_T(t)$ denote, respectively, the actual rice yield and the trend rice yield in year t. The latter values were constructed by polynomial fit, as shown by the continuous smooth line in *Figure 4.1*. The yield index is closely related to the mean temperature for the July–August period in both Hokkaido and Tohoku (NIAS, 1975). These relationships (shown in *Figure 4.2*) can be well approximated by equations (4.2) and (4.3):

$$\text{Hokkaido:} \quad IY = 1.2035 - 0.04803\,(T_{7.8} - 22.5)^2 \tag{4.2}$$

$$\text{Tohoku:} \quad IY = 1.1207 - 0.03147\,(T_{7.8} - 24.5)^2 \tag{4.3}$$

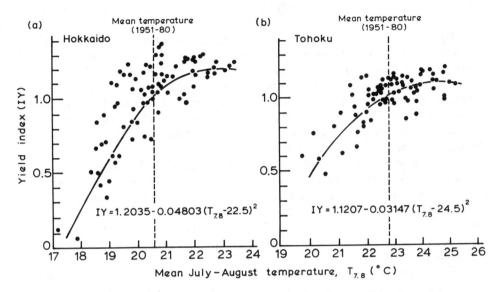

Figure 4.2. Relationships between the yield index (IY) and mean July–August temperature ($T_{7.8}$) in (*a*) Hokkaido and (*b*) Tohoku districts.

where $T_{7.8}$ is mean July–August temperature. The temperature in Hokkaido is represented by average data from two meteorological stations (Sapporo and Asahigawa), and in Tohoku by the average from six stations (Aomori, Miyako, Ishinomaki, Akita, Yamagata and Fukushima). Station locations are shown in *Figure 2.2* in Section 2.

A convenient threshold of temperature may be selected from these results that gives a yield index of 1.0 in each district (20.4 °C in Hokkaido and 22.5 °C in Tohoku), and a remarkable decrease in the yield index can be observed for the range of temperatures below these values. Employing these relationships, we can thus estimate rice yields due to a range of climatic conditions, and determine the critical temperatures required to obtain particular yield levels.

4.3. Altitudinal Shifts of the Cultivable Area

Air temperature at a given location is related to its latitude and altitude. In Japan, the lapse rate of monthly mean temperature with increasing latitude is between 0.59 and 1.19 °C/degree, while the lapse rate with increasing altitude is between 0.51 and 0.64 °C/100m. Combining these characteristics, July–August mean normal temperatures at any location in Japan can be approximated by:

$$T_{7.8} = 29.3 - 0.62(L - 30) - 0.0063 H \tag{4.4}$$

where L is the latitude (degrees) and H is the altitude (meters) (JMA, 1960). Thus, for example, it can be shown from equation (4.4) that at a given latitude the altitude of a given isotherm will be shifted by 100 m for a mean temperature change of 0.63 °C.

By substituting this altitudinal lapse rate into equations (4.2) and (4.3) it is possible to relate the altitudinal shift of temperature isotherms induced by a temperature anomaly to a change in yield index at a fixed location (*Figure 4.3*). A positive temperature anomaly leads to an altitudinal shift of thermal resources uphill, with yields at any fixed point in both districts increased. In contrast, there is a rapid decrease of the yield index for equivalent altitudinal shifts downslope. The response of the yield index is greater in Hokkaido than in Tohoku because of its more northerly location.

The long-term mean yield index in each district is 1.0, and this represents yields for an average elevation in the cultivated area. Towards the upper limits of cultivation the mean yield index clearly decreases and the probability of failing to achieve satisfactorily high yields (arbitrarily defined) increases. Most of the rice area in recent years has been distributed below about 250 m in Hokkaido and below about 350 m in Tohoku (*Figure 4.4*). If we assume that these altitudes correspond to the mean long-term upper limits of the cultivated zone, for which the mean yield index is 1.0, then this zone can be matched to mean July–August temperatures that give the yield index 1.0. Year-to-year variations in temperature relative to the mean can thus be translated into shifts in the altitudinal level of temperature isotherms and of the "safe" cultivation zone.

Figure 4.5 shows these interannual shifts in limits of potential cultivation in Hokkaido and Tohoku, during the period 1890–1983. There are noticeable falls in the cultivable limit during the 1890s, 1910s, 1930s, 1950s and 1980s in both districts, with the greater variations occurring in Hokkaido, situated at a higher latitude than Tohoku.

For any arbitrary elevation, for example the average altitude of present-day cultivation (dashed line B in *Figure 4.5*), the number of years in which the potential cultivation limit is at a lower level can be used as an indicator of "failure-frequency" (i.e., failure to achieve the threshold temperature requirement). At altitude B the proportion of failure-years is 22.4% in Hokkaido, and 7.0% in Tohoku. These proportions are almost equal to the proportions of cool summer damage years recorded in these districts.

The relationships between frequency of these failure-years and altitude in Tohoku district is illustrated in *Figure 4.6*. At higher altitudes the risk of failure evidently increases, and a zone of high risk is plausible where rice cultivation can be considered only marginally viable. Unless supported by empirical field evidence, the delineation of such a marginal zone is essentially arbitrary. Here we assume that the zone lies in the failure probability range of 0.15–0.35 (i.e., between 1 year-in-7 and 1-in-3), which would be found at elevations of between about 200–330 m.

Clearly, shifts in the altitudinal limits of cultivable area give rise to changes in the area of safe cultivation. Using the estimates of the distribution of cultivated area (*Figure 4.4*), changes in safely cultivable area have been evaluated for certain specified temperature conditions relative to the period 1951–80 (*Table 4.1*). Analysis of these results shows that for a low temperature anomaly in Hokkaido occurring with a return period of 5 years, the altitudinal limit for safe

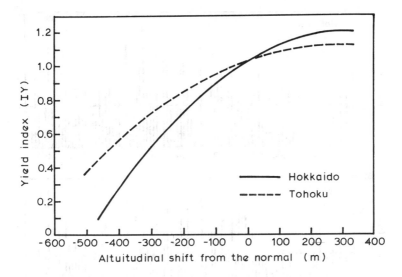

Figure 4.3. Relationships between the yield index at an average elevation in the cultivated area, and the altitudinal shift of temperature isotherms induced by anomalous temperatures.

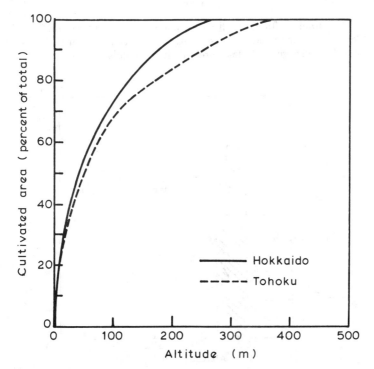

Figure 4.4. Proportion of total cultivated rice area (%) at different altitudes in Hokkaido (solid line) and Tohoku (dashed line). Data are for 1978. (Source: Ministry of Agriculture, Forestry and Fisheries, various dates.)

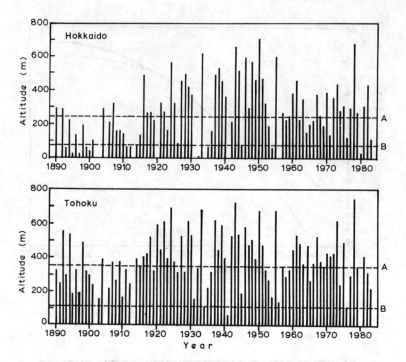

Figure 4.5. Altitudinal shifts of the limit to potential cultivation, 1890–1983, for Hokkaido and Tohoku. Dashed lines represent A: the altitudinal limit of present-day rice cultivation, and B: average altitude of present-day rice cultivation.

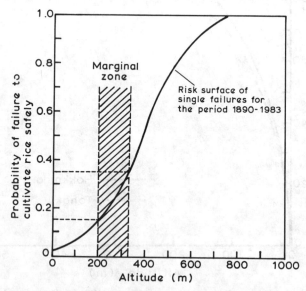

Figure 4.6. Probability of failure to cultivate rice safely at different elevations in Tohoku based on frequencies calculated from *Figure 4.5*. The shaded area depicts an arbitrary "marginal" zone.

Table 4.1. The effect of anomalies of mean July–August temperature (relative to the average for 1951–80) on altitudinal limits of potentially "safe" cultivation, the area of potentially safe cultivation, and yield index (for fixed cultivation area).

Temperature event	Temperature anomaly (°C)[a]	Altitudinal shift (m)	Cultivable area (%)	Yield index[b]
HOKKAIDO				
1951–1980 Average	0.0	0	100	1.03
Return period (years):				
5	−1.2	−190	56	0.74
10	−1.7	−270	0	0.58
20	−2.2	−349	0	0.40
30	−2.4	−381	0	0.32
TOHOKU				
1951–1980 Average	0.0	0	100	1.03
Return period (years):				
5	−0.9	−143	84	0.91
10	−1.3	−206	76	0.84
20	−1.9	−302	38	0.71
30	−2.2	−349	20	0.64

[a] Deviations from 1951–80 mean temperatures: 20.6 °C in Hokkaido, 22.8 °C in Tohoku.
[b] Assuming no change in actual cultivated area from present day.

cultivation shifts downhill by about 190 m, representing a decrease in the safely cultivable area for rice to about 56% of normal. For anomalies with return periods of 10 years or greater, the altitudinal shifts would be such that no safely cultivable area remains. In Tohoku the safely cultivable area contracts to 84% of the present-day area under low temperature anomalies corresponding to a return period of 5 years, and to only 20% in the case of the 1-in-30 year temperature anomaly. Since they are based on district averages, these estimations probably exaggerate the impact of cool temperatures because in both districts there are regions where conditions remain favorable even under the most severe (average) temperature anomalies.

The shifts depicted in *Table 4.1* are of *potentially* safe cultivable area. In reality, of course, the cultivated area does not change markedly from year to year, so the observed impact of temperature anomalies needs to be assessed in terms of changes in average yield. *Table 4.1* therefore also includes information on yield index for the temperature anomalies corresponding to the specified return periods.

4.4. Changes in Rice Yield and Cultivable Area Under Different Climatic Scenarios

4.4.1. Climatic scenarios

As described in Section 2, two types of climatic scenario were selected in this section for assessing impacts on rice cultivation in northern Japan: instrumentally derived scenarios and GISS model-derived estimates of $2 \times CO_2$ climate. To construct the scenarios, we considered only temperature, because rice yield in Japan depends chiefly upon temperature during the growing season.

The instrumental scenarios, based on the characteristics of variation in the Japanese climate, were selected from about 100 years of observed meteorological data. As a reference against which to compare anomalous individual years or periods, the period 1951–1980 was adopted as a "baseline". The scenarios chosen for this study are listed in *Table 4.2*. Note that most of these differ from the years selected in Section 3, which were based on variations in rice yield.

The $2 \times CO_2$ temperature scenario is based on the GISS model outputs for $2 \times CO_2$–$1 \times CO_2$ equilibrium conditions (*see* Subsection 2.5). Deviations from the baseline temperatures in Hokkaido and Tohoku (*Table 4.2*) were obtained as mean July–August values averaged from four grid point locations representing each district (respectively, Hokkaido: 46°N, 140°E; 46°N, 145°E; 42°N, 140°E and 42°N, 145°N; Tohoku: 42°N, 140°E; 42°N, 145°E; 38°N, 140°E and 38°N, 145°E).

Table 4.2. Climatic scenarios for Section 4, representing variations in July–August mean temperature.

Climatic scenario	Years	Mean July–August Temperature (°C)	Deviation from baseline (°C)
		HOKKAIDO	
Warmest year	1978	23.1	+2.5
Warm year	1955	22.6	+2.0
Warm decade	1921–1930	21.0	+0.4
Baseline	1951–1980	20.6	0.0
Cool decade	1902–1911	19.5	−1.1
Cool year	1980	19.0	−1.6
Coolest year	1902	17.3	−3.3
$2 \times CO_2$	Future	24.1	+3.5
		TOHOKU	
Warmest year	1978	25.1	+2.3
Warm year	1955	24.7	+1.9
Warm decade	1921–1930	23.5	+0.7
Baseline	1951–1980	22.8	0.0
Cool decade	1902–1911	21.8	−1.0
Cool year	1980	20.2	−2.6
Coolest year	1902	19.7	−3.1
$2 \times CO_2$	Future	26.0	+3.2

4.4.2. Changes in yield and the altitudinal limits of cultivable rice

Estimates of the impact of the scenarios on rice yield and cultivable area are shown in *Table 4.3*. Under the climate represented in the "coolest year" scenario, the hypothetical altitude of the safely cultivable zone is shifted downward in both districts by about 500 m thus effectively precluding safe rice cultivation in all but the most favorable areas in either district (*see Figure 4.4*), while the average yield index for the present cultivated area would collapse to zero relative to the normal (baseline) in Hokkaido, and 38% of normal in Tohoku. Under the "cool decade" scenario, the altitudinal limit shifts downhill by more than 150 m and the yield index is 75% of the normal in Hokkaido, and 87% in Tohoku. In contrast, under the "warmest year" scenario, the altitude of the safely cultivable zone is shifted upward by nearly 400 m and the yield index increases to 115% of the normal in Hokkaido, and 108% in Tohoku. Under the "warm decade" scenario, the yield index is estimated to be about 106% relative to the baseline in each district.

The $2 \times CO_2$ scenario would cause a shift in the upper limit of potentially safe cultivation of more than 500 m in both districts, while the yield index would be about 105% and 102% of normal in Hokkaido and Tohoku, respectively. Thus rice yields under a $2 \times CO_2$ climate might well be smaller than those for the warm year and warm decade scenarios, a result of the curvilinear relationships shown in *Figure 4.2*.

Table 4.3. Effects of the climatic scenarios on altitudinal limits of potentially safe cultivation and the yield index in Hokkaido and Tohoku. Values are relative to the baseline period (1951–80).

Climatic scenario	Mean July–August temperature (°C)	Altitudinal shift (m)	Yield index (% of baseline)
HOKKAIDO			
Warmest year	23.1	397	115
Warm year	22.6	317	117
Warm decade	21.0	63	106
Cool decade	19.5	−175	75
Cool year	19.0	−254	60
Coolest year	17.3	−524	0
$2 \times CO_2$	24.1	556	105
TOHOKU			
Warmest year	25.1	365	108
Warm year	24.7	302	109
Warm decade	23.5	111	106
Cool decade	21.8	−159	87
Cool year	20.2	−413	52
Coolest year	19.7	−492	38
$2 \times CO_2$	26.0	508	102

4.4.3. Changes in the safely cultivable area for rice

As already stated above, one method of defining the safely cultivable area for rice under present-day agricultural technology is to consider those areas with an effective accumulated temperature (ΣT_{10}) greater than 2600 degree-days in Hokkaido, and above 3200 degree-days in Tohoku. Based on the regional distribution of these critical ETS values, the safely cultivable areas have been estimated for each of the climatic scenarios. Selected results are illustrated in *Figure 4.7*.

The present cultivated area of rice in Hokkaido is about 180 000 ha (*see Figure 1.9*), with the majority distributed at elevations below 200 m in the southwestern part of the island [*Figure 4.7(a)*]. Under the cool decade scenario, with mean July–August temperatures about 1 °C lower than the normal, the safely cultivable area is restricted solely to elevations below 100 m in southwestern Hokkaido, where the Sea of Japan acts to ameliorate the climate [*Figure 4.7(b)*]. Under the warm decade scenario, with temperatures about 0.4 °C higher than normal, the area would expand to include land up to about 250 m altitude in western parts, and in part of the eastern region too, up to elevations of about 100 m [*Figure 4.7(c)*].

With the conditions implied in the $2 \times CO_2$ scenario and if other factors do not intervene, the estimated temperature rise of about 3.5 °C could open up most of the land below about 500 m for rice cultivation, except for the low temperature areas in the east and in the mountains [*Figure 4.7(d)*].

In Tohoku, the area at present cultivated for rice is about 600 000 ha, with the bulk of this land distributed at elevations below about 300 m. However, northeastern regions that are liable to be influenced by cool summer weather caused by incursions of the Okhotsk air mass are unfavorable for rice cultivation under present conditions, even at low elevations [*Figure 4.7(a)*]. The cool decade scenario, giving mean temperatures some 1 °C lower than the baseline value, would cause the safely cultivable area to contract to elevations below 200 m, and the unfavorable regions in the northeast to expand southwards [*Figure 4.7(b)*]. During the warm decade scenario with temperatures about 0.7 °C higher than normal, part of the northeastern region, up to about 100 m elevation, would become suitable for cultivation [*Figure 4.7(c)*]. Finally, under the $2 \times CO_2$ scenario in Tohoku, the increase in mean temperatures of about 3 °C could open up most of the area below about 600 m altitude for rice cultivation, except in the mountainous regions [*Figure 4.7(d)*].

Interestingly, and consistent with the results recorded (using a different approach) in *Table 4.1*, no safely cultivable area would remain in Hokkaido under the coolest year scenario, and the area would be restricted to elevations below about 200 m in the southern part of Tohoku, emphasizing the vulnerability of these northern districts to low temperature episodes.

4.5. Conclusion

From the results presented, it can be concluded that under the anomalously warm conditions that have been observed historically in northern Japan, the rice

Altitudinal shift of rice yield and cultivable area

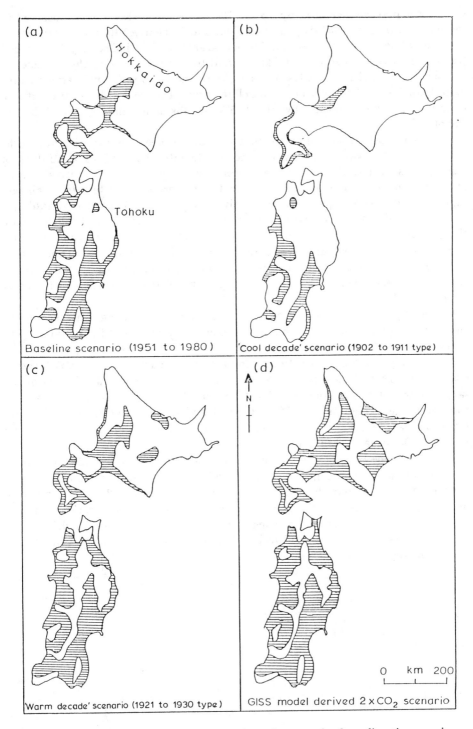

Figure 4.7. Safely cultivable rice area in northern Japan under four climatic scenarios.

yield index may be increased by up to 17% above the average levels observed in the baseline period (1951–80) in Hokkaido, and to nearly 10% above average in Tohoku, while under the cool climate scenarios the yield index may be reduced (in extreme cases) to zero in Hokkaido, and to less than 40% of the average, in Tohoku. Under the $2 \times CO_2$ scenario, with mean July–August temperatures more than 3 °C higher than in the baseline period, the yield index of present rice varieties may be increased by about 5% in Hokkaido, and by about 2% in Tohoku.

It is important to emphasize that in order to keep the experimental techniques simple it was necessary to incorporate into the analysis a number of assumptions. First, although the altitudinal shifts in potentially cultivable area may be realistic meteorologically, in practice factors such as terrain, drainage, exposure and soils may preclude rice cultivation even if the climate is suitable.

A second assumption (which is problematic in cases of significant climatic change such as under the $2 \times CO_2$ scenario) concerns the exclusion of other climatic variables, particularly solar radiation, windspeed and precipitation, from the analyses. Although temperature anomalies exert an important influence on rice yields, these other factors can also be significant (as will be demonstrated in the next section).

Thirdly, the incidence of crop diseases could well change in response to climatic changes, thus influencing both the potential yield and the suitable areas for cultivation. This was not considered in the above analyses.

Fourthly, in order to estimate impacts of scenario climates not hitherto observed in the historical record (namely, the $2 \times CO_2$ scenario), it was necessary to extrapolate statistical relationships, such as those for the yield index, *outside* the range of climatic conditions for which they were developed. This problem is discussed further in the next section, and caution is advised in interpreting the $2 \times CO_2$ scenario results presented above.

Finally, the estimations conducted in this section were for quick-maturing varieties of rice that are grown today in northern Japan. Later-maturing rice varieties that exploited the higher temperatures under the $2 \times CO_2$ scenario might well offer substantially greater yields. This problem is investigated in the next section, with experiments using a rice–weather simulation model. Other possible adjustments in agricultural technology in response to climatic change are considered in Section 7.

SECTION 5

The Effects on Rice Yields in Hokkaido

5.1. Introduction

We have seen that cool summer damage can have a significant impact on Japanese rice production, especially in the northern regions of Hokkaido and Tohoku, which produce more than a third of the national output. In this section we use a dynamic model to simulate the growth and yield of the rice crop in Hokkaido under some of the climatic scenarios described in Section 2. Results from these experiments will subsequently be used to substantiate further experimental results in the next section of this report.

The model developed here is a process-oriented dynamic model that can be used to predict the growth and yield of rice under altered climatic conditions, in contrast to empirical–statistical rice models which cannot be extrapolated with much confidence beyond the range of data upon which they were constructed (for example, see Sections 3 and 4, this report; Matsuda, 1960; Hanyu et al., 1966; Kudo, 1975; Munakata, 1976). Since the model is dynamic, it may also be applied to a real time prediction of the growth and yield of rice under changing weather conditions, providing timely information that is useful both for crop management and policy determinations.

5.2. The Model

The *SI*mulation *M*odel for *RI*ce-*W*eather relationships (SIMRIW) was constructed on the basis of the general principle that the grain yield Y_G forms a specific component of the total dry matter production W_t of a crop:

$$Y_G = hW_t \tag{5.1}$$

in which h is the harvest index.

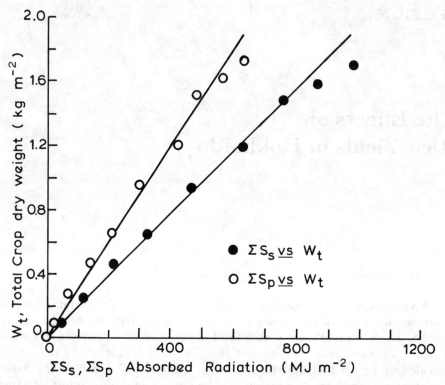

Figure 5.1. Relation between total crop dry weight (W_t) at different times during growth and absorbed shortwave radiation (S_s) or PAR (S_p) cumulated over time for "Nipponbare" rice grown at Tsukuba, Ibaraki prefecture (Horie and Sakuratani, 1985).

It has been shown that crop dry matter production is proportional to the photosynthetically active radiation (PAR) or the shortwave radiation absorbed by a crop canopy (Shibles and Weber, 1966; Monteith, 1977; Gallagher and Biscoe, 1978). Horie and Sakuratani (1985) showed that this is also true in rice and that the proportional constant, the conversion efficiency from the radiation to biomass, is constant until the middle of the ripening stage and thereafter it decreases curvilinearly (*Figure 5.1*). Moreover, they concluded from both simulations and experiments that the conversion efficiency is practically unaffected by climatic conditions in a wide range of environments.

SIMRIW is based on this general principle as follows:

$$\frac{dW_t}{dt} = c_s I_s \qquad (5.2)$$

where c_s is the conversion efficiency from the absorbed shortwave radiation to the rice biomass (g/MJ), and I_s the absorbed radiation per unit time. Since the time constant of rice growth after transplanting into the field (the reciprocal of the relative growth rate) is more than 5 days, it is sufficiently accurate to integrate equation (5.2) with a time interval of 1 day. Hence equation (5.2) can also be presented as:

$$\Delta W_t = c_s S_s \tag{5.3}$$

in which ΔW_t is the daily increment of the crop dry weight and S_s the daily total absorbed radiation. Likewise, in SIMRIW the growth in dry weight, leaf area and yield is computed day by day by inputting daily weather data.

The quantities h, c_s and S_s are functions of the environment, crop developmental stage and growth attributes. These functions and parameters have been determined by analyzing the data from field experiments conducted for major Japanese rice cultivars grown under widely different environmental conditions. Since it is not the purpose of this report to describe the details of derivation of each functional relationship and parameters of the model, only the principal functions of the model are described in the subsequent sections.

5.2.1. Crop development

The developmental processes of a rice crop, such as ear initiation, booting, heading, flowering and maturation are strongly influenced both by the environment and by the crop genotype. In SIMRIW these are represented by a continuous variable named developmental stage, DVS, in a similar way to that employed by de Wit et al. (1970). This variable is defined in such a way that DVS is zero at the onset of the crop emergence, 1.0 at heading and 2.0 at maturation. Thus the development at any moment in time is represented by a DVS value between 0 and 2.0.

The value of DVS at any moment of the crop development is given by integrating the developmental rate, DVR, with respect to time. Day length and temperature are known to be the major environmental factors determining DVR. Under Japanese geographical and climatological conditions, rice development was found to be explained by a function of temperature alone (Horie, 1984), as follows:

$$DVR = \frac{1}{G}\left[1.0 - \exp\left\{-K_D\left(T - T_{cD}\right)\right\}\right] \quad \text{for } T \geq T_{cD}$$

$$DVR = 0 \quad \text{for } T < T_{cD} \tag{5.4}$$

where T is daily mean temperature; T_{cD} minimum temperature for development; G and K_D are parameters. Since the developmental process is different between the vegetative phase ($0 < DVS < 1$) and reproductive phase ($1 < DVS < 2$), T_{cD} and the parameters in equation (5.4) are also different between these phases.

The values of parameters in equation (5.4) also differ between varieties. In general late-maturing varieties have a larger G, while cultivars with a higher sensitivity to temperature have a smaller K_D. By using experimental data, these parameters were determined for the leading rice varieties in various regions of Japan by the "simplex" method, one of the trial-and-error methods used to estimate parameters of nonlinear functions.

5.2.2. Dry matter production

To simulate crop dry matter production by equation (5.3) it is necessary to compute the absorbed radiation by the crop canopy S_s which is the function of the incident solar radiation S_{so}, leaf area index (LAI) F, and the structure and optical properties of the canopy. Using crop micrometeorological theory originating from Monsi and Saeki (1953), S_s is given by:

$$S_s = S_{so} \left[1 - r - \left(1 - r_o\right) \cdot \exp\left\{-\left(1 - m\right)k^* F\right\}\right] \tag{5.5}$$

where r and r_o are the reflectances of the canopy and the bare soil, respectively, m is the scattering coefficient, and k^* is the extinction coefficient of the canopy to daily shortwave radiation. The canopy reflectance r can be approximated well by the following equation (Research Group of Evapotranspiration, 1967):

$$r = r_f - \left(r_f - r_o\right) \cdot \exp\left(-0.5F\right) \tag{5.6}$$

in which r_f is the reflectance when the surface is completely covered by the vegetation. From Horie and Sakuratani (1985), the following values were adopted for the parameters of the above functions: $k^* = 0.6$, $m = 0.25$, $r_f = 0.22$ and $r_o = 0.1$.

In most simulation models for crop growth, leaf area growth is calculated from the growth of leaf weight by multiplying with a simple conversion factor, the specific leaf area. In SIMRIW, however, the expansion of leaf area in rice is modeled independently of the weight, for reasons outlined in Horie et al. (1978).

Under the present conditions of rice production, it is reasonable to assume that water and nutrients are not limiting factors to the expansion of the leaf area, while the main governing factor is temperature. In SIMRIW the relationship between LAI (F) and daily mean temperature for the period before heading is given as:

$$\frac{1}{F} \cdot \frac{dF}{dt} = A\left[1.0 - \exp\left\{-K_f\left(T - T_{cf}\right)\right\}\right]\left\{1.0 - \left(F/F_{as}\right)\eta\right\} \tag{5.7}$$

in which A is the maximum relative growth rate of LAI; T_{cf} is the minimum temperature for LAI growth; F_{as} is an asymptotic value of the LAI when temperature is nonlimiting, and K_f and η are parameters.

In rice, it is commonly observed that LAI attains a maximum at about the heading stage and thereafter declines gradually during the maturing stage.

However, the physiology and the environmental response of the leaves during this maturation process are quite obscure. For this reason, the change in LAI from the time just before the heading to the maturation of the crop is represented in SIMRIW by a unique function of the crop developmental stage (DVS), alone.

By use of the above equations, the crop-absorbed radiation S_s in each day can be obtained, and the dry matter production can be simulated with the appropriate value of the conversion factor c_s. As has been shown in *Figure 5.1*, c_s is constant until the middle of the grain filling stage and then it decreases gradually. These features of the change in c_s may be represented by:

$$c_s = 19.5 \text{ g/MJ} \qquad \text{for } 0 < DVS < 1.0$$
$$c_s = 19.5 \frac{1+B}{1+B \exp\{(DVS-1)/\tau\}} \qquad \text{for } 1.0 < DVS < 2.0 \tag{5.8}$$

in which B and τ are parameters.

5.2.3. Harvest index

Figure 5.2 shows that the relationship between the dry weight of the brown rice (unprocessed grain), W_y, and that of the whole crop, W_t, is linear over a wide range of W_t, indicating that the proportionality constant, the harvest index (h), is constant. However, this is not always so. The harvest index decreases if the percentage sterility of the spikelets increases, or if the crop ceases growth before completing its development, due to cool late-summer temperatures or frost. The increase in the number of sterile grains brought about by cool temperatures at the booting and flowering stages is called cool summer damage due to floral impotency, and the premature cessation of growth due to late-summer coolness is called cool summer damage due to delayed growth.

In SIMRIW, the harvest index h is represented as a function of the percentage sterility of rice spikelets, γ, and the crop developmental stage, DVS, in order to take into account both types of cool summer damage, as follows:

$$h = h_m \big(1.0 - \gamma\big)\bigg[1.0 - \exp\big\{-K_h\big(DVS - 1.22\big)\big\}\bigg] \tag{5.9}$$

where h_m is the potential harvest index of a given cultivar and K_h is a parameter. Equation (5.9) implies that the harvest index decreases as γ increases due to cool temperature at the booting and flowering stages, or according to the date of cessation of growth before full maturation ($DVS = 2.0$) due to cool late-summer temperatures.

Using the "cooling degree-day" concept (Uchijima, 1976), the relation between daily mean temperature, T, and the percentage sterility may be approximated by the following equation (*Figure 5.3*):

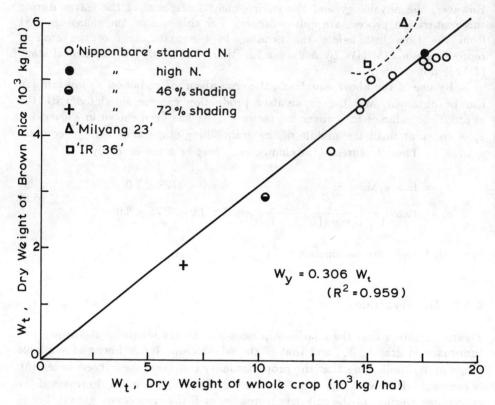

Figure 5.2. Relation between the dry weight of brown rice and that of whole crop for "Nipponbare" rice grown under different environmental conditions. Data on the high yielding cultivars "Milyang 23" and "IR-36" are also shown.

$$\gamma = \gamma_0 + K_q Q_T^a \tag{5.10}$$

where γ_0, K_q and a are empirical constants, and Q_T is the cooling degree-days, given by:

$$Q_T = \sum_i \left[22.0 - T_i \right] \tag{5.11}$$

The summation of equation (5.11) is made for the period of greatest sensitivity of the rice panicle to cool temperatures ($0.75 \leq DVS \leq 1.2$).

The terminal condition of the model simulation occurs either when DVS reaches 2.0 (full maturation), or when the number of days with $T \leq 10.0\,°C$ reaches three. In the latter case, with daily mean temperatures below about 10 °C in Hokkaido, autumn frosts are likely and cool summer damage due to delayed growth usually occurs.

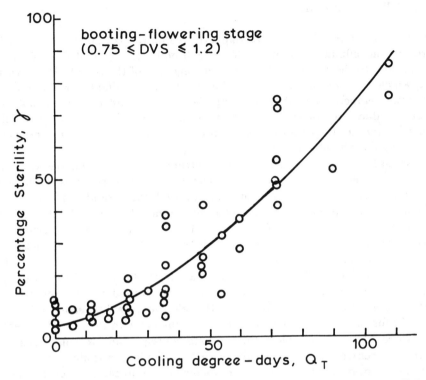

Figure 5.3. Relation between cooling degree-days and percentage sterility of spikelets of "Eiko" rice between the booting and flowering stages. (From the data of Shibata *et al.*, 1970.)

5.3. Crop Parameters and Climatic Scenarios

To simulate rice yields using the model described above, values of the crop parameters and climatic variables (temperature, and radiation or sunshine hours) are required.

5.3.1. Crop parameters

The crop parameters were determined by analyzing well specified field experimental data on the crop–weather relationships for three major rice cultivars in Japan. The cultivars are "Ishikari", a dominant variety in Hokkaido, "Koshihikari", a middle-maturing variety commonly grown in Honshu and "Nipponbare", a late-maturing variety, also grown in Honshu. To simulate the rice yield under the present level of technology, the crop parameters for "Ishikari" were used for Hokkaido.

5.3.2. Climatic scenarios

The climatic data observed at Sapporo (43°03′N, 141°20′E) were used to represent the climatic conditions for rice production in Hokkaido, since it is located close to the center of the rice producing area of the island. Daily weather data, which are required as inputs to the model, are available for about the last 40 years. Temperature data can be applied directly, but consistent daily solar radiation values are not available for the scenarios described below. As a substitute, solar radiation was estimated from daily sunshine hours using the method of Yoshida and Shinoki (1978).

As described in Section 2, both instrumental and model-based climatic scenarios were employed in this study. The instrumental scenarios comprised an anomalously cool summer year, and decades of "stable" and "unstable" yields (primarily reflecting the frequency of cool summer damage). Within the constraints of data availability, 1971 was selected at the extreme case scenario, the climatic data from 1957–66 represented the stable decade, and those from 1974–1983 the unstable decade, the latter choice differing from the unstable decade used in Section 3 because suitable data from the period 1926–35 were not available. The standard baseline period, 1951–80, is used as a reference in most of the experiments in this section, although the recent decade, 1974–83, is employed for this purpose in some of the experiments with the $2 \times CO_2$ scenario.

To represent a possible future climate corresponding to conditions under doubled concentrations of atmospheric carbon dioxide, outputs from the GISS general circulation model were utilized, following the procedures outlined in Subsection 2.5.2. Adjustments to the reference daily temperature data at Sapporo were made by adding the differences between GISS model-derived $1 \times CO_2$ and $2 \times CO_2$ equilibrium monthly temperatures, for a grid point location approximating the Sapporo station. In this application, the arithmetic mean of values at four grid points surrounding Sapporo (at 46°N, 140°E; 46°N, 145°E; 42°N, 140°E and 42°N, 145°E) were used. [Note that in Subsection 2.5.1 only a *single* grid point location (42°N, 140°E) was used to compare the GISS model outputs with data for Sapporo.] *Table 5.1* shows the computed GISS model-derived monthly temperature changes for a doubling of atmospheric CO_2 at Sapporo during the main rice growing season.

Table 5.1. Monthly temperature change, ΔT (°C), due to a doubling of atmospheric CO_2 predicted by the GISS model, for the rice growing season in Hokkaido.

	May	June	July	August	September	October
ΔT (°C)	4.0	3.7	3.5	3.5	3.4	3.3

The temperature changes given in *Table 5.1* were added to the observed daily data in each month during the recent 10-year period 1974–83 and also to the daily normals for the baseline period (1951–80). The GISS model also predicts that the precipitation and cloud cover would slightly increase with a doubling of CO_2, but since the rice crop is assumed to be fully irrigated in the model, and because information is not available for changes in solar radiation, we have assumed that all other climatic factors remain unaltered from their baseline values.

5.4. Yield Simulations, Technology Adjustment and Sensitivity Analysis

Figure 5.4 illustrates the results of a model simulation of the growth processes of a rice crop under changing weather conditions. For daily inputs of air temperature and derived solar radiation, the crop development stages, leaf area expansion and dry matter weight (of the whole crop and of the grain) are depicted along the same time axis.

Since it is assumed in SIMRIW that water, nutrients, insects and diseases are managed adequately, the model estimates are of *potential* yield achievable under the present level of rice production technology, with climate as the only constraint. To adjust the simulated yield to match the observed average farm yields, comparisons were made between the simulated average district yields and the actual yields in Hokkaido for the recent decade (1974–1983). The results are shown in *Figure 5.5*.

Since the actual yield level is a function not only of temperature and radiation, but also of typhoon, pest and disease damage, the model cannot fully explain the year-to-year changes in yield. Nevertheless, it is clear that a large part of the yield variation in Hokkaido *can* be explained by variations in temperature and sunshine hours, as modeled.

Figure 5.5 shows that actual yield in Hokkaido at the present level of technology may be obtained by multiplying the factor 0.84 by the simulated yield. Accordingly, the impact of the scenario climates on rice production under present technology has also been estimated by multiplying this factor to the simulated yield in the experiments reported below.

It was important, before conducting the scenario experiments, to evaluate first the sensitivity of the model to a range of climatic perturbations. The results of this sensitivity analysis are given in *Figure 5.6*, in which yield isopleths are depicted as a function of temperature and sunshine hours anomalies relative to the baseline period (1951–80). In this analysis, crop parameters for the cultivar "Ishikari" were used. *Figure 5.6* shows that, with a negative anomaly of the average temperature over the whole growth period of greater than $-1\,°C$, a large yield reduction is expected, being almost independent of the sunshine hours. This is due to the overriding influence of cool summer damage at these temperatures. However, as the temperature anomaly increases from $-1\,°C$, the effect of sunshine hours becomes more conspicuous.

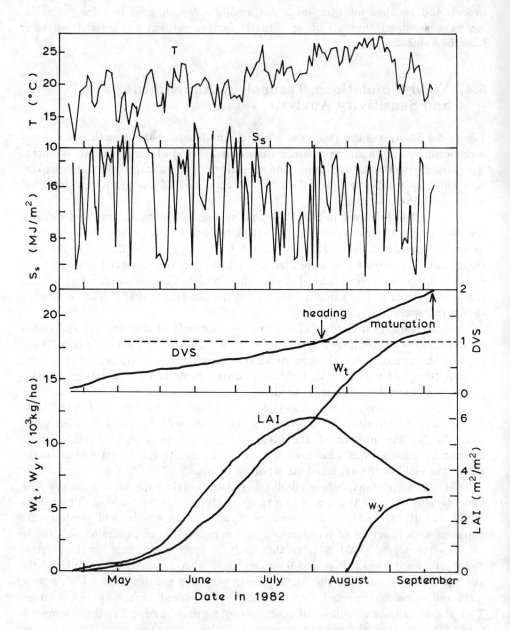

Figure 5.4. Simulated dynamics of the crop developmental stage (*DVS*), leaf area index (LAI) and dry weight of both brown rice (W_y) and the whole crop (W_t), plotted alongside the daily mean temperature (T) and shortwave radiation (S_s) conditions during an entire growth period of "Nipponbare" rice in 1982 at Tsukuba.

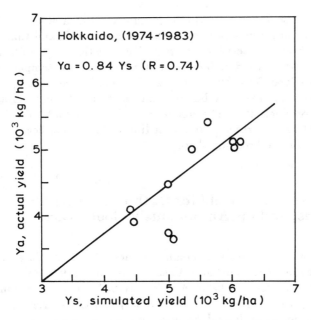

Figure 5.5. Comparison of simulated and actual rice yields in Hokkaido, 1974–1983.

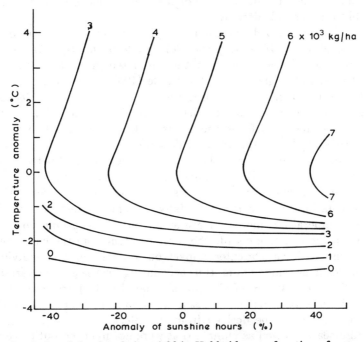

Figure 5.6. Isopleths of simulated rice yield in Hokkaido as a function of anomalies relative to the baseline (1951–80) of temperature and sunshine hours over the whole growing period.

Figure 5.6 also shows that the maximum yield at a given level of sunshine hours is obtained at a temperature anomaly of about 0 °C, and that as the anomaly deviates in both the positive and negative directions, a significant reduction in yield occurs. The yield reduction under the positive temperature anomaly conditions is derived from the fact that "Ishikari" rice develops too rapidly, reaching the reproductive stage before establishing a sufficient vegetative growth for maximum productivity. The above result implies that this rice cultivar is very well suited to the present climate in Hokkaido, having been selected during about 100 years of rice breeding there.

5.5. Estimated Rice Yield for Stable and Unstable Decades, and for Anomalously Cool Years

Figure 5.7 shows the year-to-year changes in actual and simulated rice yields for the stable decade (1957–66) and unstable decade (1974–83), together with anomalies from the 30-year baseline period (1951–80) of total sunshine hours over the main growth period (July–September) and the average air temperature over July and August. July and August temperature is used (as in Sections 3, 4 and 6) because this is the period in which cool summer damage due to floral impotency of rice usually occurs. The magnitude of the variations in these temperatures during each period was one of the criteria that influenced the selection of the stable and unstable decades.

As has been described, the simulated yield level was adjusted to the recent technology level in the 1974–83 (unstable) decade by multiplying by a factor of 0.84. Hence the differences between the simulated yield (averaging 4.48 t/ha) and actual yield (3.57 t/ha) during the stable (1957–66) decade can be ascribed to the advance in technology during the period between the two decades. Thus in Hokkaido the increase in the rice yield due to the advance of the technology is estimated to be about 25% for the 17 years from 1961 (middle of the stable decade) to 1978 (middle of the unstable decade).

The computed difference in the simulated average yield between the stable and unstable decades is negligibly small, while the coefficients of variation (CV) of the simulated yields in the two decades are 7.8% and 11.6%, respectively. The specific explanation for this result is derived from the fact that in the stable decade the temperature anomalies are small but the sunshine hour anomalies large, while in the unstable decade the reverse is true, and the large temperature anomalies tend to have a greater impact on annual yields in individual years than anomalies in sunshine hours, thus accounting for the greater variability.

Note that these values of CV (for simulated yields) are somewhat smaller than the CV of actual yields in Hokkaido (about 15% in both decades), the difference considered to be derived from the effects of the technological trend, insects and disease, as well as climate that were not modeled but are observed in the actual yields. The CV values are also considerably smaller than that of 30.7% reported for the Hokkaido in Subsection 4.1 of this report. However, this value was calculated for variations in the yield index, *IY*, over about 100 years,

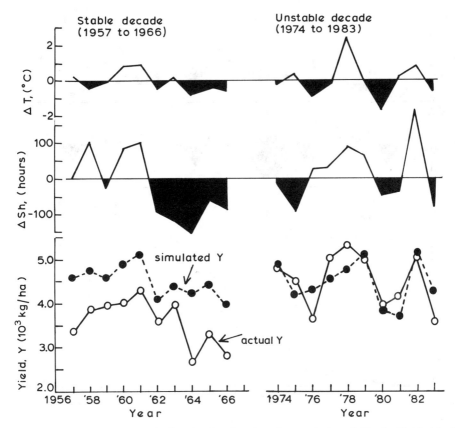

Figure 5.7. Year-to-year variation in simulated and actual rice yields in Hokkaido for the stable (1957–1966) and unstable (1974–1983) decades. ΔS_h is the deviation from the baseline of the total sunshine hours from June to September, and ΔT is the deviation of July–August mean temperature.

while the CV values given in this section are for 10-year periods in recent years with modern rice cultivation technology.

Figure 5.8 shows the simulated growth and yield of rice under the "worst-case" annual climatic conditions (1971), together with those under baseline climate. In 1971 the mean air temperature was below normal (1951–80 average) for almost the entire growth period, and sunshine hours were also lower than normal for most of the period. The anomalously low temperatures in July brought about cool summer damage due to floral impotency, and those in August and September caused cool summer damage due to delayed growth. In this year the actual rice yield in Hokkaido was 66% of the average level at that time, while the simulated yield for that year, assuming the present advanced technology, is 73% of the normal (i.e., simulated for the baseline climate). This suggests that, if the climatic conditions, comparable with those in 1971, were observed today, the yield reduction would be similar to that in 1971, even under the present advanced technology for rice cultivation.

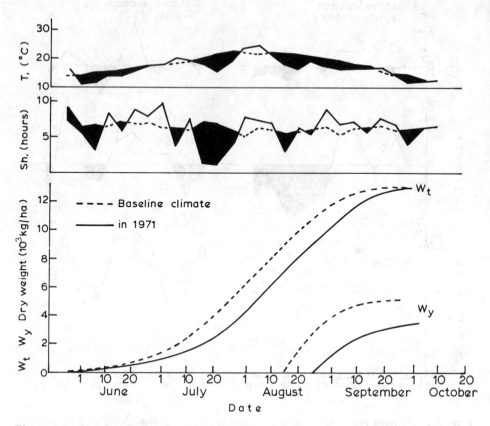

Figure 5.8. Simulated growth and yield of rice in the extreme year (1971) and in the average (baseline) climate in Hokkaido. T and S_h denote daily mean temperature and sunshine hours, respectively.

5.6. Estimated Rice Yields for the $2 \times CO_2$ Scenario under Present and Adjusted Technology

An assessment of the likely impacts of the GISS-derived $2 \times CO_2$ climatic scenario (described above) on rice in Hokkaido was conducted for two cases. The first involved estimating the impact on rice productivity under current technology (fixed cultivar and the planting date, etc.). The second evaluated the impact under an adjusted technology (changed cultivars and planting date). In both cases, the actual climatic data from the period 1974–83 and the 30-year "normal" climate (1951–80) were used as the bases, and the predicted change in climate (in fact, in monthly temperature alone) was added to these bases.

Figure 5.9 shows the simulated growth curves for total crop and grain dry weights with current and adjusted technologies under the present-day $(1 \times CO_2)$ and future $(2 \times CO_2)$ climatic conditions. The current technology means that the present dominant rice cultivar "Ishikari" is used with the current transplanting date (around 25 May) and management technologies, while the adjusted

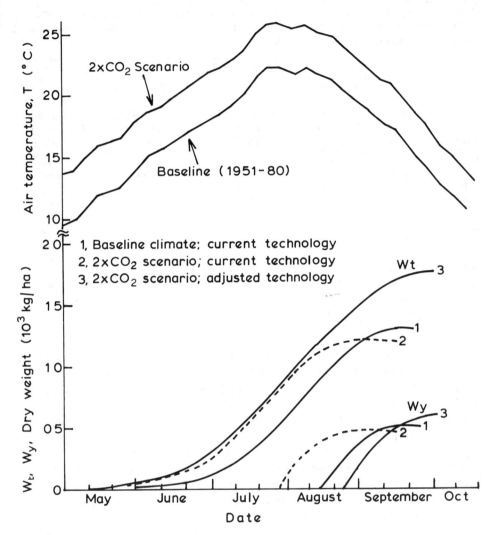

Figure 5.9. Simulated growth curves of the total crop and grain dry weight under the present (baseline) and GISS model-derived $2 \times CO_2$ climates in Hokkaido with both current and adjusted technology. Current technology indicates that the present cultivar ("Ishikari"), transplanting date, and technology are adopted; adjusted technology assumes a late-maturing cultivar ("Nipponbare") is adopted, with 25-day earlier transplanting.

technology refers to the later-maturing varieties, "Koshihikari" and "Nipponbare" (mentioned above in Subsection 5.3.1), cultivated with a transplanting date 25 days earlier than "Ishikari".

Under the $2 \times CO_2$ scenario, as predicted by the GISS model, the mean monthly temperature in Hokkaido during the rice growing period would increase by 3.3–4.0 °C (*see Table 5.1*). For the situation where this temperature rise is

Table 5.2. Actual and simulated annual rice yield variations in Hokkaido (1974–83) under observed climate and current technology, and simulated yields under the GISS 2 × CO_2 scenario with current and adjusted technologies. Also shown are the period means and coefficients of variation (CV).

	Observed climate		$2 \times CO_2$ climate/simulated yield (t/ha)		
Year	Actual rice yield (t/ha)	Simulated yield with current technology (t/ha)	Current variety (Ishikari) & current transplanting date (TD)	Mid-maturing variety (Koshihikari) & TD 25 days earlier	Late-maturing variety (Nipponbare) & TD 25 days earlier
1974	5.03	5.06	4.87	5.64	5.70
1975	4.46	4.20	3.94	4.83	4.92
1976	3.61	4.24	4.93	5.77	5.87
1977	5.04	4.53	4.67	5.41	5.48
1978	5.36	4.76	4.65	5.63	5.85
1979	5.02	5.06	4.81	5.90	6.07
1980	3.85	3.83	4.69	5.27	5.25
1981	4.13	3.70	4.32	5.13	5.43
1982	5.01	5.15	5.45	6.38	6.45
1983	3.55	4.30	4.44	5.15	5.43
Mean	4.51	4.48	4.68	5.51	5.64
CV (%)	14.9	11.6	8.5	8.1	7.8

added to the normal annual course of the temperature, the present rice cultivar in Hokkaido would develop very rapidly, reaching the reproductive stage too early, before a sufficient vegetative growth is established. For this reason, the total crop dry weight and grain yield both decrease relative to the baseline under this scenario, provided that the current rice cultivation technology is fixed (curve 2 in *Figure 5.9*).

A warmer $2 \times CO_2$ climate would extend the possible period for rice cultivation in Hokkaido to about 150 days from the present 120 days. By introducing "Nipponbare" rice, which can exploit almost fully the temperature resources under the $2 \times CO_2$ scenario, a higher dry matter production and yield could be expected (curve 3 in *Figure 5.9*).

The above results are based on 30-year average climate, which tends to smooth out the inevitable interannual variations in climate that would reduce the long-term average yield. To study the year-to-year impacts, the predicted temperature rise was added to observed temperatures during 1974–1983, and the effects on rice productivity in Hokkaido were simulated. The results are shown in *Table 5.2* and *Figure 5.10*. Under the $2 \times CO_2$ climate for current technology, yield reductions due to cool summer damage that were observed in 1976, 1980 and 1981 would almost disappear. In the favorable harvest years (1974, 1978 and 1979), however, yield reductions could be expected due to too high temperatures under the scenario climate. For this reason, substantial yield increases cannot be expected under the doubled-CO_2 climate, provided that the current rice cultivars are applied with the current cultivation technology. The model simulations show that, only when different rice cultivars are applied with an

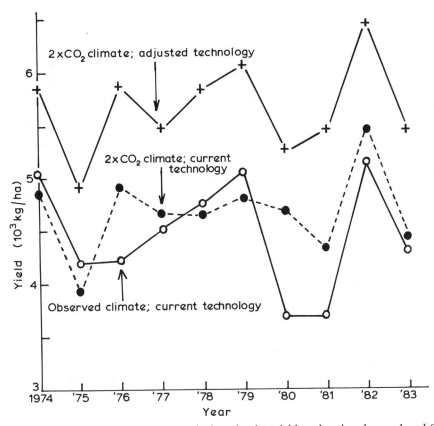

Figure 5.10. Simulated year-to-year variations in rice yield under the observed and 2 × CO_2 scenario climates for the period 1974–1983 in Hokkaido. Estimates are for current technology and adjusted technology (as defined for *Figure 5.9*).

altered transplanting date, can substantial yield increases be expected under 2 × CO_2 climate. By applying middle-maturing "Koshihikari" (not shown in *Figure 5.10*) and late-maturing "Nipponbare" cultivars from Honshu Island with transplanting advanced by 25 days, 23% and 26% yield increases are expected, respectively, in Hokkaido (*Table 5.2*).

The above assessments indicate that the likely climatic change indicated by the GISS model for a doubling of atmospheric carbon dioxide concentrations would dramatically reduce the incidence of cool summer damage of rice crops in Hokkaido, but would not increase the yield level substantially, as long as the current technology for rice cultivation were applied. (The probable effects of increased atmospheric concentrations of carbon dioxide on plant photosynthesis and water use efficiency are not considered in this report.) Substantial yield increases under a 2 × CO_2 climate would only be achieved by the adjustment of rice production technology to the changed climate, e.g., selecting alternative rice varieties, changing the planting date and procedures, and modifying the management of nutrients and water.

SECTION 6

The Effects on the Japanese Rice Market

6.1. Introduction

The delicate balance between supply and demand in Japan's rice market is affected both by government policy and by climate. The purpose of this section is, firstly, to simulate the effects of temperature variations on national rice supply and rice stocks using an integrated econometric model; and, secondly, to consider what fiscal and policy responses are most appropriate. The temperature variations considered here are similar to those adopted as scenarios in the preceding sections.

The rice production system in Japan is an interrelated structure comprising climatic, technological and political factors, economic and fiscal variables, and rice policy measures. The impact of climate on rice yield and production has been shown in the preceding sections of this study to differ considerably between regions. The dominant effect of climate is through cold summer damage to rice in the northern part of Japan, the magnitude of which is well represented by July–August temperature (Uchijima, 1981; *see also* Sections 4 and 5). In this study a model of the Japanese rice system, incorporating those variables (including climate) mentioned above, has been constructed for three regions of Japan (Hokkaido, Tohoku, and the rest of Japan), and its parameters estimated using econometric methods for the period 1966–82. The model thus represents the integration of a crop–weather relation into an economic model, the need for which has been emphasized elsewhere (Parry and Carter, 1984).

The model is based on a prototype developed for Japan and its rice trading partners for the 1950s and 1960s period (Tsujii, 1982). The prototype is a nationally aggregated rice market model with an emphasis on rice production technology and on international rice trade relations. The new model is a rice market (supply and demand) model for Japan alone, with the supply side disaggregated into the three subregions of Japan for the purpose of capturing more accurately the regional impacts of temperature changes on rice supply.

In this section the development and performance of the model is first described. Secondly, impacts of climatic variations on rice production are evaluated by model simulations for a number of specific climatic scenarios. Finally, implications regarding policies for rice prices, surplus disposal, land diversion, reserve stocks and production technology, and food and agricultural policy in general will be explored, based on the simulation results.

6.2. The Rice Production System

Rice is the staple food for the Japanese, and is grown all over the Japanese archipelago on four million farms (*Table 6.1*). Both in economic and political terms it is therefore the most important agricultural product in Japan. Because all paddy fields in Japan are irrigated (a situation that was achieved over 100 years ago), drought is not a problem. However, the recent northward extension of rice production (*see* Section 1) has exposed rice crops in the north (especially in Hokkaido and Tohoku districts) to cooler temperatures that can be damaging to production. This assessment focuses on such effects by considering the two northern districts separately from those in the rest of Japan. The contributions of each region to national rice production are shown in *Table 6.1*.

Japan's rice balance is shown in *Table 6.2* for two years, 1980 and 1983. In both years, domestic rice production (supplemented by some minor imports and previous years' stocks) was utilized for direct consumption and for other uses such as animal feed, while a small proportion was also exported.

We can illustrate the role of government in influencing rice production in Japan by using a straightforward supply–demand diagram (*Figure 6.1*). Simply expressed, supply, in this case rice production, increases as the farmers' price increases (represented by the supply curve S), while demand, the quantity of rice that is required and can be afforded by the consumers, decreases as the consumers' price increases (the demand curve D). By superimposing government pricing policies onto these basic relations, the problems of this sector of Japanese agriculture quickly become apparent.

The domestic rice market is separated from the world rice market by the government monopoly of international rice trade, and is effectively under the control of the government, which determines both the prices paid to farmers (P_F) and the price (P_C) paid by consumers [which is itself determined by the government's resale price (P_R) – *Figure 6.1*]. Thus in Japan discrepancies between the demand for and supply of rice are not adjusted by changes in the rice prices, and thus not through market mechanisms. Since the end of World War II a dominant but declining part ($S_2 - R$) of the total marketed rice ($S_2 - T$) has been purchased from farmers by the Food Agency at P_F, stored, and resold at a lower price, P_R, to wholesalers. Mainly because of the strong political power of nationally organized agricultural cooperatives, and due to the effects

The Japanese rice market

Table 6.1. Statistics of Japanese rice cultivation, 1983. (Source: Ministry of Agriculture, Forestry and Fisheries, 1985.)

	Number of rice farms (1980)	Planted area (000 ha)	Production (brown rice 000 tonnes)	Yield (kg/ha)	Paddy area withdrawn (000 ha)
Hokkaido	54 700	147	522	3551	117
Tohoku	625 600	558	2910	5215	99
Rest of Japan	3 154 000	1568	6934	4422	410
Japan total	3 834 300	2273	10 366	4560	626

Table 6.2. Balance of production, trade, stocks and consumption in 1980 and 1983. (Source: Ministry of Agriculture, Forestry and Fisheries, 1985.)

Japanese rice balance (000 tonnes) [a]	1980	1983
Production	9751	10 366
Exports	−754	−384
Imports	27	18
Stock change	−2185	−1489
Utilization (white rice[b]):		
Direct consumption	−8971	−8826
Other users	−1172	−1572
Per capita consumption (kg)	78.9	75.7

[a] Brown rice unless otherwise stated.
[b] Values expressed as the edible component of brown rice: white rice = 0.905 brown rice (by weight).

of rice shortages before 1968, the farmers' rice price (P_F) has been supported and raised to extremely high levels during the past two and a half decades (Tsujii, 1984). In recent years it has been 4–5 times higher than the world market price of rice (P_W). The resale price (P_R) has also been raised following the fast rise in producer prices in order to limit the extent of government expenditure involved in the purchasing and reselling operation. The product of these costs $(P_F - P_R)$ combined with the government marketing cost $(Q - P_F)$, and the amount of rice purchased $(S_2 - R)$, determines the total government expenditure on procurement (depicted as the area HIJK in *Figure 6.1*).

The farmers' price has remained higher than the government resale price of rice (P_R), though the difference between them $(P_F - P_R)$ has recently tended to decrease. Thus the resale price of rice (P_R), has been raised faster than P_F, which has led to a faster rise of the price of rice to the consumer (P_C) than P_F.

The potential supply is governed by the producer price, P_F (which intersects the supply curve, S, at point A and is extended to point C), and the potential demand is determined by the consumer price, P_C (which crosses the demand curve, D, at point E). In *Figure 6.1*, a *potential surplus* is represented by CE, the excess of potential supply over demand. The potential rice surplus condition

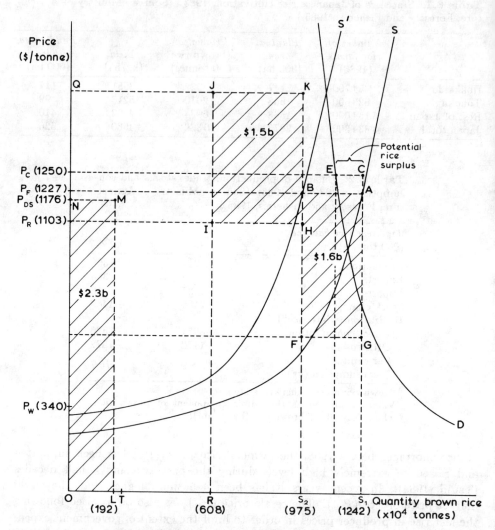

Figure 6.1. Relationship between government pricing policy and rice production in Japan. Values for 1980 are shown in parenthesis (¥240 = US$1). S potential supply curve; S' actual supply curve; D demand curve; P_W world market price; P_R resale price; P_F farmers' price; P_C consumers' price; P_{DS} disposal cost; $Q - P_F$ government marketing cost; OL quantity of rice disposal; $S_2 - T$ quantity of marketed rice; $S_2 - R$ quantity of rice purchased by Japanese Food Agency; S_1 potential quantity of rice production according to farmers' price; HIJK subsidy on government marketed rice; LMNO subsidy on rice disposal; ABFG subsidy on land diversion; $C - E$ potential rice surplus. For full explanation, see text.

(CE) has worsened considerably, and Japan has experienced two severe rice surplus periods during the last two decades.

The recent trend in the highly supported producer price, and the two recent and large government-owned surpluses are shown in *Table 6.3*. It is clear

Table 6.3. The farmers' price of rice and the rice surplus in Japan, 1960–1984. (Source: Food Agency, various dates.)

	Farmers' price, P_F [a] (yen/tonne)	Relative Price Index P_F/P_{AG} [b] (1970 = 1)	Rice surplus[c] (000 tonnes)	Surplus as percentage of total demand (%)
1960	65 033	0.93	440	3.5
1961	68 816	0.91	500	3.8
1962	76 033	0.91	100	0.8
1963	83 833	0.94	200	0.2
1964	96 200	1.03	10	0.1
1965	103 800	1.00	52	0.4
1966	115 600	1.04	210	1.7
1967	126 533	1.04	640	5.1
1968	134 800	1.09	2980	24.3
1969	134 866	1.02	5530	46.2
1970	135 866	1.00	7202	60.3
1971	141 366	1.02	5891	49.7
1972	148 000	1.01	3074	26.0
1973	170 300	0.95	1477	12.2
1974	224 850	1.03	615	5.1
1975	257 333	1.05	1142	9.5
1976	273 866	1.02	2641	22.4
1977	284 766	1.07	3761	32.8
1978	286 266	1.03	5783	51.0
1979	286 266	0.99	6517	58.1
1980	292 266	0.97	6693	59.7
1981	293 383	0.95	4415	39.0
1982	296 616	0.98	2700	22.9
1983	302 700	0.98	909	7.9
1984	311 133	–	360	–

[a] The average government procurement price.
[b] Wholesale price of agricultural products.
[c] Rice held by the government more than 1 year after its production, on 1 November each year.

from this that the large surpluses accumulated when the farmers' price level increased faster than the wholesale price of agricultural products (represented by a relative price index).

The effect of the government policy and the severe potential and actual rice surpluses has been to force the government into high levels of expenditure in the form of three types of subsidy. First is the cost of the difference between the government procurement price (P_F) and the government resale price (P_R), plus the government rice marketing cost ($Q - P_F$) multiplied by the government rice procurement ($S_2 - R$) (i.e., the shaded area, HIJK in *Figure 6.1*). The second comprises large subsidies that have been paid to many farmers during the last 15 years for transferring paddy fields to other crops. In *Figure 6.1*, the subsidy in 1980 ($593/tonne) shifts the total rice supply curve to the left, from S to S′, and the total expenditure is represented by the area ABFG. Thirdly, large sums have been spent during the last 15 years on disposal of surplus rice as feed, export goods, etc. (area LMNO). This subsidy is a product of the disposal cost per tonne (P_{DS}) and the annual amount of disposal OL. The disposal cost is the sum of the government marketing cost ($Q - P_F$) and the procurement price (P_F) minus the world market price (P_W). In addition to these large subsidies, the government-operated compulsory rice crop insurance scheme is also expensive (Tsujii, 1985). This is an important policy institution which modifies the economic impacts of climatic variations on the rice economy in Japan, and though beyond the scope of this report, it merits consideration of its efficiency and distributive effects.

The sum of the first three subsidies (1.5, 1.6 and 2.3 billion dollars, respectively – *see Figure 6.1*) amounted to about 38% of the total current agricultural expenditure by government, and about 3.2% of the total current expenditure in the fiscal year 1980. The large share of these three subsidies in total agricultural expenditure increased during the 1960s and 1970s and the slow growth in total government revenue since the first oil crisis has made the subsidies increasingly heavier burdens to the government.

In summary, since 1965 the rice market in Japan has been characterized by severe potential surplus – a potential that has been realized twice because of good weather, rapid increases in the government procurement price of rice, and sharp decreases in the paddy field diversion subsidy. Near balances between rice supply and demand in the mid-1970s and mid-1980s were reached only through heavy government subsidization of paddy field diversion and measures to dispose of rice surpluses.

The above problems are confounded by weather effects on productivity. Poor rice yields are usually a result of cool temperature damage in the northern regions of Japan. In 1984, the Japanese almost faced a severe rice shortage because of four consecutive cold damage years before 1984, a very cold spring in 1984, and trouble with rice fumigation. However, in 1984 a severe shortage was averted because of favorable, hot summer conditions. That such a severe shortage threatened even under an overall condition of large potential rice surplus demonstrates that the balance of supply and demand for rice in Japan is extremely delicate. It can easily be transformed into a severe rice surplus or into a rice shortage as a result of either inappropriate policy measures or the effects of weather.

6.3. A Model of the Japanese Rice Market

6.3.1. Description

There are three components of the Japanese rice system that are endogenous to the model used in this section: consumption, supply and stocks. These are expressed as functions of a number of explanatory variables in the following description.

Consumption

The first equation is a behavioral function for total national rice consumption in Japan, which has the standard specification of a demand function with deflated net national product and the relative price between rice and its most important substitute (i.e., bread) as its explanatory variables:

$$CF = f_1(NNP, RB) \tag{6.1}$$

where CF is the total direct consumption of rice as food (thousand tonnes), NNP is the net national product (billion yen, deflated), and RB is the consumers' rice price relative to the consumers' bread price (1980 = 1).

Supply

Three regional rice supply functions are required in the model, each having the same specification. This is the component that includes a climatic variable, temperature, which is perturbed in the scenario experiments described below. The respective functions for Hokkaido (subscript H), Tohoku (subscript T), and the rest of Japan (subscript R) are:

$$O_H = f_2\left[T_H, TS_H, P, F_H, YR, A_H\right] \tag{6.2}$$

$$O_T = f_3\left[T_T, TS_T, P, F_T, YR, A_T\right] \tag{6.3}$$

$$O_R = f_4\left[T_R, TS_R, P, F_R, YR, A_R\right] \tag{6.4}$$

where O_i is the rice supply after harvest (November 1 each year) in region i (thousand tonnes), T_i is the mean July–August temperature in region i (0.1 °C), TS_i is the square of T_i (i.e., T_i^2), P is the government procurement price of rice (yen per 60 kg, deflated), F_i is the fertilizer price in region i (yen per 22 kg mixed fertilizer, deflated), YR is the annual technological trend (indexed to the year), and A_i is the area of rice planted in region i (thousand hectares).

The form of the explanatory temperature variable, T, in the above equations is consistent with that of the statistical yield–climate relationships reported in Section 4 [see equations (4.2) and (4.3), Subsection 4.2]. It was demonstrated in Section 5 (for example, see *Figure 5.6*) that solar radiation (often substituted by hours of sunshine) also affects rice yield in Japan, especially during the grain-maturing period (August–September). This variable tends to be positively correlated with August–September temperatures, which suggests some justification for omitting it from simple regression-derived functions in the model that are applied to the range of present-day climatic conditions. However, under possible carbon dioxide-induced climatic change, this temperature–radiation relationship may not hold, so it would be useful to incorporate a separate radiation term in any future modification of the model.

For an econometric specification of the rice supply function, the planted area variable, A, can be treated separately from the other variables which are used to estimate rice yield:

$$y = 1(P_F, P_N) + \alpha_0 T + \alpha_1 T^2 + \mu_0 \tag{6.5}$$

and

$$O = y \cdot A \tag{6.6}$$

where y is the rice yield, O is the total rice supply, P_F is the rice procurement price, P_N is the input price, T is the mean July–August temperature, μ_o is a disturbance term and α_o and α_1 are coefficients.

The planted area, A, has been determined by the government since 1970 through the paddy field diversion policy, in order to attain domestic equilibrium in the rice market. A is thus exogenous and equations (6.5) and (6.6) permit a separate analysis of the effects of this policy on the Japanese rice market. However, because this is not an objective of this study, a more simple linear specification has been adopted that includes the area term, A:

$$O = \delta_o + \delta_1 T + \delta_2 T^2 + \delta_3 P_F + \delta_4 P_N + \delta_5 A + \mu \tag{6.7}$$

where δ_1, δ_2, δ_3, δ_4 and δ_5 are coefficients and μ is a disturbance term.

Equation (6.7) specifies the form of the regional supply functions [equations (6.2), (6.3) and (6.4)], which are combined to give the national rice supply:

$$O = O_H + O_T + O_R \tag{6.8}$$

Stocks

The change in rice stock is related to the supply and consumption variables defined above, as well as to trade and other uses of rice, expressed as:

$$I = O - CF - X + M - CNF \tag{6.9}$$

where I is the change in rice stocks, O is total rice supply, CF is total consumption of rice as food, X is rice exports, M is rice imports and CNF is the total non-food utilization of rice (all units in thousand tonnes).

The total accumulated stock, S, is the final effect variable in the system:

$$S = LAG + I \tag{6.10}$$

where LAG is the rice stock at the end of the preceding harvest-year (thousand tonnes).

A flow diagram of the model is shown in *Figure 6.2*, indicating the linkages between different variables included in the model and some other factors not included. Enclosed within the bold rectangle are the endogenous variables [described by equations (6.1)–(6.4) and (6.8)–(6.10)], while outside this but inside the outer rectangle are the exogenous variables. Elements outside the outer rectangle are related to the model but not included in it. Solid arrows show the linkages specified in the model, and dashed lines indicate long-run policy decisions and other connections not explicitly modeled.

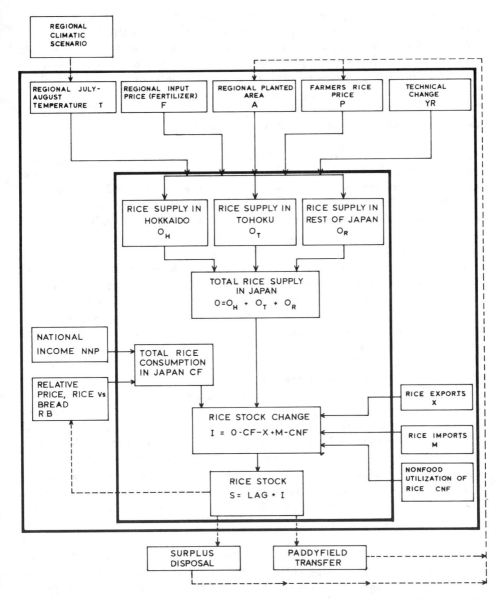

Figure 6.2. Integrated regional and national econometric model of the Japanese rice market.

As mentioned above, both the farmers' price and consumers' price of rice are fixed by the government so that the supply and demand system in reality is *not* simultaneous in the sense that a market equilibrium (supply = demand) is reached automatically by market adjustments in the prices of rice. Relationships between rice stock, consumers' price of rice, surplus disposal, paddy field transfer, farmers' price of rice and planted area are not endogenized in the model. They are, in fact, long-run feedback mechanisms in the Japanese rice system, and involve not only economic but also political, institutional and cultural factors that are not easily modeled. They are illustrated as dashed arrows in *Figure 6.2*, and in the experiments reported below they are assumed fixed at the actual annual levels recorded during the model simulation period (1966–82).

6.3.2. Model coefficients and significance testing

Fairly good annual and regional time series data for the model variables are available in Japan. Unlike the experiments reported in earlier sections of this report, however, this study employs only the years 1966–82 (rather than 1951–80) as the baseline period. This is because the basic economic structure of Japan's rice market changed considerably during the late 1960s from a rice shortage market to a rice surplus market and the whole economy grew very rapidly from 1959 to 1972. The climatic variable, regional average July–August temperature, was calculated for each of the years 1966–1982, based on a close check of consistency between the regional distribution of the rice production areas and weather stations. [The weather stations selected to represent each region were, for Hokkaido: Haboro, Rumoi, Asahigawa, Iwamizawa, Suto and Tomakomai; for Tohoku: Aomori, Hachinohe, Akita, Morioka, Sakata, Yamagata, Sendai and Fukushima; and for the rest of Japan: Niigata, Kanazawa, Kumagaya, Nagoya, Tottori, Hiroshima, Osaka, Saga, Kagoshima and Kochi. Their locations are shown in *Figure 2.2*.]

Since the model is not a simultaneous (dynamic) system, each behavioral function [equations (6.1)–(6.4)] was estimated using the Ordinary Least Squares (OLS) regression procedure. This involved relating the recorded annual values of all the explanatory variables to the corresponding values of the dependent variables (i.e., national consumption or regional supply) for the period 1966–1982 using multiple linear regression. This procedure supplies estimates of the values of coefficients required in the model equations [for example, in equation (6.7)]. These are listed in *Tables 6.4* and *6.5*, along with the standard errors and significance (T-ratios).

Signs of the estimated parameters are in accordance with theoretical expectations (for example, the signs of the temperature coefficients are consistent with those presented in Section 4), and the statistical T-ratios allow us, in most cases, to reject the statistical hypothesis that the coefficients are zero. The R-squared values show the percentage of the variation in the endogenous (dependent) variable of each behavioral function explained by the function. These are also satisfactorily high, implying that the validation of each estimated behavioral function is acceptable both on theoretical and on statistical grounds.

6.3.3. Comparison of model estimates with observed values

Another way of verifying the model is to check how well it can simulate the actual time series of values of the endogenous variables of the model, given

The Japanese rice market

Table 6.4. Magnitude and significance of the coefficients used in the rice consumption function [equation (6.1)].[a]

Estimates	Intercept	NNP	RB
Coefficients	13 902.8	−0.0132	−1204.7
Standard error	318.5	0.0007	227.9
Prob > |T|	0.0001	0.0001	0.0001

[a] Total Japanese rice consumption, CF: $R^2 = 20.967$; Prob > F = 0.0001.

Table 6.5. Magnitude and significance of the coefficients used in the regional supply functions [equations (6.2)–(6.4)].

Hokkaido rice supply, O_H	$R^2 = 0.793$	Prob > F = 0.0054					
Estimates	Intercept	T_H	TS_H	P	F_H	YR	A_H
Coefficients	−46 207	104.82	−0.234	0.0487	−0.143	17.18	4.55
Standard error	22 299	100.6	0.25	0.058	0.13	8.8	1.4
Prob > |T|	0.0650	0.3220	0.3741	0.4240	0.2795	0.0792	0.0089
Tohoku rice supply, O_T	$R^2 = 0.854$	Prob > F = 0.0011					
Estimates	Intercept	T_T	TS_T	P	F_T	YR	A_T
Coefficients	−112 628	147.82	−0.290	0.131	−0.491	46.25	6.87
Standard error	37 138	98.9	0.22	0.09	0.34	16.0	1.9
Prob > |T|	0.0126	0.1658	0.2103	0.1632	0.1760	0.0159	0.0046
Rest of Japan rice supply, O_R	$R^2 = 0.956$	Prob > F = 0.0001					
Estimates	Intercept	T_R	TS_R	P	F_R	YR	A_R
Coefficients	−350 016	254.64	−0.424	0.166	−0.852	156.26	5.51
Standard error	123 837	460.7	0.87	0.13	0.60	51.2	1.0
Prob > |T|	0.0180	0.5926	0.6383	0.2371	0.1871	0.0122	0.0002

actual time series values of the exogenous variables as inputs to the model. This is called the "final test of the model" in the discipline of econometrics. Since the model includes a variable which updates the annual rice stocks, LAG, the final test is effectively a dynamic simulation. The results are presented in *Figure 6.3*.

The plots indicate that the simulation estimates correspond very closely to the actual time series values of the endogenous model variables, representing the actual behavior of the Japanese rice system very accurately. Further indications of this accuracy can also be obtained from goodness-of-fit statistics (*Table 6.6*).

In passing, it may be noted that rice in Japan has become an "inferior good" [see *Figure 6.3(a)*] as indicated by the almost continuous decline both in total and in per capita rice consumption as the average income of the Japanese has increased during the sample period (i.e., the income elasticity of demand for rice was negative).

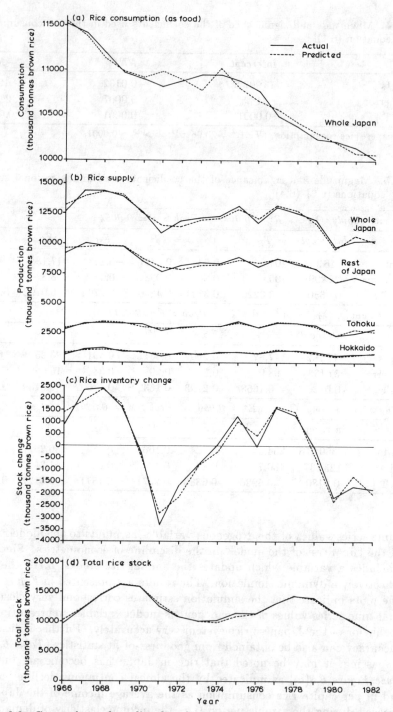

Figure 6.3. Comparison of predicted and actual values of endogenous model variables for the period 1966–1982. The graphs show: (a) rice consumption, (b) rice supply (national and regional), (c) rice inventory change, and (d) total rice stock.

Table 6.6. Goodness of fit of model estimates to observed values of variables over the simulation period (1966–82). RMS error is the standard deviation of the differences between predicted and observed values, expressed both in absolute terms and as a percentage of the actual mean value.

Variable	RMS error (000 tonnes)	RMS error (%)
CF	77.46	0.72
O_H	82.93	11.70
O_T	121.17	4.30
O_R	201.36	2.37
O	317.96	2.68
I	305.82	369.69
S	243.75	2.20

6.4. Scenario Experiments and Results

Having validated the model for conditions during the simulation period, the effects of specified temperature changes on regional rice supplies and on the whole rice market were investigated in six separate scenario experiments.

6.4.1. The climatic scenarios

The six climatic scenarios that were selected in this section have been described briefly in Section 2. They include four instrumental scenarios, based on observed historical temperature data, and two scenarios of the climate estimated for conditions of doubled atmospheric concentrations of carbon dioxide by the Goddard Institute for Space Studies (GISS) general circulation model. All scenarios represent adjustments of mean July–August temperatures from each year of the model simulation period, 1966–82 (the baseline climate employed in this section). The scenarios are described here and the results of model scenario runs are reported in the following subsection.

Stable decade (1957–66) and unstable decade (1926–35)

Mean July–August temperatures from these 10-year periods are assumed to occur in each region of Japan for the period 1973–82, replacing the baseline values. Since the model simulation period is 17 years, the years 1966–72 are retained, and the temperature values in those years remain unaltered. The stable decade is the same period employed as a scenario in Sections 3 and 5, and the unstable decade is also described in Section 3.

Ten consecutive cold damage years

1980 was a severe cold damage year for rice in much of northern Japan (*see* Sections 2, 3 and 4). Here, the regional average July–August temperatures in 1980 have been assumed to occur in the ten consecutive years 1973–82, in reality an unlikely episode, but effective as a "synthetic" scenario to illustrate the potential impacts. The 1966–72 temperatures remain unaltered.

Three repetitions of three cold damage years

Three repetitions of the temperatures recorded during the three years 1980–82 (each a cold damage year for rice) were applied to the nine-year period 1974–82. The remaining temperatures for the period 1966–73 were retained at their baseline values. As in the previous case, this is also a rather unrealistic, synthetic scenario but it does incorporate some interannual variability in climate that is lacking in the 10 consecutive "1980" events.

$2 \times CO_2$ equilibrium climate

To simulate the average climate existing under doubled concentrations of atmospheric carbon dioxide, the GISS general circulation model estimates (see Subsection 2.5) were employed for each of the three regions. Following the procedures outlined in Section 2, the changes in mean July–August temperature between the GISS model $1 \times CO_2$ and $2 \times CO_2$ equilibrium conditions were averaged for grid point locations representing each region as follows:

Hokkaido: $(GP_{46,140} + GP_{46,145} + GP_{42,135} + GP_{42,140} + GP_{42,145})/5$
Tohoku: $(GP_{42,135} + GP_{42,140} + GP_{42,145} + GP_{38,140} + GP_{38,145})/5$
Rest of Japan: $(GP_{38,135} + GP_{34,130} + GP_{34,135} + GP_{34,140} + GP_{30,130} + GP_{30,135})/6$

where GP_{ij} is the grid point value at latitude i and longitude j. The regionally averaged changes were then added to the baseline values for the period 1966–82, thus representing a new "equilibrium" $2 \times CO_2$ climate, though retaining the same year-to-year variations as the baseline climate.

Although the estimates themselves may be reasonably justified as a representative equilibrium climate under a $2 \times CO_2$ atmosphere, they are treated here as an *instantaneous* change in present-day conditions rather than as the climate occurring at that time in the future when CO_2 concentrations have actually doubled (probably towards the end of the next century – see, for example, Carbon Dioxide Assessment Committee, 1983; Bolin et al., 1986). This is because there is no appropriate method of predicting so far into the future how values of the other exogenous variables in the economic model might change. Thus this scenario represents the application of a future climate to present-day (baseline) economic conditions, which although unrealistic, is presented here for illustrative purposes.

$2 \times CO_2$ transition climate

This is a modification of the previous scenario which attempts to simulate a transition between the present-day climatic conditions and a $2 \times CO_2$ climate by adding to the baseline time series (1966–82) progressively larger increments of temperature change, increasing linearly from zero in 1966 (the assumed "present" condition) to the GISS model-derived $2 \times CO_2 - 1 \times CO_2$ equilibrium change in 1982 (the $2 \times CO_2$ climate). Of course, this represents an extremely rapid change in climate (from a $1 \times CO_2$- to a $2 \times CO_2$-type climate in only 17 years), and such changes would inevitably be accompanied, in reality, by large adjustments in other exogenous model variables that are assumed fixed in this application. However, it is probably more realistic to think of future climatic change as a transient change in the average condition rather than an abrupt, step-like change from one instantaneous mean condition to another. As such, experiments with this type of scenario represent an initial attempt at providing estimates of rates of change of impact in response to time-dependent changes in climate.

6.4.2. Results

The more extreme climatic scenarios, as already discussed, represent large, long-lasting temperature changes to which farmers would inevitably adapt their cultivation methods. The simulated impacts below, however, are calculated based on the assumption that production methods are not adjusted from their present condition. As a result, there is likely to be a bias in some of the estimates of rice supply. For large negative and persistent temperature anomalies, the model over-estimates the probable reduction of supply. For large positive and persistent temperature anomalies, however, the model under-estimates the probable increases of supply. This under-estimation may be enhanced since the parabolic relationship between temperature and supply [i.e., the functional form of equations (6.2)–(6.4), which closely resemble the curves in *Figure 4.2*] implies declining supply in the high- and low-temperature ranges of the curve.

Mean July–August temperature is included only in the supply functions as an explanatory variable, and does not have a modeled affect on rice consumption in Japan. Thus the graph for consumption [*Figure 6.3(a)*] is not presented with other results of the six scenario simulation experiments described below.

Stable decade (1957–66) and unstable decade (1926–35)

The impacts of these scenarios on Japan's rice supply and balance are quite small. Thus only the impacts on the total national rice production and on the total rice stock are described here, and the differences in the impacts between the two scenario simulations are compared. The simulated rice supply under the unstable decade [*Figure 6.5(a)*] varies slightly more than under the stable decade scenario [*Figure 6.4(a)*] relative to the actual (baseline) rice production. This is not surprising considering that the unstable decade contains five, while the stable decade includes only two (minor) cold damage years (*see* Section 3). *Figure 6.4(b)* and *Figure 6.5(b)* show the impacts on the total rice stock of the stable and unstable decade scenarios, respectively. The results indicate neither a surplus nor a shortage of rice and there is no significant difference between the

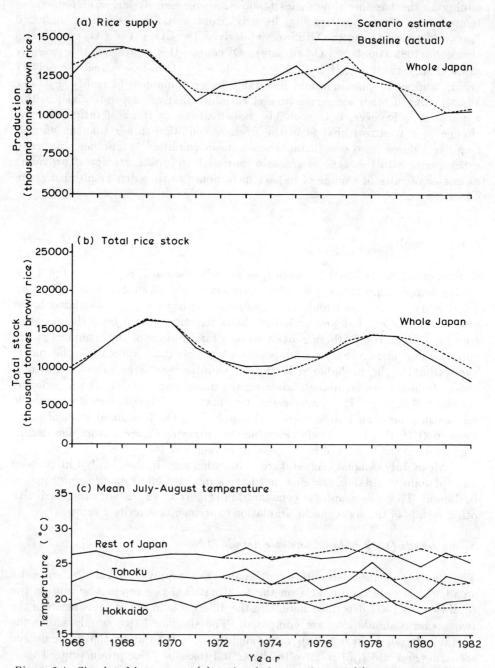

Figure 6.4. Simulated impacts on (a) national rice supply, and (b) total rice stock, as a result of (c) the stable decade scenario (1957–66), where mean July–August temperatures are adjusted for the simulation period 1973–82. Scenario estimates (dashed lines) are compared with the actual (baseline) values (solid lines). Temperatures are for the three modeled regions.

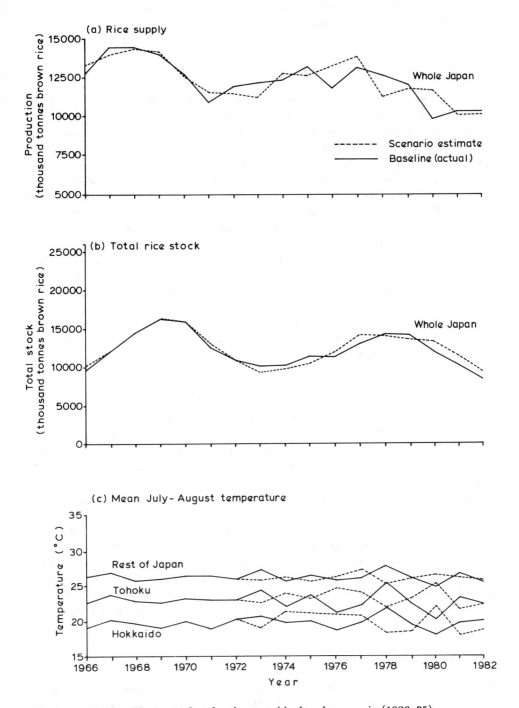

Figure 6.5. As for *Figure 6.4*, but for the unstable decade scenario (1926–35).

impacts of these two scenarios. The simulations imply that the rice supply in Japan under present technological and management conditions is quite resilient to periods of anomalously high temperature variability that have occurred over the last century.

Ten consecutive cold damage years

Figure 6.6 shows the simulated impacts of ten consecutive occurrences of the severe cold damage year of 1980. The scenario temperatures lead to a decline in rice supply relative to the baseline values, which is proportionally greater in the northernmost region of Hokkaido than further south and southwest in Tohoku and the rest of Japan. Consequently, the total rice supply also declines considerably [*Figure 6.6(a)*]. This, in turn, results in a sharp decline in the total rice stock in Japan as shown in *Figures 6.6(b)* and *6.6(c)*, such that in the last two years of the scenario decade, the rice stock is exhausted. This dramatic fall in rice stocks indicates that Japan would have to import more than 1 million tonnes of *Japonica* rice annually during the simulation decade, in order to maintain rice stocks after harvest at the annual human consumption level of about 10 million tonnes. This would be an essential measure for any government, both because rice is such an important staple food for the Japanese population, and because of the advisability of maintaining an appropriate food security level. However, this is a difficult task to achieve because, while about 10 million tonnes of rice are traded annually on the world market, this is predominantly *indica* rice which is not favored by the Japanese.

Three repetitions of three cold damage year

The estimated effects of three repetitions, (over the period 1974–82) of the three consecutive cold damage years (1980–82) are depicted in *Figure 6.7*. The impacts are considerably moderated compared with the results of the previous scenario. Regional rice supplies are reduced slightly in some of the years from 1974 to 1979, and hence the total rice supply is also reduced during the same period [*Figure 6.7(a)*]. These impacts on rice supply result in a decline in the total rice stock as indicated in *Figure 6.7(c)*. The simulated stock falls below 10 million tonnes in 1974, 1975, and after 1979, dropping to nearly 5 million tonnes by 1982. Japan would thus again face a severe rice shortage situation in these years, and 1–2 million tonnes of *Japonica* would need to be imported to feed her population. Again, this would not be easy to achieve considering the total amount and the kind of rice traded at present on the international rice market. It would also undermine the government's objective of retaining self-sufficiency in rice.

$2 \times CO_2$ equilibrium climate

The CO_2-induced increases, relative to the baseline, of over 3 °C in mean July–August temperature estimated by the GISS model for all regions of Japan, and applied to the whole simulation period [1966–82; *Figure 6.8(d)*] lead to an

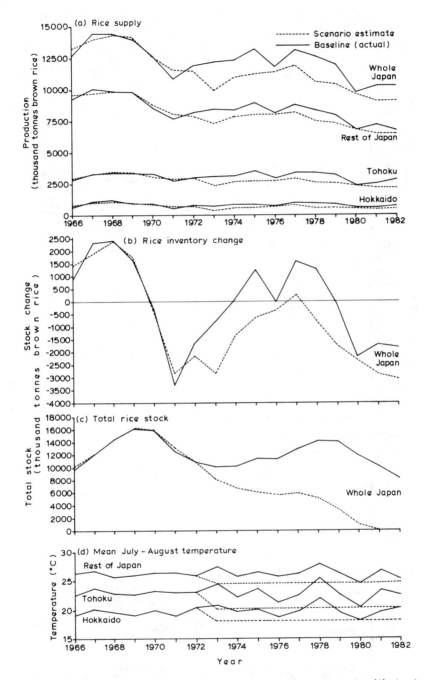

Figure 6.6. Simulated impacts on (*a*) national and regional rice supply, (*b*) rice inventory change, and (*c*) total rice stock, as a result of (*d*) ten consecutive occurrences of the 1980 cold damage year, where mean July–August temperatures are adjusted for the simulation period 1973–82. Scenario estimates (dashed lines) are compared with the actual (baseline) values (solid lines). Temperatures are for the three modeled regions.

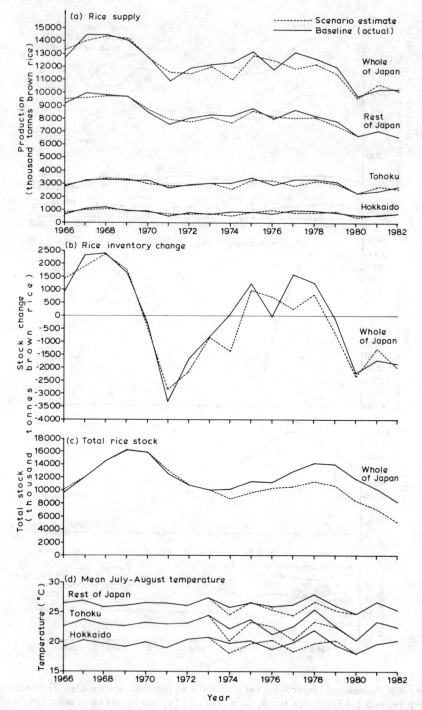

Figure 6.7. As for Figure 6.6, but for a scenario of three repetitions of the three cold damage years 1980–82, where mean July–August temperatures are adjusted for the simulation period 1974–82.

increase in rice supply in all three subregions in almost all years. This results in total rice supply increases of 1–2 million tonnes of brown rice nearly every year [*Figure 6.8(a)*]. The two severe rice surplus periods in the late 1960s and late 1970s are further worsened and the total rice stock in Japan reaches 25 million tons by the last year (1982) of the simulation period [*Figure 6.8(b,c)*]. This is more than twice the total annual rice consumption in Japan, and about 15 million tonnes more than the actual rice stock in 1982, indicating an extreme rice surplus. Moreover, if the bias in the simulated response to supply is taken into account, the actual surplus would be even greater.

$2 \times CO_2$ transition climate

The effects of this scenario of transient change in temperatures between the baseline climate and the GISS model-derived $2 \times CO_2$ climate over the 17-year simulation period [*Figure 6.9(d)*] are not as extreme as for the instantaneous climatic change assumed in the previous experiment. Nevertheless, while the simulated increases of rice production and surplus are much less than for the "equilibrium change" scenario, the total rice stock in 1982 still reaches a level of 20 million tonnes, which signifies a serious rice surplus. If we consider the likely under-estimate of the simulated supply then this surplus might well be greater.

6.5. Policy Implications

Severe rice shortages or surpluses in Japan are the simulated impacts of the two more extreme groups of scenarios (i.e., consecutive occurrences of cold damage and the $2 \times CO_2$ climate, respectively). These impacts would have serious effects on the Japanese rice system, the economy and national budget, and would reduce national food security. In this subsection we consider the policies most appropriate as responses to the changed conditions.

6.5.1. Implications of consecutive occurrences of cold damage to rice

Of the two scenarios that cause a severe rice shortage situation in Japan, ten consecutive occurrences of the very severe cold damage year (1980) is less probable than the three repetitions of the three consecutive occurrences of cold damage years (1980–82). Thus the specific policy implications that are discussed here relate to the second scenario although the general implications are the same for both cases.

As already explained, rice shortages occur from 1974 onwards under this scenario, and Japan must import 1–2 million tonnes of rice annually after that year in order to supply demand. But Japan probably faces difficulties in importing these amounts of rice. *Japonica* rice, which the Japanese like to consume, is not traded in large quantities in the *indica*-dominated world rice market. Further, the size of the world rice trade is only about 10 million tonnes per year, rather small in comparison with the simulated Japanese imported requirement.

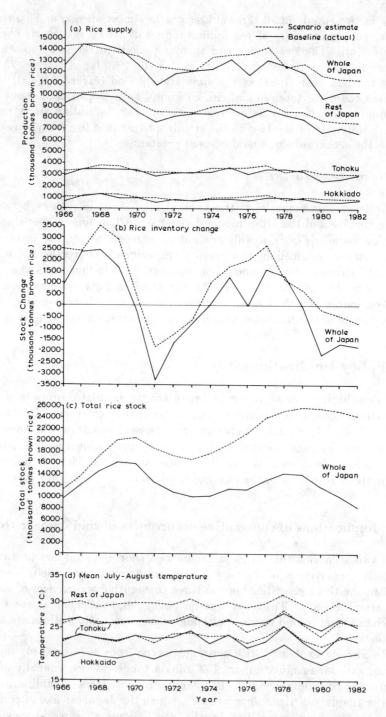

Figure 6.8. As for *Figure 6.6*, but for the $2 \times CO_2$ equilibrium climate scenario, where mean July–August temperatures are adjusted for the full simulation period, 1966–82.

The Japanese rice market

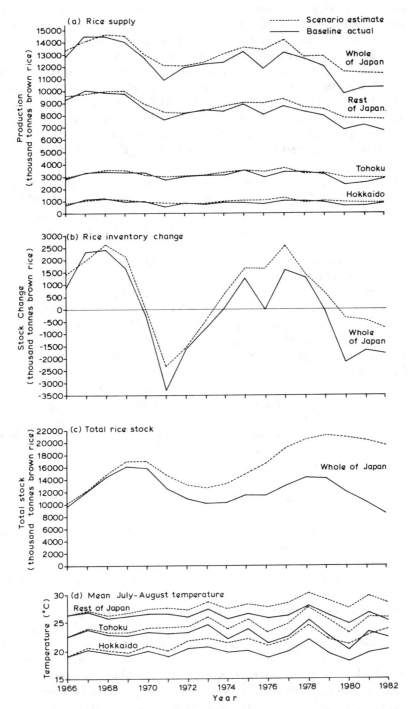

Figure 6.9. As for Figure 6.6, but for the $2 \times CO_2$ transition climate scenario, where mean July–August temperatures are adjusted for the full simulation period, 1966–82.

This and the ensuing rice shortage would be in sharp conflict with the goal of maintaining national food security. Neither the government nor the public would be likely to tolerate the occurrence of such shortages and this could lead to increased social dissatisfaction. Under present policy, various strategies can be followed in order to mitigate a rice shortage. First, the heavily subsidized paddy land transfer policy to other crops can be abandoned. This policy has been used in order to attain a subsidized artificial balance between rice supply and demand in Japan, and its abandonment would considerably enlarge the area planted under rice possibly by up to 630 000 hectares – see Table 6.1). Although this would save the government subsidy needed for the paddy land transfer, it would not, however, increase rice supply greatly.

The rice pricing policy could be adjusted, raising the producers' price to levels that increased the incentive of farmers to grow rice in the cold damage years, and increasing the consumers' price to discourage consumption, and encourage a shift in consumer demand from rice to its substitutes, especially bread.

Thirdly, increases in rice imports and changes in the policy concerning reserve stock would be a high priority.

The important issue would be to decide upon the appropriate combination of these policy measures. Systematic methods, such as the model simulation described here, can readily be used to explore possible combinations of measures given certain policy objectives and public preferences.

With respect to production technology there would occur a switch to rice varieties more resistant to cool damage, as well as a change of planting dates and other rice production technologies, although probably with some time lag. The simulated results do not take into account these likely technological responses, thus overestimating the actual rice shortage that would occur. Government policy would be important here too, in coordinating and facilitating the technological adjustments required to cope with changed climate.

6.5.2. Implications of a $2 \times CO_2$ climate

Both of the $2 \times CO_2$ scenarios represent extreme deviations from the baseline conditions, though the transition scenario is perhaps the more appropriate for model simulations in this application. Thus the effects and policy implications of this scenario are taken up here in detail, although they are similar to those pertaining to the step-wise scenario.

As shown above, an incremental warming of the climate leads to a very severe rice surplus during the simulation period. Such a situation in Japan results in heavy government expenditure in order to dispose of it. This is because the domestic prices of rice are fixed by the government at a level some five times higher than the international rice price. Thus it cannot be exported without very heavy export subsidization. In addition, there is little demand for *Japonica* rice on the international rice market. If, as an alternative, surplus rice were fed to animals, the required government subsidy would be larger still, because feed prices are even further below the world market price.

The accumulated 20 million tonnes of total rice stock by 1982 resulting from the $2 \times CO_2$ transition scenario would require extremely large additional government expenditures to dispose of it, and would waste considerable resources in its production. Paddy land transfers would need to be increased, but at a cost the exchequer could barely support.

Thus another policy approach relating to the government rice price policy would be necessary. Farmers' and consumers' prices would have to be reduced considerably by the government, resulting in increased domestic rice consumption and decreased rice supply. Rice stock accumulation would therefore slow down, as the estimated parameters of the rice consumption and regional rice supply functions presented in Subsection 6.3 above indicate. This approach does not entail subsidies, which is thus desirable to the government. Furthermore, if the consumers' price is reduced by more than the producers' price, the government expenditures for the procurement and sales of rice may even be reduced.

In the past the nationally unified agricultural cooperative would not have been in favor of price decreases. However, after the two severe rice surplus periods experienced during the last two decades, the cooperative has come to appreciate the potential for severe rice surplus, and a policy of price decrease will probably be easier to follow in the future.

Early warning of a rice surplus is also important. With slight modifications of the supply side of the rice model presented in this study (for example, by incorporating additional climatic variables or by refining the spatial and/or temporal resolution), a short-term early warning capability could be achieved.

Finally, technological adjustments in rice production would surely follow such large changes in climate, with developments at agricultural experimental stations complementing the farmers' independent adjustments. These and other policy issues are explored in the next section.

SECTION 7

The Implications for Agricultural Policies and Planning

In the preceding sections we have shown that Japanese agriculture is finely adjusted to the climatic conditions prevailing in different regions of the country. In northern Japan in particular, where thermal conditions are critical determinants of rice production, seemingly small temperature anomalies can result in wide variations in productivity and regional rice supply that have implications for the national rice market, itself a precarious balance of domestic heavily subsidized supply and demand.

Through the use of simple models we have evaluated some of these impacts for a set of specified climatic conditions (scenarios). In this section we will first summarize the experiments and their results and, secondly, discuss their implications in terms of the requirements for further research, technological developments, economic restructuring and policy responses.

7.1. Summary of Results

7.1.1. Climatic scenarios

Experiments were conducted to simulate a range of feasible future climatic conditions defined largely on the basis of temperature. As described in Section 2, listed in *Table 2.1* and summarized here in *Table 7.1*, scenarios were selected to reflect:

(1) The weather in extreme individual years (viz: 1978 warm, and 1980 cool) identified from the existing records of agroclimatic conditions over about one century up to the present.
(2) The weather in anomalously warm or cool decades recorded historically (viz: 1921-30 warm, and 1902-11 cool).

Table 7.1. Summary of results of impact experiments in Hokkaido district and in the whole of Japan, as absolute values for the baseline period and as deviations from the baseline for each climatic scenario.

Climatic scenario	Representative period[a]	Mean July-August temperature (°C)[b]	Hokkaido district						Whole of Japan		
			Altitudinal shift of potential cultivation limit (m)[c]	Rice yield index[c]	Simulated rice yield[d]				Relative total net production (vegetation)[h]	Rice yield index[h]	Rice stock (1982 equivalent)[i]
					Present technology[e]		Adjusted technology[f]				
					Mean	CV[g]	Mean	CV[g]			
Baseline	1951-80	20.6	250[j]	1.03	4.48 t/ha[k]	11.6%	4.48 t/ha[k]	11.6%	1.00	100%	8.34 M tonnes[p]
Cool year	1980	-1.6	-254	-40%	-15%	-	+17%[m]	-	-7%	-13%	-100%[q]
Cool decade	1902-11	-1.1	-175	-25%	-	-	-	-	-	-	-
Unstable decade	1926-35	-0.3	-48	-6%	+0%[l]	+0%[l]	-	-	-2%	+0%	+11%
Stable decade	1957-66	-0.1	-16	-2%	+0%	-33%	-	-	-1%	+3%	+20%
Warm decade	1921-30	+0.4	+63	+6%	-	-	-	-	-	-	-
Warm year	1978	+2.5	+397	+15%	+6%	-	+31%[m]	-	+3%[n]	+3%[n]	-
2 × CO$_2$	Future	+3.5	+556	+5%	+4%	-27%	+26%	-33%	+9%	-	+191%[r]

[a] Unless otherwise stated.
[b] Mean of two stations: Sapporo and Asahigawa (from Section 2).
[c] From Section 4.
[d] From Section 5.
[e] Present cultivar: Ishikari; present planting date.
[f] Late-maturing cultivar: Nipponbare; planting date 25 days earlier than present.
[g] Coefficient of variation (standard deviation/mean).
[h] From Section 3.
[i] From Section 6.
[j] Present-day upper limit of main cultivation area.
[k] Baseline climate is 1974-1983.
[l] Unstable decade is 1974-1983.
[m] Extreme years are adjusted to 2 × CO$_2$ climate.
[n] 1979 chosen as warm year.
[p] Baseline climate is 1966-1982.
[q] 10 consecutive occurrences of 1980 climate from 1973 to 1982.
[r] 2 × CO$_2$ equilibrium climate.

(3) The weather during decades of particularly stable or unstable agroclimatic conditions (viz: 1957–66 stable, and 1926–35 unstable).
(4) The climate derived from the Goddard Institute for Space Studies (GISS) model estimates for doubled concentrations of atmospheric carbon dioxide.

The realism of the scenarios was investigated in Section 2 by examining the synoptic mechanisms that could explain the surface weather conditions assumed under each type. Composite maps of surface air temperature and precipitation anomalies, and mid-tropospheric geopotential height anomalies for selected extreme months or seasons taken from the historical record, revealed that cooler-than-average summers in central and northern Japan are characterized by an extension of the influence of the northeasterly winds from the Okhotsk anticyclone, and an intensification of the polar frontal zone located along the southern coast of Honshu Island. In contrast, warm summers tend to be associated with weaker activity of the polar frontal zone over southwest Japan and with a westward shift of the low pressure trough over Siberia such that the warm southwesterly winds circulating around its eastern part affect much of northern Japan.

There are some similarities between the conditions observed in warm years and those derived for the $2 \times CO_2$ climate, and though the magnitude of warming in the latter scenario is much larger, it suggests that further studies utilizing historical data as analogues of possible future CO_2-induced climatic change may provide some helpful statistical information as well as verification material for use in climate impact analysis. Geographical analogues might also serve a useful purpose: for example, under the GISS model-derived $2 \times CO_2$ scenario, the climate of Hokkaido district (in the lowlands) would resemble the present climate in lowland Tohoku, or in upland Chuba districts, about 500 km and 700 km to the south, respectively. Agricultural adaptation to $2 \times CO_2$ conditions in northern Japan might require the adoption of management practices and technologies already existing in these analogue regions. These considerations are discussed further below.

7.1.2. Estimated effects on net primary productivity of vegetation

Geographical distribution maps of net primary productivity (NPP) of vegetation were constructed for each climatic scenario on the basis of estimates using the Chikugo model (*see* Section 3). It was found that the northward or southward shift of the NPP-isopleths due to climatic change is larger in the northern districts of Japan than in the southern districts. This result agreed well with the results obtained for the latitudinal shift of effective accumulated temperature (ΣT_{10}) isopleths. To quantify the effect of climatic change on NPP distribution, distribution curves of percentage area of zones with various NPP values (NPP spectrum), and the relative total net production ($RTNP$) were calculated using the NPP distribution maps for the respective climatic scenarios. $RTNP$ is the ratio of total net production (TNP_0) for the climatic scenarios to the average values, where the 30-year baseline (1951–80) gives a ratio of 1.00. The results are shown in *Table 7.1*, the shifts in the NPP spectrum under each scenario expressed in terms of $RTNP$ values: 1.03, 0.99, 0.98, 0.93 and 1.09 for the good year, stable period, unstable period, bad year and $2 \times CO_2$ climate, respectively. Thus the climatic change due to a doubling in atmospheric CO_2 concentration may lead to the increase of about 9% in total net production of vegetation in Japan, well in excess of the increases experienced in even the warmest years in recent history.

7.1.3. Estimated effects on cultivable area and rice yields in northern Japan

In addition to their effects on productivity of vegetation and crops, temperature changes can also affect the potential area of cultivation of a crop such as rice. In some regions, the change in the thermal regime may be sufficient either to permit or to preclude the "safe" cultivation of rice (i.e., where the risk of cool summer damage is within acceptable limits). For Hokkaido, the shift of potential cultivable area under different climatic scenarios has been described both as a function of latitude and of altitude in Section 4. In a cool year such as 1980, the altitudinal shift of the upper limit of safe cultivation of rice (-254 m relative to the present approximate limit of 250 m a.s.l.) would effectively demarcate the whole island (except for a few favorably sited regions) as unsuitable for rice cultivation (*Table 7.1*). In contrast, a warm year (such as 1978, or during the $2 \times CO_2$ era) would potentially open up much of upland Hokkaido for rice cultivation, given that other factors such as topography, soils and exposure are suitable.

Japanese rice yields increased, with large year-to-year fluctuations, from about 2 t/ha in the 1880s to about 5 t/ha in the 1980s, reflecting the improvement of rice cultivation techniques throughout the past 100 years. The strong year-to-year fluctuations of yield around its nonlinear trend were caused mainly by weather variations, in particular by interannual variations in thermal resources during the rice growing period. The temperature–yield relationship was investigated using both a simple empirical statistical approach (Section 4) and a process-based simulation approach (Section 5). The estimated yield responses to the different climatic scenarios in Hokkaido district are shown in *Table 7.1*. The responses in Tohoku, the northernmost district of Honshu Island, are not shown in *Table 7.1*, but are similar in direction, though slightly different in magnitude, to those in Hokkaido.

In the warm year, 1978, with a mean July–August temperature anomaly relative to the baseline period of $+2.5$ °C in Hokkaido, the yield index is 1.19 (15% above the baseline), while in the cool year, 1980, with July–August temperatures 1.6 °C below the baseline value, the corresponding yield index was 0.62 (40% lower than the baseline yield index). Assuming that the relation between the rice yield index and temperature established from historical data can be applied to a case of future climatic change due to doubling in CO_2 concentration (i.e., by extrapolating the regression equation), the estimated change in yield index is $+5%$ relative to the baseline. This is a significantly smaller increase than for the warm year, 1978, reflecting the curvilinear nature of the functional relationship and implying that the present rice cultivars would not be well adapted to an average warming of such a large magnitude in Hokkaido. This conclusion was supported in Section 5 by results generated using a process-based rice yield simulation model.

The results from that study are shown in *Table 7.1*, and the important findings may be summarized as follows:

(1) A warming of the climate due to enhanced atmospheric CO_2 concentrations may reduce the year-to-year fluctuations of rice yield in the northern part of Japan.
(2) If the current rice cultivation techniques, including the use of early-maturing rice cultivars, are retained under warm climatic conditions, the average rice yield in Hokkaido would remain close to the present level.
(3) If later-maturing cultivars (such as those cultivated in Tohoku at present) were introduced into Hokkaido, the rice yield would probably increase markedly under warmer conditions.

7.1.4. Estimated effects on national rice supply, rice stocks and pricing policies

Just as the yield of rice is highly sensitive to temperature variations in northern Japan, so the levels of supply and demand in the national rice market are balanced precariously through a system of expensive government-funded subsidies. This balance can easily be upset as a result of misguided policies or because of natural variations in rice supply including those attributable to weather effects.

The impacts of the temperature change scenarios on the Japanese rice market were investigated in Section 6 using an integrated regional econometric rice model. The model was able to simulate the effects of temperature changes on regional rice supply in the two northern districts of Japan (where cool summer damage can cause considerable production losses in some years) and in the remainder of Japan. The total national supply was then matched to demand – through relationships with trade, prices, technology and consumption – to determine the change in national rice stocks during a specified simulation period (17 years).

The scenario impacts are shown in *Table 7.1*, two of which are worth highlighting here because they illustrate two extreme opposite situations. Firstly, the warmer conditions simulated for a $2 \times CO_2$ climate, if they were to occur with present-day management under the current market system, would cause a considerable increase in regional rice supply and a large accumulation of surplus rice stocks. Since rice is priced at extremely high levels in Japan, its disposal is a serious problem, so it would be very important to implement appropriate precautionary policies to avoid a potentially acute surplus during a period of climatic warming.

In contrast, the second scenario, representing the effects on national rice supply of ten consecutive occurrences of the cool damage year 1980, is characterized by considerable reductions in regional production and a rapid decline in total rice stocks (if present policies are retained). The goal of national food security would be jeopardized under this scenario, and immediate policy adjustments would be needed to encourage reversion of land (transferred to other crops through previous policy measures) to rice cultivation, reassess the pricing structure of the rice market and increase rice imports.

An additional requirement under either type of climatic change would be for increased government support for research into appropriate technologies to cope with the changed conditions (e.g., different rice cultivars with properties adapted to the new climate, such as those experimented with in Section 5 and summarized in *Table 7.1*).

7.2. Discussion

There are several points that arise from the results presented above, particularly with respect to the $2 \times CO_2$ scenario experiments:

(1) The temperature anomalies used in the instrumental scenarios (*see Table 7.1*) are smaller (even in the extreme warm year, 1978) than those estimated by the GISS model for doubled atmospheric CO_2 concentrations. One problem in attempting to assess the effects of such large changes on agricultural productivity is to justify the application of present-day climate–yield relationships to conditions that have never before been recorded.

(2) It was concluded in Section 5 that the climatic change derived for a doubling of CO_2 would decrease the cool summer damage of paddy production in Hokkaido, but would not increase the level of rice yield substantially if the current cultivation technology remains fixed. Significant increases in yield will be achieved only by the adjustment of technology to the new conditions produced by the climatic change (for instance, the selection of different crop varieties, changed planting dates and cultivation practices, and altered nutrient and water management). Combining point (1) with this conclusion, it is clear that further studies are needed on the limits of extrapolation of the crop–weather relationships deduced from our experience, and the likely form of relationships beyond present limits. Such studies should include the problem of evaluating the rate of acclimatization of agriculture.

(3) The variability of yields is related to the variability of climate, but we have little or no knowledge about the likely climatic variability under the higher air temperature conditions predicted for a future $2 \times CO_2$ equilibrium climate. High amplitude year-to-year fluctuations could, of course, have considerable economic effects.

(4) The results of the GISS general circulation model simulation runs were interpreted in this study for the regions surrounding Japan. It is questionable, however, whether we can interpret with high accuracy the absolute values at each grid point, given that they have already been smoothed from a coarser resolution (*see* Part I, Section 3 of this volume). Generally speaking, the historical variation of temperatures and the estimated CO_2-induced

changes are greater at higher latitudes than at lower latitudes. This is also the case in Japan where they are greater in Hokkaido than in Kyushu. On the other hand, contrasts in climatic conditions (for example, annual precipitation receipt) are also great in the west–east direction in the part of East Asia that includes Japan, which is the boundary region between a vast water mass, the Pacific Ocean, and the biggest continental land mass, Eurasia. As a result, the isolines in this region often run longitudinally rather than latitudinally. Thus the absolute values of air temperature and rainfall simulated under a changed climate should be used with the greatest care, and only after comparison with the present-day regional- and local-scale climatological characteristics of a particular area.

(5) The climatic changes induced by doubled CO_2 conditions may well influence agriculture in a less direct way than discussed above. For example, if the mean air temperature under $2 \times CO_2$ conditions increased by 3 °C relative to the present during the winter half year, snow accumulation in the mountainous regions on the side of Honshu Island that faces the Sea of Japan would be much lower than at present, and as a result springtime river discharges could be expected to decrease dramatically, with likely implications for the supply of water for irrigation. Furthermore, under conditions of increasing air temperature but unchanged or decreasing precipitation amount, the water deficit would increase because of the higher evapotranspiration. If this were to occur (as simulated by the GISS model for southwest Japan) the available water for agriculture would be reduced and the potential for wind erosion might increase in some regions during the drier parts of the year.

(6) The results of the economic analysis in Section 6 raise similar issues to those discussed in points (1) to (4). The model employed there considered policies for prices, surplus disposal, land diversion, reserve stock, farm structure and production technology of rice, but the model equations are all based on the present (1966–1982) conditions. There are great problems in extrapolating these economic and social relationships to a future situation that we have never experienced. In particular, there are immense difficulties in anticipating the adjustments in technology, farm structure, trade, and markets, etc., that might occur by the time CO_2 doubling actually occurs in the future, and inserting them into the simulation model. Other countries will also experience changes in their agricultural production and structure under global climatic changes, and so the exports and imports between Japan and her trading partners may also change drastically.

(7) Though the problems are many, as mentioned above, methodologically our approach may be correct. The method of study in this report has has been a hierarchical one (see Figure 1.11), proceeding as follows: climatic model → climatic scenario → shift of climatic (mainly thermal) regime → modeled first-order impact on agricultural productivity → modeled second-order (economic) impact → response experiments. Examples of the last two of these are presented in Sections 5 and 6.

7.3. Implications for Policy and Planning

7.3.1. Policies related to weather-induced crop damage

Japan has been developing policies for several decades to minimize crop damage due to the weather. Under Japanese law, the government has a responsibility to establish countermeasures for various agricultural damages. Since 1961, it has been the responsibility (again, by law) of the local authorities to manage these problems, which are outside the jurisdiction of single settlements such as cities and villages. Local authorities therefore play an important role in deciding upon compensation for farmers, which, while additional to anything received from central government, is assessed based on local knowledge, because the weather conditions and cultivated crops differ from region to region. Indeed, some prefectures have their own loan system to assist farmers who are affected by weather-induced damage to their crops.

In the early 1950s, a low interest loan was made available to any farmer who suffered heavy damages when an extreme damage-event occurred. It was, however, difficult to legislate quickly after every event. Therefore in 1955 a general law was passed for the "Loan System in the Case of Natural Calamity" with details indicated by a government ordinance for every type of case, including weather-induced crop damage. Under this system farmers can receive a loan when their harvest is less than 70% of the normal and the loss of agricultural income by the decreased harvest is more than 10%, subject to the approval of the local authorities. Local authorities grant the payment at a fixed rate of interest, and the government subsidizes the interest paid by the local authorities.

An integrated field crop and livestock insurance scheme was established in 1947 (Tsujii, 1985). Insurance specifically for fruit cultivation was introduced tentatively in 1968, and confirmed in 1973. Also in 1973, a mutual aid system was introduced experimentally for field crops and horticultural equipment. Damage by strong wind, flood, drought, cool summer, heavy snow and other nonmeteorological causes including earthquakes and volcanic eruptions are all covered by this system.

Thus the political countermeasures have been developed step by step, by introducing or revising laws, insurance and mutual aid systems in Japan. Though they have only recently been established, revisions of these measures provide some basis for adjustment to long-term climatic change. At present, the Japanese government has no alternative policies for adapting to future climatic change apart from the encouragement of studies of possible countermeasures.

7.3.2. Economic policy

At a national level, the Japanese government has intervened very strongly in the rice market because of the importance of rice as the staple food for the population and as the single commodity relevant to food security in Japan. The government monopoly of the international rice trade means that the domestic market is effectively separated from the world market. Further, the domestic rice market is controlled by heavy government intervention in rice marketing and the government's determination of both farmers' and consumers' rice prices. The former has risen to four or five times the international price of rice in recent years, and Japan has been faced with a potential (and on two occasions, a real) rice surplus (as described in Section 6).

The government offers three large rice subsidies:

(1) For the difference (the negative cost margin) between the government procurement price and the government resale price, plus the government marketing cost.
(2) For paddy field diversion to other crops [about US$4000 per hectare ($1 = 155 yen in 1986)].
(3) For the disposal of rice surpluses as feed and for export.

In addition considerable expenditure is incurred by the government-operated compulsory rice crop insurance scheme.

The simulated results of the contrasting scenario experiments have suggested that large rice surpluses or shortages could result from variations of climate. The two 2 × CO_2 scenarios would, *ceteris paribus*, result in an enormous rice surplus. Under the present rice policy, the producers' and the consumers' prices of rice would need to be reduced considerably to decrease supply and to increase consumption. If these price adjustments alone were not sufficient (and this is highly probable because of strong fiscal and political constraints, including the strength of the agricultural lobby), then the paddy land diversion policy would need to be extended. Over the long term there should be a gradual transition to a policy of reduced government intervention, allowing the market mechanism to regulate demand and price, while retaining a government monopoly of the international rice trade.

Since potential rice surpluses have long been the norm, the distinct rice shortages predicted under the consecutive cool damage year scenarios are easier to handle under the present policy framework. The paddy land diversion policy can be reduced considerably or abandoned altogether with appropriate adjustments in the regional distribution of rice production. This step would reduce subsidy expenditures and also the fiscal deficit. Increases in the producers' and consumers' prices may also be needed but this step is limited by the already high level of prices. Even after adopting these policy measures, rice shortages may still occur in some of the more severe cold damage years. Thus some rice imports by government may be needed, and this would require a diversification of import sources, for the specific *Japonica* rice favored by the Japanese public is not cultivated widely outside Japan.

A new policy framework under these conditions could be similar to that proposed as a response to rice surplus, the rice market being expected to adjust itself in order to cope with the predicted shortage. Government rice imports might be necessary in order to smooth out extreme market adjustments over several years. An improved rice crop insurance policy could be retained.

7.3.3. Policy regarding research and technology

Finally, some general recommendations can be made in relation to research priorities that apply not only in Japan but also further afield. Modern agriculture depends heavily on five technologies: high yielding crop varieties, fertilization, mechanization, irrigation, and the chemical control of weeds, pests and diseases. Each of these technologies has made an important contribution to the population capacity of the earth, and to the rise in the nutrition levels. But, because much of the potential arable land in the world is already in crop production, one of the few ways of increasing food production will be to use more fossil energy for the intensification of agricultural management (particularly in the production of fertilizers and agricultural chemicals). Yet the increased burning of fossil fuels along with carbon dioxide release from biological sources, has been responsible for raising the level of atmospheric CO_2 concentrations throughout this century, and further increases will inevitably intensify the atmospheric greenhouse effect, probably causing a change in world climate.

The climatic change induced by increased atmospheric CO_2 will probably affect physiological and ecological activities of plants (as we have shown for Japan in this report) and alter the relationship between climate and the distribution of terrestial ecosystems, including world agricultural systems. Assessing the direct effects of increased atmospheric CO_2 and the indirect effects of CO_2-induced climatic change on food production is very difficult, because we do not fully understand the direct and indirect influences of increased CO_2 levels on crops and their surrounding environment. This is probably one of the results of the conventional practice of regarding the effects of variations of climate on crops as origins of error in field experiments that are designed to study and develop new agricultural technologies. By referring to results reported by several authors, including the findings of this study in Japan, the direct and indirect influences of increased atmospheric CO_2 in the future on crops and their environment have been summarized in *Figure 7.1*.

The diagram shows clearly that increased atmospheric CO_2 would be profitable for crops and food production according to some criteria and unprofitable according to others. With regard to these effects and their mitigation, the following scientific and technological problems need to be solved in the next few years:

(1) Assessment of the change in geographical and seasonal distributions of thermal and water resources important to crop production.
(2) Assessment of the change in microclimate within crop fields and forests associated with the change in large-scale climate.
(3) Elucidation of the direct effects of long-term CO_2 changes on dry matter production, yield formation, and water use efficiency of crop and forest plants.
(4) Evaluation of the responses of natural and controlled plant ecosystems to climatic change.

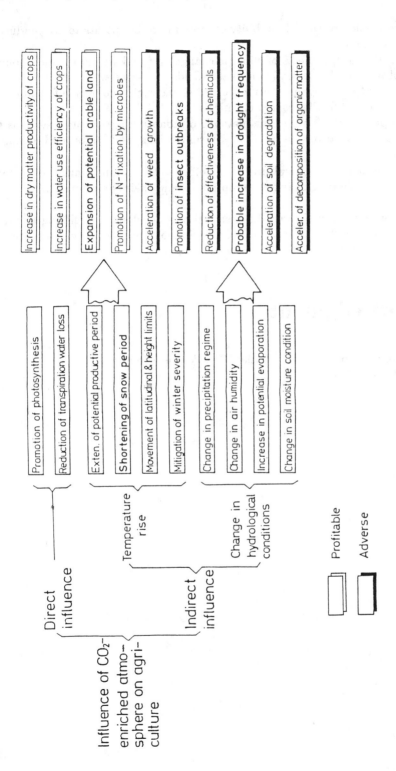

Figure 7.1. Possible effects of increased atmospheric carbon dioxide on agriculture in Japan.

(5) Determination of the adaptability of crop and forest plants to CO_2-induced climatic change.
(6) Elucidation of the effects of climatic change on the distribution and population of pests and diseases in cultivated fields and forests.
(7) Assessment of the longer-term effects of climatic change on soil fertility and effectiveness of fertilizers and chemicals.
(8) Development of management methods of adapting agricultural systems to the increased atmospheric CO_2 concentrations and associated climatic change.
(9) Establishment of agricultural technologies for minimizing the impact imposed by climatic change on crop plants and their environment.

In order to facilitate successful research and development relating to these problems, it is desirable to organize interdisciplinary national and international projects of which this study forms one contribution.

Acknowledgment

Acknowledgment is made to Mr. Atsushiro Taki for his help in compilation and computation of data in Section 6.

References

Arakawa, H. and Taga, S. (1969). Climate of Japan. In H. Arakawa (ed.) *Climates of Northern and Eastern Asia*. World Survey of Climatology, Volume 8, Elsevier, Amsterdam, pp. 119-158.

Bolin, B., Döös, B.R., Jäger, J. and Warrick, R.A. (eds.) (1986). *The Greenhouse Effect, Climatic Change and Ecosystems*, Wiley, Chichester, 542 pp.

Budyko, M.I. (1956). *The Heat Balance of the Earth's Surface* (translated into English by N.I. Stepanova). U.S. Weather Bureau, Washington, 1958.

Budyko, M.I. (1971). *Climate and Life* (translated into Japanese by Z. Uchijima and Iwakiri) (Todai Shuppan-Kai).

Cannell, M.G.R. (1982). *World Forest Biomass and Primary Production Data*. Academic Press, New York.

Carbon Dioxide Assessment Committee (1983). *Changing Climate*, National Academy of Sciences, Washington, D.C., 496 pp.

Clark, W.C. (1985). Scales of climate impacts. *Climatic Change*, 7(1), 5-27.

Food Agency (various dates). *Annual Statistical Report of Food Control*, Ministry of Agriculture, Forestry and Fisheries, Tokyo.

Gallagher, J.N. and Biscoe, P.V. (1978). Radiation absorption, growth and yield of cereals. *J. Agric. Sci.*, 91, 47-60.

Gen, M. and Ida, K. (1983). *Calculation Programs for Science and Technology*. Kogaku Tosho, Tokyo, pp. 125-140 (in Japanese).

Gifford, R.M. (1979). Growth and yield of CO_2-enriched wheat under water limited conditions. *Aust. J. Plant Physiol.*, 6, 367-378.

Golts'berg, I.A. (ed.) (1972). *World Atlas of Agroclimatic Resources*. Gidrometeoizdat. Leningrad, 115 pp.

Goudriaan, J. and de Ruiter, H.E. (1983). Plant growth in response to CO_2 enrichment at two levels of nitrogen and phosphoorun supply. 1. Dry matter, leaf area and development. *Neth. J. Agric. Sci.*, 31, 237-250.

Hansen, J., Russell, G., Rind, D., Stone, P., Lacis, A., Lebedeff, S., Ruedy, R. and Travis, L. (1983). Efficient three dimensional global models for climate studies: models I and II. *Monthly Weather Review*, 111(4), 609-662.

Hansen, J., Lacis, A., Rind, D., Russell, G., Stone, P., Fung, I., Ruedy, R. and Lerner, J. (1984). Climate sensitivity: analysis of feedback mechanisms. In J. Hansen and T. Takahashi (eds.) *Climate Processes and Climate Sensitivity*, Maurice Ewing Series, 5. American Geophysical Union, Washington, D.C., pp. 130–163.

Hanyu, J. and Ishiguro, T. (1972). Temperature and rice yield stability during the cultivation period in Hokkaido. (2). Relation of the fluctuation of rice yield to temperature during the safe cultivation period. *Res. Bull. Hokkaido Natl. Agric. Exp. Sta.*, **102**, 93–103.

Hanyu, J., Uchijima, T. and Sugawara, S. (1966). Studies on the agro-climatological method for expressing the paddy rice products. I. *Bull. Tohoku Natl. Agric. Exp. Sta.*, **34**, 27–36.

Hare, F.K. (1983). *Climate and Desertification: A Revised Analysis*, WCP-44. WMO, World Climate Applications Programme.

Horie, T. (1984). Studies on crop-weather relationship model in rice. (3). A dynamic model for simulating the crop development and the values of parameters for major cultivars. *Proc. of Ann. Meeting of Soc. Agric. Meteor., Japan, 1984*, pp. 5–6.

Horie, T. and Sakuratani, T. (1985). Studies on crop-weather relationship model in rice. (1) Relation between absorbed solar radiation by the crop and the dry matter production. *J. Agric. Meteorol.*, **40**, 331–342.

Horie, T., de Wit, C.T., Goudriaan, J. and Bensink, J. (1978). A formal template for the development of cucumber in its vegetative stage. *Proc. KNAW, Ser. C*, **82**, 433–479.

Ichikawa, T., Yamashita, Sh. and Yoshino, M.M. (1984). Climate and agricultural land use in Japan. In M.M. Yoshino (ed.) *Climate and Agricultural Land Use in Monsoon Asia*. University of Tokyo Press, Tokyo, pp. 317–332.

Ishii, M. (1980). Regional trends in the changing agrarian structure of postwar Japan. In Ass. Japanese Geographers (eds.) *Regional Geography of Japan*. Teikoku-shoin, Tokyo, pp. 199–222.

Iwaki, H. (1984). Assessment of regional distribution of phytomass and net primary productivity in Japan. *SPEY*, **4**, 167–172.

Japanese Meteorological Agency (JMA), Statistical Section (1960). A method of estimating monthly mean temperature. *Technical Report of the Japan Met. Agency*, No. **2**, 1–9.

Japanese Meteorological Agency (1981). *Monthly normals of atmospheric pressure, temperature, relative humidity, precipitation and sunshine duration, 1951–1980*, 71 pp.

Japanese Meteorological Agency (1984). *Ijokisho-hakusho* (on unusual weather).

Japanese Meteorological Agency, Observation Department (1985). *Nôgyô-kishô-tsûhoro-tebiki* (shiryo-hen) (Guidebook to Agrometeorology), 74 pp.

Kira, T. (1973). Primary production of forests. In J.P. Cooper (ed.) *Photosynthesis and Productivity in Different Environments*. Cambridge University Press, pp. 5–49.

Kira, T. (1976). *Introduction to Terrestrial Ecosystems*. Kyoritsu Shuppan. Tokyo, 166 pp. (in Japanese).

Kudo, K. (1975). Economic yield and climate. In Y. Murata (ed.) *JIBP Synthesis*, Vol. **11**. University of Tokyo Press, pp. 99–220.

Lamb, H.H. (1972). *Climate Present, Past and Future. Volume 1: Fundamentals and Climate Now*, Methuen, London, 614 pp.

Matsuda, K. (1960). Statistical studies between the rice-culture and meteorological variants in warm districts II. *J. Agric. Meteorol.*, **16**, 17–19.

Ministry of Agriculture, Forestry and Fisheries (various dates). *Crop Statistics*, MAFF, Tokyo.

Monsi, M. and Saeki, T. (1953). Über den Lichtfaktor in den Pflanzengesellschaften und seine Bedeutung für die Stoffproduktion. *Jap. J. Bot.*, **15**, 22–52.

Monteith, J.L. (1977). Climate and the efficiency of crop production in Britain. *Phil. Trans. R. Soc. Lond. B*, **281**, 277–294.

Müller, M.J. (1982). *Selected Climatic Data for a Global Set of Standard Stations for Vegetation Science*, Dr. W. Junk Publishers, The Hague, 306 pp.

Munakata, K. (1976). Effects of temperature and light on the reproductive growth and ripening of rice. In IRRI, *Climate and Rice*, Los Banos, Philippines, pp. 187–207.

National Institute of Agricultural Sciences (NIAS), Division of Agrometeorology (1975). *Climatic Change and Crop Production*, 40 pp. (in Japanese).

NIAS, Division of Agrometeorology (1976). *Crop Production and Unusual Weather*. Misc. Report, National Institute of Agricultural Sciences, Tokyo, 30 pp.

Parry, M.L. and Carter, T.R. (1984). *Assessing the Impact of Climatic Change in Cold Regions*. Summary Report SR-84-1, International Institute for Applied Systems Analysis, Laxenburg, Austria, 42 pp.

Research Group of Evapotranspiration (1967). Radiation balance of paddy field. *J. Agric. Meteorol.*, **22**, 97–102.

Sapozhnikova, S.M., Mel', M.I. and Smirnova, V.A. (1957). Agroclimatic characteristics of various regions in USSR in relation to the introduction of maize. *Try. Nauchino-Issledo., Inst. Aeroclimatologii*, **2**, 5–194.

Satake, T. (1978). Sterility-type cool injury to rice plants in Japan. In K. Takahashi and M.M. Yoshino (eds.) *Climatic Change and Food Production*. University of Tokyo Press, Tokyo, pp. 245–254.

Seino, H. and Uchijima, Z. (1985). Agroclimatic evaluation of net primary productivity of natural vegetation. (2). Assessment of total production in Japan. *J. Agric. Meteorol.*, **41**(2), 139–144.

Shibata, M., Sasaki, K., and Shimazaki, Y. (1970). Effects of air-temperature and water-temperature at each stage of the growth of lowland rice. I. Effect of air-temperature and water-temperature on the percentage of sterile grains. *Proc. Crop Sci. Soc. Japan*, **39**, 401–408.

Shibles, R.M. and Weber, C.R. (1966). Interception of solar radiation and dry matter production by various soybean planting patterns. *Crop Sci.*, **6**, 55–59.

Tabayashi, A. (1983). Development of rice cultivation in the beech forest belt, Japan. *Tsukuba Studies in Human Geography*, **7**, 233–256 (in Japanese with English abstract).

Tadaki, R. and Hachiya, K. (1968). *Forest Ecosystems and their Biomass Production*, Ringyokagaku Shinkojo, Tokyo, 64 pp. (in Japanese).

Tani, N. (1978). Agricultural crop damage by weather hazards and the countermeasures in Japan. In K. Takahashi and M.M. Yoshino (eds.) *Climatic Change and Food Production*. University of Tokyo Press, Tokyo, pp. 197–215.

Toriyama, K. (1978). Rice breeding for tolerance to climatic injury in Japan. In K. Takahashi and M.M. Yoshino (eds.) *Climatic Change and Food Production*. University of Tokyo Press, Tokyo, pp. 237–243.

Tsujii, H. (1982). A quantitative model of the international rice market and analysis of the national rice policies, with special reference to Thailand, Indonesia, Japan, and the United States. In M.R. Langham and R.H. Retzlaff (eds.) *Agricultural Sector Analysis in Asia*. Singapore University Press for the Agricultural Development Council, pp. 291–321.

Tsujii, H. (1984). Agricultural Cooperatives and the Farmers' Price of Rice – A Comparison of Integration in Rural Development Policy between Japan and Thailand. Presented at the *International Seminar on the Role of Agricultural Cooperatives in Integrated Rural Development, Seoul, South Korea, July 3–6, 1984*.

Tsujii, H. (1985). An economic analysis of Japanese rice insurance policy: estimation of farmers' risk response behavior by risk premium function and rice supply function. In A. Valdés and P.B.R. Hazell (eds.) *Risk and Crop Insurance in the World*. Johns Hopkins University Press, Baltimore, USA.

Uchijima, T. (1976). Some aspects of the relation between low air temperature and sterile spikelets in rice plants. *J. Agric. Meteorol.*, **31**, 199–202.

Uchijima, T. (1983). Agrometeorological studies on the safety cropping season of paddy rice in Hokkaido and Tohoku Districts. *Bull. Natl. Inst. Agric. Sci.*, Series A, **31**, 23–113 (In Japanese with English summary).

Uchijima, Z. (1976a). Long-term change and variability of sum of air temperature during period with daily mean above 10 °C. *J. Agric. Meteorol.*, **31**, 185–194.

Uchijima, Z. (1976b). Return period of duration with daily mean above 10 °C. *J. Agric. Meteorol.*, **32**, 19–21.

Uchijima, Z. (1976c). Long-term change and variability of air temperature above 10 °C in relation to crop production. In Takahashi and M.M. Yoshino (eds.) *Climatic Change and Food Production*. University of Tokyo Press, pp. 217–229.

Uchijima, Z. (1981). Yield variability of crops in Japan. *Geojournal*, **5**(2), 151–164.

Uchijima, Z. and Horibe, Y. (1977). Long-term change and variability of warmth index and coldness index. *J. Agric. Meteorol.*, **33**, 137–148

Uchijima, Z. and Seino, H. (1985a). Agroclimatic evaluation of net primary productivity of natural vegetations. (1). Chikugo model for evaluating net primary productivity. *J. Agric. Meteorol.*, **40**, 343–352

Uchijima, Z. and Seino, H. (1985b). *Distribution Map of the Net Primary Productivity of Natural Vegetations in Japan*, BCP-85-I-1-1. National Institute of Agro-Environmental Science, and Kyushu National Agricultural Experimental Station.

Wit, C.T. de, Brouwer, R. and Penning de Vries, F.W.T. (1970). The simulation of photosynthetic systems. In *Proceedings of the IBP/PP, Technical Meeting, Trebon (1969)*, pp. 47–60.

Yoshida, S. and Shinoki, S. (1978). Preparation of the maps of monthly mean global solar radiation and its year to year variability for the Japanese islands. *Tenki*, **25**, 375–389.

Yoshino, M.M. (1975). *Climate in a Small Area*. University of Tokyo Press, Tokyo.

Yoshino, M.M. (1978a). Altitudinal vegetation belts of Japan with special reference to climatic conditions. *Arctic and Alpine Research*, **10**(2), 449–456

Yoshino, M.M. (1978b). Regionality of climatic change in monsoon Asia. In K. Takahashi and M.M. Yoshino (eds.) *Climatic Change and Food Production*. University of Tokyo Press, Tokyo, pp. 331–342.

Yoshino, M.M. (1983). Time-space scales of biolevel and human life in relation to the climatic phenomena. In *Cross-Section of Modern Ecology*, Kyoritsu-shuppan, Tokyo, (in Japanese), pp. 16–23.

Yoshino, M.M. and Yasunari, T. (1986). Climatic anomalies of El Niño and anti-El Niño years and their socio-economic impacts in Japan. *Sci. Rept., Inst. Geosci., Univ. Tsukuba. A*, **7**, 41–53.

Index

Editors' note: We refer to place names and authors in this index only where this is necessary to identify a climatic phenomenon, a method of impact assessment or an analytical model.

Accumulated temperature, *see* Effective accumulated temperature, Effective temperature sum
Acid deposition, 18
Adjoint method, *see* Climate impact assessment, adjoint method
Adjustments (at farm level) to climatic variations, 18, 25, 44–50, 71–72, 486, 499, 701–705, 744–747; adjustment experiments 20–21, 114; *see also* Agricultural policy, Crop varieties, Farm operations, Food purchases, Irrigation, Soil drainage
Afforestation, 405, 491, 501
Agricultural cooperatives, 828
Agricultural expenditure, *see* Farm expenditure
Agricultural output, *see* Food production, Yield
Agricultural policy, 18, 20, 27, 50–51, 72, 237, 241, 345–348, 360–363, 365, 487, 490–493, 498–501, 531, 544, 605–607, 677–693, 695–706, 853–864; at regional level, 50–51, 680–681; at national level, 51, 72, 681, 692, 715–717; regional model of, 682–684, 684–691, 710–714; *see also* Adjustments, Exports, Food prices, Imports, Set-aside, Subsidies
Agricultural potential, 29, 53, 70, 279, 415–444; climatic index of, 52, 54, 82, 84, 105, 255, 264–270, 369; spatial pattern of, 36–39, 70; spatial shift of, 38, 773–795; *see also* Agroclimatic index, Agroecological potential

Agricultural research, 364, 365, 470, 501–503, 862–864
Agrocenose, 664
Agroclimatic index, 24, 104–105, 110; *see also* Agricultural potential, climatic index of
Agroclimatic models, *see* Agroclimatic index, Crop–climate models
Agroecological potential, 412–413
Analogue regions, *see* Regional analogues

Barley, 391, 441, 472, 498–499, 500, 526, 547–563, 559–594, 582, 631, 632–633, 686, 739; crop–climate simulation model of, 245, 690–691; potential for cultivation, 469; potential area of successful ripening, 35; yields, 26, 42, 56–67, 88–89, 91, 326, 331, 527–528, 533, 547–563, 600–602, 685, 700, 745
Baseline climate, *see* Climatic scenarios
Base temperature, *see* Threshold temperature
Beef, production of, 235, 501, 531; *see also* Livestock
Bioclimatic potential, 633
Biomass, 16–17; net production of, 61, 120; potential, 26, 353, 358, 360; productivity of, 256, 265–266, 269, 352, 359; *see also* Chikugo model, Net primary productivity of vegetation
Biomes, *see* Vegetation zones
Biophysical models, *see* Agroclimatic index, Crop–climate models
Birch woodland, 390, 439, 440, 441

Blight, *see* Disease
Boreal forest, 55, 86, 183–195, 629, 738; growth rates of, 44; limits of, 121, 187, 208, 210, 439–442; physical productivity of, 43, 183–195; potential economic output from, 43, 197–218; *see also* Afforestation, Forest sector model, Timber, Tree growth

Canola yields, 326, 331
Canopy reflectance, 812
Capital, investment of, 677–693; stock of, 683, 687
Carbon dioxide, in atmosphere, 11, 23, 81, 99, 125, 143, 150, 172, 241, 626, 637, 647; other trace gases, 11, 648; direct effects of, 24, 48, 115, 184, 194, 208, 365, 577, 673, 675, 862; CO_2-induced warming, 33, 206, 207, 398, 405, 545; *see also* Climatic scenarios, GISS
Carcass weight, 26, 64; of lambs, 435–437, 475–487, 494, 496, 497; of meat, 624; of sheep, 51, 54–55, 67, 87, 405–406, 442
Carrying capacity, 26, 32, 50, 64, 389, 473; of sheep on improved grassland, 54, 87, 431–435, 442; of sheep on rangelands, 54, 67, 87, 431–436, 442, 444, 475–487, 496, 497; *see also* Livestock, Stocking level
Carry-over effects, 63, 69, 448, 484
Cattle, *see* Livestock
Centrally planned economies, *see* Agricultural policy
Chikugo model, of net primary productivity, 104–105, 116, 782–786, 788, 790–791
Climatic change, *see* Climatic scenarios, Climatic variations
Climatic fluctuations, *see* Climatic variations
Climate impact assessment, adjoint method, 19–20, 73; direct method, 18–20, 73; impact approach, 13–14; integrated approach, 15–17, 73, 229; interaction approach, 14–15; iterative approach, 73; spatial approach, 17, 21, 38, 73, 121–122, 290–291, 364, 412–413, 773–795
Climatic index of agricultural potential, *see* Agricultural potential, climatic index of
Climatic optimum (hypsithermal), 752
Climatic scenarios, 24–25, 125–157, 245–252, 357–358, 544, 636–637, 748, 751–772, 786–788, 815–817, 835, 839–841, 853–855; instrumentally based scenarios, 25, 77–81, 559–562, 754–757, 786–788, 804; instrumentally based scenarios, single years, 24, 52–53, 59, 60, 61, 175–179, 229, 246–247, 269, 296, 301–305, 315–316, 330–338, 351, 407–409, 425–426, 462–464, 469, 484–487, 493–498, 754–757, 804–805, 816, 820–821; instrumentally based scenarios, periods, 24, 53, 54, 56, 60, 174–179, 229, 246–247, 261, 269, 291–292, 296, 301–305, 311–315, 330–338, 351, 389, 407–409, 425–426, 462–464, 468–469, 495–498, 534–537, 569–572, 589–596, 600–601, 754–757, 804– 805, 816, 820–821, 839; instrumentally based scenarios using Vinnikov–Groisman method, 79, 81, 635, 637, 647–649, 672, 696; baseline climate, 29, 52, 69, 117–118, 141, 142, 249, 407, 411, 537, 544, 572, 637, 655–656, 696, 755; analogue scenarios, 763– 767; synthetic scenarios, 25, 59, 77–81, 115, 637, 685–687; based on GCM outputs, 24, 77–81, 115, 351, 354–356, 364, 390; transient, 77–81, 118, 150, 700; *see also* General circulation models; GISS
Climatic variables, *see* Precipitation, Solar radiation, Temperature, Windspeed
Climatic variations and variability, 12–13, 69, 159, 360; anomalies (extremes), 11, 31–32, 66–68, 69, 100, 107, 407, 463, 793; return periods, 276–278; cool periods, 408, 443–444, 756, 767, 786–788, 804, 806; warm periods, 408, 444, 756, 767, 786–788, 804, 805; stable periods, 786, 816, 820, 839; unstable periods, 786, 816, 820, 839; *see also* Climatic scenarios, Frequency, Precipitation, Probability, Temperature
Climatic warming, *see* Carbon dioxide, Climatic scenarios
Cloud seeding, 704
Coldness index, 103, 774, 775
Collective farms, 626, 627
Collinearity, 104, 110
Comparative advantage, 36, 70, 75, 603, 699
Competition, between land uses, 76; between regions, 75
Construction sector, 338; impact on, 327, 345–347
Cool periods, *see* Climatic variations
Cooling degree-day, 103, 813–814, 815
Cool summer damage, 731, 740, 742, 744, 746, 749, 752, 786, 813, 817, 821, 824, 858
Cooperatives, agricultural, 828
Corn for silage, crop–climate simulation model of, 690–691; yields, 91, 685, 700
Cornell University Soil–Plant–Atmosphere model, 298
Cotton, 624, 627
Cretaceous period, 140
Crop allocation, *see* Land allocation
Crop breeding, *see* Plant breeding
Crop certainty, 577; *see also* Probability
Crop calendar, for spring wheat, 99
Crop–climate models, 105–107, 252–257; assumptions in, 109–110; baseline specification for, 117–118; data requirements for, 109; extrapolation from, 110; missing variables for, 116; sensitivity analysis of, 111, 298–301; spatial resolution of, 117; temporal resolution of, 16; time-dependent

Index 871

changes in, 118; validation of, 106, 108, 111–112, 252–253, 297–298, 352, 364, 424–425, 427, 455–458, 557–558, 569, 571, 586–589, 649–655, 653, 668; *see also* Agroclimatic index, Empirical–statistical models, Simulation models
Crop combinations, 597
Crop dispersion, 702, 704
Crop drying, 567
Crop failure, 592, 800; frequency of, 800; probability of, 802
Crop receipts, *see* Farm income
Crop rotation, 324
Crop storage, 237
Crop varieties, changes in, 27, 28, 36, 44–46, 361, 701–703; changes in varieties with high thermal refinements, 45, 67; changes in varieties giving less variable yields, 45–46; changes from spring to winter wheat, 342–345; cold-resistant varieties, 738; early-maturing varieties, 551, 565, 568, 857; high-yielding varieties, 471, 562; late-maturing varieties, 48, 92, 565, 570, 823, 825, 857; middle-maturing varieties, 48, 92; of rice, 39; quick-maturing varieties, 101; *see also* Plant breeding, Thermal requirements
Cumulative effects, 31, 69
Cultivation limits, *see* Limits

Dairy farming, 235, 464–467; dairy cattle, 432; milk production, 87, 393, 462, 465–466, 531, 624; milk products, 606
Data, availability of, 160–161, 534, 567–568; homogeneity of, 161; requirements, 417–420; spatial resolution of, 74; temporal resolution of, 74
Day length, 39, 811
Degree-days, *see* Effective accumulated temperature; Effective temperature sum
Depopulation, 491–492
Desert pavement, 284
Deterministic functions, 106
Development rate, of plants, 811
Development stage, of plants, 811, 818
Diagnostic approach, to scenario development, 78
Disease, 40, 71, 75, 282, 472, 563, 808, 817, 864; blight, 40, 472; control of, 677, 702; rust, 41; ryegrass mosaic virus, 472
Drainage, *see* Soil drainage
Drought, 12, 237, 254, 416, 549, 604, 624, 634, 740; frequency of, 26, 53, 82, 84, 250, 254–255, 270–279, 353, 355, 360, 863; index of, 254, 256; in 1930s, 14, 238–239, 246, 275, 291, 304, 311–315; 1936-type, 343–345; Palmer Drought Index, 104, 255, 270–278, 355–356, 359, 369–370; relief payments after, 238

Dwarf shrubs, 392

Economic models, 17; for regional economic planning, 119; input–output models, 27, 43, 322, 328; in Saskatchewan Drought Study, 325–328; of employment, 17; of farm production, 17, 42, 43; of farm profitability, 17, 118–119, 322; of farm purchases, 17; of Japan rice market, 827–851, 857–858; optimization models, 692, 715–717; validation of, 364, 836
Effective accumulated temperature, 640–641, 739, 774–775, 777–778, 781, 797, 806; *see also* Effective temperature sum
Effective temperature sum, 26, 28–29, 36–37, 52, 54–56, 59, 82, 84, 88–89, 92, 102, 104, 184–193 *passim*, 209, 255–256, 260–262, 279, 303, 353, 359, 406, 437, 524–525, 530, 535–536, 549, 554, 556, 568, 584–597, 629, 631, 633, 640–641, 675; definition of, 82; Thom's method of calculation, 260; *see also* Effective accumulated temperature
Egg production, 235, 531, 624; *see also* Poultry
El Niño, 12
Empirical–statistical models, of crop–climate relationships, 13, 24, 27, 243–244, 253, 493, 664; of barley, 108, 544, 547–563, 584–586; of carcass weight of lambs, 483–484; of hay, 108, 416, 420–424, 445–474; of oats, 108, 120, 584–586; of pasture, 108; of rice, 108, 748, 749; of spring wheat, 108, 649–665, 653–657; of winter rye, 108, 649–665
Employment, 43, 52–54, 64, 71, 85, 321–349, 338, 342; at farm level, 26, 28, 339, 352–353; model of, 322, 329, 338; regional, 67, 255
Erosion, *see* Soil erosion
Evapotranspiration, *see* Climatic scenarios, Climatic variations
Expenditure, *see* Farm expenditure
Exports, 829, 850
Extreme events, *see* Frequency, Probability

Fallow, 233–234, 528
Famine, 391–392, 536, 752
Farm abandonment, 238, 281, 468, 469; *see also* Depopulation, Population, migration of
Farm employment, *see* Employment
Farm expenditure, 27, 49–50, 72, 322, 346, 353, 687–692
Farm income, 26, 28, 51–53, 56–57, 75, 332, 338–339, 342, 346–347, 492, 533, 544, 605–607; household level, 85, 332; national level, 64–68, 89; regional level, 42, 71; regional equity of, 51, 72, 605; stability of, 45; *see also* Farm profitability
Farm Income Act (Finland), 606

Farming operations, timing of, 28, 677
Farm inputs, 51
Farmlab program (Canada), 362
Farm models, see Economic models
Farm production, 322, 605–607
Farm profitability, 27, 33, 41, 71, 75, 544, 579–597, 603–604
Farm purchases, 50, 72
Farm size, 234
Farm wages, 579, 593, 606
Feed grains, see Fodder
Feed requirements, see Livestock
Fertilizer, use of, 27–28, 36, 40, 46–49, 63, 65, 72, 75, 87, 117, 324, 406, 418–420, 424, 428, 443, 460–461, 465, 468, 469, 471, 497, 623, 633–634, 653–654, 657–659, 672, 703, 862, 864; costs of, 579–580; experiments with, 446–448, 454, 458, 473; fertilizer index, 422–423, 426; fertilizer pollution index, 58; see also Nitrate pollution
Financial sector, effects on, 338
First-order interactions, 15, 25, 32–36, 43, 70– 71, 97–123, 229, see also Disease, Droughts, Livestock, Second-order interactions, Soils, Third-order interactions, Waterlogging, Weeds, Yields
Fishing, 327, 345–347, 391, 393
Flax yields, 331
Fodder, crops, 576; purchase of, 491, 492; reserves of, 491; shortage of, 492
Föhn wind, 416
Food consumption and demand, 827–851, 857–858
Food prices, 684, 827–851, 857–858, 860; support of, 28–29, 51; see also Subsidies
Food production, 75; national, 41, 71; optimization of, 50–51, 72; regional, 41, 71; stabilization of, 51, 72
Food production costs, 41, 71, 677–693; barley, 91; corn, 41, 91; grain, 49; hay, 91; oats, 91; potatoes, 91; vegetables, 91; winter wheat, 41, 91
Food purchases, 27
Food security, 51, 72, 844, 850, 857–858, 860
Food stocks, 26, 28, 607; of rice, 31, 60, 64, 68, 92, 827–851, 857–858
Food supply, 26; of rice, 31, 56, 61, 64, 92, 749, 827–851, 857–858; district, 61; national, 61, 92
Food surplus, 42, 71, 493, 532, 577, 606, 747; of rice, 827–851, 857–858, 860
Forage production, see Grass, Hay, Yields
Forest, see Birch woodland, Boreal forest, Broadleaved woodland, Norwegian spruce
Forest fires, 12, 390
Forest Growth Units, 103–104
Forest sector model, 198–207, 214
Freeze-thaw cycles, 284

Frequency, 21; of cold summer damage, 34; of crop failure, 583, 802; of dry spells, 21, 40, 84, 126; of extreme events, 31, 69, 75, 126; of flooding, 40, 126; see also Probability
Frost, 237, 548–549, 588, 592, 634, 641, 643, 814; probability of, 577, 608; date of, 631; first autumn frost, 437; night frost, 592; frost-free period, 104, 231, 294, 398, 400, 629–631, 685
Fuel, 332, 339; fuel and oil purchases, 332; effects on fuel mines sector, 327

General circulation models, 17, 24–25, 125–157, 354–356, 364, 637; Chervin significance test, 143, 152; climate inversion problem, 148; grid points/spatial resolution, 130, 131, 523, 537, 630; parameterization, 129, 139, 149, 150; time steps/ temporal resolution, 130, 131; see also GFDL, GISS, NCAR, OSU, UKMO
Geopotential height anomaly, 765, 766
Geostrophic wind, 168
GFDL (Geophysical Fluid Dynamics Laboratory) general circulation model, 131, 132, 134, 135, 189
GIMMS cartographical program, 585
GISS (Goddard Institute for Space Studies) general circulation model, 24, 77, 78, 113, 115–118, 133–148 passim, 184, 256, 354–356, 364, 404, 519, 569–572, 635–636, 639, 732; GISS-based transient scenario, 647, 655–657, 696; GISS 2 × CO_2 scenario, 29, 32–34, 38–68, 77–81, 113, 120, 143–148, 183– 188, 191, 197, 201, 208, 210, 229, 246–252, 257, 260–261, 268, 271, 289, 291, 296, 301–308, 316–320, 330–338, 348, 354, 357–358, 407, 409–412, 425–426, 462–464, 471–473, 484–485, 493–498, 523, 537–542, 559–561, 572–575, 581–596, 600–601, 636, 639, 644–647, 655, 673–675, 696, 757–763, 769–772, 788, 804, 816, 822–825, 840–841; grid point coverage, 78; interpolation between grid points, 78, 151–152, 538; spatial resolution of, 137–138, 143; validation of, 78, 139–143, 152–153, 247–248, 409, 538–539, 645, 674, 758–759
Glaciers, 392, 393
Glacial Maximum, 140
Grapes, 627
Grass, 526, 631; common bent (*Agrostis capillaris*), 446; digestibility of, 36; meadow fox-tail (*Alopecurus pratensis*), 446, red fescue (*Festuca rubra*), 446; ryegrass (*Lolium perenne*), 471–472; smooth meadow grass (*Poa pratensis*), 446; timothy (*Phleum pratense*) 459, 462; see also Carrying capacity, Hay

Index 873

Grass growth index, 480, 481
Grass margin, see Farm profitability
Grazing intensity, see Stocking rate
Grazing season, 31, 45; see also Growing season
Greenhouse effect, see Carbon dioxide
Greenhouses, 703
Gross Domestic Product, 205; regional, 326, 346, 347
Groundwater, depletion of, 18; level of, 659, 660, 700; supply of, 40
Growing-degree days, see Effective accumulated temperature, Effective temperature sum
Growing season, 44, 183, 456, 471, 565; length of, 33, 70, 301, 303, 458–460, 520, 524, 562, 631, 774; onset of, 26, 28, 54, 86, 496; see also Grazing season
Growth unit, 439, 440, 441

Hailstorms, 704
Harvest index, 809, 813–815
Harvest procurement system, 627–628
Harvest quality, see Yield quality
Heliothermic factor, 105, 264, 266–267, 369
Hay, 429–431; crop-climate simulation model of, 690–691; harvest date, 449–454; harvest failure, 493; herbage quality, 461, 465, 469, 473, 477, 483–484; yields, 26, 31–32, 39, 54–55, 62, 67, 86, 91, 113, 406, 415–474 passim, 492, 494–496, 528, 685, 700; see also Empirical-statistical models, Grass, Livestock
Herbicides, see Weed control
Hierarchies, see Model hierarchies
Holdridge classification of vegetation zones, 189
Holocene epoch, 140
Household consumption, 326
Household income, see Farm income
Household purchasing power, 52, 54, 82, 85; see also Farm income

Impact assessment, see Climate impact assessment
Impacts, see First-order interactions, Second-order interactions, Third-order interactions
Imports, 829, 850
Income, see Farm income
Income elasticity of demand, 837
Industrial sector, effects on, 16
Insurance, 332, 339, 345–347, 491, 704, 747, 860; of crops, 76, 324, 499
Interactions, see First-order interactions, Second-order interactions, Third-order interactions
International Biological Programme (IBP), 782

Irrigation, 238, 627, 703, 859, 862, see also Water resources

Köppen climatic classification, 232, 398

Labor, see Employment
Lambs, growth rate of, 476, 483; see also Livestock
Land abandonment, see Farm abandonment
Land allocation, changes in, 27–28, 46, 50, 72, 75, 468, 626, 677–693, 700, 702
Land management, changes in, 703
Limits, altitudinal, 28, 797–808, 856–857; of barley, 437–438, 442; of cereal cropping, 31; of crop zones, 542–543; of cultivable area, 38, 64, 468, 474, 519, 526, 529; of rice, 59–61, 92, 773–795; of spring wheat, 529; see also Boreal forest, limits of
Livestock, 28; feed requirements for, 51, 54–55, 86, 393, 429–431, 465, 494; health of, 17; numbers of, 392–394, 432–435, 441, 469, 531; production of, 395; rearing of, 429–431; yield of, 17; see also Carcass weight, Carrying capacity, Lambs
Lodging, 101

Maize, production of, 36, 262, 686; yields, 12, 26, 62
Management, 14, 39, 44–50, 71–72, 117, 320, 457, 490; changes in, 446; of experiments, 456; trend, 114; see also Adjustments, Agricultural policy, Technology
Marginality, 21–22, 391, 415, 542, 731; marginal areas, 63, 472, 485, 501, 532, 800; marginal cropland, 51
Marketable yield, 579, 582, 584–586, 588–589, 594; see also Yields
Meat, 627; consumption of, 606; see also Beef production, Lambs, Carcass weight, Livestock
Mechanistic schemes, see Simulation models
Mechanization, 580, 623, 862; cost of, 580; machine hire, 579; machine repair, 332, 339
Methane, see Carbon dioxide (and other trace gases)
Milk production, see Dairy farming
Mining, 327
Model hierarchies, 17, 24, 118–120, 183, 197, 216, 255, 348, 358–360, 364, 732, 747, 827, 859
Models, see Economic models, Empirical-statistical models, General circulation models, Model hierarchies, Simulation models
Moisture stress, 36, 231, 263, 460–461
Monsoon, 734
Mortality, see Famine, Population
Mulberries, 739

Mulching, 704
Multiplicative effects, 31, 69

National agricultural policy, see Agricultural policy
NCAR (National Center for Atmospheric Research) general circulation model, 131, 132, 134
Net primary productivity of vegetation, 749, 750, 773–795, 855–856; see also Biomass, Chikugo model
Nitrate pollution, 49, 657–658, 661, 700
Nitrogen, see Fertilizers, Soil fertility
Nitrous oxide, see Carbon dioxide (and other trace gases)
Nonlinear effects, 31, 69, 73
Norwegian spruce, 439–441
Nutrients, cycling of, 208; requirements for, 99; supply of, 18, 75, 448; transport of, 98; uptake of, 98
Nutrient depletion, see Soil degradation

Oats, 472, 526, 579–597, 582, 631–633, 686, 739; crop–climate simulation model of, 245, 690–691; yields, 26, 33, 36, 42, 57, 64–65, 88–89, 91, 120, 527–528, 530, 600–601, 685, 700
Obukhov model, of winter rye yields, 649, 651–659, 709–710
Ocean currents, 732
Oil, see Fuel
Oil seed yields, 528
Optimal land use, 50; see also Land allocation
Optimization models, see Economic models
OSU (Oregon State University) general circulation model, 131–132, 134–135, 139
Over-production, see Food surplus
Ozone, see Carbon dioxide (and other trace gases)

Palmer Drought Index, 104, 255, 270–278, 355–356, 359, 369–370; see also Drought
Pasture yields, 26, 54–55, 62, 64–65; see also Carrying capacity, Grass, Livestock, Yield
Peat, 189, 392
Permafrost, 652
Pests, 40, 71, 75, 208, 282, 563, 817, 863–864; control of, 36, 51, 99, 100, 702, 862
Phenological stages, 100, 300, 429–430, 568, 570, 572, 811
Phosphate, see Fertilizers, Soil fertility
Photoperiods, 39, 471–472
Photosynthesis, 98, 472, 577
Plant breeding, 39, 76, 101, 361, 563, 607, 627, 675, 773; see also Crop varieties
Policy responses, see Agricultural policy
Population, change of, 391; migration of, 238; see also Depopulation

Pork production, 531
Potato, 393, 501, 627, 686, 739; crop–climate simulation model of, 690–691; limits of cultivation, 444; yields, 26, 33, 91, 528, 685, 700
Poultry, 235, 531
Prairie Farm Rehabilitation Administration (PFRA), 238, 254, 286, 362
Precipitation, 161, 170–172, 401–404; decreases in, 179; increases in, 40, 143, 170, 179; variability of, 34, 169, 232, 250, 278, 304, 438; see also Climatic scenarios, Climatic variations
Precipitation effectiveness, 26, 28, 52, 54, 82, 84, 255–256, 279, 353, 359; definition of, 82; see also Precipitation index
Precipitation index, 262–264, 289
Pressure, 167–168, 397; anomalies of, 171
Prices, see Food prices
Probability, 364; of barley ripening, 438, 469, 496; of cool summer damage, 66; of crop failure, 469, 802; of extreme events, 706; of frost, 208, 577; of maturation/ripening, 26, 28; of winter kill, 458; of yields, 34–35, 70; see also Frequency
Procurement prices, 627; see also Food prices
Producer prices, 582; see also Food prices
Production costs, see Food production costs
Profit maximization, assumption of, 200
Prognostic approach, to scenario development, 78
Proxy data, 397
Pulses, 739

Radiative dryness index (RDI), 782–784
Rainfall, see Precipitation
Rangeland, see Carring capacity, Grass, Livestock, Yield
Red clover, 631
Regional analogues, 38–39, 70, 73, 472, 540, 542, 762–763, 855
Regional Compensation Effect, 623
Regional policy, see Agricultural policy
Regression models, see Empirical–statistical models
Rice, 738–740, 746, 808, 827–851; crop–climate simulation model of, 34, 60, 108, 116, 118, 749, 809–825; cultivation area of, 797–808, 856–857; production of, 41, 731, 741, 797–825; stocks of, 31, 60, 64, 68, 92, 827–851, 857–858; yield index of, 59–61, 92, 113, 778–779, 781, 786–787, 793, 797–808, 856–857; yields, 12, 26, 32, 37, 62–64, 66, 120, 740, 779–781, 787, 797–825 passim, 856–857
Risk, 21, 75, see also Frequency, Probability
Root crops, 472, 498
Rye, 526, 582, 631–633, 642, 643–644, 739; yields, 64, 527–528, 530; winter rye yields,

26, 56, 58–59, 65, 68, 90, 118, 642, 649–659 *passim*, 698

Saltation, 283
Sand dunes, 284
Saskatchewan Wheat Pool (Canada), 291
Sea ice, 131, 134, 392, 394–396
Second-order interactions, 15, 25, 27, 41–44, 71, 119; *see also* Employment, Farm expenditure, Farm income, Farm production, Farm profitability, Farm purchases, First-order interactions, Third-order interactions
Seed cost, 579, 580
Self-sufficiency, 532, 604, 606–607, 844
Sensitivity, 23, 28, 61, 70, 237, 364, 389, 415, 470, 772, 817–820; analysis of, 36, 44, 227
Service sector, effects on, 16, 327, 338, 345, 346, 347
Set-aside, of farmland, 42, 51, 607, 747, 835, 858
Severe winters, 391, 415, 417
Sheep, 429–437, 475–487; *see also* Carrying capacity, Lambs, Livestock
Shelter, for livestock, 439
Shelterbelts, 281, 502, 704
Shift of margins, *see* Climate Impact Assessment (spatial approach), Limits
Silage yields, 528
Simulation models, of crop–climate relationships, 24, 27, 45, 107, 245, 253, 664, 680, 714–715; of barley, 245, 690–691; of corn for silage, 690–691; of hay, 690–691; of oats, 245, 690–691; of potatoes, 690–691; of rice, 34, 60, 108, 116, 118, 749, 809–825; of spring wheat, 108, 115, 293–320, 322, 329–330, 663–675; of vegetables, 690–691; of winter wheat, 690–691
Snow cover, 416, 484, 520, 524, 576, 631, 632, 643; variability of, 398
Snow depth, 735, 737
Soils, 231, 522, 629–630, 634, 642, 808; acidity of, 631; classification of, 229; conservation of, 99–100, 240; degradation of, 227, 234–235, 238, 240, 255, 357, 362, 364, 863; drainage of, 27, 46–49, 98, 72, 473, 633–634, 659, 661, 703, 808; erosion of, 18, 40, 49, 51, 71–72, 75, 98, 239, 357, 362, 390, 651–652, 659; frost in, 629; heating of, 502; leaching of, 98; properties of, 101, 114; structure of, 18; *see also* Waterlogging
Soil fertility, 40, 66, 71, 864; index of, 26, 40, 90, 652, 656–657, 700
Soil moisture, 98, 143, 270–278, 449, 652
Soil salinity, 651–652; salinization, 18, 75, 98, 240, 357, 362
Solar radiation, 98, 99, 100, 125, 265, 525, 642, 811, 813, 818
Spring wheat, *see* Wheat

Stable period, *see* Climatic variations
State farms, 626, 627
Stocking level, 51, 63, 477, 483; *see also* Carrying capacity
Storm tracks, 397
Subsidies, 531, 606, 747, 850
Sugar beets, 67, 528, 624, 627
Sunflowers, 627
Sweet potatoes, 739

Taiga, *see* Boreal forest
Tea, 739
Technology, 14, 39, 117, 495, 798, 822–825, 862–864; change in, 44, 69, 243, 835; trend of, 110, 114, 243, 244, 656, 833; *see also* Adjustments, Agricultural policy, Crop varieties, Irrigation, Management, Mechanization, Plant breeding, Soil drainage
Temperature, of northern hemisphere, 160–175; global mean, 162; sea-surface, 131, 134, 161; *see also* Climatic scenarios, Climatic variations, Effective accumulated temperature, Effective temperature sum, Threshold temperature
Temperature index, 421–422, 424, 435–436; *see also* Warmth index
Thermal requirements, of crops, 33, 45, 48, 72; of barley, 437–438
Third-order interactions, 15, 25, 41–44, 71, 229; *see also* Agricultural policy, Employment, First-order interactions, Food prices, Food production, Food security, Food stocks, Food supply, Food surplus, Second-order interactions
Threshold, 30, 69, 73; of temperature, 104, 184–185, 192, 209, 260, 437, 797
Timber, carrying capacity for, 207; demand for, 201, 205, 212; prices of, 198–199, 214–215; stocks of, 206–207, 211, 213–214; supply of, 198–199, 201, 206, 212; supply models for, 260; trade of, 199, 202–203, 215
Timberline, *see* Boreal forest, Tree growth
Timing of operations, *see* Farming operations
Trace gases, *see* Carbon dioxide (and other trace gases)
Trade sector, 327, 338, 347
Transient response, of climate, *see* Climatic scenarios, GISS
Transportation sector, 338, 345–347, 681
Tree growth, potential for, 54, 82; potential tree zone, 398, 413, 496; *see also* Boreal forest
Tundra, 193, 210
Typhoons, 731, 734–735, 817

UKMO (United Kingdom Meteorological Office) general circulation model, 131, 133–135

Unstable period, *see* Climatic variations
Urbanization effects, 556

Vegetables, 67, 395, 624, 686, 739; crop–climate simulation model of, 690–691; yields, 26, 91, 685, 700
Vegetation zones, 737–738, 783, 784
VNIISI (All-Union Research Institute for Systems Studies, USSR) ecological model, 637, 652–653, 655–659
Volcanic eruptions, 391, 392
Vulnerability, 14, 21, 28, 806; *see also* Sensitivity

Warm periods, *see* Climatic variations
Warmth index, 102, 774, 775; *see also* Temperature index
Waterlogging, 66, 68, 644, 651, 657
Water requirements, of plants, 99, 100
Water resources, 16, 40, 71; management of, 51, 72; *see also* Irrigation
Water pollution, 90; *see also* Nitrate pollution
Water-use efficiency, 24, 184, 863; *see also* Carbon dioxide
Weeds, 282; control of, 99–100, 324, 862, 863
Western Grain Stabilization Act (Canada), 343, 345–346
Wheat, 41, 52, 54, 739; production of, 64, 67, 85, 233, 236; spring wheat, 353, 526, 565–577, 582, 632–633; spring wheat, crop–climate simulation model of, 108, 115, 293–320, 322, 329–330, 663– 675; spring wheat maturation time, 300, 303, 306–308, 317; winter wheat, 237, 342–343, 632, 686; winter wheat, crop–climate simulation model of, 690–691; *see also* Crop varieties, Empirical–statistical models, Yields
Wind erosion, 26, 237, 262, 360, 362, 859; climatic potential for, 40, 52, 53, 83, 84, 255, 256, 281–292, 352, 353, 359; Chepil model of potential, 283, 286; days of blowing dust, 20, 286–288
Winter housing, for livestock, 476
Winter kill, 44, 417, 456–458, 468, 473, 479, 489; probability of, 458
Winter wheat, *see* Wheat

Yield index, of rice, 59–61, 92, 113, 778–779, 781, 786–787, 793, 797–808, 856–857
Yields, 33–34, 527; barley, 26, 42, 56–67, 88–89, 91, 326, 331, 527–528, 533, 547–563, 600–602, 685, 700, 745; canola, 326, 331; corn for silage, 91, 685, 700; flax, 331; hay, 26, 31–32, 39, 54–55, 62, 64–65, 67, 86, 91, 113, 406, 415–474 *passim*, 492, 494–496, 528, 685, 700; maize, 12, 26, 62; oats, 26, 33, 36, 42, 57, 64–65, 88–89, 91, 120, 527–528, 530, 600–601, 685, 700; oil seed, 528; pasture, 26, 54–55, 62, 64–65, 86; potatoes, 26, 33, 91, 528, 685, 700; rice, 12, 26, 32, 37, 62–64, 66, 120, 740, 779–781, 787, 797–808, 809–825, 856–857; rye, 64, 527–528, 530; silage, 528; spring wheat, 26–27, 29, 33–34, 46, 52–54, 56–59, 63–66, 84, 88–90, 118–119, 255–256, 293–320, 326, 331, 359, 526–528, 565–577, 600–601, 663–675, 698; sugar beets, 528; vegetables, 26, 91, 685, 700; wheat, 32, 62, 27, 243–244, 745; winter wheat, 26, 33, 36, 91, 255, 528, 685, 688, 700; winter rye, 26, 56–59 *passim*, 65, 68, 90, 118, 642, 649–659 *passim*, 698
Yield quality, 70, 565; *see also* Marketable yield
Yield stability/variability, 26, 34, 47, 63, 66, 70, 122, 559, 602, 623, 642, 677, 779–781, 797, 816